Lecture Notes in Computer Science 13407

More information about this series at https://link.springer.com/bookseries/558

Cristina Alcaraz · Liqun Chen · Shujun Li ·
Pierangela Samarati (Eds.)

Information and Communications Security

24th International Conference, ICICS 2022
Canterbury, UK, September 5–8, 2022
Proceedings

Editors
Cristina Alcaraz
University of Malaga
Malaga, Spain

Liqun Chen
University of Surrey
Guildford, UK

Shujun Li
University of Kent
Canterbury, UK

Pierangela Samarati
University of Milan
Milan, Italy

ISSN 0302-9743 ISSN 1611-3349 (electronic)
Lecture Notes in Computer Science
ISBN 978-3-031-15776-9 ISBN 978-3-031-15777-6 (eBook)
https://doi.org/10.1007/978-3-031-15777-6

This Springer imprint is published by the registered company Springer Nature Switzerland AG
The registered company address is: Gewerbestrasse 11, 6330 Cham, Switzerland

Preface

This volume contains the papers that were selected for presentation and publication at the 24th International Conference on Information and Communications Security (ICICS 2022), which was jointly organized by the University of Kent (UK), theUniversità degli Studi di Milano (Italy), the University of Surrey (UK), and the University of Malaga (Spain). The conference was held at the main campus of the University of Kent, Canterbury, UK, during September 5–8, 2022. Due to post-pandemic conditions and travel limitations in some countries, the conference was held as a hybrid event, offering both in-person and remote participation options for attendees.

ICICS is one of the mainstream security conferences with the longest history. It started in 1997 and aims at bringing together leading researchers and practitioners from both academia and industry to discuss and exchange their experiences, lessons learned, and insights related to computer and communication security. This year's Program Committee (PC) consisted of 114 members with diverse backgrounds and broad research interests. A total of 150 valid paper submissions were received. The review process was double blind, and the papers were evaluated on the basis of their significance, novelty, and technical quality. Practically all the papers were reviewed by four or more PC members and then discussed among the Program Committee. The discussions were held online intensively over more than three weeks. Finally, 34 papers were selected for presentation at the conference giving an acceptance rate of 22.7%.

Following the reviews, two papers were selected for the Best Paper Award and the Best Student Paper Award, respectively. Both awards were generously sponsored by Springer. The conference also selected winners of four additional awards, a Best Presentation Award, a Best Artifact Award, a Best Poster Award and a Best Demo Award, all sponsored by the Institute of Cyber Security for Society (iCSS), University of Kent. Additionally, ICICS 2022 was honored to offer three outstanding keynote talks by Ross Anderson, University of Cambridge (UK), Nicholas Carlini, Google (USA), and Guang Gong, University of Waterloo (Canada). Our deepest and sincere thanks to Ross, Nicholas, and Guang for sharing their knowledge and experience during the conference. The conference also called for posters and demo presentations of "already published/accepted work", which were presented at a Poster/Demo session. In addition, a panel discussion was also organized at the conference.

For the success of ICICS 2022, we would like to first thank the authors of all submissions and the PC members for their great effort in selecting the papers. We also thank all the external reviewers for assisting the reviewing process. For the conference organization, we would like to thank the ICICS Steering Committee, the Publicity Chairs, Kalikinkar Mandal and Ding Wang, the Local Arrangement Co-Chairs, Budi Arief and Sanjay Bhattacherjee, the Poster/Demo Chairs, Özgür Kafalı and Vineet Rajani, the Panel Chair Zonghua Zhang, the Local Award Judging Chair Keenan Jones, the Sponsorship

Chairs, Fauzia Idrees and Clare Patterson, and the VI (Visual Identity) Designers, Yinglong He and Zhonghai Liu. Finally, we thank everyone else, speakers, session chairs, and volunteer helpers, for their contributions to the program of ICICS 2022.

September 2022

Cristina Alcaraz
Liqun Chen
Shujun Li
Pierangela Samarati

Organization

Steering Committee

Jianying Zhou	Singapore University of Technology and Design, Singapore
Robert Deng	Singapore Management University, Singapore
Dieter Gollmann	Hamburg University of Technology, Germany
Javier Lopez	University of Malaga, Spain
Qingni Shen	Peking University, China
Zhen Xu	Institute of Information Engineering, Chinese Academy of Sciences, China

General Chairs

Shujun Li	University of Kent, UK
Pierangela Samarati	Università degli Studi di Milano, Italy

Program Chairs

Cristina Alcaraz	University of Malaga, Spain
Liqun Chen	University of Surrey, UK

Local Arrangement Chairs

Budi Arief	University of Kent, UK
Sanjay Bhattacherjee	University of Kent, UK

Publicity Chairs

Kalikinkar Mandal	University of New Brunswick, Canada
Ding Wang	Nankai University, China

Poster/Demo Chairs

Özgür Kafalı	University of Kent, UK
Vineet Rajani	University of Kent, UK

Panel Chair

Zonghua Zhang	Huawei Paris Research Center, Huawei Technologies France S.A.S.U, France

Sponsorship Chairs

Fauzia Idrees	Royal Holloway, University of London, UK
Clare Patterson	University of Kent, UK

Local Award Judging Chair

Keenan Jones	University of Kent, UK

VI (Visual Identity) Designers

Yinglong He	University of Birmingham, UK
Zhonghai Liu	Guangdong Vgreen Intelligent Home Technology Co., Ltd., China

Program Committee

Chuadhry Mujeeb Ahmed	Singapore University of Technology and Design, Singapore
Man Ho Au	University of Hong Kong, Hong Kong
Zhongjie Ba	Zhejiang University, China
Joonsang Baek	University of Wollongong, Australia
Guangdong Bai	University of Queensland, Australia
Jia-Ju Bai	Tsinghua University, China
Diogo Barradas	University of Waterloo, Canada
Yinzhi Cao	Johns Hopkins University, USA
Guangke Chen	ShanghaiTech University, China
Rongmao Chen	National University of Defense Technology, China
Ting Chen	University of Electronic Science and Technology of China, China
Xiaofeng Chen	Xidian University, China
Xun Chen	Samsung Research America, USA
Sherman S. M. Chow	Chinese University of Hong Kong, Hong Kong
Mauro Conti	University of Padua, Italy
Nora Cuppens-Boulahia	Polytechnique Montréal, Canada
Jose Maria de Fuentes	Universidad Carlos III de Madrid, Spain
Roberto Di Pietro	Hamad Bin Khalifa University, Qatar
Wenrui Diao	Shandong University, China
Changyu Dong	Newcastle University, UK
Constantin Catalin Dragan	University of Surrey, UK
François Dupressoir	University of Bristol, UK
Afonso Ferreira	CNRS - Institut de Recherches en Informatique de Toulouse, France
Debin Gao	Singapore Management University, Singapore
Fei Gao	Beijing University of Posts and Telecommunications, China
Xing Gao	University of Delaware, USA
Joaquin Garcia-Alfaro	Institut Polytechnique de Paris, France
Amrita Ghosal	University of Limerick, Ireland
Dieter Gollmann	Hamburg University of Technology, Germany
Stefanos Gritzalis	University of Piraeus, Greece
Le Guan	University of Georgia, USA
Fuchun Guo	University of Wollongong, Australia
Shuai Hao	Old Dominion University, USA
Jiaqi Hong	Singapore Management University, Singapore
Hongxin Hu	University at Buffalo, SUNY, USA
Pengfei Hu	Shandong University, China

Jun Huang	City University of Hong Kong, Hong Kong
Xinyi Huang	Fujian Normal University, China
Jinyuan Jia	Duke University, USA
Chenglu Jin	CWI Amsterdam, The Netherlands
Sokratis Katsikas	Norwegian University of Science and Technology, Norway
Doowon Kim	University of Tennessee, USA
Hyoungshick Kim	Sungkyunkwan University, South Korea
Costas Lambrinoudakis	University of Piraeus, Greece
Wenjuan Li	Hong Kong Polytechnic University, Hong Kong
Kaitai Liang	Delft University of Technology, The Netherlands
Feng Lin	Zhejiang University, China
Jingqiang Lin	University of Science and Technology of China, China
Xiangyu Liu	Alibaba Inc., China
Zhuotao Liu	Tsinghua University, China
Javier Lopez	University of Malaga, Spain
Kangjie Lu	University of Minnesota, USA
Rongxing Lu	University of New Brunswick, Canada
Bo Luo	University of Kansas, USA
Xiapu Luo	Hong Kong Polytechnic University, Hong Kong
Haoyu Ma	Xidian University, China
Christian Mainka	Ruhr University Bochum, Germany
Daisuke Mashima	Advanced Digital Sciences Center, Singapore
Jake Massimo	Royal Holloway, University of London, UK
Weizhi Meng	Technical University of Denmark, Denmark
Jiang Ming	University of Texas at Arlington, USA
Yuhong Nan	Sun Yat-sen University, China
Siaw-Lynn Ng	Royal Holloway, University of London, UK
Jianbing Ni	Queen's University, Canada
Jianting Ning	Singapore Management University, Singapore
Liang Niu	New York University, USA
Rolf Oppliger	eSECURITY Technologies, Switzerland
Manos Panousis	University of Greenwich, UK
Günther Pernul	Universität Regensburg, Germany
Joachim Posegga	University of Passau, Germany
Elizabeth Quaglia	Royal Holloway, University of London, UK
Giovanni Russello	University of Auckland, New Zealand
Nitesh Saxena	Texas A&M University, USA
Shawn Shan	University of Chicago, USA
Vishal Sharma	Queen's University Belfast, UK
Qingni Shen	Peking University, China

Wenbo Shen	Zhejiang University, China
Purui Su	Institute of Software, Chinese Academy of Sciences, China
Hung-Min Sun	National Tsing Hua University, Taiwan
Kun Sun	George Mason University, USA
Willy Susilo	University of Wollongong, Australia
Qiang Tang	Luxembourg Institute of Science and Technology, Luxembourg
Yuzhe Tang	Syracuse University, USA
Luca Viganò	King's College London, UK
Ding Wang	Nankai University, China
Haoyu Wang	Huazhong University of Science and Technology, China
Lingyu Wang	Concordia University, Canada
Ting Wang	East China Normal University, China
Xiuhua Wang	Huazhong University of Science and Technology, China
Zhe Wang	Institute of Computing Technology, Chinese Academy of Sciences, China
Jinpeng Wei	University of North Carolina at Charlotte, USA
Weiping Wen	Peking University, China
Zhe Xia	Wuhan University of Technology, China
Dongpeng Xu	University of New Hampshire, USA
Jia Xu	NUS-Singtel Cyber Security R&D Lab, Singapore
Toshihiro Yamauchi	Okayama University, Japan
Guomin Yang	University of Wollongong, Australia
Kang Yang	State Key Laboratory of Cryptology, China
Zheng Yang	Southwest University, China
Xun Yi	RMIT University, Australia
Qilei Yin	Tsinghua University, China
Meng Yu	Roosevelt University, USA
Xingliang Yuan	Monash University, Australia
Chuan Yue	Colorado School of Mines, USA
Fan Zhang	Zhejiang University, China
Jiang Zhang	Institute of Software, Chinese Academy of Sciences, China
Kehuan Zhang	Chinese University of Hong Kong, Hong Kong
Tianwei Zhang	Amazon Web Services, USA
Yuan Zhang	Fudan University, China
Liang Zhao	Sichuan University, China
Qingchuan Zhao	City University of Hong Kong, Hong Kong
Yongjun Zhao	Nanyang Technological University, Singapore

Yunlei Zhao	Fudan University, China
Yongbin Zhou	Nanjing University of Science and Technology, China

Additional Reviewers

Bai, Weiheng
Biswas, Partha
Cao, Nhat Quang
Chen, Chenyang
Chen, Jinrong
Chen, Tianyang
Lin, Chengjun
Cui, Hongrui
Du, Minxin
Ehsanpour, Maryam
Eichhammer, Philipp
Empl, Philip
Feng, Qi
Fernandez, Carmen
Fouque, Pierre-Alain
Friedl, Sabrina
Gao, Yiwen
Gholipour, Mahmood
Glas, Magdalena
Gong, Borui
Guo, Hui
Guo, Xiaojie
He, Xu
Jia, Xiangkun
Jiang, Anqi
Jin, Renjie
Kabir, Mohammad Ekramul
Kailun, Yan
Kelarev, Andrei
Knittel, Lukas
Kumar, Gulshan
Lai, Qiqi
Lee, Moon Sung
Lepore, Cristian
Li, Bingyu
Li, Rui
Li, Xinyu
Li, Yannan
Li, Yongqiang
Limniotis, Konstantinos
Lin, Chao
Little, Rachael
Liu, Lin
Liu, Xiaoning
Liu, Yuejun
Lou, Xin
Lu, Xingye
Luo, Junwei
Lv, Chunyang
Ma, Mimi
Mladenov, Vladislav
Mui, William H. Y.
Muñoz, Antonio
Nissenbaum, Olga
Nowroozi, Ehsan
Pakki, Aditya
Pei, Weiping
Pöhls, Henrich C.
Rios, Ruben
Schlette, Daniel
Shen, Jun
Shi, Wenhao
Song, Qiyang
Song, Shang
Spielvogel, Korbinian
Spolaor, Riccardo
Tao, Yang
Tefek, Utku
Tian, Guangwei
Tian, Guohua
Torabi, Sadegh
Tricomi, Pier Paolo
Tsohou, Aggeliki
Wang, Jiafan
Wang, Shu
Wang, Tianyu
Wang, Xinda
Wang, Yi

Contents

Cryptography

Authentication

Network Security and Forensics

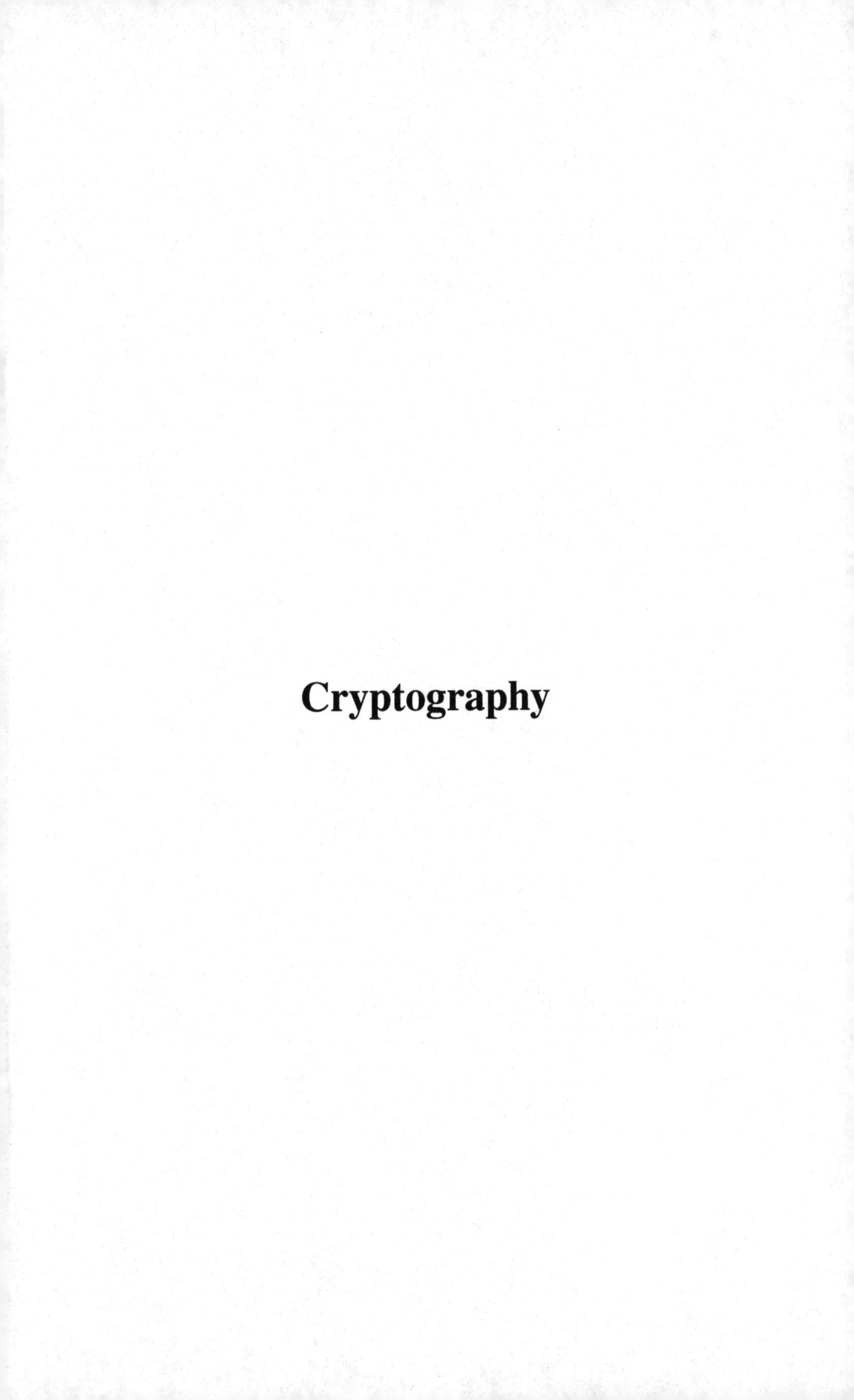

Cryptography

BS: Blockwise Sieve Algorithm for Finding Short Vectors from Sublattices

Jinzheng Cao[1], Qingfeng Cheng[1(✉)], Xinghua Li[2], and Yanbin Pan[3]

[1] Strategic Support Force Information Engineering University, Zhengzhou 450001, China
qingfengc2008@sina.com
[2] School of Cyber Engineering, Xidian University, Xi'an 710071, China
[3] Key Laboratory of Mathematics Mechanization, Academy of Mathematics and Systems Science, Chinese Academy of Sciences, Beijing 100190, China

Abstract. The Shortest Vector Problem is a crucial part of the lattice theory and a central lattice problem in analyzing lattice-based cryptography. This work provides a new algorithm that finds a short vector by calling the sieve oracle in projected sublattices orthogonal to each other. We first propose the Block Sieve algorithm. With blockwise sieving, proper insertion and reduction, the coordinates of the right end of vector v can be recovered. The algorithm moves the block to recover the other coordinates. We continue to optimize the algorithm and propose the Progressive Block Sieve algorithm, employing techniques such as skipping to accelerate the procedure. In a d-dimensional lattice, smaller sieve calls in $(d - \Theta(d/\ln d))$-dimensional sublattices are enough to find a short vector. We compare the experimental results on different lattices to test the performance of the new approach. On challenge lattices, our algorithm takes less time and fewer tours than original reduction algorithms to reach a similar outcome. As an application of the new algorithm, we test the performance of solving Learning With Errors problem. Our algorithm is able to solve the instances about 5% faster than sieving.

Keywords: Lattice · Shortest vector problem · Blockwise reduction · BKZ · Sieving

1 Introduction

Lattice-based cryptography is one of the most promising candidates for the post-quantum cryptography standard. A lattice $\mathcal{L}$ in $\mathbb{R}^m$ is a discrete subgroup of $\mathbb{R}^m$, defined as the set of all integer linear combinations of d linearly independent vectors $b_1, \ldots, b_d \in \mathbb{R}^m$. The matrix $B = [b_1, \ldots, b_d]$ form a basis of the lattice. The hardness of the Shortest Vector Problem (SVP) is at the center of estimating the security of lattice-based cryptosystems. Given a lattice basis, SVP asks to find the shortest non-zero vector in the lattice. Such a problem

C. Alcaraz et al. (Eds.): ICICS 2022, LNCS 13407, pp. 3–18, 2022.
https://doi.org/10.1007/978-3-031-15777-6_1

is proved to be NP-hard under certain assumptions. The extended applications of SVP lead to a list of variants, such as approximate SVP and unique SVP. To find the shortest vector, two classes of algorithms are proposed. One is the *accurate* approach, namely enumeration [23] and sieving. The algorithms aim to directly find a vector shorter than a certain upper bound, and are usually assumed to be able to recover the shortest vector. Enumeration was first initiated by Pohst, who proved that the shortest vector could be recovered by the algorithm with time complexity $2^{\Theta(d \log d)}$ and polynomial memory. The sieve algorithm was initiated in 2001, and is faster in theory at the cost of exponential memory. Both algorithms have been closely reviewed and are accelerated with a list of practical techniques. The other, *approximate*, approach is referred to as basis reduction, e.g. Lenstra-Lenstra-Lovász (LLL) algorithm [18] or Blockwise Korkine-Zolotarev (BKZ) algorithm [26]. Such algorithms output a reduced basis, with basis vectors short and nearly orthogonal to each other. Sieving and enumeration have an exponential time complexity, while basis reduction algorithms usually have lower complexity determined by parameters.

Most lattice-based cryptosystems do not explicitly rely on SVP to design their algorithms and parameters, but other hard lattice problems [2,16] instead. Among the many hard problems in lattice, the Learning With Errors (LWE) problem has been used to construct a series of KEMs and digital signatures. Introduced by Regev in 2005 [24], the problem makes it possible to build LWE-based primitives. In the third round of NIST PQC project, the security of several candidates such as Kyber and FrodoKEM is based on variants of LWE. Recent study has been focused on using lattice reduction and SVP solvers to implement and optimize the primal strategy.

1.1 Related Work

Most hard problems in lattice can be reduced to large instances of SVP. Thus, the algorithms for SVP lie at the core of the security analysis of lattice cryptosystems and have been closely studied over decades. The approximate approach to SVP mainly refers to lattice reduction algorithms. Such algorithms take in a basis as input and output another basis satisfying certain qualities. The first practical lattice reduction algorithm, LLL, was first proposed in 1982. The algorithm is still in active use today. In polynomial time the algorithm is able to produce a reasonably short lattice vector, but usually far from the shortest in high dimensions. Schnorr and Euchner proposed BKZ, which has a time complexity $\beta^{\beta/(2e)+o(\beta)}$, but outputs a better lattice basis and consequently a shorter lattice vector. While BKZ is the main-stream algorithm for lattice reduction, its actual behavior is a more complicated matter [15].

A line of studies and improvements on lattice reduction algorithms concentrated on the running time and the optimization of strategies and parameters. The work of Chen and Nguyen [10] analyzed the behavior of blockwise reduction by simulation and proposed a new variant of blockwise reduction algorithm, BKZ 2.0. The algorithm employed techniques such as extreme pruning and auto-abort, along with other fixes. This variant performs well in practice, and the algorithm has been the most recognized variant of lattice reduction algorithm. A variant of progressive reduction was proposed in 2016 [7], suggesting that BKZ-like reduction algorithms may achieve a better output when the block size is properly updated according to simulation result. Slide reduction [12] was initiated in 2008 by Aggarwal et al., and recently reviewed in 2020 [1]. The algorithm is based on Mordell's inequality, reaching a rather small approximation factor in theory. With techniques such as relaxation, a new enumeration-based reduction algorithm was proposed in 2020 and analyzed in 2021 by Albrecht et al., achieving a small Hermite factor with a lower-dimension SVP solver [3,4]. They suggest a new aspect that the context in which a reduction solver is called may be decoupled from the block, where the SVP solver is called.

Enumeration and sieving are two categories of accurate SVP algorithms. Enumeration is known to have an asymptotical complexity $d^{d/(2e)+o(d)}$ [20] and polynomial memory cost. The enumeration can be optimized with pruning [14], which accelerates the speed while lowering the probability of success. Sieving, however, has a time complexity ranging from $(4/3)^{d+o(d)}$ to $(3/2)^{d/2+o(d)}$ considering different techniques and strategies [8,17]. However, sieving has a $(4/3)^{d/2+o(d)}$ memory complexity, making it hard to solve larger instances. Variants of sieving have been proposed to improve the practical performance of sieving. In 2018, Ducas proposed Subsieve [11], combining sieve and Babai's Nearest Plane algorithm. The algorithm can find the shortest vector by sieving in a context of rank $\Theta(d/\log d)$. In 2019, Albrecht et al. gave an implementation of the algorithm, along with a general kernel for sieving [5], and achieved several new records of SVP challenges. With the development of new sieving variants, it is possible to use sieve algorithm as the SVP solver in certain cases.

1.2 Our Contribution

We summarize this paper's contribution in this subsection. Three main contributions are concluded as following.

- We extend Albrecht et al.'s approach of solving uSVP to the random lattices and propose the Block Sieve algorithm (BS). For a BKZ-preprocessed basis, we heuristically show that some projection of a short vector v can be found by a sieve call. With a projected vector found, the block can be moved leftwards recursively, until recovering all coordinates with sieving.
- In addition to the baseline algorithm described above, we adopt several other techniques and propose the Progressive Block Sieve algorithm (PBS). The improvements include Babai's algorithm to find coordinates in an extended context and combining progressive reduction, as well as skipping.
- We illustrate the performance of Block Sieve algorithm by experiments on various lattice challenges and comparing it to other algorithms (such as BKZ and Progressive BKZ). On lattices of SVP challenge, the progressive variant of the algorithm takes less time than BKZ to reach a similar norm and requires fewer tours and a smaller β at preprocessing stage. When applied to LWE, PBS is able to solve the instances about 5% faster than sieving.

1.3 Organization of the Paper

Preliminaries about lattice and hard problems are described in Sect. 2. In Sect. 3, we briefly review some important results of solving uSVP from previous literature and generalize the strategy to propose Block Sieve algorithm. Section 4 compares the performance computing costs of different algorithms in solving approximate SVP and LWE. We summarize the paper in Sect. 5.

2 Preliminaries

In this section, we introduce the basic notions concerned in the following sections. We provide the basic concepts about lattice and necessary assumptions. More details about lattice are introduced in [22].

2.1 Lattice

Definition 1 (Lattice). *Let* $b_1, b_2, \ldots, b_d \in \mathbb{R}^d$ *be linearly independent vectors, the basis matrix* $B = [b_1, b_2, \ldots, b_d]$ *and the lattice generated by* B *is* $\mathcal{L}(B) = \{\sum_{i=1}^{d} x_i b_i : x_i \in \mathbb{Z}\}$.

The 2-norm $\|\cdot\|$ of lattice vectors is called the length. The determinant of $\mathcal{L}$ denotes the volume of the fundamental area $\det \mathcal{L} = \det B = \|b_1^*\| \|b_2^*\| \ldots \|b_d^*\|$.

Definition 2 (Projection). *For a given basis* B *of lattice* $\mathcal{L}$, $\pi_i(v)$ *is the projections of vector* v *orthogonal to the span of* $b_1, b_2, \ldots, b_{i-1}$. *Further, the Gram-Schmidt orthogonalization of basis* B *is* $B^* = [b_1^*, b_2^*, \ldots, b_d^*]$, *where* $b_i^* = \pi_i(b_i)$.

The projected lattice $\mathcal{L}_{[l:r]}$, where $1 \leq l < r \leq d$ is defined as the lattice with basis $B_{[l:r]} = [\pi_l(b_l), \ldots, \pi_l(b_r)]$. We refer to $\mathcal{L}$ as the full lattice compared with the sublattice generated by the submatrix of full basis B. We use $\lambda_1(\mathcal{L})$ to denote the length of the shortest non-zero vector in $\mathcal{L}$, and $\lambda_2(\mathcal{L})$ the length of the second-shortest non-zero vector that is linearly independent of the first shortest vectors.

Most hard lattice problems can be reduced to the Shortest Vector Problem. The problem is to find a non-zero lattice vector of minimal length.

Definition 3 (Shortest Vector Problem, SVP). *Given a basis* B *of a lattice* $\mathcal{L}$, *find a non-zero lattice vector* $v \in \mathcal{L}$ *of minimal length* $\lambda_1(\mathcal{L}) = \min_{0 \neq w \in L} \|w\|$.

In practice, approximate versions of SVP are usually used, where we aim to find a vector longer than the shortest vector by a polynomial factor. Another widely used variant is unique-SVP, where the shortest vector has a small norm.

Definition 4 (unique SVP, uSVP). *For a lattice* $\mathcal{L}$ *satisfying* $\lambda_1(\mathcal{L}) \ll \lambda_2(\mathcal{L})$, *finding the shortest vector in* $\mathcal{L}$ *is called unique SVP (uSVP). The* γ-*unique SVP (*γ*-uSVP) is to find the shortest vector in* $\mathcal{L}$, *where* $\gamma\lambda_1(\mathcal{L}) < \lambda_2(\mathcal{L})$. *If the gap* $\lambda_2(\mathcal{L})/\lambda_1(\mathcal{L})$ *is bigger, it is easier to find the shortest vector.*

To analyze the problems on lattice, heuristic assumptions are necessary to estimate the quality of a lattice. We will rely on the following Gaussian heuristic [21] to explain our analysis in the rest of the paper.

Heuristic 1 (Gaussian). Let $K \subset \mathbb{R}^d$ be a measurable body, then there is $|K \cap \mathcal{L}| \approx \mathrm{vol}(K)/\det(\mathcal{L})$. When applying the heuristic to a d-dimension ball of volume $\det(\mathcal{L})$ we obtain that

$$\lambda_1(\mathcal{L}) \approx \frac{\Gamma\left(\frac{d}{2}+1\right)^{\frac{1}{d}} \det(\mathcal{L})^{\frac{1}{d}}}{\sqrt{\pi}} \approx \sqrt{\frac{d}{2\pi e}} \det(\mathcal{L})^{1/d}.$$

We denote the length as $gh(\mathcal{L})$ or $\mathrm{GH}(\mathcal{L})$ in short. In a random lattice $\mathcal{L}$, we assume the shortest vector will have norm $\mathrm{GH}(\mathcal{L})$.

2.2 Lattice Reduction Algorithms

The first widely applied reduction algorithm is LLL. Given the basis of a lattice $\mathcal{L}$, LLL outputs a basis $B = [b_1, b_2, \ldots, b_d]$ with the following statements hold:

(1) $\forall 1 \le i \le d, j < i, |\mu_{i,j}| \le 1/2$.
(2) $\forall 1 \le i < d, \|\delta b_i^*\|^2 \le \|\mu_{i+1,i} b_i^* + b_{i+1}^*\|^2$.

LLL has been widely used in attacks on several public-key cryptosystems. The algorithm has a polynomial time complexity, but the quality of output basis is limited. BKZ algorithm is commonly used to get a basis better than LLL. A BKZ-reduced lattice basis B for block size $\beta \ge 2$ satisfies $b_i^* = \lambda_1(\mathcal{L}_{[i:\min(i+\beta,d)-1]}), i = 1, \ldots, d-1$. The BKZ algorithm takes the lattice basis B and block size β as input. In its process, BKZ calls an SVP oracle on every projected block of dimension β. The BKZ algorithm achieves a good balance between the quality of reduced basis and running-time, and is the most commonly used lattice reduction algorithm to analyze the lattice. Hermite Factor (HF) is adopted to measure the quality of a reduced lattice basis [13]. The Hermite Factor has the form

$$\mathrm{HF}(b_1, \ldots, b_d) = \|b_1\|/\det(\mathcal{L})^{1/d}. \quad (1)$$

To obtain a basis of better quality, we expect the Hermite factor to be as small as possible. This is done by stronger reduction algorithms. To analysis the quality of the basis independent of dimension n, the Root-Hermite factor (RHF) is defined.

$$\delta = \mathrm{RHF}(b_1, \ldots, b_d) = (\|b_1\|/\det(\mathcal{L})^{1/d})^{1/d}. \quad (2)$$

The RHF describes the relation of the given basis and the short vector output by the lattice reduction algorithm. From the definition, the first vector of some reduced basis has norm $\|v\| = \delta^d \cdot \det(\mathcal{L})^{1/d}$. For LLL algorithm, $\delta \approx 1.0219$. For BKZ-β, $\delta = \left(\frac{\beta}{2\pi e}(\pi\beta)^{\frac{1}{\beta}}\right)^{\frac{1}{2(\beta-1)}}$ [9]. The problem of finding a non-zero lattice vector of length $\le \gamma \cdot \det(\mathcal{L})^{1/d}$ is called Hermite-SVP with parameter γ (γ-HSVP). To analyze the behavior of BKZ algorithm and predict the quality of the reduced basis, Schnorr's Geometric Series Assumption (GSA) is introduced to describe some reduced lattice basis [27].

Heuristic 2 (Geometric Series Assumption). For a BKZ-reduced basis of lattice $\mathcal{L}$, $\|b_{i+1}^*\|/\|b_i^*\| \approx \alpha, i = 1, \ldots, d-1$, where $\frac{3}{4} \leq \alpha^2 < 1$.

From GSA, the norms of the Gram-Schmidt vectors satisfy $\|b_i^*\| = \alpha^{i-1}\|b_1\|$. The definition of the Root-Hermite Factor implies $\|b_1\| = \delta^d \det(\mathcal{L})^{1/d}$. The determinant of the lattice $\det(\mathcal{L}) = \prod_{i=1}^d \|b_i^*\|$. Therefore, we can get $\alpha = \delta(\beta)^{-2d/(d-1)} \approx \delta(\beta)^{-2}$ [19].

2.3 Learning with Errors

Definition 5 (Learning with Errors, LWE). *Let $n, q \in \mathbb{Z}$, χ be a discrete Gaussian distribution on $\mathbb{Z}$ of standard deviation σ. Matrix $A \in \mathbb{Z}_q^{m\times n}$ and secret vector $s \in \mathbb{Z}_q^n$ are uniformly sampled. Vector e is sampled from χ. Given $A \in \mathbb{Z}_q^{m\times n}$ and secret $b \equiv As + e \mod q \in \mathbb{Z}_q^n$, the search-LWE is to compute s and e. The decision-LWE is to decide whether e is sampled from χ or from the uniform distribution.*

For an LWE instance (A, b), the row vectors correspond to samples (A_i, b_i), where $A_i s + e_i \equiv b_i \mod q$. The amount of samples is denoted as m. LWE can be solved via reducing it to other lattice problems. The *primal* strategy views LWE as a Bounded Distance Decoding (BDD) instance on a q-ary lattice, and reduce it to the unique-shortest vector problem on a basis defined by Kannan's embedding technique [6].

3 Block Sieve Algorithm

In this section, we introduce our main work, Block Sieve algorithm (BS). The goal is to find short vectors in $\mathcal{L}(B)$. Combined with BKZ preprocessing, the algorithm is able to solve the approximate SVP. In the first subsection, we introduce the basic version of BS algorithm. A modified variant, Progressive Block Sieve algorithm (PBS), is introduced in the second subsection.

3.1 Basic Block Sieve Algorithm

We describe our algorithm to find short vectors in a lattice. Suppose the unique shortest vector v is drawn from a spherical distribution, not skewed to any particular subspace. Thus, when $v \in \mathbb{R}^d$ is projected to a subspace $\mathbb{R}^k$, the projection has expected length $\sqrt{k/d}\|v\|$. The problem of searching for original v's coordinates can be performed by the size-reduction subroutine of LLL. According to Gaussian heuristic the norm of the shortest vector v is estimated as $\|v\| = \lambda_1(\mathcal{L}) \approx \sqrt{\frac{d}{2\pi e}} \det(\mathcal{L})^{1/d}$. We also assume that v's projection to $\mathcal{L}_{[d-\eta+1,d]}$ has length $\|\pi_{d-\eta+1}(v)\| = \sqrt{\eta/d}\|v\| \approx \sqrt{\frac{\eta}{2\pi e}} \det(\mathcal{L})^{1/d}$. Instead of running one sieve call on a block $\mathcal{L}_{[d-\eta+1,d]}$, we construct different projected blocks, and update the basis by calling sieve procedures on each of them.

Assuming we have found the projection of the shortest vector v in $\mathcal{L}_{[d-\eta+1,d]}$, the last η coefficients of v over basis $[b_1, b_2, \ldots, b_d]$ are also recovered. We go on to recover the rest of the coefficients of v. Instead of simply using Babai's method, we project the vectors on orthogonal blocks, and find the coefficients by finding the shortest projected vectors in each of them.

The algorithm requires that $\pi_{d-\eta+1}(v)$ is the shortest vector (or up to an approximate factor α) in $\mathcal{L}_{[d-\eta+1,d]}$, so that it can be found by a call of sieve, and inserted at $d-\eta+1$. Then we continue to move the block to $[d-2\eta+2, d-\eta+1]$. Since

$$\pi_{d-2\eta+2}(v) = \sum\nolimits_{i=d-2\eta+2}^{d} \nu_i \pi_{d-2\eta+2}(b_i) \in \mathcal{L}_{d-2\eta+2},$$

where $\nu_i \in \mathbb{Z}$ with $v = \sum_{i=1}^{d} \nu_i b_i$, and $b_{d-\eta+1}^{\text{new}}$ is recovered and inserted, it holds that

$$\pi_{d-2\eta+2}(v) \in \mathcal{L}(\pi_{d-2\eta+2}(b_{d-2\eta+2}, \ldots, b_{d-\eta+1})).$$

Therefore, v's coefficients $\nu_{d-2\eta+2,\ldots,d-\eta}$ can be recovered via another call of SVP solver on the projected sublattice $\mathcal{L}(\pi_{d-2\eta+2}(b_{d-2\eta+2}, \ldots, b_{d-\eta}))$. Continue this process until the block reaches the left end, indicating a vector in $\mathcal{L}$ is recovered. Operations in the next block will not affect the previous block. In fact, projected basis vectors $[b_1^*, \ldots, b_{d-\eta}^*]$ and $[b_{d-\eta+1}^*, \ldots, b_d^*]$ are orthogonal, so sieving and lifting in $\mathcal{L}_{[1,d-\eta]}$ will only change the coefficients of the first $d-\eta$ elements.

Particularly, an SVP solver can find a short vector within a much smaller context, which is useful in terms of updating the basis at the early stages of lattice reduction. For a BKZ reduced basis, the sublattice generated by the first few vectors may contain a short vector that can be used to update the basis. For $k \in \{1, 2, \ldots, d-1\}$, $\text{vol}(\mathcal{L}_{[1,k]}) = \|b_1^*\|\|b_2^*\| \ldots \|b_k^*\| = \delta(\beta)^{\frac{dk(d-k)}{d-1}} \text{vol}(\mathcal{L})^{k/d} \approx \delta(\beta)^{k(d-k)} \text{vol}(\mathcal{L})^{k/d}$, where $\delta(\beta)$ is the Root-Hermite Factor of the reduced basis by BKZ-β. Following the Gaussian heuristic, this sublattice contains a vector of length $\lambda(\mathcal{L}_{[1,k]}) = \sqrt{\frac{k}{2\pi e}} \delta(\beta)^{d-k} \text{vol}(\mathcal{L})^{1/d}$. By sieving in the k dimensional block, we obtain a vector of length

$$\lambda(\mathcal{L}_{[1,k]}) = \sqrt{\frac{k}{d}} \delta(\beta)^{d-k} \lambda(\mathcal{L}).$$

For example, set $k = \lceil 2d\delta(\beta)^{2(k-d)} \rceil$, then it is possible to recover a vector of norm $\approx \sqrt{2}\lambda_1(\mathcal{L})$. Even if the sieve call in $\mathcal{L}_{[1,k]}$ does not produce the shortest vector of $\mathcal{L}$, it is still possible to update the basis and increase δ. When the sieve produces a vector with norm

$$\|v\| \approx \lambda(\mathcal{L}_{[1,k]}) = \sqrt{\frac{k}{2\pi e}} \delta(\beta)^{d-k} \text{vol}(\mathcal{L})^{1/d} = \delta^d \text{vol}(\mathcal{L})^{1/d}, \tag{3}$$

the δ value is updated as

$$\delta = \left(\frac{k}{2\pi e}\right)^{\frac{1}{2d}} \delta(\beta)^{\frac{d-k}{d}}. \tag{4}$$

Therefore, after every sieve call on a block, shorter vectors are produced, and the quality of the basis is updated. The algorithm continues to reduce the basis by calling sieve on projected blocks, until a vector short enough is found.

Based on the discussion about recovering vectors via blockwise sieving, we present a new algorithm to find a short vector. The algorithm takes a BKZ-reduced basis as input. Sieve procedures are recursively performed on projected blocks $\mathcal{L}_{[r-\eta+1,r]}$. Insertion is applied following each sieving call to update the basis. Continue the process until all coordinates are found for a short lattice vector.

Algorithm 1 Block Sieve, BS

Require: $\mathcal{L}$ with BKZ reduced basis B of dimension d, parameters η and k
Ensure: short vector v
set $r = d$
while $r > k$ **do**
  Sieve($\mathcal{L}_{[\max(r-\eta+1,1),r]}$)
  insert output vector v' into index $r - \eta$
  LLL($\mathcal{L}$)
  $r = \max(r - \eta+1, 1)$
end while
Sieve($\mathcal{L}_{[1,k]}$), insert output vector v into index 1

Remark 1. Unlike previous algorithms like [6, 11], which rely on one SVP call on a projected block, the new algorithm projects the lattice over a list of orthogonal sublattices and use sieve calls on the blocks consecutively. Heuristically, the shortest vector is still a short vector when projected on a sublattice. Thus, by calling sieve on each block, we are able to obtain a better local quality of the basis, and progressively reach a shorter vector in the full lattice.

3.2 Progressive Block Sieve Algorithm

Algorithm 1 illustrates the basic procedure of BS. There is a list of possible improvements to the approach. As a result of the improvements, we will propose a practical variant of BS.

Progressive Algorithm. One single call of Algorithm 1 may not recover a vector short enough, partly because of a small δ. We BKZ-reduce the basis with a larger block size β in every tour. With an updated δ, we are able to recover a shorter lattice vector by sieving.

Simplified Tour. The expected behavior of BS is to run a sieve call in every projected sublattice $\mathcal{L}_{[l,r]}$. However, the local sieve is too expensive and can hardly recover a shorter vector. To simplify the algorithm, we merge the several sublattices in the middle into a few blocks and use Subsieve [11] to run sieve to lower the cost.

Skipping. The analysis of the sublattices and norm of vectors rely on the quality of the lattice basis. In fact, we assume the input basis follows GSA after BKZ reduced. After sieving, however, with short vectors inserted back into the basis, the structure of basis is different from expected. To maintain the structure of the basis between sieve calls, we skip some sieve calls and use stronger BKZ instead to ensure the quality of the lattice basis.

Improving on the basic BS algorithm with the techniques, we present the progressive version of Block Sieve algorithm.

Algorithm 2 Progressive Block Sieve, PBS

Require: $\mathcal{L}$ with basis B of dimension d, list of block sizes L, free dimension f, skipping index I
Ensure: short vector v
  for all $\beta \in L$ **do**
    BKZ-β reduce B
    if $\beta \notin I$ **then**
      calculate η and k
      set $r = d$
      while $r > k$ **do**
        Sieve($\mathcal{L}_{[r-\eta+1,r]}$)
        size-reduction($\mathcal{L}_{[r-\eta-f+1,r]}$)
        insert output vector v' into index $r - \eta - f$
        LLL($\mathcal{L}$)
        $r = \max(r - \eta+1, 1)$
      end while
      Sieve($\mathcal{L}_{[1,k]}$), insert output vector v into index 1
    end if
  end for

Remark 2. In the PBS algorithm, we see the BKZ reduction as a part of the preprocessing, and let the block sizes in L increase by an interval, for example up to 60 when $d \leq 200$. In our experiments, we determine f according to the analysis in [5], and set I so that a sieve procedure is called every 3 blocks. The size of the sieving context η affects the possibility and efficiency of recovering a short lattice vector. Generally, sieving has an exponential time complexity, so a smaller η is preferred. However, the conditions may not hold for a smaller η. In fact, for solving uSVP, η is affected by $\delta(\beta)$, as the Gaussian heuristic for $\mathcal{L}_{d-\eta+1}$ is determined by the quality of reduced basis. The desired condition is $\sqrt{\eta/d}\|v\| < \mathrm{GH}(\mathcal{L}_{d-\eta+1}) \approx \delta(\beta)^{2\beta-d}\mathrm{vol}(\mathcal{L})^{1/d}$.

4 Analysis of BS and PBS

4.1 Complexity Analysis

We start from the choice of η, which is the size of sieve calls in BS. For an average-case basis which follows the Gaussian heuristic, the norm of $\pi_{d-\eta+1}(v)$ has the estimated value $\sqrt{\frac{\eta}{2\pi e}}\det(\mathcal{L})^{1/d}$. Choosing η is then about comparing the projected norm to the shortest projected vector $\mathrm{GH}(\mathcal{L}_{d-\eta+1}) \approx \delta(\beta)^{2\eta-d}\det(\mathcal{L})^{1/d}$. Then, we find the maximum t that satisfies the inequality $\frac{4}{3}\delta(\beta)^{2t-d}\mathrm{vol}(\mathcal{L})^{1/d} > \sqrt{\frac{t}{2\pi e}}\det(\mathcal{L})^{1/d}$ as the η value. The condition indicates that the shortest projected vector is shorter than the Gaussian heuristic of sublattice, thus can be found via a sieving call. Note that with every updated basis B, parameters η and k should be adjusted according to the updated $\delta(\beta)$, in order to acquire better performance. The time cost of calculating the parameters can be neglected, so the total running time of the algorithm will rely mainly on reduction and sieving.

For a certain basis, we assume BS calls sieve oracle on $\Theta(d/\eta)$ sublattices. Further, PBS calls sieving for $\mathbf{poly}(d/\eta)$ times. According to the work of Ducas et al. [11], when the basis is BKZ-$d/2$ reduced, then $d-\eta = \Theta(d/\ln d)$ is a sufficient condition for solving SVP. In that case, our strategy solves the problem for at most $\mathbf{poly}(d/\eta)$ times the cost of Sieving in dimension $\Theta(d - d/\ln d)$.

Using GSA we calculate the volume of $\mathcal{L}_{d-\eta+1,d}$:

$$\mathrm{vol}(\mathcal{L}_{d-\eta+1,d}) = \prod_{i=d-\eta+1}^{d} \alpha^{\frac{d-1}{2}-i} = \alpha^{-\eta(d-\eta)/2}.$$

Recalling that $\pi_{d-\eta+1}(v) = \sqrt{\frac{\eta}{2\pi e}}\det(\mathcal{L})^{1/d}$, then the condition of algorithm is reorganized as

$$\sqrt{\frac{\eta}{2\pi e}} \leq \sqrt{\frac{4}{3}} \cdot \sqrt{\frac{\eta}{2\pi e}}\alpha^{-(d-\eta)/2}.$$

Taking logarithms on both sides, the condition is rewritten as $(d-\eta)\ln\alpha \leq \ln\frac{4}{3}$. Following the analysis of Subsieve, we choose $\beta = d/2$, to make sure a negligible cost of preprocessing. Therefore, according to definition, $\ln\alpha = \Theta(\beta/\ln\beta) = \Theta(d/\ln d)$. For some $\eta = d - \Theta(d/\ln d)$, the condition is satisfied. Thus, we heuristically claim that BS has the time complexity $\Theta(d/\eta)$ times of sieve in dimension $\Theta(d - \Theta(d/\ln d))$, and the PBS algorithm will find the shortest vector for a cost of $\mathbf{poly}(d/\eta)$ calls of sieve in dimension $\Theta(d - \Theta(d/\ln d))$. In particular, to analyze the sieve in $\mathcal{L}_{[1,k]}$, we assume the dimension $k = d/2$. Compared with BKZ, which outputs a vector of length $\approx (\beta^{1/2\beta})^d \cdot \det(\mathcal{L})^{1/d}$, a sieve call on a sublattice will give a vector of norm $\sqrt{\frac{k}{2\pi e}}\delta(\beta)^{d-k}\mathrm{vol}(\mathcal{L})^{1/d} \approx \sqrt{\frac{k}{2\pi e}}(\beta^{1/2\beta})^{d-k}\det(\mathcal{L})^{1/d}$. Therefore, a shorter vector is produced by BS when $k > \frac{\log(\sqrt{k/2\pi e})}{\log(\beta^{1/2\beta})}$.

4.2 Performance on Challenge Lattices

In this subsection, we implement the BS algorithm and show with experimental data how it actually performs. We implement the Block Sieve algorithm in Python. Namely, we implement the lattice operations based on FPLLL and

FPYLLL packages [28,29], and sieve algorithms based on G6K package [30]. For each parameter set, we run multiple experiments and get the average result. In the implementation of the algorithm, we preprocess the basis with BKZ of one tour in order to save time. For every tour, we let the BKZ block size β increase progressively.

Testing Basic BS on Challenge Lattices. We first analyze the performance of basic BS on challenge lattices. The lattice instances are selected from the TU Darmstadt SVP challenge [25]. The dimensions of the lattices are 140, 160, and 180. We compare the average norm of the first output basis vectors to test the performance of the algorithms, and compare the average running time to test the efficiency. We take the 140-dimension lattice as an example. When preprocessed with BKZ with block $\beta = 30$, BS will output a vector of average norm 10706.8, while the average norm after preprocessing is 14995.1. That is a 28% improvement, illustrating the effectiveness of our algorithm. In the experiments, we set the size of sieve subroutine the same with BKZ block size β. The d represents the dimension. The L_1, T_1 in the table represent the average norm and time cost of the algorithm after BKZ preprocessing but before sieve. The L_2, T_2 represent the average norm and time cost after sieve is called. The results are listed below in Table 1.

Table 1. Performance of basic BS

d	β	L_1	T_1	L_2	T_2
140	30	14995	0.2 s	10706	1.0 s
140	40	12503	0.4 s	9411	1.3 s
140	60	5062	15.8 s	4994	16.4 s
160	20	24007	0.3 s	17211	1.6 s
160	40	22186	1.1 s	11584	2.6 s
160	60	6284	29.2 s	5897	30.2 s
180	40	35358	1.6 s	26908	3.0 s
180	50	12502	9.4 s	10540	10.6 s

Remark. When a single call of BS is concerned, the running time is relatively longer than BKZ. In fact, the BS algorithm includes BKZ preprocessing, so the time for BKZ reduction is also counted. We also notice that with a larger block size β, the time cost of BKZ reduction is not negligible. In that case, the running time of BS and BKZ are relatively close. On average, BS will output a vector 30% shorter than BKZ, with time cost 2.4 times of BKZ.

Testing PBS on Challenge Lattices. Previous experiments are conducted with relatively small blocks and only take account of one tour of the algorithm.

We move on to run PBS and compare its performance with Progressive BKZ (PBKZ) [7]. Take the 160-dimensional lattice as an example. Let β increase progressively in [40, 60], PBS outputs a vector of norm 5797.06 in 164 s. The Progressive BKZ with the same β setting, however, outputs a vector of norm ≈6300 in 4 min. In fact, when β is up to 68, the Progressive BKZ will finally reach the norm ≈6000 even in 13 min (Fig. 1).

Table 2. Performance of PBS and Progressive BKZ (PBKZ) on SVP challenges

d	Method	β	Norm	Time	d	Method	β	Norm	Time
140	PBS	30–50	5171	14 s	160	PBS	40–60	5797	164 s
140	PBKZ	30–58	5187	36 s	160	PBKZ	40–68	5987	795 s
140	PBKZ	30–50	5895	7 s	160	PBKZ	40–60	6284	276 s
160	PBS	20–40	7813	18 s	180	PBS	30–50	9918	36 s
160	PBKZ	20–50	7888	27 s	180	PBKZ	30–54	10450	44 s
160	PBKZ	20–40	10452	6 s	180	PBKZ	30–50	12100	27 s
180	PBS	40–58	8615	111 s	180	PBS	40–60	7636	201 s
180	PBKZ	40–58	9142	97 s	180	PBKZ	40–60	8639	127 s

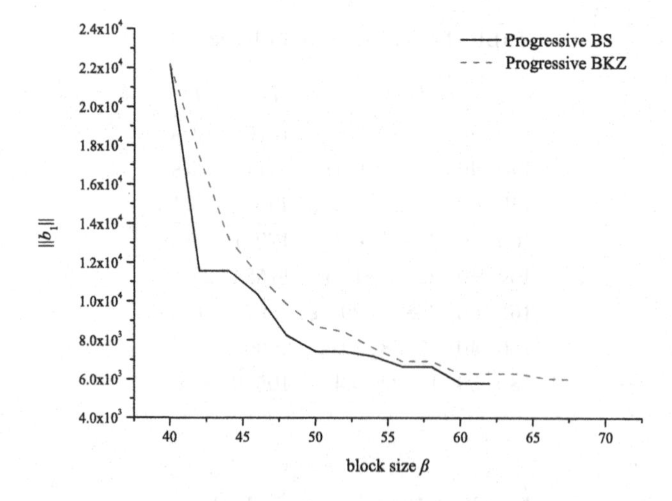

Fig. 1. PBS compared with PBKZ on 160-dimensional lattice

We tested the performance of Progressive BS and Progressive BKZ on lattices of different sizes, the result is listed in Table 2. For the same β, BS returns a vector 15% shorter than Progressive BKZ, with time cost 15% more on average. To reach the same output norm, Progressive BKZ has to run several more tours of reduction with larger blocks, thus costing more time. Note that in the process of BS, most computation is still done in BKZ reduction. Set $\beta = 50$ and call BS on a 180-dimensional lattice. Progressive BKZ takes 7.6 s while sieving takes

1.00 s. With a relatively small cost compared with Progressive BKZ reduction, BS is able to recover a considerably short vector.

4.3 Performance of PBS on LWE Instances

In this subsection, we use PBS to solve LWE instances, in order to put the algorithm into practical context. We generate different instances, with $q = 256, \sigma = \sqrt{8}$ and $n \in [60, 80]$. The matrix A is uniformly distributed. The instance's error rate $\sigma/q \approx 0.011$ indicates the hardness of the instances. To solve the instances, we adopt the primal strategy, reducing LWE to a BDD instance on a q-array basis, and implement the embedding technique to convert it to solving uSVP on a $(m+1) \times (m+1)$ lattice basis. We compare the performance of our approach with the sieving implementation from [5] and BKZ from FPYLLL.

To solve the LWE instances in practice, we run a subroutine to let η run from 0, to get the largest possible value. To be exact, we check the inequality $\sqrt{\eta} \cdot \sigma \leq \sqrt{4/3} \cdot \mathrm{GH}(\mathcal{L}_{[d-\eta+1,d]})$ until maximizing η. We also set an upper bound for η to let sieving end at an acceptable time cost.

Table 3. Average time cost of solving LWE

n	60	65	65	70	70	75	80
m	130	150	170	170	190	190	200
Subsieve	75 s	194 s	131 s	939 s	901 s	2107 s	27154 s
PBS	68 s	189 s	127 s	804 s	645 s	2072 s	25780 s

Remark. We also tried to solve the instances with BKZ. BKZ successfully solved the 60-dimensional instance in 131.69 s, slower than both sieving and PBS. For larger instances, we did not manage to solve them at an acceptable time cost. In fact, these instances require larger β, adding to the overhead. From the experimental results listed in Table 3, PBS succeeded in solving the instances. On average, PBS is about 5.5% faster than sieving (preprocessing time included). The advantage of PBS is more oblivious when a larger instance is concerned. We heuristically explain that a smaller instance allows for only a small sieving context size η, reducing the effectiveness of the sieve subroutine.

5 Conclusion

This paper reviewed the sieving approach to uSVP and adapted the strategy to solving SVP in an average case. For a BKZ-preprocessed basis of some δ, it is possible to generate projected sublattice blocks where the projections of the short vector v can be found by a sieving call. Based on the idea, we propose the basic Block Sieve algorithm and the optimized Progressive Block Sieve algorithm. To test the efficiency of our approach, we run experiments on different lattices. On lattices of SVP challenge, the basic BS is able to reach a shorter vector than BKZ with a slightly higher time cost. When the progressive variant is concerned, PBS takes less time than BKZ to reach a similar norm and requires fewer tours and a smaller β as preprocessing. As an application of the new approach, we go on to accelerate the primal strategy of solving LWE with PBS. Our algorithm is able to solve the instances about 5% faster than sieving.

Acknowledgements. This work was supported by National Natural Science Foundation of China (Grant Nos. 61872449, 62125205).

References

1. Aggarwal, D., Li, J., Nguyen, P.Q., Stephens-Davidowitz, N.: Slide reduction, revisited—filling the gaps in SVP approximation. In: Micciancio, D., Ristenpart, T. (eds.) CRYPTO 2020. LNCS, vol. 12171, pp. 274–295. Springer, Cham (2020). https://doi.org/10.1007/978-3-030-56880-1_10
2. Ajtai, M.: Generating hard instances of the short basis problem. In: Wiedermann, J., van Emde Boas, P., Nielsen, M. (eds.) ICALP 1999. LNCS, vol. 1644, pp. 1–9. Springer, Heidelberg (1999). https://doi.org/10.1007/3-540-48523-6_1
3. Albrecht, M.R., Bai, S., Fouque, P.-A., Kirchner, P., Stehlé, D., Wen, W.: Faster enumeration-based lattice reduction: root Hermite factor $k^{1/(2k)}$ time $k^{k/8+o(k)}$. In: Micciancio, D., Ristenpart, T. (eds.) CRYPTO 2020. LNCS, vol. 12171, pp. 186–212. Springer, Cham (2020). https://doi.org/10.1007/978-3-030-56880-1_7
4. Albrecht, M.R., Bai, S., Li, J., Rowell, J.: Lattice reduction with approximate enumeration oracles: practical algorithms and concrete performance. Cryptology ePrint Archive, Report 2020/1260 (2020). https://eprint.iacr.org/2020/1260
5. Albrecht, M.R., Ducas, L., Herold, G., Kirshanova, E., Postlethwaite, E.W., Stevens, M.: The general sieve kernel and new records in lattice reduction. In: Ishai, Y., Rijmen, V. (eds.) EUROCRYPT 2019. LNCS, vol. 11477, pp. 717–746. Springer, Cham (2019). https://doi.org/10.1007/978-3-030-17656-3_25
6. Albrecht, M.R., Göpfert, F., Virdia, F., Wunderer, T.: Revisiting the expected cost of solving uSVP and applications to LWE. In: Takagi, T., Peyrin, T. (eds.) ASIACRYPT 2017. LNCS, vol. 10624, pp. 297–322. Springer, Cham (2017). https://doi.org/10.1007/978-3-319-70694-8_11
7. Aono, Y., Wang, Y., Hayashi, T., Takagi, T.: Improved progressive BKZ algorithms and their precise cost estimation by sharp simulator. In: Fischlin, M., Coron, J.-S. (eds.) EUROCRYPT 2016. LNCS, vol. 9665, pp. 789–819. Springer, Heidelberg (2016). https://doi.org/10.1007/978-3-662-49890-3_30
8. Becker, A., Ducas, L., Gama, N., Laarhoven, T.: New directions in nearest neighbor searching with applications to lattice sieving. In: Proceedings of the Twenty-Seventh Annual ACM-SIAM Symposium on Discrete Algorithms, SODA 2016, pp. 10–24. Society for Industrial and Applied Mathematics (2016)
9. Chen, Y.: Réduction de réseau et sécurité concrète du chiffrement complètement homomorphe. Ph.D. thesis, Higher Normal School - PSL (2013). https://www.theses.fr/2013PA077242, thèse de doctorat dirigée par Nguyen, Phong-Quang Informatique Paris 7 2013

10. Chen, Y., Nguyen, P.Q.: BKZ 2.0: better lattice security estimates. In: Lee, D.H., Wang, X. (eds.) ASIACRYPT 2011. LNCS, vol. 7073, pp. 1–20. Springer, Heidelberg (2011). https://doi.org/10.1007/978-3-642-25385-0_1
11. Ducas, L.: Shortest vector from lattice sieving: a few dimensions for free. In: Nielsen, J.B., Rijmen, V. (eds.) EUROCRYPT 2018. LNCS, vol. 10820, pp. 125–145. Springer, Cham (2018). https://doi.org/10.1007/978-3-319-78381-9_5
12. Gama, N., Nguyen, P.Q.: Finding short lattice vectors within Mordell's inequality. In: Proceedings of the Fortieth Annual ACM Symposium on Theory of Computing, STOC 2008, pp. 207–216. Association for Computing Machinery, New York (2008). https://doi.org/10.1145/1374376.1374408
13. Gama, N., Nguyen, P.Q.: Predicting lattice reduction. In: Smart, N. (ed.) EUROCRYPT 2008. LNCS, vol. 4965, pp. 31–51. Springer, Heidelberg (2008). https://doi.org/10.1007/978-3-540-78967-3_3
14. Gama, N., Nguyen, P.Q., Regev, O.: Lattice enumeration using extreme pruning. In: Gilbert, H. (ed.) EUROCRYPT 2010. LNCS, vol. 6110, pp. 257–278. Springer, Heidelberg (2010). https://doi.org/10.1007/978-3-642-13190-5_13
15. Hanrot, G., Pujol, X., Stehlé, D.: Analyzing blockwise lattice algorithms using dynamical systems. In: Rogaway, P. (ed.) CRYPTO 2011. LNCS, vol. 6841, pp. 447–464. Springer, Heidelberg (2011). https://doi.org/10.1007/978-3-642-22792-9_25
16. Hoffstein, J., Pipher, J., Silverman, J.H.: NTRU: a ring-based public key cryptosystem. In: Buhler, J.P. (ed.) ANTS 1998. LNCS, vol. 1423, pp. 267–288. Springer, Heidelberg (1998). https://doi.org/10.1007/BFb0054868
17. Laarhoven, T.: Sieving for shortest vectors in lattices using angular locality-sensitive hashing. In: Gennaro, R., Robshaw, M. (eds.) CRYPTO 2015. LNCS, vol. 9215, pp. 3–22. Springer, Heidelberg (2015). https://doi.org/10.1007/978-3-662-47989-6_1
18. Lenstra, A.K., Lenstra, H.W., Lovasz, L.: Factoring polynomials with rational coefficients. Math. Ann. **261**, 515–534 (1982). https://doi.org/10.1007/BF01457454
19. Lindner, R., Peikert, C.: Better key sizes (and attacks) for LWE-based encryption. In: Kiayias, A. (ed.) CT-RSA 2011. LNCS, vol. 6558, pp. 319–339. Springer, Heidelberg (2011). https://doi.org/10.1007/978-3-642-19074-2_21
20. Micciancio, D., Walter, M.: Fast lattice point enumeration with minimal overhead. In: Proceedings of the Twenty-Sixth Annual ACM-SIAM Symposium on Discrete Algorithms, SODA 2015, pp. 276–294. Society for Industrial and Applied Mathematics (2015)
21. Micciancio, D., Walter, M.: Practical, predictable lattice basis reduction. In: Fischlin, M., Coron, J.-S. (eds.) EUROCRYPT 2016. LNCS, vol. 9665, pp. 820–849. Springer, Heidelberg (2016). https://doi.org/10.1007/978-3-662-49890-3_31
22. Nguyen, P., Valle, B. (eds.): The LLL Algorithm. Springer, Berlin (2010). https://doi.org/10.1007/978-3-642-02295-1
23. Pohst, M.: On the computation of lattice vectors of minimal length, successive minima and reduced bases with applications. SIGSAM Bull. **15**(1), 37–44 (1981). https://doi.org/10.1145/1089242.1089247
24. Regev, O.: On lattices, learning with errors, random linear codes, and cryptography. J. ACM **56**(6) (2009). https://doi.org/10.1145/1568318.1568324
25. Schneider, M., Gama, N.: SVP challenge. [EB/OL]. https://www.latticechallenge.org/svp-challenge. Accessed 25 June 2021
26. Schnorr, C.P., Euchner, M.: Lattice basis reduction: improved practical algorithms and solving subset sum problems. In: Budach, L. (ed.) FCT 1991. LNCS, vol. 529, pp. 68–85. Springer, Heidelberg (1991). https://doi.org/10.1007/3-540-54458-5_51

27. Schnorr, C.P.: Lattice reduction by random sampling and birthday methods. In: Alt, H., Habib, M. (eds.) STACS 2003. LNCS, vol. 2607, pp. 145–156. Springer, Heidelberg (2003). https://doi.org/10.1007/3-540-36494-3_14
28. T. F. Development Team: FPLLL, a lattice reduction library, Version: 5.4.1 (2021). https://github.com/fplll/fplll
29. T. F. Development Team: FPYLLL, a Python wraper for the FPLLL lattice reduction library, Version: 0.5.6 (2021). https://github.com/fplll/fpylll
30. T. F. Development Team: The general sieve kernel (g6k) (2021). https://github.com/fplll/fpylll

Calibrating Learning Parity with Noise Authentication for Low-Resource Devices

Teik Guan Tan(✉), De Wen Soh, and Jianying Zhou

Singapore University of Technology and Design, Singapore, Singapore
teikguan_tan@mymail.sutd.edu.sg

Abstract. Learning Parity with Noise (LPN) is an attractive post-quantum cryptosystem for low-resource devices due to its simplicity. Communicating parties only require the use of AND and XOR gates to generate or verify LPN cryptogram samples exchanged between the parties. However, the LPN setup is complicated by different parameter choices including key length, noise rate, sample size, and verification window which can determine the usability and security of the implementation. To address advances in LPN cryptanalysis, recommendations for ever increasing key lengths have made LPN no longer feasible for low-resource devices. In this paper, we use a series of experiments to simulate and cryptanalyze LPN authentication under different parameter values to arrive at recommended values suitable for low-resource devices. We also examine the impact of limiting the key lifespan of the LPN secret vector as a means to balance security while keeping key lengths relatively short.

Keywords: Learning Parity with Noise (LPN) · Cryptanalysis · Machine learning · Post quantum cryptography

1 Introduction

Learning Parity with Noise (LPN) [4] is already a decades-old problem. However, it has received renewed interest by the research community of late due to awareness and activities related to post-quantum cryptography. LPN has its roots in code-based cryptography and can also be seen as a special case of a lattice-based Learning-with-Error problem, both of which are still being evaluated by National Institute of Science and Technology (NIST) as possible candidates for post-quantum cryptography standardization [27]. What is attractive about LPN is the simplicity in its computation, requiring only AND and XOR gates to compute the cryptogram, which makes it a quantum-secure alternative for authenticating communications in low-resource devices such as sensors, wearables and radio-frequency RFID tags.

LPN, however, is not a typical cryptosystem. The security of most cryptosystems is solely proportional to the size of the secret key[1]. In LPN, a higher key

[1] In the case of LPN, the key is also referred to as the secret vector. For this paper, we will use key and secret vector interchangeably for readability purposes.

C. Alcaraz et al. (Eds.): ICICS 2022, LNCS 13407, pp. 19–36, 2022.
https://doi.org/10.1007/978-3-031-15777-6_2

length reduces the chance of key recovery, but it is not the only factor that determines the security of an LPN implementation. Esser et al. [9] describes LPN to be a "two-parameter problem" whose hardness is determined at setup by the size of secret vector and the noise rate. Higher noise rates at constant key lengths decrease the chance of an adversary being able to guess the secret vector while lower noise rates work vice versa. In the extreme case of zero noise, the LPN secret vector can be recovered in polynomial time using Gaussian elimination.

The security of an LPN implementation is also affected by the number of samples obtained during runtime. Much research [9,24,28] on the security strength of LPN has assumed an unbounded number of samples available for cryptanalysis, thus recommending larger key lengths and even larger number of samples per cycle in order to thwart the ever-growing cryptanalysis capacity of the adversary's machine. This inadvertently increases the compute and bandwidth overheads of LPN implementations beyond the reach of low-resource devices, a common concern echoed by other researchers [12,25]. We ask if restricting the key lifespan by putting a limit on the total number of samples generated is a feasible means to mitigate against key recovery attacks.

Just as how authentication systems can be adjusted (e.g. different password combinations and lengths, number of features in a facial biometric minutiae, etc.) according to the application's needs for security versus convenience, we want to calibrate the values of these three (or more) LPN parameters: key length, noise rate, and number of samples per cycle along the same purpose. In this paper, we will use statistical bounds, algorithmic analysis and computational simulations to examine the inter-dependencies among the parameters, and attempt to arrive at optimal recommendations for deciding the parameter values while balancing practical constraints against security requirements. Our contributions are:

- Execution and analysis of over 70,000 node-hours of LPN simulation and cryptanalysis on a high-performance computational cluster.
- Use of machine-learning algorithms, specifically Extremely Randomized Trees and Genetic Algorithms, to attempt LPN key recovery.
- Introduction of a key lifespan parameter as the mitigating means to keep key lengths and sample sizes sufficiently low while maintaining security.
- Validated list of recommended LPN parameters for low-resource devices.

The organization of the paper is as follows. Section 2 covers LPN basics and assumptions. Section 3 uses the authentication concept of "precision" to model the LPN parameters for determining false-acceptance rates (FAR) and false-rejection rates (FRR). Section 4 explores how limiting the key lifespan could mitigate the need to increase key length against key recovery attacks. Section 5 summarizes the findings and concludes the paper.

2 Preliminaries

2.1 Notation

For positive integers $i, j \in \mathbb{N}$, we denote $v \in \mathbb{Z}_2^i$ as a vector with i binary elements and $M \in \mathbb{Z}_2^{i \times j}$ as a two-dimensional matrix with i rows and j columns of binary elements. v_i represents the ith element in vector v, and M_i is a vector representing the ith row in matrix M. The hamming weight of vector v is denoted by $\|v\|_1$ and represents the number of "1" elements in the vector. log denotes the binary logarithm, while ln denotes the natural logarithm.

For a real number $\tau \in [0,1]$, we use $v \leftarrow Ber_{\tau}^{i}$ and $M \leftarrow Ber_{\tau}^{i \times j}$ to denote a vector v of i binary elements and a two-dimensional matrix M of i by j binary elements respectively where each of the vector and matrix elements are randomly chosen to follow a Bernoulli distribution where $Pr[v_i == 1] = \tau$ and $Pr[M_{i,j} == 1] = \tau$.

2.2 LPN Basics

Definition 1 (LPN). *When given access to a LPN oracle $\mathbb{O}_{\tau}^{k}$ which possesses a secret vector $s \in \mathbb{Z}_2^k$ and returns n pairs of samples in the form $\{A_i, b_i = A_i \cdot s \oplus e_i\}$ where $A_i \leftarrow Ber_{0.5}^{k}$, $b_i, e_i \in \{0,1\}$, $Pr[e_i = 1] = \tau$, $0 \leq \tau < 0.5$ and $i = \{1, 2, ..., n\}$, the ϵ-hardness of the LPN (search) problem is defined as the probability that a polynomial running-time algorithm $\mathcal{S}$ can return s.*

$$Pr[\mathcal{S}(A, b) \rightarrow s] \leq \epsilon \tag{1}$$

The HB [17] protocol is a vanilla construction of the LPN problem embodied in an identification protocol between a Verifier and Prover.

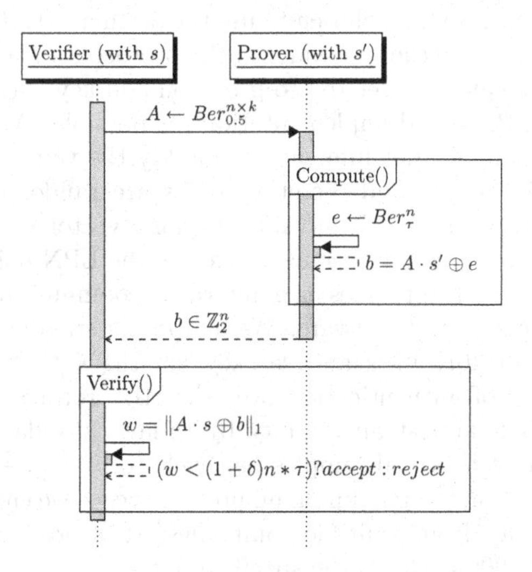

Fig. 1. The HB protocol [17]

Figure 1 shows the interaction between the Verifier and Prover where the Verifier is assured to a degree that the Prover has knowledge of the shared secret vector s of length k. It goes as follows:

1. Noise rate τ is pre-agreed. Verifier possesses s and Prover possesses s', both vectors of length k.

2. Verifier sends a n by k binary challenge matrix, A, to the Prover. A is randomly generated and each element has equal probability of being 1 or 0.
3. Prover generates an error vector $e \in Ber_\tau^n$ where $Pr[e_i = 1] = \tau$ and computes the response vector $b = A \cdot s' \oplus e$. Prover sends b to the Verifier.
4. Verifier has a verification window parameter $\delta \in \mathbb{R}$ where $0 \leq \delta \leq 1$. Verifier computes the hamming weight $w = \|A \cdot s \oplus b\|_1$, and accepts that $s = s'$ if w does not exceed $(1 + \delta) * n * \tau$. And rejects otherwise.

We note that over and above the LPN parameters of key length k, noise rate τ and sample size per cycle n, there is an additional verification window δ that affects the authentication outcome. Assuming c represents the total number of authentication cycles exchanged between the Prover and Verifier, we define the lifespan of the key $N = n * c$ which represents total number of samples generated in the life of the key. HB is only secure against passive attacks where the values of A_i are random and not under the influence or control of any adversary. Subsequent protocols such as HB^+ [18], HB^{++} [8], HB# [12] and many more evolve from HB and attempt to address active attacks and man-in-the-middle attacks. Other constructions on improving LPN security include Ring-LPN [15] which replaces the binary field with a ring, using a subspace of the secret key [20] on which to build a MAC scheme, and moving to a 3-round LPN [25].

2.3 Assumptions

For our investigations, we have chosen to use the original HB [17] protocol as it is a common denominator to the subsequent LPN protocols and derived schemes. It allows our findings which are related to precision and key lengths to remain relevant to as many LPN-based implementations as possible. We therefore assume a passive attacker who cannot influence or modify the values of matrix A. The binary values in matrix A and secret vector s are randomly generated with equal probability to be 1 or 0. The values in noise vector e follows a Bernoulli distribution where $Pr[e_i == 1] = \tau$ in which τ is the LPN noise rate.

ISO 18000-6:2013 [1] provides the interface communication standard for RFID-based equipment and software. We adopt the standard's ultra-high frequency data rate of 40Kbits/s and set key lengths k to start from 80 bits. Assuming one cycle of authentication over the HB protocol happens in under one second (which is almost an eternity by today's standards), this gives an approximate range of maximal samples per cycle to be $n \leq \frac{40000}{80} = 500$. While the standard is silent on the maximum number of cycles, we checked for available product specifications from multiple companies[2] that produce RFID products and found that 100,000 cycles is the specified norm.

3 Exploring Precision

Since LPN, by design, can be made statistically secure [19] by proper choice of key length, noise and number of samples, we want to create an applicable framework for LPN projects to suitably choose these parameters.

[2] Including 3M, EM Microelectronic, Fujitsu, NXP and Rockwell Automation.

Definition 2 (Precision). *Generalizing from Definition 1 in Hopper and Blum [17] and referring to Fig. 1, we envision an identification system of two parties, Verifier and Prover, where the Verifier has in possession a secret vector* $s \in \mathbb{Z}_2^k$ *and the Prover has in possession a secret vector* $s' \in \mathbb{Z}_2^k$. *The Verifier is required to accept the identity of the Prover if* $s == s'$ *and reject otherwise. We define precision as the measure of two limits:*

False-Acceptance-Rate (FAR) represents the probability that a Verifier will accept a Prover with a different secret vector:

$$FAR = Pr\left[Verify() \rightarrow accept | s \neq s'\right] \tag{2}$$

False-Rejection-Rate (FRR) represents the probability that a Verifier will reject a Prover with the same secret vector:

$$FRR = Pr\left[Verify() \rightarrow reject | s == s'\right] \tag{3}$$

FAR represents the authentication security of the system where a very low FAR represents a system which adversaries are unable to spoof. FRR represents the usability of the system where higher FRR rates cause unhappiness to valid users, who are forced to retry the authentication.

3.1 Statistical Bounds

Levieil and Fouque [24] attempted to address precision by using a completeness error P_c (defined similarly to FRR) and soundness error P_s (defined similarly to FAR). The computations of P_c and P_s using Stirling's formula, where $u = (1+\delta)\tau$, are shown in Eq. 4.

$$\begin{aligned} g(x,y) &= \left(\frac{x}{y}\right)^x \left(\frac{1-x}{1-y}\right)^{1-x} \\ FRR = P_c &\sim g(u,\tau)^{-n} \\ FAR = P_s &\sim g(u,\frac{1}{2})^{-n} \end{aligned} \tag{4}$$

The approach in Eq. 4 considers the set of all possible answers consisting of correct and incorrect answers, and then accounts for the probability P_c that an answer from a valid Prover may not be accepted due to noise and the probability P_s that an adversary could consecutively guess the correct answer for each sample. Levieil and Fouque then proposed that for $P_c < 2^{-40}$ and $P_s < 2^{-80}$, several pairs of [*noise,samples*] are listed: {[0.01,159], [0.05,249], [0.125,441], [0.25,1164], [0.4,7622], [0.49,554360]}. Note that the proposed number of samples for $\tau \geq 0.25$ already exceeds what is feasible in ISO 18000-6 (see Sect. 2.3).

On the other hand, Kübler in his PhD thesis [22] defines α and β in the same definition as FRR and FAR respectively, and uses Chernoff Bounds and Piling-Up Lemma to propose a different hypothesis test for verifying LPN. Following

Theorem 2 and Lemma 7 of [22], and using $t = (1+\delta)n\tau$ we have:

$$\begin{aligned} FRR = \alpha = Pr[wt(As' + b) > t] \leq e^{-\frac{1}{2}\delta \ln(1+\delta)n\tau} \\ FAR = \beta = Pr[wt(As' + b) \leq t] \leq e^{-\frac{1}{2}\delta^2 n\tau} \end{aligned} \tag{5}$$

By setting $\alpha = \beta = e^{-k}$, Kübler's recommendations are for $n = 4k(\frac{1}{2} - \tau)^{-2}$ and $t = n * \tau + \sqrt{k * m}$. From a configuration standpoint, the Kübler's system is rigid as it fixes both false error rates to be below e^{-k}. Furthermore, relying solely on the key length k to determine the precision means that the FAR/FRR for such LPN systems cannot be controlled dynamically after deployment.

We argue that fixing both FAR and FRR before computing number of samples per cycle n may not be the optimal approach since there are inter-dependencies observed between them. Using n = 500 and δ on the x-axis, Fig. 2 shows plots using www.desmos.com for Eq. 4 on the left and Eq. 5 on the right. FRR in depicted blue and dotted-green, and FAR in depicted red and dotted-orange with noise rate $\tau = 0.125$ and 0.4 respectively.

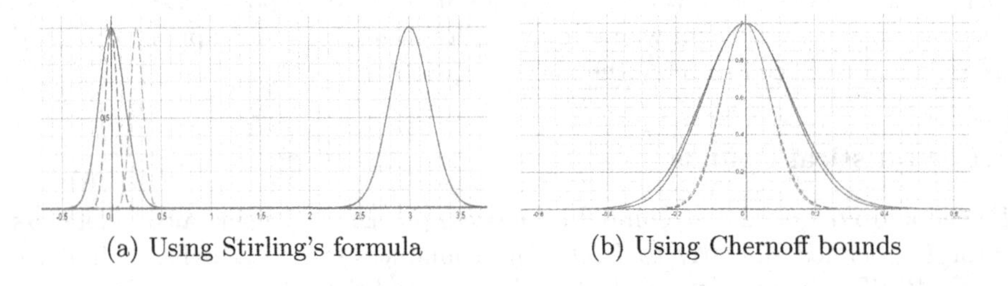

(a) Using Stirling's formula (b) Using Chernoff bounds

Fig. 2. FAR and FRR for n = 500 (Color figure online)

Referring to Fig. 2a, for $\tau = 0.125$, the cross-over range where FRR and FAR are both below 0.01 happens for $0.43 < \delta < 2.38$. However, for $\tau = 0.4$, the lowest cross-over range for FRR and FAR is 0.135 when $\delta = 0.124$. It will require $n > 900$ before the cross-over range falls below 0.01. While in Fig. 2b, both FRR and FAR have very similar values for the same noise rate τ. For $\tau = 0.125$, FRR and FAR are below 0.01 if $\delta > 0.42$. Increasing τ to 0.4 causes the curves to narrow, thus allowing δ to be at least 0.226 in order to keep FRR and FAR below 0.01. The effect of increasing sample size n from 500 to larger values causes the curves to narrow, allowing for an even smaller δ to keep FRR and FAR low. These bounds seem to suggest that FRR and FAR can be kept low if $0.5 < \delta < 1$, which we want to confirm in our Experiment 1 below.

3.2 Computational Simulations

Experiment 1. *We design a HB protocol experiment to explore how close the statistical bounds of FAR and FRR are to actual execution. The parameters used are as follows:*

- *Values of noise rate* τ *= {0.01, 0.05, 0.125, 0.25, 0.4, 0.45, 0.49}*
- *Values of sample size per cycle* n *= {50, 100, 200, 350, 500, 900.}*
- *Values of key length* k *= {100, 200, 300, 400, 500}*
- *Values of verification window* δ *= 0.0 to 1.0 in increments of 0.1.*

The FAR/FRR measurement algorithm is found in Algorithm 1 in Appendix A. It takes in four parameters—noise rate τ, number of samples n, key length k and verification window δ. The FAR/FRR measurement algorithm performs the HB protocol to generate A, e and compute b as per Fig. 1. We ran 1,000 simulations for each of the parameter values to measure FAR and FRR, and the results are plotted in Fig. 3 and Fig. 4 respectively. The y-axis represents the number of errors (either false-accept or false-reject) that happened out of 1,000 simulations which ideally should be zero. The x-axis represents the range of sample sizes tested for each of the noise rates indicated. The different δ values are depicted by family of colours where $\{colour = \delta\}$ = {maroon = 0.0, red = 0.1, orange = 0.2, yellow = 0.3, light-green = 0.4, green = 0.5, light-blue = 0.6, blue = 0.7, indigo = 0.8, purple = 0.9, grey = 1.0} respectively. The observations are:

- Key lengths k do not affect FAR/FRR.
- For noise rate $\tau \leq 0.125$, FAR can be kept low with practical sample sizes regardless of δ. For $\tau \geq 0.25$, far higher values of sample sizes with $\delta = 0$ is needed to keep FAR low.
- It is not possible to find low FAR for $\tau > 0.4$ for practical sample sizes.
- $\delta = 0$ cannot be used as $FRR \sim 50\%$ regardless of sample size.
- A combination of higher δ, higher noise rate τ and much larger sample sizes n is needed to keep FRR low.

3.3 Summary of Precision Results

With both statistical bounds and computational simulations analysed, we are now ready to recommend appropriate $[\tau, n, \delta]$ combinations with low FAR and FRR. The methodology used is:

Step 1: From Fig. 3, extract the combinations of $[\tau, n, \delta]$ with $FAR < 1$.
Step 2: From Fig. 4, extract the combinations of $[\tau, n, \delta]$ with $FRR < 1$.
Step 3: From the intersection of the two sets from Step 1 and Step 2, compute the statistical FAR and FRR using Eq. 4 and Eq. 5.
Step 4: Remove combinations whose FAR and FRR are above acceptable statistical bounds.

Assuming the acceptable statistical bounds for FAR and FRR are set at 2^{-40}, the final set of recommended $[\tau, n, \delta]$ combinations is found in Table 1.

Table 1. Verification window δ values for acceptable $[\tau, n]$ combinations.

τ	$n = 200$	350	500	900
0.05	–	0.9–1.0	0.8–1.0	0.6–1.0
0.125	0.8–0.9	0.6–1.0	0.5–1.0	0.4–1.0

4 Exploring Key Lengths

While we have established that key lengths k does not impact precision, it is directly correlated to the hardness (or difficulty) of key recovery of a LPN problem. This is a well-studied topic where related research has provided recommen-

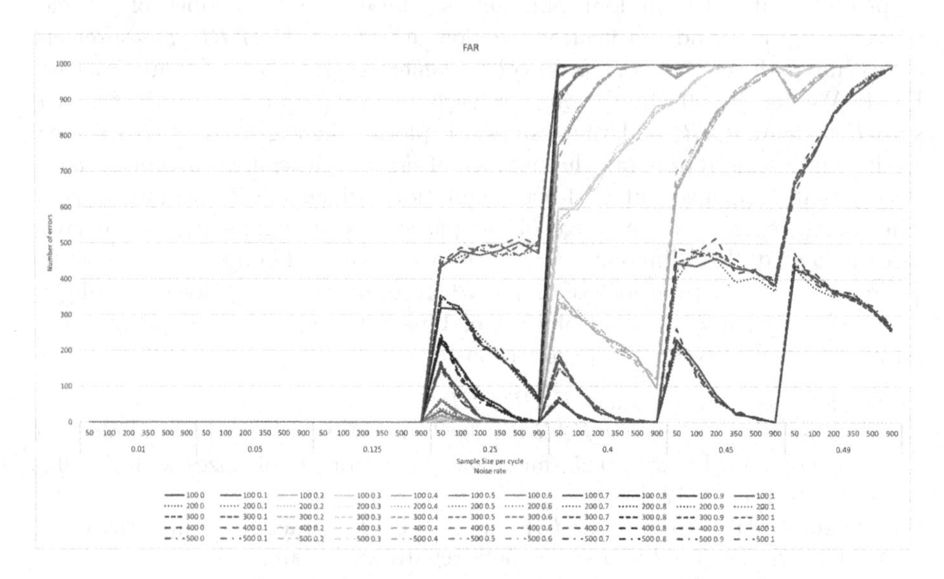

Fig. 3. (Experiment 1) computational simulation of FAR

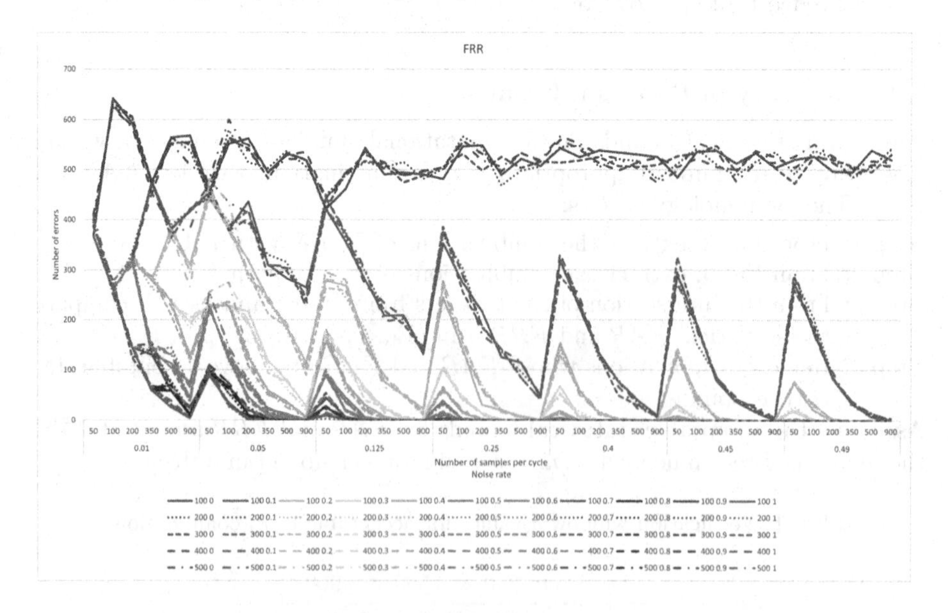

Fig. 4. (Experiment 1) computational simulation of FRR

dations into the choice of key length versus the intended security bit-strength. For 80-bit security, key length recommendations range from $k = 512, \tau = \frac{1}{8}$ by Levieil and Fonque [24], $k > 1090$ by Bogos et al. [6] and $k \geq 2048$ by Esser et al. [9]. Since these results are based on cryptanalysis techniques with access to large quantities of collected samples (A, b), we continue our research into how limiting the key lifespan can be used as a mitigation tool to maintain LPN security for small key lengths such as $k = 80$.

We take direction from the key length selection framework by Lenstra and Verheul [23] which uses the *Wassenaar Arrangement* to evaluate the appropriate LPN key length. The topics explored are the key length recommendation, the effectiveness of important known (i.e., published) attacks, guessing and incomplete attacks, and any expected cryptanalytic progress.

4.1 Key Length Recommendation

Since the HB protocol is symmetric-key based, the *Wassenaar Arrangement* asks for key-lengths not exceeding 64-bits for retail or mass-market sold products. This is less than the 80-bit size in our assumptions (equivalent to RSA-1024), and thus susceptible to brute-force attacks.

4.2 Effectiveness of Known Attacks

The noise rate τ, in addition to key length k, impacts the outcome of LPN cryptanalysis. When $\tau = 0$, the LPN problem becomes polynomial-time solvable using Gaussian Elimination when the number of samples $n \geq k$ and are linearly independent. However, when $\tau > 0$, the errors in b increases quadratically (by the Piling-up Lemma [26]) with each round of elimination, thus making the problem intractable in polynomial-time.

Blum et al. [5] describes the $\mathcal{BKW}$ algorithm that works similarly to Guassian Elimination, but reduces the cumulative errors by searching for and eliminating partial sequences instead of individual elements. The total number of samples needed is shown computationally to be $2^{k/\log \frac{k}{\tau}}$ by Esser et al. [9]. Levieil and Fouque [24] improved on $\mathcal{BKW}$ by applying the Walsh-Hadamard Transform (WHT) in the solving phase of their algorithm $\mathcal{LF}1$. This has an effect of increasing the time and memory complexity while reducing the total samples needed which is shown to be $\frac{4k \log 2}{(1-2\tau)^2}$ [28]. Guo et al. [14] introduced the use of covering code, subspace distinguishing [3] and WHT in a five-step $\mathcal{CC}$ algorithm which has an effect of reducing the query complexity to $\frac{4k \log 2}{\tau^2}$. The only algorithm we found that starts from the basis of a limited pool of samples is a hybrid-$\mathcal{BKW}$ by Belaïd et al. [2] which uses a birthday-paradox heuristic to select edge-case samples with assumptions before using WHT to solve a partial solution. In another approach, Esser et al. [9] proposes a Well-pooled Gaussian algorithm ($\mathcal{WPG}$) by analyzing the samples and pooling their commonalities to solve smaller subsets of the secret vector. The query complexity of $\mathcal{WPG}$ is $O(\frac{1}{(1-\tau)^k})^{\overline{\frac{1}{1+\log(\frac{1}{1-\tau})}}}$.

Table 2. Selected known total samples used for cryptanalysis algorithms.

Algorithm	Key length k, noise τ				
	$64, 0.25$	$80, 0.1$	$135, 0.25$	$200, 0.125$	$512, 0.125$
$\mathcal{BKW}$ [5]	–	$2^{27.3}$ [6][a]	–	–	–
$\mathcal{LF}1$ [24]	$2^{20.7}$ [6][a]	$2^{17.98}$ [6][a]	–	–	$2^{75.2}$ [14]
$\mathcal{CC}$ [14]	–	–	2^{33} [9]	$2^{31.2}$ [28]	2^{65} [14]
$\mathcal{WPG}$ [9]	$2^{32.56}$ [6][a]	–	2^{33} [9]	–	–

[a]Based on sparse secret.

We tabulate selected known total samples used from related research discussed above in Table 2.

The effectiveness of the attacks arise from the availability of large-enough samples which allow the adversary to perform dimension reduction to retrieve a subset of the key. We therefore replicate the first reduction step from $\mathcal{BKW}$ as the basis for checking the boundaries of limited sample sizes.

Experiment 2. *We perform an experiment to find blocks of size* $\frac{k}{\log \frac{k}{\tau}}$ *(representing a good probability of cryptanalysis success) for different sample sizes using codes provided by Esser et al.* [9] *hosted at https://github.com/Memphisd/LPN-decoded. We vary the following parameters:*

- *Values of noise rate* $\tau = \{0.05,\ 0.125,\ 0.25, 0.4\}$
- *Values of total sample size* $n = \{1k,\ 5k,\ 10K,\ 50K,\ 100K,\ 500K,\ 1M,\ 5M,\ 10M,\ 50M\}$ *where* $K =$ *thousand,* $M =$ *million.*
- *Values of key length* $k = 16$ *to* 52 *in steps of* 4.

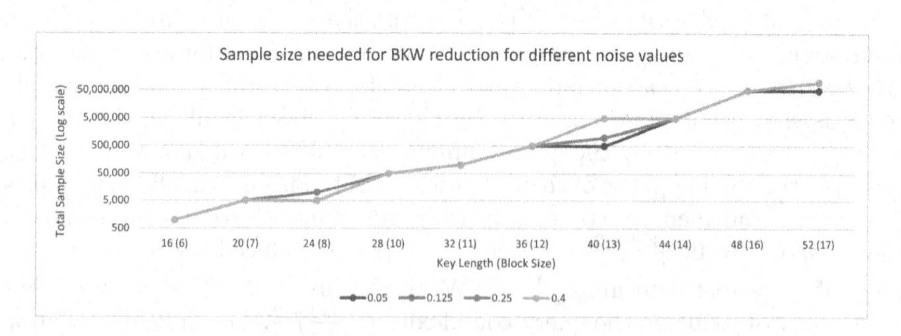

Fig. 5. (Experiment 2) BKW reduction for limited sample sizes

Experiment 2 is repeated for each key length & noise rate, starting with the lowest total sample size. The total sample size is then increased if no vector survives the reduction after ten independent attempts. Figure 5 shows the graph for the minimum total sample size needed for at least one vector to survive the reduction for each of the key length values. The required total sample size grows exponentially with key length and exceeds 50 million after $k > 52$ while the noise rate τ does not materially affect the total number of samples needed to complete the first reduction step.

4.3 Effectiveness of Guessing

We envision that the modern adversary has more tools in the arsenal compared to when Lenstra and Verheul wrote the article [23] some 20 years ago. Beyond random-guessing which the adversary has a $\frac{1}{2^k}$ chance in guessing the right secret vector, we want to know if an adversary can make use of machine-learning to predictably guess the secret vector. Kübler [21] provided an experimental walk-through using Extremely Randomized Trees ($\mathcal{ERT}$) [11], an ensemble learning technique, to successfully predict the value of s for a LPN implementation with parameters $k = 16, \tau = 0.125, n = 100000$. Informally, $\mathcal{ERT}$ works by randomly splitting samples into independent tree structures based on a subset (instead of all) of the data attributes. Such a method has a possible effect of ignoring the noise impact on some of the computed responses in vector b and allow the predictor to guess a correct secret vector s.

Experiment 3. *We design a HB protocol experiment (see Algorithm 2 in Appendix A) around Kübler's $\mathcal{ERT}$ implementation with the following parameters:*

- *Values of noise rate τ = {0.05, 0.125, 0.25, 0.4}*
- *Values of total sample size n = {10K, 50K, 100K, 500K, 1M, 5M, 10M, 50M} where K = thousand, M = million.*
- *Values of key length k = {16, 20, 24, 28, 32}*

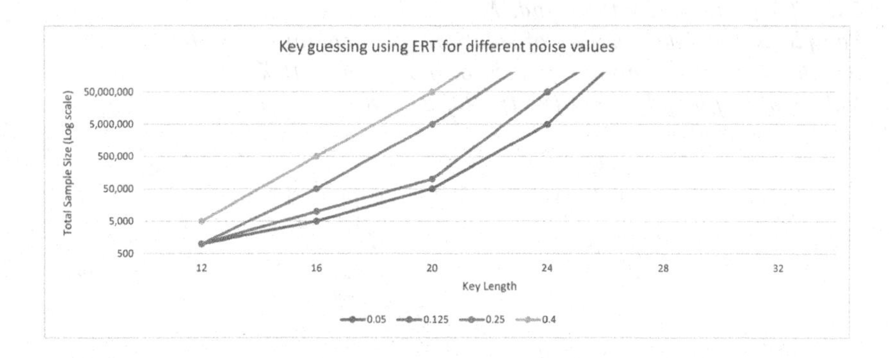

Fig. 6. (Experiment 3) Key guessing using $\mathcal{ERT}$ machine-learning

Experiment 3 is repeated for each key & noise rate, starting with the lowest total sample size. The total sample size is then increased if $\mathcal{ERT}$ fails to guess the correct key after 100 attempts. We chose the ceiling of 50 million samples as it represents the total of 100,000 cycles with 500 samples in each cycle as per the assumptions in Sect. 2.3. Figure 6 shows the total number of samples needed for $\mathcal{ERT}$ to correctly guess the correct key. When $k = 16$, only 10,000 samples (or $\frac{10000}{500} = 20$ cycles) will allow an adversary to guess the secret vector. However, the difficultly increases exponentially with each small increase in key length. As expected, $\mathcal{ERT}$ performs better for lower noise rates. Hence, if a secret vector s is not used over to compute over 50,000,000 samples, then key lengths of $k > 28$ are safe from machine-learning guessing.

4.4 Effectiveness of Incomplete Attacks

We use Genetic Algorithm ($\mathcal{GA}$) [16], a meta-heuristic method of improving a partial solution through multiple cycles of evolution, to answer the question if the noise parameter τ can change LPN's susceptibility to incomplete attacks where an adversary is able to obtain a partially-correct secret vector s'.

Intuitively, the existence of partial solutions would mean the adversary has some form of indication about the proximity of s' to s. This can be mapped into $\mathcal{GA}$'s fitness function and allow an adversary to arrive at the correct secret vector faster than a brute-force search. The fitness function (see Algorithm 3 in Appendix A) is designed to mimic the $Verify()$ function in the HB protocol (see Fig. 1) which returns 1 if the response vector b is within the verification window and the reciprocal of the difference otherwise.

Experiment 4. *We design a HB-protocol experiment using $\mathcal{GA}$ to attempt to recover the LPN secret vector. We build on the python library provided by pygad [10] and use the following parameters with Algorithm 3 as the fitness function:*

- *Values of noise rate τ = {0.05, 0.125, 0.25, 0.4}*
- *Number of generations = {10, 50, 100, 500, 1K, 10K, 50K, 100K, 500K, 1M, 5M, 10M} where K = thousand, M = million.*
- *Sample size = 500, number of solutions per generation = 10*
- *Number of mating parents = 4, % of mutation = 10%*
- *Values of key length k = {12, 16, 20, 24, 28}*

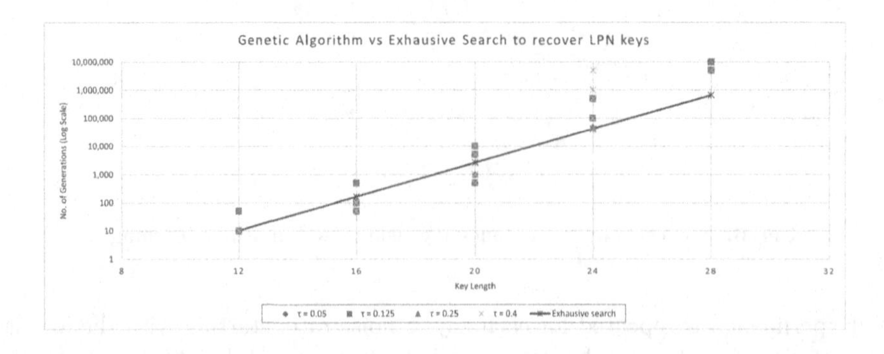

Fig. 7. (Experiment 4) Scatter graph results of key recovery using $\mathcal{GA}$

Experiment 4 is run five times for each key length & noise rate, starting with the lowest generation. The number of generations is then increased if $\mathcal{GA}$ fails to find the correct key after 20 attempts. Figure 7 shows the results of the experiment plotted in a scatter-graph where the y-axis shows the minimum number of generations needed for each of the five runs for each of the key length and noise values, while the x-axis is the key length used. As a reference, we include a line function representing the statistically-computed number of generations needed if brute-force, instead of $\mathcal{GA}$, is used to guess the secret vector. Our experiment confirms that LPN is not susceptible to incomplete key attacks. In fact, $\mathcal{GA}$ increasingly performs poorly as the size of key increases which we suspect is due to repetitions in the mutated solutions during $\mathcal{GA}$ evolution.

4.5 Cryptanalytic Progress

We first examine the possibility of an adversary carrying out a brute-force attack to recover the secret vector. Taking a conservative approach where each LPN sample takes one clock cycle to compute and the adversary has access to a super-computing cluster with 10,000 nodes running at 3GHz each, we give an estimated breakdown of time needed T to brute-force ($\mathcal{BF}$) a LPN setup with sample size of $n = 500$ per cycle in Eq. 6.

$$\begin{aligned} k = 64 : T &= \frac{2^{63} * 500}{10000 * 3000000000 * 3600 * 24 * 365} = 4.87 \text{ years} \\ k = 80 : T &= \frac{2^{79} * 500}{10000 * 3000000000 * 3600 * 24 * 365 * 1000} = 319 \text{ millennia} \end{aligned} \tag{6}$$

Expected Developments. Since we are preparing to use LPN in the post-quantum era, we cannot discount the possibility of attacks using quantum computers. The LPN problem is closely related to decoding random linear codes and solving worst-case hardness nearest-codeword problem can be reduced to LPN [7]. Esser et al. [9] propose the use of quantum-based Grover algorithm within their ($\mathcal{WPG}$) algorithm to improve the execution time and query complexity by a factor of $\frac{2}{3}$. Grilo et al. [13] application of quantum-based Bernstein-Vazirani algorithm on a Q2 LPN oracle could yield results with exponentially less queries, but the Q2 oracle counts as an active attack and falls outside the scope of our assumptions.

4.6 Summary of Key Length Results

We summarize in Table 3 the various key length evaluation results from this section, and apply a 50% margin to take into account any improvements, especially in the area of quantum computation.

Table 3. Recommended settings for smaller key lengths.

Algorithm	$k = 64$	$k = 80$	Remarks
$\mathcal{BKW}$[a]	500,000 samples	50,000,000 samples	Experiment 2
$\mathcal{ERT}$[a]	>50,000,000 samples	>50,000,000 samples	Experiment 3
$\mathcal{GA}$[b]	$=\mathcal{BF}$	$=\mathcal{BF}$	Experiment 4
$\mathcal{BF}$	<2.4 years	>10 years	Eq. 6

[a]These results are based on actual cryptanalysis on full sample sizes.
[b]These results are extrapolated from lower key sizes.

For $k = 64$, we recommend that it be used only for systems with short usage period of less than one year and for < 1000 authentication cycles. For $k = 80$, it can be used for five years but limited to a maximum of $100,000$ cycles (i.e. $50,000,000$ samples) before the secret vector needs to be changed.

5 Conclusion

We have used a series of experiments to calibrate LPN authentication for low-resource devices using the HB protocol. Such devices can only handle small sample sizes per authentication cycle and small key lengths due to bandwidth and computational restrictions. By using the authentication concept of precision, we have identified appropriate noise rate and verification window values for sample size per cycle $\leq$500. To mitigate against key recovery attacks, we have proposed to limit the lifetime of a secret vector to less than 100,000 cycles (or a total of 50,000,000 samples) for key length $k = 80$.

Our next steps will be to apply the recommended values on derivative HB protocols such as HB$^+$, HB^{++}, HB#, etc. and provide a usability study on a real-world implementation of LPN authentication using RFID devices.

Acknowledgement. This project is supported by the Ministry of Education, Singapore, under its MOE AcRF Tier 2 grant (MOE2018-T2-1-111). The computational work for this article was partially performed on resources of the National Supercomputing Centre, Singapore (https://www.nscc.sg).

The work is also supported by A*STAR under its RIE2020 Advanced Manufacturing and Engineering (AME) Industry Alignment Fund - Pre Positioning (IAF-PP) Award A19D6a0053. Any opinions, findings and conclusions or recommendations expressed in this material are those of the author(s) and do not reflect the views of A*STAR.

A Algorithm Pseudocode

Algorithm 1: *FAR/FRR* measurement algorithm.

```
1  begin
2      τ ← noise rate; n ← number of samples per cycle;
3      k ← key length; δ ← verification window;
4      Generate matrix A = Random(n * k, 0.5);
5      Generate secret vector s = Random(k, 0.5);
6      Generate noise vector e = Random(n, τ);
7      Compute b = A · s ⊕ e;
8      Create s' = s with random 2 bits changed;
9      if ||A · s' ⊕ b||_1 < (1 + δ)τn then
10         FAR_error = TRUE;
11     end
12     if ||A · s ⊕ b||_1 > (1 + δ)τn then
13         FRR_error = TRUE;
14     end
15 end
```

We assume the existence of a function $Random(n, p)$ that returns a binary matrix/vector of size n where each element has a probability p to be 1. The secret key s is randomly generated.

Algorithm 2: $\mathcal{ERT}$ algorithm for LPN Key Recovery

```
1  begin
2      τ ← noise rate; n ← number of samples; k ← key length;;
3      Generate matrix A = Random(n * k, 0.5);
4      Generate secret vector s = Random(k, 0.5);
5      Generate noise vector e = Random(n, τ);
6      Compute b = A · s ⊕ e;
7      Create ERT = ExtRaTree classifier with n/k estimators ;
8      Call ERT.fit(A, b);
9      Create I = Identity Matrix of size k by k;
10     s' = ERT.predict(I);
11     if s' == s then
12         Key_Recovery = TRUE;
13     end
14 end
```

Algorithm 3: Fitness function for $\mathcal{GA}$

```
1  begin
2      τ ← noise rate; n ← number of samples; δ ← verification window;
3      s' ← vector to be tested; A ← challenge matrix; b ← response vector;
4      Compute w = ||A · s' ⊕ b||_1;
5      if w ≤ (1 + δ)τn then
6          return 1
7      else
8          return 1/(w−(1+δ)τn)
9      end
10 end
```

We performed a sub-experiment to measure the efficacy of the fitness function by varying the number of erroneous bits in s' and noise rate to find any advantage that adversaries may be able to uncover.

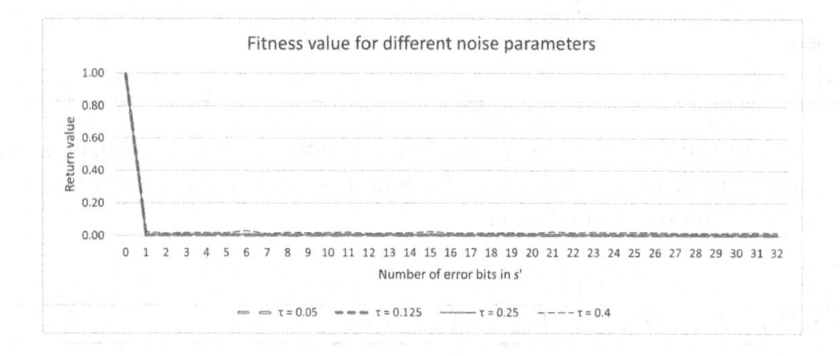

Fig. 8. Return values for simulated fitness function for $k = 64, \delta = 0.5$

Figure 8 shows the graph which plots the return values of the fitness function for error bits in s' from 0 to $\frac{k}{2}$ in increments of 1 and for noise rate $\tau = \{0.05, 0.125, 0.25, 0.4\}$. For clarity purposes, we have fixed $k = 64, \delta = 0.5, n = 500$. It clearly shows that the fitness function is unable to tell the difference in the number of error bits for partial solutions since the fitness values become close to zero once there is at least one error bit in s'.

References

1. 2013, I...: Information technology-radio frequency identification for item management-part 6: Parameters for air interface communications at 860 MHz to 960 MHz general (2013)
2. Belaïd, S., Coron, J.-S., Fouque, P.-A., Gérard, B., Kammerer, J.-G., Prouff, E.: Improved side-channel analysis of finite-field multiplication. In: Güneysu, T., Handschuh, H. (eds.) CHES 2015. LNCS, vol. 9293, pp. 395–415. Springer, Heidelberg (2015). https://doi.org/10.1007/978-3-662-48324-4_20

3. Bernstein, D.J., Lange, T.: Never trust a bunny. In: Hoepman, J.-H., Verbauwhede, I. (eds.) RFIDSec 2012. LNCS, vol. 7739, pp. 137–148. Springer, Heidelberg (2013). https://doi.org/10.1007/978-3-642-36140-1_10
4. Blum, A., Furst, M., Kearns, M., Lipton, R.J.: Cryptographic primitives based on hard learning problems. In: Stinson, D.R. (ed.) CRYPTO 1993. LNCS, vol. 773, pp. 278–291. Springer, Heidelberg (1994). https://doi.org/10.1007/3-540-48329-2_24
5. Blum, A., Kalai, A., Wasserman, H.: Noise-tolerant learning, the parity problem, and the statistical query model. J. ACM (JACM) **50**(4), 506–519 (2003)
6. Bogos, S., Tramer, F., Vaudenay, S.: On solving LPN using BKW and variants. Cryptogr. Commun. **8**(3), 331–369 (2016)
7. Brakerski, Z., Lyubashevsky, V., Vaikuntanathan, V., Wichs, D.: Worst-case hardness for LPN and cryptographic hashing via code smoothing. In: Ishai, Y., Rijmen, V. (eds.) EUROCRYPT 2019. LNCS, vol. 11478, pp. 619–635. Springer, Cham (2019). https://doi.org/10.1007/978-3-030-17659-4_21
8. Bringer, J., Chabanne, H., Dottax, E.: HB^{++}: a lightweight authentication protocol secure against some attacks. In: Second international Workshop on Security, Privacy and Trust in Pervasive and Ubiquitous Computing (SecPerU 2006), pp. 28–33. IEEE (2006)
9. Esser, A., Kübler, R., May, A.: LPN decoded. In: Katz, J., Shacham, H. (eds.) CRYPTO 2017. LNCS, vol. 10402, pp. 486–514. Springer, Cham (2017). https://doi.org/10.1007/978-3-319-63715-0_17
10. Gad, A.F.: PyGAD: An Intuitive Genetic Algorithm Python Library (2021)
11. Geurts, P., Ernst, D., Wehenkel, L.: Extremely randomized trees. Mach. Learn. **63**(1), 3–42 (2006)
12. Gilbert, H., Robshaw, M.J.B., Seurin, Y.: HB#: increasing the security and efficiency of HB^{+}. In: Smart, N. (ed.) EUROCRYPT 2008. LNCS, vol. 4965, pp. 361–378. Springer, Heidelberg (2008). https://doi.org/10.1007/978-3-540-78967-3_21
13. Grilo, A.B., Kerenidis, I., Zijlstra, T.: Learning-with-errors problem is easy with quantum samples. Phys. Rev. A **99**(3), 032314 (2019)
14. Guo, Q., Johansson, T., Löndahl, C.: Solving LPN using covering codes. In: Sarkar, P., Iwata, T. (eds.) ASIACRYPT 2014. LNCS, vol. 8873, pp. 1–20. Springer, Heidelberg (2014). https://doi.org/10.1007/978-3-662-45611-8_1
15. Heyse, S., Kiltz, E., Lyubashevsky, V., Paar, C., Pietrzak, K.: Lapin: an efficient authentication protocol based on ring-LPN. In: Canteaut, A. (ed.) FSE 2012. LNCS, vol. 7549, pp. 346–365. Springer, Heidelberg (2012). https://doi.org/10.1007/978-3-642-34047-5_20
16. Holland, J.H.: Genetic algorithms. Sci. Am. **267**(1), 66–73 (1992)
17. Hopper, N.J., Blum, M.: Secure human identification protocols. In: Boyd, C. (ed.) ASIACRYPT 2001. LNCS, vol. 2248, pp. 52–66. Springer, Heidelberg (2001). https://doi.org/10.1007/3-540-45682-1_4
18. Juels, A., Weis, S.A.: Authenticating pervasive devices with human protocols. In: Shoup, V. (ed.) CRYPTO 2005. LNCS, vol. 3621, pp. 293–308. Springer, Heidelberg (2005). https://doi.org/10.1007/11535218_18
19. Kearns, M.: Efficient noise-tolerant learning from statistical queries. J. ACM (JACM) **45**(6), 983–1006 (1998)
20. Kiltz, E., Pietrzak, K., Venturi, D., Cash, D., Jain, A.: Efficient authentication from hard learning problems. J. Cryptol. **30**(4), 1238–1275 (2017)
21. Kübler, R.: Where Machine Learning meets Cryptography (2020). https://towardsdatascience.com/where-machine-learning-meets-cryptography-b4a23ef54c9e. Accessed Mar 2022

22. Kübler, R.J.: Time-memory trade-offs for the learning parity with noise problem. Ph.D. thesis, Ruhr University Bochum, Germany (2018)
23. Lenstra, A.K., Verheul, E.R.: Selecting cryptographic key sizes. J. Cryptol. **14**(4), 255–293 (2001)
24. Levieil, É., Fouque, P.-A.: An improved LPN algorithm. In: De Prisco, R., Yung, M. (eds.) SCN 2006. LNCS, vol. 4116, pp. 348–359. Springer, Heidelberg (2006). https://doi.org/10.1007/11832072_24
25. Lyubashevsky, V., Masny, D.: Man-in-the-middle secure authentication schemes from LPN and weak PRFs. In: Canetti, R., Garay, J.A. (eds.) CRYPTO 2013. LNCS, vol. 8043, pp. 308–325. Springer, Heidelberg (2013). https://doi.org/10.1007/978-3-642-40084-1_18
26. Matsui, M.: Linear cryptanalysis method for DES cipher. In: Helleseth, T. (ed.) EUROCRYPT 1993. LNCS, vol. 765, pp. 386–397. Springer, Heidelberg (1994). https://doi.org/10.1007/3-540-48285-7_33
27. NIST: Post-Quantum Cryptography: Round 3 Submissions (2019). https://csrc.nist.gov/projects/post-quantum-cryptography/round-3-submissions. Accessed Mar 2022
28. Wiggers, T., Samardjiska, S.: Practically solving LPN. In: 2021 IEEE International Symposium on Information Theory (ISIT), pp. 2399–2404. IEEE (2021)

New Results of Breaking the CLS Scheme from ACM-CCS 2014

Jing Gao[1,2], Jun Xu[1,2](✉), Tianyu Wang[1,2], and Lei Hu[1,2]

[1] State Key Laboratory of Information Security, Institute of Information Engineering, Chinese Academy of Sciences, Beijing 100093, China
xujun@iie.ac.cn

[2] School of Cyber Security, University of Chinese Academy of Sciences, Beijing 100093, China

Abstract. At ACM-CCS 2014, Cheon, Lee and Seo introduced a particularly fast additively homomorphic encryption scheme (CLS scheme) based on a new number theoretic assumption, the co-Approximate Common Divisor (co-ACD) assumption. However, at Crypto 2015, Fouque et al. presented several lattice-based attacks that effectively devastated this scheme. They proved that a few known plaintexts are sufficient to break both the symmetric-key and the public-key variants, and they gave a heuristic lattice method for solving the search co-ACD problem.

In this paper, we mainly improve in terms of the number of samples, and propose a new key-retrieval attack. We first give an effective attack by Coppersmith's method to break the co-ACD problem with $N = p_1 \cdots p_n$ is known. If n is within a certain range, our work is theoretically valid for a wider range of parameters. When $n = 2$, we can successfully solve it with only two samples, that is the smallest number of needed samples to the best of our knowledge. A known plaintext attack on the CLS scheme can be simply converted to solving the co-ACD problem with a known N, again requiring fewer samples than before to retrieve the private key. Finally, we show a ciphertext-only attack with a hybrid approach of direct lattice and Coppersmith's method that can recover the key with a smaller number of ciphertexts and without any restriction on the plaintext size, but N is needed. All of our attacks are heuristic, but we have experimentally verified that these attacks work efficiently for the parameters proposed in the CLS scheme, which can be broken in seconds by experiments.

Keywords: co-ACD problem · Lattice · LLL algorithm · Coppersmith's method

The work of this paper was supported in part by the National Natural Science Foundation of China (No.61732021).

C. Alcaraz et al. (Eds.): ICICS 2022, LNCS 13407, pp. 37–54, 2022.
https://doi.org/10.1007/978-3-031-15777-6_3

1 Introduction

The approximate common divisor problem (ACD) is firstly presented by Howgrave-Graham [21], and its variants are utilized to prove the security of the proposed homomorphic encryption schemes over the integers [4,7,12,13,16]. There are many analyses of the ACD problem, such as [3,6,14,18].

At ACM-CCS 2014, Cheon et al. [5] first presented the co-approximate common divisor problem (co-ACD) and a new additive homomorphic encryption scheme (CLS scheme), which includes both symmetric key version and public key version. This scheme is based on the inverse function of Chinese remainder theorem under a co-ACD assumption that is divided into decisional co-ACD assumption and search co-ACD assumption. They claimed that, due to the simplicity of the operations in encryption and decryption, their scheme was the most efficient additive homomorphic encryption scheme compared to those of the same type. In this scheme, a plaintext $M \in \mathbb{Z}_Q$ (Q is the size of the plaintext space) is encrypted as $(C_1, C_2) = ((M + eQ) \bmod p_1, (M + eQ) \bmod p_2)$, where the prime numbers p_1 and p_2 are large enough to form the secret key.

In the latter part of CLS, they also took into account the hardness of their assumptions and presented a few attacks to validate these assumptions. They utilized Chen-Nguyen's attack [2], orthogonal lattice [31,32] and Coppersmith's method [6,8–10] to show that these attacks cannot break CLS scheme with their parameters.

At Crypto 2015, Fouque et al. [17] presented three attacks that completely and devastatingly broke the CLS scheme. The first attack is a known-plaintext attack. If only one plaintext is unknown and a few other plaintexts are all known, then this attack can retrieve the unknown plaintext. It breaks both the symmetric key and public key versions of CLS, as well as the underlying decisional co-ACD assumption with an orthogonal lattice attack [31,32]. The second attack is a ciphertext-only attack on the symmetric key CLS scheme. If the plaintexts are small, then even if any plaintext is unknown, the attack can also retrieve the plaintexts by doubly orthogonal lattice technique [30]. The third attack combines direct lattice and Coppersmith's method together to break the public key scheme and solve the search co-ACD assumption by factoring the known modulus N and revealing the entire private key p_i. In these attacks, only the first one is rigorous, while the other two are heuristic, but they are all effective in practice.

Our Contributions. In this paper, we mainly attack the CLS scheme and its underlying co-ACD problem by Coppersmith's method, which reduces the number of required samples and improves on some restrictions compared to the previous attacks.

First, we describe how to solve the search co-ACD problem in Sect. 3. That is, the factorization of modulo N can be obtained with a small number of samples. We construct the equations about some unknown noise e_j and use Coppersmith's method to solve small roots to find e_j. Then we can obtain the factors of N by a simple calculation of the greatest common divisor. Compared with previous work, our attack only requires two samples for $N = p_1p_2$, and when $N = p_1 \cdots p_n$, we can theoretically complete the attack for larger ρ if n is within a certain range. For the known plaintext attack, if we substitute the known messages into the equation, the attack is converted to solving the search co-ACD problem.

Next, we present a ciphertext-only attack on the CLS scheme in Sect. 4. We propose a method combining the direct lattice and Coppersmith's method [17] to attack the case that only the ciphertexts and N are known without knowing any corresponding plaintexts. The main idea is to construct a relevant matrix to get the approximate short target vector, and then further get the exact target vector by Coppersmith's method. This attack also needs a smaller number of samples than the previous work, and there are no other size requirements for the plaintexts.

Our attack algorithm is a heuristic polynomial time algorithm. We have experimentally verified that the co-ACD problem and the CLS scheme can be broken with a smaller number of samples by recovering the secret key when N is known and $n = 2$. Although the CLS scheme has been completely broken [17], we present attacks that can be done with a smaller number of samples, and recover the key with any sized plaintext.

In this paper, we mainly use Coppersmith's method, which was originally proposed by Coppersmith in 1996 to solve univariate modulo polynomial equations [9] and bivariate integer polynomial equations based on lattice basis reduction [8] and was summarized in 1997 [10]. After years of continuous improvement [1,20,22,26], Coppersmith's method is now widely used in cryptanalysis, such as RSA [19,24,28,33], CRT-RSA [11,15,23,29] and hidden number problem [34].

Organization. The paper is organized as follows. In Sect. 2, we introduce preliminary information related to basic knowledge. Section 3 describes the detailed attack to break the search co-ACD assumption and retrieve p_i. We give only ciphertext attack in Sect. 4. Section 5 is a conclusion.

2 Preliminaries

2.1 Notation

For $x \in \mathbb{R}$, rounding down the number x is denoted by $\lfloor x \rfloor$. For an integer n, $\mathbb{Z}_n$ is the ring of integers modulo n. We use the letter $\vec{v}$ to denote the vector $(v_1, \cdots, v_n) \in \mathbb{Z}^n$. The ℓ_2-norm of vector $\vec{v}$ is denoted by $\|\vec{v}\|$, and the ℓ_2-norm of the polynomial coefficient vector is denoted by $\|f(x_1, \cdots, x_n)\|$. In addition, $a \xleftarrow{\$} A$ denotes uniformly sampling an element a from a finite set A. When $\mathcal{A}$ is a distribution, $a \leftarrow \mathcal{A}$ means to sample a according to $\mathcal{A}$, and $\vec{v} \leftarrow \mathcal{A}^n$ means that in the vector $\vec{v}$ each element v_i is sampled from the distribution $\mathcal{A}$.

2.2 co-ACD Problem

Definition 1 (co-ACD problem [17]**).** *Let $n, Q, \eta, \rho \geq 1$ and denote π the uniform distribution over the η-bit prime integers. The co-ACD distribution for* *a given* $\vec{p} = (p_1, \cdots, p_n) \in \mathbb{Z}^n$ *is the set of tuples* ($eQ \bmod p_1, \cdots, eQ \bmod p_n$) *where* $e \xleftarrow{\$} (-2^\rho, 2^\rho) \cap \mathbb{Z}$.

The search co-ACD problem is: for a vector $\vec{p} \leftarrow \pi^n$ *and given arbitrarily many samples from the co-ACD distribution for* $\vec{p}$, *to compute* $\vec{p}$.

The decisional co-ACD problem is: for some fixed vector $\vec{p} \leftarrow \pi^n$ *and given arbitrarily many samples from* $\mathbb{Z}_{p_1} \times \cdots \times \mathbb{Z}_{p_n}$, *to distinguish whether the samples are distributed uniformly or whether they are distributed as the co-ACD distribution for* $\vec{p}$.

The original paper [5] assumed that solving the decisional co-ACD problem was hard, i.e., it is not solvable in polynomial time, and proposed an efficient additive homomorphic encryption scheme based on this assumption.

2.3 CLS Additive Homomorphic Encryption Scheme

The CLS scheme encrypts a message by adding an error and performing modular reductions with two hidden primes respectively. Then the ciphertext is decrypted by using CRT and removing the error using a modular reduction. The parameters used in the scheme show that η is the bit-length of p_i's, ρ is the bit-length of the noise e and Q is the size of the plaintext space $\mathbb{Z}_Q$ in Table 1. The specific symmetry-key scheme is as follows.

Key Generation: Generate two random distinct prime integers p_1, p_2 of η bits and a positive integer Q for the plaintext space which satisfies $\gcd(Q, p_i)$=1. Set $N = p_1 p_2$. Output the private key sk= $\{p_1, p_2\}$.

Encryption: For any plaintext $M \in \mathbb{Z}_Q$, the sender generates a random noise $e \leftarrow (2^{-\rho}, 2^\rho) \cap \mathbb{Z}$, and outputs the ciphertext vector $\vec{C} = (C_1, C_2) = (M + eQ \bmod p_1, M + eQ \bmod p_2)$.

Decryption: The receiver decomposes $\vec{C}$ into (C_1, C_2) and calculates $e' = C_1\overline{p}_1 + C_2\overline{p}_2 \bmod N$, where $\overline{p}_1 = p_2(p_2^{-1} \bmod p_1)$ and $\overline{p}_2 = p_1(p_1^{-1} \bmod p_2)$ are the CRT coefficients. Then, outputs $e' \bmod Q = M$.

Table 1. Parameters in the CLS scheme for $\lambda = 128$ bits of security

Parameters	λ	η	ρ	$\log Q$
Set-I	128	1536	1792	256
Set-II	128	2194	2450	256
Set-III	128	2706	2962	256

2.4 Lattice

A lattice $\mathcal{L}$ is a discrete subgroup of $\mathbb{R}^m$. Given n linearly independent vectors $\vec{b}_1, \vec{b}_2, \cdots, \vec{b}_n \in \mathbb{R}^m$, the lattice generated by them is defined as

$$\mathcal{L}(\vec{b}_1, \vec{b}_2, \cdots, \vec{b}_n) = \{\sum x_i \vec{b}_i | x_i \in \mathbb{Z}\}.$$

The set $\{\vec{b}_1, \vec{b}_2, \cdots, \vec{b}_n\}$ is referred to as a basis of the lattice $\mathcal{L}$. Equivalently, define B as the $n \times m$ basis matrix whose rows are the basis vectors $\vec{b}_1, \vec{b}_2, \cdots, \vec{b}_n$ which can be written as $B = [\vec{b}_1^T, \cdots, \vec{b}_n^T]^T$. Then the lattice generated by B is

$$\mathcal{L}(B) = \{\vec{x}B | \vec{x} \in \mathbb{Z}^n\}.$$

The dimension and determinant of $\mathcal{L}$ when $n < m$ are respectively

$$\dim \mathcal{L} = n, \det \mathcal{L} = \sqrt{\det BB^T}.$$

When $n = m$, the lattice is called full rank and $\det \mathcal{L} = |\det B|$. λ_1 denotes the length of the shortest nonzero vector in the lattice, and λ_i is the i-th successive minimum. The celebrated LLL lattice reduction algorithm [25] can output a reduced basis satisfying the following property (please see e.g. [27] for proof).

Lemma 1 (LLL). *Let $\mathcal{L}$ be a n-dimensional lattice. Within polynomial time, the LLL algorithm outputs reduced basis vectors $\vec{v}_1, \ldots, \vec{v}_n$ that satisfy*

$$\|\vec{v}_1\| \le \|\vec{v}_2\| \le \cdots \le \|\vec{v}_i\| \le 2^{\frac{n(n-1)}{4(n+1-i)}} (\det \mathcal{L})^{\frac{1}{n+1-i}}, 1 \le i \le n.$$

In practice, LLL algorithm tends to output the vectors whose norms are much smaller than theoretically predicted. The Gaussian heuristic gave the approximate norm of the shortest vector in $\mathcal{L}$.

Assumption 1 (Gaussian heuristic). *Let $\mathcal{L}$ be a random n-dimensional lattice of $\mathbb{Z}^m$. Then, with overwhelming probability, the length of the shortest non-zero vectors in $\mathcal{L}$ is asymptotically close to:*

$$\text{GH}(\mathcal{L}) = \sqrt{\frac{n}{2\pi e}} \det(\mathcal{L})^{\frac{1}{n}}.$$

2.5 Coppersmith's Method

Coppersmith's method can be used to solve modular polynomial equations with small roots. The first step is to construct more modular polynomials with the same desired roots. The second step is to construct a lattice with coefficients of these polynomials. Finally, utilize the lattice reduction algorithms such as the LLL algorithm to obtain integer polynomials over $\mathbb{Z}$ with the desired roots. In these processes, the following lemma, reformulated by Howgrave-Graham [20], is needed.

Lemma 2 (Howgrave-Graham). *Let $f(x_1,\ldots,x_m)$ be an integer polynomial that consists of at most ω monomials. Let t be a positive integer and the X_i be the upper bound of $|x_i|$ for $i=1,\cdots,m$. Suppose that*

1. $f(x_1,\ldots,x_m)=0 \pmod{p^t}$,
2. $\|f(x_1X_1,\ldots,x_mX_m)\| < \frac{p^t}{\sqrt{\omega}}$,

then $f(x_1,\ldots,x_m)=0$ holds over $\mathbb{Z}$.

Note that $\|f(x_1X_1,\ldots,x_mX_m)\|$ in Lemma 2 also stands for the ℓ_2-norm of the corresponding row vector in the involved lattice. Therefore, according to Lemma 1 and Lemma 2, in order to obtain at least m polynomials with the common desired root $(x_1,\ldots,x_m)$, there is the following condition:

$$2^{\frac{\omega(\omega-1)}{4(\omega+1-m)}} \cdot (\det L)^{\frac{1}{\omega+1-m}} < \frac{p^t}{\sqrt{\omega}} \text{ where } \omega = \dim L. \tag{1}$$

Finally, in order to find the desired root $(x_1,\ldots,x_m)$ by utilizing the resultant method or the Gröbner basis technique, we expect that the obtained integer polynomials are algebraically independent. However, we cannot prove such an argument. Thus, the following assumption is necessary and often used by Coppersmith-type cryptanalysis [22].

Assumption 2. *Let $g_1,\cdots,g_m \in \mathbb{Z}[x_1,\cdots,x_m]$ be the polynomials that are found by Coppersmith's method. Then the variety of the ideal generated by $g_1(x_1,\cdots,x_m),\cdots,g_m(x_1,\cdots,x_m)$ is zero-dimensional.*

3 Strategy for Solving Search co-ACD Problem

In this section, we break the search co-ACD problem by using Coppersmith's method when $N=p_1\cdots p_n$ is public and $\gcd(Q,N)=1$. Given m co-ACD samples $(Qe_j \bmod p_1,\cdots,Qe_j \bmod p_n)$, where $j=1,\cdots,m$ and $m\geq n$, our goal is to retrieve the specific p_i's with these m samples, in other words, fully factorize the modulus N.

3.1 Solution for $N = p_1\cdots p_n$

Construct Modular Polynomials with Same Roots

In the case of $N=p_1\cdots p_n$, each sample is denoted as $(C_{j1},\cdots,C_{jn}) = (Qe_j \bmod p_1,\cdots,Qe_j \bmod p_n)$ for $j=1,\cdots,m$ and $m\geq n$.

The first type of polynomials $f_j(x_j)$ for $1\leq j\leq n-1$ are collected as follows. To begin, multiply these n equations $Qe_j - C_{ji}\equiv 0 \bmod p_i$ together and obtain a new equation $(Qe_j-C_{j1})\cdots(Qe_j-C_{jn})\equiv 0 \bmod N$. Rearrange this equation, then acquire

$$\sum_{i=0}^{n}(-1)^i\sigma_i(C_{j1},\cdots,C_{jn})Q^{n-i}e_j^{n-i}\equiv 0 \bmod N.$$

Here, $\sigma_i(x_1, \cdots, x_n) = \sum_{1 \le s_1 < s_2 < \cdots < s_i \le n} x_{s_1} x_{s_2} \cdots x_{s_i}$ is the i-th elementary symmetric polynomial on variables $x_1, \cdots, x_n$, such as $\sigma_1(x_1, \cdots, x_n) = x_1 + \cdots + x_n$ and $\sigma_n(x_1, \cdots, x_n) = x_1 \cdots x_n$. Let the coefficient of e_j^n be 1, then the following equation can be obtained

$$e_j^n + (-1)\sigma_1(C_{j1}, \cdots, C_{jn})Q^{-1}e_j^{n-1} + \cdots + (-1)^n \sigma_n(C_{j1}, \cdots, C_{jn})Q^{-n} \equiv 0 \bmod N.$$

We construct the following $n-1$ polynomials

$$f_j(x_j) := a_{j0} + a_{j1}x_j + \cdots + a_{j,n-1}x_j^{n-1} + x_j^n \tag{2}$$

where $j = 1, \cdots, n-1$ and $a_{ji} = (-1)^{n-i}\sigma_{n-i}(C_{j1}, \cdots, C_{jn})Q^{i-n} \bmod N$ for $0 \le i \le n-1$. Apparently, e_j is the root of $f_j(x_j) \bmod N$.

Next, we construct the second type of polynomials $f_k(x_1, \cdots, x_{n-1}, x_k)$ for $n \le k \le m$. Denote $\overline{p}_i := (p_1 \cdots p_{i-1}p_{i+1} \cdots p_n) \cdot ((p_1 \cdots p_{i-1}p_{i+1} \cdots p_n)^{-1} \bmod p_i) \bmod N$ for $i = 1, \cdots, n$. From

$$Qe_j \equiv C_{j1} \bmod p_1, \cdots, Qe_j \equiv C_{jn} \bmod p_n,$$

utilizing the Chinese remainder theorem, we arrive at the following modular equation:

$$Q\begin{pmatrix} e_1 \\ \vdots \\ e_{n-1} \\ e_k \end{pmatrix} \equiv \overline{p}_1 \begin{pmatrix} C_{11} \\ \vdots \\ C_{n-1,1} \\ C_{k1} \end{pmatrix} + \cdots + \overline{p}_n \begin{pmatrix} C_{1n} \\ \vdots \\ C_{n-1,n} \\ C_{kn} \end{pmatrix} \bmod N.$$

For simplicity, denote $\mathbf{C}_{n,k}$ as a $n \times n$ matrix $\begin{pmatrix} C_{11} & C_{12} & \cdots & C_{1n} \\ \vdots & \vdots & \ddots & \vdots \\ C_{n-1,1} & C_{n-1,2} & \cdots & C_{n-1,n} \\ C_{k1} & C_{k2} & \cdots & C_{kn} \end{pmatrix}$. We can rewrite the above equation as

$$Q\begin{pmatrix} e_1 \\ \vdots \\ e_{n-1} \\ e_k \end{pmatrix} \equiv \mathbf{C}_{n,k} \begin{pmatrix} \overline{p}_1 \\ \vdots \\ \overline{p}_n \end{pmatrix} \bmod N. \tag{3}$$

Assume that matrices $\mathbf{C}_{n,k}$ are invertible in $\mathbb{Z}_N$; otherwise, a non-trivial factor of N can be easily obtained. Because $-nN < \det(\mathbf{C}_{n,k}) < nN$ according to the standard determinant calculation method, the probability that $gcd(\det(\mathbf{C}_{n,k}), N) = N$ holds is extremely low. If $\det(\mathbf{C}_{n,k})$ is a multiple of N, reselect a set of samples. Left multiply $\mathbf{C}_{n,k}^{-1}$ by both sides of (3) and get

$$Q\mathbf{C}_{n,k}^{-1}\begin{pmatrix} e_1 \\ \vdots \\ e_{n-1} \\ e_k \end{pmatrix} \equiv \begin{pmatrix} \overline{p}_1 \\ \vdots \\ \overline{p}_n \end{pmatrix} \bmod N. \tag{4}$$

Note that $\overline{p}_i \equiv 1 \pmod{p_i}$ and $\overline{p}_j \equiv 0 \pmod{p_i}$ for $j \neq i$. Then the congruence $\overline{p}_1 + \cdots + \overline{p}_n \equiv 1 \pmod{N}$ holds. Thus, left multiply the n-dimensional row vector $(1, \cdots, 1)$ by both sides of (4) and get

$$Q(1, \cdots, 1, 1)\mathbf{C}_{n,k}^{-1} \begin{pmatrix} e_1 \\ \vdots \\ e_{n-1} \\ e_k \end{pmatrix} \equiv 1 \bmod N.$$

Let $(b_{k1}, \cdots, b_{k,n-1}, b_{kn}) := Q(1, \cdots, 1, 1)\mathbf{C}_{n,k}^{-1} \bmod N$, then we can rewrite the above equation as

$$b_{k1}e_1 + \cdots + b_{k,n-1}e_{n-1} + b_{kn}e_k \equiv 1 \bmod N.$$

Multiplying the equation by $b_{k,n}^{-1}$ produces the following polynomials:

$$f_k(x_1, \cdots, x_{n-1}, x_k) := a_{k0} + a_{k1}x_1 + \cdots + a_{k,n-1}x_{n-1} + x_k \tag{5}$$

where $a_{k0} = -b_{kn}^{-1} \bmod N$ and $a_{ki} = b_{ki}b_{kn}^{-1} \bmod N$ for $1 \leq i \leq n-1$. Clearly, $(e_1, \cdots, e_{n-1}, e_k)$ is a root of $f_k(x_1, \cdots, x_{n-1}, x_k) \bmod N$.

Solve e_i by Coppersmith's Method

Given some positive integer t, which is the power of modulo N^t, we generate the following polynomials for any integers $s_1, \cdots, s_m$ satisfying $0 \leq s_1 + \ldots + s_m \leq t$:

$$g_{s_1,\ldots,s_m}(x_1, \ldots, x_m) := N^{(t - \sum_{j=1}^{n-1} \lfloor \frac{s_j}{n} \rfloor - \sum_{k=n}^{m} s_k)} \prod_{j=1}^{n-1} x_j^{(s_j - n\lfloor \frac{s_j}{n} \rfloor)} \prod_{j=1}^{n-1} f_j^{\lfloor \frac{s_j}{n} \rfloor} \prod_{k=n}^{m} f_k^{s_k}.$$

It is obvious that $g_{s_1,\ldots,s_m}(e_1, \ldots, e_m) \equiv 0 \bmod N^t$. Then we construct the lattice $L_n(m, t)$ spanned by the coefficient vectors of the polynomials

$$g_{s_1,\cdots,s_m}(Xx_1, \cdots, Xx_m),\ 0 \leq s_1 + \ldots + s_m \leq t.$$

In order to make the involved basis matrix be lower triangular, we define the following orders. First, variables $x_1, \cdots, x_m$ are arranged according to $x_1 < \cdots < x_m$. The monomials $x_1^{s_1} \cdots x_m^{s_m}$ and $x_1^{s'_1} \cdots x_m^{s'_m}$ are arranged by terms of

$$x_1^{s_1} \cdots x_m^{s_m} < x_1^{s'_1} \cdots x_m^{s'_m} \Leftrightarrow (s_1, \cdots, s_m) \prec_{lex} (s'_1, \cdots, s'_m),$$

where $(s_1, \cdots, s_m) \prec_{lex} (s'_1, \cdots, s'_m)$ denotes that the lexicographic order of $(s_1, \cdots, s_m)$ is lower than that of $(s'_1, \cdots, s'_m)$. In the case of $m = 3$ and $t = 2$, we have

$$x_1 < x_1^2 < x_2 < x_1x_2 < x_2^2 < x_3 < x_1x_3 < x_2x_3 < x_3^2.$$

Therefore, $x_1^{s_1} \cdots x_m^{s_m}$ is obviously the leading monomial of $g_{s_1,\cdots,s_m}(Xx_1, \cdots, Xx_m)$.

Lemma 3. *The basis matrix of lattice $L_n(m,t)$ spanned by the coefficient vectors of the polynomials $g_{s_1,\cdots,s_m}(Xx_1,\cdots,Xx_m)$ is triangular if the monomials corresponding to the coefficient vectors are arranged according to the monomials order defined above.*

Proof. The polynomials $g_{s_1,\cdots,s_m}(Xx_1,\cdots,Xx_m)$ and $g_{s_1,\cdots,s_m}(x_1,\cdots,x_m)$ are in one-to-one correspondence, so we can prove this lemma by illustrating that the coefficients of $g_{s_1,\cdots,s_m}(x_1,\cdots,x_m)$ form a triangular matrix. The proof relies on mathematical induction.

First, when $s_2 = \cdots = s_m = 0$ and s_1 goes from 0 to t, the leading monomial of the polynomial $g_{s_1,\cdots,s_m}(x_1,\cdots,x_m)$ corresponds to $1, x_1, \cdots, x_1^{t-1}, x_1^t$ respectively. It is easy to see that the matrix formed by these $t+1$ polynomials is lower triangular.

$$\begin{array}{c} \\ g_{0,0\cdots,0} \\ g_{1,0,\cdots,0} \\ g_{2,0,\cdots,0} \\ \vdots \\ g_{t-1,0,\cdots,0} \\ g_{t,0,\cdots,0} \end{array} \begin{array}{c} \begin{array}{cccccc} 1 & x_1 & x_1^2 & \cdots & x_1^{t-1} & x_1^t \end{array} \\ \begin{bmatrix} N^t & 0 & 0 & \cdots & 0 & 0 \\ 0 & N^t & 0 & \cdots & 0 & 0 \\ 0 & 0 & N^t & \cdots & 0 & 0 \\ \vdots & \vdots & \vdots & \ddots & \vdots & \vdots \\ * & * & * & \cdots & N^{t-\lfloor \frac{t-1}{n} \rfloor} & 0 \\ * & * & * & \cdots & * & N^{t-\lfloor \frac{t}{n} \rfloor} \end{bmatrix} \end{array}$$

The symbol $*$ indicates some value about N, a_{10} and a_{11}.

Assume that for some $s_1, \cdots, s_m$, all polynomials $g_{t_1,\cdots,t_m}(x_1,\cdots,x_m)$ whose $t_1,\cdots,t_m$ satisfy $x_1^{t_1}\cdots x_m^{t_m} \leq x_1^{s_1}\cdots x_m^{s_m}$ form a lower triangular matrix. Then, in the order we defined above, add a new polynomial $g_{s_1',\cdots,s_m'}(x_1,\cdots,x_m)$ such that $(s_1',\cdots,s_m')$ is the next sequence of $(s_1,\cdots,s_m)$. The only new monomial is $x_1^{s_1'}\cdots x_m^{s_m'}$, and all other terms $x_1^{t_1}\cdots x_m^{t_m}$ in the new polynomial that satisfy $x_1^{t_1}\cdots x_m^{t_m} \leq x_1^{s_1}\cdots x_m^{s_m} < x_1^{s_1'}\cdots x_m^{s_m'}$ are the leading monomials of $g_{t_1,\cdots,t_m}(x_1,\cdots,x_m)$, which is part of the previously constructed triangular matrix. Therefore, after adding the new polynomial, the matrix is still lower triangular.

In conclusion, the lemma is proved.

Next, we analyze the dimension and determinant of $L_n(m,t)$. The dimension is the number of combinations of $(s_1,\cdots,s_m)$ satisfying $0 \leq s_1 + \ldots + s_m \leq t$. Therefore, the dimension of $L_n(m,t)$ is equal to

$$\dim L_n(m,t) = \sum_{0 \leq s_1+\ldots+s_m \leq t} 1 = \binom{m+t}{m}.$$

From the entries on the diagonal

$$N^{(t-\sum_{j=1}^{n-1}\lfloor\frac{s_j}{n}\rfloor - \sum_{k=n}^{m} s_k)} X^{\sum_{k=1}^{m} s_k},$$

we can get the determinant of $L_n(m,t)$ is

$$\det L_n(m,t) = N^{w_N} X^{w_X},$$

where

$$w_N = \sum_{0\le s_1+\ldots+s_m\le t} \left(t - \sum_{j=1}^{n-1} \lfloor \frac{s_j}{n} \rfloor - \sum_{k=n}^{m} s_k\right),$$

$$w_X = \sum_{0\le s_1+\ldots+s_m\le t} \sum_{k=1}^{m} s_k.$$

The specific calculation of w_N and w_X is shown in Appendix A. Then the upper bound of $\det L_n(m,t)$ can be expressed as follows:

$$\det L_n(m,t) < N^{\frac{-mn-2n+n^2+1}{n}\binom{m+t}{m+1}+(t+n-1)\binom{m+t}{m}} X^{m\binom{m+t}{m+1}}. \tag{6}$$

Substituting $\omega = \dim L_n(m,t)$, $\det L_n(m,t)$ and $\binom{m+t}{m+1} = \frac{t}{m+1}\omega$ into the inequality (1), the result is:

$$X < 2^{-\frac{\omega-1}{4t}} \omega^{-\frac{\omega+1-m}{2\omega t}} N^{1-\frac{(n-1)^2}{mn}-\frac{(n-1)(m-1)}{mt}-\frac{m^2-1}{mw}}.$$

For a sufficiently large N, the powers of 2 and ω are negligible, and thus we only consider the exponent of N:

$$1 - \frac{(n-1)^2}{mn} - \frac{(n-1)(m-1)}{mt} - \frac{m^2-1}{mw},$$

where $\omega = \binom{m+t}{m} = \frac{(m+t)!}{m!t!} = \frac{\prod_{i=1}^{m}(t+i)}{m!} > mt$, which is proved easily by induction. Thus, above formula can be lower bound by

$$1 - \frac{(n-1)^2}{mn} - \frac{(mn+1)(m-1)}{m^2 t}.$$

Therefore, we obtain

$$X < N^{1-\frac{(n-1)^2}{mn}-\frac{(mn+1)(m-1)}{m^2 t}}.$$

By plugging $\log_2 X = \rho$ and $\log_2 N \ge n(\eta - 1)$ into the inequality above with $\frac{(mn+1)(m-1)}{m^2 t}$ as an error term ϵ, we obtain

$$\rho < n(\eta - 1)\left(1 - \frac{(n-1)^2}{mn} - \epsilon\right),$$

where the error term $\epsilon > 0$ is any positive number that satisfies $t \ge \frac{(mn+1)(m-1)}{m^2 \epsilon}$.

In summary, we get the result as follows.

Result 1. *Given an integer $N = p_1 \cdots p_n$ with unknown factors p_i, and m co-ACD samples $(e_j Q \bmod p_1, \cdots, e_j Q \bmod p_n)$. Under Assumption 2, as long as $m \ge n$ and $\epsilon > 0$, co-ACD problem can be solved in polynomial time when*

$$\rho < n(\eta - 1)\left(1 - \frac{(n-1)^2}{mn} - \epsilon\right). \tag{7}$$

Remark 1. When $n = 2$, we can solve the co-ACD problem with two samples, whereas [17] requires $m > \frac{3\eta-\rho}{2\eta-\rho}$. If $m = 2$, this will cause $\eta > \rho$, which does not meet the conditions for the selection of the parameters of the co-ACD problem. Therefore, they are unable to solve the problem when $m = 2$. As m tends to infinity, we approximately need to satisfy $\rho < 2\eta$, which is the same as the bound $\rho < (2 - \frac{1}{m-1})\eta$ in [17].

When $n \geq 3$, we can solve it when $m > (n-1)^2$. In the asymptotic case, we only need satisfy $\rho < n(\eta-1)$, which corresponds to $\rho < (n-1+1/n)\eta$ in Sect. 5.2 of [17]. As a result, when $n(\eta-1) > (n-1+1/n)\eta$, that is $3 \leq n < \frac{\eta+\sqrt{\eta^2-4\eta}}{2}$, the theoretical bound of ρ in our work outperforms previous work.

Remark 2. If $M_1, M_2, ..., M_m$ are m known messages, their ciphertexts are $\vec{C}_1, ...,$ $\vec{C}_m$ that satisfy $\vec{C}_j = (M_j + Qe_j \bmod p_1, \cdots, M_j + Qe_j \bmod p_n)$, for $j = 1, ..., m$. As we already know M_j, we can obtain $Qe_j - C'_{ji} \equiv 0 \bmod p_i$, which is the same as the co-ACD problem.

3.2 Experimental Results

We ran the above attacks over 500 times in SageMath 9.0 on a PC with an Intel(R) Core(TM) i7-9750H CPU @ 2.60 GHz, 16 GB RAM, and Windows 10 to ensure success for $n = 2$ which is the parameter of the CLS scheme. The attack works efficiently since the main cost of the attack is LLL algorithm, and they need only six lattice dimensions, which can run very fast with only a few tenths of a second. We compare our work with [17] in the number of samples m and running time. Please see Table 2 for details, and the symbol – indicates that the corresponding experiments were not given in [17].

Table 2. Attack of the search co-ACD problem for $N = p_1 p_2$

Parameters		Minimal m	Running time	Success rate
This work	Set-I	2	0.16 s	100%
	Set-II	2	0.19 s	100%
	Set-III	2	0.20 s	100%
[17]	Set-I	3	0.31 s	100%
	Set-II	3	0.57 s	100%
	Set-III	–	–	–

4 Ciphertext-Only Attack

In this section, we attack the scheme with unknown plaintexts, and there is no restriction to the plaintexts. That is, any unknown plaintext $M_i \in \mathbb{Z}_Q$ can be attacked. The main idea is to construct the lattice using known equations, hoping that information about the unknown variables can be obtained by finding the short vector in the lattice.

4.1 Solution for $N = p_1 p_2$

For the first two samples,

$$\begin{cases} M_1 + e_1 Q - C_{11} \equiv 0 \bmod p_1 \\ M_1 + e_1 Q - C_{12} \equiv 0 \bmod p_2 \end{cases} \text{and} \begin{cases} M_2 + e_2 Q - C_{21} \equiv 0 \bmod p_2 \\ M_2 + e_2 Q - C_{22} \equiv 0 \bmod p_1 \end{cases},$$

let $x_i = M_i + e_i Q$ for $i = 1, \cdots, m$, we can multiply the two equations in each row correspondingly to obtain:

$$\begin{cases} x_1 x_2 - C_{22} x_1 - C_{11} x_2 + C_{11} C_{22} \equiv 0 \bmod N \\ x_1 x_2 - C_{21} x_1 - C_{12} x_2 + C_{12} C_{21} \equiv 0 \bmod N \end{cases}.$$

If we eliminate the same variable $x_1 x_2$, these two equations can be combined into one equation:

$$(C_{21} - C_{22}) x_1 + (C_{12} - C_{11}) x_2 + C_{11} C_{22} - C_{12} C_{21} \equiv 0 \bmod N.$$

Normalize the coefficient of the variable x_2, then

$$(C_{12} - C_{11})^{-1} (C_{21} - C_{22}) x_1 + x_2 + (C_{12} - C_{11})^{-1} (C_{11} C_{22} - C_{12} C_{21}) \equiv 0 \bmod N.$$

Then we multiply the remaining $m-2$ samples with the first sample, respectively, by the same operation as above, and we can get $m - 1$ equations,

$$A_i x_1 + x_i + D_i \equiv 0 \bmod N,$$

where $i = 2, \cdots, m$, $A_i = (C_{12} - C_{11})^{-1} (C_{i1} - C_{i2}) \bmod N$ and $D_i = (C_{12} - C_{11})^{-1} (C_{11} C_{i2} - C_{12} C_{i1}) \bmod N$.

Next, we construct a lattice $\mathcal{B}$ which is spanned by the row vectors of the matrix:

$$B = \begin{pmatrix} 1 & A_2 & \cdots & A_m \\ 0 & N & \cdots & 0 \\ \vdots & \vdots & \ddots & \vdots \\ 0 & 0 & \cdots & N \end{pmatrix},$$

whose determinant is N^{m-1}. The lattice $\mathcal{B}$ contains two short vectors,

$$\vec{v}_1 = (C_{12} - C_{11}, -(C_{22} - C_{21}), \cdots, -(C_{m2} - C_{m1})),$$

$$\vec{v}_2 = (x_1 - C_{11}, -(x_2 - C_{21}), \cdots, -(x_m - C_{m1})).$$

It is easy to see that $\vec{v}_1 = (C_{12} - C_{11}) \times \vec{z}_1 \bmod N$ and $\vec{v}_2 = (x_1 - C_{11}) \times \vec{z}_1 \bmod N$ where $\vec{z}_1$ is the first row vector of B. So the ℓ_2-norm of $\vec{v}_1$ and $\vec{v}_2$ is $\|\vec{v}_1\| \approx 2^{\eta}$ and $\|\vec{v}_2\| \approx 2^{\rho+256}$. In addition, the target vector $\vec{v}_2$ satisfies the following equation:

$$\vec{v}_2 = \alpha \vec{b}_1 + \beta \vec{b}_2,$$

where $\vec{b}_1$ and $\vec{b}_2$ are the first two vectors of the LLL-reduced basis of B. The details are as follows.

Let $\vec{u}_i$ denote the i-th shortest vector in lattice $\mathcal{B}$. When m is large enough, there is $\|\vec{v}_1\| \ll \|\vec{v}_2\| \ll \mathrm{GH}(\mathcal{B})$, which illustrates that $\vec{v}_1$ and $\vec{v}_2$ are much shorter than other independent vectors in $\mathcal{B}$ according to Gaussian heuristic assumption. At this point, $\vec{v}_1 = \gamma\vec{u}_1$ and $\vec{v}_2 = \beta\vec{u}_2 + \alpha\vec{u}_1$ hold with high possibility. Since γ is the greatest common factor of the coefficients $(C_{12} - C_{11}, k_2, \cdots, k_m)$ of $\vec{v}_1 = (C_{12} - C_{11}, k_2, \cdots, k_m)B$, the shortest non-zero vector $\vec{u}_1 = (\frac{C_{12}-C_{11}}{\gamma}, \frac{k_2}{\gamma}, \cdots, \frac{k_m}{\gamma})B = \frac{1}{\gamma}\vec{v}_1$, and $\gamma = \gcd(C_{12} - C_{11}, k_2, \cdots, k_m)$ is very small with high probability. Then $\|\vec{u}_1\| \approx \|\vec{v}_1\| \approx 2^{\eta}$.

Under the average sense, $\|\vec{u}_2\| \approx \frac{\|\vec{v}_2\|}{\beta}$, because $\|\vec{v}_2\| \geq \beta\|\vec{u}_2\| \sin(\vec{u}_2, \vec{u}_1)$ and the distribution of the angle between vectors $\vec{u}_1$ and $\vec{u}_2$ converges to a normal distribution with mean $\frac{\pi}{2}$ and variance proportional to $\frac{1}{\sqrt{m}}$, that is, $\|\vec{v}_2\| \approx \beta\|\vec{u}_2\|$. According to the property of LLL algorithm, when $\|\vec{u}_1\| \ll \mathrm{GH}(\mathcal{B})$, $\vec{b}_1 = \vec{u}_1$. When $\pi_1(\vec{u}_2) \ll \mathrm{GH}(\pi_1(\mathcal{B}))$[1], $\vec{b}_2 = \vec{u}_2$. To sum up, if $\pi_1(\vec{u}_2) \ll \mathrm{GH}(\pi_1(\mathcal{B}))$ is satisfied, then $\vec{v}_2 = \alpha\vec{b}_1 + \beta\vec{b}_2$, that is, $\sqrt{\frac{m}{m+1}}\|\vec{u}_2\| \ll \sqrt{\frac{m}{2\pi e}}(\frac{\det(\mathcal{B})}{2^{\eta}})^{\frac{1}{m}}$. If we omit the constant term, then $\|\vec{v}_2\| = 2^{\rho+256} < (\frac{\det(\mathcal{B})}{2^{\eta}})^{\frac{1}{m}}$ approximately.

Therefore, when we apply the LLL algorithm to matrix B, we can get an equation about variables α and β, according to $v_{21} = x_1 - C_{11} \equiv 0 \bmod p_1$:

$$v_{21} = \alpha b_{11} + \beta b_{21} \equiv 0 \bmod p_1.$$

Due to $\alpha b_{11} = v_{21} - \beta b_{21}$, there is approximately $\alpha < \frac{v_{21}}{v_{11}} \approx 2^{\rho-\eta+256} = X_\alpha$ and the bound of β is $X_\beta = O(1)$ for the same reasons as γ. Then the bounds of α and β satisfy condition $\log_N|\alpha| + \log_N|\beta| \approx \frac{\rho-\eta+256}{2\eta} = \frac{256}{\eta} < (\frac{1}{2})^2 - \epsilon$, where modulo $p_1 \geq N^{\frac{1}{2}-\epsilon'}$. So we can solve α and β using Coppersmith's method by Theorem 7 in [26], and p_1 is $\gcd(\alpha b_{11} + \beta b_{21}, N)$.

Finally, in order to meet the above condition, we need to have $\|\vec{v}_2\| = 2^{\rho+256} < (\frac{\det(\mathcal{B})}{2^{\eta}})^{\frac{1}{m}}$, where $\det(\mathcal{B}) = N^{m-1}$. Then the condition of m is as follows:

$$m > \frac{3\eta}{2\eta - \rho - 256}.$$

Result 2. *Given the integer $N = p_1p_2$ with unknown factors p_1 and p_2, and m ciphertexts $(M_i + e_iQ \bmod p_1, M_i + e_iQ \bmod p_2)$. Under Assumption 2, the secret key p_1 and p_2 can be retrieved in polynomial time when*

$$m > \frac{3\eta}{2\eta - \rho - 256}.$$

Remark 3. For the ciphertext-only attack, we can retrieve the secret key p_1 and p_2 as long as the number of samples m satisfies $m > \frac{3\eta}{2\eta-\rho-256}$ with known N, while [17] requires the plaintexts to be small $\|\vec{M}\| \approx 2^{\mu}m^{1/2}$ and $m \gtrsim 3 + \frac{\log Q+\rho+3}{\log Q-\mu} > 3 + \frac{\log Q+\rho+3}{\log Q}$, but it is not necessary to know the value of N. When $\rho > 1065$, their number of samples is much larger than ours.

[1] π_1 denotes the projection onto $\langle\vec{b}_1\rangle^{\perp}$.

4.2 Experimental Results

We carried out the attack described in Sect. 4. Compared with [17], our attack has no restrictions on plaintexts; that is, any $M_i \in [0, Q)$ can be attacked successfully, but N is required. We still conducted 500 experiments on each set of parameters, and the specific experimental results are in Table 3.

Table 3. Ciphertext-only attack of the CLS scheme

Parameters	m	Running time	Success rate
Set-I	3	0.05 s	100%
Set-II	3	0.22 s	100%
Set-III	3	0.27 s	100%

5 Conclusion

In the case where N is known, we study the co-ACD problem and key recovery attacks against CLS scheme, which can be accomplished with a smaller number of samples, and the arbitrary size of plaintext is unknown. We work with a better theoretical result on the range of ρ when n is in a certain range. However, for the case where N is unknown, how to complete the key recovery attack or solve the co-ACD problem is the research direction of our future work.

Acknowledgements. The authors would like to thank anonymous reviewers for their helpful comments and suggestions. The work of this paper was supported by the National Natural Science Foundation of China (No.61732021) and the National Key R&D Program of China (No.2018YFB0803801 and No.2018YFA0704704).

A Calculation of w_N and w_X

First, we compute w_X:

$$\begin{aligned} w_X = & \sum_{0 \le s_1+\ldots+s_m \le t} \sum_{k=1}^{m} s_k = \sum_{0 \le s_1+\ldots+s_m \le t} (s_1 + \cdots + s_m) \\ = & \sum_{l=0}^{t} \sum_{s_1+\cdots+s_m=l} l = \sum_{l=0}^{t} l \binom{m+l-1}{m-1} \\ = & \, m \binom{m+t}{m+1}. \end{aligned}$$

Since that $s_1, \cdots, s_m$ are identical in terms of value and weight, then$\sum_{0\le s_1+\ldots+s_m\le t} s_1 = \cdots = \sum_{0\le s_1+\cdots+s_m\le t} s_m$. This way, we can get

$$\sum_{0\le s_1+\ldots+s_m\le t} s_i = \frac{1}{m} \sum_{0\le s_1+\ldots+s_m\le t} \sum_{k=1}^{m} s_k = \binom{m+t}{m+1}$$

Next, we compute $w_N = \sum\limits_{0\le s_1+\ldots+s_m\le t} (t - \sum\limits_{j=1}^{n-1} \lfloor \frac{s_j}{n} \rfloor - \sum\limits_{k=n}^{m} s_k)$. Clearly,

$$\sum_{0\le s_1+\cdots+s_m\le t} t = t \sum_{0\le s_1+\cdots+s_m\le t} 1 = t\binom{m+t}{m}.$$

Denote $\lfloor \frac{s_j}{n} \rfloor = \frac{s_j}{n} - \delta_j$ where $0 \le \delta_j < 1$, then

$$\begin{aligned}
- \sum_{0\le s_1+\ldots+s_m\le t} \sum_{j=1}^{n-1} \lfloor \frac{s_j}{n} \rfloor &= \sum_{0\le s_1+\ldots+s_m\le t} \left(-\frac{s_1+\cdots+s_{n-1}}{n} + \delta_1 + \cdots + \delta_{n-1} \right) \\
&= -\frac{n-1}{n}\binom{m+t}{m+1} + \sum_{0\le s_1+\ldots+s_m\le t} (\delta_1 + \cdots + \delta_{n-1}) \\
&< -\frac{n-1}{n}\binom{m+t}{m+1} + (n-1)\binom{m+t}{m}.
\end{aligned}$$

Moreover,

$$- \sum_{0\le s_1+\ldots+s_m\le t} \sum_{k=n}^{m} s_k = -(m-n+1) \sum_{0\le s_1+\ldots+s_m\le t} s_k = -(m-n+1)\binom{m+t}{m+1}.$$

Summarizing the above analysis, we find

$$\begin{aligned}
w_N &< t\binom{m+t}{m} - \frac{n-1}{n}\binom{m+t}{m+1} + (n-1)\binom{m+t}{m} - (m-n+1)\binom{m+t}{m+1} \\
&= \frac{-mn-2n+n^2+1}{n}\binom{m+t}{m+1} + (t+n-1)\binom{m+t}{m}.
\end{aligned}$$

References

1. Bauer, A., Joux, A.: Toward a rigorous variation of Coppersmith's algorithm on three variables. In: Naor, M. (ed.) EUROCRYPT 2007. LNCS, vol. 4515, pp. 361–378. Springer, Heidelberg (2007). https://doi.org/10.1007/978-3-540-72540-4_21
2. Chen, Y., Nguyen, P.Q.: Faster algorithms for approximate common divisors: breaking fully-homomorphic-encryption challenges over the integers. In: Pointcheval, D., Johansson, T. (eds.) EUROCRYPT 2012. LNCS, vol. 7237, pp. 502–519. Springer, Heidelberg (2012). https://doi.org/10.1007/978-3-642-29011-4_30

3. Cheon, J.H., Cho, W., Hhan, M., Kim, J., Lee, C.: Algorithms for CRT-variant of approximate greatest common divisor problem. J. Math. Cryptol. **14**(1), 397–413 (2020)
4. Cheon, J.H., Coron, J.-S., Kim, J., Lee, M.S., Lepoint, T., Tibouchi, M., Yun, A.: Batch fully homomorphic encryption over the integers. In: Johansson, T., Nguyen, P.Q. (eds.) EUROCRYPT 2013. LNCS, vol. 7881, pp. 315–335. Springer, Heidelberg (2013). https://doi.org/10.1007/978-3-642-38348-9_20
5. Cheon, J.H., Lee, H.T., Seo, J.H.: A new additive homomorphic encryption based on the co-ACD problem. In: Ahn, G., Yung, M., Li, N. (eds.) ACM SIGSAC Conference on Computer and Communications Security, pp. 287–298. ACM (2014)
6. Cohn, H., Heninger, N.: Approximate common divisors via lattices. CoRR abs/1108.2714 (2011)
7. Cominetti, E.L., Jr., Simplicio, M.A.: Fast additive partially homomorphic encryption from the approximate common divisor problem. IEEE Trans. Inf. Forensics Secur. **15**, 2988–2998 (2020)
8. Coppersmith, D.: Finding a small root of a bivariate integer equation; factoring with high bits known. In: Maurer, U. (ed.) EUROCRYPT 1996. LNCS, vol. 1070, pp. 178–189. Springer, Heidelberg (1996). https://doi.org/10.1007/3-540-68339-9_16
9. Coppersmith, D.: Finding a small root of a univariate modular equation. In: Maurer, U. (ed.) EUROCRYPT 1996. LNCS, vol. 1070, pp. 155–165. Springer, Heidelberg (1996). https://doi.org/10.1007/3-540-68339-9_14
10. Coppersmith, D.: Small solutions to polynomial equations, and low exponent RSA vulnerabilities. J. Cryptol. **10**(4), 233–260 (1997)
11. Coron, J.-S., Faugère, J.-C., Renault, G., Zeitoun, R.: Factoring $N = p^r q^s$ for large r and s. In: Sako, K. (ed.) CT-RSA 2016. LNCS, vol. 9610, pp. 448–464. Springer, Cham (2016). https://doi.org/10.1007/978-3-319-29485-8_26
12. Coron, J.-S., Mandal, A., Naccache, D., Tibouchi, M.: Fully homomorphic encryption over the integers with shorter public keys. In: Rogaway, P. (ed.) CRYPTO 2011. LNCS, vol. 6841, pp. 487–504. Springer, Heidelberg (2011). https://doi.org/10.1007/978-3-642-22792-9_28
13. Coron, J.-S., Naccache, D., Tibouchi, M.: Public key compression and modulus switching for fully homomorphic encryption over the integers. In: Pointcheval, D., Johansson, T. (eds.) EUROCRYPT 2012. LNCS, vol. 7237, pp. 446–464. Springer, Heidelberg (2012). https://doi.org/10.1007/978-3-642-29011-4_27
14. Coron, J., Notarnicola, L., Wiese, G.: Simultaneous diagonalization of incomplete matrices and applications. CoRR abs/2005.13629 (2020)
15. Coron, J.-S., Zeitoun, R.: Improved factorization of $N = p^r q^s$. In: Smart, N.P. (ed.) CT-RSA 2018. LNCS, vol. 10808, pp. 65–79. Springer, Cham (2018). https://doi.org/10.1007/978-3-319-76953-0_4
16. van Dijk, M., Gentry, C., Halevi, S., Vaikuntanathan, V.: Fully homomorphic encryption over the integers. In: Gilbert, H. (ed.) EUROCRYPT 2010. LNCS, vol. 6110, pp. 24–43. Springer, Heidelberg (2010). https://doi.org/10.1007/978-3-642-13190-5_2
17. Fouque, P.-A., Lee, M.S., Lepoint, T., Tibouchi, M.: Cryptanalysis of the co-ACD assumption. In: Gennaro, R., Robshaw, M. (eds.) CRYPTO 2015. LNCS, vol. 9215, pp. 561–580. Springer, Heidelberg (2015). https://doi.org/10.1007/978-3-662-47989-6_27

18. Galbraith, S.D., Gebregiyorgis, S.W., Murphy, S.: Algorithms for the approximate common divisor problem. IACR Cryptology ePrint Archive, p. 215 (2016)
19. Herrmann, M., May, A.: Maximizing small root bounds by linearization and applications to small secret exponent RSA. In: Nguyen, P.Q., Pointcheval, D. (eds.) PKC 2010. LNCS, vol. 6056, pp. 53–69. Springer, Heidelberg (2010). https://doi.org/10.1007/978-3-642-13013-7_4
20. Howgrave-Graham, N.: Finding small roots of univariate modular equations revisited. In: Darnell, M. (ed.) Cryptography and Coding 1997. LNCS, vol. 1355, pp. 131–142. Springer, Heidelberg (1997). https://doi.org/10.1007/BFb0024458
21. Howgrave-Graham, N.: Approximate integer common divisors. In: Silverman, J.H. (ed.) CaLC 2001. LNCS, vol. 2146, pp. 51–66. Springer, Heidelberg (2001). https://doi.org/10.1007/3-540-44670-2_6
22. Jochemsz, E., May, A.: A strategy for finding roots of multivariate polynomials with new applications in attacking RSA variants. In: Lai, X., Chen, K. (eds.) ASIACRYPT 2006. LNCS, vol. 4284, pp. 267–282. Springer, Heidelberg (2006). https://doi.org/10.1007/11935230_18
23. Jochemsz, E., May, A.: A polynomial time attack on RSA with private CRT-exponents smaller than $N^{0.073}$. In: Menezes, A. (ed.) CRYPTO 2007. LNCS, vol. 4622, pp. 395–411. Springer, Heidelberg (2007). https://doi.org/10.1007/978-3-540-74143-5_22
24. Kakvi, S.A., Kiltz, E., May, A.: Certifying RSA. In: Wang, X., Sako, K. (eds.) ASIACRYPT 2012. LNCS, vol. 7658, pp. 404–414. Springer, Heidelberg (2012). https://doi.org/10.1007/978-3-642-34961-4_25
25. Lenstra, A.: Factoring polynomial with rational coefficients. Mathematiche Annalen **261**, 515–534 (1982)
26. Lu, Y., Zhang, R., Peng, L., Lin, D.: Solving linear equations modulo unknown divisors: revisited. In: Iwata, T., Cheon, J.H. (eds.) ASIACRYPT 2015. LNCS, vol. 9452, pp. 189–213. Springer, Heidelberg (2015). https://doi.org/10.1007/978-3-662-48797-6_9
27. May, A.: New RSA vulnerabilities using lattice reduction methods. Ph.D. thesis, University of Paderborn (2003). http://ubdata.uni-paderborn.de/ediss/17/2003/may/disserta.pdf
28. May, A.: Using LLL-reduction for solving RSA and factorization problems. In: Nguyen, P.Q., Vallée, B. (eds.) The LLL Algorithm - Survey and Applications. ISC, pp. 315–348. Springer, Heidelberg (2010). https://doi.org/10.1007/978-3-642-02295-1_10
29. May, A., Nowakowski, J., Sarkar, S.: Partial key exposure attack on short secret exponent CRT-RSA. IACR Cryptology ePrint Archive, p. 972 (2021)
30. Nguyen, P., Stern, J.: Merkle-Hellman revisited: a cryptanalysis of the Qu-Vanstone cryptosystem based on group factorizations. In: Kaliski, B.S. (ed.) CRYPTO 1997. LNCS, vol. 1294, pp. 198–212. Springer, Heidelberg (1997). https://doi.org/10.1007/BFb0052236
31. Nguyen, P., Stern, J.: Cryptanalysis of a fast public key cryptosystem presented at SAC '97. In: Tavares, S., Meijer, H. (eds.) SAC 1998. LNCS, vol. 1556, pp. 213–218. Springer, Heidelberg (1999). https://doi.org/10.1007/3-540-48892-8_17

32. Nguyen, P., Stern, J.: The hardness of the hidden subset sum problem and its cryptographic implications. In: Wiener, M. (ed.) CRYPTO 1999. LNCS, vol. 1666, pp. 31–46. Springer, Heidelberg (1999). https://doi.org/10.1007/3-540-48405-1_3
33. Suzuki, K., Takayasu, A., Kunihiro, N.: Extended partial key exposure attacks on RSA: improvement up to full size decryption exponents. Theor. Comput. Sci. **841**, 62–83 (2020)
34. Xu, J., Sarkar, S., Hu, L., Wang, H., Pan, Y.: New results on modular inversion hidden number problem and inversive congruential generator. In: Boldyreva, A., Micciancio, D. (eds.) CRYPTO 2019. LNCS, vol. 11692, pp. 297–321. Springer, Cham (2019). https://doi.org/10.1007/978-3-030-26948-7_11

A Note on the Security Framework of Two-key DbHtS MACs

Tingting Guo[1,2] and Peng Wang[1,2](✉)

[1] SKLOIS, Institute of Information Engineering, CAS, Beijing, China
guotingting@iie.ac.cn, w.rocking@gmail.com

[2] School of Cyber Security, University of Chinese Academy of Sciences, Beijing, China

Abstract. Double-block Hash-then-Sum (DbHtS) MACs are a class of MACs that achieve beyond-birthday-bound (BBB) security, including SUM-ECBC, PMAC_Plus, 3kf9, LightMAC_Plus, etc. Recently, Shen et al. (CRYPTO 2021) proposed a security framework for two-key DbHtS MACs in the multi-user setting, stating that when the underlying block cipher is ideal and the universal hash functions are regular and almost universal, the two-key DbHtS MACs achieve $2n/3$-bit security. Unfortunately, the regular and universal properties can not guarantee the BBB security of two-key DbHtS MACs. We propose three counter-examples which are proved to be $2n/3$-bit secure in the multi-user setting by the framework, but can be broken with probability 1 using only $\mathcal{O}(2^{n/2})$ queries even in the single-user setting. We also point out the miscalculation in their proof leading to such a flaw. However, we haven't found attacks against 2k-SUM-ECBC, 2k-PMAC_Plus, and 2k-LightMAC_Plus proved $2n/3$-bit security in their paper.

Keywords: MAC · DbHtS · Beyond-birthday-bound security · Multi-user security

1 Introduction

Message Authentication Code (MAC). MAC is a symmetric-key crypto primitive to ensure integrity of messages. Most of their security proofs, including XCBC [4], PMAC [5,13], HMAC [2], and NMAC [2], follow the Hash-then-(Fixed-Input-Length) Function (HtF) framework:

$$\mathrm{HtF}[H,E](K_h,K,M) = E_K(H_{K_h}(M)).$$

When H is an almost universal (AU) hash function and E is a fixed-input-length PRF (often instantiated as a block cipher), HtF is a variable-input-length PRF [16] with birthday bound security, i.e. they are secure up to $\mathcal{O}(2^{n/2})$ queries where n is the input size of E. However, birthday-bound security is always not enough for modes of lightweight block ciphers, such as PRESENT [6], GIFT [1], etc. whose block size is 64-bit. In these cases, the securities are only 32-bit

C. Alcaraz et al. (Eds.): ICICS 2022, LNCS 13407, pp. 55–68, 2022.
https://doi.org/10.1007/978-3-031-15777-6_4

(i.e., security up to 2^{32} queries), which are practically vulnerable. Therefore, researchers make great efforts to construct MACs with better security bounds.

Birthday-Birthday-Bound (BBB) MACs. Plenty of MACs with BBB security have been put forward, such as SUM-ECBC [17], PMAC_Plus [18], 3kf9 [19], LightMAC_Plus [11], and so on. Their primary proofs [11,17–19] gave $2n/3$-bit security (ignoring the maximum message length). At FSE 2019, Datta et al. showed that these MACs all follow the three-key Double-block Hash-then-Sum (DbHtS) framework [8]:

$$\text{DbHtS}[H, E]\,(K_h, K_1, K_2, M) = E_{K_1}(H^1_{K_{h,1}}(M)) \oplus E_{K_2}(H^2_{K_{h,2}}(M)),$$

where M is the massage, $K_h = (K_{h,1}, K_{h,2})$ is the key for two universal hash functions H^1 and H^2. K_1, K_2 are keys for block cipher E. In the following, we treat the block cipher as a Pseudorandom Permutation (PRP). E_{K_1} and E_{K_2} mean two independent PRPs. Let $H_{K_h} = (H^1_{K_{h,1}}, H^2_{K_{h,2}})$. Datta et al. proved that when H is weak-cover-free and weak-block-wise universal, the three-key DbHtS is $2n/3$-bit secure. Later, Leurent et al. [10] showed the best attacks to them cost $\mathcal{O}(2^{3n/4})$ queries. Recently at EUROCRYPT 2020, Kim et al. [9] have proved the tight $3n/4$-bit security (ignoring the maximum message length) if H^1 and H^2 are only almost universal.

Datta et al. [8] also found that the two-key DbHtS, that is to say, $K_1 = K_2$ in the above framework, is still $2n/3$-bit security when H is cover-free, block-wise universal, and colliding.

Two-key DbHtS in the Multi-user Setting. All the above MAC frameworks only considered single-user settings. In practice, the adversary can attack multiple users. For instance, MACs are core elements of real-world security protocols such as TLS, SSH, and IPsec, which are used by lots of websites with plenty of daily active users. However, by a generic reduction, all above BBB results degrade to (or even worse than) the birthday bound in the multi-user setting [14].

At CRYPTO 2021, Shen et al. [14] revisited the security of the two-key DbHtS framework in the multi-user setting elaborately. Their framework (Theorem 1 in [14]) states that when the underlying block cipher is an ideal cipher and the two independent universal hash functions $H^1_{K_{h,1}}$ and $H^2_{K_{h,2}}$ are both regular and almost universal, the two-key DbHtS MACs, including 2k-SUM-ECBC, achieve $2n/3$-bit security. They adjusted the proof of the framework for adapting to 2k-PMAC_Plus and 2k-LightMAC_Plus based on two dependent universal hash functions, stating they achieve $2n/3$-bit security, too.

We summarise the above security frameworks for MACs in Table 1.

Our Contributions. We show that *Theorem* 1 *in Shen et al.'s paper* [14], *giving the security of the two-key DbHtS framework, has a critical flaw* by three counter-examples. According to their framework, these counter-examples are proved $2n/3$-bit security (ignoring the maximum message length) in the multi-user setting. However, they are all attacked successfully with only $\mathcal{O}(2^{n/2})$ queries even in the single-user setting. We also show clearly the miscalculation in their proof leading to such a flaw.

Table 1. Summary of security frameworks for MACs. n is the input size of E. 'SU' means single-user setting. 'MU' means multi-user setting. The security of MACs ignores the maximum message length.

Framework	Property of H	Property of E	Setting	Security
HtF	AU	PRF	SU	$n/2$ [16]
Three-key DbHtS	Weak-cover-free Weak-block-wise universal	PRP	SU	$2n/3$ [8]
Three-key DbHtS	AU	PRP	SU	$3n/4$ [9]
Two-key DbHtS	Cover-free Block-wise universal Colliding	PRP	SU	$2n/3$ [8]
Two-key DbHtS	Regular AU	Ideal cipher	MU	$2n/3$ [14]

2 Preliminaries

Notation. For a finite set $\mathcal{X}$, let $X \xleftarrow{\$} \mathcal{X}$ denote sampling X from $\mathcal{X}$ uniformly and randomly. Let $|\mathcal{X}|$ be the size of the set $\mathcal{X}$. For a domain $\mathcal{X}$ and a range $\mathcal{Y}$, let $\mathsf{Func}(\mathcal{X}, \mathcal{Y})$ denote the set of all functions from $\mathcal{X}$ to $\mathcal{Y}$.

Multi-user Pseudorandom Function (PRF). Let $F : \mathcal{K} \times \mathcal{X} \to \mathcal{Y}$ be a function. For u users, the game $\mathbf{G}_F^{\text{prf}}(\mathscr{A})$ about adversary $\mathscr{A}$ is defined as follows.

1. Initialize $K_1, K_2, \ldots, K_u \xleftarrow{\$} \mathcal{K}, f_1, f_2, \ldots, f_u \xleftarrow{\$} \mathsf{Func}(\mathcal{X}, \mathcal{Y})$, and $b \xleftarrow{\$} \{0, 1\}$;
2. $\mathscr{A}$ queries Eval function with (i, X) and get $\mathsf{Eval}(i, X)$, where $i \in \{1, 2, \ldots, u\}$, $X \in \mathcal{X}$, and
$$\mathsf{Eval}(i, X) = \begin{cases} F(K_i, X), & \text{if } b = 1, \\ f_i(X), & \text{if } b = 0; \end{cases}$$
3. $\mathscr{A}$ output b'.

Then the advantage of the adversary $\mathscr{A}$ against the multi-user PRF security of F is
$$\begin{aligned} \text{Adv}_F^{\text{prf}}(\mathscr{A}) &= |2\Pr[b' = b] - 1| \\ &= |\Pr[b' = 1 | b = 1] - \Pr[b' = 1 | b = 0]| \,. \end{aligned}$$

We call a function F is a multi-user secure PRF if the advantage $\text{Adv}_F^{\text{prf}}(\mathscr{A})$ is negligible for any adversary $\mathscr{A}$. When $u = 1$, we call F is a single-user secure PRF.

The H-Coefficient Technique. When considering interactions between an adversary $\mathscr{A}$ and an abstract system $\mathbf{S}$ which answers $\mathscr{A}$'s queries, let X_i denote the query from $\mathscr{A}$ to $\mathbf{S}$ and Y_i denote the response of X_i from $\mathbf{S}$ to $\mathscr{A}$. Then the resulting interaction can be recorded with a transcript $\tau = ((X_1, Y_1), \ldots, (X_q, Y_q))$. Let $p_{\mathbf{S}}(\tau)$ denote the probability that $\mathbf{S}$ produces τ. In fact, $p_{\mathbf{S}}(\tau)$ is the description of $\mathbf{S}$ and independent of the adversary $\mathscr{A}$. Then we describe the H-coefficient technique [7,12]. Generically, it considers an adversary that aims at distinguishing a "real" system $\mathbf{S}_1$ from an"ideal" system $\mathbf{S}_0$. The interactions of the adversary with those two systems induce two transcript distributions D_1 and D_0 respectively. It is well known that the statistical distance $\mathsf{SD}(D_0, D_1)$ is an upper bound on the distinguishing advantage of $\mathscr{A}$. That is to say,

$$\mathrm{Adv}_F^{\mathrm{prf}}(\mathscr{A}) \leq \mathsf{SD}(D_0, D_1).$$

Lemma 1. [7,12] *Suppose that the set of attainable transcripts for the ideal system can be partitioned into good and bad ones. If there exists $\epsilon \geq 0$ such that $\frac{\mathrm{ps}_1(\tau)}{\mathrm{ps}_0(\tau)} \geq 1 - \epsilon$ for any good transcript τ, then*

$$\mathsf{SD}(D_0, D_1) \leq \epsilon + \Pr[D_0 \textit{ is bad}].$$

This lemma shows that $\epsilon + \Pr[D_0$ is bad$]$ is the upper bound of $\mathrm{Adv}_F^{\mathrm{prf}}(\mathscr{A})$.

Regular and AU. Let $H : \mathcal{K}_h \times \mathcal{X} \rightarrow \mathcal{Y}$ be a hash function where $\mathcal{K}_h$ is the key space, $\mathcal{X}$ is the domain and $\mathcal{Y}$ is the range. Hash function H is said to be ϵ_1-regular if for any $X \in \mathcal{X}$, $Y \in \mathcal{Y}$,

$$\Pr[K_h \xleftarrow{\$} \mathcal{K}_h : H_{K_h}(X) = Y] \leq \epsilon_1.$$

And hash function H is said to be ϵ_2-AU if for any two distinct strings $X, X' \in \mathcal{X}$,

$$\Pr[K_h \xleftarrow{\$} \mathcal{K}_h : H_{K_h}(X) = H_{K_h}(X')] \leq \epsilon_2.$$

3 BBB-Security Framework in [14]

Let $\mathcal{M}$ be the message space and $\mathcal{K}_h \times \mathcal{K}$ be the key space. Let block cipher $E : \mathcal{K} \times \{0,1\}^n \rightarrow \{0,1\}^n$ and $\mathcal{K} = \{0,1\}^k$. Let hash function $H : \mathcal{K}_h \times \mathcal{M} \rightarrow \{0,1\}^n \times \{0,1\}^n$. The function H is consist of two n-bit hash functions H^1 and H^2, i,e., $H_{K_h}(M) = (H^1_{K_{h,1}}(M), H^2_{K_{h,2}}(M))$ where $K_h = (K_{h,1}, K_{h,2}) \in \mathcal{K}_{h,1} \times \mathcal{K}_{h,2}$ and $K_{h,1}, K_{h,2}$ are two independent keys. Then the two-key DbHtS framework in paper [14] (see Fig. 1) is

$$\mathrm{DbHtS}[H, E](K_h, K, M) = E_K\left(H^1_{K_{h,1}}(M)\right) \oplus E_K\left(H^2_{K_{h,2}}(M)\right).$$

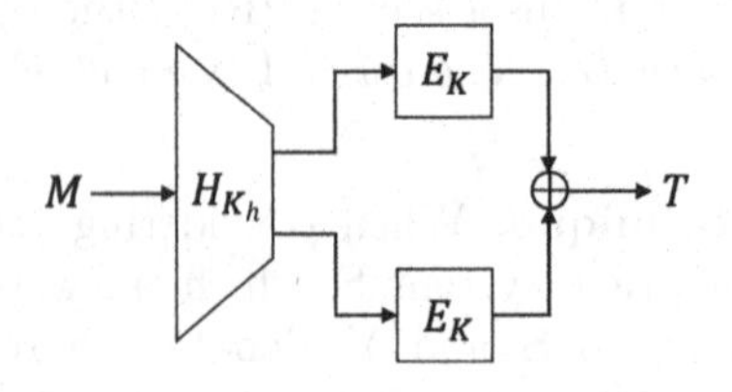

Fig. 1. The two-key DbHtS construction. Here H is a $2n$-bit hash function from $\mathcal{K}_h \times \mathcal{M}$ to $\{0,1\}^n \times \{0,1\}^n$, and E is a n-bit block cipher from $\mathcal{K} \times \{0,1\}^n$ to $\{0,1\}^n$.

Shen et al. [14] considered the multi-user security of this framework in the ideal-cipher model, where they regarded the keyed block cipher as ideal cipher. So they gave the additional ideal-cipher oracles expect for the evaluation oracles when considering the multi-user security of the two-key DbHtS framework. Let $\mathbf{S}_1$ be "real" system and $\mathbf{S}_0$ be "ideal" system. That is to say, the system $\mathbf{S}_b$ with $b \xleftarrow{\$} \{0,1\}$ about adversary $\mathscr{A}$ against the two-key DbHtS frameworks or random functions in the u-user setting performs the following procedure.

1. Initialize $(K_h^1, K_1), \ldots, (K_h^u, K_u) \xleftarrow{\$} \mathcal{K}_h \times \mathcal{K}$ if $b = 1$; otherwise, initialize $f_1, \ldots, f_u \xleftarrow{\$} \mathsf{Func}(\mathcal{M}, \{0,1\}^n)$;
2. If an adversary $\mathscr{A}$ queries Eval function with (i, M), where $i \in \{1, 2, \ldots\}$, $M \in \mathcal{M}$, return

$$\mathsf{Eval}(i, M) = \begin{cases} \mathrm{DbHtS}[H, E](K_h^i, K_i, M), & \text{if } b = 1, \\ f_i(M), & \text{if } b = 0; \end{cases}$$

3. If an adversary $\mathscr{A}$ queries Prim function with (J, X), where $J \in \mathcal{K}, X \in \{+,-\} \times \{0,1\}^n$, return

$$\mathsf{Prim}(J, X) = \begin{cases} E_J(x), & \text{if } X = \{+, x\}, \\ E_J^{-1}(y), & \text{if } X = \{-, y\}. \end{cases}$$

After the procedure in system $\mathbf{S}_b$, adversary $\mathscr{A}$ outputs a bit $b' \in \{0,1\}$. The advantage of the adversary $\mathscr{A}$ against the multi-user PRF security of two-key DbHtS framework is still $\mathrm{Adv}_{\mathrm{DbHtS}}^{\mathrm{prf}}(\mathscr{A}) = \left|\Pr[b' = 1 | b = 1] - \Pr[b' = 1 | b = 0]\right|$. When $u = 1$, the adversary $\mathscr{A}$ is against the single-user PRF security. They called the query to Eval evaluation query and the query to Prim ideal-cipher query.

Theorem 1 in [14]**.** Let E be modeled as an ideal cipher. Let H^1 and H^2 both satisfy ϵ_1-regular and ϵ_2-AU. Then Shen et al. [14] proved the security of two-key DbHtS in the multi-user setting as following, which is the core of their paper and they named it Theorem 1. For any adversary $\mathscr{A}$ that makes at most q evaluation queries and p ideal-cipher queries,

$$\begin{aligned} \mathrm{Adv}_{\mathrm{DbHtS}}^{\mathrm{prf}}(\mathscr{A}) \leq \frac{2q}{2^k} + \frac{q(3q+p)(6q+2p)}{2^{2k}} + \frac{2qp\ell}{2^{n+k}} + \frac{2qp\epsilon_1}{2^k} + \frac{4qp}{2^{n+k}} \\ + \frac{4q^2\epsilon_1}{2^k} + \frac{2q^2\ell\epsilon_1}{2^k} + 2q^3(\epsilon_1 + \epsilon_2)^2 + \frac{8q^3(\epsilon_1 + \epsilon_2)}{2^n} + \frac{6q^3}{2^{2n}} \end{aligned} \tag{1}$$

where ℓ is the maximal block length among these evaluation queries and assuming $p + q\ell \leq 2^{n-1}$.

An Overview of the Proof of Theorem 1 in [14]**.** They used H-coefficient technique to get the upper bound of $\mathrm{Adv}_{\mathrm{DbHtS}}^{\mathrm{prf}}(\mathscr{A})$. For each query $T \leftarrow \mathsf{Eval}(i, M)$, they associated it with an entry $(eval, i, M, T)$. The query to Prim is similar to it. These two entries are included in transcript τ. Then they defined bad transcripts, including fourteen cases. If a transcript is not bad then they said it's good. Let D_1 and D_0 be the random variables for the transcript distributions in the system $\mathbf{S}_1$ and $\mathbf{S}_0$ respectively. They firstly bounded the probability that D_0 is bad as follows. Let Bad_i be the event that the i-th case of bad transcripts happens. They calculated the probability $\Pr[\mathrm{Bad}_1], \ldots, \Pr[\mathrm{Bad}_{14}]$ in sequence. After summing up, they got

$$\begin{aligned}\Pr\left[D_0 \text{ is bad }\right] &\leq \sum_{i=1}^{14} \Pr\left[\mathrm{Bad}_i\right] \\ &\leq \frac{2q}{2^k} + \frac{q(3q+p)(6q+2p)}{2^{2k}} + \frac{2qp\ell}{2^{k+n}} + \frac{2qp\epsilon_1}{2^k} + \frac{4qp}{2^{n+k}} \\ &\quad + \frac{4q^2\epsilon_1}{2^k} + \frac{2q^2\ell\epsilon_1}{2^k} + 2q^3\left(\epsilon_1+\epsilon_2\right)^2 + \frac{8q^3\left(\epsilon_1+\epsilon_2\right)}{2^n}.\end{aligned}$$

Besides, they proved the transcript ratio $\frac{\mathrm{p}_{\mathbf{S}_1}(\tau)}{\mathrm{p}_{\mathbf{S}_0}(\tau)} \geq 1 - \frac{6q^3}{2^{2n}}$ for any good transcript τ. Thus they concluded Theorem 1 in [14] by Lemma 1.

4 Counter-Examples

We show that the regular and universal properties of the hash functions can not guarantee the BBB security of two-key DbHtS MACs. We construct three universal hash functions which satisfy the properties, leading to $2n/3$-bit security of two-key DbHtS MACs by Theorem 1 in [14]. But all the instantiations can be broken with probability 1 using only $\mathcal{O}(2^{n/2})$ queries even in the single-user setting.

4.1 Counter-Example 1

Our first counter-example uses universal hash functions with fixed input length. Let function

$$H_{K_h}(M) = (H^1_{K_1}(M), H^2_{K_2}(M)) = (M \oplus K_1, M \oplus K_2),$$

where M is the message from massage space $\{0,1\}^n$, $K_h = (K_1, K_2)$ and $K_1, K_2 \xleftarrow{\$} \{0,1\}^n$. Let block cipher $E_K : \{0,1\}^n \times \{0,1\}^n \to \{0,1\}^n$. Then the derived two-key DbHtS MAC is $F : \{0,1\}^{2n} \times \{0,1\}^n \times \{0,1\}^n \to \{0,1\}^n$ as

$$F[H,E](K_h, K, M) = E_K(H^1_{K_1}(M)) \oplus E_K(H^2_{K_2}(M)).$$

H^1 and H^2 are $\frac{1}{2^n}$-Regular and 0-AU. It is easy to know that for any $M \in \{0,1\}^n, Y \in \{0,1\}^n$ and $i \in \{1,2\}$,

$$\Pr[K_i \xleftarrow{\$} \{0,1\}^n : M \oplus K_i = Y] \leq \frac{1}{2^n}.$$

And for any two distinct strings $M, M' \in \{0,1\}^n$ and $i \in \{1,2\}$,

$$\Pr[K_i \xleftarrow{\$} \{0,1\}^n : M \oplus K_i = M' \oplus K_i] = 0.$$

So hash functions H^1 and H^2 are both $\frac{1}{2^n}$-regular and 0-AU.

$2n/3$-Bit Security. According to Theorem 1 [14], function F is secure up to $\mathcal{O}(2^{2n/3})$ evaluation queries assuming ideal-cipher queries is $\mathcal{O}(1)$ in the multi-user setting.

Attack with $\mathcal{O}(2^{n/2})$ Query Complexity. Assume adversary $\mathscr{A}$ is an adversary against the single-user PRF security of F. Then $\mathscr{A}$ will output $b' \in \{0,1\}$ for system $\mathbf{S}_b$ where $u = 1$ (see Sect. 3). In the following, we will construct an adversary $\mathscr{A}$ such that $\mathrm{Adv}_F^{\mathrm{prf}}(\mathscr{A}) = |\Pr[b'=1|b=1] - \Pr[b'=1|b=0]| \approx 1$ with only $\mathcal{O}(2^{n/2})$ evaluation queries. It means there is an adversary that can distinguish F from random function f with only $\mathcal{O}(2^{n/2})$ evaluation queries, which is contradictory to Theorem 1 [14].

It is easy to know that for all keys in keyspace and messages in message space,

$$\begin{aligned} F[H,E](K_h, K, M \oplus K_1 \oplus K_2) &= E_K(M \oplus K_2) \oplus E_K(M \oplus K_1) \\ &= F[H,E](K_h, K, M). \end{aligned}$$

It means F has a period $s := K_1 \oplus K_2$. Based on this, we construct adversary $\mathscr{A}$ as follows.

1. $\mathscr{A}$ firstly makes $\mathcal{O}(2^{n/2})$ evaluation queries of distinct massages $M_1, M_2, \ldots$ chosen uniformly and randomly, and get $T_1, T_2, \ldots$;
2. $\mathscr{A}$ outputs 1 if it searches a message pair (M_i, M_j) for $M_i \neq M_j, M_i, M_j \in \{M_1, M_2, \ldots\}$ which makes (i) and (ii) hold.
   (i) $T_i = T_j$;
   (ii) After make another two evaluation queries with massages M' and $M' \oplus M_i \oplus M_j$ for $M' \notin \{M_i, M_j\}$, $\mathscr{A}$ gets two identical answers.
   Else, $\mathscr{A}$ outputs 0.

If it is in system $\mathbf{S}_1$, the evaluation query is to F. One can expect on average that there exists one message pair (M_i, M_j) among $\mathcal{O}(2^{n/2})$ massages such that $M_i = M_j \oplus s$. Conditions (i) and (ii) in the second step of $\mathscr{A}$ filter out such a pair.

If it is in system $\mathbf{S}_0$, the evaluation query is to random function f. On average there exists one message pair (M_i, M_j) among $\mathcal{O}(2^{n/2})$ massages such that $T_i = T_j$. However, random function f has no period. So the probability of $f(M') = f(M' \oplus M_i \oplus M_j)$ for any $M' \notin \{M_i, M_j\}$ is only $1/2^n$. Thus $\mathscr{A}$ finds a pair (M_i, M_j) satisfying conditions (i) and (ii) with $1/2^n$.

From above, we get that such $\mathscr{A}$ distinguishes F from random function with probability $1 - 1/2^n \approx 1$.

4.2 Counter-Example 2

Compared with the first counter-example with fixed input length, our second counter-example can handle variable-length input. We construct two hash functions H^1 and H^2 dealing with messages from $(\{0,1\}^n)^*$:

$$H^i_{K_i}(M) = M[1] \oplus M[2]K_i \oplus M[3]K_i^2 \oplus \ldots \oplus M[m]K_i^{m-1} \oplus |M|K_i^m, i = 1, 2.$$

where $M = M[1] \parallel M[2] \parallel \ldots \parallel M[m]$ and every message block is n-bit. This example is a variant of PolyMAC [9].

H^1 and H^2 are $\frac{\ell}{2^n}$-Regular and $\frac{\ell}{2^n}$-AU. Assume the maximal block length of all evaluation queries is ℓ. Any equation of at most ℓ degree has at most ℓ roots. So it is easy to know that for any $M \in (\{0,1\}^n)^*, Y \in \{0,1\}^n$ and $i \in \{1,2\}$,

$$\Pr[K_i \xleftarrow{\$} \{0,1\}^n : H^i_{K_i}(M) = Y] \leq \frac{\ell}{2^n}.$$

And for any two distinct strings $M, M' \in (\{0,1\}^n)^*$ and $i \in \{1,2\}$,

$$\Pr[K_i \xleftarrow{\$} \{0,1\}^n : H^i_{K_i}(M) = H^i_{K_i}(M')] \leq \frac{\ell}{2^n}.$$

It means H^1 and H^2 are both $\frac{\ell}{2^n}$-regular and $\frac{\ell}{2^n}$-AU.

$2n/3$-Bit Security. According to Theorem 1 [14], function F is secure up to $\mathcal{O}(2^{2n/3})$ evaluation queries assuming ideal-cipher queries is $\mathcal{O}(1)$ and $\ell = \mathcal{O}(1)$ in the multi-user setting.

Attack with $\mathcal{O}(2^{n/2})$ Query Complexity. Similar to counter-example 1, there is an adversary $\mathscr{A}$ who can distinguish F from random function f with only $\mathcal{O}(2^{n/2})$ evaluation queries in single-user setting, which is contradictory to Theorem 1 [14].

Fix any arbitrary string

$$M_{fix} := M[2] \| M[3] \| \ldots \| M[m] \in (\{0,1\}^n)^{m-1},$$

where $2 \leq m \leq \ell = O(1)$. Let

$$K_i' := M[2]K_i \oplus M[3]K_i^2 \oplus \ldots M[m]K_i^{m-1} \oplus nmK_i^m, i = 1, 2.$$

Then it is easy to obtain for any keys in key space and $M[1] \in \{0,1\}^n$,

$$\begin{aligned} &F[H,E](K_h, K, (M[1] \oplus K_1' \oplus K_2') \parallel M_{fix}) \\ =\, &E_K(M[1] \oplus K_2') \oplus E_K(M[1] \oplus K_1') \\ =\, &F[H,E](K_h, K, M[1] \parallel M_{fix}). \end{aligned}$$

It means F has a period $s := (K_1' \oplus K_2') \parallel 0^{n(m-1)}$ for any $M \in \{0,1\}^n \times \{M_{fix}\}$. Based on this, we construct adversary $\mathscr{A}$ as follows.

1. $\mathscr{A}$ firstly makes $\mathcal{O}(2^{n/2})$ evaluation queries with distinct massages $M_1 \parallel M_{fix}$, $M_2 \parallel M_{fix}, \ldots$ where $M_1, M_2, \ldots \xleftarrow{\$} \{0,1\}^n$, and get $T_1, T_2, \ldots$;
2. $\mathscr{A}$ outputs 1 if it searches a pair (M_i, M_j) for $M_i \neq M_j, M_i, M_j \in \{M_1, M_2, \ldots\}$ which makes (i) and (ii) hold.
   (i) $T_i = T_j$;
   (ii) After make another two evaluation queries with massages $M' \parallel M_{fix}$ and $(M' \oplus M_i \oplus M_j) \parallel M_{fix}$ for $M' \notin \{M_i, M_j\}$, $\mathscr{A}$ gets two identical answers.
   Else, $\mathscr{A}$ outputs 0.

The same as counter-example 1, $\mathscr{A}$ distinguishes F from f with the probability of almost 1.

4.3 Counter-Example 3

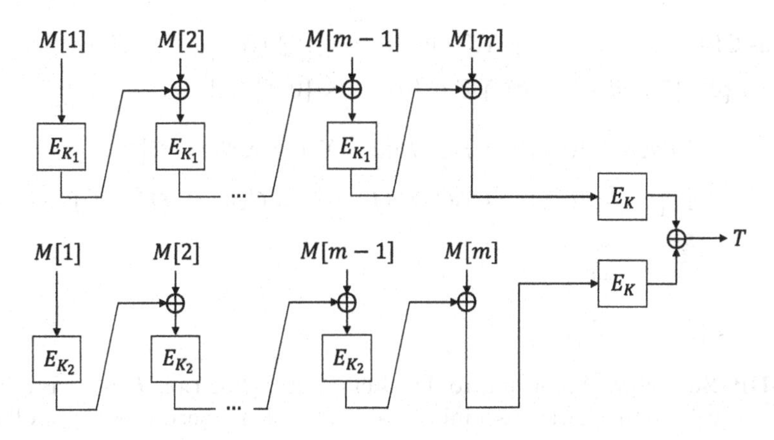

Fig. 2. The variant of 2k-SUM-ECBC. K_1, K_2, K_3 are three independent keys in $\{0,1\}^n$. E is a n-bit block cipher from $\{0,1\}^n \times \{0,1\}^n$ to $\{0,1\}^n$.

Unlike counter-examples 1 and 2, the third counter-example is based on block ciphers. Let $E : \{0,1\}^n \times \{0,1\}^n \to \{0,1\}^n$ be a block cipher with key $K \in \{0,1\}^n$. The two n-bit hash functions used in this function are two CBC MACs without the last block cipher call, which we name CBC$'$. They are keyed with two independent keys K_1 and K_2 respectively. And they deal with at least two message blocks respectively. For a message $M = M[1] \parallel M[2] \parallel \ldots \parallel M[m]$ where every message block is n-bit and $m \geq 2$, the CBC$'$ algorithm $\mathrm{CBC}'[E](K, M)$ is defined as Y_m, where

$$Y_1 = M[1],$$
$$Y_j = E_K(Y_{j-1}) \oplus M[j], j = 2, \ldots, m.$$

Let $K_h = (K_1, K_2)$. Then we define the function as

$$F[\mathrm{CBC}'[E], E](K_h, K, M) = E_K(\mathrm{CBC}'[E](K_1, M)) \oplus E_K(\mathrm{CBC}'[E](K_2, M)).$$

F (see Fig. 2) can be seen as a variant of 2k-SUM-ECBC [14].

$\text{CBC}'[E]$ is $\left(\frac{2\ell}{2^n}+\frac{16\ell^4}{2^{2n}}\right)$-Regular and $\left(\frac{2\ell}{2^n}+\frac{16\ell^4}{2^{2n}}\right)$-AU. For any two different message $M, M' \in (\{0,1\}^n)^*$ with at most ℓ blocks and the block cipher E_K being a random permutation, Ballare et al. [3] showed that for $i \in \{1,2\}$,

$$\Pr[E_K(\text{CBC}'[E](K_i, M)) = E_K(\text{CBC}'[E](K_i, M'))] \le \frac{2\ell}{2^n} + \frac{16\ell^4}{2^{2n}}.$$

block cipher E_K is a permutation. So

$$\Pr[\text{CBC}'[E](K_i, M) = \text{CBC}'[E](K_i, M')] \le \frac{2\ell}{2^n} + \frac{16\ell^4}{2^{2n}}.$$

Thus for ideal block cipher E we get

$$\Pr[K_i \xleftarrow{\$} \{0,1\}^n : \text{CBC}'[E](K_i, M) = \text{CBC}'[E](K_i, M')] \le \frac{2\ell}{2^n} + \frac{16\ell^4}{2^{2n}}.$$

It means CBC' is $\left(\frac{2\ell}{2^n}+\frac{16\ell^4}{2^{2n}}\right)$-AU. Let $M = X[1] \parallel (X[2] \oplus Y) \parallel Z \in (\{0,1\}^n)^* \times \{0,1\}^n \times \{0,1\}^n$ and $M' = 0^n \parallel Z \in \{0,1\}^n \times \{0,1\}^n$. Then

$$\begin{aligned} &\Pr[K_i \xleftarrow{\$} \{0,1\}^n : \text{CBC}'[E](K_i, X[1] \parallel X[2]) = Y] \\ =\ &\Pr[K_i \xleftarrow{\$} \{0,1\}^n : \text{CBC}'[E](K_i, M) = \text{CBC}'[E](K_i, M')] \\ \le\ &\frac{2\ell}{2^n} + \frac{16\ell^4}{2^{2n}}. \end{aligned}$$

So CBC' is $\left(\frac{2\ell}{2^n}+\frac{16\ell^4}{2^{2n}}\right)$-regular.

$2n/3$-Bit Security. According to Theorem 1 [14], function F is secure up to $\mathcal{O}(2^{2n/3})$ evaluation queries assuming no ideal-cipher queries and $\ell = \mathcal{O}(1)$ in the multi-user setting.

Attack with $\mathcal{O}(2^{n/2})$ Query Complexity. Fix any arbitrary string $M_{fix} \in (\{0,1\}^n)^{m-1}$ where $2 \le m \le \ell = O(1)$. Let

$$s' = \text{CBC}'[E](K_1, M_{fix} \parallel 0^n) \oplus \text{CBC}'[E](K_2, M_{fix} \parallel 0^n)).$$

Then it is easy to obtain for any keys in key space and $M[m] \in \{0,1\}^n$,

$$\begin{aligned} &F[\text{CBC}'[E], E](K_h, K, M_{fix} \parallel (M[m] \oplus s')) \\ =\ &E_K(\text{CBC}'[E](K_2, M_{fix} \parallel 0^n) \oplus M[m]) \oplus \\ &E_K(\text{CBC}'[E](K_1, M_{fix} \parallel 0^n) \oplus M[m]) \\ =\ &E_K(\text{CBC}'[E](K_2, M_{fix} \parallel M[m])) \oplus E_K(\text{CBC}'[E](K_1, M_{fix} \parallel M[m])) \\ =\ &F[\text{CBC}'[E], E](K_h, K, M_{fix} \parallel M[m]). \end{aligned}$$

It means F has a period $s := 0^{n(m-1)} \parallel s'$ for any $M \in \{M_{fix}\} \times \{0,1\}^n$. So there is an adversary $\mathscr{A}$ distinguishes F from random function with only $\mathcal{O}(2^{n/2})$ evaluation queries when considering a single user similar to counter-example 2.

5 The Flaw of the Proof of Theorem 1 in [14]

In Sect. 3, we have given an overview of how Shen et al. [14] proved Theorem 1 based on the H-coefficient technique. However, we find they made a critical flaw when they were calculating $\Pr[\mathrm{Bad}_9]$ in their proof, which leads to our counter-examples.

We firstly introduce some preliminaries in their proof for understanding the flaw. Assume there are u users and the adversary make q_i evaluation queries to the i-th user in all. Let $(eval, i, M_a^i, T_a^i)$ be the entry obtained when the adversary makes the a-th query to user i. In the "real" system $\mathbf{S}_1$, during the computation of entry $(eval, i, M_a^i, T_a^i)$, let Σ_a^i and Λ_a^i be the internal outputs of hash function H, namely $\Sigma_a^i = H^1_{K_{h,1}}(M_a^i)$ and $\Lambda_a^i = H^2_{K_{h,2}}(M_a^i)$ respectively, and denote by U_a^i and V_a^i the outputs of block cipher E with inputs Σ_a^i and Λ_a^i respectively, namely $U_a^i = E(K_i, \Sigma_a^i)$ and $V_a^i = E(K_i, \Lambda_a^i)$ respectively. The adversary also obtain the entries of ideal-cipher queries. After all queries, they further gave it: (i) the keys (K_h^i, K_i) where $K_h^i = (K_{h,1}^i, K_{h,2}^i)$ and (ii) the internal values U_a^i and V_a^i. In the "ideal" system $\mathbf{S}_0$, they instead gave the adversary truly random strings $(K_h^i, K_i) \xleftarrow{\$} \mathcal{K}_h \times \mathcal{K}$, independent of its queries. In addition, they gave the adversary dummy values U_a^i and V_a^i computed by an simulation oracle. These additional information can only help the adversary. Thus a transcript consists of the revealed keys (K_h^i, K_i), the internal values U_a^i and V_a^i, the ideal-cipher queries and evaluation queries.

The ninth bad event is

"There is an entry $(eval, i, M_a^i, T_a^i)$ such that either $\Sigma_a^i = \Sigma_b^i$ or $\Sigma_a^i = \Lambda_b^i$, and either $\Lambda_a^i = \Lambda_b^i$ or $\Lambda_a^i = \Sigma_b^i$ for some entry $(eval, i, M_a^i, T_a^i)$."

They defined this event as bad for the reason that the appearance of such bad event is easily used to distinguish systems $\mathbf{S}_1$ and $\mathbf{S}_0$. We call the event of either $\Sigma_a^i = \Sigma_b^i$ or $\Sigma_a^i = \Lambda_b^i$ as event 1, and the event of either $\Lambda_a^i = \Lambda_b^i$ or $\Lambda_a^i = \Sigma_b^i$ as event 2. Then we can regard the simultaneous events 1 and 2 as one of the following 4 events:

- Event 3: $\Sigma_a^i = \Sigma_b^i \wedge \Lambda_a^i = \Lambda_b^i$;
- Event 4: $\Sigma_a^i = \Sigma_b^i \wedge \Lambda_a^i = \Sigma_b^i$;
- Event 5: $\Sigma_a^i = \Lambda_b^i \wedge \Lambda_a^i = \Lambda_b^i$;
- Event 6: $\Sigma_a^i = \Lambda_b^i \wedge \Lambda_a^i = \Sigma_b^i$.

In the "real" system $\mathbf{S}_1$, event 4 or 5 leads to $T_a^i = 0^n$; event 3 or 6 leads to $T_a^i = T_b^i$. However in the "ideal" system $\mathbf{S}_0$ these happen with negligible probability by the randomness of random function f_i. And in the "real" system $\mathbf{S}_1$, $U_a^i = E(K_i, \Sigma_a^i)$ and $V_a^i = E(K_i, \Lambda_a^i)$. However in the "ideal" system $\mathbf{S}_0$, $(U_a^i, V_a^i) = \bot$ by the simulation oracle defined by them. Thus it is easy distinguish these two systems.

When calculating $\Pr[\mathrm{Bad}_9]$ in the "ideal" system $\mathbf{S}_0$, Shen et al. [14] regarded that the event 1 is independent from event 2 when $K_{h,1}^i, K_{h,2}^i$ are independent from each other. So by H^1, H^2 are both ϵ_1-regular and ϵ_2-AU, they thought the probability of event 1 (resp. event 2) is at most $\epsilon_1 + \epsilon_2$. Note that for each user, there are at most q_i^2 pairs of (a, b). So they summed among u users and got

$$\Pr[\text{Bad}_9] \leq \Sigma_{i=1}^{u} q_i^2 (\epsilon_1 + \epsilon_2)^2 \leq q^2 (\epsilon_1 + \epsilon_2)^2.$$

In fact, even if $K_{h,1}^i, K_{h,2}^i$ are independent of each other, the event 1 and event 2 may not be independent, which has been shown in counter-examples 1–3. We regard the ninth event as the union set of events 3, 4, 5 and 6. Event 3 holds with probability at most ϵ_2^2 by the assumption that H^1 and H^2 are ϵ_2-AU. Event 4 holds with probability at most $\epsilon_1\epsilon_2$ by the assumption that H^1 is ϵ_2-AU and H^2 is ϵ_1-regular. Event 5 holds with probability at most $\epsilon_1\epsilon_2$ by the assumption that H^1 is ϵ_1-regular and H^2 is ϵ_2-AU. For event 6,

$$\begin{aligned}
&\Pr[K_{h,1}^i \xleftarrow{\$} \mathcal{K}_{h,1}, K_{h,2}^i \xleftarrow{\$} \mathcal{K}_{h,2} : \Sigma_a^i = \Lambda_b^i \wedge \Lambda_a^i = \Sigma_b^i] \\
&= \Pr[K_{h,1}^i \xleftarrow{\$} \mathcal{K}_{h,1}, K_{h,2}^i \xleftarrow{\$} \mathcal{K}_{h,2} : \Sigma_a^i = \Lambda_b^i | \Lambda_a^i = \Sigma_b^i] \\
&\quad \cdot \Pr[K_{h,1}^i \xleftarrow{\$} \mathcal{K}_{h,2}, K_{h,2}^i \xleftarrow{\$} \mathcal{K}_{h,1} : \Lambda_a^i = \Sigma_b^i] \\
&\leq \epsilon_3\epsilon_1
\end{aligned}$$

by the assumption that H^2 is ϵ_1-regular and let

$$\epsilon_3 = \Pr[K_{h,1}^i \xleftarrow{\$} \mathcal{K}_{h,1}, K_{h,2}^i \xleftarrow{\$} \mathcal{K}_{h,2} : \Sigma_a^i = \Lambda_b^i | \Lambda_a^i = \Sigma_b^i].$$

So we sum among u users and get

$$\Pr[\text{Bad}_9] \leq \Sigma_{i=1}^{u} q_i^2 (\epsilon_2^2 + 2\epsilon_1\epsilon_2 + \epsilon_3\epsilon_1) \leq q^2 (\epsilon_2^2 + 2\epsilon_1\epsilon_2 + \epsilon_3\epsilon_1).$$

For counter-examples 1–3, it is easy to get $\epsilon_3 = 1$. So for these cases, $\Pr[\text{Bad}_9] \leq q^2(\epsilon_2^2 + 2\epsilon_1\epsilon_2 + \epsilon_1)$. If we substitute our $\Pr[\text{Bad}_9]$ for that in paper [14], we get the security of proofs of counter-examples 1–3 should be up to $\mathcal{O}(2^{n/2})$ evaluation queries assuming ideal-cipher queries are $\mathcal{O}(1)$ and the maximal block length of all evaluation queries is $\mathcal{O}(1)$, which is consistent with attacks.

6 Conclusion

In this paper, we point out a flaw of the security framework for two-key DbHtS in the multi-user setting proposed by Shen et al. [14] by constructing three counter-examples. We also analyze how the flaw happens in their proof. This is due to the fact that the authors overlooked the dependence of $\Sigma_a^i = \Lambda_b^i$ and $\Lambda_a^i = \Sigma_b^i$ in the proof of Theorem 1 [14]. In their paper, they also stated that 2k-SUM-ECBC, 2k-PMAC_Plus, and 2k-LightMAC_Plus all achieve $2n/3$-bit security. For 2k-SUM-ECBC based on two independent CBC MACs, the probability ϵ_3 is about $\frac{1}{2^n}$. So if we substitute our $\Pr[\text{Bad}_9]$ for that in paper [14], 2k-SUM-ECBC still achieves $2n/3$ security. The two universal hash functions of 2k-PMAC_Plus or 2k-LightMAC_Plus are dependent, they adjusted the concrete proof of these two MACs from the framework. We haven't found attacks against these three MACs.

Recently, Shen et al. refined their paper [15] because of what we have found in this paper. Their new framework for two-key DbHtS is not universal. Because the $2n/3$-bit security of two-key DbHtS MACs doesn't only come from the regular and AU properties of the hash functions anymore. To be specific, they added two variables to capture the probabilities of two subcases '$\Sigma_a^i = \Lambda_b^i \wedge \Lambda_a^i = \Sigma_b^i$' and '$\Sigma_a^i = \Lambda_b^i \wedge \Lambda_a^i = \Sigma_c^i$', the values of which will be clear until in the analysis of concrete MAC. In fact, these two added subcases have been included in the cover-free property of H by Datta et al. [8] when they considered the framework of two-key DbHtS in the single-user setting.

Acknowledgments. The authors thank the anonymous reviewers for many helpful comments. This paper was supported by the NSFC of China (61732021) and the National Key R&D Program of China (2018YFB0803801 and 2018YFA0704704).

References

1. Banik, S., Pandey, S.K., Peyrin, T., Sasaki, Y., Sim, S.M., Todo, Y.: GIFT: a small present. In: Fischer, W., Homma, N. (eds.) CHES 2017. LNCS, vol. 10529, pp. 321–345. Springer, Cham (2017). https://doi.org/10.1007/978-3-319-66787-4_16
2. Bellare, M.: New proofs for NMAC and HMAC: security without collision-resistance. In: Dwork, C. (ed.) CRYPTO 2006. LNCS, vol. 4117, pp. 602–619. Springer, Heidelberg (2006). https://doi.org/10.1007/11818175_36
3. Bellare, M., Pietrzak, K., Rogaway, P.: Improved security analyses for CBC MACs. In: Shoup, V. (ed.) CRYPTO 2005. LNCS, vol. 3621, pp. 527–545. Springer, Heidelberg (2005). https://doi.org/10.1007/11535218_32
4. Black, J., Rogaway, P.: CBC MACs for arbitrary-length messages: the three-key constructions. In: Bellare, M. (ed.) CRYPTO 2000. LNCS, vol. 1880, pp. 197–215. Springer, Heidelberg (2000). https://doi.org/10.1007/3-540-44598-6_12
5. Black, J., Rogaway, P.: A block-cipher mode of operation for parallelizable message authentication. In: Knudsen, L.R. (ed.) EUROCRYPT 2002. LNCS, vol. 2332, pp. 384–397. Springer, Heidelberg (2002). https://doi.org/10.1007/3-540-46035-7_25
6. Bogdanov, A., Knudsen, L.R., Leander, G., Paar, C., Poschmann, A., Robshaw, M.J.B., Seurin, Y., Vikkelsoe, C.: PRESENT: an ultra-lightweight block cipher. In: Paillier, P., Verbauwhede, I. (eds.) CHES 2007. LNCS, vol. 4727, pp. 450–466. Springer, Heidelberg (2007). https://doi.org/10.1007/978-3-540-74735-2_31
7. Chen, S., Steinberger, J.: Tight security bounds for key-alternating ciphers. In: Nguyen, P.Q., Oswald, E. (eds.) EUROCRYPT 2014. LNCS, vol. 8441, pp. 327–350. Springer, Heidelberg (2014). https://doi.org/10.1007/978-3-642-55220-5_19
8. Datta, N., Dutta, A., Nandi, M., Paul, G.: Double-block hash-then-sum: a paradigm for constructing BBB secure PRF. IACR Trans. Symmetric Cryptol. **2018**(3), 36–92 (2018). https://doi.org/10.13154/tosc.v2018.i3.36-92
9. Kim, S., Lee, B., Lee, J.: Tight security bounds for double-block hash-then-sum MACs. In: Canteaut, A., Ishai, Y. (eds.) EUROCRYPT 2020. LNCS, vol. 12105, pp. 435–465. Springer, Cham (2020). https://doi.org/10.1007/978-3-030-45721-1_16
10. Leurent, G., Nandi, M., Sibleyras, F.: Generic attacks against beyond-birthday-bound MACs. In: Shacham, H., Boldyreva, A. (eds.) CRYPTO 2018. LNCS, vol. 10991, pp. 306–336. Springer, Cham (2018). https://doi.org/10.1007/978-3-319-96884-1_11

11. Naito, Y.: Blockcipher-based MACs: beyond the birthday bound without message length. In: Takagi, T., Peyrin, T. (eds.) ASIACRYPT 2017. LNCS, vol. 10626, pp. 446–470. Springer, Cham (2017). https://doi.org/10.1007/978-3-319-70700-6_16
12. Patarin, J.: The "Coefficients H" technique. In: Avanzi, R.M., Keliher, L., Sica, F. (eds.) SAC 2008. LNCS, vol. 5381, pp. 328–345. Springer, Heidelberg (2009). https://doi.org/10.1007/978-3-642-04159-4_21
13. Rogaway, P.: Efficient instantiations of tweakable blockciphers and refinements to modes OCB and PMAC. In: Lee, P.J. (ed.) ASIACRYPT 2004. LNCS, vol. 3329, pp. 16–31. Springer, Heidelberg (2004). https://doi.org/10.1007/978-3-540-30539-2_2
14. Shen, Y., Wang, L., Gu, D., Weng, J.: Revisiting the security of DbHtS MACs: beyond-birthday-bound in the multi-user setting. In: Malkin, T., Peikert, C. (eds.) CRYPTO 2021. LNCS, vol. 12827, pp. 309–336. Springer, Cham (2021). https://doi.org/10.1007/978-3-030-84252-9_11
15. Shen, Y., Wang, L., Weng, J.: Revisiting the security of DbHtS MACs: beyond-birthday-bound in the multi-user setting. IACR Cryptology ePrint Archive, p. 1523 (2020). https://eprint.iacr.org/2020/1523
16. Shoup, V.: Sequences of games: a tool for taming complexity in security proofs. IACR Cryptology ePrint Archive, p. 332 (2004). http://eprint.iacr.org/2004/332
17. Yasuda, K.: The sum of CBC MACs is a secure PRF. In: Pieprzyk, J. (ed.) CT-RSA 2010. LNCS, vol. 5985, pp. 366–381. Springer, Heidelberg (2010). https://doi.org/10.1007/978-3-642-11925-5_25
18. Yasuda, K.: A new variant of PMAC: beyond the birthday bound. In: Rogaway, P. (ed.) CRYPTO 2011. LNCS, vol. 6841, pp. 596–609. Springer, Heidelberg (2011). https://doi.org/10.1007/978-3-642-22792-9_34
19. Zhang, L., Wu, W., Sui, H., Wang, P.: 3kf9: enhancing 3GPP-MAC beyond the birthday bound. In: Wang, X., Sako, K. (eds.) ASIACRYPT 2012. LNCS, vol. 7658, pp. 296–312. Springer, Heidelberg (2012). https://doi.org/10.1007/978-3-642-34961-4_19

Maliciously Secure Multi-party PSI with Lower Bandwidth and Faster Computation

Zhi Qiu[1], Kang Yang[2], Yu Yu[1(✉)], and Lijing Zhou[3]

[1] Shanghai Jiao Tong University, Shanghai, China
{chonps,yyuu}@sjtu.edu.cn
[2] State Key Laboratory of Cryptology, Beijing, China
yangk@sklc.org
[3] Huawei Technology, Shanghai, China
zhoulijing@huawei.com

Abstract. Private Set Intersection (PSI) allows a set of mutually distrustful parties, each holds a private data set, to compute the intersection of all sets, such that no information is revealed except for the intersection. The state-of-the-art PSI protocol (Garimella et al., CRYPTO'21) in the multi-party setting tolerating any number of malicious corruptions requires the communication bandwidth of $O(n\ell|\mathbb{F}|)$ bits for the central party P_0 due to the star architecture, where n is the number of parties, ℓ is the size of each set and $|\mathbb{F}|$ is the size of an exponentially large field $\mathbb{F}$. When n and ℓ are large, this forms an efficiency bottleneck (especially for networks with restricted bandwidthes). In this paper, we present a new multi-party PSI protocol in dishonest-majority malicious setting, which reduces the communication bandwidth of the central party P_0 from $O(n\ell|\mathbb{F}|)$ bits to $O(\ell|\mathbb{F}|)$ bits using a tree architecture. Furthermore, our PSI protocol reduces the expensive LPN encoding operations performed by P_0 by a factor of n as well as the computational cost by $2n\ell$ hash operations in total. Additionally, while the multi-party PSI protocol (Garimella et al., CRYPTO'21) with a single output is secure, we present a simple attack against its multi-output extension, which allows an adversary to learn more information on the sets of honest parties beyond the intersection of all sets.

1 Introduction

Private Set Intersection (PSI) allows a set of mutually distrustful parties, where each holds a private set, to compute the intersection of all sets, without revealing anything beyond the intersection. PSI and its variants have found a wide variety of applications, including measuring the effectiveness of online advertising [18, 19], private contact discovery [10,20] and more. In the two-party setting, PSI protocols has been extensively studied and become truly practical with extremely fast implementations (see the recent work [7,14,15,26,27,29,31] and references therein). While two-party PSI is interesting for many applications, there are a lot

C. Alcaraz et al. (Eds.): ICICS 2022, LNCS 13407, pp. 69–88, 2022.
https://doi.org/10.1007/978-3-031-15777-6_5

of applications which are better suitable for the multi-party setting. For example, a) several companies intend to combine their data sets to find a target audience for an ad campaign [18]; b) a variant of multi-party PSI was recently used for cache sharing in edge computing, which allows multiple network operators to obtain a set of common data items with the highest access frequencies in the privacy-preserving way [25]. We refer the reader to [24] for more interesting examples that are suited to the multi-party case.

The problem of multi-party PSI was first introduced in [13]. The previous work [8,13,15,16,22,32,33] constructed theoretical multi-party PSI protocols, where all of these protocols are not concretely efficient (especially for large sets). The first practical PSI protocol in the multi-party setting was proposed by Kolesnikov et al. [23]. This protocol is secure against semi-honest adversaries in the dishonest-majority setting (i.e., the adversary can corrupt up to $n-1$ parties of the n parties but must follow the protocol specification). For semi-honest security, efficient multi-party PSI protocols were further developed, including [17,21] in the dishonest-majority setting based on Garbled Bloom Filter (GBF), and [6] in the honest-majority setting (i.e., the adversary can corrupt up to less than a half of parties). In this work, we focus on multi-party PSI protocols in the dishonest-majority setting in the presence of malicious adversaries (i.e., the adversary allows to run an arbitrary attack strategy in its attempt to break the protocol). This is the strongest adversary model that was considered in the previous PSI protocols.

In the dishonest-majority malicious setting, several concretely efficient PSI protocols [12,14,24,37] have been proposed. Among these protocols, the multi-party PSI protocol by Garimella et al. [14] achieves the best efficiency in this setting. This protocol builds on the multi-party PSI protocol with augmented semi-honest security [23], and instantiates the underlying Oblivious Programmable Pseudo-Random Function (OPPRF) primitive with the Oblivious Key-Value Store (OKVS) scheme [14] and Oblivious Pseudo-Random Function (OPRF). Garimella et al. [14] use a 3-hash garbled cuckoo table to design the OKVS scheme, which is an optimized version of the PaXoS construction [27] and can achieve much better communication efficiency than GBF. The state-of-the-art protocol [31] to realize OPRF adopts the recent LPN-based Vector Oblivious Linear-function Evaluation (VOLE) protocol with *sublinear* communication [1–4,9,34,35]. Garimella et al. [14] modified the augmented semi-honest PSI protocol [23] into a maliciously secure protocol by adding a random oracle to wrap the OPRF output,[1] and proved that the modified multi-party PSI protocol is secure against malicious adversaries. Although the state-of-the-art multi-party PSI protocol [14] is concretely efficient, we observed the following two aspects that need to be further improved and are addressed by this work.

- The multi-party PSI protocol [14] adopts the *star* architecture for communication. Specifically, the central party P_0 will interact with parties $P_1, \ldots, P_n$ to compute n OPRF outputs that will include n VOLE protocol executions, and then receives n OKVS from the n parties. When n and the size ℓ of each set

[1] A similar observation was also made by Nevo et al. [24].

are large, the communication bandwidth of P_0 is $O(n\ell|\mathbb{F}|)$ bits, which forms an efficiency bottleneck (especially for networks with restricted bandwidthes), where $|\mathbb{F}|$ is the size of an exponentially large field $\mathbb{F}$.
- The n VOLE executions for computing n OPRF outputs will require n encoding operations of Learning Parity with Noise (LPN) for the central party P_0, where the LPN encoding is computationally expensive [36] and forms a computational efficiency bottleneck of the state-of-the-art VOLE protocol with malicious security [35].

1.1 Our Contributions

In this paper, we propose a new multi-party PSI protocol in the dishonest-majority malicious setting, which improves the state-of-the-art multi-party PSI protocol [14] in the following two aspects.

- We reduce the computation cost of LPN encoding for the central party P_0 implied in the PSI protocol [14] by a factor of n. Meanwhile, we further reduce the total computation cost of their PSI protocol by $2n\ell$ hash operations. To achieve the efficiency gain, we construct the PSI protocol by directly using VOLE in the multi-party setting (instead of calling OPRF) and integrating the repetitive operations (see Sect. 3.2 for a technical overview).
- Building on the above technique to improve computation, we use a *tree* architecture to reduce the communication bandwidth of the central party P_0. In particular, we reduce the bandwidth complexity of P_0 from $O(n\ell)$ field elements to $O(\ell)$ field elements by amortizing the communication among all parties in a tree network architecture. We present two types of tree architectures: one is used to make P_0 send the same message to parties $P_1, \ldots, P_n$; and the other is used to let $P_1, \ldots, P_n$ send the sum of n different messages to P_0 in an aggregation way (see Sect. 3.3 for a technical overview).

Our PSI protocol requires the underlying OKVS scheme to be *linear*, which is satisfied by the recent efficient constructions [14,27,29]. We prove security of our protocol in the Universal Composability (UC) model and random oracle model.

In addition, Garimella et al. [14] proposed a multi-output extension of their multi-party PSI protocol such that all parties can obtain the output instead of only the central party P_0. While their multi-party PSI protocol with a single output is provably secure, we present a simple attack against the multi-output extension. This attack allows an adversary to leak the information of the sets held by honest parties more than that allowing to be obtained from the intersection of the sets of all parties, even if the adversary behaves semi-honestly. In particular, if only P_0 is honest, then the adversary can leak some secret items in the private set of P_0.

Comparison of Communication, Bandwidth and Rounds. In Table 1, we compare our protocol with two recent multi-party PSI protocols in the dishonest-majority malicious setting (where the adversary can corrupt up to n parties of the $n + 1$ parties). For the sake of simplicity, this table does not compare

Table 1. Comparison between our protocol and recent multi-party PSI protocols tolerating all-but-one malicious corruptions. $n+1$ is the number of parties, ℓ is the size of each set, and κ and ρ are the computational and statistical security parameters respectively. The bandwidth column denotes the maximum communication sent or received by each party.

MP-PSI	Topology	Communication	Bandwidth	Rounds
[12]	Star	$O(n\ell\kappa^2 + n\ell\kappa\log(\ell\kappa))$	$O(n\ell\kappa^2 + n\ell\kappa\log(\ell\kappa))$	4
[14]	Star	$O(n^2\kappa + n\ell(\kappa + \rho + \log\ell))$	$O(n\ell(\kappa + \rho + \log\ell))$	2
This work	Tree	$O(n^2\kappa + n\ell(\kappa + \rho + \log\ell))$	$O(\ell(\kappa + \rho + \log\ell))$	$2\lfloor\log(n+1)\rfloor$

the recent PSI protocol [24], as it has the same complexities as [14] in this setting. All costs are counted in the ROT/VOLE-hybrid model, where ROT represents random oblivious transfer. For the state-of-the-art ROT or VOLE protocols, the communication of these protocols is small compared to the whole communication. From this table, we can see that our protocol has the lowest bandwidth complexity while keeping the same communication complexity. In particular, our tree architecture does not increase the total communication cost, compared to the state-of-the-art multi-party PSI protocol [14]. As a trade-off of lower communication bandwidth, the round complexity of our PSI protocol is increased to $O(\log n)$, compared to the multi-party PSI protocol [14,24] with the round complexity of $O(1)$.

2 Preliminaries

2.1 Notation

We use κ and λ to denote the computational and statistical security parameters, respectively. For $a, b \in \mathbb{N}$ and $a \leq b$, we use $[a, b]$ to denote the set $\{a, \dots, b\}$. For a finite set S, we use $x \leftarrow S$ to denote sampling x uniformly at random from S. For a vector $\boldsymbol{x}$, we denote by x_i the i-th component of $\boldsymbol{x}$ with x_1 the first entry. For a set S, we use $|S|$ to denote the size of set S.

2.2 Security Model and Functionalities

Security Model. We use the Universal Composability (UC) framework [5] to prove security in the presence of a static, malicious adversary. We say that a protocol Π *UC-realizes* an ideal functionality $\mathcal{F}$ if for any Probabilistic Polynomial Time (PPT) adversary $\mathcal{A}$, there exists a PPT simulator $\mathcal{S}$, such that for any PPT environment $\mathcal{Z}$, the output distribution of $\mathcal{Z}$ in the *real-world* execution where the parties interact with $\mathcal{A}$ and execute Π is computationally indistinguishable from the output distribution of $\mathcal{Z}$ in the *ideal-world* execution where the parties interact with $\mathcal{S}$ and $\mathcal{F}$.

Functionality $\mathcal{F}_{\mathsf{mpsi}}$

Parameters: The size ℓ of the input set for every honest party. The maximal allowed size $\ell' \geq \ell$ of the input set for any malicious party. Let $\mathcal{C}$ be the set of corrupted parties.

Execution: This functionality runs with $P_0, \ldots, P_n$. Upon receiving (input, i, X_i) from every party P_i for $i \in [0, n]$, this functionality executes as follows:

1. If there exists some $i \in \mathcal{C}$ such that $|X_i| > \ell'$, send abort to all parties and then abort.
2. If there exists some $i \notin \mathcal{C}$ such that $|X_i| > \ell$, send abort to all parties and then abort.
3. Otherwise, compute $Y := \bigcap_{i\in[0,n]} X_i$, and then send Y to P_0.

Fig. 1. Functionality for multi-party private set intersection.

Functionality for Multi-party PSI. In a PSI protocol, parties $P_0, P_1, \ldots, P_n$ with each having an input set X_i of size ℓ compute the intersection of their input sets, i.e., $\bigcap_{i\in[0,n]} X_i$. As a result of the protocol execution, P_0 obtains the intersection output, and all other parties learn nothing. The multi-party PSI functionality is shown in Fig. 1. Following prior work such as [12,27,30,31], we allow the adversary who corrupts a party P_i to input a set X_i with $\ell \leq |X_i| \leq \ell'$.

Functionality for VOLE. Following the previous definition [2,35], the functionality for Vector Oblivious Linear-function Evaluation (VOLE) is given in Fig. 2. Two parties run the initialization procedure only once, and then can repeatedly call the extend procedure to obtain multiple batches of VOLE correlations. The VOLE functionality can be securely realized by the recent LPN-based protocols with *sublinear* communication [1–4,9,34,35].

2.3 Oblivious Key-Value Stores

We recall the definitions of an Oblivious Key-Value Store (OKVS) proposed by Garimella et al. [14]. OKVS is a general notion, and captures the functionality and security property of existing constructions, including polynomials, dense matrix, garbled Bloom filter [11] and PaXoS [27]. Then, we roughly discuss the state-of-the-art OKVS construction as well as its communication and computation complexities.

Definition 1 ([14]). *A key-value store is parameterized by a set $\mathcal{K}$ of keys, a set $\mathcal{V}$ of values, and a set $\mathcal{H}$ of functions, and consists of the following two algorithms:*

- $\mathsf{Encode}_{\mathcal{H}}(\{(k_i, v_i)\}_{i\in[1,\ell]})$ *takes as input a set of key-value pairs $\{(k_i, v_i)\}_{i\in[1,\ell]}$, and outputs an object $\mathbf{S}$ (or an error indicator $\perp$ with statistically small probability).*

Functionality $\mathcal{F}_{\mathsf{vole}}$

This functionality runs with a sender P_S, a receiver P_R and an adversary, and operates as follows:

Initialize: Upon receiving (init) from P_S and P_R, if P_S is honest, then sample $\Delta \leftarrow \mathbb{F}$, otherwise receive $\Delta \in \mathbb{F}$ from the adversary. Store Δ and output Δ to P_S, and ignore all subsequent (init) commands.

Extend: Upon receiving (extend, m) from P_S and P_R, do the following:

1. If P_S is honest, sample $\boldsymbol{v} \leftarrow \mathbb{F}^m$. Otherwise, receive $\boldsymbol{v} \in \mathbb{F}^m$ from the adversary.
2. If P_R is honest, sample $\boldsymbol{u} \leftarrow \mathbb{F}^m$ and compute $\boldsymbol{w} := \boldsymbol{v} + \boldsymbol{u} \cdot \Delta$. Otherwise, receive $\boldsymbol{u}, \boldsymbol{w} \in \mathbb{F}^m$ from the adversary, and then recompute $\boldsymbol{v} := \boldsymbol{w} - \boldsymbol{u} \cdot \Delta$.
3. Output $\boldsymbol{v}$ to P_S and $(\boldsymbol{u}, \boldsymbol{w})$ to P_R.

Fig. 2. Functionality for vector oblivious linear-function evaluation.

- $\mathsf{Decode}_{\mathcal{H}}(\mathbf{S}, k)$ *takes as input an object* $\mathbf{S}$, *a key* k, *and outputs a value* v.

A KVS is correct, if for all $A \subseteq \mathcal{K} \times \mathcal{V}$ with distinct keys:

$$(k, v) \in A \text{ and } \bot \neq \mathbf{S} \leftarrow \mathsf{Encode}_{\mathcal{H}}(A) \Rightarrow \mathsf{Decode}_{\mathcal{H}}(\mathbf{S}, k) = v.$$

For the sake of simplicity, we choose to omit the underlying parameter $\mathcal{H}$ in the rest of exposition, as long as the context is clear. In the known OKVS constructions, we always have that the decision whether Encode outputs $\bot$ depends on the functions $\mathcal{H}$ and the keys $\{k_i\}_{i \in [\ell]}$, and is independent of the values $\{v_i\}_{i \in [\ell]}$. In the following, we define the security property guaranteeing that one cannot decide whether a key k was used to generate $\mathbf{S}$ or not.

Definition 2 ([14]). *A KVS is an Oblivious KVS (OKVS), if for all distinct keys* $\{k_1^0, ..., k_\ell^0\}$ *and all distinct keys* $\{k_1^1, ..., k_\ell^1\}$, *when* Encode *does not output* $\bot$ *for both* $(k_1^0, ..., k_\ell^0)$ *and* $(k_1^1, ..., k_\ell^1)$, *the output of* $\mathcal{R}(k_1^0, ..., k_\ell^0)$ *is computationally indistinguishable from that of* $\mathcal{R}(k_1^1, ..., k_\ell^1)$, *where* $\mathcal{R}(k_1, ..., k_\ell)$ *is defined as:*

- *For* $i \in [1, \ell]$, *sample* $v_i \leftarrow \mathcal{V}$.
- *Output* $\mathsf{Encode}(\{(k_i, v_i)\}_{i \in [1,\ell]})$.

From the above definition, we have that if the OKVS encodes random values, for any two sets of keys K^0, K^1, it is infeasible to distinguish the OKVS encoding of the keys of K^0 from that of the keys of K^1.

Our multi-party PSI protocol requires that an OKVS has some kind of additively homomorphic property. In particular, $\mathsf{Decode}(\cdot, k)$ is a *linear function* for all $k \in \mathcal{K}$. The formal definition is recalled as follows.

Definition 3 ([14]). *An OKVS is linear over a field* $\mathbb{F}$ *if* $\mathcal{V} = \mathbb{F}$ *(i.e., "values" are elements of* $\mathbb{F}$*), then the output of* Encode *is a vector* $\mathbf{S}$ *in* $\mathbb{F}^m$, *and the* Decode *function is defined as:*

$$\mathsf{Decode}(\mathbf{S}, k) = \langle d(k), \mathbf{S} \rangle \stackrel{\text{def}}{=} \sum_{i=1}^{m} d(k)_i \cdot S_i$$

for some function $d : \mathcal{K} \to \mathbb{F}^m$, where d is typically defined by hash functions. Thus, $\mathsf{Decode}(\cdot, k)$ *is a linear map from $\mathbb{F}^m$ to $\mathbb{F}$.*

The idea of constructing a linear OKVS is that generating a solution to the linear system of equations:

$$\begin{bmatrix} - d(k_1) - \\ - d(k_2) - \\ \vdots \\ - d(k_\ell) - \end{bmatrix} \cdot \mathbf{S}^\top = \begin{bmatrix} v_1 \\ v_2 \\ \vdots \\ v_\ell \end{bmatrix}.$$

If Encode chooses uniformly from the set of solutions to the linear system and the values are uniform, then the output $\mathbf{S}$ is uniformly distributed (and thus independent of the keys). That is, a linear OKVS satisfies the obliviousness property. The recent linear OKVS scheme [14] (building on the PaXoS technique [27]) has the computation complexity *linear* to the number of key-value pairs, and achieves the rate of $0.81 - o(1)$, where the rate of an OKVS that encodes ℓ items from $\mathbb{F}$ is the ratio between $\ell \cdot |\mathbb{F}|$ and the size of the OKVS. Very recently, Rindal and Raghuraman [29] significantly improved the computation efficiency of the linear OKVS scheme [14], and achieve the best concrete performance for now.

An OKVS whose parameters are chosen to encode N items may hold even more than N items, when it is generated by the adversary. In the context of PSI, this allows the adversary to encode more items than advertised. In Appendix A, we review the the definition of OKVS overfitting to bound the number of items that the adversary can "overfit" into an OKVS, which will be used in the security proof of our PSI protocol.

3 Technical Overview

In this section, we give an overview of our techniques to improve the communication bandwidth and computation cost of the state-of-the-art multi-party PSI protocol [14] tolerating any number of malicious corruptions. The PSI protocol by Garimella et al. [14] is constructed by transforming the augmented semi-honest PSI protocol [23] into a maliciously secure version by adding a random oracle to wrap the OPRF output, which is also observed by Nevo et al. [24]. Firstly, we review the multi-party PSI protocol [14] at a high level.

3.1 Overview of the Best-Known Multi-party PSI Protocol

The state-of-the-art multi-party PSI protocol [14] tolerating any number of malicious corruptions, which builds on the augmented semi-honest protocol [23], executes as follows:

Functionality $\mathcal{F}_{\text{mvole}}$

Let $\mathcal{C}$ be the set of corrupted parties. This functionality runs with parties $P_0, P_1, \ldots, P_n$ and an adversary, and operates as follows:

Initialize: For each $i \in [1,n]$, upon receiving (init, i) from P_i, if P_i is honest, then sample $\Delta_i \leftarrow \mathbb{F}$, else receive $\Delta_i \in \mathbb{F}$ from the adversary. Store $(\Delta_1, \ldots, \Delta_n)$ and output Δ_i to P_i for each $i \in [1,n]$, and ignore all subsequent (init) commands.

Extend: Upon receiving (extend, m) from all parties, execute as follows:

1. For $i \in [1,n]$, if P_i is honest, then sample $\boldsymbol{v}_i \leftarrow \mathbb{F}^m$, else receive $\boldsymbol{v}_i \in \mathbb{F}^m$ from the adversary.
2. If P_0 is honest, sample $\boldsymbol{u} \leftarrow \mathbb{F}^m$ and compute $\boldsymbol{w}_i := \boldsymbol{v}_i + \boldsymbol{u} \cdot \Delta_i$ for $i \in [1,n]$.
3. If P_0 is corrupted, receive $\boldsymbol{u}_i \in \mathbb{F}^m$ for all $i \in [1,n]$ and $\boldsymbol{w}_i \in \mathbb{F}^m$ for all $i \notin \mathcal{C}$ from the adversary, and then recompute $\boldsymbol{v}_i := \boldsymbol{w}_i - \boldsymbol{u}_i \cdot \Delta_i$ for each $i \notin \mathcal{C}$ and compute $\boldsymbol{w}_i := \boldsymbol{v}_i + \boldsymbol{u}_i \cdot \Delta_i$ for each $i \in \mathcal{C}$.
4. Output $\{(\boldsymbol{u}_i, \boldsymbol{w}_i)\}_{i \in [1,n]}$ to P_0 and $\boldsymbol{v}_i$ to P_i for $i \in [1,n]$.

Fig. 3. Functionality for multi-party VOLE.

1. $P_0, P_1, \ldots P_n$ are the $n+1$ parties who will compute the intersection of input sets $X_0, X_1, \ldots, X_n$ with $|X_i| = \ell$ for $i \in [0,n]$, where P_0 will obtain the output. For any h, every party P_i can compute a zero share s_h^i such that $\sum_{i \in [0,n]} s_h^i = 0$ by exchanging Pseudo-Random Function (PRF) keys.
2. For each $i \in [1,n]$, P_0 and P_i call an oblivious PRF functionality $\mathcal{F}_{\text{oprf}}$ with malicious security, and then P_i obtains the key denoted by PRF_i and P_0 gets the set $\{\mathsf{PRF}_i(h) \mid h \in X_0\}$.
3. Let $(\mathsf{Encode}, \mathsf{Decode})$ be an OKVS scheme which maps ℓ items to m slots, and $\mathsf{H} : \{0,1\}^* \to \mathbb{F}$ be a random oracle. For each $i \in [1,n]$, P_i computes an OKVS $\mathbf{Q}_i$ as
$$\mathbf{Q}_i = \mathsf{Encode}\left(\{(h, \mathsf{H}(\mathsf{PRF}_i(h), h) + s_h^i) \mid h \in X_i\}\right),$$
and then sends $\mathbf{Q}_i$ to P_0.
4. P_0 computes an OKVS $\mathbf{Q}_0$ as
$$\mathbf{Q}_0 = \mathsf{Encode}\Big(\Big\{(h, -\sum_{i=1}^{n} \mathsf{H}(\mathsf{PRF}_i(h), h) + s_h^0) \Big| h \in X_0\Big\}\Big).$$
5. After receiving the OKVS from other n parties, P_0 computes the output as
$$\Big\{h \in X_0 \Big| \sum_{i=0}^{n} \mathsf{Decode}(\mathbf{Q}_i, h) = 0\Big\}.$$

3.2 Our Approach to Improve Computation Efficiency

The state-of-the-art OPRF protocol [31] is constructed by combining VOLE with an OKVS scheme called PaXoS [27], where the efficiency of OKVS can be further improved using the recent constructions [14,29]. At a high level, this protocol running between P_i and P_0 works as follows:

1. P_i and P_0 call functionality $\mathcal{F}_{\mathsf{vole}}$ (shown in Fig. 2), which returns $\boldsymbol{v}_i$ to P_i and $(\boldsymbol{u}, \boldsymbol{w}_i)$ to P_0, where $\boldsymbol{w}_i = \boldsymbol{v}_i + \boldsymbol{u} \cdot \Delta_i$.
2. For an input set X_0, P_0 computes an OKVS as:

$$\mathbf{S} = \mathsf{Encode}\left(\{(h, \mathsf{H}(h)) \mid h \in X_0\}\right).$$

3. P_0 computes $\boldsymbol{d} := \mathbf{S} - \boldsymbol{u} \in \mathbb{F}^m$ and sends it to P_i who computes $\mathbf{V}_i := \boldsymbol{v}_i - \boldsymbol{d} \cdot \Delta_i$. Let $\mathbf{W}_i = \boldsymbol{w}_i$, and thus $\mathbf{W}_i = \mathbf{V}_i + \mathbf{S} \cdot \Delta_i$.
4. P_0 computes $\mathsf{PRF}_i(h) = \mathsf{H}(\mathsf{Decode}(\mathbf{W}_i, h), h)$ for $h \in X_0$, and P_i can compute $\mathsf{PRF}_i(x) = \mathsf{H}(\mathsf{Decode}(\mathbf{V}_i, x) + \mathsf{H}(x) \cdot \Delta_i, x)$ for any x in the domain.

If integrating the above OPRF protocol into the state-of-the-art multi-party PSI protocol with malicious security shown in the previous section, P_0 needs to call $\mathcal{F}_{\mathsf{vole}}$ with n different parties. Note that the LPN encoding is the computational efficiency bottleneck for the state-of-the-art VOLE protocol [35] in the malicious setting. When instantiating functionality $\mathcal{F}_{\mathsf{vole}}$, the multi-party PSI protocol requires P_0 to perform n operations of LPN encoding, which is computationally expensive.

Our Solution. Our approach reduces the cost of LPN encoding associated with vector $\boldsymbol{u}$ by a factor of n. We make an important observation that a single random vector $\boldsymbol{u}$ is sufficient to mask the OKVS $\mathbf{S}$, and thus it is unnecessary to compute n random vectors $\boldsymbol{u}_1, \ldots, \boldsymbol{u}_n$. Thus, P_0 needs to generate n VOLE correlations with the same vector $\boldsymbol{u}$ by running a protocol with n different parties. We model this as a multi-party VOLE functionality $\mathcal{F}_{\mathsf{mvole}}$ shown in Fig. 3. Note that when P_0 is corrupted, the adversary allows to choose different vectors $\boldsymbol{u}$. That is, we do not require the consistency check of vector $\boldsymbol{u}$. Furthermore, we do *not* require P_0 to broadcast the vector $\boldsymbol{d} = \mathbf{S} - \boldsymbol{u}$. These are sufficient to design multi-party PSI protocols in the malicious setting (see the security proof of our protocol given in Sect. 4.3). Such a functionality $\mathcal{F}_{\mathsf{mvole}}$ can be UC-realized by calling the standard VOLE functionality (shown in Fig. 2) n times and programming the input/output of P_0 to keep the consistency of $\boldsymbol{u}$ in the honest case (see [3] for more details).

Furthermore, we further reduce the computation cost of the multi-party PSI protocol described as above by $2n\ell$ H operations. In particular, we construct the PSI protocol by directly using $\mathcal{F}_{\mathsf{mvole}}$ as the underlying primitive instead of the OPRF primitive. We adopt $\mathcal{F}_{\mathsf{mvole}}$ and an OKVS to obtain an OPRF, but simplify the computation of $\mathsf{H}(\mathsf{PRF}_i(h), h) = \mathsf{H}\big(\mathsf{H}(\mathsf{Decode}(\mathbf{V}_i, h) + \mathsf{H}(h) \cdot \Delta_i, h), h\big)$ performed by P_i for $i \in [1, n], h \in X_i$ in the previous construction as

$$\mathsf{H}(\mathsf{Decode}(\mathbf{V}_i, h) + \mathsf{H}(h) \cdot \Delta_i, h).$$

As such, the computation of

$$\sum_{i=1}^{n} \mathsf{H}(\mathsf{PRF}_i(h), h) = \sum_{i=1}^{n} \mathsf{H}\big(\mathsf{H}(\mathsf{Decode}(\mathbf{W}_i, h), h), h\big)$$

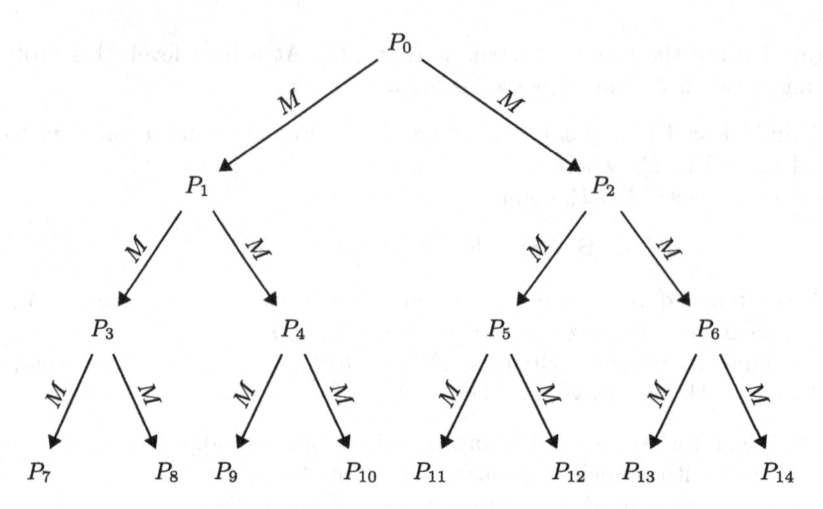

Fig. 4. Transmission of a message M using a binary tree when $n = 14$.

performed by P_0 for $h \in X_0$ can be simplified as

$$\sum_{i=1}^{n} \mathsf{H}(\mathsf{Decode}(\mathbf{W}_i, h), h).$$

In other words, we combine two hash operations into one, which removes the redundant hash operations and significantly improves the computation efficiency.

3.3 Our Approach to Reduce Communication Bandwidth

The state-of-the-art multi-party PSI protocol [14] adopts the *star* network architecture. In particular, the central party P_0 will have to send the messages $\{\boldsymbol{d}_i = \mathbf{S} - \boldsymbol{u}_i \in \mathbb{F}^m\}_{i \in [1,n]}$ to n parties for computing OPRF values and receive the OKVSs $\mathbf{Q}_1, \ldots, \mathbf{Q}_n \in \mathbb{F}^m$ from n parties. These make the communication bandwidth of P_0 be $nm|\mathbb{F}|$ bits per protocol execution, where $m = \ell/(0.81 - o(1))$ for the recent OKVS scheme [14] and ℓ is the size of a set. For a large number n of parties and a large size ℓ of input sets, this forms an efficiency bottleneck of the PSI protocol (especially for networks with restricted bandwidthes). For example, when $\ell = 2^{24}$, $n = 20$ and $|\mathbb{F}| = 128$, P_0 needs about 53 gigabits of communication bandwidth. To reduce the communication bandwidth, one may consider using the Cuckoo hashing approach [23] to divide a large set into multiple bins. However, this approach is *not* secure against malicious adversaries [24].

Our Solution. In the following, we present an efficient approach to reduce the communication bandwidth of P_0 to $2m|\mathbb{F}|$ bits by amortizing the communication among all parties in a *binary-tree* architecture. Specifically, we first apply the optimized approach described in the previous section to the multi-party PSI protocol. In this case, P_0 needs to send the same message $\boldsymbol{d} = \mathbf{S} - \boldsymbol{u}$ to all other

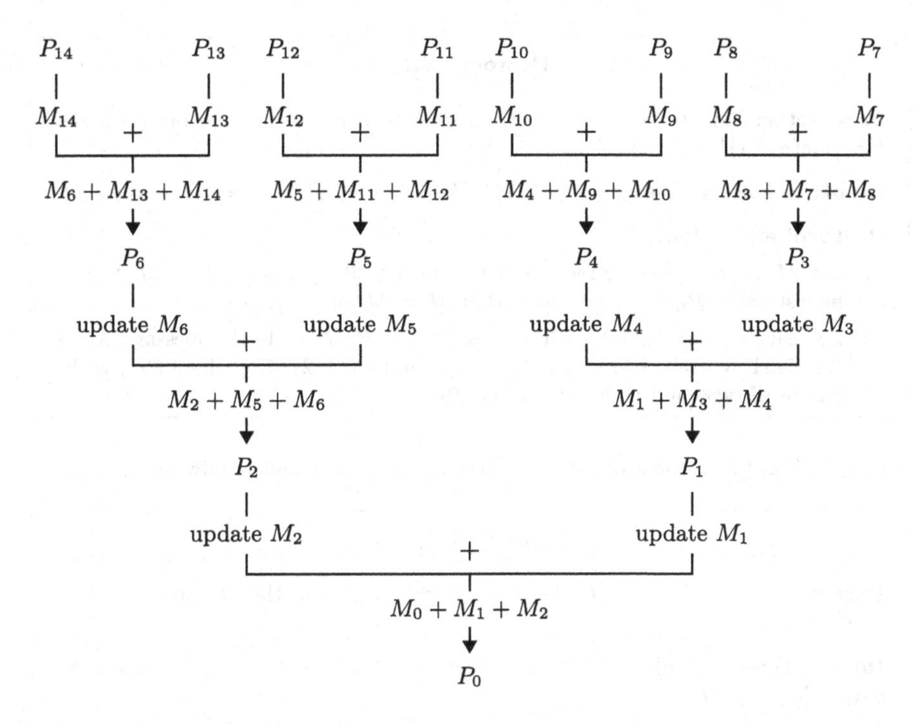

Fig. 5. Message addition aggregation using a binary tree when $n = 14$.

parties. Therefore, we can use a binary-tree structure to send $\boldsymbol{d}$. The parties $P_0, P_1, \ldots, P_n$ constitute a binary tree with the root P_0. Party P_0 can only send $\boldsymbol{d}$ to its two child nodes P_1 and P_2. Then, for each level of the tree, every parent node P_{j-1} sends $\boldsymbol{d}$ to its children P_{2j-1} and P_{2j}. See Fig. 4 for an example of $n = 14$ where the message $M = \boldsymbol{d}$. Note that here we do *not* require to broadcast vector $\boldsymbol{d}$.

To reduce the bandwidth that communicates the OKVSs $\mathbf{Q}_1, \ldots, \mathbf{Q}_n$, we require that the OKVS scheme is *linear*, which is satisfied by the recent highly-efficient constructions [14,29,31]. In this case, we can aggregate these OKVSs into one OKVS through a binary tree. For each level of the tree (from bottom to top), two child nodes P_{2j-1} and P_{2j} respectively send $\mathbf{Q}_{2j-1}$ and $\mathbf{Q}_{2j}$ to their parent node P_{j-1} who updates its OKVS as $\mathbf{Q}_{j-1} := \mathbf{Q}_{j-1} + \mathbf{Q}_{2j-1} + \mathbf{Q}_{2j}$. Finally, the root node P_0 obtains the OKVS $\mathbf{Q}_0 := \sum_{i=0}^{n} \mathbf{Q}_i$. See Fig. 5 for an example of $n = 14$, where P_i holds an OKVS $M_i = \mathbf{Q}_i \in \mathbb{F}^m$ for $i \in [0, n]$. According to the linearity of the OKVS scheme, we have $\sum_{i=0}^{n} \mathsf{Decode}(\mathbf{Q}_i, h) = \mathsf{Decode}(\sum_{i=0}^{n} \mathbf{Q}_i, h)$. Therefore, the OKVS $\sum_{i=0}^{n} \mathbf{Q}_i$ that has been aggregated allows P_0 to obtain the correct output. The message-aggregation approach can also be used for the recent multi-party PSI protocols based on garbled Bloom filters [12,21], where the approach can reduce the communication bandwidth of the central party P_0 in the protocol [12] by a factor of $O(n)$ and the rounds in the protocol [21] from $O(n)$ to $O(\log n)$. In addition, our tree-architecture approach

Protocol $\Pi^{\text{tree}}_{\text{send}}$

Parameter: Let $P_0, P_1, ..., P_n$ be $n+1$ parties that execute the protocol. Let $h = \lfloor \log(n+1) \rfloor$.

Input: P_0 holds a message $M \in \{0,1\}^*$. For $i \in [1, n]$, P_i will receive message M.

Protocol execution:

1. Let M_i be the message received by P_i for $i \in [0, n]$, where $M_0 = M$. For an honest party P_0, we always have that $M_i = M$ for $i \in [1, n]$.
2. For each $i \in [1, h]$, for each $j \in [2^{i-1}, 2^i)$, P_{j-1} sends the message M_{j-1} (received by itself) to P_{2j-1} if $2j - 1 \leq n$ and P_{2j} if $2j \leq n$, where P_{j-1} is the parent of two child nodes P_{2j-1} and P_{2j}.

Fig. 6. Protocol for sending messages with constant bandwidth.

Protocol $\Pi^{\text{tree}}_{\text{aggregate}}$

Parameter: Let $P_0, P_1, ..., P_n$ be $n+1$ parties that run the protocol. Let $h = \lfloor \log(n+1) \rfloor$.

Inputs: For $i \in [1, n]$, P_i holds a message $M_i \in \mathbb{F}^m$. Party P_0 will obtain the output $\sum_{i \in [1,n]} M_i$.

Protocol execution:

1. P_0 sets $M_0 := 0$.
2. From $i = h$ to $i = 1$, for each $j \in [2^{i-1}, 2^i)$, P_{2j-1} sends M_{2j-1} to P_{j-1} (if $2j - 1 \leq n$) and P_{2j} sends M_{2j} to P_{j-1} (if $2j \leq n$), and then P_{j-1} computes $M'_{j-1} := M_{j-1} + M_{2j-1} + M_{2j}$ and updates $M_{j-1} := M'_{j-1}$.
3. P_0 outputs the final message M_0.

Fig. 7. Protocol for aggregating messages with constant bandwidth.

is able to be applied in the multi-party PSI protocol [24] for any corruption threshold $t \leq n$.

4 Maliciously Secure Multi-party PSI Protocol

We first describe two sub-protocols which are used to send and aggregate messages respectively in a binary-tree architecture. Then, we present the detailed construction of our multi-party PSI protocol with malicious security. Finally, we give a formal proof of security for our PSI protocol.

4.1 Sub-protocols for Sending and Aggregating Messages

We first describe the sub-protocol $\Pi^{\text{tree}}_{\text{send}}$ shown in Fig. 6, which allows the central party P_0 to send a message M to all other parties $P_1, \ldots, P_n$. The message is transmitted in a binary-tree architecture, which enables us to obtain $O(1)$ communication bandwidth instead of $O(n)$. As a trade-off, the round complexity is increased from $O(1)$ to $O(\log n)$. A malicious party P_i for $i \in [1, n]$ may send $M' \neq M$ to its left child and $M'' \neq M$ to its right child. In the next subsection, we will show such malicious behavior is harmless for the security of our protocol.

In Fig. 7, we describe the sub-protocol $\Pi^{\mathsf{tree}}_{\mathsf{aggregate}}$, which allows n parties to send the sum of their messages to the central party P_0. These messages are aggregated by addition operations in the binary-tree network architecture, which reduces the communication bandwidth from $O(n)$ to $O(1)$. Similarly, the round complexity is increased from $O(1)$ to $O(\log n)$.

In both of two sub-protocols, the message is always transmitted between a parent node and two child nodes. While the parent node sends the same message to two child nodes in the protocol $\Pi^{\mathsf{tree}}_{\mathsf{send}}$, two child nodes send two different messages to the parent node who sums the two messages and its message as an aggregation message in the protocol $\Pi^{\mathsf{tree}}_{\mathsf{aggregate}}$.

Extending to k-Ary Trees. Although we describe our protocols in the binary-tree architecture, they are easy to be extended to work in the k-ary tree architecture with any $2 \leq k \leq n$. For a k-ary tree, while the communication bandwidth will be increased by a factor of $k/2$, the round complexity will be reduced from $O(\log n)$ to $O(\log_k n)$. In particular, the star architecture used in prior work can be considered as the special case of n-ary trees.

4.2 Our PSI Protocol with Efficient Bandwidth and Computation

In Fig. 8, we describe the details of our multi-party PSI protocol in the dishonest-majority malicious setting. The communication bandwidth of the central party P_0 is $O(\ell|\mathbb{F}|)$ bits, where the state-of-the-art protocol [14] that builds on the technique [23] requires the bandwidth of $O(n\ell|\mathbb{F}|)$ bits for P_0. This protocol works in the $\mathcal{F}_{\mathsf{mvole}}$-hybrid model, and is executed by $n+1$ parties $P_0, P_1, \ldots, P_n$. We separate the protocol into two phases: preprocessing phase where the input sets are unknown, and online phase in which the sets are known. In this protocol, only the central party P_0 obtains the output.

Correctness. To see that the PSI protocol as described in Fig. 8 is correct in the honest case, we first note that for any h,

$$\sum_{i=0}^{n} s_h^i = \sum_{i=0}^{n} \left(\sum_{j<i} \mathsf{PRF}(k_{i,j}, h) - \sum_{j>i} \mathsf{PRF}(k_{j,i}, h) \right) = 0.$$

Then, we observe that for each $h \in \bigcap_{i\in[0,n]} X_i, i \in [1, n]$, we have $z_{i,h} - z'_{i,h} =$

$$\begin{aligned}
&= \mathsf{H}(\mathsf{Decode}(\mathbf{V}_i, h) + \mathsf{H}(h) \cdot \Delta_i, h) - \mathsf{H}(\mathsf{Decode}(\mathbf{W}_i, h), h) \\
&= \mathsf{H}(\mathsf{Decode}(\mathbf{V}_i, h) + \mathsf{H}(h) \cdot \Delta_i, h) - \mathsf{H}(\mathsf{Decode}(\mathbf{V}_i + \mathbf{S} \cdot \Delta_i, h), h) \\
&= \mathsf{H}(\mathsf{Decode}(\mathbf{V}_i, h) + \mathsf{H}(h) \cdot \Delta_i, h) - \mathsf{H}(\mathsf{Decode}(\mathbf{V}_i, h) + \mathsf{Decode}(\mathbf{S}, h) \cdot \Delta_i, h) \\
&= \mathsf{H}(\mathsf{Decode}(\mathbf{V}_i, h) + \mathsf{H}(h) \cdot \Delta_i, h) - \mathsf{H}(\mathsf{Decode}(\mathbf{V}_i, h) + \mathsf{H}(h) \cdot \Delta_i, h) = 0.
\end{aligned}$$

Protocol Π_{mpsi}

Parameters:
- A large finite field $\mathbb{F}$ with $|\mathbb{F}| \geq 2^{\kappa}$.
- Linear OKVS scheme (Encode, Decode) that maps ℓ items to m slots.
- A cryptographic hash function $\mathsf{H} : \{0,1\}^* \to \mathbb{F}$ modeled as a random oracle.
- Let $\mathsf{PRF} : \{0,1\}^{\kappa} \times \{0,1\}^{\kappa} \to \mathbb{F}$ be a pseudo-random function.

Inputs: For parties $P_0, P_1, ..., P_n$, every party P_i holds a set X_i such that $X_i \in \mathbb{F}$ and $|X_i| = \ell$. Let P_0 be the designated party to receive the output $\bigcap_{i\in[0,n]} X_i$.

Preprocessing: This procedure can be executed when the sets are *unknown.*

1. Every party P_i samples $k_{i,j} \leftarrow \{0,1\}^{\kappa}$ as a random PRF key, and then sends it to P_j over a private channel for each $j < i$. (The zero-sharing setup phase is run only once.)
2. For each $i \in [1,n]$, P_i sends (init, i) to the multi-party VOLE functionality $\mathcal{F}_{\mathsf{mvole}}$ (shown in Figure 3), which returns $\Delta_i \in \mathbb{F}$ to P_i.
3. The parties $P_0, P_1, ..., P_n$ send (extend, m) to functionality $\mathcal{F}_{\mathsf{mvole}}$, which returns $\boldsymbol{u} \in \mathbb{F}^m$ and $\{\boldsymbol{w}_i\}_{i\in[1,n]}$ to P_0, and outputs $\boldsymbol{v}_i \in \mathbb{F}^m$ to P_i for each $i \in [1,n]$, such that $\boldsymbol{w}_i = \boldsymbol{v}_i + \boldsymbol{u} \cdot \Delta_i$ for $i \in [1,n]$.

Online: This procedure is run when the sets are *known.*

4. P_0 computes an OKVS
$$\mathbf{S} := \mathsf{Encode}\left(\{(h, \mathsf{H}(h)) \mid h \in X_0\}\right)$$
and $\boldsymbol{d} := \mathbf{S} - \boldsymbol{u} \in \mathbb{F}^m$. Then P_0 sends $\boldsymbol{d}$ to P_i for $i \in [1,n]$ by running sub-protocol $\Pi^{\mathsf{tree}}_{\mathsf{send}}$ (shown in Figure 6).
5. For each $i \in [1,n]$, P_i computes $\mathbf{V}_i := \boldsymbol{v}_i - \boldsymbol{d} \cdot \Delta_i$. P_0 sets $\mathbf{W}_i := \boldsymbol{w}_i$ for $i \in [1,n]$ where $\mathbf{W}_i = \mathbf{V}_i + \mathbf{S} \cdot \Delta_i$.
6. For each $i \in [0,n]$, for each $h \in X_i$, every party P_i computes the following:
$$s_h^i := \sum_{j<i} \mathsf{PRF}(k_{i,j}, h) - \sum_{j>i} \mathsf{PRF}(k_{j,i}, h).$$
7. For each $i \in [1,n]$, P_i computes $z_{i,h} := \mathsf{H}(\mathsf{Decode}(\mathbf{V}_i, h) + \mathsf{H}(h) \cdot \Delta_i, h)$ for each $h \in X_i$, and then computes an OKVS $\mathbf{Q}_i$ via
$$\mathbf{Q}_i := \mathsf{Encode}\left(\left\{(h, z_{i,h} + s_h^i) \mid h \in X_i\right\}\right).$$
8. Parties $P_1, ..., P_n$ run the sub-protocol $\Pi^{\mathsf{tree}}_{\mathsf{aggregate}}$ (shown in Figure 7) to send $\mathbf{Q} = \sum_{i=1}^{n} \mathbf{Q}_i \in \mathbb{F}^m$ to P_0.
9. P_0 computes $z'_{i,h} = \mathsf{H}(\mathsf{Decode}(\mathbf{W}_i, h), h)$ for each $h \in X_0$, and then outputs
$$\left\{h \in X_0 \;\middle|\; \mathsf{Decode}(\mathbf{Q}, h) + s_h^0 - \sum_{i=1}^{n} z'_{i,h} = 0\right\}.$$

Fig. 8. Maliciously secure multi-party PSI in the $\mathcal{F}_{\mathsf{mvole}}$-hybrid model.

Therefore, for each $h \in \bigcap_{i\in[0,n]} X_i$, we have

$$\begin{aligned}
\mathsf{Decode}(\mathbf{Q}, h) + s_h^0 - \sum_{i=1}^{n} z'_{i,h} &= \mathsf{Decode}(\sum_{i=1}^{n} \mathbf{Q}_i, h) + s_h^0 - \sum_{i=1}^{n} z'_{i,h} \\
&= s_h^0 + \sum_{i=1}^{n} \mathsf{Decode}(\mathbf{Q}_i, h) - \sum_{i=1}^{n} z'_{i,h} \\
&= s_h^0 + \sum_{i=1}^{n} (z_{i,h} + s_h^i) - \sum_{i=1}^{n} z'_{i,h} \\
&= \sum_{i=0}^{n} s_h^i + \sum_{i=1}^{n} (z_{i,h} - z'_{i,h}) = 0.
\end{aligned}$$

4.3 Proof of Security

In the following, we prove the security of our PSI protocol in the multi-party malicious setting. Our proof of security will use the following lemma, which has been proven in [14].

Lemma 1 ([14]). *Given a set of parties that run the zero-sharing setup (i.e., the step 1 in Fig. 8) such that a pair of parties P_i, P_j are honest and the adversary's view is independent of the P_i's share s_h^i, then the P_j's share s_h^j is computationally indistinguishable from a uniform value.*

Based on the above lemma, we prove the following theorem.

Theorem 1. *Let* H *be a random oracle and* $(\mathsf{Encode}, \mathsf{Decode})$ *be a linear OKVS scheme. Let* PRF *be a pseudo-random function. Then protocol Π_{mpsi} shown in Fig. 8 UC-realizes functionality $\mathcal{F}_{\mathsf{mpsi}}$ in the presence of a malicious adversary corrupting up to n of the $n+1$ parties in the $\mathcal{F}_{\mathsf{mvole}}$-hybrid model.*

Due to space limit, the formal proof of the above theorem can be found in the full version of this paper [28].

5 An Attack Against Multi-output Extension of PSI

In the malicious setting, it is a non-trivial task to extend a multi-party PSI protocol to support multiple outputs where every party (instead of only party P_0) will obtain the output. This is because the parties cannot be trusted to deliver the intersection output faithfully. In the multi-party malicious setting, Garimella et al. [14] extended their PSI protocol with a single output to a protocol supporting multiple outputs that achieves the best efficiency. In this section, we present a simple and practical attack for the multi-output extension [14], which allows the attacker to reveal more information of the sets of honest parties than that obtained from the intersection of the sets of all parties. Specifically, we first review the multi-output extension by Garimella et al. [14]. This extension modifies the multi-party PSI protocol shown in Sect. 3.1 to realize that all parties obtain the output in the following procedure:

- All parties $P_0, P_1, \ldots, P_n$ publicly commit to their OKVSs $\{\mathbf{Q}_i\}_{i\in[0,n]}$. That is, every party P_i broadcasts a commitment $com_i = \mathsf{Commit}(\mathbf{Q}_i; r_i)$ to all other parties where r_i is a randomness.
- After all the commitments that have been made, the parties open these commitments. That is, for $i \in [0, n]$, every party P_i sends $(\mathbf{Q}_i, r_i)$ to all other parties, and then verifies the correctness of all the openings (i.e., checking that $com_j = \mathsf{Commit}(\mathbf{Q}_j; r_j)$ for all $j \neq i$).
- Every party P_i with $i \in [0, n]$ computes the output as

$$\left\{h \in X_i \,\middle|\, \sum_{j=0}^{n} \mathsf{Decode}(\mathbf{Q}_j, h) = 0\right\}.$$

An Attack to Leak the Information of the Sets Held by Honest Parties. Below, we show a simple but practical attack against the above multi-output extension to reveal more information of the sets held by honest parties beyond the intersection of all sets. Suppose that P_0 is honest. Let $\mathcal{C} \subseteq [1, n]$ denote the set of corrupted parties. In particular, an adversary $\mathcal{A}$ (who even behaves semi-honestly) is able to perform the following attack:

1. $\mathcal{A}$ receives the OKVS $\mathbf{Q}_i$ for all $i \notin \mathcal{C}$ that contain $\mathbf{Q}_0$. Recall that the OKVS $\mathbf{Q}_i$ for $i \notin \mathcal{C}, i \neq 0$ is defined as follows:

$$\mathbf{Q}_i = \mathsf{Encode}\Big(\Big\{\big(h, \mathsf{H}(\mathsf{PRF}_i(h), h) + s_h^i\big) \,\Big|\, h \in X_i\Big\}\Big).$$

   The OKVS $\mathbf{Q}_0$ is computed as

$$\mathbf{Q}_0 = \mathsf{Encode}\Big(\Big\{\big(h, s_h^0 - \sum_{i=1}^{n} \mathsf{H}(\mathsf{PRF}_i(h), h) \,\big|\, h \in X_0\big)\Big\}\Big).$$

2. $\mathcal{A}$ computes $\mathbf{Q} := \sum_{i\notin\mathcal{C}} \mathbf{Q}_i$. According to the definition of zero sharings, we know that $\sum_{i\in[0,n]} s_h^i = 0$ for any h. Therefore, for each $h \in \bigcap_{i\notin\mathcal{C}} X_i$, we obtain the following:

$$\sum_{i\notin\mathcal{C}} \mathsf{Decode}(\mathbf{Q}_i, h) = -\sum_{i\in\mathcal{C}} s_h^i - \sum_{i\in\mathcal{C}} \mathsf{H}(\mathsf{PRF}_i(h), h). \tag{1}$$

3. $\mathcal{A}$ who corrupts P_i for $i \in \mathcal{C}$ can compute the shares s_h^i and values $\mathsf{PRF}_i(h)$ for all $i \in \mathcal{C}$ when any h is known. Then, for any $h \notin \bigcap_{i\in[0,n]} X_i$, $\mathcal{A}$ is able to decide whether $h \in \bigcap_{i\notin\mathcal{C}} X_i$ by checking if the Eq. (1) holds. For the items in the set $\bigcap_{i\notin\mathcal{C}} X_i$ that have a low entropy, $\mathcal{A}$ can directly reveal these items by enumerating the items and then checking their correctness, and thus obtain the secret data included in $\bigcap_{i\notin\mathcal{C}} X_i$.

In the special case that only P_0 is honest and all other parties are corrupted (i.e., $|\mathcal{C}| = n$), the adversary can leak some secret items in the set X_0 by performing the above attack. The reason behind the successful attack is that the OKVS $\mathbf{Q}_0$ is revealed by the honest central party P_0. This does not occur in the original multi-party PSI protocol [14] that only P_0 obtains the output, where P_0 only uses the OKVS $\mathbf{Q}_0$ *locally* in the original protocol. To prevent the attack as describe above, one may need to change the way computing the OKVS $\mathbf{Q}_0$ and introduce some high-entropy secrets into $\mathbf{Q}_0$. We leave it as an interesting future work to design a concretely efficient multi-party PSI protocol supporting multiple outputs in the malicious setting.

Acknowledgements. Work of Kang Yang is supported by the National Natural Science Foundation of China (Grant Nos. 62102037, 61932019). Work of Yu Yu is supported by the National Key Research and Development Program of China (Grant Nos. 2020YFA0309705 and 2018YFA0704701) and the National Natural Science Foundation of China (Grant Nos. 62125204 and 61872236). Yu Yu also acknowledges the support from the XPLORER PRIZE. We thank anonymous reviewers for their helpful comments.

A OKVS Overfitting

For the security proof of a maliciously secure PSI protocol, the simulator obtains an OKVS from a corrupted party, and needs to extract the keys that are encoded in the OKVS. In general, this is done by defining $v_i = \mathsf{H}(k_i)$ for $i \in [1, \ell]$ where H is a random oracle. Then, the simulator can observe the queries to H made by the adversary, and then check which of the keys k satisfy $\mathsf{Decode}(\mathbf{S}, k) = \mathsf{H}(k)$. An OKVS whose parameters are chosen to encode ℓ keys may often hold even more than ℓ keys, when it is generated by the adversary. In the context of PSI, this allows the adversary to encode more keys than advertised. Therefore, we need to bound the number of keys that the adversary can "overfit" into an OKVS. Following the previous work [14], we model the property in the following definition.

Definition 4 ([14]). *The (ℓ, ℓ')-OKVS overfitting game is defined as follows.*

- *Let* $(\mathsf{Encode}, \mathsf{Decode})$ *be an OKVS with parameters chosen to support ℓ items, and $\mathcal{A}$ be any PPT adversary. Let $\mathsf{H} : \mathcal{K} \to \mathcal{V}$ be a random oracle.*
- *Run* $\mathbf{S} \leftarrow \mathcal{A}^{\mathsf{H}(\cdot)}(1^\kappa)$.
- *Define the following set:*

$$X = \{k \mid \mathcal{A} \textit{ made a query } (k) \textit{ to } \mathsf{H} \textit{ and } \mathsf{Decode}(\mathbf{S}, k) = \mathsf{H}(k)\}.$$

- *If $|X| > \ell'$, then the adversary $\mathcal{A}$ wins.*

We say that the (ℓ, ℓ')-OKVS overfitting problem is hard for an OKVS, if no PPT adversary wins this game except with negligible probability.

For $\kappa = 128$ and $\lambda = 40$, according to the analysis [27], when $\mathsf{H} : \mathcal{K} \to \mathbb{F}$ is used to define the values, a linear OKVS with a field size $|\mathbb{F}| = 128$ can guarantee that the successful probability of the adversary in the above overfitting game is less than $1/2^{40}$, even though the adversary is allowed to make 2^{80} queries to H.

References

1. Boyle, E., Couteau, G., Gilboa, N., Ishai, Y.: Compressing vector OLE. In: ACM Conference on Computer and Communications Security (CCS) 2018, pp. 896–912. ACM Press (2018). https://doi.org/10.1145/3243734.3243868
2. Boyle, E., et al.: Efficient two-round OT extension and silent non-interactive secure computation. In: ACM Conference on Computer and Communications Security (CCS) 2019, pp. 291–308. ACM Press (2019). https://doi.org/10.1145/3319535.3354255
3. Boyle, E., Couteau, G., Gilboa, N., Ishai, Y., Kohl, L., Scholl, P.: Efficient pseudorandom correlation generators: silent OT extension and more. In: Boldyreva, A., Micciancio, D. (eds.) CRYPTO 2019, Part III. LNCS, vol. 11694, pp. 489–518. Springer, Cham (2019). https://doi.org/10.1007/978-3-030-26954-8_16
4. Boyle, E., Couteau, G., Gilboa, N., Ishai, Y., Kohl, L., Scholl, P.: Correlated pseudorandom functions from variable-density LPN, pp. 1069–1080. IEEE (2020). https://doi.org/10.1109/FOCS46700.2020.00103
5. Canetti, R.: Universally composable security: a new paradigm for cryptographic protocols. In: 42nd Annual Symposium on Foundations of Computer Science (FOCS), pp. 136–145. IEEE (2001). https://doi.org/10.1109/SFCS.2001.959888
6. Chandran, N., Dasgupta, N., Gupta, D., Obbattu, S.L.B., Sekar, S., Shah, A.: Efficient linear multiparty PSI and extensions to circuit/quorum PSI. In: Proceedings of the 2021 ACM SIGSAC Conference on Computer and Communications Security, CCS 2021, pp. 1182–1204. Association for Computing Machinery (2021)
7. Chase, M., Miao, P.: Private set intersection in the Internet setting from lightweight oblivious PRF. In: Micciancio, D., Ristenpart, T. (eds.) CRYPTO 2020, Part III. LNCS, vol. 12172, pp. 34–63. Springer, Cham (2020). https://doi.org/10.1007/978-3-030-56877-1_2
8. Cheon, J.H., Jarecki, S., Seo, J.H.: Multi-party privacy-preserving set intersection with quasi-linear complexity. Cryptology ePrint Archive, Report 2010/512 (2010). https://eprint.iacr.org/2010/512
9. Couteau, G., Rindal, P., Raghuraman, S.: Silver: silent VOLE and oblivious transfer from hardness of decoding structured LDPC codes. In: Malkin, T., Peikert, C. (eds.) CRYPTO 2021. LNCS, vol. 12827, pp. 502–534. Springer, Cham (2021). https://doi.org/10.1007/978-3-030-84252-9_17
10. Demmler, D., Rindal, P., Rosulek, M., Trieu, N.: PIR-PSI: scaling private contact discovery. Proc. Priv. Enhancing Technol. **2018**(4), 159–178 (2018)
11. Dong, C., Chen, L., Wen, Z.: When private set intersection meets big data: an efficient and scalable protocol. In: ACM Conference on Computer and Communications Security (CCS) 2013, pp. 789–800. ACM Press (2013). https://doi.org/10.1145/2508859.2516701
12. Efraim, A.B., Nissenbaum, O., Omri, E., Paskin-Cherniavsky, A.: PSimple: practical multiparty maliciously-secure private set intersection. Cryptology ePrint Archive, Report 2021/122 (2021). https://ia.cr/2021/122
13. Freedman, M.J., Nissim, K., Pinkas, B.: Efficient private matching and set intersection. In: Cachin, C., Camenisch, J.L. (eds.) EUROCRYPT 2004. LNCS, vol. 3027, pp. 1–19. Springer, Heidelberg (2004). https://doi.org/10.1007/978-3-540-24676-3_1
14. Garimella, G., Pinkas, B., Rosulek, M., Trieu, N., Yanai, A.: Oblivious key-value stores and amplification for private set intersection. In: Malkin, T., Peikert, C. (eds.) CRYPTO 2021. LNCS, vol. 12826, pp. 395–425. Springer, Cham (2021). https://doi.org/10.1007/978-3-030-84245-1_14

15. Ghosh, S., Nilges, T.: An algebraic approach to maliciously secure private set intersection. In: Ishai, Y., Rijmen, V. (eds.) EUROCRYPT 2019, Part III. LNCS, vol. 11478, pp. 154–185. Springer, Cham (2019). https://doi.org/10.1007/978-3-030-17659-4_6
16. Hazay, C., Venkitasubramaniam, M.: Scalable multi-party private set-intersection. In: Fehr, S. (ed.) PKC 2017, Part I. LNCS, vol. 10174, pp. 175–203. Springer, Heidelberg (2017). https://doi.org/10.1007/978-3-662-54365-8_8
17. Inbar, R., Omri, E., Pinkas, B.: Efficient scalable multiparty private set-intersection via garbled bloom filters. In: Catalano, D., De Prisco, R. (eds.) SCN 2018. LNCS, vol. 11035, pp. 235–252. Springer, Cham (2018). https://doi.org/10.1007/978-3-319-98113-0_13
18. Ion, M., et al.: On deploying secure computing: private intersection-sum-with-cardinality. In: 2020 IEEE European Symposium on Security and Privacy (EuroS&P), pp. 370–389 (2020)
19. Ion, M., et al.: Private intersection-sum protocol with applications to attributing aggregate ad conversions. Cryptology ePrint Archive, Report 2017/738 (2017). https://eprint.iacr.org/2017/738
20. Kales, D., Rechberger, C., Schneider, T., Senker, M., Weinert, C.: Mobile private contact discovery at scale. In: 28th USENIX Security Symposium (USENIX Security 2019), pp. 1447–1464 (2019)
21. Kavousi, A., Mohajeri, J., Salmasizadeh, M.: Efficient scalable multi-party private set intersection using oblivious PRF. Cryptology ePrint Archive, Report 2021/484 (2021). https://ia.cr/2021/484
22. Kissner, L., Song, D.: Privacy-preserving set operations. In: Shoup, V. (ed.) CRYPTO 2005. LNCS, vol. 3621, pp. 241–257. Springer, Heidelberg (2005). https://doi.org/10.1007/11535218_15
23. Kolesnikov, V., Matania, N., Pinkas, B., Rosulek, M., Trieu, N.: Practical multi-party private set intersection from symmetric-key techniques. In: ACM Conference on Computer and Communications Security (CCS) 2017, pp. 1257–1272. ACM Press (2017). https://doi.org/10.1145/3133956.3134065
24. Nevo, O., Trieu, N., Yanai, A.: Simple, fast malicious multiparty private set intersection. In: Proceedings of the 2021 ACM SIGSAC Conference on Computer and Communications Security, CCS 2021, pp. 1151–1165. Association for Computing Machinery (2021)
25. Nguyen, D.T., Trieu, N.: MPCCache: privacy-preserving multi-party cooperative cache sharing at the edge. Cryptology ePrint Archive, Report 2021/317 (2021). https://ia.cr/2021/317
26. Pinkas, B., Rosulek, M., Trieu, N., Yanai, A.: SpOT-light: lightweight private set intersection from sparse OT extension. In: Boldyreva, A., Micciancio, D. (eds.) CRYPTO 2019, Part III. LNCS, vol. 11694, pp. 401–431. Springer, Cham (2019). https://doi.org/10.1007/978-3-030-26954-8_13
27. Pinkas, B., Rosulek, M., Trieu, N., Yanai, A.: PSI from PaXoS: fast, malicious private set intersection. In: Canteaut, A., Ishai, Y. (eds.) EUROCRYPT 2020, Part II. LNCS, vol. 12106, pp. 739–767. Springer, Cham (2020). https://doi.org/10.1007/978-3-030-45724-2_25
28. Qiu, Z., Yang, K., Yu, Y., Zhou, L.: Maliciously secure multi-party PSI with lower bandwidth and faster computation. Cryptology ePrint Archive, Paper 2022/772 (2022). https://eprint.iacr.org/2022/772
29. Rindal, P., Raghuraman, S.: Blazing fast PSI from improved OKVS and subfield vole. Cryptology ePrint Archive, Report 2022/320 (2022). https://ia.cr/2022/320

30. Rindal, P., Rosulek, M.: Improved private set intersection against malicious adversaries. In: Coron, J.-S., Nielsen, J.B. (eds.) EUROCRYPT 2017, Part I. LNCS, vol. 10210, pp. 235–259. Springer, Cham (2017). https://doi.org/10.1007/978-3-319-56620-7_9
31. Rindal, P., Schoppmann, P.: VOLE-PSI: fast OPRF and circuit-PSI from vector-OLE. In: Canteaut, A., Standaert, F.-X. (eds.) EUROCRYPT 2021. LNCS, vol. 12697, pp. 901–930. Springer, Cham (2021). https://doi.org/10.1007/978-3-030-77886-6_31
32. Sang, Y., Shen, H.: Privacy preserving set intersection protocol secure against malicious behaviors. In: Proceedings of the Eighth International Conference on Parallel and Distributed Computing, Applications and Technologies, PDCAT 2007, pp. 461–468. IEEE Computer Society (2007)
33. Sang, Y., Shen, H.: Privacy preserving set intersection based on bilinear groups. In: The 31th Australasian Computer Science Conference, ACSC 2008, vol. 74, pp. 47–54. Australian Computer Society (2008)
34. Schoppmann, P., Gascón, A., Reichert, L., Raykova, M.: Distributed vector-OLE: improved constructions and implementation. In: ACM Conference on Computer and Communications Security (CCS) 2019, pp. 1055–1072. ACM Press (2019). https://doi.org/10.1145/3319535.3363228
35. Weng, C., Yang, K., Katz, J., Wang, X.: Wolverine: fast, scalable, and communication-efficient zero-knowledge proofs for Boolean and arithmetic circuits. In: 2021 IEEE Symposium on Security and Privacy (SP), pp. 1074–1091 (2021)
36. Yang, K., Weng, C., Lan, X., Zhang, J., Wang, X.: Ferret: fast extension for correlated OT with small communication. In: ACM Conference on Computer and Communications Security (CCS) 2020, pp. 1607–1626. ACM Press (2020). https://doi.org/10.1145/3372297.3417276
37. Zhang, E., Liu, F.H., Lai, Q., Jin, G., Li, Y.: Efficient multi-party private set intersection against malicious adversaries. In: Proceedings of the 2019 ACM SIGSAC Conference on Cloud Computing Security Workshop, CCSW 2019, pp. 93–104. Association for Computing Machinery (2019)

Conditional Cube Attacks on Full Members of KNOT-AEAD Family

Siwei Chen, Zejun Xiang(✉), Xiangyong Zeng, and Shasha Zhang

Faculty of Mathematics and Statistics, Hubei Key Laboratory of Applied Mathematics, Hubei University, Wuhan, China
{xiangzejun,xzeng}@hubu.edu.cn

Abstract. KNOT is a family of permutation-based lightweight AEAD and hashing algorithms, which is submitted to the NIST Lightweight Cryptography Standardization process and becomes one of the 32 candidates in the second round. In this paper, we focus on the security of the initialization phase of full members of KNOT-AEAD family against conditional cube attacks in the nonce-respecting setting. To be specific, we introduce a conditional cube attack framework by exploiting linearization technique and division property based degree evaluation. With this framework, we can use a conditional cube set including only one public variable to construct a conditional cube distinguisher, and then to recovery one key bit with a negligible time and data. As a result, we present the 8-, 8-, 8-, and 9-round practical full key-recovery attacks on KNOT-AEAD (128, 256, 64), KNOT-AEAD (128, 384, 192), KNOT-AEAD (192, 384, 96) and KNOT-AEAD (256, 512, 128), and a 9-round theoretical full key-recovery attack on KNOT-AEAD (128, 384, 192) with a complexity of $2^{51.72}$ and a negligible data complexity. To the best of our knowledge, this is the first time to achieve the practical full key-recovery attacks on all members of round-reduced KNOT-AEAD.

Keywords: KNOT-AEAD · Conditional cube attack · Division property · Full key-recovery · Practical attack

1 Introduction

In secure information systems, it is necessary to achieve the confidentiality of source data and the integrity of encrypted/decrypted data. In general, the source data is handled by encryption schemes like block ciphers and the encrypted/decrypted data is verified by message authentication code (MAC). In recent years, a new type of symmetric-key primitive called authenticated encryption with associated data (AEAD) [11] has come into sight. The AEAD primitives can provide confidentiality and integrity simultaneously such that performance loss caused by encryption and authentication in separate algorithms can be avoided. Thus the design and cryptanalysis of AEAD have attracted

C. Alcaraz et al. (Eds.): ICICS 2022, LNCS 13407, pp. 89–108, 2022.
https://doi.org/10.1007/978-3-031-15777-6_6

attentions of cryptographers, especially with the beginning of CAESAR competition[1], which aimed to solicit high-performance and strong-security AEAD schemes. With the development of lightweight cryptography, the National Institute of Standards and Technology (NIST) started the LWC project[2] in 2013 to solicit lightweight AEAD and hashing schemes suitable for highly constrained computing environments.

Our target in this paper is the AEAD algorithm of KNOT [18] suite, called KNOT-AEAD, which is designed by Zhang et al. and becomes one of the 32 candidates in the second round of NIST LWC. There are four members of KNOT-AEAD with the difference on the size of the key, rate and internal state. Besides the security announced in the official design document [18] and the supplementary materials[3] by designers, there are several researches [1,3,5,16] on the security of KNOT. All of them [1,3,5] aimed to search for better differential and integral distinguishers for the underlying permutation of KNOT. In addition, Wang et al. [16] presented key-recovery attacks for KNOT-AEAD (128, 256, 64) and KNOT-AEAD (128, 384, 192) by differential-linear attacks, which supplemented the security of this two members against differential-linear attacks. As far as we know, there is no third-party cryptanalysis on KNOT-AEAD from the perspective of algebraic attacks such as cube-like attacks [4,6], and the designers in [18] only give a simple statement as follows:
"The KNOT S-box does not exhibit any special algebraic structure. Furthermore, it seems that successful applications of algebraic attacks on block ciphers/permuta-tions can only reach a very limited number of rounds. Therefore, we do not expect that algebraic attacks form a danger for any KNOT member". Therefore, it is necessary and significant to fill this gap for a better understanding of the security of KNOT-AEAD.

Our Contributions. In this paper, we evaluate the security of the initialization phase of KNOT-AEAD family against cube-like attacks in the nonce-respecting setting. Specifically, we firstly propose a framework of conditional cube attacks for KNOT-AEAD using the linearization technique and division property based degree evaluation. On the one hand, the ANF of internal states have a special form after linearization phase, based on which we can choose conditional key bits to form the corresponding conditional equations. On the other hand, we exploit the division property to estimate the maximal algebraic degree of monomials only including the given public variables in two cases where the conditional equations hold or not. Thus a gap on the estimated degrees in two cases can be easily detected and we then form a cube set that contains the fewest public variables to construct a cube distinguisher. In addition, we give the

[1] The Competition for Authenticated Encryption: Security, Applicability, and Robustness. https://competitions.cr.yp.to/caesar-submissions.html.
[2] Lightweight Cryptography Standardization process. https://csrc.nist.gov/projects/lightweight-cryptography.
[3] Update on Security Analysis and Implementations of KNOT. https://csrc.nist.gov/CSRC/media/Projects/lightweight-cryptography/documents/round-2/status-update-sep2020/KNOT_Update.pdf.

concept of *conditional S-box* in which some input bits are fixed to constants, and re-describe the division property propagation of conditional S-boxes to avoid the influence of constants on the development of algebraic degrees of Sboxes. Because the conventional description on the division property propagation of original S-box will indeed introduce some invalid division trails for conditional S-boxes, which leads to a non-precise MILP model and an inaccurate estimation of the degree. Benefiting from these optimized methods, the cube distinguisher constructed due to the gap on estimated degrees is reliable enough and can be exploited to achieve a key-recovery attack with a high probability. As a result, we presented 8-, 8-, 8- and 9-round practical key-recovery attacks on KNOT-AEAD (128, 256, 64), KNOT-AEAD (128, 384, 192), KNOT-AEAD (192, 384, 96) and KNOT-AEAD (256, 512, 128). Moreover, we presented a 9-round key-recovery attack on KNOT-AEAD (128, 384, 192) with a time complexity of $2^{51.72}$ and a negligible data complexity. To the best of our knowledge, this is the first time to evaluate the security from cube-like attack perspectives and achieve practical full key-recovery attacks on full versions of KNOT-AEAD. Our attacks cannot threaten the security of KNOT-AEAD family but give a better understanding on the resistance against cube-like attacks. The key-recovery attacks of this paper and [16] are summarized in Table 1.

Table 1. Summary of key-recovery attacks on KNOT-AEAD

Member†	#Round	#Key bit	Method	Time	Data	Ref
(128, 256, 64)	15	128/128	Differential-linear	$2^{48.8}$	$2^{47.5}$	[16]
	8	**128**/128	**Conditional cube**	$\mathbf{2^{18.55}}$	$\mathbf{2^{18.55}}$	Sect. 4.2
(128, 384, 192)	17	1/128	Differential-linear	$2^{59.2}$	$2^{58.2}$	[16]
	8	**128**/128	**Conditional cube**	$\mathbf{2^{33}}$	$\mathbf{2^{18.12}}$	Sect. 4.3
	9	**128**/128	**Conditional cube**	$\mathbf{2^{51.72}}$	$\mathbf{2^{18.12}}$	Sect. 4.3
(192, 384, 96)	**8**	**192**/192	**Conditional cube**	$\mathbf{2^{18.13}}$	$\mathbf{2^{18.13}}$	Sect. 4.2
(256, 512, 128)	**9**	**256**/256	**Conditional cube**	$\mathbf{2^{18.55}}$	$\mathbf{2^{18.55}}$	Sect. 4.2

† The KNOT-AEAD (k, b, r) is written as (k, b, r) for short. For example, the member $(128, 256, 64)$ denotes the KNOT-AEAD $(128, 256, 64)$.

Organization of This Paper. In Sect. 2, we give a brief introduction on the algebraic degree evaluation by division property, conditional cube attack and KNOT-AEAD family. In Sect. 3, we propose a framework of conditional cube attacks for KNOT-AEAD, and based on the proposed framework, we will present full key-recovery attacks on all members of round-reduced KNOT-AEAD family in Sect. 4. Finally we conclude this paper in Sect. 5.

2 Preliminaries

We firstly introduce some notations used throughout this paper. Let $\mathbb{F}_2$ denote the finite field with two elements (0 and 1) and $\boldsymbol{a} \in \mathbb{F}_2^n$ be an n-bit vector where a_i denotes the ith bit of $\boldsymbol{a}$. A unit vector where the ith element is 1 and the others are 0 is denoted by $\boldsymbol{e}_i$. Especially, an n-bit vector whose all elements are 0 (or 1) is denoted by 0^n (or 1^n). The concatenation of two vectors $\boldsymbol{a} \in \mathbb{F}_2^n$ and $\boldsymbol{b} \in \mathbb{F}_2^m$ is an $(n+m)$-bit vector, which is denoted by $\boldsymbol{a} \| \boldsymbol{b}$.

2.1 Algebraic Degree Evaluation by Division Property

In this paper, we use the *two-subset division property* [15] to estimate the algebraic degree, which has been proved to be the most optimal non-tight method in terms of the accuracy in [2].

Division Property. At EUROCRYPT 2015, Todo [14] proposed a generalized integral property called division property to search for longer integral distinguishers for block ciphers. Soon after this, Todo and Morii [15] extended this method to more accurate variants: two-subset and three-subset bit-based division property. At ASIACRYPT 2016, Xiang et al. [17] first adopted Mixed Integer Linear Programming (MILP), which has been widely applied to cryptanalysis in recent years [10,13], to further extend the application of division property.

Degree Evaluation Based on Division Property Using MILP. For an iterative cipher, let $\boldsymbol{s}^r = (s^r_{m-1}, ..., s^r_0)$ denote the r-round internal state and $(a^0_{m-1}, ..., a^0_0) \rightarrow \cdots \rightarrow (a^r_{m-1}, ..., a^r_0)$ denote an r-round division trail. In order to evaluate the algebraic degree of s^r_i w.r.t. (s^0_i)'s, we introduce an MILP model $\mathcal{M}$ to describe the division property propagation through the r-round encryption. Moreover, we fix $(a^r_{m-1}, ..., a^r_0)$ to $\boldsymbol{e}_i$ and maximize the objective function $\sum_{j=0}^{m-1} a^0_j$. The solution of $\mathcal{M}$ returned by MILP solvers like `Gurobi`[4] is regarded as the estimated degree of s^r_i. The procedure is illustrated in Algorithm 1.

2.2 Conditional Cube Attack

Assuming that $f(\boldsymbol{x}, \boldsymbol{v})$ is a Boolean function defined over the variables $\boldsymbol{x} = (x_{m-1}, ..., x_0)$ and $\boldsymbol{v} = (v_{n-1}, ..., v_0)$. Given a set of variables $I = \{v_{i_1}, v_{i_2}, ..., v_{i_{|I|}}\} \subset \{v_0, ..., v_{n-1}\}$ (we call *cube set*), and denote the monomial $v_{i_1} v_{i_2} \cdots v_{i_{|I|}}$ by t_I. Then $f(\boldsymbol{x}, \boldsymbol{v})$ can be rewritten as

$$f(\boldsymbol{x}, \boldsymbol{v}) = t_I \cdot pp_{S(I)} + q(\boldsymbol{x}, \boldsymbol{v}),$$

where $p_{S(I)}$ called *superpoly* does not contain any variable in I, and $q(\boldsymbol{x}, \boldsymbol{v})$ misses at least one variable from I. Moreover, the cube set I defines an $|I|$-dimensional space $\boldsymbol{C}_I$ (we call *cube*) of $2^{|I|}$ vectors, in which we assign all the possible combinations of 0/1 values to variables in I and fix remaining variables in $\{v_0, ..., v_{n-1}\} \setminus I$ to constant values.

[4] Gurobi optimization, https://www.gurobi.com/.

Algorithm 1. Algebraic degree evaluation by division property

Require: The model $\mathcal{M}$ describing the division property propagations of r-round cipher, and output bit index i.
Ensure: The algebraic degree d of the fixed output position.
1: $\mathcal{M}.obj \leftarrow$ maximize $\{a_0^0 + \cdots + a_{m-1}^0\}$; * Setting the objective function of $\mathcal{M}$. *\
2: $\mathcal{M}.con \leftarrow a_i^r = 1$;
3: $\mathcal{M}.con \leftarrow \sum_j a_j^r = 0$ for all $j \in \{0, 1, ..., m-1\} \setminus \{i\}$; * Focusing on the i-th bit of output, thus let others' division property be 0. *\
4: $\mathcal{M}.optimize()$; * Solving model $\mathcal{M}$ by GUROBI optimization. *\
5: $obj = \mathcal{M}.getObjective()$;
6: $d = obj.getValue()$; * Getting the solution of model $\mathcal{M}$. *\
7: **return** d

Theorem 1 ([4]). *For any polynomial $f(\boldsymbol{x}, \boldsymbol{v})$ and cube set I, the Xor-sum of $f(\boldsymbol{x}, \boldsymbol{v})$ over $\boldsymbol{C}_I$ (we call cube sum) is just the superpoly $p_{S(I)}$, i.e.,*

$$\bigoplus_{v \in \boldsymbol{C}_I} f(\boldsymbol{x}, \boldsymbol{v}) = \bigoplus_{v \in \boldsymbol{C}_I} t_I \cdot p_{S(I)} + \bigoplus_{v \in \boldsymbol{C}_I} q(\boldsymbol{x}, \boldsymbol{v}) \equiv p_{S(I)}.$$

At EUROCRYPT 2017, Huang et al. [6] proposed a variant of cube attack [4] called conditional cube attack. It aims to impose conditions on a series of key bits or equations w.r.t. key bits (collectively called conditional equations) and then detects the non-randomness of the cube sum to recover some key bits. There are two phases in the conditional cube attack summarized as follows:

- **Preprocessing phase.** In this phase, the adversary needs to prepare a set of conditional equations and a cube set such that the corresponding cube sum is always 0 if the conditional equations hold and is unknown otherwise, i.e. constructing the following distinguisher:

$$\bigoplus_{v \in \boldsymbol{C}_I} f(\boldsymbol{x}, \boldsymbol{v}) = \begin{cases} 0, & \text{if the conditional equations hold,} \\ \text{unknown,} & \text{otherwise.} \end{cases}$$

- **Online phase.** In this phase, the key of the cipher is unknown but fixed, the adversary can choose the value of public variables and has access to the corresponding outputs. Therefore, the adversary can compute the cube sum using the prepared conditional cube set and then judges whether the conditional equations hold or not according to the property of the cube sum. As a result, some key bits included in the conditional equations can be recovered.

2.3 KNOT-AEAD Family

KNOT [18] is a lightweight cipher suite designed by Zhang et al., which became one of the 32 candidates in the second round of NIST LWC. It contains the authenticated encryption scheme KNOT-AEAD and the hash function KNOT-Hash. This paper focuses on the security of KNOT-AEAD, thus we just introduce the KONT-AEAD here and omit the KNOT-Hash.

KNOT Permutations. All the four members of KNOT-AEAD family are designed on the underlying permutation KNOT-b, which is an SPN primitive with size of $b\,(b = 256, 384, 512)$. The b-bit internal state S can be divided into four $\frac{b}{4}$-bit words as $S = S_3\|S_2\|S_1\|S_0$ and described as the following $4 \times \frac{b}{4}$ array:

$$\begin{bmatrix} S_0[\frac{b}{4}-1], \cdots S_0[1], S_0[0] \\ S_1[\frac{b}{4}-1], \cdots S_1[1], S_1[0] \\ S_2[\frac{b}{4}-1], \cdots S_2[1], S_2[0] \\ S_3[\frac{b}{4}-1], \cdots S_3[1], S_3[0] \end{bmatrix}. \tag{1}$$

The round function of KNOT-b, denoted by p_b, is composed of three steps:

1. The $AddRoundConstant_b$ transformation. A simple bitwise Xor of a d-bit constant to the lowest d bits of the intermediate state, where $d = 6$ if $b = 128$ otherwise $d = 7$. We omit the detail to generate the constant set $CONST_d$, which can be referred to [18].
2. The $SubColumn_b$ transformation. This is a nonlinear layer by parallel applications of $\frac{b}{4}$ same S-boxes of size 4 to the columns. The truth table of the S-box is illustrated in Table 2.
3. The $ShiftRow_b$ transformation. The word S_0 is not rotated and S_i $(i \in \{1, 2, 3\})$ is left rotated over c_i bits, where the rotation parameters (c_1, c_2, c_3) are $(1, 8, 25)$, $(1, 8, 55)$ and $(1, 16, 25)$ for $b = 256, 384$ and 512 respectively.

Table 2. The truth table of KNOT-b S-box

Input	0x0	0x1	0x2	0x3	0x4	0x5	0x6	0x7	0x8	0x9	0xa	0xb	0xc	0xd	0xe	0xf
Output	0x4	0x0	0xa	0x7	0xb	0xe	0x1	0xd	0x9	0xf	0x6	0x8	0x5	0x2	0xc	0x3

KNOT-AEAD Family. The KNOT-AEAD family is based on the MonkeyDuplex mode of permutation KNOT-b. Let KNOT-AEAD (k, b, r) be the member that has the k-bit key and nonce, b-bit state and r-bit rate. Let K and N denote the key and nonce of KNOT-AEAD (k, b, r), where $K[i]$ and $N[i]$ are the ith bits of K and N respectively. The initial state of KNOT-AEAD (k, b, r) is loaded by key, nonce and constant as

$$S_3\|S_2\|S_1\|S_0 = \begin{cases} (0^{128}\|K\|N) \oplus (1\|0^{383}), & \text{if}\,(k, n, b) = (128, 384, 192), \\ K\|N, & \text{otherwise}. \end{cases} \tag{2}$$

in the initialization phase. Let P_i and C_i denote the ith block of plaintext and ciphertext, then the r-bit ciphertext C_i is generated as

$$C_i = \begin{cases} P_i \oplus (S_1\|S_0), & \text{if}\,(k, b, r) = (128, 384, 192), \\ P_i \oplus S_0, & \text{otherwise}. \end{cases} \tag{3}$$

in the encryption phase. Here we omit the processing of associated data and finalization phases, because our attacks are launched under the scenario where there is no associated data and only the first block of ciphertext C_0 is utilized.

3 A Framework of Conditional Cube Attacks for KNOT-AEAD

In conditional cube attacks, cube sums can be experimentally computed and distinguished when the cube set is small. But when the cube set is too large to practically compute the corresponding cube sums, the common approach to distinguish cube sums is to estimate the algebraic degrees by trivial bound, i.e. $\deg(G \circ F) \leq \deg(G) \cdot \deg(F)$ where G and F are vectorial Boolean functions. The difference of estimated degrees can be utilized to check whether the conditional equations hold or not. Apparently, estimating degrees by trivial bound is quite efficient but not accurate enough. In the worst situation, the estimated degrees are nearly equal and both greatly far from the exact degrees. This will lead to an invalid key-recovery attack since cube sums are always 0 and actually cannot be distinguished at all. For example, the authors of [9] points out that Liu et al.'s key-recovery attack in [8] indeed degraded to a distinguishing attack because the exact algebraic degrees are always lower than 64 whether the conditional equations hold or not.

In this section, we will introduce an improved framework for conditional cube attacks, which exploits the linearization technique and division property. Benefiting from the accuracy of division property for degree estimation, the estimated degree is very close to the exact degree and the gap on degrees estimated in two cases (conditional equations holds or not) is quite distinct. Therefore, we can select a more optimal cube set to achieve key-recover attacks. Before introducing this framework, we first give some concepts frequently used in the later content.

Definition 1 (CPV and FPV). *Let g be an $(m+n)$-variable function over $\boldsymbol{x}$ and $\boldsymbol{v}$. Considering a set consisting of I public variables $\boldsymbol{V}_c := \{v_{i_1}, v_{i_2}, ..., v_{i_I}\}$ and a set $\boldsymbol{V}_f := \{v_0, v_1, ..., v_{n-1}\} \setminus \boldsymbol{V}_c$, if g degenerates to a linear function w.r.t. the variables in $\boldsymbol{V}_f$ after imposing condition on every variable in $\boldsymbol{V}_c$ to a constant (0 or 1), then we call $v_c \in \boldsymbol{V}_c$ a conditional public variable (CPV) and call $v_f \in \boldsymbol{V}_f$ a free public variable (FPV).*

Example 1. Assuming that f is a function from $\mathbb{F}_2^4 \times \mathbb{F}_2^4$ to $\mathbb{F}_2$ as

$$f(\boldsymbol{v}, \boldsymbol{x}) = v_0 v_1 v_2 x_0 x_1 + v_0 v_2 x_3 + v_1 v_3 x_2 + v_1 v_2 + v_0 v_2 v_3 + x_0 x_1 + v_0 x_3,$$

where $\boldsymbol{v}$ and $\boldsymbol{x}$ are public and secret variables, respectively. Let $v_1 = 1$ and $v_2 = 0$, then we have

$$f(v_0, 1, 0, v_3, \boldsymbol{x}) = v_3 x_2 + x_0 x_1 + v_0 x_3,$$

the algebraic degree of f on the public variable is decreased from three to one when $\{v_1, v_2\}$ are CPV and $\{v_0, v_3\}$ are FPV. Similarly, choosing $\{v_0, v_1\}$ as CPV and setting $(v_0, v_1) = (0, 1)$ can also linearize f.

Definition 2 (CRPV, CSV). *In a conditional cube attack, assuming that there is a state bit of a cipher can be represented as*

$$f(\boldsymbol{v}, \boldsymbol{x}) = v_i \cdot L(\boldsymbol{x}) + P(\boldsymbol{v}) \cdot Q(\boldsymbol{x}), \tag{4}$$

where $\boldsymbol{x}$ and $\boldsymbol{v}$ are the secret variable and public variable respectively. Moreover, $P(\boldsymbol{v})$ is a linear function w.r.t. $\boldsymbol{v}$ but excluding v_i, $L(\boldsymbol{x})$ and $Q(\boldsymbol{x})$ are linear and nonlinear functions only w.r.t. $\boldsymbol{x}$ respectively. If we impose conditions on some bits of $\boldsymbol{x}$ involved in $L(\boldsymbol{x})$ such that $L(\boldsymbol{x}) = c$ where c is a constant in $\mathbb{F}_2$, then we call v_i a conditional equation related public variable (CRPV) and call the involved bits of $\boldsymbol{x}$ in $L(\boldsymbol{x})$ conditional secret variables (CSV).

We now introduce our framework of conditional cube attacks. Let E be an iterative cipher and E_i $(i \geq 1)$ denote the i-round encryption of E. Denote $\boldsymbol{s}^0$ the initial state of E loaded with the public variable $\boldsymbol{v}$ and the secret variable $\boldsymbol{x}$. We use $\boldsymbol{s}^i$ $(i > 0)$ to denote the i-round state, i.e. $\boldsymbol{s}^i = E_i(\boldsymbol{s}^0)$. Our framework to construct a conditional cube attack on E is composed of four steps. In particular, they are linearization of state bits, construction of conditional equations, degree evaluation of output bits and construction of conditional cube sets.

Linearization of State Bits. In this step, we aim to linearize the state bits by choosing appropriate public variables as CPV. On the one hand, it can extend the number of attacked rounds. On the other hand, it will be more convenient for us to construct conditional equations from these linearized state bits. In general, the monomials and the algebraic degree of state bits will grow exponentially with round increasing. Thus the round number of linearization cannot be too large, otherwise it will be quite difficult to achieve linearization and construct conditional cube sets using FPV. We can determine the number of linearized rounds by analyzing the distribution of the algebraic degree and the ANF's structural property of the short-round state bits. Assuming that for the r_f-round state $\boldsymbol{s}^{r_f}$, the most bits are linear or there are quite fewer nonlinear monomials in the nonlinear bits, then we are supposed to linearize $\boldsymbol{s}^{r_f}$.

In [7], Li et al. utilized linearization technique in conditional cube attacks to eliminate all quadratic terms except a particular one in the first round. It is different from our framework since our goal is to eliminate all nonlinear terms.

Construction of Conditional Equations. After linearization on the r_f-round state, all the bits of $\boldsymbol{s}^{r_f}$ are linear functions over the FPV with the form as Eq. (4). Assuming that the ith bit $s_i^{r_f}$ can be represented as

$$s_i^{r_f} = v_{i_0} \cdot L(\boldsymbol{x}) + P(\boldsymbol{v}) \cdot Q(\boldsymbol{x}).$$

Now we impose conditions on the CSV to construct a conditional equation $L(\boldsymbol{x}) = c_i$, where c_i is a constant. Thus v_{i_0} becomes the CRPV. Note that the specific value of c_i (0 or 1) can possibly change the diffusion of v_{i_0}. This might lead to a conditional distinguisher on the cube sum by choosing v_{i_0} as one of the cube variables. In the next step we will explain how to detect the effect of c_i on the diffusion of v_{i_0}.

Degree Evaluation of Output Bits. Evaluating algebraic degrees of output bits is an effective approach to detect the effect on diffusion of v_{i_0} when fixing c_i to different values. In the previous works [6–8], the cube set including v_{i_0} is prepared in advance and the algebraic degrees of output bits on cube variables are estimated by the trivial bound. Nevertheless, we here only focus on the maximal degree of monomials that contain v_{i_0} and utilize the MILP-aided division property to achieve the degree evaluation. Note that in the corresponding MILP model, the division property of v_{i_0} and CPV are 1 and 0, respectively. Moreover, the objective is to maximize the sum of division property of FPV including v_{i_0}. Compared with the previous works, our method only focuses on the degree of monomials that contain v_{i_0} and can obtain a more accurate degree. This is of great significance to construct cube sets with fewer variables and recover CSV.

Construction of Conditional Cube Sets. Let us focus on the kth bit of $\boldsymbol{s}^r$, denoted by s_k^r. Assuming that the estimated degree as described in the last step is d_0 for $c_i = 0$ and d_1 for $c_i = 1$, then there are two following cases:

- There is a huge gap between d_0 and d_1. Without loss of generality, we suppose that d_0 is less than d_1 and one of the d_1-degree monomials is $v_{i_0} v_{i_1} \cdots v_{i_{d_1}}$. Now we can arbitrarily select d variables from the set $\{v_{i_0}, v_{i_1}, ..., v_{i_{d_1}}\}$ to form a cube set I that must include v_{i_0}. In theory, the following distinguisher can be constructed as long as $d \in (d_0, d_1]$,

$$\bigoplus_{\boldsymbol{v} \in \boldsymbol{C}_I} s_k^r(\boldsymbol{x}, \boldsymbol{v}) = \begin{cases} 0, & \text{if}\, L(\boldsymbol{x}) = c_0 = 0, \\ \text{unknown}, & \text{otherwise}, \end{cases}$$

  where $\boldsymbol{C}_I$ is the cube of I. Note that the estimated degree by division property is an upper bound on the exact degree, thus the above cube sum is possibly always 0 if d is equal or very close to d_1. To ensure that the cube sum can be distinguished as well as to decrease the complexity of computation, $d = d_0 + 1$ will be the best choice.
- d_0 is equal or very close to d_1. In this case, the real degree is probably lower than both d_0 and d_1, thus the cube sum will be always 0 even if we choose a cube set of size $d_0 + 1$.

As a result, if d_0 is far from d_1 then we can construct a cube set as illustrated in the above first case and recover one bit of CSV with 2^d computations. Otherwise we need to go back to the **construction of conditional equations** step and consider other bits of $\boldsymbol{s}^{r_f}$ or go back to the **degree evaluation of output bits** step and consider other bits of $\boldsymbol{s}^r$.

It should be emphasized that the above steps are general for conditional cube attacks. When applied to a concrete cipher, we can explore some good properties of the cipher to achieve a better conditional cube attack.

4 Conditional Cube Attacks on Full Members of KNOT-AEAD Family

In this section, we will apply the proposed framework to all members of KNOT-AEAD family. Note that KNOT-AEAD (k, $2k$, $\frac{k}{2}$) ($k \in \{128, 192, 256\}$) follow the same style to load the initial state and generate ciphertexts as illustrated in Eqs. (2) and (3), but KNOT-AEAD (128, 384, 192) does not. Therefore we focus on KNOT-AEAD (128, 256, 64) and KNOT-AEAD (128, 384, 192) to demonstrate our conditional cube attacks in detail and we only list the results of KNOT-AEAD (192, 384, 96) and KNOT-AEAD (256, 512, 128).

Our attacks are launched at the initialization phase in nonce-respecting setting under the scenario where there is no associated data. In particular, after r_R-round initialization, the r-bit intermediate state is Xored with the first block of plaintext P_0 to get the corresponding ciphertext block C_0, which will be utilized to compute cube sums. Our attack model is depicted in Fig. 1. All experiments in this section are implemented on the following platform: Intel(R) Core(TM) i7-8700 CPU @ 3.20 GHz, 8.00 GB RAM, 64-bit Windows 10 system.

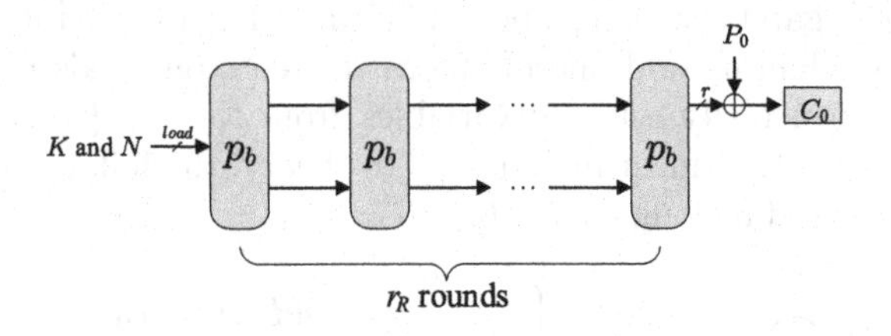

Fig. 1. Attack model on KNOT-AEAD

4.1 Modeling the Division Property Propagation of Conditional S-boxes

In our framework for conditional cube attacks, we exploit MILP-aided division property to estimate degrees of output bits, thus we need to describe the propagation of division property by MILP language.

We use Xiang et al.'s algorithm [17] to enumerate all valid division trails of the KNOT S-box as shown in Table 6 of Appendix A. Then we adopt Sasaki and Todo's strategy [12] to select **12** inequalities from the H-representation of the convex hull of division trails to describe the division property propagation as illustrated in Eq. (12) of Appendix A, where (a_3, a_2, a_1, a_0) and (b_3, b_2, b_1, b_0) denote the input and output division property, respectively.

In our attack model, there are many incomplete S-boxes in the first round, some input bits of which are assigned to constants. We call these incomplete S-boxes conditional S-boxes as the following definition.

Definition 3 (Conditional S-box). *Given an S-box from $\mathbb{F}_2^n$ to $\mathbb{F}_2^n$. A new S-box from $\mathbb{F}_2^{n-n_c}$ to $\mathbb{F}_2^n$ can be derived by imposing conditions on n_c $(0 < n_c < n)$ input bits of the original S-box, we call this derived S-box an n_c-bit conditional S-box.*

Example 2. Consider the KNOT S-box, if we assign the middle two input bits to 0 i.e. $x_1 = x_2 = 0$, then we can obtain a 2-bit conditional S-box: 0x0 $\xrightarrow{S}$ 0x4, 0x1 $\xrightarrow{S}$ 0x0, 0x8 $\xrightarrow{S}$ 0x9 and 0x9 $\xrightarrow{S}$ 0xf. Its ANF is

$$y_0 = x_3,\ y_1 = x_0x_3,\ y_2 = 1 + x_0 + x_3,\ y_3 = x_3. \tag{5}$$

The 12 inequalities listed in Eq. (12) are still suitable but not accurate enough to model conditional S-boxes. In order to obtain a more accurate MILP model to estimate a more tight bound on the degree, we need to consider the influence of constants on division trails of conditional S-boxes. Here we utilize the KNOT 2-bit conditional S-box, as described in Example 2, to illustrate the re-description on the division property propagation of conditional S-boxes.

Similarly, the division trails are enumerated by Xiang et al.'s algorithm as listed in Table 6 of Appendix A from the derived ANF (see Eq. (5)). Then the corresponding inequalities can be described as in Eq. (11) of Appendix A, where (a_3, a_2, a_1, a_0) and (b_3, b_2, b_1, b_0) denote the input and output division property respectively and $a_1 = a_2 = 0$ in this case.

From Table 6, we can see that the original S-box has two extra division trails 0x1 $\rightarrow$ 0x1 and 0x1 $\rightarrow$ 0x9 in the case of $a_2 = a_1 = 0$. However, both are **invalid** division trails for this 2-bit conditional S-box. Moreover, the trail 0x9 $\rightarrow$ 0x1 is also **invalid**. It indicates that the inequality system $\mathcal{L}_c$ is more accurate than $\mathcal{L}$ to describe the 2-bit conditional S-box. Thus we should consider the actual values of inputs and generate division trails using the new ANF when modeling conditional S-boxes.

4.2 Key-Recovery Attack on KNOT-AEAD (128, 256, 64)

We reuse the notations used in Sect. 2.3. In addition, S^r denotes the r-round intermediate state, especially S^0 denotes the initial state. From Eq. (2), we know $S_1^0 \| S_0^0 = N$ and $S_2^0 \| S_3^0 = K$, which indicates that S_0^1 and S_3^1 are quadratic w.r.t. nonce variables according to the ANF of KNOT S-box. Here we choose the higher 64 bits of nonce as CPV and set all of them to 0, then the first bit of S_0^1 can be linearized and expressed as

$$\begin{aligned} S_0^1[0] &= K[0] + (N[0] + 1) \cdot K[0] + K[64] + K[0] \cdot K[64] \\ &= \boldsymbol{N[0]} \cdot \boldsymbol{K[0]} + K[0] \cdot K[64] + K[64]. \end{aligned}$$

We next impose a condition on $K[0]$ and construct the conditional equation $K[0] = c_0$, then $N[0]$ becomes a CPV naturally. Note that $c_0 = 0$ can eliminate the existence of $N[0]$ in $S_0^1[0]$, thus we expect to construct a distinguisher on the ciphertext C_0 by detecting the maximal degree of monomials that contain $N[0]$. Because all the bits of S_1^0 are fixed to 0 and $K[0] = c_0$ is the conditional equation, there are sixty-three **1**-bit conditional S-boxes and only one **2**-bit conditional S-box. Thus when we utilize the MILP-aided division property to estimate degrees, it is necessary to rebuild the description of division property propagation through the $SubColumn_{256}$ transformation in the first round. The whole procedure to model conditional S-boxes has been illustrated in Sect. 4.1 so we omit the details both here and later. We use Algorithm 1 to estimate the algebraic degree of $C_O[0]$ that denotes the first bit of C_O, and list the estimated degrees from the second round to the ninth round in Table 3. From this table we know that the

Table 3. The estimated degrees of monomials including $N[0]$ in $C_O[0]$

#Round	2	3	4	5	6	7	8	9
$K[0] = \mathbf{0}$	0	0	0	0	0	0	**0**	64
$K[0] = \mathbf{1}$	2	6	12	23	36	49	**60**	64

gap on the degree in different cases ($K[0] = 0$ holds or not) is pretty significant. To be more specific, the output bit $C_O^r[0]$ $(2 \leq r \leq 8)$ actually does not contain the variable $N[0]$ if $K[0] = 0$ holds. Therefore, we can construct a cube set $I_0 = \{N[0]\}$ and obtain the following distinguisher:

$$\bigoplus_{N \in C_{I_0}} C_O^8[0](K, N) = \begin{cases} 0, & \text{if } K[0] = 0, \\ \text{unknown}, & \text{otherwise}, \end{cases} \tag{6}$$

where $\boldsymbol{C}_{I_0}$ is the cube of I_0 and $C_O^8[0]$ is the first bit of C_O in the case where the round number of initialization is reduced to 8. Note that there are only two elements in $\boldsymbol{C}_{I_0}$, thus it is practical and very efficient to verify the correctness of the distinguisher by experiments. We randomly choose 2^{10} cubes by setting the non-cube free variables to constants and then compute cube sums through 2nd to 9th rounds under 2^{10} random keys. In other words, there are 2^{20} cube sums. Our experiment shows that the cube sums are always equal to 0 on $C_O^i[0]$ $(2 \leq i \leq 8)$ and uncertain on $C_O^9[0]$ when modifying $K[0]$ to 0, and the cube sums on $C_O^i[0]$ $(2 \leq i \leq 9)$ are always uncertain when modifying $K[0]$ to 1. We repeat this experiment ten times to calculate the average probability that $K[0] = 0$ does not hold but the cube sum is 0. The probability is summarized in Table 4.

Table 4. Probability that cube sum on $C_O[0]$ equals to 0

#Round	2	3	4	5	6	7	8	9
$K[0] = \mathbf{0}$	100%	100%	100%	100%	100%	100%	**100%**	84.02%
$K[0] = \mathbf{1}$	49.98%	75.01%	87.49%	93.79%	96.86%	98.43%	**99.22%**	83.23%

The experimental results on the correctness of the above disitinguisher (see Eq. (6)) are actually consistent with our expectations. Thus we can recover the key bit $K[0]$ by an 8-round conditional cube attack. Note that the cube sum on $C_O^8[0]$ is 0 with a probability of **99.22%** when $K[0]$ equals to 1, which means that the recovered $K[0]$ being 0 (in fact it is 1) is **false** with a probability of 99.22% by computing the cube sum only one time in the online phase. Assuming that we repeat the computation T times, then the probability that there exists at least one cube sum of being 1 is $Pr_0 = 1 - (99.22\%)^T$ when the value of $K[0]$ is 1. It is easy to calculate that $Pr_0 \geq \mathbf{99.99\%}$ if $T \geq \mathbf{1177}$ and Pr_0 is almost equal to **100%** if $T \geq \mathbf{1500}$. Consequently, we can recover $K[0]$ with success rate of **100%** with time and data complexity of $\mathbf{1500 \times 2^1 \approx 2^{11.55}}$.

The $ShiftRows_b$ transformation is *translation-invariant* in each row, which leads to the fact that every bit in the same row has a similar algebraic structure. Therefore, we can construct the conditional equation $K[i] = 0$ and cube set $I_i = \{N[i]\}$ to generate the following distinguisher:

$$\bigoplus_{N \in \boldsymbol{C}_{I_i}} C_0^8[i](K, N) = \begin{cases} 0, & \text{if } K[i] = 0, \\ \text{unknown}, & \text{otherwise}, \end{cases}$$

where $\boldsymbol{C}_{I_i}$ is the corresponding cube and $i \in \{0, 1, ..., 63\}$. Hence, we can recovery the **64** key bits $(K[63], ..., K[0])$ with time and data complexity of $\mathbf{64 \times 1500 \times 2^1 \approx 2^{17.55}}$.

Additionally, we choose the lower 64 bits of the nonce as CPV and fix them to constants as follows:

$$N[i] = \begin{cases} 1, & \text{if } i = 0, \\ 0, & \text{if } i \in \{1, 2, ..., 63\}. \end{cases}$$

After one round encryption, the first bit of S_0 can be represented as

$$S_0^1[0] = \boldsymbol{N[64]} \cdot \boldsymbol{K[64]} + K[0] + K[64] + K[0] \cdot K[64].$$

Let $K[64] = 0$ be a conditional equation, then $N[64]$ becomes the corresponding CRPV. Thus by estimating the maximal algebraic degree of monomials including $N[64]$ in C_0, we find that the first bit of C_0 has a distinct gap on the estimated degree, which is shown in Table 7 of Appendix B. We can construct a cube set $I_{64} = \{N[64]\}$ and get the following distinguisher:

$$\bigoplus_{N \in \boldsymbol{C}_{I_{64}}} C_0^8[0](K, N) = \begin{cases} 0, & \text{if } K[64] = 0, \\ \text{unknown}, & \text{otherwise}, \end{cases}$$

where $\boldsymbol{C}_{I_{64}}$ is the cube of I_{64}. Moreover, on the basis of *translation-invariant* property of $ShiftRow_b$, we can construct the cube set $I_i = \{N[i]\}$ and get the distinguisher

$$\bigoplus_{N \in \boldsymbol{C}_{I_i}} C_0^8[i - 64](K, N) = \begin{cases} 0, & \text{if } K[i] = 0, \\ \text{unknown}, & \text{otherwise}, \end{cases}$$

for any $i \in \{64, 65, ..., 127\}$. As a result, we can recover the **64** key bits $(K[127], ..., K[64])$ with time and complexity of $\mathbf{64} \times T \times \mathbf{2^1} = T \times \mathbf{2^7}$, where T

is the repetitive time of computation on cube sums. Similarly, our experiments indicate that the success rate to recover correct key bit is nearly equal to **100%** when setting $T = 1500$. In conclusion, we can launch the 8-round conditional cube attack on KNOT-AEAD (128, 256, 64) to recover the full **128** key bits, and the total time complexity T_C and data complexity D_C are calculated as

$$T_C = D_C = \mathbf{128 \times 1500 \times 2^1 \approx 2^{18.55}}.$$

The full key-recovery attack proposed in [16] using differential-linear method is 7 rounds longer than ours, but what we want to state is that the complexity of our attack is negligible and it indeed takes no more than one second to implement our full key-recovery attack on a PC.

In the same way, we can launch conditional cube attacks to achieve full key-recovery on **8**-round KNOT-AEAD (192, 384, 96) and **9**-round KNOT-AEAD (256, 512, 128). The details are omitted here and only the positions of the ciphertext, conditional equations, cube sets and complexities are summarized in Table 5, where C.E. is the abbreviation of conditional equation.

Table 5. Attacks on KNOT-AEAD $(k, 2k, \frac{k}{2})$ $(k \in \{128, 192, 384\})$

Member	#Round	Ciphertext	C.E.	Cube set	Index	Time	Data
(192,384,96)	8	$C_0^8[(i+55) \bmod 96]$	$K[i]=0$	$\{N[i]\}$	$i \in [0, 95]$	$2^{18.13}$	$2^{18.13}$
		$C_0^8[(i+43) \bmod 96]$	$K[i+96]=0$				
(256,512,128)	9	$C_0^9[(i+25) \bmod 128]$	$K[i]=0$	$\{N[i]\}$	$i \in [0, 127]$	$2^{18.55}$	$2^{18.55}$
		$C_0^9[(i+24) \bmod 128]$	$K[i+128]=0$				

4.3 Key-recovery Attacks on KNOT-AEAD (128, 384, 192)

For this version, we split the 128-bit nonce and key to four 32-bit parts as: $N = N_3\|N_2\|N_1\|N_0$, $K = K_3\|K_2\|K_1\|K_0$. Thus according to Eq. (2), the initial state can be represented as

$$\begin{bmatrix} N_2 & N_1 & N_0 \\ K_1 & K_0 & N_3 \\ 0^{32} & K_3 & K_2 \\ 1\|0^{31} & 0^{32} & 0^{32} \end{bmatrix}. \tag{7}$$

Obviously, there are three types of S-boxes in the $SubColumn_{384}$ transformation of the first round corresponding to the three columns of Eq. (7). Next, we will discuss the three columns separately.

The Right Column. We choose N_3 as CPV and fix N_3 to 0^{32}, thus after the $SubColumn_{384}$ transformation, the intermediate state is

$$\begin{aligned}S_0[i] &= \boldsymbol{N_0[i]} \cdot \boldsymbol{K_2[i]} + K_2[i],\\ S_1[i] &= K_2[i],\\ S_2[i] &= N_0[i] + K_2[i] + 1,\\ S_3[i] &= 0,\end{aligned}$$

for $i \in \{0, ..., 31\}$. Naturally, let $K_2[i] = 0$ be a conditional equation, then $N_0[i]$ becomes the corresponding CRPV. We estimate the maximal algebraic degree of monomials including $N_0[i]$ in C_0, and find that $C_0[i+2]$ has a distinct gap on the algebraic degree as shown in Table 8 of Appendix B. It indicates that the estimated degree of $C_0^9[i+2]$ is 0 if $K_3[i] = 0$ holds, otherwise it is 96. Therefore, we can use the cube set $I_i = \{N_0[i]\}$ to construct the 9-round distinguisher as follows:

$$\bigoplus_{N \in \boldsymbol{C}_{I_i}} C_0^9[i+2](K, N) = \begin{cases} 0, & \text{if } K_3[i] = 0,\\ \text{unknown}, & \text{otherwise},\end{cases}$$

where $\boldsymbol{C}_{I_i}$ is the cube of I_i for $i \in \{0, ..., 31\}$. Our experiments indicate that when repeating 1500 times, the key bit $K_3[i]$ can be recovered with a success rate of 100% by the conditional cube distinguisher. As a result, we can recover the **32**-bit key K_3 for **9**-round KNOT-AEAD (128, 384, 192), and the corresponding time complexity $T_{C_9}^R$ and data complexity $D_{C_9}^R$ are

$$T_{C_9}^R = D_{C_9}^R = \mathbf{32 \times 1500 \times 2^1 \approx 2^{16.55}}. \tag{8}$$

Note that the **8**-round key-recovery attack has the same time and data complexity as that of the 9-round version, here we denote them by $T_{C_8}^R$ and $D_{C_8}^R$.

The Left Column. The intermediate state after the $SubColumn_{384}$ transformation is linear as follows

$$\begin{aligned}S_0[i+64] &= \boldsymbol{N_2[i]} \cdot \boldsymbol{K_1[i]},\\ S_1[i+64] &= K_1[i],\\ S_2[i+64] &= N_2[i] + K_1[i] + 1,\\ S_3[i+64] &= \boldsymbol{N_2[i]} \cdot \boldsymbol{K_1[i]} + K_1[i],\end{aligned}$$

for $i \in \{0, 1, ..., 30\}$, and the case of $i = 31$ will be discussed later. Let $K_1[i] = 0$ be the conditional equation and we estimate the maximal algebraic degree of monomials including $N_2[i]$. We find the output bit $C_0[i+27]$ ($i \in \{0, ..., 30\}$) has a distinct gap on the estimated degree as shown in Table 9 of Appendix B. The following 9-round distinguishers can be constructed using the cube set $I_i = \{N_2[i]\}$:

$$\bigoplus_{N \in \boldsymbol{C}_{I_i}} C_0^9[i+27](K, N) = \begin{cases} 0, & \text{if } K_1[i] = 0,\\ \text{unknown}, & \text{otherwise},\end{cases}$$

where C_{I_i} is the cube of I_i for $i \in \{0, ..., 30\}$. Thus we can recover the **31**-bit key $(K_1[30], ..., K_1[0])$ for **9**-round KNOT-AEAD (128, 384, 192), and the time complexity $T^L_{C_9}$ and data complexity $D^L_{C_9}$ are calculated as

$$T^L_{C_9} = D^L_{C_9} = \mathbf{31} \times \mathbf{1500} \times \mathbf{2^1} \approx \mathbf{2^{16.50}}. \tag{9}$$

In addition, the remaining key bit $K_1[31]$ needs to be guessed with a time complexity of **2**. The above distinguisher can be also used to launch an **8**-round key-recovery attack, which has the same time and data complexity as that of 9-round key-recovery attack. We denote them by $T^L_{C_8}$ and $D^L_{C_8}$.

The Middle Column. The intermediate state after the *SubColumn*$_{384}$ transformation is also linear as follows

$$\begin{aligned}
S_0[i+32] &= \boldsymbol{N_1[i]} \cdot (\boldsymbol{K_0[i]} + \boldsymbol{K_3[i]}) + K_3[i],\\
S_1[i+32] &= K_0[i] + K_3[i],\\
S_2[i+32] &= N_1[i] + K_0[i] \cdot K_3[i] + K_0[i] + K_3[i] + 1,\\
S_3[i+32] &= \boldsymbol{N_1[i]} \cdot \boldsymbol{K_0[i]} + K_0[i] + K_3[i],
\end{aligned}$$

for $i \in \{0, ..., 31\}$. We choose $K_0[i] + K_3[i] = 0$ as the conditional equation and do not consider the concrete values of $K_0[i]$ and $K_3[i]$, then we find the output bit $C_0[i+34]$ has a distinct gap on the estimated degree as illustrated in Table 10 of Appendix B. Thus we can construct the 8-round distinguisher using cube set $I_i = \{N_1[i]\}$ as

$$\bigoplus_{N \in C_{I_i}} C_0^8[i+34](K, N) = \begin{cases} 0, & \text{if } K_0[i] + K_3[i] = 0,\\ \text{unknown}, & \text{otherwise}, \end{cases}$$

for $i \in \{0, ..., 31\}$. However, we can only recover one bit information of $K_0[i]$ and $K_3[i]$. As a result, we can recover **32** bits information by 8-round conditional cube attack, the corresponding time complexity $T^M_{C_8}$ and data complexity $D^M_{C_8}$ are calculated as

$$T^M_{C_8} = D^M_{C_8} = \mathbf{32} \times \mathbf{1500} \times \mathbf{2^1} \approx \mathbf{2^{16.55}}. \tag{10}$$

The remaining **32** bits information needs to be guessed with a time complexity of $\mathbf{2^{32}}$.

Besides, if we consider the concrete values of $K_0[i]$ and $K_3[i]$, it is potential to launch a 9-round key-recovery attack. To be specific, we find that the estimated degree of monomials including $N_1[i]$ in $C_0^9[i+34]$ is **0** if $K_0[i] = K_3[i] = 0$ otherwise it is **128**, which is listed in Table 10 of Appendix B. Thus we can construct the following 9-round distinguisher using cube set $I_i = \{N_1[i]\}$:

$$\bigoplus_{N \in C_{I_i}} C_0^9[i+34](K, N) = \begin{cases} 0, & \text{if } K_0[i] = K_3[i] = 0,\\ \text{unknown}, & \text{otherwise}, \end{cases}$$

for $i \in \{0, ..., 31\}$. This is a weak distinguisher since there are three possible cases for $(K_0[i], K_3[i])$: $(1,0)$, $(0,1)$ and $(1,1)$ when the condition does not hold. If we obtain a cube sum of 1, then we need to guess $(K_0[i], K_3[i])$ with a time complexity of 3. Thus in the worst situation, we need a time complexity of $\mathbf{3^{32} \approx 2^{50.72}} < 2^{64}$ to recover the **64**-bit key $K_0 \| K_3$.

Combination of Three Columns. In summary, by combining the complexities of three parts (see Eqs. (8), (9) and (10)), we can achieve the full key-recovery attack on 8-round KNOT-AEAD $(128, 384, 192)$, the time complexity T_{C_8} is

$$T_{C_8} = T_{C_8}^R + T_{C_8}^L + T_{C_8}^M + \mathbf{2 \times 2^{32} \approx 2^{33}},$$

and the data complexity D_{C_8} is

$$D_{C_8} = D_{C_8}^R + D_{C_8}^L + D_{C_8}^M \approx \mathbf{2^{18.12}}.$$

For the full key-recovery attack on 9-round KNOT-AEAD $(128, 384, 192)$, the worst time complexity T_{C_9} is

$$T_{C_9} = T_{C_9}^R + T_{C_9}^L + \mathbf{32 \times 1500 \times 2^1 + 2^1 \times 3^{32} \approx 2^{51.72}},$$

where $\mathbf{2^1 \times 3^{32}}$ is the time to guess the **65** key bits: ($K_1[31]$, $K_0[0], ..., K_0[31]$, $K_3[0], ..., K_3[31]$), and the data complexity D_{C_9} is

$$D_{C_9} = D_{C_9}^R + D_{C_9}^L + \mathbf{32 \times 1500 \times 2^1 \approx 2^{18.12}}.$$

Note that a 17-round key-recovery attack is proposed by Wang et al. [16], but we can recover the full key bits whereas Wang et al. recovered only one bit. Moreover, our 8-round attacks are practical that can be efficiently launched.

5 Conclusion

In this paper, we evaluate the security of KNOT-AEAD family against cube-like attacks. We combine the linearization technique and division property to propose a framework for conditional cube attacks. In particular, we linearize the state bit to a special form, from which it is convenient for us to construct conditional equations. Moreover, we utilize division property to detect the gap on the algebraic degree caused by conditional equations and further choose cube sets to construct conditional cube distinguishers. We apply it to KNOT-AEAD family and obtain the first practical full key-recovery on all round-reduced members.

In addition, our experiments show that the probability of cube sum being 0 in the situation where the conditional equation does not hold is larger than 99%. It is very odd but interesting, we guess the reason is that the high-degree monomials in output bits are very sparse or output bits have a very special algebraic structure, which is worthy of further discussion.

Acknowledgement. We would like to thank all the anonymous reviewers for their helpful comments. This work was supported by the Research Foundation of Department of Education of Hubei Province, China (No. D2020104) and the National Natural Science Foundation of China (No. 61802119).

A Division Trails and Linear Descriptions of the KNOT S-box

Table 6. The division trails of KNOT S-box

The original S-box			
Input	Output	Input	Output
0x0	0x0	0x8	0x1, 0x2, 0x4, 0x8
0x1	0x1, 0x2, 0x4, 0x8	0x9	0x1, 0x2, 0xc
0x2	0x1, 0x2, 0x4, 0x8	0xa	0x1, 0x2, 0xc
0x3	0x1, 0x6, 0x8	0xb	0x1, 0x6, 0xc
0x4	0x1, 0x2, 0x4, 0x8	0xc	0x1, 0x2, 0xc
0x5	0x1, 0x6, 0xa, 0xc	0xd	0x7, 0x9, 0xa
0x6	0x2, 0x4, 0x9	0xe	0x2, 0x5, 0xc
0x7	0x7, 0xa, 0xd	0xf	0xf
The 2-bit conditional S-box			
Input	Output	Input	Output
0x0	0x0	0x8	0x1, 0x2, 0x4, 0x8
0x1	0x2, 0x4	0x9	0x2, 0x5, 0xc

$$\mathcal{L}_c = \begin{cases} a_3 + a_0 - b_3 - b_2 - b_1 - b_0 \geq 0, \\ -a_0 + b_2 + b_1 \geq 0, \\ -a_3 - a_0 + b_3 + b_2 + 2b_1 + b_0 \geq 0, \\ -b_3 - b_1 - b_0 + 1 \geq 0, \\ -b_2 - b_1 + 1 \geq 0, \\ a_3, a_0, b_3, b_2, b_1, b_0 \in \{0, 1\}. \end{cases} \tag{11}$$

$$\mathcal{L} = \begin{cases} -a_3 - 4a_2 - 3a_1 - 2a_0 + b_3 + 2b_2 + b_1 - b_0 + 7 \geq 0, \\ a_3 + a_2 + 2a_1 + a_0 - 2b_3 - b_2 - b_1 - 2b_0 + 1 \geq 0, \\ a_3 + a_2 + a_1 + 3a_0 - b_3 - 2b_2 - 2b_1 - 2b_0 + 1 \geq 0, \\ 3a_2 + a_1 + a_0 - b_3 - 2b_2 - 2b_1 - 2b_0 + 2 \geq 0, \\ -2a_3 - a_2 - a_1 + 2b_3 + 2b_2 + 4b_1 + 3b_0 \geq 0, \\ -a_3 - a_2 - a_0 + b_3 + b_2 + 2b_1 + 2b_0 \geq 0, \\ -a_2 - a_0 + b_3 + b_2 - b_1 + 2 \geq 0, \\ a_3 - a_2 - a_1 - b_3 + b_1 + b_0 + 2 \geq 0, \\ -a_3 - 2a_2 - a_1 - 3a_0 + 2b_3 - b_2 + 2b_1 + b_0 + 4 \geq 0, \\ 2a_2 - a_1 - a_0 - b_3 + b_2 - 2b_1 - b_0 + 3 \geq 0, \\ a_3 + 3a_2 - 2b_3 - b_2 - b_1 - 2b_0 + 2 \geq 0, \\ a_0 - b_3 - b_2 - 2b_1 + b_0 + 2 \geq 0, \\ a_3, a_2, a_1, a_0, b_3, b_2, b_1, b_0 \in \{0, 1\}. \end{cases} \tag{12}$$

B Some Tables about Estimated Algebraic Degrees

Table 7. The estimated degrees of monomials including $N[64]$ in $C_0[0]$

#Round	2	3	4	5	6	7	8	9
$K[64] = \mathbf{0}$	0	0	0	0	0	0	**0**	64
$K[64] = \mathbf{1}$	2	6	12	23	36	49	**60**	64

Table 8. The estimated degrees of monomials including $N_0[i]$ in $C_0[i+2]$

#Round	2	3	4	5	6	7	8	9	10
$K_2[i] = \mathbf{0}$	0	0	0	0	0	0	0	**0**	96
$K_2[i] = \mathbf{1}$	0	0	12	22	38	58	80	**96**	96

Table 9. The estimated degrees of monomials including $N_2[i]$ in $C_0[i+27]$

#Round	2	3	4	5	6	7	8	9	10
$K_1[i] = \mathbf{0}$	0	0	0	0	0	0	0	**0**	128
$K_1[i] = \mathbf{1}$	0	0	0	0	57	83	112	**128**	128

Table 10. The estimated degrees of monomials including $N_1[i]$ in $C_0[i+34]$

#Round	2	3	4	5	6	7	8	9	10
$(K_0[i], K_3[i]) = (\mathbf{0}, \mathbf{0})$	0	0	0	0	0	0	**0**	**0**	128
$(K_0[i], K_3[i]) = (\mathbf{1}, \mathbf{0})$	0	0	15	32	58	86	**113**	**128**	128
$(K_0[i], K_3[i]) = (\mathbf{0}, \mathbf{1})$	0	0	15	32	58	86	**113**	**128**	128
$(K_0[i], K_3[i]) = (\mathbf{1}, \mathbf{1})$	0	0	0	0	0	0	**0**	**128**	128

References

1. Baksi, A., Breier, J., Chen, Y., Dong, X.: Machine learning assisted differential distinguishers for lightweight ciphers. In: Design, Automation & Test in Europe Conference & Exhibition, DATE 2021, Grenoble, France, 1–5 February 2021, pp. 176–181. IEEE (2021). https://doi.org/10.23919/DATE51398.2021.9474092
2. Chen, S., Xiang, Z., Zeng, X., Zhang, S.: On the relationships between different methods for degree evaluation. IACR Trans. Symmetric Cryptol. **2021**(1), 411–442 (2021). https://doi.org/10.46586/tosc.v2021.i1.411-442
3. Ding, T., Zhang, W., Zhou, C., Ji, F.: An automatic search tool for iterative trails and its application to estimation of differentials and linear hulls. Cryptology ePrint Archive, Report 2020/1152 (2020). https://ia.cr/2020/1152
4. Dinur, I., Shamir, A.: Cube attacks on tweakable black box polynomials. In: Joux, A. (ed.) EUROCRYPT 2009. LNCS, vol. 5479, pp. 278–299. Springer, Heidelberg (2009). https://doi.org/10.1007/978-3-642-01001-9_16

5. Ghosh, S., Dunkelman, O.: Automatic search for bit-based division property. In: Longa, P., Ràfols, C. (eds.) LATINCRYPT 2021. LNCS, vol. 12912, pp. 254–274. Springer, Cham (2021). https://doi.org/10.1007/978-3-030-88238-9_13
6. Huang, S., Wang, X., Xu, G., Wang, M., Zhao, J.: Conditional cube attack on reduced-round Keccak sponge function. In: Coron, J.-S., Nielsen, J.B. (eds.) EUROCRYPT 2017. LNCS, vol. 10211, pp. 259–288. Springer, Cham (2017). https://doi.org/10.1007/978-3-319-56614-6_9
7. Li, Z., Dong, X., Wang, X.: Conditional cube attack on round-reduced ASCON. IACR Trans. Symmetric Cryptol. **2017**(1), 175–202 (2017). https://doi.org/10.13154/tosc.v2017.i1.175-202
8. Liu, F., Isobe, T., Meier, W.: Cube-based cryptanalysis of subterranean-SAE. IACR Trans. Symmetric Cryptol. **2019**(4), 192–222 (2019). https://doi.org/10.13154/tosc.v2019.i4.192-222
9. Liu, Y., Chen, S., Zhang, S., Xiang, Z., Zeng, X.: Conditional cube attacks on Subterranean-SAE (In Chinese). J. Cryptol. Res. **9**(1), 45 (2022). https://doi.org/10.13868/j.cnki.jcr.000502
10. Mouha, N., Wang, Q., Gu, D., Preneel, B.: Differential and linear cryptanalysis using mixed-integer linear programming. In: Wu, C.-K., Yung, M., Lin, D. (eds.) Inscrypt 2011. LNCS, vol. 7537, pp. 57–76. Springer, Heidelberg (2012). https://doi.org/10.1007/978-3-642-34704-7_5
11. Rogaway, P.: Authenticated-encryption with associated-data. In: Atluri, V. (ed.) CCS 2002, pp. 98–107. ACM (2002). https://doi.org/10.1145/586110.586125
12. Sasaki, Yu., Todo, Y.: New algorithm for modeling S-box in MILP based differential and division trail search. In: Farshim, P., Simion, E. (eds.) SecITC 2017. LNCS, vol. 10543, pp. 150–165. Springer, Cham (2017). https://doi.org/10.1007/978-3-319-69284-5_11
13. Sun, S., Hu, L., Wang, P., Qiao, K., Ma, X., Song, L.: Automatic security evaluation and (related-key) differential characteristic search: application to SIMON, PRESENT, LBlock, DES (L) and other bit-oriented block ciphers. In: Sarkar, P., Iwata, T. (eds.) ASIACRYPT 2014. LNCS, vol. 8873, pp. 158–178. Springer, Heidelberg (2014). https://doi.org/10.1007/978-3-662-45611-8_9
14. Todo, Y.: Structural evaluation by generalized integral property. In: Oswald, E., Fischlin, M. (eds.) EUROCRYPT 2015. LNCS, vol. 9056, pp. 287–314. Springer, Heidelberg (2015). https://doi.org/10.1007/978-3-662-46800-5_12
15. Todo, Y., Morii, M.: Bit-based division property and application to SIMON family. In: Peyrin, T. (ed.) FSE 2016. LNCS, vol. 9783, pp. 357–377. Springer, Heidelberg (2016). https://doi.org/10.1007/978-3-662-52993-5_18
16. Wang, S., Hou, S., Liu, M., Lin, D.: Differential-linear cryptanalysis of the lightweight crytographic algorithm KNOT. In: Yu, Yu., Yung, M. (eds.) Inscrypt 2021. LNCS, vol. 13007, pp. 171–190. Springer, Cham (2021). https://doi.org/10.1007/978-3-030-88323-2_9
17. Xiang, Z., Zhang, W., Bao, Z., Lin, D.: Applying MILP method to searching integral distinguishers based on division property for 6 lightweight block ciphers. In: Cheon, J.H., Takagi, T. (eds.) ASIACRYPT 2016. LNCS, vol. 10031, pp. 648–678. Springer, Heidelberg (2016). https://doi.org/10.1007/978-3-662-53887-6_24
18. Zhang, W., et al.: KNOT: algorithm specifications and supporting document. submission to NIST (Round 2) (2019). https://csrc.nist.gov/CSRC/media/Projects/lightweight-cryptography/documents/round-2/spec-doc-rnd2/knot-spec-round.pdf

Fast Fourier Orthogonalization over NTRU Lattices

Shuo Sun[1,2], Yongbin Zhou[1,2,3](✉), Rui Zhang[1,2], Yang Tao[1], Zehua Qiao[1,2], and Jingdian Ming[1,2]

[1] Institute of Information Engineering, Chinese Academy of Sciences, Beijing, China
{sunshuo,zhouyongbin,r-zhang,taoyang,qiaozehua,mingjingdian}@iie.ac.cn

[2] School of Cyber Security, University of Chinese Academy of Sciences, Beijing, China

[3] School of Cyber Security and Engineering, Nanjing University of Science and Technology, Nanjing, China

Abstract. FALCON is an efficient and compact lattice-based signature scheme. It is also one of the round 3 finalists in the NIST PQC standardization process. The core of FALCON is a trapdoor sampling algorithm, which has found numerous applications in lattice-based cryptography. It needs the fast Fourier orthogonalization algorithm to build an LDL tree. But the LDL tree needs much RAM to store, which may limit the application of FALCON on memory-constrained devices. On the other hand, if building the LDL tree dynamically, the signature cost will almost double.

In this work, we discover the LDL tree of FALCON has some symmetric structure, and prove why this phenomenon occurs. With this property, we can reduce the generation time and storage of the LDL tree by almost half without affecting the efficiency of FALCON. We verify the correctness and validity of our way in the implementations of FALCON. In addition, the result applies to the cyclotomic field $\mathbb{Q}[x]/(x^n - x^{n/2} + 1)$ with $n = 3 \cdot 2^\kappa$. But we can not apply it to NTRU module lattices so far.

Keywords: fast Fourier orthogonalization · lattice-based cryptography · NTRU · FALCON · trapdoor sampling

1 Introduction

Lattice-based cryptography has attracted much attention because it can construct many advanced cryptographic objects, such as fully homomorphic encryption [12], and practical post-quantum secure public-key encryptions and signatures. In the NIST post-quantum cryptography (PQC) standardization process, a substantial proportion of candidates is based on lattices. In the round 3 finalists, there are two lattice-based digital signature schemes. FALCON [23] is one of them. Among the finalists and alternate candidates of round 3, FALCON has the least sum of public key and signature length, which means it's the most bandwidth-efficient signature. Its signature verification is very fast and

C. Alcaraz et al. (Eds.): ICICS 2022, LNCS 13407, pp. 109–127, 2022.
https://doi.org/10.1007/978-3-031-15777-6_7

RAM efficient. The trapdoor sampling algorithm of FALCON can be used as a building block to construct (hierarchical) identity-based encryption schemes [1,4,7,13,25], attribute-based encryption [3], ring signature [16], group signature [20], public-key encryption with keyword search [2], and many other schemes.

FALCON is an instantiation of the generic framework proposed by Gentry, Peikert, and Vaikuntanathan [13]. The framework can build provably secure hash-and-sign lattice-based signature schemes. FALCON chooses NTRU lattices to obtain the compact public key and signature [7,14,24]. To speed up the signature generation, it uses the fast Fourier orthogonalization algorithm and the fast Fourier sampling algorithm [9] to reduce the signing time from $O(n^2)$ to $O(n \log n)$.

The signature generation of FALCON involves two main steps: build an LDL tree of the secret key (i.e. the short NTRU lattice basis) by the fast Fourier orthogonalization algorithm, and then sample a short vector through the LDL tree and the fast Fourier sampling algorithm. Because the LDL tree depends only on the secret key, not the message to be signed, it can be built in advance and be a part of the secret key. However, the signature generation algorithm will require a lot of RAM to store the bulky LDL tree. For the ring degree of NTRU lattices being 1024, the LDL tree needs 88 KB of memory. It may limit the application of FALCON on memory-constrained devices. To reduce RAM usage, FALCON offers the reference implementation regenerating the LDL tree dynamically, and then only a path from tree root to the current leaf needs to be present in RAM. But this will almost double the computational overhead of the signature, thus reducing the competitiveness of FALCON. Consider the implementations without AVX2 vector instructions at NIST security level 5. The signature cost of Dilithium, another lattice-based signature of round 3 finalists, is about 2.38×10^6 cycles [17]. For FALCON, the signature cost with the LDL tree is 1.51×10^6 cycles. If building the LDL tree dynamically, the signature cost is 2.69×10^6 cycles [21]. Therefore, it's meaningful to reduce the storage of the LDL tree without affecting the signature efficiency of FALCON.

1.1 Our Contributions

Let $\mathcal{Q} = \mathbb{Q}[x]/(x^n + 1)$, where n is a power-of-2. For any self-adjoint $d(x) \in \mathcal{Q}$, the fast Fourier orthogonalization algorithm makes full use of the field tower of $\mathcal{Q}$ to construct the LDL tree of $d(x)$ in a recursive way. The secret key of FALCON is a short lattice basis $\mathbf{B}_{f,g} = \begin{bmatrix} f & g \\ F & G \end{bmatrix} \in \mathcal{Q}^{2\times 2}$, which verifies the NTRU equation $fG - gF = q$ for a given positive integer q. To compute the LDL tree of the secret key, FALCON first computes the LDL* decomposition of the Gram matrix $\mathbf{G}_{f,g} = \mathbf{B}_{f,g} \times \mathbf{B}_{f,g}^{\star}$ over $\mathcal{Q}$:

$$\mathbf{G}_{f,g} = \begin{bmatrix} 1 & 0 \\ \frac{Ff^\star + Gg^\star}{ff^\star + gg^\star} & 1 \end{bmatrix} \times \begin{bmatrix} ff^\star + gg^\star & 0 \\ 0 & \frac{q^2}{ff^\star + gg^\star} \end{bmatrix} \times \begin{bmatrix} 1 & \frac{fF^\star + gG^\star}{ff^\star + gg^\star} \\ 0 & 1 \end{bmatrix},$$

then builds the LDL trees of $ff^\star + gg^\star$ and $\frac{q^2}{ff^\star+gg^\star}$ by the fast Fourier orthogonalization algorithm. Finally, it recursively obtains the LDL tree of the secret key, where the polynomial $\frac{Ff^\star+Gg^\star}{ff^\star+gg^\star}$ is the root node and the LDL trees of $ff^\star + gg^\star$ and $\frac{q^2}{ff^\star+gg^\star}$ are the left and right subtrees respectively.

Our observation is that the LDL trees of $ff^\star + gg^\star$ and $\frac{q^2}{ff^\star+gg^\star}$ are very similar such that the LDL tree of the secret key is symmetrical to some extent. We give a toy example of the LDL tree in Fig. 1. Therefore, it's possible to reduce the LDL tree storage of the secret key without affecting the signature efficiency of FALCON. We outline our contributions as follows:

- We discover that the LDL tree of the secret key in FALCON is symmetric to some extent and prove why this phenomenon occurs. The core of our proof is to show that for any self-adjoint $a(x)$, $\frac{1}{a(x)} \in \mathcal{Q}$, the reciprocal relationship between the diagonal elements always exists during their LDL tree constructions.
- Not storing the LDL tree of $\frac{q^2}{ff^\star+gg^\star}$ hardly affects the computation complexity of the fast Fourier sampling over NTRU lattices, so FALCON can almost halve the storage and generation time of the LDL tree of the secret key without affecting its signature efficiency.
- The result can be extended to the cyclotomic field $\mathbb{Q}[x]/(x^n - x^{n/2} + 1)$ with $n = 3 \cdot 2^\kappa$. Because, for any self-adjoint $a(x)$, $\frac{1}{a(x)} \in \mathbb{Q}[x]/(x^n - x^{n/2} + 1)$, there is also the reciprocal relationship between the diagonal elements during their LDL tree constructions. However, we can not reduce the computation and storage of the LDL tree of NTRU module lattices [5] so far. For any self-adjoint $a(x), b(x)$, $\frac{1}{a(x)b(x)} \in \mathcal{Q}$, we can not establish some invariable relationship between the diagonal elements during their LDL tree constructions. Therefore, there does not seem to be an *efficient* method to construct the LDL tree of $\frac{1}{a(x)b(x)}$ by the LDL trees of $a(x)$ and $b(x)$.
- We further verify the correctness and validity of our technique in the round 3 implementations of FALCON, and perform the benchmark on two different architectures, Intel i7-4790 CPU and ARM Cortex M4.

1.2 Related Works

There are two ways to sample from Gaussian distributions over lattices. They are derived from Babai's nearest plane and Babai's round-off algorithms. The sampler [13] based on the nearest plane algorithm usually has compact parameters and needs the Gram-Schmidt orthogonalization (or the equivalent LDL* decomposition). The sampler [19] based on the round-off algorithm can thoroughly avoid floating-point operations with an integral matrix Gram root [6], and it is easy to parallelize. FALCON belongs to the former.

In the seminal work [13], for generic unstructured lattices, it takes $O(n^3)$ time and $O(n^2)$ storage to compute the Gram-Schmidt orthogonalization of the basis. For ideal lattices, Lyubashevsky and Prest reduced the time to $O(n^2)$ through the isometry property of the basis [8]. In 2016, Ducas and Prest further reduced the time and storage to $O(n \log n)$ using the field tower structure of the cyclotomic field [9].

There are some follow-up works of FALCON. ModFalcon extends NTRU lattices of FALCON to NTRU module lattices to provide more modularity in parameter selection [5]. MITAKA [10] and Zalcon [11] are also hash-and-sign signature schemes. It's easier to provide provably secure masking implementations of them. The idea of MITAKA comes from the hybrid Gaussian sampler [22] that combines the Gaussian samplers in [13] and [19]. Zalcon is based on the technique in [6] to avoid floating-point operations. However, both of their parameters are worse than FALCON.

2 Preliminaries

2.1 Notations

Let $\mathbb{Z}, \mathbb{Q}, \mathbb{R}, \mathbb{C}$ be the ring of integers and the fields of rationals, real and complex numbers respectively. We denote by $\mathbb{N}^\star$ the set of positive integers. For an integer $r > 0$, we denote by $\mathbb{Z}_r$ the ring of integers modulo r. We denote matrices in bold uppercase (e.g. $\mathbf{B}$) and vectors in bold lowercase (e.g. $\mathbf{v}$). We use the row convention for vectors.

2.2 Polynomial Rings and Fields

Let $\phi(x)$ be a monic irreducible polynomial of degree $n \geq 1$ in $\mathbb{Z}[x]$. We can define the ring $\mathbb{Z}[x]/(\phi(x))$ and the field $\mathbb{Q}[x]/(\phi(x))$. For a positive integer q, $\phi(x)$ may be factorable over $\mathbb{Z}_q$. We can define the ring $\mathbb{Z}_q[x]/(\phi(x))$. For any polynomial $f(x)$, $c(f)$ represents its coefficient vector. If $f(x) \in \mathbb{Q}[x]/(\phi(x))$, we denote by $\mathcal{C}(f)$ the $n \times n$ matrix whose j-th row is the coefficient vector $c(x^{j-1}f \bmod \phi)$. For any $f(x), g(x) \in \mathbb{Q}[x]/(\phi(x))$, we have:

$$c(fg) = c(f)\mathcal{C}(g), \qquad \mathcal{C}(fg) = \mathcal{C}(f)\mathcal{C}(g).$$

We extend the definition to matrices: for $\mathbf{B} = (b_{ij})_{ij}$ in $(\mathbb{Q}[x]/(\phi(x)))^{m \times n}$, $\mathcal{C}(\mathbf{B}) = (\mathcal{C}(b_{ij}))_{ij}$.

Most lattice-based cryptographic algorithms using polynomial rings to represent structured lattices rely on cyclotomic polynomials. The m-th cyclotomic polynomial $\Phi_m(x)$ is defined as:

$$\Phi_m(x) = \prod_{k \in \mathbb{Z}_m^\times} \left(x - e^{2i\pi k/m}\right).$$

Cyclotomic polynomials are in $\mathbb{Z}[x]$ and irreducible over $\mathbb{Q}$. So $\mathbb{Q}[x]/(\Phi_m(x))$ are the fields for all $m \geq 1$ and we call them *cyclotomic fields*. The degree n of $\Phi_m(x)$ is $\varphi(m)$, where φ denotes Euler's function: $\varphi(m) = |\mathbb{Z}_m^\times|$. In this paper, we are mainly interested in two common types of cyclotomic polynomials: (1)

$\Phi_m(x) = x^n + 1$ with $n = 2^\kappa$ and $m = 2n$; (2) $\Phi_m(x) = x^n - x^{n/2} + 1$ with $n = 3 \cdot 2^\kappa$ and $m = 3n$.

Let $f(x) = \sum_{i=0}^{n-1} f_i x^i$ and $g(x) = \sum_{i=0}^{n-1} g_i x^i$ be arbitrary elements in $\mathcal{Q} = \mathbb{Q}[x]/(\Phi_m(x))$. We denote by $f^*(x)$ the (Hermitian) adjoint of $f(x)$, which means that $f^*(\zeta) = \overline{f(\zeta)}$ for any root ζ of $\Phi_m(x)$, where $\bar{\cdot}$ is the usual complex conjugation over $\mathbb{C}$. For a matrix $\mathbf{B} \in \mathcal{Q}^{m \times n}$, its adjoint $\mathbf{B}^\star$ is the component-wise adjoint of the transpose of $\mathbf{B}$.

The inner product over $\mathcal{Q}$ and its associated norm $\|\cdot\|$ are

$$\langle f, g \rangle = \frac{1}{\varphi(m)} \sum_{\Phi_m(\zeta)=0} f(\zeta) \cdot \overline{g(\zeta)}, \qquad \|f\| = \sqrt{\langle f, f \rangle}.$$

If $\Phi_m(x) = x^n + 1$ with $n = 2^\kappa$, then

$$\langle f, g \rangle = \sum_{0 \le i < n} f_i g_i.$$

If $\Phi_m(x) = x^n - x^{n/2} + 1$ with $n = 3 \cdot 2^\kappa$, then

$$\langle f, g \rangle = \sum_{0 \le i < n/2} (f_i g_i + f_{i+n/2} g_{i+n/2} + \frac{1}{2} f_i g_{i+n/2} + \frac{1}{2} f_{i+n/2} g_i).$$

We extend the definition to vectors: for $\mathbf{u} = (u_i)_i$ and $\mathbf{v} = (v_i)_i$ in $\mathcal{Q}^m$, $\langle \mathbf{u}, \mathbf{v} \rangle = \sum_i \langle u_i, v_i \rangle$.

2.3 The Field Norm

Definition 1. *Let $\mathbb{K}$ be a number field and $\mathbb{L}$ be a Galois extension of $\mathbb{K}$. We denote by $\mathrm{Gal}(\mathbb{L}/\mathbb{K})$ the Galois group of the field extension $\mathbb{L}/\mathbb{K}$. The field norm $\mathrm{N}_{\mathbb{L}/\mathbb{K}} : \mathbb{L} \to \mathbb{K}$ is a map defined for any $f \in \mathbb{L}$ by the product of the Galois conjugates of f:*

$$\mathrm{N}_{\mathbb{L}/\mathbb{K}}(f) = \prod_{\tau \in \mathrm{Gal}(\mathbb{L}/\mathbb{K})} \tau(f).$$

Equivalently, $\mathrm{N}_{\mathbb{L}/\mathbb{K}}(f)$ can be defined as the determinant of the $\mathbb{K}$-linear map $\psi_f : a \in \mathbb{L} \longmapsto fa$.

One can check that the field norm is a multiplicative morphism. When $f \in \mathbb{L}$ and $\mathbb{K}$ is the unique largest proper subfield of $\mathbb{L}$, we denote $\mathrm{N}(f) = \mathrm{N}_{\mathbb{L}/\mathbb{K}}(f)$.

For the cyclotomic field $\mathbb{Q}[x]/(x^n + 1)$ with $n = 2^\kappa$, we have the following field tower:

$$\mathbb{Q} \subseteq \mathbb{Q}[x]/(x^2 + 1) \subseteq \ldots \subseteq \mathbb{Q}[x]/(x^{n/2} + 1) \subseteq \mathbb{Q}[x]/(x^n + 1).$$

Let $\mathbb{L} = \mathbb{Q}[x]/(\Phi_{2m})$ and $\mathbb{K} = \mathbb{Q}[y]/(\Phi_m)$ with $m = 2^\kappa$, the field norm is particularly simple to express. Any $f(x) \in \mathbb{L}$ can be split into its coefficients of even and odd degress:

$$f(x) = f_0(x^2) + x f_1(x^2)$$

with $f_0(y), f_1(y) \in \mathbb{K}$. Noting $\psi_f : a \in \mathbb{L} \longmapsto fa$, we have

$$\mathrm{N}_{\mathbb{L}/\mathbb{K}}(f) = \det_{\mathbb{K}}(\psi_f) = \det \begin{bmatrix} f_0 & f_1 \\ yf_1 & f_0 \end{bmatrix} = f_0^2 - yf_1^2.$$

Similarly, for the cyclotomic field $\mathbb{Q}[x]/(x^n - x^{n/2} + 1)$ with $n = 3 \cdot 2^\kappa$, we have the following field tower:

$$\mathbb{Q} \subseteq \mathbb{Q}[x]/(x^2 - x + 1) \subseteq \mathbb{Q}[x]/(x^6 - x^3 + 1) \subseteq \mathbb{Q}[x]/(x^{12} - x^6 + 1) \subseteq ...$$
$$\subseteq \mathbb{Q}[x]/(x^{n/2} - x^{n/4} + 1) \subseteq \mathbb{Q}[x]/(x^n - x^{n/2} + 1).$$

For $\mathbb{L} = \mathbb{Q}[x]/(\Phi_{2m})$ and $\mathbb{K} = \mathbb{Q}[y]/(\Phi_m)$ with $m = 3^2 \cdot 2^\kappa$, the field norm $\mathrm{N}_{\mathbb{L}/\mathbb{K}}$ is also expressed by splitting the polynomial with respect to its even or odd coefficients. Let $\mathbb{L} = \mathbb{Q}[x]/(x^6 - x^3 + 1)$ and $\mathbb{K} = \mathbb{Q}[y]/(y^2 - y + 1)$, any $f(x) \in \mathbb{L}$ can be split into three parts:

$$f(x) = f_0(x^3) + xf_1(x^3) + x^2 f_2(x^3)$$

with $f_0(y), f_1(y), f_2(y) \in \mathbb{K}$. Then wen have

$$\mathrm{N}_{\mathbb{L}/\mathbb{K}}(f) = \det \begin{bmatrix} f_0 & f_1 & f_2 \\ yf_2 & f_0 & f_1 \\ yf_1 & yf_2 & f_0 \end{bmatrix} = f_0^3 + yf_1^3 + y^2 f_2^3 - 3yf_0 f_1 f_2.$$

If $\mathbb{L} = \mathbb{Q}[x]/(x^2 - x + 1)$, for any $f(x) = f_0 + xf_1 \in \mathbb{L}$, we have

$$\mathrm{N}_{\mathbb{L}/\mathbb{Q}}(f) = \det \begin{bmatrix} f_0 & f_1 \\ -f_1 & f_0 + f_1 \end{bmatrix} = f_0^2 + f_0 f_1 + f_1^2.$$

2.4 The GSO and LDL* Decomposition

For a number field $\mathcal{Q}$, a full-rank matrix $\mathbf{B} \in \mathcal{Q}^{m \times n}$ can be uniquely decomposed as follows:

$$\mathbf{B} = \mathbf{L} \times \widetilde{\mathbf{B}},$$

where $\mathbf{L}$ is lower triangular with 1's on the diagonal, the rows $\widetilde{\mathbf{b}}_i$'s of $\widetilde{\mathbf{B}}$ verify $\langle \mathbf{b}_i, \mathbf{b}_j \rangle = 0$ for $i \neq j$. It is called the Gram-Schmidt orthogonalization (or GSO).

The LDL* decomposition writes any full-rank Gram matrix as a product $\mathbf{LDL}^\star$, where $\mathbf{L} \in \mathcal{Q}^{m \times m}$ is lower triangular with 1's on the diagonal, and $\mathbf{D} \in \mathcal{Q}^{m \times m}$ is diagonal. If the Gram matrix $\mathbf{G} = \mathbf{BB}^\star$, then $\mathbf{G} = \mathbf{L} \cdot (\widetilde{\mathbf{B}}\widetilde{\mathbf{B}}^\star) \cdot \mathbf{L}^\star$ is the unique LDL* decomposition of $\mathbf{G}$, which means the GSO of $\mathbf{B}$ is equivalent to the LDL* decomposition of $\mathbf{G}$.

The GSO is a more familiar concept in lattice-based cryptography, whereas the use of LDL* decomposition is faster and therefore makes more sense from an algorithmic point of view.

2.5 The Fast Fourier Orthogonalization and LDL Tree

For any non-zero $b(x) \in \mathbb{Q}[x]/(x^n+1)$ with $n = 2^\kappa$, $b(x), xb(x), ..., x^{n-1}b(x)$ generate the ideal lattice $\mathcal{L}(\mathbf{B})$ and the lattice basis $\mathbf{B}$ is $\mathcal{C}(b)$. Let $d(x) = b(x)b^*(x)$. If we want the GSO of $\mathcal{C}(b)$, we can compute the LDL* decomposition of $\mathcal{C}(d)$ by the fast Fourier orthogonalization algorithm. According to the field tower for $\mathbb{Q}[x]/(x^n+1)$, we consider the associated endomorphism $\psi_d : a \in \mathbb{Q}[x]/(x^n+1) \longmapsto da$ and compute the LDL* decomposition of the transformation matrix $\mathbf{D}$ over the smaller field $\mathbb{Q}[y]/(y^{n/2}+1)$:

$$\mathbf{D} = \begin{bmatrix} d_0 & d_1 \\ yd_1 & d_0 \end{bmatrix} = \begin{bmatrix} 1 & 0 \\ l_{10} & 1 \end{bmatrix} \times \begin{bmatrix} d_0 & 0 \\ 0 & d_0' \end{bmatrix} \times \begin{bmatrix} 1 & l_{10}^* \\ 0 & 1 \end{bmatrix},$$

where $d(x) = d_0(x^2) + xd_1(x^2)$. The matrix $\mathbf{D}$ is self-adjoint, which means $d_0^\star(y) = d_0(y)$ and $d_1^\star(y) = yd_1(y)$. The matrix $\mathcal{C}(\mathbf{D})$ permutes the rows and columns compared with $\mathcal{C}(d)$, which doesn't change the generated lattice up to an isometry [9]. Then we break the diagonal elements d_0 and d_0' into the matrices over $\mathbb{Q}[z]/(z^{n/4}+1)$ and recursively compute the LDL* decompositions. We continue the recursion until the diagonal elements are in $\mathbb{Q}$. It's worth noting that the diagonal elements in $\mathbb{Q}[x]/(x^2+1)$ are also in $\mathbb{Q}$ as they are self-adjoint. So the recursion terminates with the diagonal elements in $\mathbb{Q}[x]/(x^2+1)$ in practice. We need to store all polynomials l_{10}'s computed in the LDL* decompositions and the final diagonal elements in $\mathbb{Q}$. They constitute an *LDL tree*, where the values of internal nodes are the polynomials l_{10}'s, and the values of leaf nodes are the diagonal elements.

If the non-zero $b(x)$ belongs to $\mathbb{Q}[x]/(x^n - x^{n/2}+1)$ with $n = 3 \cdot 2^\kappa$, we can construct a similar LDL tree based on the field tower of $\mathbb{Q}[x]/(x^n - x^{n/2}+1)$. The only difference is that we have to break the diagonal element into three terms when it belongs to $\mathbb{Q}[x]/(x^6 - x^3 + 1)$, then compute the following LDL* decomposition over the smaller field $\mathbb{Q}[y]/(y^2 - y + 1)$:

$$\mathbf{D} = \begin{bmatrix} d_0 & d_1 & d_2 \\ yd_2 & d_0 & d_1 \\ yd_1 & yd_2 & d_0 \end{bmatrix} = \begin{bmatrix} 1 & 0 & 0 \\ l_{10} & 1 & 0 \\ l_{20} & l_{21} & 1 \end{bmatrix} \times \begin{bmatrix} d_0 & 0 & 0 \\ 0 & d_0' & 0 \\ 0 & 0 & d_0'' \end{bmatrix} \times \begin{bmatrix} 1 & l_{10}^* & l_{20}^* \\ 0 & 1 & l_{21}^* \\ 0 & 0 & 1 \end{bmatrix},$$

where $d(x) = d_0(x^3) + xd_1(x^3) + x^2d_2(x^3)$. The matrix $\mathbf{D}$ is also self-adjoint, which means $d_0^\star(y) = d_0(y)$, $d_1^\star(y) = yd_2(y)$ and $d_2^\star(y) = yd_1(y)$. We need to store three polynomials l_{10}, l_{20} and l_{21}. In practice, the recursion terminates with the diagonal elements in $\mathbb{Q}[x]/(x^2 - x + 1)$.

2.6 NTRU Lattices

Let $\mathcal{Q} = \mathbb{Q}[x]/(\Phi_m(x))$, $q \in \mathbb{N}^\star$, and $f, g \in \mathcal{Q}$. Let $h = f^{-1}g \bmod q$. The NTRU lattice associated to h and q is $\mathcal{L}_{\text{NTRU}} = \{(u,v) \in \mathcal{Q}^2 \mid uh - v = 0 \bmod q\}$. There are two bases of the NTRU lattice:

$$\mathbf{B}_h = \begin{bmatrix} 1 & h \\ 0 & q \end{bmatrix} \text{ and } \mathbf{B}_{f,g} = \begin{bmatrix} f & g \\ F & G \end{bmatrix},$$

where $F, G \in \mathcal{Q}$ are such that $fG - gF = q$. Typically, h will be a public key, whereas f, g, F, G will be secret keys in the cryptographic schemes based on NTRU lattices.

As described in Sect. 1.1, when building the LDL tree of $\mathbf{B}_{f,g}$, we can first compute the LDL* decomposition of the Gram matrix $\mathbf{G}_{f,g} = \mathbf{B}_{f,g} \times \mathbf{B}^\star_{f,g}$, then build the LDL trees of the diagonal elements by the fast Fourier orthogonalization algorithm.

2.7 Discrete Gaussians

The n-dimensional Gaussian function $\rho_{\sigma,\mathbf{c}}(\mathbf{x})$ on $\mathbb{R}^n$ centered at $\mathbf{c}$ with the standard deviation σ is defined by $\rho_{\sigma,\mathbf{c}}(\mathbf{x}) = \exp\left(-\frac{\|\mathbf{x}-\mathbf{c}\|^2}{2\sigma^2}\right)$. For any lattice $\Lambda \subset \mathbb{R}^n$, let $\rho_{\sigma,\mathbf{c}}(\Lambda) = \sum_{x \in \mathcal{L}} \rho_{\sigma,\mathbf{c}}(\mathbf{x})$. Then we can define the discrete Gaussian distribution $D_{\Lambda,\sigma,\mathbf{c}}$ over Λ by $D_{\Lambda,\sigma,\mathbf{c}}(\mathbf{x}) = \frac{\rho_{\sigma,\mathbf{c}}(\mathbf{x})}{\rho_{\sigma,\mathbf{c}}(\Lambda)}$.

3 Fast Fourier Orthogonalization over NTRU Lattices

In this section, we prove why we can reduce about half the computation and storage of the LDL tree of the NTRU lattice basis when the field is $\mathbb{Q}[x]/(x^n+1)$ or $\mathbb{Q}[x]/(x^n - x^{n/2} + 1)$. We also briefly explain why the idea does not apply to NTRU module lattices [5] in Sect. 3.1.

3.1 The Cyclotomic Field $\mathbb{Q}[x]/(x^n + 1)$

Although the storage complexity of the LDL tree is $O(n \log n)$, in practice, it's still a little bulky. In addition, if we generate the LDL tree on the fly, the computational cost of the cryptographic schemes will increase significantly. For example, in the FALCON signature [23], the LDL tree needs about 88 KB memory for $n = 1024$ and building the tree on the fly almost doubles the signature generation time. Therefore reducing the storage of LDL trees is practical for the Gaussian sampling over lattices.

When building the LDL tree of the NTRU lattice with the lattice basis $\mathbf{B}_{f,g}$, we have to compute the LDL trees of $ff^\star + gg^\star$ and $\frac{q^2}{ff^\star+gg^\star}$. In Theorem 1, we prove that in the LDL trees of the self-adjoint polynomial $a(x) \in \mathbb{Q}[x]/(x^n+1)$ and its multiplicative inverse $b(x)$, the value of each leaf node of $a(x)$ is the reciprocal of that of some leaf node of $b(x)$, and the polynomials of their internal nodes differ by a negative sign. Because the difference between the LDL trees of $\frac{q^2}{ff^\star+gg^\star}$ and $\frac{1}{ff^\star+gg^\star}$ is that the values of former leaf nodes increase by a factor q^2, there is no need to compute the LDL tree of $\frac{q^2}{ff^\star+gg^\star}$ if we have the LDL tree of $ff^\star + gg^\star$. As a result, we can almost halve the generation time and storage of the LDL tree of the NTRU lattice.

Theorem 1. *Let $a(x), b(x) \in \mathbb{Q}[x]/(x^n + 1)$ with $n = 2^\kappa$, and $a(x) = a^\star(x)$, $b(x) = b^\star(x)$. If $a(x)b(x) = 1$, then we can build the LDL tree of $b(x)$ according to the LDL tree of $a(x)$.*

Proof. Let $\mathbb{L} = \mathbb{Q}[x]/(x^n + 1)$ and $\mathbb{K} = \mathbb{Q}[y]/(y^{n/2} + 1)$. We can split $a(x)$ and $b(x)$ into their coefficients of even and odd degrees respectively: $a(x) = a_0(x^2) + xa_1(x^2)$, $b(x) = b_0(x^2) + xb_1(x^2)$, where $a_0(y), a_1(y), b_0(y), b_1(y) \in \mathbb{K}$. As $a^\star(x) = a(x)$ and $b^\star(x) = b(x)$, we have $a_0^\star(y) = a_0(y)$, $a_1^\star(y) = ya_1(y)$, $b_0^\star(y) = b_0(y)$ and $b_1^\star(y) = yb_1(y)$.

When we build the LDL tree of $a(x)$ by the fast Fourier orthogonalization algorithm, the first step is to write the transformation matrix $\mathbf{D}_a$ over $\mathbb{K}$ with the $\mathbb{K}$-linear map $\psi_a : c \in \mathbb{L} \longmapsto ac$, and compute its LDL* decomposition:

$$\mathbf{D}_a = \begin{bmatrix} a_0 & a_1 \\ a_1^\star & a_0 \end{bmatrix} = \begin{bmatrix} 1 & 0 \\ \frac{a_1^\star}{a_0} & 1 \end{bmatrix} \times \begin{bmatrix} a_0 & 0 \\ 0 & \frac{N_{\mathbb{L}/\mathbb{K}}(a)}{a_0} \end{bmatrix} \times \begin{bmatrix} 1 & \frac{a_1}{a_0} \\ 0 & 1 \end{bmatrix}.$$

Then the algorithm recursively computes the LDL trees of the diagonal elements. As $N_{\mathbb{L}/\mathbb{K}}(a) = a_0^2 - a_1a_1^\star$, the LDL* decomposition of $\mathbf{D}_a$ only involves $a_0(y)$, $a_1(y)$, and $a_1^\star(y)$.

Let $\mathbf{D}_b$ be the transformation matrix of $b(x)$ over $\mathbb{K}$. Our goal is to figure out the relation between the LDL* decompositions of $\mathbf{D}_a$ and $\mathbf{D}_b$, so it's necessary to express $b_0(y)$ and $b_1(y)$ with $a_0(y)$, $a_1(y)$, and $a_1^\star$. As $a(x)b(x) = 1$, we have

$$\begin{cases} a_0(y)b_0(y) + ya_1(y)b_1(y) & = 1 \\ a_0(y)b_1(y) + a_1(y)b_0(y) & = 0 \end{cases} \Rightarrow \begin{cases} b_0(y) = \dfrac{a_0(y)}{N_{\mathbb{L}/\mathbb{K}}(a)(y)} \\ b_1(y) = -\dfrac{a_1(y)}{N_{\mathbb{L}/\mathbb{K}}(a)(y)} \end{cases}.$$

Now we can replace $b_0(y)$, $b_1(y)$ and $b_1^\star(y)$ in the LDL* decomposition of $\mathbf{D}_b$ with $a_0(y)$, $a_1(y)$ and $a_1^\star(y)$:

$$\begin{aligned} \mathbf{D}_b = \begin{bmatrix} b_0 & b_1 \\ b_1^\star & b_0 \end{bmatrix} &= \begin{bmatrix} 1 & 0 \\ \frac{b_1^\star}{b_0} & 1 \end{bmatrix} \times \begin{bmatrix} b_0 & 0 \\ 0 & \frac{N_{\mathbb{L}/\mathbb{K}}(b)}{b_0} \end{bmatrix} \times \begin{bmatrix} 1 & \frac{b_1}{b_0} \\ 0 & 1 \end{bmatrix} \\ &= \begin{bmatrix} 1 & 0 \\ -\frac{a_1^\star}{a_0} & 1 \end{bmatrix} \times \begin{bmatrix} \frac{a_0}{N_{\mathbb{L}/\mathbb{K}}(a)} & 0 \\ 0 & \frac{1}{a_0} \end{bmatrix} \times \begin{bmatrix} 1 & -\frac{a_1}{a_0} \\ 0 & 1 \end{bmatrix}. \end{aligned}$$

In the LDL* decompositions of $\mathbf{D}_a$ and $\mathbf{D}_b$, the polynomials to be stored differ only by a negative sign, and the diagonal elements are pairwise reciprocals of each other in reverse order. Because there always exists the reciprocal relationship during the whole recursion, we can build the complete LDL tree of $b(x)$ by the LDL tree of $a(x)$. The computational costs are negating the polynomials of internal nodes and computing the reciprocals of the values of leaf nodes. □

We give an example of the LDL tree of the NTRU lattice in Fig. 1, where the field is $\mathbb{Q}[x]/(x^8 + 1)$ and the modulus is q. It's clear from the figure that the right subtree of the root node l_0 can be obtained from its left subtree.

Next, we evaluate the impact of our technique on the fast Fourier sampling over NTRU lattices while the field is $\mathbb{Q}[x]/(x^n + 1)$. We can divide the time-consuming operations in the fast Fourier sampling algorithm into two types. The

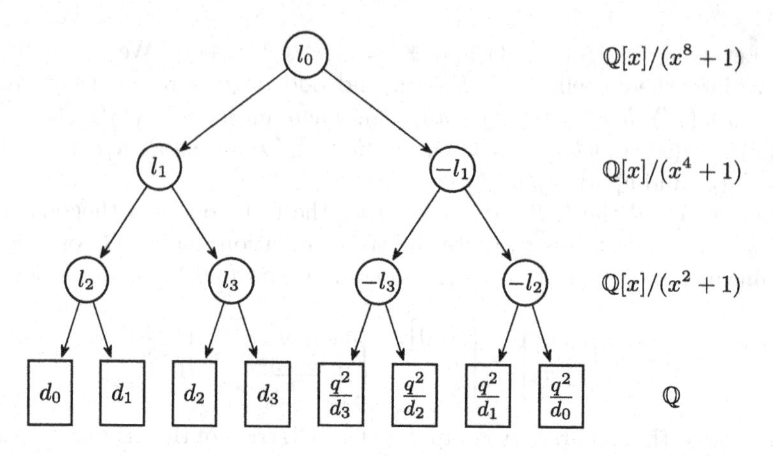

Fig. 1. The LDL tree of the NTRU lattice with the modulus q over $\mathbb{Q}[x]/(x^8+1)$

first is visiting the leaf nodes of the LDL tree from right to left to sample from the Gaussian distributions over the integers based on the values of leaf nodes. The second is combining the sampled integers and the polynomials of internal nodes to compute the polynomials as sampling centers.[1] When we replace the LDL tree of $\frac{1}{ff^\star+gg^\star}$ with that of $ff^\star+gg^\star$, there are three differences in the fast Fourier sampling algorithm. The first two differences are that the algorithm has to visit the leaf nodes of $ff^\star+gg^\star$ from left to right and recompute the values of leaf nodes before sampling from the Gaussian distributions over the integers. The third is negating the polynomials of internal nodes before computing the polynomials as sampling centers. These differences don't need any precomputation and only take a little computational cost. Therefore, not storing the LDL tree of $\frac{1}{ff^\star+gg^\star}$ hardly affects the computation complexity of the fast Fourier sampling over NTRU lattices.

At the end of this section, we briefly discuss if this idea works on NTRU module lattices [5]. The ModFalcon extends NTRU lattices to NTRU module lattices. It has extra flexibility in the choice of parameters and allows an intermediate security level. The field in [5] is also $\mathbb{Q}[x]/(x^n+1)$ with $n=2^\kappa$. Let q be the modulus. For the parameter set chosen in [5], a natural extension of our idea is constructing the LDL tree of $\frac{q^2}{a(x)b(x)}$ by the LDL trees of $a(x)$ and $b(x)$, where $a(x)$ and $b(x)$ are independent random and self-adjoint polynomials. According to the proof of Theorem 1, we have to analyze the first LDL* decompositions of $a(x)$, $b(x)$ and $\frac{1}{a(x)b(x)}$ in building their LDL trees over $\mathbb{Q}[x]/(x^n+1)$ by the fast Fourier orthogonalization algorithm. Let $a(x)=a_0(x^2)+xa_1(x^2)$ and $b(x)=b_0(x^2)+xb_1(x^2)$, there are the following LDL* decompositions over $\mathbb{Q}[y]/(y^{n/2}+1)$:

[1] The two types of operations are performed in turn. The interested readers may refer to Algorithm 11 in [23] for more details.

$$\mathbf{D}_a = \begin{bmatrix} a_0 & a_1 \\ a_1^\star & a_0 \end{bmatrix} = \begin{bmatrix} 1 & 0 \\ \frac{a_1^\star}{a_0} & 1 \end{bmatrix} \times \begin{bmatrix} a_0 & 0 \\ 0 & \frac{N_{\mathbb{L}/\mathbb{K}}(a)}{a_0} \end{bmatrix} \times \begin{bmatrix} 1 & \frac{a_1}{a_0} \\ 0 & 1 \end{bmatrix},$$

$$\mathbf{D}_b = \begin{bmatrix} b_0 & b_1 \\ b_1^\star & b_0 \end{bmatrix} = \begin{bmatrix} 1 & 0 \\ \frac{b_1^\star}{b_0} & 1 \end{bmatrix} \times \begin{bmatrix} b_0 & 0 \\ 0 & \frac{N_{\mathbb{L}/\mathbb{K}}(b)}{b_0} \end{bmatrix} \times \begin{bmatrix} 1 & \frac{b_1}{b_0} \\ 0 & 1 \end{bmatrix},$$

$$\mathbf{D}_{\frac{1}{ab}} = \begin{bmatrix} 1 & 0 \\ -\frac{a_0 b_1^\star + a_1^\star b_0}{a_0 b_0 + y a_1 b_1} & 1 \end{bmatrix} \times \begin{bmatrix} \frac{a_0 b_0 + y a_1 b_1}{N_{\mathbb{L}/\mathbb{K}}(ab)} & 0 \\ 0 & \frac{1}{a_0 b_0 + y a_1 b_1} \end{bmatrix} \times \begin{bmatrix} 1 & -\frac{a_0 b_1 + a_1 b_0}{a_0 b_0 + y a_1 b_1} \\ 0 & 1 \end{bmatrix}.$$

We hope the product of the diagonal elements $a_0, b_0, \frac{1}{a_0 b_0 + y a_1 b_1}$ equals 1, then there may be an invariable relationship between the diagonal elements during the LDL tree constructions. But the product equals 1 only if $y a_1 b_1 = 0$. We can not find an *efficient* method to construct the LDL tree of $\frac{1}{a(x)b(x)}$ by the LDL trees of $a(x)$ and $b(x)$, although the LDL tree of $\frac{1}{a(x)b(x)}$ is determined by $a(x), b(x)$. Therefore, we can not reduce the computation and storage of the LDL tree of NTRU module lattices so far.

3.2 The Cyclotomic Field $\mathbb{Q}[x]/(x^n - x^{n/2} + 1)$

The cyclotomic polynomial most commonly used in the cryptographic schemes is $x^n + 1$ with $n = 2^\kappa$. For well-chosen q, the rings $\mathbb{Z}_q[x]/(x^n + 1)$ can support the Number Theory Transform (NTT) to perform efficient multiplication and division. In addition, the expansion factor of these rings controlling the growth of polynomial products is the minimal of all rings [18]. One disadvantage of these rings is that they are sparse, so one may not be able to select an appropriate ring for the desired security level. In practice, obtaining 192-bit hardness requires taking the dimension n somewhere between 512 and 1024. Since n is a power of 2, one has to take $n = 1024$ which usually reaches about 256-bit hardness and incurs the significant expansion of parameters. One way to overcome this issue is using the cyclotomic polynomial $x^n - x^{n/2} + 1$ with $n = 3 \cdot 2^\kappa$. The rings $\mathbb{Z}_q[x]/(x^n - x^{n/2} + 1)$ also support the efficient NTT for well-chosen q, and the hardness with $n = 768$ is close to 192-bit.

When building the LDL tree of the NTRU lattice with the lattice basis $\mathbf{B}_{f,g}$ over $\mathbb{Q}[x]/(x^n - x^{n/2} + 1)$, we also need to compute the LDL trees of $ff^\star + gg^\star$ and $\frac{q^2}{ff^\star + gg^\star}$. As described in Sect. 2.5, we will split the polynomial into two or three terms according to the chosen subfield to construct the transformation matrix. In Theorem 2, we prove that for a self-adjoint polynomial $a(x) \in \mathbb{Q}[x]/(x^n - x^{n/2} + 1)$ and its multiplicative inverse $b(x)$, we can construct the LDL tree of $b(x)$ by that of $a(x)$. The detailed proof is given in Appendix A.

Theorem 2. *Let $a(x), b(x) \in \mathbb{Q}[x]/(x^n - x^{n/2} + 1)$ with $n = 3 \cdot 2^\kappa$, and $a(x) = a^\star(x)$, $b(x) = b^\star(x)$. If $a(x)b(x) = 1$, then we can build the LDL tree of $b(x)$ according to the LDL tree of $a(x)$.*

We give an example of the LDL tree of the NTRU lattice in Fig. 2, where the field is $\mathbb{Q}[x]/(x^{12} - x^6 + 1)$ and the modulus is q.

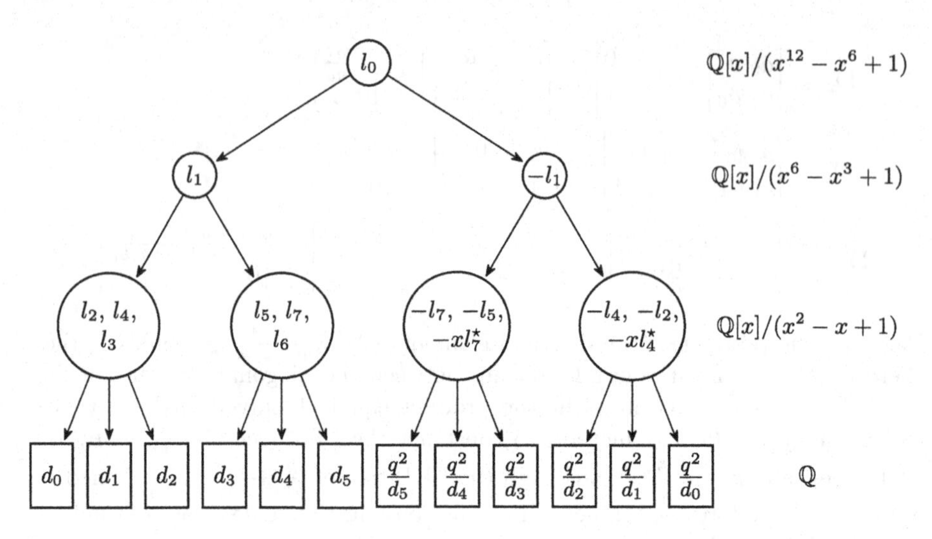

Fig. 2. The LDL tree of the NTRU lattice with the modulus q over $\mathbb{Q}[x]/(x^{12} - x^6 + 1)$

Let's evaluate the impact of our technique on the fast Fourier sampling over NTRU lattices while the field is $\mathbb{Q}[x]/(x^n - x^{n/2} + 1)$. Similar to the case of $\mathbb{Q}[x]/(x^n + 1)$, there are also three differences in the fast Fourier sampling algorithm when we replace the LDL tree of $\frac{1}{ff^\star + gg^\star}$ with that of $ff^\star + gg^\star$. The first two differences are that the algorithm has to visit the leaf nodes of $ff^\star + gg^\star$ from left to right and recompute the values of leaf nodes before sampling from the Gaussian distributions over the integers. The third is recomputing the polynomials of internal nodes before computing the polynomials as sampling centers. If the diagonal element is broken into two terms, the algorithm only needs to negate the polynomial of the internal node. If it's broken into three, the algorithm needs to compute the adjoint of the polynomial and the polynomial multiplication in addition. In practice the polynomials are all in fast Fourier transformation (FFT) representation, it's easy to compute their adjoints and products. Due to the complex conjugate, the polynomial in FFT representation over $\mathbb{Q}[x]/(x^n - x^{n/2} + 1)$ has $n/2$ complex numbers, and one polynomial multiplication requires $2n$ floating-point multiplications. For $n = 3 \cdot 2^\kappa$, all polynomial multiplications require $2n/3$ floating-point multiplications while computing the polynomials of internal nodes of $\frac{1}{ff^\star + gg^\star}$ by that of $ff^\star + gg^\star$. Therefore, building the LDL tree of $\frac{1}{ff^\star + gg^\star}$ by the LDL tree of $ff^\star + gg^\star$ is more efficient than building the LDL tree of $\frac{1}{ff^\star + gg^\star}$ by the fast Fourier sampling algorithm.

It's worth noting that the proof of Theorem 2 works on $n = 2^{\kappa_1} 3^{\kappa_2}$ since the degree of the extension in the field tower of $\mathbb{Q}[x]/(x^n - x^{n/2} + 1)$ is either two or three, which means the diagonal element is broken into two or three terms. However, for the larger κ_2, building the LDL tree of $\frac{1}{ff^\star + gg^\star}$ by the LDL tree of $ff^\star + gg^\star$ will require more polynomial multiplications. One may consider the impact of our technique on the efficiency carefully for $n = 2^{\kappa_1} 3^{\kappa_2}$.

4 Application to FALCON

In this section, we confirm the practicability of our technique through experiments. We applied it to FALCON, a lattice-based digital signature of NIST round 3 finalists. In the NIST round 1 version of FALCON, the package contains an implementation over the cyclotomic field $\mathbb{Q}[x]/(x^n - x^{n/2} + 1)$ with $n = 768$. But it was considered way too technique and was therefore removed from round 2. Our experiments were based on the implementations in round 3 and only considered the cyclotomic field $\mathbb{Q}[x]/(x^n + 1)$ with $n = 512$ or 1024. In the round 3 NIST package of FALCON, we chose the codes in the folder /falcon-round3/Extra/c to experiment, as it's convenient to test the correctness and performance of the modified implementations. The new implementations have been successfully tested on two different architectures, Intel i7-4790 CPU and ARM Cortex M4. Therefore, the benchmark results consist of two parts.

4.1 Intel i7-4790

We performed the experiments on Ubuntu 18.04 with a single Intel Core i7-4790 CPU core at 3.60 GHz and 4 GB RAM. The benchmark doesn't consider AVX2 acceleration and uses the C double type for all floating-point operations. We compiled the codes with clang 6.0.0 and the default optimization flag -O3.

We first tested the impact of the technique on the LDL tree generation of FALCON. The experimental results are shown in Table 1. In the implementations of FALCON, the LDL tree generation mainly consists of three steps: transforming the NTRU basis $\mathbf{B}_{f,g}$ into its FFT representation $\text{FFT}(\mathbf{B}_{f,g})$, computing $\text{FFT}(\mathbf{G}_{f,g}) = \text{FFT}(\mathbf{B}_{f,g}) \times \text{FFT}(\mathbf{B}^{\star}_{f,g})$, and building the LDL tree with $\text{FFT}(\mathbf{G}_{f,g})$. Therefore, we presented the running times of the LDL tree generation with $\mathbf{B}_{f,g}$. The generation times are reduced by about 28% while not computing the LDL tree of $\frac{1}{ff^{\star}+gg^{\star}}$. We also presented the running times of the LDL tree generation with $\text{FFT}(\mathbf{G}_{f,g})$, which excluded the first two steps. The generation times are reduced by about 48% while not computing the LDL tree of $\frac{1}{ff^{\star}+gg^{\star}}$. The implementations use the C double type for all floating-point operations, and the LDL tree over $\mathbb{Q}[x]/(x^n + 1)$ in FALCON needs $64n(\log n + 1)$ bits of memory to store. While not keeping the LDL tree of $\frac{1}{ff^{\star}+gg^{\star}}$, the LDL tree of the NTRU lattice only needs $32n(\log n + 2)$ bits of memory, which saves about 45% memory for $n = 512$ or 1024.

We also confirmed that not storing the LDL tree of $\frac{1}{ff^{\star}+gg^{\star}}$ hardly affects the performance of the fast Fourier sampling over NTRU lattices. The benchmark is shown in Table 2. As described in Sect. 3.1, while replacing the LDL tree of $\frac{1}{ff^{\star}+gg^{\star}}$ with the LDL tree of $ff^{\star} + gg^{\star}$, the fast Fourier sampling algorithm needs to recompute the values of leaf nodes and negate the polynomials of internal nodes. The timing-consuming operation is recomputing the values of leaf nodes. In FALCON's implementations, the values of the LDL tree leaf nodes are the reciprocals of the standard deviations σ_d of the Gaussian distributions over the integers, rather than the diagonal elements d obtained by

Table 1. Generation time and storage of the LDL tree on Intel i7-4790

	$n = 512$		$n = 1024$	
	FALCON	Our work	FALCON	Our work
Generation time with $\mathbf{B}_{f,g}$ (μs)	35.28	25.38	76.02	54.81
Generation time with FFT($\mathbf{G}_{f,g}$) (μs)	20.12	10.28	43.40	22.53
Memory (KB)	40	22	88	48

the fast Fourier orthogonalization algorithm. Let σ be the standard deviation of the Gaussian distribution over NTRU lattices, and we have $\sigma_d = \sigma/\sqrt{d}$. Let q be the modulus of FALCON. The value stored in the LDL tree leaf node of $ff^\star + gg^\star$ is $1/\sigma_d = \sqrt{d}/\sigma$, and the required value for the LDL tree of $\frac{1}{ff^\star+gg^\star}$ is $q/(\sigma\sqrt{d})$. While precomputing $1/\sigma^2$, for the number field $\mathbb{Q}[x]/(x^n+1)$, our technique introduces additional n floating-point multiplications and n floating-point divisions to the signing process. As shown in Table 2, these floating-point multiplications and divisions have little impact on the efficiency of signature generation. However, building the LDL tree on the fly will increase the signature generation time by 70%–80%.

Table 2. Signature generation time on Intel i7-4790

Degree	Signature generation without LDL tree (μs)	Signature generation with LDL tree (μs)	Our work (μs)
512	318.43	183.30	184.38
1024	649.21	363.68	365.17

4.2 ARM Cortex M4

It's more important for memory-constrained devices to reduce RAM usage without sacrificing the signature efficiency, so we used an STM32F4 development board to benchmark the implementations. The board is the same as the one used by the pqm4 project [15] and Pornin [21]. It provides an ARM Cortex M4 core that can run at up to 168 MHz, along with 192 kB of RAM (in two separate chunks of 128 and 64 kB, respectively). The code for our benchmark utilizes uint64_t to emulate the floating-point number and does not contain dedicated inline assembly routines for the core floating-point operations. The implementation with the inline assembly routines is more than twice faster as the generic implementation, but the inline assembly routines are compatible with our technique and have no effect on our comparisons.

Because both uint64_t and double need 64 bits memory, not keeping the LDL tree of $\frac{1}{ff^\star+gg^\star}$ can also save about 45% memory for the code using uint64_t. As shown in Table 3, if not keeping the LDL tree of $\frac{1}{ff^\star+gg^\star}$, we can reduce the LDL tree generation times with $\mathbf{B}_{f,g}$ by about 28% and reduce the LDL tree generation times with $\mathrm{FFT}(\mathbf{G}_{f,g})$ by about 47%. From Table 4, we can conclude that not keeping the LDL tree of $\frac{1}{ff^\star+gg^\star}$ hardly affects the signing time on the ARM Cortex M4, and building the LDL tree on the fly will increase the signing time by 117%–120%.

Table 3. Generation time and storage of the LDL tree on ARM Cortex M4

	$n = 512$		$n = 1024$	
	FALCON	Our work	FALCON	Our work
Generation time with $\mathbf{B}_{f,g}$ (ms)	185.72	134.05	410.61	295.31
Generation time with $\mathrm{FFT}(\mathbf{G}_{f,g})$ (ms)	111.29	59.62	246.47	131.17
Memory (KB)	40	22	88	48

Table 4. Signature generation time on ARM Cortex M4

Degree	Signature generation without LDL tree (ms)	Signature generation with LDL tree (ms)	Our work (ms)
512	470.24	216.97	219.04
1024	1028.07	466.86	471.00

5 Conclusion

In this paper, we utilized the symmetric structure of the LDL tree of FALCON to reduce the generation time and storage of the LDL tree by almost half without affecting the signature efficiency. We further confirmed the technique in the implementations of FALCON. Therefore, we could mitigate the RAM requirement of the FALCON signature algorithm. The result is true not only for the cyclotomic field $\mathbb{Q}[x]/(x^n+1)$ with $n = 2^\kappa$, but also for the cyclotomic field $\mathbb{Q}[x]/(x^n - x^{n/2} + 1)$ with $n = 3 \cdot 2^\kappa$. Although we can not extend the result to NTRU module lattices, the positive results of this paper show that we should pay more attention to the influence of NTRU lattice structure when applying the algorithm for ideal lattices to NTRU lattices.

Acknowledgements. The authors would like to thank the anonymous reviewers for their helpful comments. This work is supported in part by National Natural Science Foundation of China (No. U1936209 and No. 62002353), China Postdoctoral Science Foundation (No. 2021M701726) and Yunnan Provincial Major Science and Technology Special Plan Projects (No. 202103AA080015).

A Proof of Theorem 2

Proof. If we build the LDL trees of $a(x)$ and $b(x)$ according to the field tower described in Sect. 2.3, we split the diagonal element into three terms only when it's in $\mathbb{Q}[x]/(x^6-x^3+1)$, otherwise, we split it into its coefficients of even and odd degrees. In fact, we can split the diagonal element into three terms at any time by selecting an appropriate subfield, as long as n is a multiple of 3. When we break the polynomial into two, the transformation matrix and its LDL* decomposition have the same form as that in Theorem 1, so we only need to prove the case that the polynomial is broken into three terms.

Let $\mathbb{L} = \mathbb{Q}[x]/(x^n - x^{n/2} + 1)$ and $\mathbb{K} = \mathbb{Q}[y]/(y^{n/3} - y^{n/6} + 1)$. We write $a(x)$ and $b(x)$ as $a(x) = a_0(x^3) + xa_1(x^3) + x^2a_2(x^3)$ and $b(x) = b_0(x^3) + xb_1(x^3) + x^2b_2(x^3)$, where $a_i(y), b_i(y) \in \mathbb{K}$ for $0 \le i \le 2$. As $a^\star(x) = a(x)$ and $b^\star(x) = b(x)$, we have $a_0^\star(y) = a_0(y)$, $a_1^\star(y) = ya_2(y)$, $a_2^\star(y) = ya_1(y)$, $b_0^\star(y) = b_0(y)$, $b_1^\star(y) = yb_2(y)$, and $b_2^\star(y) = yb_1(y)$.

Let $\mathbf{D}_a$ be the transformantion matrix of the $\mathbb{K}$-linear map $\psi_a : c \in \mathbb{L} \longmapsto ac$. We can compute its LDL* decomposition over $\mathbb{K}$:

$$\mathbf{D}_a = \begin{bmatrix} a_0 & a_1 & a_2 \\ a_1^\star & a_0 & a_1 \\ a_2^\star & a_1^\star & a_0 \end{bmatrix} = \begin{bmatrix} 1 & 0 & 0 \\ l_{10a} & 1 & 0 \\ l_{20a} & l_{21a} & 1 \end{bmatrix} \times \begin{bmatrix} a_0 & 0 & 0 \\ 0 & \frac{a_0^2 - a_1a_1^\star}{a_0} & 0 \\ 0 & 0 & \frac{N_{\mathbb{L}/\mathbb{K}}(a)}{a_0^2 - a_1a_1^\star} \end{bmatrix} \times \begin{bmatrix} 1 & l_{10a}^\star & l_{20a}^\star \\ 0 & 1 & l_{21a}^\star \\ 0 & 0 & 1 \end{bmatrix},$$

where $l_{10a} = \frac{a_1^\star}{a_0}$, $l_{20a} = \frac{a_2^\star}{a_0}$, and $l_{21a} = \frac{a_0a_1^\star - a_1a_2^\star}{a_0^2 - a_1a_1^\star}$. As $N_{\mathbb{L}/\mathbb{K}}(a) = a_0^3 + a_1a_1a_2^\star + a_1^\star a_1^\star a_2 - 2a_0a_1a_1^\star$, the LDL* decomposition of $\mathbf{D}_a$ only involves $a_0(y)$, $a_1(y)$, $a_2(y)$, $a_1^\star(y)$ and $a_2^\star(y)$.

Let $\mathbf{D}_b$ be the transformation matrix of $b(x)$ over $\mathbb{K}$. Before giving the LDL* decomposition of $\mathbf{D}_b$, we need to express $b_0(y)$, $b_1(y)$ and $b_2(y)$ with $a_0(y)$, $a_1(y)$, $a_2(y)$, $a_1^\star(y)$ and $a_2^\star(y)$. As $a(x)b(x) = 1$, we have

$$\begin{cases} a_0(y)b_0(y) + ya_1(y)b_2(y) + ya_2(y)b_1(y) & = 1 \\ a_0(y)b_1(y) + a_1(y)b_0(y) + ya_2(y)b_2(y) & = 0 \\ a_0(y)b_2(y) + a_1(y)b_1(y) + a_2(y)b_0(y) & = 0 \end{cases}.$$

A tedious but easy Gaussian elimination then gives us the desired result:

$$b_0(y) = \frac{a_0(y)a_0(y) - a_1(y)a_1^\star(y)}{N_{\mathbb{L}/\mathbb{K}}(a)(y)}, \qquad b_1(y) = \frac{a_1^\star(y)a_2(y) - a_0(y)a_1(y)}{N_{\mathbb{L}/\mathbb{K}}(a)(y)},$$

$$b_2(y) = \frac{a_1(y)a_1(y) - a_0(y)a_2(y)}{N_{\mathbb{L}/\mathbb{K}}(a)(y)}.$$

Now we can express the LDL* decomposition of $\mathbf{D}_b$ with $a_0(y)$, $a_1(y)$, $a_2(y)$, $a_1^\star(y)$ and $a_2^\star(y)$ to figure out the relation between the LDL* decompositions of $\mathbf{D}_a$ and $\mathbf{D}_b$:

$$\mathbf{D}_b = \begin{bmatrix} b_0 & b_1 & b_2 \\ b_1^\star & b_0 & b_1 \\ b_2^\star & b_1^\star & b_0 \end{bmatrix} = \begin{bmatrix} 1 & 0 & 0 \\ l_{10b} & 1 & 0 \\ l_{20b} & l_{21b} & 1 \end{bmatrix} \times \begin{bmatrix} b_0 & 0 & 0 \\ 0 & \frac{b_0^2 - b_1 b_1^\star}{b_0} & 0 \\ 0 & 0 & \frac{N_{\mathbb{L}/\mathbb{K}}(b)}{b_0^2 - b_1 b_1^\star} \end{bmatrix} \times \begin{bmatrix} 1 & l_{10b}^\star & l_{20b}^\star \\ 0 & 1 & l_{21b}^\star \\ 0 & 0 & 1 \end{bmatrix}$$

$$= \begin{bmatrix} 1 & 0 & 0 \\ l_{10b} & 1 & 0 \\ l_{20b} & l_{21b} & 1 \end{bmatrix} \times \begin{bmatrix} \frac{a_0^2 - a_1 a_1^\star}{N_{\mathbb{L}/\mathbb{K}}(a)} & 0 & 0 \\ 0 & \frac{a_0}{a_0^2 - a_1 a_1^\star} & 0 \\ 0 & 0 & \frac{1}{a_0} \end{bmatrix} \times \begin{bmatrix} 1 & l_{10b}^\star & l_{20b}^\star \\ 0 & 1 & l_{21b}^\star \\ 0 & 0 & 1 \end{bmatrix},$$

where l_{10b}, l_{20b} and l_{21b} satisfy the following equalities:

$$\begin{cases} l_{10b}(y) = \dfrac{b_1^\star(y)}{b_0(y)} = -\dfrac{a_0(y)a_1^\star(y) - a_1(y)a_2^\star(y)}{a_0(y)a_0(y) - a_1(y)a_1^\star(y)} = -l_{21a}(y) \\ l_{20b}(y) = \dfrac{b_2^\star(y)}{b_0(y)} = \dfrac{y\left(a_1^\star(y)a_2(y) - a_0(y)a_1(y)\right)}{a_0(y)a_0(y) - a_1(y)a_1^\star(y)} = -y l_{21a}^\star(y) \\ l_{21b}(y) = \dfrac{b_0(y)b_1^\star(y) - b_1(y)b_2^\star(y)}{b_0(y)b_0(y) - b_1(y)b_1^\star(y)} = -\dfrac{a_1^\star(y)}{a_0(y)} = -l_{10a}(y) \end{cases}.$$

In the LDL* decompositions of $\mathbf{D}_a$ and $\mathbf{D}_b$, the diagonal elements are pairwise reciprocals of each other in reverse order. For the polynomials to be stored, we can calculate the polynomials $l_{10b}(y)$, $l_{20b}(y)$ and $l_{21b}(y)$ in the LDL* decomposition of $\mathbf{D}_b$ with the polynomials $l_{10a}(y)$ and $l_{21a}(y)$ in the LDL* decomposition of $\mathbf{D}_a$. Because there always exists the reciprocal relationship during the whole recursion, we can build the complete LDL tree of $b(x)$ by the LDL tree of $a(x)$. The computational costs include computing the adjoints of the polynomials and the polynomial multiplications, negating the polynomials of some internal nodes and computing the reciprocals of the values of leaf nodes. □

References

1. Agrawal, S., Boneh, D., Boyen, X.: Efficient lattice (H)IBE in the standard model. In: Gilbert, H. (ed.) EUROCRYPT 2010. LNCS, vol. 6110, pp. 553–572. Springer, Heidelberg (2010). https://doi.org/10.1007/978-3-642-13190-5_28
2. Behnia, R., Ozmen, M.O., Yavuz, A.A.: Lattice-based public key searchable encryption from experimental perspectives. IEEE Trans. Dependable Secur. Comput. **17**(6), 1269–1282 (2020)
3. Boneh, D., et al.: Fully key-homomorphic encryption, arithmetic circuit ABE and compact garbled circuits. In: Nguyen, P.Q., Oswald, E. (eds.) EUROCRYPT 2014. LNCS, vol. 8441, pp. 533–556. Springer, Heidelberg (2014). https://doi.org/10.1007/978-3-642-55220-5_30
4. Cash, D., Hofheinz, D., Kiltz, E., Peikert, C.: Bonsai trees, or how to delegate a lattice basis. In: Gilbert, H. (ed.) EUROCRYPT 2010. LNCS, vol. 6110, pp. 523–552. Springer, Heidelberg (2010). https://doi.org/10.1007/978-3-642-13190-5_27

5. Chuengsatiansup, C., Prest, T., Stehlé, D., Wallet, A., Xagawa, K.: ModFalcon: compact signatures based on module-NTRU lattices. In: Sun, H., Shieh, S., Gu, G., Ateniese, G. (eds.) ASIA CCS, pp. 853–866. ACM (2020)
6. Ducas, L., Galbraith, S., Prest, T., Yu, Y.: Integral matrix gram root and lattice gaussian sampling without floats. In: Canteaut, A., Ishai, Y. (eds.) EUROCRYPT 2020. LNCS, vol. 12106, pp. 608–637. Springer, Cham (2020). https://doi.org/10.1007/978-3-030-45724-2_21
7. Ducas, L., Lyubashevsky, V., Prest, T.: Efficient identity-based encryption over NTRU lattices. In: Sarkar, P., Iwata, T. (eds.) ASIACRYPT 2014. LNCS, vol. 8874, pp. 22–41. Springer, Heidelberg (2014). https://doi.org/10.1007/978-3-662-45608-8_2
8. Lyubashevsky, V., Prest, T.: Quadratic time, linear space algorithms for Gram-Schmidt orthogonalization and Gaussian sampling in structured lattices. In: Oswald, E., Fischlin, M. (eds.) EUROCRYPT 2015. LNCS, vol. 9056, pp. 789–815. Springer, Heidelberg (2015). https://doi.org/10.1007/978-3-662-46800-5_30
9. Ducas, L., Prest, T.: Fast Fourier orthogonalization. In: Abramov, S.A., Zima, E.V., Gao, X. (eds.) ISSAC, pp. 191–198. ACM (2016)
10. Espitau, T.: Mitaka: faster, simpler, parallelizable and maskable hash-and-sign signatures on NTRU lattices. In: Emura, K., Wang, Y. (eds.) APKC@AsiaCCS, p. 1. ACM (2021)
11. Fouque, P.A., Gérard, F., Rossi, M., Yu, Y.: Zalcon: an alternative FPA-free NTRU sampler for Falcon. Technical report, National Institute of Standards and Technology (2021). https://csrc.nist.gov/events/2021/third-pqc-standardization-conference
12. Gentry, C.: Fully homomorphic encryption using ideal lattices. In: Mitzenmacher, M. (ed.) STOC, pp. 169–178. ACM (2009)
13. Gentry, C., Peikert, C., Vaikuntanathan, V.: Trapdoors for hard lattices and new cryptographic constructions. In: Dwork, C. (ed.) STOC, pp. 197–206. ACM (2008)
14. Hoffstein, J., Pipher, J., Silverman, J.H.: NTRU: a ring-based public key cryptosystem. In: Buhler, J.P. (ed.) ANTS 1998. LNCS, vol. 1423, pp. 267–288. Springer, Heidelberg (1998). https://doi.org/10.1007/BFb0054868
15. Kannwischer, M.J., Rijneveld, J., Schwabe, P., Stoffelen, K.: pqm4: Testing and benchmarking NIST PQC on ARM cortex-m4. IACR Cryptol. ePrint Arch. 844 (2019)
16. Lu, X., Au, M.H., Zhang, Z.: Raptor: a practical lattice-based (linkable) ring signature. In: Deng, R.H., Gauthier-Umaña, V., Ochoa, M., Yung, M. (eds.) ACNS 2019. LNCS, vol. 11464, pp. 110–130. Springer, Cham (2019). https://doi.org/10.1007/978-3-030-21568-2_6
17. Lyubashevsky, V., et al.: CRYSTALS-DILITHIUM. Technical report, National Institute of Standards and Technology (2020). https://csrc.nist.gov/Projects/post-quantum-cryptography/round-3-submissions
18. Lyubashevsky, V., Micciancio, D.: Generalized compact knapsacks are collision resistant. In: Bugliesi, M., Preneel, B., Sassone, V., Wegener, I. (eds.) ICALP 2006. LNCS, vol. 4052, pp. 144–155. Springer, Heidelberg (2006). https://doi.org/10.1007/11787006_13
19. Peikert, C.: An efficient and parallel gaussian sampler for lattices. In: Rabin, T. (ed.) CRYPTO 2010. LNCS, vol. 6223, pp. 80–97. Springer, Heidelberg (2010). https://doi.org/10.1007/978-3-642-14623-7_5
20. del Pino, R., Lyubashevsky, V., Seiler, G.: Lattice-based group signatures and zero-knowledge proofs of automorphism stability. In: Lie, D., Mannan, M., Backes, M., Wang, X. (eds.) CCS, pp. 574–591. ACM (2018)

21. Pornin, T.: New efficient, constant-time implementations of falcon. IACR Cryptol. ePrint Arch. 893 (2019)
22. Prest, T.: Gaussian sampling in lattice-based cryptography. Ph.D. thesis, École Normale Supérieure, Paris, France (2015)
23. Prest, T., et al.: FALCON. Technical report, National Institute of Standards and Technology (2020). https://csrc.nist.gov/Projects/post-quantum-cryptography/round-3-submissions
24. Stehlé, D., Steinfeld, R.: Making NTRU as secure as worst-case problems over ideal lattices. In: Paterson, K.G. (ed.) EUROCRYPT 2011. LNCS, vol. 6632, pp. 27–47. Springer, Heidelberg (2011). https://doi.org/10.1007/978-3-642-20465-4_4
25. Zhao, R.K., McCarthy, S., Steinfeld, R., Sakzad, A., O'Neill, M.: Quantum-safe HIBE: does it cost a latte? IACR Cryptol. ePrint Arch. 222 (2021)

Secure Sketch and Fuzzy Extractor with Imperfect Randomness: An Information-Theoretic Study

Kaini Chen[1,2], Peisong Shen[1(✉)], Kewei Lv[1,2], and Chi Chen[1,2]

[1] State Key Laboratory of Information Security, Institute of Information Engineering, Chinese Academy of Sciences, Beijing 100093, China
{chenkaini,shenpeisong,lvkewei,chenchi}@iie.ac.cn
[2] School of Cyber Security, University of Chinese Academy of Sciences, Beijing 100049, China

Abstract. Fuzzy extractor retrieves the same cryptographic key from samples of one noisy source. Information-theoretically secure fuzzy extractors are usually constructed by secure sketches and seeded extractors. However, current constructions rely heavily on perfect randomness, which is difficult to obtain in the real world. In this paper, we analyze the security deficiency of current secure sketches and fuzzy extractors in the *imperfect randomness* setting where no perfect randomness is available. Furthermore, we discuss the information-theoretic security of several explicit fuzzy extractors based on two-source extractors in the *imperfect randomness* setting.

Keywords: Fuzzy extractor · Secure sketch · Imperfect randomness · Information-theoretic security

1 Introduction

Uniformly distributed randomness is essential to the modern cryptography. It is widely used by cryptographic primitives like encryption, signature and authentication, etc. Randomness extractors are proposed to generate uniformly distributed strings from entropy sources in the real world. However, in reality many high-entropy sources are noisy in nature, such as biometrics [10,19], physically unclonable functions [16,22] and quantum information generated from quantum devices [3]. Thus, two homologous samples of a noisy source are similar (or close in certain distance metric) but not identical.

In order to extract stable cryptographic system usable randomness from these noisy sources, Dodis et al. [12] proposed fuzzy extractor. A fuzzy extractor consists of two phases: Generate ("Gen") and Reproduce ("Rep"). Gen converts an input sample w into a publicly accessible help data P and a nearly uniformly distributed string R. Rep takes the help data P and a new data sample w' as inputs, if w' is close to w (their distance is within an error threshold t), then the

C. Alcaraz et al. (Eds.): ICICS 2022, LNCS 13407, pp. 128–147, 2022.
https://doi.org/10.1007/978-3-031-15777-6_8

string R is reproduced. The definition of fuzzy extractor in [12] is an information-theoretic one, which requires the statistical distance between output key R and a uniform distribution is negligible. Figure 1 illustrates the functionality of fuzzy extractor.

Besides, Dodis et al. [12] proposed secure sketch. It is an information reconciliation component. That is, secure sketch allows one to produce public information from an enrolled sample of noisy sources and to recover the sample given another sample that is close enough. A secure sketch consists of two phases: Sketch ("SS") and Recover ("Rec"). SS takes an enrolled sample w as input to output a public sketch data ss. Rec reconstructs the original sample w from sketch data ss and a new data sample w' if w' is close to w. Figure 2 illustrates the functionality of secure sketch. The security of secure sketch poses a requirement on lower bound of the average min-entropy of noisy source conditioned on public sketch data. Dodis et al. [12] also showed that secure sketch can be used to construct fuzzy extractor, i.e. combining a secure sketch with a strong seeded randomness extractor. This architecture is abbreviated as "sketch-then-extract".

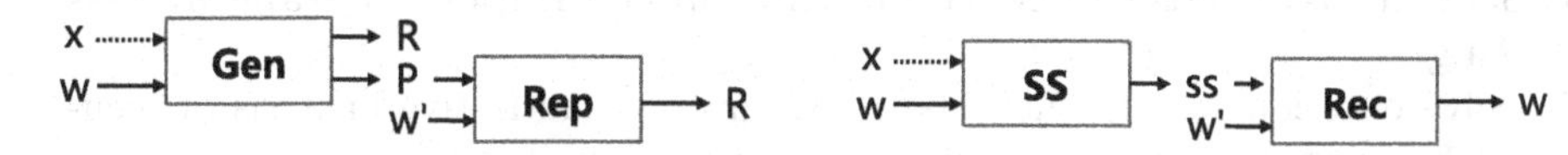

Fig. 1. Fuzzy Extractor, private randomness x is explicitly marked with a dotted arrow.

Fig. 2. Secure Sketch, private randomness x is explicitly marked with a dotted arrow.

It's worth noting that current fuzzy extractor constructions rely heavily on perfect private randomness. Some kinds of secure sketches (e.g. code offset-based, permutation-based, fuzzy vault-based) are randomized and need private randomness for sketch generation. For example, in the code-offset construction, SS process randomly picks a codeword of an error-correcting code using randomness from a private random source. Permutation-based secure sketch also relies on private random source to select a random codeword and a permutation function satisfying a certain requirement. Besides, for sketch-then-extract kind of fuzzy extractors, a truly random seed is used by a seeded extractor to extract the key from an enrolled sample. In short, perfect private randomness is vital to the security of secure sketches and fuzzy extractors.

However, it's difficult to find a perfect random source in reality. A less restrictive and more realistic assumption on the random source is that the source only contains some entropy. We call it *imperfect randomness* setting. In the imperfect randomness setting, secure sketches do not have access to a truly random source any more, but are provided with a random source with high entropy, and fuzzy extractors can no longer use a truly random seed to extract key from noisy samples. Unfortunately, most works of secure sketches and fuzzy extractors do not consider this imperfect randomness setting. Therefore, two research questions arise naturally:

Q1: How the security of secure sketch and fuzzy extractor is affected by the imperfect randomness?
Q2: How to build fuzzy extractor which remains secure in the imperfect randomness setting?

Up to now, no research works are concerned with question 1. As for question 2, Cui et al. [9] proposed a reusable and robust computational fuzzy extractor based on DDH/LPN assumptions in the CRS model. In their construction, a deterministic syndrome-based secure sketch [12] is used and the imperfect randomness is used as one input of a two-source extractor. However, whether or not secure sketch and fuzzy extractor with imperfect randomness exists under the information-theoretic setting, remains unclear.

1.1 Our Contributions

In this paper, we focus on solving above questions. Our contributions are summarized as follows.
1. Security Analysis of Secure Sketches in the Imperfect Randomness Setting
To solve question 1 mentioned above, we give a quantitative analysis of security deficiencies for three types of randomized secure sketches, in the imperfect randomness setting. In this work, the imperfectness of private entropy source is measured by entropy deficiency, which equals to the difference between string length and min-entropy of imperfect source.

For code offset-based secure sketch, the private randomness is only used to select a random codeword. We prove that the extra entropy loss incurred by imperfect randomness is just equal to the entropy deficiency of the imperfect source.

For permutation-based secure sketch, the imperfect randomness is used to select a random codeword and a permutation. We prove that the entropy loss is related to the number of all permutations, the min-entropy of the random source, and the minimum min-entropy of permutations satisfying the condition that converts a noisy sample to a random codeword.

For fuzzy vault-based secure sketch, the imperfect randomness is used in three steps: (1) selecting polynomial coefficients; (2) selecting x-coordinates of chaff points; (3) selecting y-coordinates of chaff points, we give an upper bound on the entropy loss of fuzzy vault-based secure sketches based on combinatorial analysis.
2. Security Analysis of Fuzzy Extractors in the Imperfect Randomness Setting
For fuzzy extractor of "sketch-then-extract" paradigm, the imperfect private randomness is used both in its secure sketch and seeded extractor. Intuitively, the security deficiency of fuzzy extractor is related to the extra entropy loss of secure sketch and the entropy deficiency of random seed used in seeded extractor. We give an upper bound on the statistical distance between output key and a uniform distribution in the imperfect randomness setting. The analysis result verifies our intuition.

3. Information-Theoretic Security Discussion of Fuzzy Extractors based on Two-source Extractors in the Imperfect Randomness Setting
For question 2, we simplify the fuzzy extractor construction in [9] then propose an information-theoretically secure fuzzy extractor in the imperfect randomness setting. We analyze the security of this simplified construction. The simplified construction is similar to traditional "sketch-then-extract" construction, the only difference is replacing the strong seeded extractor with a two-source extractor. The imperfect private random source is used both in randomized secure sketch and two-source extractor. Thus, we model this imperfect source as a (ρ, n, k)-correlated source [9], which assumes a sample has some amount of residual entropy conditioned on previous samples from the same source. We instantiate our construction with explicit two-source extractors for different secure sketches, and analyze the security in information-theoretic view. Table 1 demonstrates the differences between our simplified construction and "sketch-then-extractor" constructions in [9,12].

Table 1. Comparison with traditional "sketch-then-extractor" fuzzy extractor constructions ([12] and [9]). "Security model" measures the security of construction in information-theoretic(IT) view or computational view. For "Sketch construction", "all" represents all current secure sketches are suitable for construction. For "Metric space", "all" represents Hamming metric, setting difference metric and edit metric.

FE constructions	Dodis et al. [12]	Cui et al. [9]	Our approach Construction 5
Sketch construction	All	Only syndrome-based	All
Extractor construction	Seeded extractor	Two-source extractor	Two-source extractor
Metric space	All	Hamming	All
Security model	IT	Computational	IT
Imperfect randomness	×	✓	✓

1.2 Related Work

Fuzzy Extractors and Secure Sketches
For fuzzy extractor of "sketch-then-extract" paradigm, the output key length of extractor is determined by the residual entropy of noisy source conditioned on the public sketch. As an information reconciliation component, secure sketch causes certain amount of entropy loss. In order to overcome this limitation, Fuller et al. [14] proposed *computational fuzzy extractor* of which the output key is computationally indistinguishable from a uniform distribution. They found computational secure sketch has a lower bound of entropy loss similar to information-theoretic ones. Thus they proposed a LWE-based computational fuzzy extractor construction which does not use secure sketch any more, and their construction derives much longer key than the information-theoretic ones.

In recent years, researchers proposed several variants of fuzzy extractors with enhanced security properties, e.g. reusability [4,5] and robustness [6]. A reusable fuzzy extractor remains secure even if adversary knows multiple instances of help data and key for the same noisy source. A robust fuzzy extractor can detect the change of help data. Most of state-of-art reusable and robust fuzzy extractors base their security on computational assumptions [1,24,25].

Cryptographic Primitives with Imperfect Randomness
There are substantial results for cryptographic primitives with imperfect randomness.

Dodis and Yu [13] considered the security of several cryptographic primitives on imperfect random secret keys, i.e. keys are not uniformly distributed. They classify cryptographic primitives into "unpredictability applications" and "indistinguishability applications" according to the perspective of security definitions. They measure security losses of cryptographic primitives on imperfect random secret keys by entropy deficiency (the difference between length of random source and the amount of entropy it has) in terms of min-entropy and collision entropy. However their result still requires perfect private randomness, that is, although the key is imperfect, the private randomness is still perfect. Later, Yao and Li [26] extended Dodis and Yu's work to the general Rényi entropy and computational entropy setting. Backes et al. [2] considered the setting where key and private random source are both imperfect. They relax cryptographic indistinguishability to (ϵ, γ)-differential indistinguishability which means one distinguisher's probability of a certain output ($\Pr[\mathcal{D} = b]$) is within ϵ times of the probability of another output plus γ (e.g. $\Pr[\mathcal{D} = 1 - b] \leq 2^{\epsilon} \cdot \Pr[\mathcal{D} = b] + \gamma$).

In general, no previous work consider secure sketch and fuzzy extractor with imperfect private randomness, except for [9]. Worse still, the analysis technique in [13] is not applicable to secure sketch and fuzzy extractor. The reason is two-fold: (1) Although secure sketch can be modeled as a unpredictability application as in [13], Dodis and Yu's methods are mainly designed for imperfect keys, not for imperfect private randomness used in cryptographic primitives. (2) fuzzy extractor can be regarded as an indistinguishability application in view of [13]. But, the security definition of fuzzy extractor does not have phase of query to apply "double-run trick" method proposed by [13]. Besides, differential indistinguishability [2] is not suitable for the security of fuzzy extractor.

1.3 Paper Organization

The rest of this paper is organized as follows. We introduce some basic notations, definitions and primitives in Sect. 2. We present security analysis results of secure sketches in the imperfect randomness setting in Sect. 3. We present security analysis results of fuzzy extractors in the imperfect randomness setting in Sect. 4. We present our upgraded "sketch-then-extract" fuzzy extractor with two-source extractor in the imperfect randomness setting in Sect. 5. We conclude our work in Sect. 6.

2 Preliminaries

In this paper, we use uppercase letters to denote random variables as well as their distributions, use lowercase letters to denote samples. For example, X is the random variable, $x \leftarrow_r X$ denotes sampling x from X randomly, indicating that X is a perfect random source that is uniformly distributed; otherwise, $x \leftarrow X$ denotes sampling x according to distribution X. We use U_ℓ to denote uniformly distributed string on $\{0,1\}^\ell$. For any integer $t > 0$, $[t]$ denotes the set $\{1, \cdots, t\}$. Unless explicitly states, all logarithms below are base 2.

Entropy and Source
The "min-entropy" of a random variable X is $\mathrm{H}_\infty(X) := -\log(\max_x \Pr[X = x])$. The "average min-entropy" of a random variable X conditioned on Y is $\tilde{\mathrm{H}}_\infty(X|Y) := -\log(\mathbb{E}_{y \leftarrow Y}[2^{-\mathrm{H}_\infty(X|Y=y)}])$. The "max-entropy" of a random variable X is $\mathrm{H}_{\max}(X) := -\log(\min_x \Pr[X = x])$. The "entropy deficiency" of a random variable X is $d := \log|X| - \mathrm{H}_\infty(X)$. A "$(\rho, n, k)$-correlated source" is a random source X over a set of size 2^n, that satisfies $\tilde{\mathrm{H}}_\infty(X_i|\{X_j\}_{j\in[i-1]}) \geq k$ when $X_i \leftarrow X$ and $i \in [\rho]$. If $\rho = 1$, (ρ, n, k)-correlated source can abbreviation to "(n, k)-source" with min-entropy k.

Metric Space
A metric space is a set $\mathcal{M}$ with a distance function dis: $\mathcal{M} \times \mathcal{M} \to R^+ = [0, \infty)$. We mainly consider three different metric spaces:

1. Hamming metric. $\mathcal{M} = \mathcal{F}^n$ for some alphabet $\mathcal{F}$, and $\mathrm{dis}(w, w')$ is the number of positions in which the strings w and w' differ.
2. Set difference metric. Here $\mathcal{M}$ consists of all subsets of a universe $\mathcal{U}$. For two sets w and w', their symmetric difference $w\Delta w' = \{x \in w \cup w' | x \notin w \cap w'\}$. The set difference distance between two sets w and w' is $|w\Delta w'|$.
3. Edit metric. Here $\mathcal{M} = \mathcal{F}^*$, the distance between w and w' is the smallest number of character insertions and deletions needed to transform w into w'.

Statistic Distance
The statistic distance between two probability distributions X and Y is defined as $\mathrm{SD}(X, Y) := \frac{1}{2}\sum_v |\Pr[X = v] - \Pr[Y = v]|$.
Error-Correcting Codes
An $(\mathcal{M}, K, t)$-code C is an error-correcting code which contains K elements over $\mathcal{M}$, with minimum distance $d \geq 2t + 1$, and can correct up to t errors. If $\mathcal{M}$ is Hamming metric over $\mathcal{F}^n$, error-correcting code C is a $k = \log_{|\mathcal{F}|} K$ dimension subspace of $\mathcal{M}$ denoted as $[n, k, d]_\mathcal{F}$-code.

2.1 Secure Sketch

Definition 1 (Secure Sketch [12]). *Let $\mathcal{M}$ be a metric space with distance function* $\mathrm{dis}(x, y)$*. An $(\mathcal{M}, m, \tilde{m}, t)$-secure sketch is a pair of randomized procedures, "sketch"* (SS) *and "recover"* (Rec)*:*

1. *The sketching procedure* SS*: on input $w \in \mathcal{M}$ returns a sketch ss in output space SS that $ss \in SS$.*
2. *The recovery procedure* Rec*: takes element $w' \in \mathcal{M}$ and a sketch $ss \in SS$, outputs the recovered $w'' \in \mathcal{M}$.*

It has the following properties:

1. *Correctness: if* $\mathrm{dis}(w, w') \leq t$, *then* $\mathrm{Rec}(w', \mathrm{SS}(w)) = w$. *If* $\mathrm{dis}(w, w') > t$, *there is no guarantee provided about the output of* Rec.
2. *Security: for any distribution W over $\mathcal{M}$ with min-entropy m, the value of W can be recovered by an adversary who observes ss with probability no greater than $2^{-\tilde{m}}$. That is, $\tilde{\mathrm{H}}_\infty(W|\mathrm{SS}(W)) \geq \tilde{m}$.*
3. *Efficiency:* SS *and* Rec *run in expected polynomial time.*

Usually, secure sketch employs an error-correcting code to overcome noises of different samples from one source, and recover the original sample with the help of public sketch data SS. Its security is defined in an information-theoretic fashion. In secure sketch scenario, we call $\tilde{\mathrm{H}}_\infty(W|\mathrm{SS}(W))$ the residual entropy of W over a sketch. The quantity $L = \mathrm{H}_\infty(W) - \tilde{\mathrm{H}}_\infty(W|\mathrm{SS}(W)) = m - \tilde{m}$ is called the entropy loss and indicates the amount of information that a sketch leaks about the input.

2.2 Fuzzy Extractor

Definition 2 (Fuzzy Extractor [12]). *An $(\mathcal{M}, m, \ell, t, \epsilon)$-fuzzy extractor is a pair of randomized procedures, "generate"* (Gen) *and "reproduce"* (Rep).

1. *The generation procedure* Gen*: on input $w \in \mathcal{M}$ outputs an extracted string $R \in \{0,1\}^\ell$ and a helper data P within its space that $P \in \mathcal{P}$.*
2. *The reproduction procedure* Rep*: takes an element $w' \in \mathcal{M}$ and a help data $P \in \mathcal{P}$ as inputs, outputs a reproduced string R'.*

It has the following properties:

1. *Correctness: if* $\mathrm{dis}(w, w') \leq t$ *and* R, P *were generated by* $(R, P) \leftarrow \mathrm{Gen}(w)$, *then* $\mathrm{Rep}(w', P) = R$. *If* $\mathrm{dis}(w, w') > t$, *there is no guarantee provided about the output of* Rep.
2. *Security: for any distribution W on $\mathcal{M}$ of min-entropy m ($\mathrm{H}_\infty(W) \geq m$), the string R is nearly uniform even for those who observe P: $(R, P) \leftarrow \mathrm{Gen}(w)$, then $\mathrm{SD}((R, P), (U_\ell, P)) \leq \epsilon$.*
3. *Efficiency:* Gen *and* Rep *run in expected polynomial time.*

Fuzzy extractor extracts a random string R from a noisy source W. An information-theoretic fuzzy extractor guarantees the statistic distance between the output string R and U_ℓ is negligible.

2.3 Randomness Extractor

Randomness extractors are widely used in cryptography. Seeded extractor distills nearly uniform string R from a random source with the help of a short uniform seed. Two-source extractor replaces truly random seed by another independent random source. Here we recall some definitions and lemmas of randomness extractors.

Definition 3 (Strong Seeded EXT [20]**).** *A function* Ext: $\{0,1\}^n \times \{0,1\}^d \rightarrow \{0,1\}^\ell$ *is an* (n,k,ℓ,ϵ)*-strong seed extractor if for any source* X *of min-entropy* k *and uniformly distributed seed* U_d, $\mathrm{SD}(\mathrm{Ext}(X,U_d),U_d),(U_\ell,U_d)) \leq \epsilon$.

Definition 4 (Average-case Strong Seeded EXT [18]**).** *A function* Ext: $\{0,1\}^n \times \{0,1\}^d \rightarrow \{0,1\}^\ell$ *is an average-case* (n,k,ℓ,ϵ)*-strong seed extractor if for any source* X *with* $\tilde{\mathrm{H}}_\infty(X|I) \geq k$ *and uniformly distributed seed* U_d, $\mathrm{SD}(\mathrm{Ext}(X,U_d),U_d,I),(U_\ell,U_d,I)) \leq \epsilon$, *where* I *is an auxiliary variable.*

Lemma 1 ([12])**.** *For any* $\delta > 0$, *if* Ext *is an* $(n,k-\log(1/\delta),\ell,\epsilon)$*-strong seeded extractor, then* Ext *is also an average-case* $(n,k,\ell,\epsilon+\delta)$*-strong seeded extractor.*

Definition 5 (Strong Two-source EXT [18]**).** *A function* TExt: $\{0,1\}^{n_1} \times \{0,1\}^{n_2} \rightarrow \{0,1\}^\ell$ *is an* $((n_1,k_1),(n_2,k_2),\epsilon)$*-strong two-source extractor for min-entropy* k_1,k_2 *and error* ϵ *if for every independent* (n_1,k_1)*-source* X *and* (n_2,k_2)*-source* Y, $\mathrm{SD}((\mathrm{TExt}(X,Y),X),(U_\ell,X)) \leq \epsilon$ *and* $\mathrm{SD}((\mathrm{TExt}(X,Y),Y),(U_\ell,Y)) \leq \epsilon$.

Definition 6 (Average-case Strong Two-source EXT [18]**).** *A function* TExt: $\{0,1\}^{n_1} \times \{0,1\}^{n_2} \rightarrow \{0,1\}^\ell$ *is an average-case* $((n_1,k_1),(n_2,k_2),\epsilon)$*-strong two-source extractor for min-entropy* k_1,k_2 *and error* ϵ *if for every independent source* X *with* $\tilde{\mathrm{H}}_\infty(X|I) \geq k_1$ *and* (n_2,k_2)*-source* Y, $\mathrm{SD}((\mathrm{TExt}(X,Y),Y,I),(U_\ell,Y,I)) \leq \epsilon$, *where* I *is an auxiliary variable.*

Lemma 2 ([9])**.** *For any* $\delta > 0$, *if* TExt *is an* $((n_1,k_1-\log(1/\delta)),(n_2,k_2),\epsilon)$*-strong two-source extractor, then* TExt *is also an average-case* $((n_1,k_1),(n_2,k_2),$ $\epsilon+\delta)$*-strong two-source extractor.*

The following lemma indicates that randomness extractor with a non-uniform seed leads to expansion of statistic distance.

Lemma 3 (Strong Seeded Extractor with Imperfect Seed [7]**).** *Let* Ext: $\{0,1\}^n \times \{0,1\}^d \rightarrow \{0,1\}^\ell$ *be an* (n,k,ℓ,ϵ)*-strong seeded extractor. Let* X *be an* (n,k)*-source and* Y *be a* $(d,d-\lambda)$*-source, then* $\mathrm{SD}((\mathrm{Ext}(X,Y),Y),(U_\ell,Y)) \leq 2^\lambda\epsilon$.

A series of works are related to explicit (efficient) constructions of randomness extractors and two-source extractors [8,18,23]. Leftover hash lemma [15] shows that universal hash function is a good strong seeded extractor, and it is used in "sketch-then-extract" fuzzy extractor [12].

3 Security Analysis of Existing Sketch Sketches with Imperfect Randomness

In this section, we will analyze the security deficiency (residual entropy loss for secure sketch) of existing secure sketches where only imperfect private randomness is available. Quantitative security deficiency results are given on three randomized secure sketches in the imperfect randomness setting: code offset-based, permutation-based, fuzzy vault-based. Although these secure sketch constructions [12,17] are mainly concerned with Hamming metric and set difference metric, one can use the technique called "metric space embedding" [12] to convert secure sketches in Hamming metric or set difference metric to secure sketches in edit distance metric.

3.1 Code Offset-Based Construction

Construction 1 (Code Offset-Based Sketch [12]**).** *A code offset-based sketch is a secure sketch* <SS,Rec> *where*
Sketch Phase: $\mathrm{SS}(w) = w - c = ss$. SS *outputs the shift ss between input $w \in \mathcal{M}$ and a randomly chosen codeword $c \in C$, with C is an $(\mathcal{M}, K, t)$-code.*
Recover Phase: $\mathrm{Rec}(w) = \mathrm{Dec}(w' - ss) + ss$. Rec *recover w on inputs w' and sketch ss, with* Dec *the decoding procedure of C that output $c' = \mathrm{Dec}(w' - ss)$ and recover by compute $w = c' + ss$.*

Code offset-based secure sketch is suitable for Hamming metric. The error correcting code C is an $[n, k, 2t+1]_{\mathcal{F}}$-code with $n = \log_{|\mathcal{F}|} |\mathcal{M}|$ and $k = \log_{|\mathcal{F}|} K$. SS uses private randomness to select a random codeword c from C (this is equivalent to choosing a random $x \leftarrow \mathcal{F}^k$ and compute encoding function $C(x)$). Dodis et al. [12] showed that entropy loss of code offset-based sketch with $[n, k, 2t+1]_{\mathcal{F}}$-code is at most $(n-k)\log|\mathcal{F}|$. In the imperfect randomness setting, we assume random codeword is sampled from an imperfect random source X over $\mathcal{F}^k$ with $\tilde{\mathrm{H}}_\infty(X) = k\log(|\mathcal{F}|) - d$, where d is entropy deficiency. Thus, we find that the extra entropy loss of code offset-based secure sketch in the imperfect randomness setting is equal to the entropy deficiency of the imperfect source. Theorem 1 demonstrates our result.

Theorem 1. *An $(\mathcal{M}, m, \tilde{m}, t)$ code offset-based secure sketch loses at most d bits of residual entropy if it takes randomness from an imperfect source with entropy deficiency d.*

Proof. Briefly, the only private randomness in code offset-based secure sketch is the chosen of codeword c. We lose at most d bits randomness when choosing codeword from an imperfect random source. Then this entropy loss will transmit to the final residual entropy of secure sketch. The full proof is given in Appendix A.1.

3.2 Permutation-Based Secure Sketch

Construction 2 (Permutation-based Sketch [12]**).** *A permutation-based sketch is a secure sketch* <SS,Rec> *where*

Sketch Phase: SS *outputs the specification of a transitive isometric permutation* π_P *in* $\mathcal{M}$ *such that* $\pi_P[w] = c \in C$*, with* C *is an* $(\mathcal{M}, K, t)$*-code.*

Recover Phase: Rec *outputs* $(\pi_P^{-1} \circ Dec \circ \pi_P)[w']$ *on inputs* w' *and sketch* π_P*, with* Dec *the decoding procedure of* C *that maps* $\pi_P[w']$ *to* c *if* $\mathrm{dis}(w, w') \leq t$*. Recover* w *by compute* $w = \pi_P^{-1}[c]$*.*

A family of permutations $\Pi = \{\pi : \mathcal{M} \to \mathcal{M}\}$ is a transitive group if for any two elements $a, b \in \mathcal{M}$ there exists a permutation $\pi \in \Pi$ such that $\pi[a] = b$. A permutation is isometric if for any two elements $a, b \in \mathcal{M}$ that $\mathrm{dis}(a, b) = \mathrm{dis}(\pi[a], \pi[b])$.

Permutation-based secure sketch is suitable for both Hamming metric and set difference metric. We use a similar technique raised by Dodis et al. [12] to give the security analysis of permutation-based secure sketch in the imperfect randomness setting by modifying some definitions. Their analysis is conducted by adding entropy brought by the random selected codeword and the random chosen permutation, and reducing entropy by publishing the sketch. Simply, they defined Γ to be the number of elements $\pi \in \Pi$ such that $\min_{w,c} |\{\pi | \pi(w) = c\}| \geq \Gamma$. That is, for each w and c, there are at least Γ choices for isometric permutation π that satisfies $\pi(w) = c$.

We modify the definition of Γ to makes it accommodated to the imperfect randomness setting where distribution of codeword $c \leftarrow C$ is not uniform. We use $\Gamma_{w,c}$ to denote the distribution of transitive isometric permutation $\pi \leftarrow_X \Pi$ with condition $\pi(w) = c$. We use $\min_{w,c}(\mathrm{H}_\infty(\Gamma_{w,c}))$ to denote the minimum value of $\Gamma_{w,c}$'s min-entropy depending on w and c. Then in Theorem 2, we present our result of the entropy loss for permutation-based secure sketch in the imperfect randomness setting.

Theorem 2. *Assume a permutation-based secure sketch on input* W *is constructed with an error-correcting code* C*. Then, entropy loss of this secure sketch is at most* $\log |\Pi| - \mathrm{H}_\infty(X) - \min_{w,c}(\mathrm{H}_\infty(\Gamma_{w,c}))$ *if it takes randomness from an imperfect random source* X*.*

Proof. We compute the entropy loss of permutation-based secure sketch by analyzing how many bits of entropy it adds through random operation in sketch phase and it loses when public the sketch. Permutation-based secure sketch public $\pi \in \Pi$ after SS phase and loses at most $\log |\Pi|$ bits of entropy. For every chosen codeword c, the chosen operation add at least $\mathrm{H}_\infty(X)$ bits of entropy. For every w and c, $\Gamma_{w,c}$ indicates the distribution of transitive isometric permutation π that satisfies $\pi(w) = c$, adding at least $\mathrm{H}_\infty(\Gamma_{w,c})$ bits of entropy. Considering all w and c, the chosen operation on π add at least $\min_{w,c}(\mathrm{H}_\infty(\Gamma_{w,c}))$ bits of entropy. Then we get our result. $\square$

Remark. Code offset construction can be seen as a special case of permutation-based sketch since addition function in $\mathcal{F}^n$ is bijective and transitive isometric. When we use the above result of permutation-based sketch to analyze code offset-based secure sketch, only one transitive isometric permutation can convert w to c. So that $\min_{w,c}(\mathrm{H}_\infty(\Gamma_{w,c})) = 0$. For random source X which holds $\mathrm{H}_\infty(X) = k - d$, we have following formula that is in agreement with Theorem 1.

$$\textit{entropy loss} = \log|\Pi| - \mathrm{H}_\infty(X) - \min_{w,c}(\mathrm{H}_\infty(\Gamma_{w,c}) \leq n - (k - d)$$

$$\textit{Residual entropy} = \mathrm{H}_\infty(W) - \textit{entropy loss} \geq m + k - n - d.$$

3.3 Fuzzy Vault-Based Secure Sketch

Fuzzy vault-based secure sketch is suitable for set difference metric where w and w' are both in set form. The distance of w and w' is the weight of symmetric difference between w and w'. Let $(\mathcal{U}, n, r, s)$ denotes parameters of the fuzzy vault-based secure sketch where $\mathcal{U}$ is the universe of the set, n denotes the number of elements in $\mathcal{U}$. Assume that $n = |\mathcal{U}|$ is a prime power, fuzzy vault-based secure sketch works over the field $\mathcal{F} = GF(n)$. S denotes the set of elements in w, $s = |S|$ denotes the number of elements in w, and r denotes the number of all elements in the final vault. On input set w, the output vault of Juels-Sudan sketch [17] is r pairs of points (x_i, y_i), for some parameter r, and $s < r \leq n$. Lemma 4 was proposed by Dodis et al. [12] showing the entropy loss of fuzzy vault-based secure sketch in the perfect randomness setting.

Construction 3 (Juels-Sudan Fuzzy Vault-Based Secure Sketch [17]**).** *A $(\mathcal{U}, n, r, s)$-Juels-Sudan fuzzy vault is a secure sketch* <SS,Rec> *where*
Sketch Phase: *Input a set $w = \{x_1, \cdots, x_s\}$ of size s.*

1. *Choose $p(\cdot)$ at random from the set of polynomials of degree at most $k = s - t - 1$ over $\mathcal{F}$.*
2. *For $i = 1, 2, \cdots, s$, calculate $y_i = p(x_i)$.*
3. *Choose $r - s$ distinct points $x_{s+1}, \cdots, x_r$ at random from $\mathcal{U} \backslash S$.*
4. *For $i = s + 1, \cdots r$, choose $y_i \in \mathcal{F}$ at random such that $y_i \neq p(x_i)$.*
5. *Output $\mathrm{SS}(w) = \{(x_1, y_1), \cdots, (x_r, y_r)\}$ (in lexicographic order of x_i).*

Recover Phase: *Input a set w' of size s.*
1. *Pick out (x_i, y_i) where x-coordinates overlap between w and w'.*
2. *Rebuild function $p(\cdot)$ if $|w \Delta w'| \leq 2t$.*
3. *Recover w refer to $p(\cdot)$.*

Lemma 4 ([12])**.** *The entropy loss of $(\mathcal{U}, n, r, s)$-fuzzy vault scheme is at most* $\log(\binom{n}{r} n^r) - (s - t)\log n - \log\binom{n-s}{r-s} - (r - s)\log(n - 1)$.

Now we analyze what will happen if private randomness used in Juels-Sudan fuzzy vault is no longer perfect. Totally, private randomness is used in three steps: (1) selecting secret coefficient of p; (2) selecting x-coordinates x_i of chaff points; (3) selecting y-coordinates y_i of chaff points. We model this imperfect random source as a probability distribution on the universe, where each element in $\mathcal{U}$ is assigned with a probability value $w_i > 0$. Obviously, $\sum_{i=1}^{n} w_i = 1$. For convenience, we explicitly define $\mathcal{U} = \{x_1, x_2, \cdots, x_n\}$ and sort the elements in $\mathcal{U}$ according to their probabilities in descending order, i.e. $w_1 \geq w_2 \geq \cdots \geq w_n$. Thus, picking $r - s$ chaff point from $\mathcal{U}\backslash S$ can be viewed as drawing $r - s$ samples from $\mathcal{U}\backslash S$ without replacement using unequal probabilities.

In Lemma 5, we analysis the influence of imperfect randomness on selecting x-coordinates of the chaff points. The full proof is given in Appendix A.2. In Theorem 3, we analyze the whole entropy losses for fuzzy vault construction in the imperfect randomness setting. Our result shows that a $(\mathcal{U}, n, r, s)$-fuzzy vault-based secure sketch will additionally lose $(r - t)(\log n - a) + (r - s)\log \frac{n-r}{2^a - r}$ bits of entropy if it uses a private randomness from an imperfect $(\log(n), a)$-source.

Lemma 5 (Entropy Loss in Selecting x-coordinates X_i). *A $(\mathcal{U}, n, r, s)$-fuzzy vault will additionally lose at most $(r - s)\log \frac{n-r}{2^a - r}$ bits residual entropy if $2^a > r$ when the x-coordinates of the chaff points are no longer selected from a uniform distribution but from an imperfect $(\log(n), a)$-source.*

Theorem 3 (Entropy Loss of Fuzzy Vault in imperfect randomness). *The entropy loss of a $(\mathcal{U}, n, r, s)$-fuzzy vault with imperfect randomness from $(\log(n), a)$-source is at most $\log(\binom{n}{r} n^r) - (s - t)a - \log\binom{n-s}{r-s} + (r - s)\log \frac{n-r}{2^a - r} - (r - s) \cdot a$ bits if $2^a > r$.*

Proof. As mentioned before, the private randomness is independently used in three steps of selection for the fuzzy vault-based secure sketch. All the selections take randomness from universe $\mathcal{U}$.

Thus, the theorem follows easily from Lemma 4 and Lemma 5. □

4 Security Analysis of Existing Fuzzy Extractors with Imperfect Randomness

In this section, we will analyze the security deficiency (expansion of statistic distance between the output string R and U_ℓ) of fuzzy extractor in "sketch-then-extract" framework where only imperfect randomness is available. This imperfect random source is used in randomized secure sketch and seeded extractor.

The "sketch-then-extract" framework proposed by Dodis et al. [12] is the most common way to construct fuzzy extractors. It consists of an information-theoretic $(\mathcal{M}, m, \tilde{m}, t)$-secure sketch and an information-theoretic average-case $(n, \tilde{m}, \ell, \epsilon)$-strong seeded extractor, as shown in Fig. 3 with explicit private randomness x_1 and x_2. In the Gen phase, fuzzy extractor apply an extractor to w with seed x_1 to obtain R; apply a secure sketch to w with the explicit private randomness x_2 to obtain a sketch ss; store (x_1, ss) as help data P of the fuzzy extractor. In the Rep phase, fuzzy extractor apply Rec to ss with a close w' to recover original input w; apple extractor to the recovered w and x_1 in help data to reproduce R.

We can see from Fig. 3 that randomness is used in secure sketch and seeded extractor. There are two options about the relationship between two random

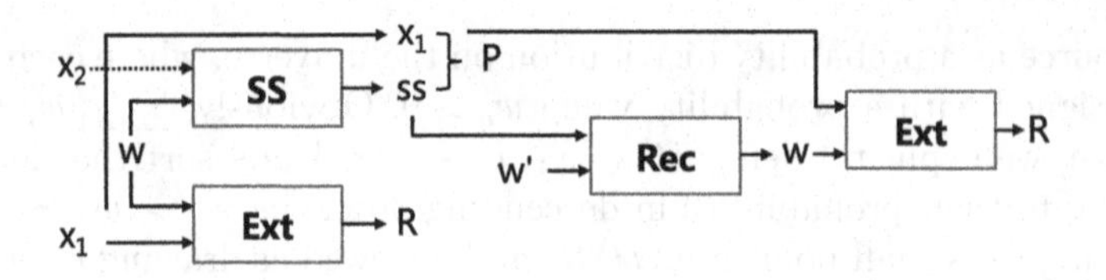

Fig. 3. Construction 4 [12]: sketch-then-extract fuzzy extractor with seeded extractor

sources used in sketch-then-extract construction: (1) X_1 and X_2 are sampled from a $(2, n, k)$-correlated source; (2) X_1 and X_2 are two independent sources. We focus on the prior one since two independent random sources is a special case of $(2, n, k)$-correlated source such that $\mathrm{H}_\infty(X_2|X_1) = \mathrm{H}_\infty(X_2) = k$. Thus, discussing sketch-then-extract construction when all randomness of the system come from $(2, n, k)$-correlated source is sufficient.

In this section we model imperfect source as a $(2, n, k)$-correlated source which remains certain amount of min-entropy after some samples. In Theorem 4, we illustrate how uniformity of output key changes in the imperfect randomness setting. Generally, the security deficiency is related to entropy deficiency of imperfect random source and extra entropy loss of secure sketch with imperfect randomness.

Theorem 4. *For any* $\delta > 0$*, assume* <SS,Rec> *is an* $(\mathcal{M}, m, \tilde{m}, t)$*-secure sketch,* Ext *is an* $(n, \tilde{m} - \delta, \ell, \epsilon)$*-strong seeded extractor, and Construction 4 is an* $(\mathcal{M}, m, \ell, t, \epsilon + 2^{-\delta})$*-fuzzy extractor in perfect randomness setting. Then, Construction 4 is also an* $(\mathcal{M}, m, \ell, t, 2^d\epsilon + 2^{-(\delta-\lambda)})$*-fuzzy extractor in the imperfect randomness setting where imperfect random source is a* $(2, n, n-d)$*-correlated source and* λ *is the extra entropy loss of secure sketch.*

Proof. In order to prove our result in the imperfect randomness setting, we firstly review the technique used in corollary 4.2 of [12], showing that secure sketch and strong seeded extractor make a fuzzy extractor in the perfect randomness setting. This proof relies on the fact of Lemma 1 that strong seeded extractor is also an average-case strong extractor.

Suppose we have (W, I) that $\tilde{\mathrm{H}}_\infty(W|I) \geq \tilde{m}$. Let $W_i = (W|I = i)$. We define the value i is "bad" if $\mathrm{H}_\infty(W_i) \leq \tilde{m} - \delta$, otherwise that i is "good". So we have:

$$\begin{aligned}\Pr[i \textit{ is bad}] &= \Pr[\mathrm{H}_\infty(W_i) \leq \tilde{m} - \delta] = \Pr[2^{-\mathrm{H}_\infty(W_i)} \geq 2^{-\tilde{m}+\delta}] \\ &= \Pr[\max_w \Pr[W = w|I = i] \geq 2^{-\tilde{m}+\delta}] \\ &\leq \frac{\mathbb{E}_i(\max_w \Pr[W = w|I = i])}{2^{-\tilde{m}+\delta}} \leq \frac{2^{-\tilde{m}}}{2^{-\tilde{m}+\delta}} = 2^{-\delta}\end{aligned}$$

The inequality in the above derivation holds by Markov inequality. For any "good" i, $(n, \tilde{m} - \delta, \ell, \epsilon)$-strong seeded extractor extracts ℓ bits that are ϵ-close to U_ℓ. Thus, we have:

$$\begin{aligned}
&\mathrm{SD}((\mathrm{Ext}(W,X),X,I),(U_\ell,X,I))\\
=&\sum_i \Pr[I=i]\cdot \mathrm{SD}((\mathrm{Ext}(W,X),X),(U_\ell,X))\\
\leq&\Pr[i\ is\ bad]\cdot 1+\sum_{i\ is\ good}\Pr[I=i]\cdot \mathrm{SD}((\mathrm{Ext}(W,X),X),(U_\ell,X))\\
\leq&2^{-\delta}+\epsilon
\end{aligned}$$

Taking I as sketch output $SS(W)$, we get that $(\mathcal{M},m,\tilde{m},t)$-secure sketch and $(n,\tilde{m}-\delta,\ell,\epsilon)$-strong seeded extractor in the perfect randomness setting construct a $(\mathcal{M},m,\ell,t,2^{-\delta}+\epsilon)$-fuzzy extractor.

Now, we consider the imperfect randomness setting that $\tilde{\mathrm{H}}_\infty^{imperfect}(W|I)\geq \tilde{m}-\lambda$.

We define the value i "bad in imperfect randomness" if $\mathrm{H}_\infty(W_i)\leq\tilde{m}-\delta$.

$$\begin{aligned}
&\Pr[i\ is\ bad\ in\ imperfect\ randomness]\\
\leq&\frac{\mathbb{E}_i^{imperfect}(\max_w \Pr[W=w|I=i])}{2^{-\tilde{m}+\delta}}\\
=&\frac{2^{-\tilde{m}-\lambda}}{2^{-\tilde{m}+\delta}}=2^{-(\delta-\lambda)}
\end{aligned}$$

Considering that the seed of the strong seeded extractor X_{im} is an $(n,n-d)$-source conditioned on the sample used in the sketch phrase, the statistic distance of R and U_ℓ in the imperfect randomness setting is $2^{-(\delta-\lambda)}+2^d\epsilon$.

$$\begin{aligned}
&\mathrm{SD}((\mathrm{Ext}(W,X_{im}),X_{im},I),(U_\ell,X_{im},I))\\
\leq&\Pr[i\ is\ bad\ in\ imperfect\ randomness]\cdot 1+\\
&\sum_{i\ is\ good}\Pr[I=i]\cdot \mathrm{SD}((\mathrm{Ext}(W,X_{im}),X_{im}),(U_\ell,X_{im}))\\
\leq&2^{-(\delta-\lambda)}+2^d\epsilon
\end{aligned}$$

The last inequality follows from Lemma 3. So we get that Construction 4 is also an $(\mathcal{M},m,\ell,t,2^d\epsilon+2^{-(\delta-\lambda)})$-fuzzy extractor in the imperfect randomness setting. □

5 Further Discussions on Fuzzy Extractors with Imperfect Randomness Based on Two-source Extractor

In order to accommodate the imperfect random source, Cui et al. [9] introduce two-source extractor to build a reusable and robust fuzzy extractor based on DDH/LPN assumptions in the CRS model. Note that in order to use two-source extractors to extract uniformly distributed key from imperfect private random source and the noisy source, one should make the assumption that imperfect private random source and the noisy source are independent. In this section, we simplify the fuzzy extractor construction in [9] by omitting its reusability and robustness functionalities, and discuss the information-theoretic security of this simplified construction. The simplified construction is similar to traditional "sketch-then-extract" construction we introduced in Sect. 4, the only difference is replacing the strong seeded extractor with a two-source extractor, as shown in Fig. 4. In Theorem 5, we present existence and security of this simplified construction in the imperfect randomness setting.

Theorem 5. *Assume* <SS,Rec> *is an* $(\mathcal{M}, m, \tilde{m}, t)$*-secure sketch in perfect randomness, it has an extra entropy loss* λ *in imperfect randomness. Let* TExt: $\{0,1\}^{n_1} \times \{0,1\}^{n_2} \rightarrow \{0,1\}^{\ell}$ *be an average-case* $((n_1, \tilde{m} - \lambda), (n_2, n_2 - d), \epsilon)$*-strong two-source extractor. Construction 5 is an* $(\mathcal{M}, m, \ell, t, \epsilon)$*-fuzzy extractor with imperfect randomness from* $(2, n_2, n_2 - d)$*-correlated source.*

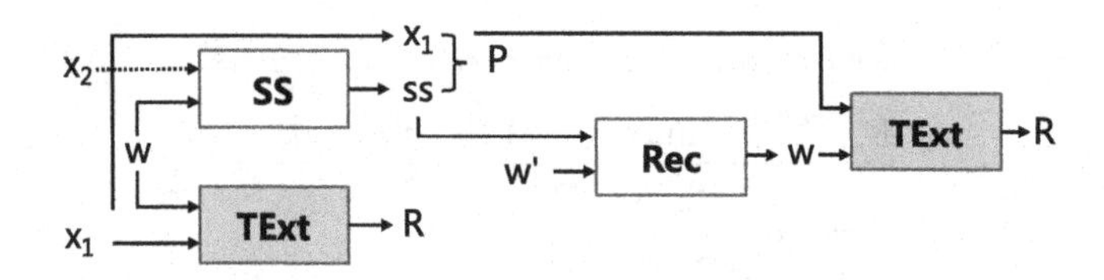

Fig. 4. Construction 5: sketch-then-extract fuzzy extractor with two-source extractor. We replace the strong seeded extractor with a two-source extractor shown as gray boxes.

Proof. Based on the security of secure sketch, we know that $\tilde{\mathrm{H}}_{\infty}(W|\mathrm{SS}(W)) \geq \tilde{m} - \lambda$ in imperfect randomness. The imperfect source X has residual min-entropy $n_2 - d$ after first sample used in sketch phase. By the definition of average-case two-source extractor, we have:

$$\mathrm{SD}((R, P), (U_\ell, P)) = \mathrm{SD}((\mathrm{TExt}(W, X), X, \mathrm{SS}(W)), (U_\ell, X, \mathrm{SS}(W))) \leq \epsilon$$

□

As regard to the efficiency of Construction 5, randomized secure sketches mentioned above work in polynomial-time, so the main goal is to find efficient constructions for two-source extractor.

In the following subsection, we instantiate our fuzzy extractor with two efficient two-source extractors: inner-product function based two-source extractor [8] for the length-consistent setting and Ramsey-Graph based [21] two-source extractor for length-inconsistent setting. Here, length-consistent means the length of noisy source is the same as length of imperfect random source. This is the case for code offset based secure sketch, we call it "length-consistent secure sketch". On the other hand, permutation-based secure sketch and fuzzy vault based secure sketch are called "length-inconsistent secure sketches".

5.1 Fuzzy Extractor Based on Length-Consistent Secure Sketch and Two-Source Extractor

For length-consistent secure sketch, such as code offset-based sketch, inner-product based two-source extractor [8] is suitable. The following theorem demonstrates the information-theoretic security of fuzzy extractor relying on inner-product function based two-source extractor and code offset-based secure sketch in the imperfect randomness setting. It reveals the security requirement on parameters $\epsilon, \tilde{m}, \delta, \ell, n$, and d.

Theorem 6. *For any $\delta > 0$, suppose that Construction 5 takes input from a $(\{0,1\}^n, m)$-noisy source W and uses a private $(2, \{0,1\}^n, n-d)$-correlated source X. In detail, Construction 5 uses a $(\{0,1\}^n, m, \tilde{m}, t)$-code offset-based secure sketch in the imperfect randomness setting, and a $((\{0,1\}^n, \tilde{m} - \delta), (\{0,1\}^n, n-d), \epsilon)$-explicit two-source extractor in* [11]. *Then Construction 5 is a $(\{0,1\}^n, m, \ell, t, \epsilon + 2^{-\delta})$-fuzzy extractor with $\log(\frac{1}{\epsilon}) = \frac{\tilde{m}-\delta-d+2-\ell}{2}$.*

Proof. The Theorem easily follows from definition of sketch-then-extract fuzzy extractor, correlated source, Lemma 2 and Theorem 1 in [11]. □

5.2 Fuzzy Extractor Based on Length-Inconsistent Secure Sketch and Two-Source Extractor

For length-inconsistent secure sketch, such as permutation-based and fuzzy vault-based secure sketch, Ramsey-Graph based [21] two-source extractor is suitable. The following theorem demonstrates the information-theoretic security of fuzzy extractor relying on Ramsey-Graph based two-source extractor and length-inconsistent secure sketch with imperfect private randomness. (Note that in following theorem, W and X can exchange their roles in two-source extractor).

Theorem 7. *For any $\delta > 0$, suppose that Construction 5 takes input from a $(\{0,1\}^{n_1}, m)$-noisy source W and uses a private $(2, \{0,1\}^{n_2}, n_2 - d)$-correlated source X. In detail, Construction 5 uses a $(\{0,1\}^{n_1}, m, \tilde{m}, t)$-secure sketch in the imperfect randomness setting, and a $((\{0,1\}^{n_1}, \tilde{m} - \delta), (\{0,1\}^{n_2}, n_2 - d), \epsilon)$-explicit two source extractor in* [21]. *Then, Construction 5 is a $(\{0,1\}^{n_1}, m, \ell, t, \epsilon + 2^{-\delta})$-fuzzy extractor if for any $0 < \alpha < \frac{1}{2}$ with $n_1 \geq 6\log n_1 + 2\log n_2$, $\tilde{m} - \delta \geq (\frac{1}{2} + \alpha)n_1 + 3\log n_1 + \log n_2$, $n_2 - d \geq 5\log(n_1 - \tilde{m} + \delta)$, $\ell \leq \alpha \min\{n_1/8, (n_2 - d)/40\} - 1$, $\epsilon = 2^{-1.5\ell}$.*

Proof. The theorem easily follows from definition of fuzzy extractor, correlated source, Lemma 2 and Theorem 2 in [21]. □

6 Conclusion

In this paper, we give an information-theoretic security analysis of current secure sketch and fuzzy extractor constructions where only imperfect randomness is available. Concretely, we analyze the extra entropy loss for three types of randomized secure sketch (code offset-based, permutation-based, fuzzy vault-based), then we analyze the security deficiency of "sketch-then-extract" fuzzy extractor in imperfect randomness.

Besides, we propose a simplified fuzzy extractor construction in the imperfect randomness setting and instantiate it with two explicit two-source extractors. Finally, we discuss the information-theoretic security of the construction.

A Appendix

A.1 Proof of Theorem 1

Proof. We directly calculate residual entropy of the secure sketch and define:

$$\begin{aligned}
&A_{ss} := \{a = (w,x) | w + x = ss\} \quad |A_{ss}| \leq |X| \\
&\Phi(w,x) := \Pr[X = x] \cdot \Pr[W = w] \\
&w_{ss} := \underset{w \in W}{argmax} \Pr[W = w | SS = ss] \\
&(\hat{w}_{ss}, \hat{x}_{ss}) := \underset{a \in A_{ss}}{argmax}(\Phi(a))
\end{aligned}$$

A_{ss} denotes the set of tuple (w,x) that can output sketch ss. $\Phi(w,x)$ denotes the joint probability of X and W. Since X and W are independent, $\Phi(w,x)$ is the multiplication of $\Pr[X = x]$ and $\Pr[W = w]$. w_{ss} denotes the $w \in W$ that make the heaviest conditional probability of $\Pr[W = w|SS = ss]$. $(\hat{w}_{ss}, \hat{x}_{ss})$ denotes a tuple of $(w,x) \in A_{ss}$ that output sketch ss and get the heaviest joint probability.

Here we can first fix $\mathcal{F} = \{0,1\}$ for the most common case, using error correcting code of $[n,k,2t+1]_2$ and then extend to $\mathcal{F}_q$.

By definition of residual entropy, we have:

$$\begin{aligned}
2^{-\tilde{H}_\infty(W|\mathrm{SS}(W))} &= \underset{ss}{\mathbb{E}} \max_{w} \Pr[W = w | SS = ss] \\
&= \sum_{ss} \Pr[SS = ss] \max_{w} \Pr[W = w | SS = ss] \\
&= \sum_{ss} \Pr[SS = ss] \Pr[W = w_{ss} | SS = ss] \\
&= \sum_{ss} \Pr[W = w_{ss}, SS = ss] \\
&= \sum_{ss} \Pr[W = w_{ss}, X = x_{ss}] \\
&\leq \sum_{ss} \Pr[W = \hat{w}_{ss}] \cdot \Pr[X = \hat{x}_{ss}] \\
&\leq 2^{-(k-d)} \cdot \sum_{ss} \Pr[W = \hat{w}_{ss}] \\
&\leq 2^{-(k-d)} \cdot 2^{-m} \cdot 2^{n}
\end{aligned}$$

The first inequality in the above derivation holds since the definition of $(\hat{w}_{ss}, \hat{x}_{ss})$. The second inequality in the above derivation holds since taking the maximum probability of randomness X. The third holds since taking the maximum probability of input variable W and $|SS| = 2^n$.

When randomness X in code offset construction is imperfectly random, we have $\tilde{\mathrm{H}}_\infty(W|\mathrm{SS}(W)) \geq m + k - n - d$, losing d bits of entropy compared to perfect randomness. When alphabet become $\mathcal{F}_q$, from the definition of entropy deficiency, we have $\max_x \Pr[X = x] \geq 2^{-(k \log |\mathcal{F}| - d)} = |\mathcal{F}|^{-k} \cdot 2^d$, which also leads to d bits additional loss in residual entropy. □

A.2 Proof of Lemma 5

Proof. Let $X = (X_1, X_2, \cdots, X_{r-s})$ be the set of x-coordinates that are selected out with size $r - s$, and in order X_1 to X_{r-s}. P_1 denotes the probabilities of a certain selection from imperfect random source, P^* denotes the probabilities of a certain selection from perfect random source. We use $\hat{w}$ to denote probability of set $\mathcal{U} \backslash S$ such that $\hat{w} = \sum_{x_i \in \mathcal{U} \backslash S} w_i$. We denote $q = \frac{\hat{w}}{w_1}$.

It is obvious that P_1 gets its maximum when the selection contain $r - s$ most heavy x-coordinates points. So we get:

$$\begin{aligned}\max P_1 &= \Pr[X = (x_1, x_2, \cdots, x_{r-s})] \leq \Pr[X = (x_1, x_1, \cdots, x_1)] \\ &= (r-s)! \cdot \frac{w_1}{\hat{w} - w_1} \cdot \frac{w_1}{\hat{w} - 2w_1} \cdots \frac{w_1}{\hat{w} - (r-s)w_1} \\ &= (r-s)! \cdot \frac{1}{(q-1)(q-2)\cdots(q-r+s)}\end{aligned}$$

Note that $(r - s)!$ means elements in set X is unordered.

Since all points are selected from $(\log(n), a)$-source, and suppose max-entropy of the imperfect random source is b, we have:

$$2^{-b} \leq w_i \leq 2^{-a} \quad w_1 = 2^{-a} \geq \frac{1}{n}$$

Equal when a=b, which means a perfect randomness.

Substituting into q, we get:

$$\frac{1 - s \cdot 2^{-b}}{\frac{1}{n}} \geq q \geq \frac{1 - s \cdot 2^{-a}}{2^{-a}} = 2^a - s > r - s$$

Now we discuss P^* in the perfect randomness case and have:

$$P^* = (r-s)! \cdot \frac{1}{(n-s-1)(n-s-2)\cdots(n-s-r+s)}$$

The extra entropy loss Δ is:

$$\begin{aligned}\Delta &= -\log(P^*) - (-\log(P_1)) \leq \log \frac{\max P_1}{P^*} \\ &= \log \frac{(n-s-1)(n-s-2)\cdots(n-s-r+s)}{(q-1)(q-2)\cdots(q-r+s)}\end{aligned}$$

We focus on the case when $\Delta \geq 0$, so $n - s \geq q$:

$$\Delta \leq \log(\frac{n-r}{q-r+s})^{r-s} \leq (r-s)\log\frac{n-r}{2^a-s-r+s}$$
$$= (r-s)\log\frac{n-r}{2^a-r}$$

□

References

1. Apon, D., Cho, C., Eldefrawy, K., Katz, J.: Efficient, reusable fuzzy extractors from LWE. In: Dolev, S., Lodha, S. (eds.) CSCML 2017. LNCS, vol. 10332, pp. 1–18. Springer, Cham (2017). https://doi.org/10.1007/978-3-319-60080-2_1
2. Backes, M., Kate, A., Meiser, S., Ruffing, T.: Secrecy without perfect randomness: cryptography with (bounded) weak sources. In: Malkin, T., Kolesnikov, V., Lewko, A.B., Polychronakis, M. (eds.) ACNS 2015. LNCS, vol. 9092, pp. 675–695. Springer, Cham (2015). https://doi.org/10.1007/978-3-319-28166-7_33
3. Bennett, C.H., Shor, P.W.: Quantum information theory. IEEE Trans. Inf. Theory **44**(6), 2724–2742 (1998). https://doi.org/10.1109/18.720553
4. Blanton, M., Aliasgari, M.: Analysis of reusability of secure sketches and fuzzy extractors. IEEE Trans. Inf. Forensics Secur. **8**(9), 1433–1445 (2013). https://doi.org/10.1109/TIFS.2013.2272786
5. Boyen, X.: Reusable cryptographic fuzzy extractors. In: Atluri, V., Pfitzmann, B., McDaniel, P.D. (eds.) Proceedings of the 11th ACM Conference on Computer and Communications Security, CCS 2004, Washington, DC, USA, 25–29 October 2004, pp. 82–91. ACM (2004). https://doi.org/10.1145/1030083.1030096
6. Boyen, X., Dodis, Y., Katz, J., Ostrovsky, R., Smith, A.: Secure remote authentication using biometric data. In: Cramer, R. (ed.) EUROCRYPT 2005. LNCS, vol. 3494, pp. 147–163. Springer, Heidelberg (2005). https://doi.org/10.1007/11426639_9
7. Chattopadhyay, E., et al.: Explicit two-source extractors and more. Ph.D. thesis (2016)
8. Chor, B., Goldreich, O.: Unbiased bits from sources of weak randomness and probabilistic communication complexity. SIAM J. Comput. **17**(2), 230–261 (1988). https://doi.org/10.1137/0217015
9. Cui, N., Liu, S., Gu, D., Weng, J.: Robustly reusable fuzzy extractor with imperfect randomness. Des. Codes Crypt. **89**(5), 1017–1059 (2021). https://doi.org/10.1007/s10623-021-00843-1
10. Daugman, J.: How iris recognition works. IEEE Trans. Circuits Syst. Video Technol. **14**(1), 21–30 (2004). https://doi.org/10.1109/TCSVT.2003.818350
11. Dodis, Y., Elbaz, A., Oliveira, R., Raz, R.: Improved randomness extraction from two independent sources. In: Jansen, K., Khanna, S., Rolim, J.D.P., Ron, D. (eds.) APPROX/RANDOM 2004. LNCS, vol. 3122, pp. 334–344. Springer, Heidelberg (2004). https://doi.org/10.1007/978-3-540-27821-4_30
12. Dodis, Y., Ostrovsky, R., Reyzin, L., Smith, A.D.: Fuzzy extractors: how to generate strong keys from biometrics and other noisy data. SIAM J. Comput. **38**(1), 97–139 (2008). https://doi.org/10.1137/060651380

13. Dodis, Y., Yu, Yu.: Overcoming weak expectations. In: Sahai, A. (ed.) TCC 2013. LNCS, vol. 7785, pp. 1–22. Springer, Heidelberg (2013). https://doi.org/10.1007/978-3-642-36594-2_1
14. Fuller, B., Meng, X., Reyzin, L.: Computational fuzzy extractors. In: Sako, K., Sarkar, P. (eds.) ASIACRYPT 2013, Part I. LNCS, vol. 8269, pp. 174–193. Springer, Heidelberg (2013). https://doi.org/10.1007/978-3-642-42033-7_10
15. Håstad, J., Impagliazzo, R., Levin, L.A., Luby, M.: A pseudorandom generator from any one-way function. SIAM J. Comput. **28**(4), 1364–1396 (1999). https://doi.org/10.1137/S0097539793244708
16. Herder, C., Yu, M.M., Koushanfar, F., Devadas, S.: Physical unclonable functions and applications: a tutorial. Proc. IEEE **102**(8), 1126–1141 (2014). https://doi.org/10.1109/JPROC.2014.2320516
17. Juels, A., Sudan, M.: A fuzzy vault scheme. Des. Codes Cryptogr. **38**(2), 237–257 (2006). https://doi.org/10.1007/s10623-005-6343-z
18. Li, X.: Non-malleable extractors, two-source extractors and privacy amplification. In: 53rd Annual IEEE Symposium on Foundations of Computer Science, FOCS 2012, New Brunswick, NJ, USA, 20–23 October 2012, pp. 688–697. IEEE Computer Society (2012). https://doi.org/10.1109/FOCS.2012.26
19. Marasco, E., Ross, A.: A survey on antispoofing schemes for fingerprint recognition systems. ACM Comput. Surv. **47**(2), 28:1–28:36 (2014). https://doi.org/10.1145/2617756
20. Nisan, N., Zuckerman, D.: Randomness is linear in space. J. Comput. Syst. Sci. **52**(1), 43–52 (1996). https://doi.org/10.1006/jcss.1996.0004
21. Raz, R.: Extractors with weak random seeds. In: Gabow, H.N., Fagin, R. (eds.) Proceedings of the 37th Annual ACM Symposium on Theory of Computing, Baltimore, MD, USA, 22–24 May 2005, pp. 11–20. ACM (2005). https://doi.org/10.1145/1060590.1060593
22. Suh, G.E., Devadas, S.: Physical unclonable functions for device authentication and secret key generation. In: Proceedings of the 44th Design Automation Conference, DAC 2007, San Diego, CA, USA, 4–8 June 2007, pp. 9–14. IEEE (2007). https://doi.org/10.1145/1278480.1278484
23. Vadhan, S.P.: Pseudorandomness. Found. Trends Theor. Comput. Sci. **7**(1–3), 1–336 (2012). https://doi.org/10.1561/0400000010
24. Wen, Y., Liu, S.: Robustly reusable fuzzy extractor from standard assumptions. In: Peyrin, T., Galbraith, S. (eds.) ASIACRYPT 2018, Part III. LNCS, vol. 11274, pp. 459–489. Springer, Cham (2018). https://doi.org/10.1007/978-3-030-03332-3_17
25. Wen, Y., Liu, S., Han, S.: Reusable fuzzy extractor from the decisional Diffie–Hellman assumption. Des. Codes Crypt. **86**(11), 2495–2512 (2018). https://doi.org/10.1007/s10623-018-0459-4
26. Yao, Y., Li, Z.: Overcoming weak expectations via the Réenyi entropy and the expanded computational entropy. In: Padró, C. (ed.) ICITS 2013. LNCS, vol. 8317, pp. 162–178. Springer, Cham (2014). https://doi.org/10.1007/978-3-319-04268-8_10

Tight Analysis of Decryption Failure Probability of Kyber in Reality

Boyue Fang[1], Weize Wang[2], and Yunlei Zhao[1](✉)

[1] Department of Computer Science, Fudan University, Shanghai 200082, China
ylzhao@fudan.edu.cn

[2] Department of Mathematics, Sun Yat-sen University, Guangzhou 510275, China

Abstract. Kyber is a candidate in the third round of the National Institute of Standards and Technology (NIST) Post-Quantum Cryptography (PQC) Standardization. However, because of the protocol's independence assumption, the bound on the decapsulation failure probability resulting from the original analysis is not tight. In this work, we give a rigorous mathematical analysis of the actual failure probability calculation, and provides the Kyber security estimation in reality rather than only in a statistical sense. Our analysis does not make independency assumptions on errors, and is with respect to concrete public keys in reality. Through sample test and experiments, we also illustrate the difference between the actual failure probability and the result given in the proposal of Kyber. The experiments show that, for Kyber-512 and 768, the failure probability resulting from the original paper is relatively conservative, but for Kyber-1024, the failure probability of some public keys is worse than claimed. This failure probability calculation for concrete public keys can also guide the selection of public keys in the actual application scenarios. What's more, we measure the gap between the upper bound of the failure probability and the actual failure probability, then give a tight estimate. Our work can also re-evaluate the traditional $1 - \delta$ correctness in the literature, which will help re-evaluate some candidates' security in NIST post-quantum cryptographic standardization.

Keywords: Post-quantum cryptography · Learning with errors · Key encapsulation mechanism · Decryption failure

1 Introduction

Cryptographic systems based on learning with errors (LWE) and related problems are the central topics of recent cryptographic research. Factorization and discrete logarithm problems have always been the basis of modern cryptography, but due to the development of quantum computing, cryptographic schemes based on these problems are no longer secure in the post-quantum era. Lattice-based cryptography makes it possible to implement a rich set of cryptographic primitives, including key exchange, key encapsulation, encryption, and digital signatures, and more advanced structures such as fully homomorphic encryption.

C. Alcaraz et al. (Eds.): ICICS 2022, LNCS 13407, pp. 148–160, 2022.
https://doi.org/10.1007/978-3-031-15777-6_9

Therefore, post-quantum cryptography (PQC) should be developed to avoid security problems in future systems to replace the existing public-key algorithms. The National Institute of Standards and Technology (NIST) is running a PQC standardization project. One type of candidate is designed based on learning with errors (LWE) and related problems such as Module-LWE. Unlike traditional public-key schemes, LWE-based schemes have the possibility of decryption failures.

In 2020, the third round of the NIST PQC project began, and only 15 candidates remained. Seven of them are finalists, and eight alternative algorithms also moved to the third round of the process. Kyber is a promising candidate for the key encapsulation mechanism (KEM), which is based on module-LWE (MLWE). For the recommended parameter sets, the failures of the Kyber decryption procedure are pretty small. For Kyber-512/768/1024 according to the NIST security categories I, III and V, the decryption failures claimed are about $2^{-139}, 2^{-164}, 2^{-174}$ respectively [2]. These upper bounds are low enough to discourage reaction attacks. However, in the current estimation technology of decryption failure probability for KEM schemes based on LWE and its variants, it assumes the failure independence in individual bits of the transmitted message. It then calculates the overall failure probability of the scheme. However, it is difficult to estimate the gap between this assumption and the actual situation, and it may cause unpredictable consequences when applying this type of encryption scheme. Therefore, this paper considers the upper bound of the actual decryption failure probability of this type of encryption scheme. This paper takes the Kyber scheme as an example to provide an analysis more in line with the actual situation. The failure probability estimation method proposed in this paper will also impact the effect of failure-boosting technology based on the independence of failure probability assumption.

Contribution. Our main contribution is to give a rigorous mathematical analysis of the actual failure probability for Kyber in reality, and then discuss the traditional $1-\delta$ correctness analysis. Our analysis of the Kyber decryption failure probability mainly focuses on the impact of the non-independence of the random vectors, and shows that the independence assumption in the traditional $1-\delta$ correctness analysis may affect the security of the encryption schemes in reality. This impact is not only related to Kyber. In the failure probability analysis, we need to consider the probability of $||e_1 s_1 + e_2 s_2 + e_3 ||_\infty < t$, where e_1, s_1, e_2, s_2, e_3 obey certain distributions, and t is some threshold. This formula is the cornerstone of error rate analysis. The rigorous mathematical analysis of this basic problem is worthy of in-depth study, which also greatly eliminates the gap between theoretical error rate and the error rate in actual applications for KEM schemes based on LWE and its variants. For the samples we selected, the difference between the upper and lower bounds in the power of 2 is usually less than 40, which means that the upper bound of the error given in this work can approximately represent its actual error rate.

Through the analysis method proposed in this paper, the failure probability of Kyber512 and Kyber768 can be considered as an overestimation of the actual situation, while Kyber1024 has a certain underestimation. The number of public keys that make the corresponding failure probability higher than the probability claimed in [2] is not negligible, which means that the security level of Kyber-1024 under certain public keys will be reduced after using technologies such as directional failure boosting.

Our analysis method can estimate the gap between the theoretical error rate and the actual error rate, quantitatively analyzes the error rate of a given public key, and provides the Kyber security estimation in actual applications rather than only in a statistical sense. Our work can also re-evaluate the traditional $1-\delta$ correctness in the literature, which will help re-evaluate some candidates' security in the third round of NIST post-quantum cryptographic standardization.

Related Work. D'Anvers et al. rejected the assumption that the failure independence in individual bits of the transmitted message theoretically and practically [7]. They provided a method to estimate the probability of decryption failure, taking the correlation of bit failures into account. Therefore, Kyber, as a KEM scheme based on the MLWE problem, the deviation between its actual decryption failure rate and the theoretical decryption failure rate given in the NIST proposal is also worthy of attention. This paper proposes an estimation method to calculate the tight failure probability upper bound of Kyber, which does not make the assumption of independence of errors. It effectively avoids the problem that the gap (between the theoretical failure probability and the actual failure probability) caused by the independence assumption is difficult to measure.

The impact of the failure probability on the encryption scheme is also reflected in the security of it. Guo et al. proposed a key recovery attack against LWE-based KEM schemes that use error correction codes to lower error probabilities. When their method is applied to LAC256-v2, the pre-computation complexity is 2^{-171}, and the success probability is 2^{-64} [8]. Bindel et al. showed that the adversary could use the first successful decryption information to increase the probability of getting the subsequent successful decryption. They also re-evaluated some candidates' security for the NIST PQC standardization [1]. When the side information about decryption failure is available, Dachman-Soled et al. proposed a cryptanalysis framework for lattice-based schemes [4]. This framework summarizes the primitive reduction attack, and allows for the gradual integration of prompts before running the final reduction step. This technique includes the sparsity of the grid, projection onto the hyperplane, and the distribution of the vector corresponding to the secret key that intersects the hyperplane. Their main contribution is to propose a toolbox and a method that can integrate this information into grid reduction attacks and can use side information to predict these grid reduction attacks' performance. They provided several end-to-end applications, such as the improvement of Frodo's single-track attack proposed by Bos et al. [3]. In particular, even with little side information, this study can also perform security loss estimation, bringing a smooth calculation trade-off for side-channel attacks. D'Anvers et al. studied the effect of decryption failure on the security of lattice-based encryption schemes, and attacked some

NIST candidate encryption schemes [6]. The results show that the attack will significantly reduce the security of the lattice-based encryption schemes with a relatively high failure rate. After applying their model to some NIST candidate cryptographic schemes, they believe that the actual security level is lower than their declared security level. Therefore, it is essential to give a failure rate analysis method that the deviation can be estimated and the theoretical bound is tight. Especially in the actual application scenarios, the public key is fixed once and for all rather than randomly selected each time, which means the failure rate in the sense of mathematical expectation cannot precisely indicate the error rate of a given public key. The failure probability estimation method proposed in this paper will provide the upper bound of the actual error rate, and conducts simulation experiments for concrete public keys. The analysis of the number of public keys corresponding to different error rates will be able to provide guidance for the public key selection.

Besides, the failure-boosting attack showed that the first decryption failure requires special attention. For example, D'Anvers et al. expanded their technology proposed in 2019 [6] and called it the "directional failure boosting" technology [5], which can speed up the search for the next decryption error. They also made an in-depth discussion on the quotient ring of the polynomial ring modulus $\langle x^N + 1\rangle$ over the finite field, and used the Kyber/Saber schemes based on module lattices to test the technology. They showed that after the decryption fails once, it can speed up the finding of the subsequent failed decryption. They proved that for such a single-target key model, the cryptographic algorithm design needs to make the first decryption failure difficult, while for the multi-target key model, the attack method is more effective. We noticed that the error analysis of D'Anvers et al. [5] is based on the analysis results given in the Kyber proposal where the independence of errors is assumed.

2 Preliminaries

2.1 Kyber

The complete description of CRYSTALS-Kyber can be found in [1]. Here we mainly focus on the failure probability analysis part of it. The system parameters are a ring R, positive integer k, d_t, d_u, d_v, and $n = 256$. The ciphertexts are of the form $(\mathbf{u}, v) \in \{0,1\}^{256 \cdot kd_u} \times \{0,1\}^{256 \cdot d_v}$. The public-key encryption scheme Kyber.CPA = KeyGen, Enc, Dec as described in Algorithms 1, 2 and 3.

Algorithm 1. Kyber.CPA.KeyGen(): key generation

1: $\rho, \sigma \leftarrow \{0,1\}^{256}$
2: $\mathbf{A} \sim R_q^{k\times k} := Sam(\rho)$
3: $(s, e) \sim \beta_\eta^k \times \beta_\eta^k := Sam(\rho)$
4: $\mathbf{t} := Compress_q(\mathbf{As} + \mathbf{e}, d_t)$
5: **return** $(pk := (\mathbf{t}, \rho), sk := s)$

Algorithm 2. Kyber.CPA.Enc($pk = (\mathbf{t}, \rho), m \in \mathcal{M}$): encryption

1: $r \leftarrow \{0,1\}^{256}$
2: $t := Decompress_q(\mathbf{t}, d_t)$
3: $\mathbf{A} \sim R_q^{k\times k} := Sam(\rho)$
4: $(\mathbf{r}, \mathbf{e}_1, e_2) \sim \beta_\eta^k \times \beta_\eta^k \times \beta_\eta := Sam(\tau)$
5: $\mathbf{u} := Compress_q(\mathbf{A}^\mathbf{T}\mathbf{r} + \mathbf{e}_1, d_u)$
6: $v := Compress_q(\mathbf{t}^\mathbf{T}\mathbf{r} + e_2 + \lfloor \frac{q}{2} \rceil \cdot m, d_v)$
7: **return** $c := (\mathbf{u}, v)$

Algorithm 3. Kyber.CPA.Dec($sk = s, c = (\mathbf{u}, v)$): decryption

1: $u := Decompress_q(\mathbf{u}, d_u)$
2: $v := Decompress_q(v, d_v)$
3: **return** $Compress_q(v - s^T\mathbf{u}, 1)$

The compression and decompression function are defined as:

$$Compress_q(x, d) = \lfloor \frac{2^d}{q} \cdot x \rceil \mod {}^{+}2^d,$$

$$Decompress_q(x, d) = \lfloor \frac{q}{2^d} \cdot x \rceil.$$

2.2 Distributions on R

Notation. For a finite set S, $|S|$ denotes its cardinality, and we write $s \leftarrow S$ to say that s is sampled uniformly from S. Denote with $\mathbb{Z}_q$ the ring of integers modulo q, represented in $(-\frac{q}{2}, \frac{q}{2}]$. Let R_q be the ring $\mathbb{Z}_q(X)/(X^N + 1)$, with N a power of two. For a vector V (or matrix A), we denote by v^T (or A^T) its transpose.

Denote with $\langle.,.\rangle$ the Eulidean inner product, and with $\lfloor x \rceil$ the nearest integer function. Let $|\cdot|$ denote taking the absolute value. These notations can be naturally extended to vectors, matrices and polynomials element wise. For an element $x \in \mathbb{Z}_q$, we write $||x||_\infty$ to mean $|x \mod \pm q|$ and $||x||_2$ to mean $|x|$. Elements of $R = \mathbb{Z}[x]/(x^n + 1)$ can be viewed as vectors in $\mathbb{R}^n$ by identifying the power basis $\{1, x, x^2, \ldots, x^{n-1}\}$ of R as an orthonormal basis of $\mathbb{R}^n$, so for $\mathbf{x} = (x_0, \ldots, x_{n-1})$, we define l_∞ norm and l_2 norm as following:

$$||\mathbf{x}||_\infty = \max_i ||x_i||_\infty$$

$$||\mathbf{x}||_2 = \sqrt{\sum_{i=0}^{n-1} ||x_i||_2}.$$

Denote with $\mathbb{P}[E]$ the probability of an event E, with $\mathbb{E}[\varepsilon]$ the expectation of the random variable ε.

The centered binomial distribution B_η are defined as follows:

Sample $\{(a_i, b_i)\}_{i=1}^{\eta} \leftarrow (\{0,1\}^2)^\eta$ and output $\sum_{i=1}^{\eta}(a_i - b_i)$.

If v is an element of R, we write $v \leftarrow \beta_\eta$ to mean that $v \in R$ is generated from a distribution where each of its coefficients is generated according to B_η. Similarly, a k-dimensional vector of polynomials $v \in R^k$ can be generated according to the distribution β_η^k.

3 Analysis of Decryption Failure Probability

Decryption failure refers to an event in which the correct ciphertext cannot be successfully restored during decryption after the decryption steps described in the algorithm are performed. The probability of decryption failure usually depends on the functions of the secret terms, denoted as s_1, s_2, e_1, e_2, e_3. Take Kyber as an example, when

$$||\langle e + c_t, r\rangle - \langle s, e_1 + c_u\rangle + e_2 + c_v||_\infty \leq B = \lfloor \frac{q}{4} \rceil,$$

the ciphertext can be decrypted successfully. And in the theorem of $1-\delta$ correctness,

$$||\langle e + c_t, r\rangle - \langle s, e_1 + c_u\rangle + e_2 + c_v||_\infty \geq B = \lfloor \frac{q}{4} \rceil$$

is usually defined as the error rate of decryption failure. In general, the error rate analysis mainly discusses the probability that

$$||\langle e^{'}, s^{''}\rangle + \langle e^{''}, s^{'}\rangle + e^{'''}|| \leq B,$$

where $e^{'}, e^{''}, e^{'''}, s^{'}, s^{''}$ obey a certain distribution, B is the threshold.

In the work of Kyber et al. [2], after expressing the failure probability problem in the above form, they adopted the independence assumption, that is, $e^{'}, e^{''}, e^{'''}, s^{'}, s^{''}$ are regarded as independent distributions for error rate calculation. However, it will bring deviations that are difficult to evaluate in the error rate calculation. Here we give a simple example to explain. It does not mean that the central binomial distribution and parameters used by Kyber are consistent with the parameters' distribution in the example. This example is to show that the independence assumption will bring massive deviations. Assume $\varepsilon_1, \varepsilon_2, \varepsilon_3$ are independent, $\varepsilon_1, \varepsilon_2$ obeys the Gaussian distribution $\varepsilon_1, \varepsilon_2 \sim \mathcal{N}(0,4)$, then define $c_1 = \frac{\varepsilon_1+\varepsilon_2}{2}, c_2 = \frac{\varepsilon_2-\varepsilon_1}{2}$, so c_1, c_2 also obey the Gaussian distribution $c_1, c_2 \sim \mathcal{N}(0,4)$. Consider the probability of

$$||\varepsilon_1 c_1 + \varepsilon_2 c_2 + \varepsilon_3|| \leq B,$$

Since we are mainly concerned with the influence of the independence assumption here, we might as well set ε_3 to 0. At this time, we consider the difference in probability calculated before and after adopting the assumption of independence. When the independence assumption is not taken, the probability is

$$\mathbb{P}(||\varepsilon_1^2 + \varepsilon_2^2|| \leq 2B),$$

and $\varepsilon_1^2 + \varepsilon_2^2 \sim \chi^2(2)$, where $\chi^2(2)$ is the chi-square distribution with 2 degrees of freedom. The probability density function is

$$f_1(x) = \frac{1}{2}e^{-\frac{x}{2}}, x \geq 0.$$

Then consider to adopt the independence assumption, the distribution can be regarded as the sum of two independent and identically distributed random variables, each of which is the product of two normally distributed variables. The probability density function of the product is $f(u) = \frac{B(0, \frac{\sqrt{u^2}}{4})}{4\pi}$, where B is the second kind of modified Bessel function, $B(0, x) = \int_0^\infty cos(x \sinh t)dt$. In this case, the probability density function is

$$f_2(x) = \int_{\mathbb{R}} f(x-u)f(u)du = \frac{1}{16\pi^2}\int_{\mathbb{R}}(\int_0^\infty cos((x-u)\sinh t)dt \int_0^\infty cos(u \sinh t)dt)du.$$

By comparing $f_1(x)$ and $f_2(x)$, it can be seen that the distribution of the corresponding random variables before and after the independence assumption is very different. Therefore, it is very important to consider the distribution of the real situation.

3.1 Decryption Failures

The Kyber key generation procedure involves ring elements s, e and matrix A. Key encapsulation involves ring elements r, e_1, e_2. The condition that decryption failure probability is less than δ is the following formula holds:

$$\mathbb{P}(||\langle e + c_t, r\rangle - \langle s, e_1 + c_u\rangle + e_2 + c_v||_\infty \geq B) \leq \delta,$$

where

$$c_u = \lfloor \frac{q}{2^{d_u}}(\lfloor \frac{2^{d_u}}{q}(A^T r + e_1)\rceil) \mod {}^+ 2^{d_u}\rceil - (A^T r + e_1)$$

$$c_t = \lfloor \frac{q}{2^{d_t}}(\lfloor \frac{2^{d_t}}{q}(As + e)\rceil) \mod {}^+ 2^{d_t}\rceil - (As + e)$$

$$c_v = \lfloor \frac{q}{2^{d_v}}(\lfloor \frac{2^{d_v}}{q}(t^T r + e_2 + \lfloor \frac{q}{2}\rceil m)\rceil) \mod {}^+ 2^{d_v}\rceil - (t^T r + e_2 + \lfloor \frac{q}{2}\rceil m)$$

and $t = As + e + c_t$. The distribution of s, e, r, e_1, e_2, A is introduced in Sect. 3.1.

3.2 Formula Derivation

Define $a_i = \frac{2^{d_i}}{q}$, $f(i,x) = x - \lfloor \frac{1}{a_i} \lfloor a_i x \rceil \rceil$.

Then

$$||\langle e + c_t, r\rangle - \langle s, e_1 + c_u\rangle + e_2 + c_v||_\infty \leq B = \lfloor \frac{q}{4} \rceil$$

can be written as

$$||\langle e, r\rangle - \langle \lfloor \frac{1}{a_t} \lfloor a_t(As+e)\rceil\rceil, r\rangle - \langle s, e_1\rangle + \langle s, \lfloor \frac{1}{a_u} \lfloor a_u(A^T r + e_1)\rceil\rceil\rangle$$
$$+ e_2 - \lfloor \frac{1}{a_v} (\lfloor a_v((As + e + c_t)^T r + e_2)\rceil\rceil ||_\infty \leq B$$

Which is equal to

$$||\langle e, r\rangle - \langle As + e, r\rangle + \langle f(t, As+e), r\rangle - \langle s, e_1\rangle + \langle s, A^T r + e_1\rangle$$
$$- \langle s, f(u, A^T r + e_1)\rangle + e_2 - \langle f(t, As+e), r\rangle - e_2 + f(v, \langle f(t, As+e), r\rangle + e_2)||_\infty \leq B$$

It's equal to

$$||f(v, \langle f(t, As+e), r\rangle + e_2) - \langle f(u, A^T r + e_1), s\rangle||_\infty \leq B.$$

We found that

$$||f(i,x)||_\infty \leq \frac{1}{2a_i} + \frac{1}{2},$$

Consider the triangular inequality of the norm, a sufficient condition is

$$||\langle f(u, A^T r + e_1), s\rangle||_\infty \leq B' = \lfloor \frac{q}{4} \rceil - \frac{1}{2a_v} - \frac{1}{2}.$$

Now we discuss about $||\langle f(u, A^T r + e_1), s\rangle||_\infty$.

Firstly,

$$A \leftarrow R_q^{k\times k}, r \leftarrow \beta_\eta^k, e_1 \leftarrow \beta_\eta^k, s \leftarrow \beta_\eta^k$$

If we write A, r, e_1, s as $A = (a^{(ij)})_{1\leq i,j\leq k}, e_1 = (e^{(i)})_{i=1}^k, r_1 = (r^{(i)})_{i=1}^k, s_1 = (s^{(i)})_{i=1}^k$, Then the formula can be written as

$$||\sum_{j=1}^{k} (s_j(\sum_{i=1}^{k} a^{(ji)} r^{(i)} + e^{(j)} - \lfloor \frac{1}{a_u} \lfloor a_u(\sum_{i=1}^{k} a^{(ji)} r^{(i)} + e^{(j)})\rceil\rceil))||_\infty \leq B',$$

where $a^{(ji)} = \sum_{m=0}^{n-1} a_m^{(ji)} x^m, r^{(i)} = \sum_{m=0}^{n-1} r_m^{(i)} x^m, e^{(j)} = \sum_{m=0}^{n-1} e_m^{(j)} x^m$.

Consider to define

$$a^{(ji)} r^{(i)} = \sum_{m=0}^{n-1} b_m x^m$$

Then

$$b_{n-1} = \sum_{k=0}^{n-1} a_k^{(ji)} r_{n-1-k}^{(i)}$$

and for $m < n-1$,

$$b_m = \sum_{k=0}^{m} a_k^{(ji)} r_{m-k}^{(i)} - \sum_{k=m+1}^{n-1} a_k^{(ji)} r_{m+n-k}^{(i)}$$

3.3 The Deviation Between the Theoretical Failure Probability and the Actual Failure Probability

Consider the following theorem in Kyber [2].

Theorem 1. Let k be a positive integer parameter. Let s, e, r, e_1, e_2 be random variables that have the same distribution. Also, let $c_t \leftarrow \varphi_{d_t}^k, c_u \leftarrow \varphi_{d_u}^k, c_v \leftarrow \varphi_{d_v}$ be distributed according to the distribution φ defined as follows:

Let φ_d^k be the following distribution over R:

- Choose uniformly-random $y \leftarrow R^k$
- **return** $(y - Decompress_q(Compress_q(y, d), d)) \mod {}^{\pm}q$.

Denote

$$\delta = Pr[||\langle e, r\rangle + e_2 + c_v - \langle s, e_1\rangle + \langle c_t, r\rangle - \langle s, c_u\rangle||_\infty \geq \lfloor \frac{q}{4} \rceil].$$

Then Kyber. CPA is $(1-\delta)$-correct.
Review our previous analysis, i.e.

$$||f(v, \langle f(t, As+e), r\rangle + e_2) - \langle f(u, A^T r + e_1), s\rangle||_\infty \leq B.$$

Because

$$|f(i, x)| \leq \frac{1}{2a_i} + \frac{1}{2},$$

The bound of the actual failure probability can be restricted between the sufficient condition and the necessary condition.

The sufficient condition can be written as:

$$||\langle f(u, A^T r + e_1), s\rangle||_\infty \geq B' = \lfloor \frac{q}{4} \rceil - \frac{1}{2a_v} - \frac{1}{2},$$

which is the upper bound of the failure probability. And the necessary condition is:

$$||\langle f(u, A^T r + e_1), s\rangle||_\infty \geq B'' = \lfloor \frac{q}{4} \rceil + \frac{1}{2a_v} + \frac{1}{2},$$

which can be seen as the lower bound of the failure probability.

For a given public key A, the actual failure probability is between these two, and the bound is tight. We draw the graph with the upper and lower bounds corresponding to the selected range of random variables to illustrate (Fig. 1):

For the 900 samples we selected, the difference between the upper and lower bounds in the power of 2 is usually less than 40, which means that the upper bound of the error given can approximately represent its actual error rate. The deviation is 12 orders of magnitude.

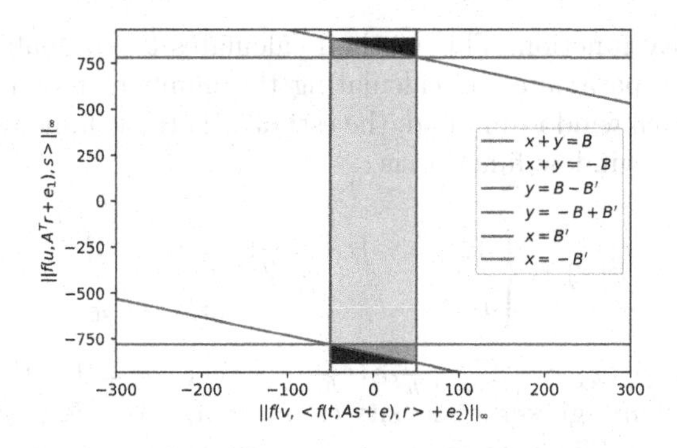

Fig. 1. The orange part corresponds to the sufficient condition we gave. The light blue part is the gap between the actual failure probability and the sufficient condition. The dark blue part is the gap between the necessary condition and the actual failure probability. (Color figure online)

4 Experiment and Sample Test

The parameter set for KYBER can be seen in the following Table 1.

Table 1. Parameter set for KYBER

	n	k	q	η_1	η_2	d_u, d_v	δ
KYBER512	256	2	3329	3	2	(10, 4)	2^{-139}
KYBER768	256	3	3329	2	2	(10, 4)	2^{-164}
KYBER1024	256	4	3329	2	2	(11, 5)	2^{-174}

In practical applications, the public key is usually fixed rather than randomly selected every time. The error rate of the mathematical expectation given in the traditional analysis cannot give the actual decryption failure probability of the fixed public key. Therefore, we give the error rate analysis for a given public key, consider the number of public keys corresponding to different error rates, and guide public key selection. The failure probability corresponding to the public key can be further discussed. This paper studies more about the probability distribution of the decryption failure. This paper uses the non-parametric estimation method proposed by Rosenblatt [10] and Parzen [9] to estimate the probability density function. This method calculates the probability density of the corresponding parameter by calculating the number of samples in a given area. For the Parzen window method, the estimated area volume used in different areas is fixed. The window function is

$$\phi(x) = \begin{cases} 1, & |x_i| \leq \frac{1}{2}; i = 1, \ldots, d \\ 0, & otherwise \end{cases}$$

By calculating $p_n(x) = \frac{1}{n} \sum_{i=1}^{n} \frac{1}{h^d} \phi(\frac{x-x_i}{h_n})$. Where $h^d = V_n$, the density function is obtained. Through kernel density estimation, it is able to give the probability that a given public key's failure probability is greater than the threshold. Therefore, this method will help the higher-order moment analysis of the failure probability, and thus have a deeper understanding of the moment characteristics of the public keys. The corresponding kernel density estimation curve is shown in Fig. 2, 3, 4.

The calculation process is quite time-consuming. It takes several hours to calculate the decryption failure probability for a given public key. As a consequence, we only testes 300 samples for each parameter set. The differences in the decryption failure probability before and after the independence assumption are given in Fig. 2, 3, 4, respectively.

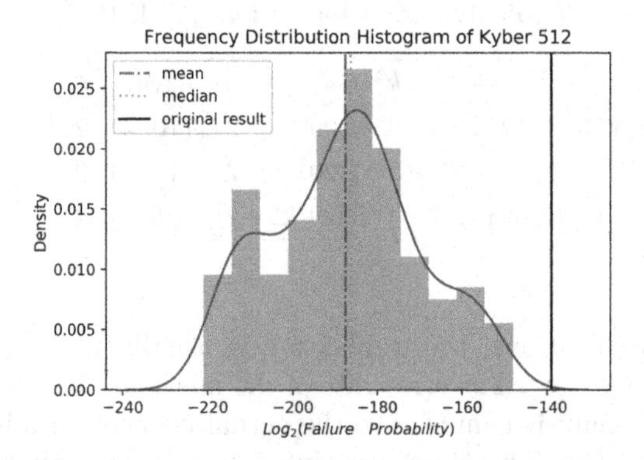

Fig. 2. Frequency histograms of $Kyber512$. The red line is the mean of all samples, the orange line is the median of all samples, and the black line is the failure probability provided in the original paper. The blue line is the kernel density estimation curve. (Color figure online)

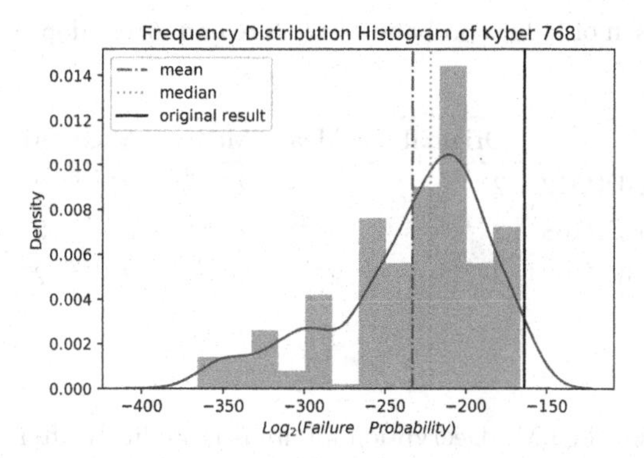

Fig. 3. Frequency histograms of *Kyber*768. The red line is the mean of all samples, the orange line is the median of all samples, and the black line is the failure probability provided in the original paper. The blue line is the kernel density estimation curve. (Color figure online)

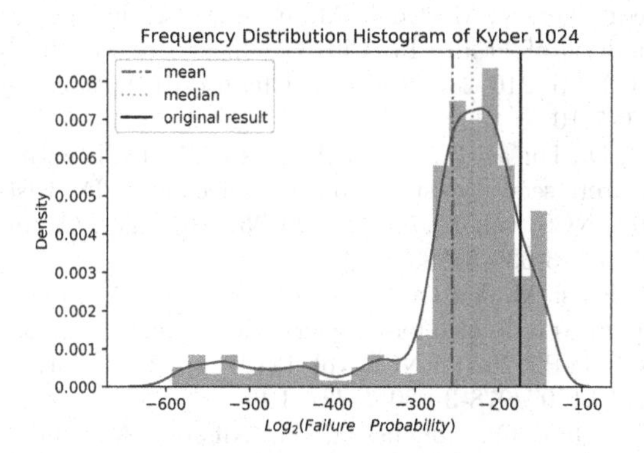

Fig. 4. Frequency histograms of *Kyber*1024. The red line is the mean of all samples, the orange line is the median of all samples, and the black line is the failure probability provided in the original paper. The blue line is the kernel density estimation curve. (Color figure online)

We select 300 random matrices A for each Kyber parameter set for testing and then draw their corresponding frequency histograms. It can be seen that the failure probability of most samples is lower than the probability given in the original paper, and the mean of the sample is less than the median of the sample (Table 2).

Table 2. Comparison of failure probabilities before and after adopting the assumption of independence.

	Original [2]	Mean	Median	Max	Min
KYBER512	2^{-139}	2^{-188}	2^{-186}	2^{-148}	2^{-221}
KYBER768	2^{-164}	2^{-233}	2^{-222}	2^{-166}	2^{-366}
KYBER1024	2^{-174}	2^{-255}	2^{-232}	2^{-142}	2^{-592}

References

1. Bindel, N., Schanck, J.M.: Decryption failure is more likely after success. In: Ding, J., Tillich, J.-P. (eds.) PQCrypto 2020. LNCS, vol. 12100, pp. 206–225. Springer, Cham (2020). https://doi.org/10.1007/978-3-030-44223-1_12
2. Bos, J., Ducas, L., Kiltz, E., et al.: CRYSTALS-Kyber: a CCA-secure module-lattice-based KEM. In: 2018 IEEE European Symposium on Security and Privacy (EuroS&P), pp. 353–367 (2018). https://doi.org/10.1109/EuroSP.2018.00032
3. Bos, J.W., Friedberger, S., Martinoli, M., et al.: Assessing the feasibility of single trace power analysis of Frodo. In: Cid, C., Jacobson, M., Jr. (eds.) SAC 2018. LNCS, vol. 11349, pp. 216–234. Springer, Cham (2019). https://doi.org/10.1007/978-3-030-10970-7_10
4. Dachman-Soled, D., Ducas, L., Gong, H., Rossi, M.: LWE with side information: attacks and concrete security estimation. In: Micciancio, D., Ristenpart, T. (eds.) CRYPTO 2020. LNCS, vol. 12171, pp. 329–358. Springer, Cham (2020). https://doi.org/10.1007/978-3-030-56880-1_12
5. D'Anvers, J.-P., Rossi, M., Virdia, F.: *(One) failure is not an option*: bootstrapping the search for failures in lattice-based encryption schemes. In: Canteaut, A., Ishai, Y. (eds.) EUROCRYPT 2020. LNCS, vol. 12107, pp. 3–33. Springer, Cham (2020). https://doi.org/10.1007/978-3-030-45727-3_1
6. D'Anvers, J.-P., Guo, Q., Johansson, T., Nilsson, A., Vercauteren, F., Verbauwhede, I.: Decryption failure attacks on IND-CCA secure lattice-based schemes. In: Lin, D., Sako, K. (eds.) PKC 2019. LNCS, vol. 11443, pp. 565–598. Springer, Cham (2019). https://doi.org/10.1007/978-3-030-17259-6_19
7. D'Anvers, J.-P., Vercauteren, F., Verbauwhede, I.: The impact of error dependencies on Ring/Mod-LWE/LWR based schemes. In: Ding, J., Steinwandt, R. (eds.) PQCrypto 2019. LNCS, vol. 11505, pp. 103–115. Springer, Cham (2019). https://doi.org/10.1007/978-3-030-25510-7_6
8. Guo, Q., Johansson, T., Yang, J.: A novel CCA attack using decryption errors against LAC. In: Galbraith, S.D., Moriai, S. (eds.) ASIACRYPT 2019. LNCS, vol. 11921, pp. 82–111. Springer, Cham (2019). https://doi.org/10.1007/978-3-030-34578-5_4
9. Parzen, E.: On estimation of a probability density function and mode. Ann. Math. Stat. **33**(3), 1065 (1962). https://doi.org/10.1214/aoms/1177704472
10. Rosenblatt, M.: Remarks on some nonparametric estimates of a density function. In: Davis, R., Lii, KS., Politis, D. (eds.) Selected Works of Murray Rosenblatt. Selected Works in Probability and Statistics. pp. 832–837. Springer, New York (1956). https://doi.org/10.1007/978-1-4419-8339-8_13
11. Wishart, J., Bartlett, M.S.: The distribution of second order moment statistics in a normal system. In: Mathematical Proceedings of the Cambridge Philosophical Society, vol. 28, no. 4, pp. 455–459. Cambridge University Press, Cambridge (1932). https://doi.org/10.1017/S0305004100010690

Authentication

Improving Deep Learning Based Password Guessing Models Using Pre-processing

Yuxuan Wu[1], Ding Wang[2](✉), Yunkai Zou[2], and Ziyi Huang[3]

[1] College of Computer Science, Nankai University, Tianjin 300350, China
[2] College of Cyber Science, Nankai University, Tianjin 300381, China
wangding@nankai.edu.cn
[3] College of Software, Nankai University, Tianjin 300457, China

Abstract. Passwords are the most widely used authentication method and play an important role in users' digital lives. Password guessing models are generally used to understand password security, yet statistic-based password models (like the Markov model and probabilistic context-free grammars (PCFG)) are subject to the inherent limitations of overfitting and sparsity. With the improvement of computing power, deep-learning based models with higher crack rates are emerging. Since neural networks are generally used as black boxes for learning password features, a key challenge for deep-learning based password guessing models is to *choose the appropriate preprocessing methods to learn more effective features*.

To fill the gap, this paper explores three new preprocessing methods and makes an attempt to apply them to two promising deep-learning networks, i.e., Long Short-Term Memory (LSTM) neural networks and Generative Adversarial Networks (GAN). First, we propose a character-feature based method for encoding to replace the canonical one-hot encoding. Second, we add so far the most comprehensive recognition rules of words, keyboard patterns, years, and website names into the basic PCFG, and find that the frequency distribution of extracted segments follows the Zipf's law. Third, we adopt Xu et al.'s PCFG improvement with chunk segmentation at CCS'21, and study the performance of the Chunk+PCFG preprocessing method when applied to LSTM and GAN.

Extensive experiments on six large real-world password datasets show the effectiveness of our preprocessing methods. Results show that within 50 million guesses: 1) When we apply the PCFG preprocessing method to PassGAN (a GAN-based password model proposed by Hitja et al. at ACNS'19), 13.83%–38.81% (26.79% on average) more passwords can be cracked; 2) Our LSTM based model using PCFG for preprocessing (short for PL) outperforms Wang et al.'s original PL model by 0.35%–3.94% (1.36% on average). Overall, our preprocessing methods can improve the attacking rates in four over seven tested cases. We believe this work provides new feasible directions for guessing optimization, and contributes to a better understanding of deep-learning based models.

Keywords: Password · Deep learning · Preprocessing · Generative Adversarial Networks · Long Short-Term Memory neural networks

C. Alcaraz et al. (Eds.): ICICS 2022, LNCS 13407, pp. 163–183, 2022.
https://doi.org/10.1007/978-3-031-15777-6_10

1 Introduction

Although passwords have some security problems and a variety of new authentication methods are constantly proposed, passwords promise to be the dominant authentication method in the foreseeable future due to their simplicity to deploy, easiness to change [2,3]. Thus, it is of great importance to understand password security, and a number of guessing algorithms have successively been proposed, such as statistical-based ones (e.g., probabilistic context free grammars (short for PCFG) [18] and Markov [10,12]) and deep-learning based ones (i.e., PassGAN [5] and FLA [11]). Password guessing algorithms study password security from the perspective of attackers who focus on the vulnerability of passwords, and they in turn can be used to build protection countermeasures, such as constructing the password strength meter (PSM) to evaluate password strength.

Password guessing attacks can be divided into targeted guessing attacks and trawling guessing attacks [16]. The former is to crack the password of a given user as quickly as possible [10], and the latter is to crack as many passwords as possible in a given password set under the limited guess number [16,19]. This paper focuses on trawling guessing attacks. A key challenge for trawling guessing attacks is to extract the password features effectively, and data preprocessing is a feasible method to improve the effect of the deep learning based models.

1.1 Related Work

Major Password guessing models based on statistical probability include PCFG [18] and the Markov model [10,12]. The main idea of PCFG is to divide a password into several segments according to the character types, and these segments can be regarded as the password features. For instance, the password `abc123` is parsed into the letter segment "abc" and digit segment "123". PCFG can also be integrated with other guessing models as a data preprocessing method [9,17,20]. The Markov model records the frequency of different characters after the password substring in the training phase, and then generates the guessing password character by character according to the statistical frequency distribution. With various improvements made on the base of PCFG, Xie et al. [19] focused on the targeted guessing attack and added the recognition of special dates and names. Wang et al. [15] added the recognition of Chinese pinyin and the six-digit dates (e.g., 201862) to further exploit the features of Chinese passwords. Houshmand et al. [6] added the recognition of keyboard patterns. Yang et al. [21] also studied keyboard patterns and analyzed the frequency distribution of the keyboard patterns. However, these PCFG-based methods are not optimal because most of these added recognition rules only consider the targeted guessing, and are not comprehensively considered in trawling guessing. In this paper, we focus on the password features, and add the most comprehensive recognition rules of keyboard patterns, words, website names, years for trawling guessing.

Recently, deep learning technology provides a new way for password attacking, and the model based on supervised learning was first used. In 2016, Melicher et al. [11] used Recurrent Neural Network (RNN) to build a password guessing model (i.e., FLA) which can be considered as a character-level model because the smallest unit it handles is each character in the password. In 2018, Liu et al. [9] proposed a multi-source PCFG+LSTM model (The LSTM based model using PCFG for preprocessing, short for PL) with the adversarial generation which can maintain high accuracy for different datasets. The password guessing model using PCFG for preprocessing can be regarded as a segment-level model, which divides passwords into different segments. In 2021, Wang et al. [17] found that the PL model could crack more passwords than PCFG and Markov within 50 million guesses. However, they only use the original PCFG which can not comprehensively extract password features. Xu et al. [20] proposed a new preprocessing method based on the Byte-Pair-Encoding (BPE) algorithm to divide passwords into chunks that consist of frequently occurring characters, and then built three models: the Markov based model using the chunk based preprocessing method, the model using the Chunk+PCFG preprocessing method, and the LSTM based model using chunk based preprocessing method. Password guessing models using the chunk-based preprocessing method are considered chunk-level. In this paper, we integrate the Chunk+PCFG preprocessing method with neural networks.

Unsupervised learning methods are also used in password cracking. In 2019, Hitaj et al. [5] first proposed a password guessing model based on Generative Adversarial Networks (GAN) and named it PassGAN. However, the cracking result of PassGAN is not ideal, and even lower than the traditional methods. We apply three preprocessing methods to PassGAN model, and find that using the basic PCFG for preprocessing can dramatically improve the cracking rate.

1.2 Our Contributions

In this work, we make the following key contributions:

(1) **Character feature based encoding method.** Character-level models usually adopt the canonical one-hot encoding method which can not fully utilize the character features. Therefore, we propose a new approach based on the type of characters and the corresponding keyboard positions, where each character is represented as a 4-dimensional vector: (character type, character serial number, keyboard row number, keyboard column number). Although this encoding method does not improve the effect, it still provides a new feasible direction for password guessing.
(2) **Refined PCFG.** Existing PCFG [6] divides passwords into four segments (letters, digits, special characters, and keyboard). We propose a refined PCFG based preprocessing method which adds the recognition rules of words, website names, and years to enable a more comprehensive password feature extraction. Inspired by Wang et al.'s work that the distribution of passwords follows the Zipf's law [14], we find that the frequency distribution of extracted segments also follows PDF-Zipf.
(3) **An extensive evaluation.** We perform a series of experiments on nine models, including two baseline ones (i.e., the LSTM based model using one-hot encoding method in Wang's work [17] and the original PassGAN model in Hitaj's work [5]), and seven models using preprocessing(i.e., the

LSTM based model using our new encoding method, the LSTM based model using basic PCFG for preprocessing [17], the LSTM based model using our refined PCFG for preprocessing, the LSTM based model using Chunk+PCFG for preprocessing, the PassGAN model using basic PCFG for preprocessing, the PassGAN model using our refined PCFG for preprocessing, and the PassGAN model using Chunk+PCFG for preprocessing). Our empirical results show that character-level models can improve their effect by using PCFG based preprocessing methods. In particular, the PassGAN model using PCFG for preprocessing can improve the success rates drastically (avg. 26.79 %) compared to the original PassGAN model.

2 Background

In this section, we briefly introduce the background on deep learning based password guessing models (i.e., LSTM based models and GAN based models).

2.1 LSTM Based Models

Recurrent Neural Network (RNN) and its Variants, such as Long Short-Term Memory neural networks (LSTM), can all be used in password guessing models [11,17]. To avoid gradient vanishing problems [8], we use LSTM instead of RNN. The one-hot encoding is usually performed on each character to convert a password string to a matrix. Moreover, it is necessary to construct the corresponding label y for the input password x due to the supervised learning method. We use an example to illustrate this training process (see in Fig. 1). Suppose the character set that contains all the characters appearing in the dataset is {a, b, c, Bos, Eos}, where Bos represents the beginning of a password and Eos represents the end of a password. Then the password `abc` is converted to a matrix: [[0, 0, 0, 1, 0], [1, 0, 0, 0, 0], [0, 1, 0, 0, 0], [0, 0, 1, 0, 0]]. The corresponding label y is: [[1, 0, 0, 0, 0], [0, 1, 0, 0, 0], [0, 0, 1, 0, 0], [0, 0, 0, 0, 1]].

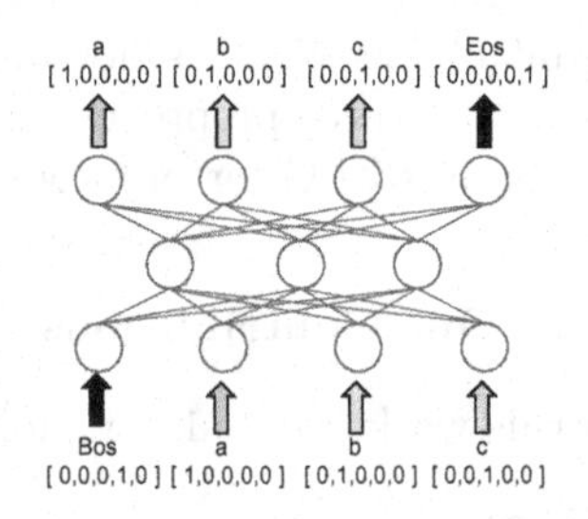

Fig. 1. The training process of LSTM based models using one-hot encoding method. The character set is {a, b, c, Bos, Eos}, where Bos represents the beginning of a password and Eos repreasents the end of a password.

The LSTM based model is a probability model which assigns probabilities to the guessing passwords. The probability of the next character is obtained by entering the prefix of the password string into LSTM. For example, a password generation process could be: `B`→`Ba`→`Bab`→ `Babc`→`BabcE`, where each time we select the character with the highest probability as the next character. The LSTM based model with this training and generating method is character-level.

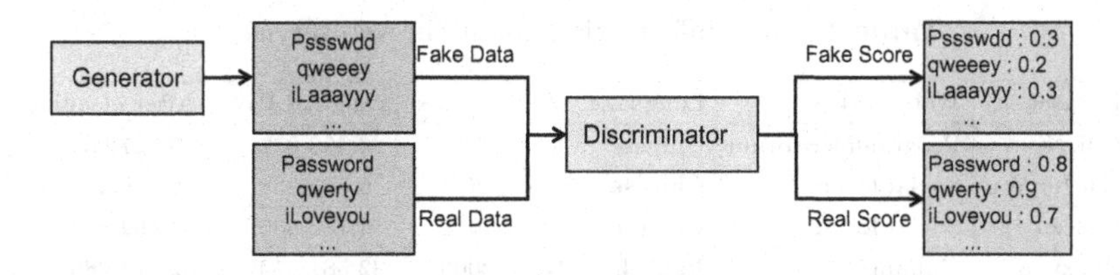

Fig. 2. The training process of PassGAN model. Passwords are first converted into matrices and then input to the discriminator. The score is the output of the discriminator which will be used to calculate the gradient.

2.2 PassGAN

PassGAN, proposed by hitaj [5], is based on WGAN-GP [4] due to the difficulty of training original GAN. PassGAN consists of a generator and a discriminator. The generator captures the real data distribution by building a mapping function from prior noise distribution to real data space and generate fake samples. The discriminator learns to determine whether a sample comes from the fake samples or the real data. PassGAN are trained adversarially in this way until the discriminator cannot identify the source of the data. The main structure of the PassGAN model is shown in Fig. 2.

Original PassGAN is character-level, but not as effective as LSTM based character-level models [17]. Another problem of original PassGAN is that it can not assign probabilities to the generated guessing passwords, so we can not obtain the priority of different guessing passwords. These problems will all be solved by using preprocessing methods in this paper.

3 Preliminaries

In this section, we first explicate the datasets used in this paper, including the basic information, the length distribution, the character composition, and the top-10 passwords. In addition, since passwords are confidential data related to personal privacy, ethical considerations are also explained.

3.1 Datasets

We compare the password guessing models with different preprocessing methods based on six large password datasets (see in Table 1) with a total of 57 million passwords. These datasets are different in terms of service, size, user localization, and language, which suggests that our models can be used to well characterize different user-chosen passwords. We name each dataset according to its website's domain name. The first three datasets, namely CSDN, YueJunYou, Renren, are all from Chinese websites. CSDN is a well-known community website of Chinese programmers, founded in 1999. Renren is a real-name social networking platform, founded in 2005. YueJunYou is a website for making friends and traveling,

Table 1. Basic information about six web services.

Dataset	Web service	Language	When leaked	Original PWs	After cleaning
CSDN	Programmer forum	Chinese	Dec., 2011	6,428,277	6,427,538
YueJunYou	Social forum	Chinese	May., 2006	5,365,338	5,286,494
Renren	Social forum	Chinese	Oct., 2011	4,733,366	4,662,654
Rockyou	Gaming	English	Dec., 2009	32,581,870	32,573,986
Yahoo	Email	English	Jul., 2012	5,737,797	5,605,985
Youporn	Video	English	Oct., 2017	2,677,951	2,105,452

founded in 2014. The last three datasets, namely Rockyou, Yahoo, Youporn, are all from English websites. Rockyou is a game website that contains 320 million passwords, which is the largest dataset among six datasets. Yahoo is a famous internet portal site in the United States, founded in 1995. Youporn is a video website only for adults, founded in 2006.

Datasets Cleaning. We note that these original datasets contain some abnormal passwords that are either too long (>40) or too short (<4), which are unlikely to be user-chosen passwords or simply junk information. Thus, we launch the work of dataset cleaning before any experiment. We first remove the passwords that contain symbols beyond the 95 printable ASCII characters, and then we also remove the passwords with length <4 or length >30, because these passwords do not comply with the password policy of most websites or may not be considered by the attackers who care about cracking efficiency [1]. Generally, the removed passwords are less than 1% for each dataset.

Here we also provide a concrete grasp of user-chosen passwords: 1) The length of most passwords is between 6 and 9, accounting for 62.08%–83.84% of each web service (see details in Table 7 of Appendix A); 2) Chinese users love to use digits (avg. 54.64%) and this figure for English users is 18.62%, while English users love to use characters (avg. 42.79%) and this figure for Chinese users is 15.43% (see details in Table 8); 3) Top-10 passwords account for 7.18% 10.43% of Chinese users, and this figure for English users is 2.05% 5.29%, indicating Chinese passwords are more concentrated, as found in [15] (see in Table 9).

3.2 Ethical Considerations

Although these datasets are widely used in the literature [5,10,11,20,21], they are still private data. Therefore, we only report the aggregated statistical information and treat each individual account as confidential, so that using them in our research will not increase the risk to the corresponding victim. Furthermore, these datasets may be utilized by attackers as cracking dictionaries, while our use is both beneficial for the academic community to understand the strength of users' password choices, and for security administrators to secure their passwords. In addition, we have consulted privacy experts a number of times. Since our datasets are all available from the Internet, the results in this work are reproducible.

4 Preprocessing Methods

In this section, we describe different preprocessing methods to improve the effect of the deep-learning based password guessing models.

4.1 Important Abbreviations

To facilitate the reading process, we introduce the important abbreviations used in this article. GAN means Generative Adversarial Networks; PassGAN is short for the password guessing model based on GAN; LSTM is short for Long Short-Term Memory neural networks; PCFG is short for probabilistic context-free grammars; LSTM/PassGAN+X means the LSTM/PassGAN based password gurssing model using X for preprocessing.

4.2 Character Feature Based Encoding Method

Canonical one-hot encoding method as used in [11,17,20] only classifies different characters, but can not fully reflect other character features. Moreover, the matrices converted by the one-hot encoding method are sparse [13]. The main challenge for character encoding is how to distinguish characters and reflect their features with as little space as possible. Therefore, we comprehensively consider different kinds of character features and then propose a new character encoding method that greatly reduces the occupied space.

The password character has two important features, one is the type, and the other one is the keyboard location since keyboard pattern is also a popular way in password creation [16]. Thus, we represent each character in four dimensions. The first dimension represents the type of characters, where we use 1, 2, 3, and 4 to represent digits, uppercase letters, lowercase letters, and special characters. The second dimension is the serial number of the characters in each type. For example, a–z can be represented by 1–26 according to the dictionary order, and the digits can be represented by themselves. The third dimension represents the keyboard row number, and the fourth dimension represents the keyboard column number. The row number of the keyboard increases from top to bottom, and the column number increases from left to right. For example, the string "1234567890-=" is in the first row, and the string "1qaz" is in the first column according to keyboard coordinates. Using our new encoding method, the password `1234` can be converted to a matrix [[1, 1, 1, 1], [1, 2, 1, 2], [1, 3, 1, 3], [1, 4, 1, 4]].

4.3 Refined PCFG

The basic PCFG [18] only divides passwords into letters, digits, and special characters, which may destroy the integrity of some segments and ignore user's habit of creating passwords. For example, the string "1!2@3#" should be considered as a complete segment due to the adjacency of characters on the keyboard, while it would be converted to $D_1S_1D_1S_1D_1S_1$ in the basic PCFG. Therefore, we add the recognition rules of 4 important password features.

Table 2. The proportion of top-10 mostly used years (1900–2100) for each web service*.

Dataset	CSDN	Renren	YunJunYou	Rockyou	Yahoo	Youporn
Proportion	58.68%	54.74%	64.16%	33.94%	26.19%	32.68%
Unique†	201	201	201	201	201	201

* We record the proportion of top-10 year segments in all year segments.
† Unique represents the number of unique years in the web service.

Year Recognition: Years have been found popular in passwords [7,15]. We count the number of the years (from 1900 to 2100, a total of 201), and record the proportion of top-10 most widely used years in Table 2. Results show that some years occupy a large proportion, which indicates users may focus on some years. However, the basic PCFG would extract years into digits. Thus, we extract digit segments with length of four and value between 1900–2100 from the passwords.

Table 3. The proportion of top-10 mostly used websites for each web service.

Dataset	CSDN	Renren	YunJunYou	Rockyou	Yahoo	Youporn
Proportion	15.15%	20.93%	25.25%	15.35%	17.23%	25.44%
Unique†	3,348	3,080	202	6,415	1,109	255

† Unique represents the number of unique websites in the web service.

Website Name Recognition: We for the first time count the number of the website names in each dataset, and record the proportion of top-10 most widely used website names in Table 3. Although the website names are less concentrated than the years, some website names still account for a large proportion, and the passwords with website name segments may be considered strong passwords in the basic PCFG. For instance, the password "csdn.net" is converted to the base structure $L_4S_1L_3$, and may be assigned a low probability of being cracked by basic PCFG. To address this problem, we add the recognition of website names. First, common website name suffixes, such as ".com", ".net", are used to construct a suffix list, and then the complete website segments are extracted from the passwords according to the suffix list. For example, the website name segment "csdn.net" is extracted from the password "123csdn.net123".

Keyboard Pattern Recognition. Since keyboard pattern is also a popular way in password creation [6,16], we add the corresponding recognition rule. Furthermore, keyboard patterns with only one character type can be extracted completely by basic PCFG, so our refined PCFG (i.e., basic PCFG with the additional recognition rules) focus on the keyboard patterns with multiple character types. For example, the password `q1w2e3` is converted to the base structure $L_1D_1L_1D_1L_1D_1$ by basic PCFG, which destroys the integrity of the keyboard pattern, while it is converted to K_6 by our refined PCFG (k represents the keyboard pattern and 6 represents the length of the segment).

Algorithm 1: Dictionary contruction algorithm.

```
Input: Password set S
Output: Word dictionary D.
1  for pwd in S do
2      letters_list = extract_letters(pwd);/* extract letter segments in pwd and store
       them in a list. */
3      for seg in letters_list do
4          if len(seg) > len_min then
5              /* len_min is the minimum word length. */
6              D[seg]+ = 1;/* D is the initial dictionary to record the frequency of
               different words. */
7  for seg in D do
8      if D[seg] < threshold then
9          /* threshold is the minimum word frequency */
10         delete D[seg];
11 return D;
```

Word Recognition: The keyboard pattern has inherent limitations that may extract wrong segments [21]. For instance, the segment "password" should be regarded as a complete segment, but "assw" would be recognized as a keyboard pattern. Although our keyboard pattern recognition method avoids most of the error cases, wrong results may still occur. Our solution is to use a dictionary that contains common words, and extract segments that appear in the dictionary from the passwords before the keyboard pattern extraction. However, the effect of this method depends on the quality of the dictionary. Once the dictionary does not contain the corresponding word, keyboard pattern extraction will still cause errors. Therefore, we choose to construct the word dictionary through the training set. This process is described in Algorithm 1.

Frequency Distribution: Wang et al. [14] found that the distribution of popular passwords follows PDF-Zipf:

$$f_r = \frac{C}{r^s} \tag{1}$$

where f_r is the frequency of the password, r is the rank of the password, C and s are constants depending on the datasets. To verify that the data conforms to this distribution, we use the following equation:

$$log(f_r) = logC - s \cdot log(r) \tag{2}$$

where $log(f_r)$ and $log(r)$ are linearly related. We verify that the extracted segment meets the Zipf distribution based on Eq. 2. The extracted results are sorted in descending order of frequency on six datasets and the $log(f_r)$-$log(r)$ graphs are shown in Fig. 3. Moreover, all the coefficients of determination (R^2) which can measure the fitting degree of the regression line to the sample data are shown in Table 4. The closer the determination coefficient is to 1, the better the fitting effect is. The results indicate that the frequency distribution of the keyboard patterns, words, website names, and years with a frequency of more than five can meet the PDF-Zipf model well.

Table 4. The coefficient of determination (R^2) for fitting different extracted segments.

R^2	Keyboard pattern	Word	Website name	Year
CSDN	0.9667	0.9974	0.9791	0.9698
YueJunYou	0.9763	0.9895	0.9697	0.9650
Renren	0.9757	0.9935	0.9764	0.9772
Rockyou	0.9843	0.9973	0.9894	0.9358
Yahoo	0.9687	0.9932	0.9767	0.8710
Youporn	0.9795	0.9949	0.9661	0.9033

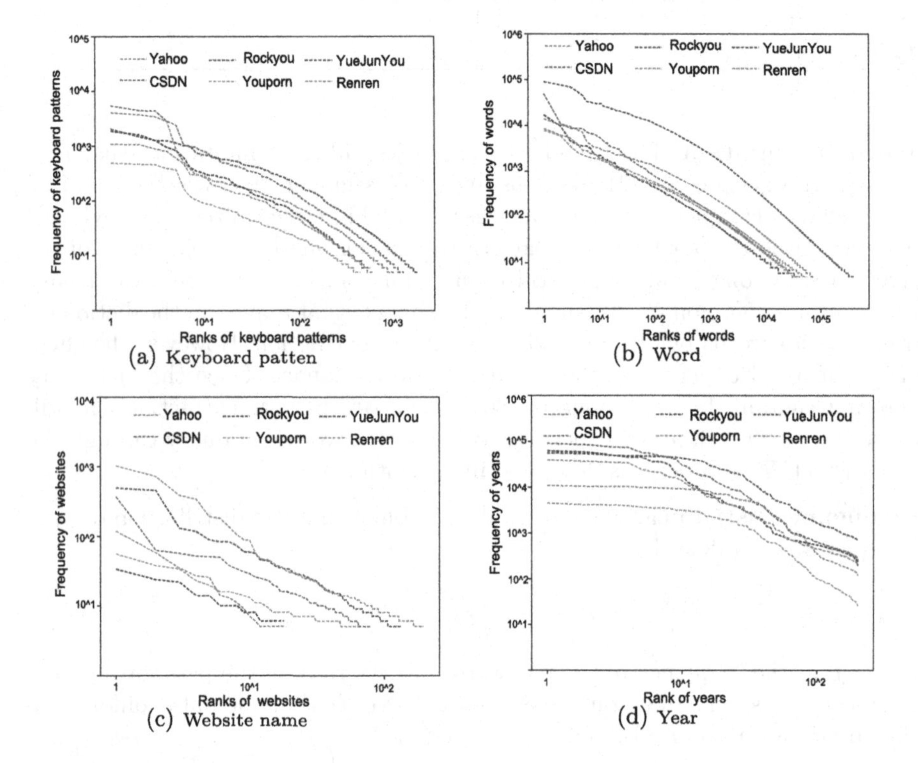

Fig. 3. Frequency distribution of different types of extracted segments.

Recognition Order: Some recognition rules may conflict with each other. For example, "989" in "1989" is recognized as a keyboard pattern, and "csdn" in "csdn.net" may be recognized as a word. Therefore, it is essential to set a reasonable extraction order and begin with the least conflict one. Since the year recognition may conflict with the keyboard pattern recognition, and the website name recognition may conflict with the word recognition. Our recognition order is: year→website name→word→keyboard pattern→basic PCFG recogni-

Algorithm 2: Extraction algorithm for our refined PCFG.

```
   Input: Password pw, word dictionary D, website name suffix list web_list
   Output: handled segment list seg_list.
1  seg_list=[pw]; /* initial list taking the whole password as a unhandled segment. */
2  kp_min=3;/* the minimum length of the keyboard pattern. */
3  for seg in seg_List do
4      if type(seg) == string and a year occurs in seg then
5          begin, end = index_year(seg);
6          divide_seg(seg, begin, end);/* divide seg into
           seg[0 : begin], ('Y' + str(end − begin), seg[begin : end]), seg[end :]. */
7  for seg in seg_List do
8      if type(seg) == string and a website suffix from web_list occurs in seg then
9          begin, end = index_website(seg);
10         divide_seg(seg, begin, end);/* divide seg into
           seg[0 : begin], ('E' + str(end − begin), seg[begin : end]), seg[end :]. */
11 for seg in seg_list do
12     if type(seg) == string and a word from D occurs in seg then
13         begin, end = index_word(seg);
14         divide_seg(seg, begin, end);/* divide seg into
           seg[0 : begin], ('W' + str(end − begin), seg[begin : end]), seg[end :]. */
15 for seg in seg_list do
16     if type(seg) == string then
17         begin, end = index_keyboard(seg);
18         if end − begin >= kp_min and seg[begin : end] contains more than one
           character type then
19             divide_seg(seg, begin, end);/* divide seg into
               seg[0 : begin], ('K' + str(end − begin), seg[begin : end]), seg[end :]. */
20 merge_unhandled(seg_list);/* merge successive unhandled segments in seg_list. */
21 PCFG_extraction(seg_list);/* use original PCFG for extraction. */
22 return seg_list;
```

tion. Overall, our refined PCFG is different from the basic PCFG on the tags of base structures, where some tags are added, such as K (keyboard patterns), W (words), E (website names), and Y (years). The complete extraction process of our refined PCFG is described in Algorithm 2. The structure of the neural networks using the PCFG based preprocessing method is shown in Fig. 4.

4.4 PassGAN Using PCFG for Preprocessing

The result of PassGAN [5] is worse than the LSTM based models within 10^7 guessing passwords, so we infer that the ability of GAN to learn text features is weaker than LSTM. Character-level passwords are relatively complex for GAN due to the length and the multiple character types. Therefore, we use PCFG based preprocessing method to simplify the data, and train the model with the base structures obtained from PCFG. In the generation process, PassGAN would generate duplicate base structures without probability, so we count the number of different base structures until the number of unique base structures reaches the target value. Then we assign each base structure with the probability $f_i/total$, where f_i represents the frequency of the corresponding base structures, and $total$ represents the total number of all base structures.

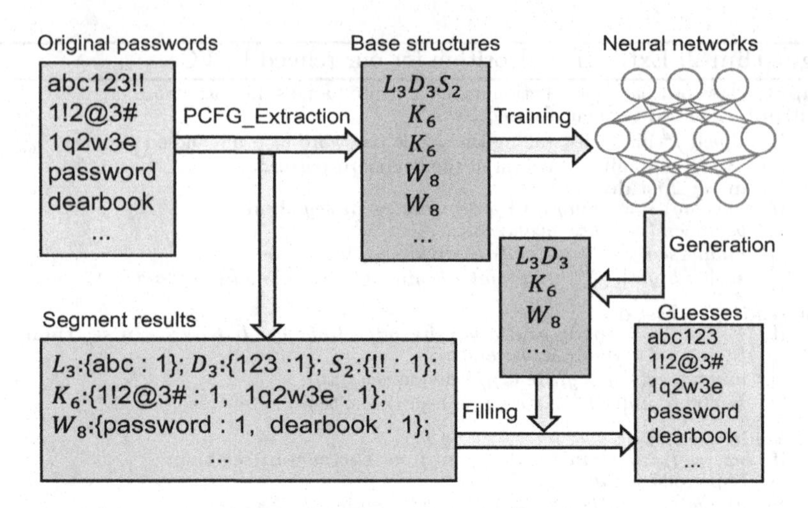

Fig. 4. An illustration of deep learning based model using PCFG based preprocessing method. K represents the keyboard pattern and W represents the word.

Algorithm 3: The process of Chunk+PCFG preprocessing method

Input: Password dictionary with the corresponding frequency *pwd_dict*
Output: Processed Dictionary *pwd_dict*.

1 **while true do**
2   (*Pairs, avg_len*)=*get_pairs*(*pwd_dict*);/*Take two consecutive chunks as a pair and record the frequency of pairs in *Pairs*, *avg_len* is the avg len of chunks.*/
3   **if** *avg_len* > *threshold* **then**
4   /**threshold* stands for the minimum average-length of chunks. */
5   *break*;
6     $best\_pair = max(Pairs, key = Pairs.get)$; /*find the most frequnt pair. */
7     $pwd\_dict = merge\_chunk(best_pair, pwd_dict)$;
8 $PCFG\_extraction(pwd_dict)$;/* perform PCFG on each chunk of passwords.*/
9 **return** *pwd_dict*;

4.5 Chunk+PCFG Preprocessing Method

We adopt Xu et al.'s PCFG improvement with chunk segmentation at CCS'21 [20], and integrate the Chunk+PCFG preprocessing method with LSTM and PassGAN separately. Byte-Pair-Encoding (BPE) algorithm is used to divide passwords into chunks, and then the chunk-level passwords are converted to the base structures by performing PCFG on each chunk of the passwords. Since chunk-level passwords are fine-grained enough, we only use the basic PCFG and a chunk can be represented as L (with only letters), D (with only digits), S (with only special characters), Two (with two character types), $Three$ (with three character types). For example, the password `iloveu4ever` can be firstly converted to the chunk-level password ["iloveu", "4ever"], and then converted to the base structure L_6Two_5. The process of the Chunk+PCFG preprocessing method is shown in Algorithm 3. The remaining training and generation process is the same as that of the LSTM based models using PCFG for preprocessing.

5 Experiments

In this section, we first describe the attacking strategies, and then evaluate the result of five different preprocessing methods combined with neural networks on multiple datasets. The details of datasets are described in Sect. 3.1.

5.1 Attacking Strategies Design

The combination of five different preprocessing methods with two leading deep-learning models gives rise to a total of ten guessing strategies, and we focus on seven promising ones: the LSTM based model using one-hot encoding method [17], the LSTM based model using our new encoding method, the LSTM based model using the basic PCFG for preprocessing (short for PL) [17], the LSTM based model using our refined PCFG for preprocessing, the LSTM based model using Chunk+PCFG for preprocessing, PassGAN using the basic PCFG for preprocessing, PassGAN using our refined PCFG for preprocessing, and PassGAN using Chunk+PCFG for preprocessing. Here LSTM includes a hidden layer that contains 128 neurons and a softmax layer, and the structure of GAN is the same as the original PassGAN model which is described in Sect. 2.2. For chunk-level models, the minimum average length of chunks is 3.0. All models are trained and tested on six datasets, which are divided into the training set, test set, validation set according to 8:1:1 as recommended in [17].

5.2 Evaluation Results

To extensively evaluate the effect of five different preprocessing methods, 50 million guessing passwords are generated from each model and sorted by probability in order to obtain a smooth curve of the result. We use the guess-number-graph and record cracking results to intuitively reflect the effects of different models.

Overall Analysis. Table 5 and 6 and Fig. 5 and 6 show that, with 50 million guessing passwords: (1) The LSTM based model using one-hot encoding outperforms using our new encoding by 4.22% on average; (2) The LSTM based model using the basic PCFG for preprocessing (short for PL) outperforms the LSTM based model using one-hot encoding by 7.85% on average; (3) Refined PL model outperforms the basic PL model by 1.36% on average; (4) The basic PL model outperforms the LSTM based model using Chunk+PCFG for preprocessing (short for CKPL) by 6.49% on average; (5) The PassGAN model using the basic PCFG for preprocessing (short for PCFG+PassGAN) outperforms the original PassGAN model by 26.79% on average; (6) The PCFG+PassGAN model outperforms refined PCFG+PassGAN model by 1.41% on average; (7) The original PassGAN model outperforms the PassGAN model using Chunk+PCFG for preprocessing (short for CKP+PassGAN) by 13.85% on average.

Table 5. Cracking results of LSTM based models with five different preprocessing methods (Guess number = $5 * 10^7$)†

Dataset	Language	LSTM one-hot [17]	LSTM 4-dim	LSTM b_PCFG [17]	LSTM r_PCFG	LSTM CK_PCFG
CSDN	Chinese	39.81%	37.73%	43.54%	43.89%	37.30%
YueJunYou	Chinese	60.72%	59.53%	78.29%	78.93%	61.78%
Renren	Chinese	50.72%	48.01%	46.11%	47.27%	53.41%
Rockyou	English	52.40%	46.71%	63.17%	67.11%	46.22%
Yahoo	English	40.66%	35.31%	44.37%	45.04%	48.62%
Youporn	English	50.42%	42.11%	66.36%	67.76%	55.53%

† LSTM 4-dim means using our new encoding method; b_PCFG means with basic PCFG for preprocessing [17]; r_PCFG means with our refined PCFG; CK_PCFG means with the Chunk+PCFG method. The results show that our refined PCFG method increases the effect by 9.21% on average for original model [17].

Encoding Method. The experimental results on six datasets show that our new encoding method does not perform well, which can be attributed to two reasons. First, the sparsity problem caused by the one-hot encoding may not have serious side effects because passwords are generally short in length. Second, since our new encoding method is only used to represent each character, its advantage which reflects multiple character features has not been fully utilized.

Table 6. Cracking results of GAN based models with four different preprocessing methods (Guess number = $5 * 10^7$)†

Dataset	Language	PassGAN [5]	PassGAN b_PCFG	PassGAN r_PCFG	PassGAN CK_PCFG
CSDN	Chinese	27.47%	42.14%	27.58%	11.87%
YueJunYou	Chinese	49.35%	78.88%	62.01%	19.72%
Renren	Chinese	36.17%	49.99%	33.20%	19.65%
Rockyou	English	31.25%	64.41%	18.35%	15.58%
Yahoo	English	24.74%	53.38%	21.04%	23.96%
Youporn	English	38.07%	76.88%	34.31%	31.11%

† PassGAN b_PCFG means with basic PCFG for preprocessing; PassGAN r_PCFG means with our refined PCFG; CK_PCFG means with the Chunk+PCFG method. The results show that the basic PCFG method increases the effect by 26.79% on average for PassGAN model [5].

PCFG Based Preprocessing Method. Using PCFG for preprocessing can improve the effect, and the LSTM based model with our refined PCFG is even better than with the basic PCFG, which indicates that the fine-grained rules

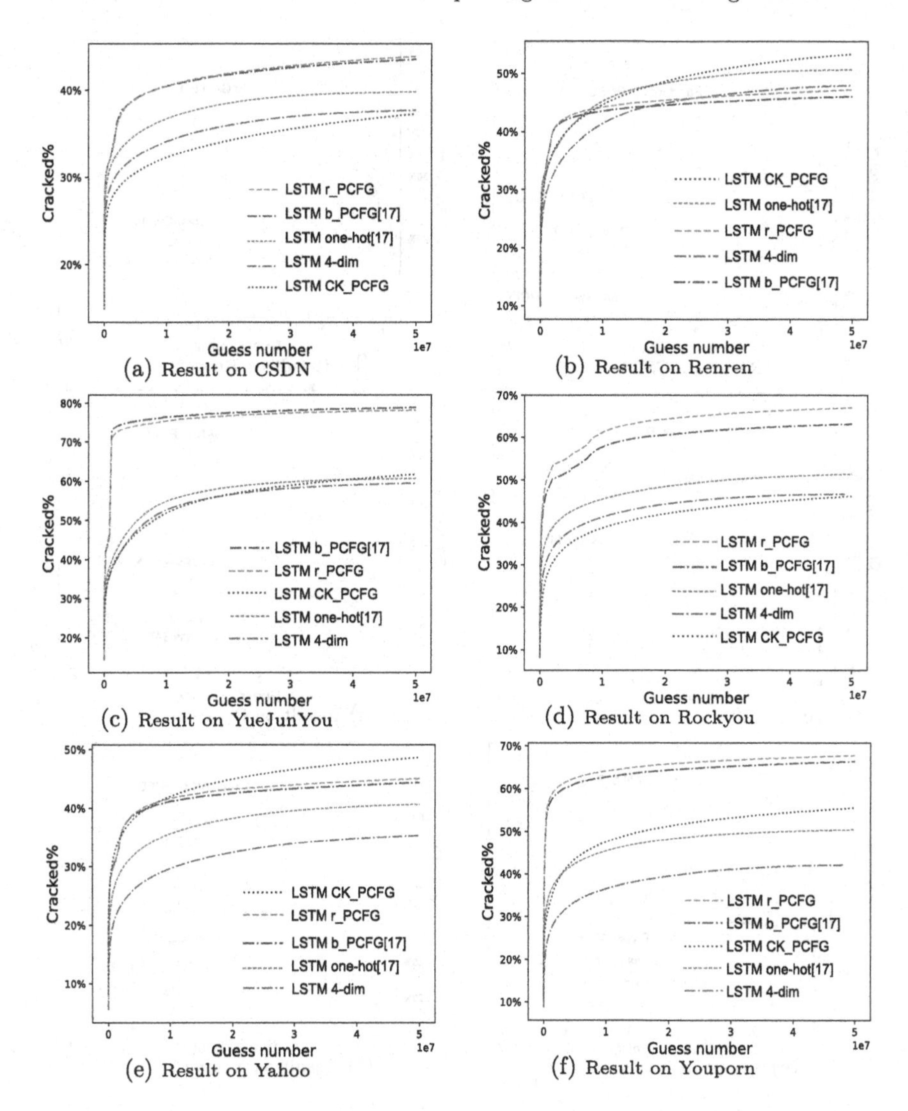

(a) Result on CSDN (b) Result on Renren (c) Result on YueJunYou (d) Result on Rockyou (e) Result on Yahoo (f) Result on Youporn

Fig. 5. Cracking results of LSTM based models with five different preprocessing methods (Guess number = $5 * 10^7$). The training set and test set are from the same dataset, with a division ratio of 8:1. LSTM one-hot represents the original model using one-hot for encoding [17]; LSTM 4-dim means using our new encoding method; b_PCFG means with basic PCFG for preprocessing [17]; r_PCFG means with our refined PCFG; CK_PCFG means with the Chunk+PCFG method. The results show that our refined PCFG method increases the effect by 9.21% on average for original LSTM model [17].

can extract more effective features. However, PassGAN with our refined PCFG decreases the effect due to the complexity of our method. Furthermore, since the Chunk+PCFG method first performs PCFG on each chunk of the passwords, it

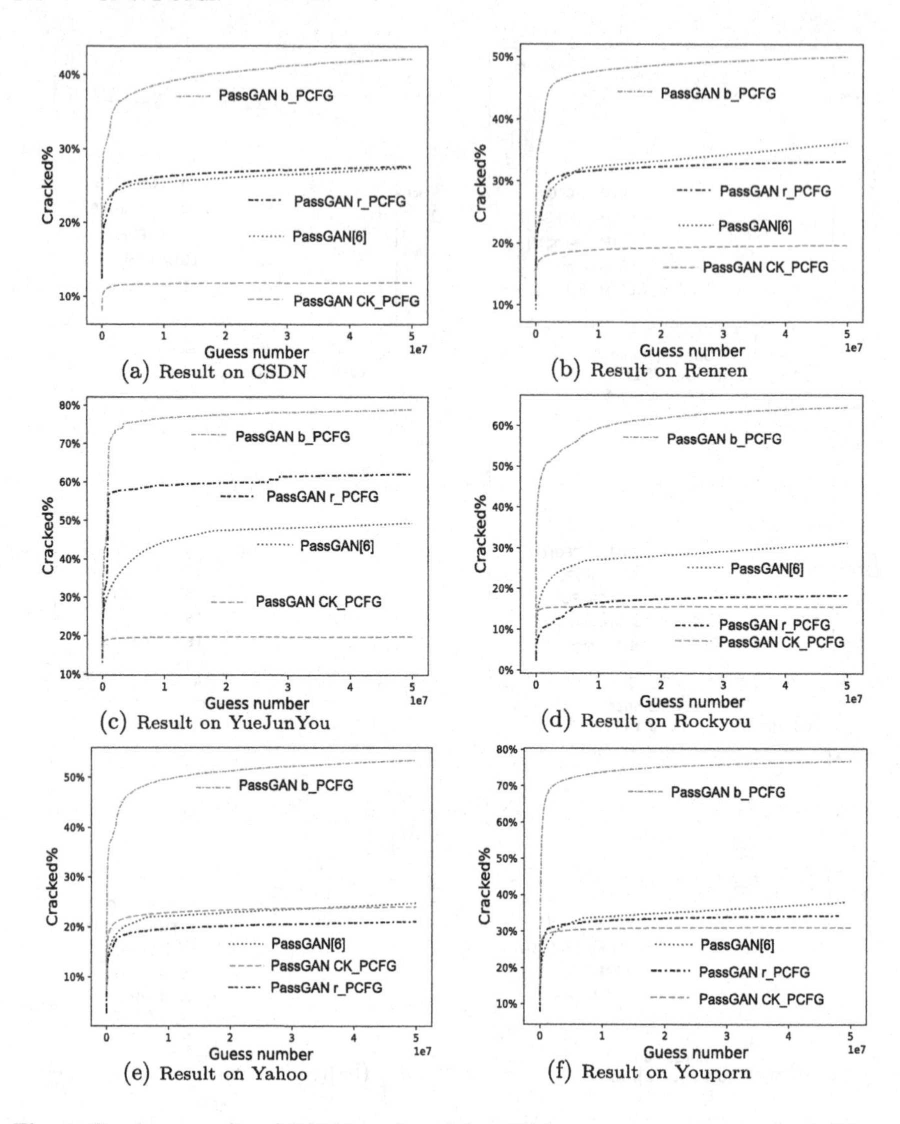

Fig. 6. Cracking results of GAN based models with four preprocessing methods (Guess number $= 5 * 10^7$). The training set and test set are from the same dataset, with a division ratio of 8:1. PassGAN represents the original model in [5]; b_PCFG means with basic PCFG for preprocessing; r_PCFG means with our refined PCFG; CK_PCFG means with the Chunk+PCFG method. The results show that using basic PCFG for preprocessing increases the effect by 26.79% on average for PassGAN model [5].

would generate even more complicated base structures than our refined PCFG, which can be the reason why all models using the Chunk+PCFG preprocessing method do not perform well compared to the original models. To explore the impact of different extraction rules added to the basic PCFG, we set up a series of experiments where only one extraction rule is added each time. The results (see deatails in Table 10 of Appendix B) indicate that word recognition has the best performance among all four rules and using separate recognition rules is not as good as our refiend PCFG (i.e., with all four rules).

Limitations. Firstly, our new encoding method can not improve the performance of the LSTM based models compared with the one-hot encoding, which may indicate that password guessing models can not be improved only by the character encoding. Secondly, we only improve the effect of PassGAN from the aspect of preprocessing, but not change the structure of PassGAN.

Future Directions. Firstly, new training methods to match up with our new encoding method can be a viable direction due to the underutilization of character features. Secondly, the research on PCFG based preprocessing method should be more fine-grained. One viable direction is to add more fine-grained recognition rules based on our refined PCFG, and another direction is to apply natural language processing(NLP) technology to extract the characteristics of passwords. Thirdly, this paper contributes to a better understanding of deep-learning based guessing models in that: preprocessing indeed can effectively enhance the use of the neural networks' learning ability, which is in turn intrinsically determined by the deep-learning model's network structure. For instance, the text learning ability of GAN is weaker than that of LSTM, which leads to the poor effect of fine-grained preprocessing methods integrated with GAN. Thus, the ability of GAN to learn text features may be improved by changing its structure, for example, using LSTM to compose the generator of GAN.

6 Conclusion

This paper studies the deep-learning based password guessing models from the aspect of preprocessing. Firstly, considering the limitations of the one-hot encoding method, we propose a new encoding method that comprehensively reflects the character features. Secondly, considering that basic PCFG does not fully extract the password features, we propose a refined PCFG with comprehensive recognition rules. Thirdly, we adopt the idea of chunk segmentation at CCS'21, and apply the chunk+PCFG preprocessing method to LSTM and GAN.

Extensive experimental results show that: 1) Our refined PCFG outperforms the basic PCFG by 1.36% on average when integrated with LSTM; 2) Using basic PCFG for preprocessing improves the effect of the PassGAN model drastically by 26.79% on average; 3) Although our new encoding method does not improve the effect compared with the one-hot encoding, it still provides a feasible new research direction; 4) The performance of Chunk+PCFG preprocessing method is not ideal due to the complexity of its base structures.

Our results suggest that using PCFG for preprocessing is an effective way to improve the deep-learning based guessing models. Still, it should be used with care: although more fine-grained PCFG (e.g., our refined PCFG and Chunk+PCFG) extracts the passwords more comprehensively, it also generates more complicated base structures, which increases the training complexity for neural networks, and may even reduce the cracking rates for these neural networks with weak text feature learning ability.

Acknowledgment. The authors are grateful to the anonymous reviewers for their invaluable comments. Ding Wang is the corresponding author. This research was in part supported by the National Natural Science Foundation of China under Grant No.62172240, and by the Natural Science Foundation of Tianjin, China under Grant No. 21JCZDJC00190. There is no competing interests.

Appendix 1 Some Statistics About User-Chosen Passwords

The length distributions of each dataset are shown in Table 7. Most passwords' length are between six and nine (avg. 73.81%). The length distribution is affected by the password policy. For example, CSDN dataset has much fewer passwords of length under eight as compared to other datasets, which may be caused by the fact that CSDN website changed the password policy to a more strict one. The character composition information is summarized in Table 8. Chinese users prefer to use digits in passwords, while English users prefer to use letters. This may be caused by cultural differences because most Chinese users use more digits in their daily lives than English words. In addition, English users prefer lowercase letters rather than uppercase letters. The top-10 passwords information is shown in Table 9. The password "123456" is the most commonly used password except for CSDN (due to its password policy). It is also interesting to see that the top-10 passwords in Chinese datasets are almost all pure digits.

Table 7. Length distribution information of each web service.

Dataset	1–5	6	7	8	9	10–16	17–30	30+
CSDN	0.63%	1.29%	0.26%	36.38%	24.15%	36.98%	0.32%	0.00%
YueJunYou	3.09%	24.00%	22.59%	24.13%	13.12%	13.05%	0.01%	0.00%
Renren	6.63%	25.36%	18.18%	20.24%	12.05%	17.20%	0.32%	0.00%
Rockyou	4.31%	26.04%	19.29%	19.98%	12.11%	17.86%	0.40%	0.01%
Yahoo	10.33%	17.86%	14.36%	25.03%	12.39%	20.04%	0.00%	0.00%
Youporn	11.44%	26.66%	16.17%	20.40%	10.82%	14.26%	0.22%	0.02%
Avg-CN†	3.45%	**16.89%**	**13.68%**	**26.92%**	**16.44%**	22.41%	0.22%	0.00%
Avg-EN	8.69%	**23.52%**	**16.60%**	**21.80%**	**11.77%**	17.39%	0.21%	0.01%
Avg-total	6.07%	20.20%	15.14%	24.36%	14.11%	19.90%	**0.21%**	**0.01%**

† Avg-X stands for the average proportion of X datasets. For example, where CN stands for three Chinese datasets and EN stands for three English datasets.

Table 8. Character composition information of each web service*.

Dataset	[a-z]+	[A-Z]+	[A-Za-z]+	[0-9]+	[a-zA-Z0-9]+	[a-z0-9]+	[a-z]+1	[0-9a-z]+
CSDN	11.64%	0.47%	12.35%	45.01%	96.31%	26.14%	0.24%	5.88%
YueJunYou	12.94%	0.23%	13.39%	65.86%	99.38%	13.10%	0.25%	2.88%
Renren	19.06%	0.64%	20.55%	53.05%	97.79%	17.83%	1.24%	2.80%
Rockyou	41.68%	1.50%	44.04%	15.93%	96.19%	27.69%	4.55%	2.53%
Yahoo	32.51%	1.70%	35.90%	19.80%	97.99%	27.14%	3.47%	3.32%
Youporn	45.94%	1.04%	48.42%	20.12%	96.50%	20.59%	2.75%	1.91%
Avg-CN†	14.55%	0.45%	**15.43%**	**54.64%**	97.83%	19.02%	0.58%	3.85%
Avg-En	40.04%	1.41%	**42.79%**	**18.62%**	96.89%	25.14%	3.59%	2.59%
Avg-total	27.30%	0.93%	29.11%	36.63%	**97.36%**	22.08%	2.08%	3.22%

* Note that the first row is written in regular expressions. For instance, [a-z]+ means passwords composed of lower-case letters; [A-Za-z]+ means passwords composed of letters; [a-z]+1 means passwords composed of lowercase letters, followed by the digit 1.
† Avg-X stands for the average proportion of X datasets, where CN stands for three Chinese datasets and EN stands for three English datasets.

Table 9. Top-10 password information of each web service.

Rank	CSDN	YueJunYou	Renren	Rockyou	Yahoo	Youporn
1	123456789	123456	123456	123456	123456	123456
2	12345678	111111	123456789	12345	123456789	123456789
3	11111111	0	111111	123456789	password	12345
4	dearbook	123456789	0	password	null	1234
5	00000000	123123	123123	iloveyou	12345	password
6	123123123	5201314	5201314	princess	12345678	qwerty
7	1234567890	wangyut2	12345	1234567	1234567	12345678
8	88888888	12345678	12345678	rockyou	iloveyou	123
9	111111111	123	123	12345678	qwerty	1234567
10	147258369	123321	123321	abc123	comeon11	111111
Top-1 %	3.66%	4.78%	3.74%	0.89%	0.86%	2.57%
Top-3 %	8.15%	7.11%	4.99%	1.37%	1.35%	3.71%
Top-10 %	**10.43%**	9.99%	**7.18%**	**2.05%**	2.13%	**5.29%**
Top-10 %	670,881	535,884	339,639	669,126	119,864	113,702
Total num	6,428,277	5,365,338	4,733,366	32,603,388	5,626,485	2,148,224

† Top-x per means the percentage of Top-x passwords, top-10 num means the total number of top-10 passwords.

Appendix 2 Exploratory Experiments

In Sect. 4.3, Probabilistic context-free grammars (i.e., PCFG) [10,18] can be used for data preprocessing when integrated with neural networks. Our refined PCFG are based on the basic PCFG with four additional recognition rules, including keyboard pattern, word, website and year. The experiment result in Sect. 5.2 has already shown that our refiend PCFG can improve the performance by 1.36% on average compared to the basic PCFG when integrated with Long Short-Term Memory neural networks (i.e., LSTM) [17]. To explore the impact of different recognition rules on the experiment results, we evaluate the performance of LSTM based models using PCFG for preprocessing, where only one recognition rule is added to basic PCFG each time.

The result in Table 10 shows that compared to the LSTM based model with basic PCFG for preprocessing: (1) Using PCFG with additional word recognition for preprocessing has a 0.26% improvement on average; (2) Using PCFG with additional keyboard recognition for preprocessing has a 0.06% improvement on average; (3) The remaining recognition rules (i.e., website and year) have little improvement on the results (less than 0.01% on average). In general, adding one recognition rule to the basic PCFG [10] alone is not as effective as adding all the rules (i.e., our refined PCFG) when integrated with LSTM. The reason why the year recognition rule has the worst performance can be can be attributed to two reasons. Firstly, years are part of birthdays and birthdays vary widely among users, which has little effect on trawling password guessing attack. Secondly, individual year segments can be replaced by digit segments. Moreover, the promotion effect of different recognition rules to some extent reflects the pattern that users tend to use when creating passwords.

Table 10. Cracking results of LSTM based models using PCFG based preprocessing methods (Guess number $= 5 * 10^7$)†

Dataset	Language	Basic [17]	Keyboard	Word	Website	Year
CSDN	Chinese	43.54%	43.68%	43.63%	43.55%	43.36%
YueJunYou	Chinese	78.29%	78.85%	78.84%	78.84%	78.34%
Renren	Chinese	46.11%	45.99%	46.58%	46.05%	46.46%
Rockyou	English	63.17%	63.10%	63.15%	63.11%	63.09%
Yahoo	English	44.37%	44.38%	44.57%	44.19%	44.53%
Youporn	English	66.36%	66.20%	66.65%	66.18%	66.13%

† Basic means the LSTM based model with basic PCFG for preprocessing [17]; Keyboard means adding keyboard recognition rule to the basic PCFG; Word means adding word recognition rule to the basic PCFG; Website means adding website recognition rule to the basic PCFG; Year means adding year recognition rule to the basic PCFG. The experiment setup is the same as Sect. 5.

References

1. Blocki, J., Harsha, B., Zhou, S.: On the economics of offline password cracking. In: Proceedings of IEEE S&P 2018, pp. 853–871 (2018)
2. Bonneau, J., Herley, C., Van Oorschot, P.C., Stajano, F.: The request to replace passwords: a framework for comparative evaluation of web authentication schemes. In: Proceedings of IEEE S&P 2012, pp. 553–567 (2012)

3. Bonneau, J., Herley, C., Van Oorschot, P.C., Stajano, F.: Passwords and the evolution of imperfect authentication. Commun. ACM **58**(7), 78–87 (2015)
4. Gulrajani, I., Ahmed, F., Arjovsky, M., Dumoulin, V., Courville, A.C.: Improved training of Wasserstein GANs. In: Proceedings of the NIPS 2017, pp. 5769–5779 (2017)
5. Hitaj, B., Gasti, P., Ateniese, G., Perez-Cruz, F.: PassGAN: a deep learning approach for password guessing. In: Proceedings of the ACNS 2019 (2019)
6. Houshmand, S., Aggarwal, S., Flood, R.: Next gen PCFG password cracking. IEEE Trans. Inf. Forensics Secur. **10**(8), 1776–1791 (2015)
7. Li, Z., Han, W., Xu, W.: A large-scale empirical analysis of Chinese web passwords. In: Proceedings of the USENIX Security 2014, pp. 559–574 (2014)
8. Lipton, Z.C., Berkowitz, J., Elkan, C.: A critical review of recurrent neural networks for sequence learning. arXiv preprint arXiv:1506.00019 (2015)
9. Liu, Y., et al.: GENPass: a general deep learning model for password guessing with PCFG rules and adversarial generation. In: Proceedings of ICC 2018, pp. 1–6 (2018)
10. Ma, J., Yang, W., Luo, M., Li, N.: A study of probabilistic password models. In: Proceedings of IEEE S&P 2014, pp. 689–704 (2014)
11. Melicher, W., Ur, B., Komanduri, S., Bauer, L., Christin, N., Cranor, L.F.: Fast, lean and accurate: modeling password guessability using neural networks. In: Proceedings of the USENIX SEC 2017, pp. 1–17 (2017)
12. Narayanan, A., Shmatikov, V.: Fast dictionary attacks on passwords using time-space tradeoff. In: Proceedings of the ACM CCS 2005, pp. 364–372 (2005)
13. Rodríguez, P., Bautista, M.A., Gonzàlez, J., Escalera, S.: Beyond one-hot encoding: lower dimensional target embedding. Image Vis. Comput. **75**, 21–31 (2018)
14. Wang, D., Cheng, H., Wang, P., Huang, X., Jian, G.: Zipf's law in passwords. IEEE Trans. Inf. Forensics Secur. **12**(11), 2776–2791 (2017)
15. Wang, D., Wang, P., He, D., Tian, Y.: Birthday, name and bifacial-security: understanding passwords of Chinese web users. In: Proceedings of the USENIX SEC 2019 (2019)
16. Wang, D., Zhang, Z., Wang, P., Yan, J., Huang, X.: Targeted online password guessing: an underestimated threat. In: Proceedings of the ACM CCS 2016, pp. 1242–1254 (2016)
17. Wang, D., Zou, Y., Tao, Y., Wang, B.: Password guessing based on recurrent neural networks and generative adversarial networks. Chin. J. Comput. 1519–1534 (2021)
18. Weir, M., Aggarwal, S., de Medeiros, B., Glodek, B.: Password cracking using probabilistic context-free grammars. In: Proceedings of the IEEE S&P 2009, pp. 391–405 (2009)
19. Xie, Z., Zhang, M., Yin, A., Li, Z.: A new targeted password guessing model. In: Liu, J.K., Cui, H. (eds.) ACISP 2020. LNCS, vol. 12248, pp. 350–368. Springer, Cham (2020). https://doi.org/10.1007/978-3-030-55304-3_18
20. Xu, M., Wang, C., Yu, J., Zhang, J., Zhang, K., Han, W.: Chunk-level password guessing: towards modeling refined password composition representations. In: Proceedings of the ACM CCS 2021, pp. 5–20 (2021)
21. Yang, K., Hu, X., Zhang, Q., Wei, J., Liu, W.: Studies of keyboard patterns in passwords: recognition, characteristics and strength evolution. In: Gao, D., Li, Q., Guan, X., Liao, X. (eds.) ICICS 2021. LNCS, vol. 12918, pp. 153–168. Springer, Cham (2021). https://doi.org/10.1007/978-3-030-86890-1_9

Exploring Phone-Based Authentication Vulnerabilities in Single Sign-On Systems

Matthew M. Tolbert, Elie M. Hess, Mattheus C. Nascimento, Yunsen Lei, and Craig A. Shue(✉)

Worcester Polytechnic Institute, Worcester, MA 01609, USA
{mmtolbert,emhess,mcnascimento,ylei3,cshue}@wpi.edu

Abstract. Phone-based authenticators (PBAs) are commonly incorporated into multi-factor authentication and passwordless login schemes for corporate networks and systems. These systems require users to prove that they possess a phone or phone number associated with an account. The out-of-band nature of PBAs and their security may not be well understood by users. Further, the frequency of PBA prompts may desensitize users and lead to increased susceptibility to phishing or social engineering. We explore such risks to PBAs by exploring PBA implementation options and two types of attacks. When employed with a real-world PBA system, we found the symptoms of such attacks were subtle. A subsequent user study revealed that none of our participants noticed the attack symptoms, highlighting the limitations and risks associated with PBAs.

1 Introduction

To authenticate users, some organizations combine traditional passwords with other verification mechanisms in a multi-factor authentication (MFA) scheme. Others eliminate passwords entirely and use passwordless authentication mechanisms. Both MFA and passwordless schemes can use a proof-of-possession authentication factor in which the end user must prove physical access to a device. Phone-based authenticator (PBA) systems are commonly used for proof-of-possession schemes since users often already have and protect smartphones. These PBA systems require a user to promptly interact with the phone associated with an account. If the user completes that interaction successfully, the system approves the authentication attempt.

PBAs are widespread in MFA schemes associated with financial institutions [31] and online account providers [5]. Experts and vendors have encouraged broader use of PBAs with the promise of reducing account compromise risks [25]. Based on industry surveys [28] and legal directives [16], PBA use is expected to grow in the future.

PBAs are commonly paired with single sign-on (SSO) systems [4] in which an identity provider authenticates users for a set of relying parties. However, if SSO implementations do not cache credentials across applications or are improperly tuned, they may frequently prompt users to authenticate using PBAs [26]. Prior

C. Alcaraz et al. (Eds.): ICICS 2022, LNCS 13407, pp. 184–200, 2022.
https://doi.org/10.1007/978-3-031-15777-6_11

work in usable security has found that repetitive warnings and confirmations can desensitize users to the importance of the security decisions they are making [6]. This may enable adversaries to deceive users into risky behavior.

In this work, we ask: *To what extent can adversaries deceive end users into authorizing malicious behavior via phone-based mechanisms (e.g., SMS OTP, email OTP, push notifications)? What phone-based authentication mechanisms have greater risk and what symptoms result? Do end users notice these symptoms? Would additional context help end users distinguish malicious phone-based authentication interactions?*

We explore some common PBA configuration options and their implications using an empirical study with a popular production SSO system. We implement techniques to undermine the PBA system and measure their effectiveness. This leads to the following contributions:

- **Exploration of PBA Settings in Two Attack Scenarios:** We explore a range of PBA implementation options and their potential vulnerabilities. We implement two attacks on PBAs: one using a malicious SSO relying party and one using network packet profiling and strategic delay. We find the malicious SSO relying party can compromise each tested PBA option. The profiling and timing attack is effective against application-based approval prompts. The observable characteristics of both attacks appear to be subtle.
- **Report of User Study on Attack Effectiveness:** We conduct an IRB-approved user study with 13 participants to determine if people notice the authenticator attacks when they occur. We found that 12 participants did not notice the attacks, while the last participant was excluded by our testing protocol before reaching the PBA attack test. Our observations and participant reports indicate only cursory review of PBA prompts and notices, providing ample opportunity for adversary deception.

2 Background and Related Work

Our work combines phone-based authentication, social engineering, and deception with computer users' perception and management of risk. To implement our tools, we use established networking techniques. Accordingly, we review background and prior work in each of these areas. To the best of our knowledge, **we are the first research work to explore attacks that undermine phone-based authenticators without compromising the user's endpoint device, their phone, or the phone's connection (e.g., SIM-swapping)**.

Multi-factor authentication (MFA) schemes often consider what the user knows, what the user possesses, and what the user is as different authentication factors [12]. Some organizations, such as Microsoft, indicate that MFA can prevent more than 97% of identity-based breaches [25]. A study on a data set of Google account authentication records found that device-based second-factor authentication blocks more than 90% of account compromise attempts [14]. However, prior work indicates that MFA has several usability challenges that affect its adoption rate [11]. Ease-of-use, required cognitive effort, and trustworthiness are three major factors that affect MFA's usability [9]. To improve the adoption of MFA, Das et al. [10] conducted a usability study on the Yubico Security Key and observed user difficulties in configuring and using the technology.

Prior research has examined mechanisms to compromise PBAs by compromising the user's endpoint device, their phone, or the communication channel with the phone. The simplest mechanism intercepts unencrypted text messages to phones by falsely registering a device (e.g., "SIM swapping attacks") or by network operators [20,24]. Alternatively, Konoth et al. [23] explore a scenario where an attacker has compromised the endpoint, including the user's browser, and synchronizes with a SMS-stealing application on the user's phone. More recent report [18] showed that scammers create a fake surveys on behalf of reputable companies to mislead users into scan QR code and falsely authenticate online services. Our work explores a simpler attack scenario that does not require a compromise of the user's phone, phone connection, or endpoint.

Phishing is a type of social engineering attack that deceives users into providing their personal information. An attacker can create a fake website to facilitate phishing. Attackers may rely on victims' lack of understanding of URL components to deceive victims with little effort to disguise the destination site [19]. In a user study, Dhamija et al. [13] asked participants to evaluate a website for symptoms of fraud. They found victims of impersonation attacks often only consider the content of a webpage to determine its legitimacy and few considered SSL indicators.

Security warnings are widely used to convey risk. However, prior research found users often ignore these warnings due to a lack understanding of the jargon [34] or habituation effects [2]. Akhawe et al. conducted a field study [1] to examine different types of browser warnings and their click-through rates. They found that malware and phishing warnings have a low click-through rate, while SSL warnings can have high click-through rates, depending on the warning's interface design. Later work [17] examined a new design for SSL warnings to improve adherence via simple, non-technical text and promoting a clear cause of action. To examine how habituation affects disregard for security warnings [33], a user study found that participants who learned to ignore warnings in one task were likely to ignore security warnings in a subsequent task. To combat such habituation, researchers proposed using polymorphic dialogues that continuously change the form of user required input [8] or interface appearance [3] to require user attention for security decisions. We explore the impact of PBA prompt messages and the user perception of these PBA prompts.

3 Understanding PBA Goals, Options, and Impacts

Phone-based authenticator (PBA) systems attempt to validate a user's identity by verifying that the user physically possesses a smartphone associated with their account. A typical SSO authentication session using PBA is illustrated in Fig. 1. In it, a user visits the relying party's website, and when the user wants to authenticate, the relying party redirects the user's browser to the identity provider site. Most identity providers have authentication APIs for relying parties to integrate the login processes into their applications. The identity provider then prompts the user for the credential. Upon validating the user's credentials, the identity provider then issues the PBA challenge to the user. The challenge is typically a task with specific instructions that can be completed only through or with the user's phone. For instance, the challenge might ask the user to input a nonce (i.e., a single-use value) that is only transmitted to the user's phone or ask the user to approve the login using an authenticator application installed on the user's phone. In some options, the response to the challenge is sent via the browser; in other options, it is sent via the user's phone. After successful completion of the PBA challenge, the identity provider redirects the user's browser back to the content provider, along with a token. The token both specifies the user and proves that user's identity.

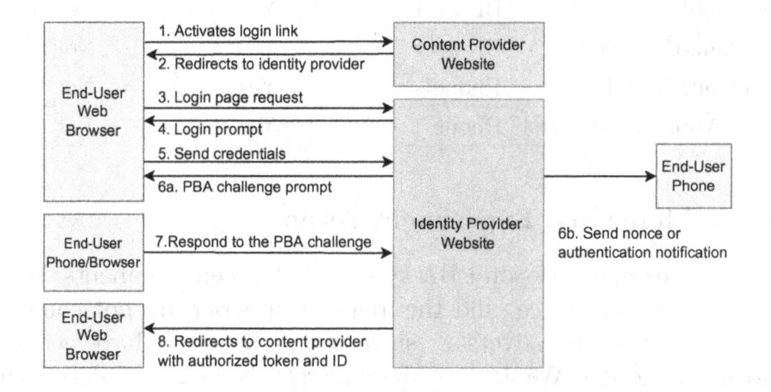

Fig. 1. A general PBA workflow without an attacker

We explored a set of phone-based authenticators as shown in Table 1. The first four entries require the end user to obtain a nonce value and to supply that nonce via the device being authenticated[1]. The underlying mechanism to share the value varies: it can be transmitted via an SMS text message, through a phone call in which an automated system verbally provides a string of numbers, through an email with a code, or through output in a phone application (often implemented via a time-based, one-time password). The next two PBA mechanisms do not require the end user to supply the nonce. In the code matching scenario, the end user selects a value in their phone application that matches what is displayed via the login prompt on the browser. This action links the approval with the active browser session. The "approve" request scenario omits the number matching requirement and simply asks the end user to press an "approve" (or similarly labeled) button on the phone to approve a request; however, this option lack an

[1] We refer to the device being authenticated as "the browser," for simplicity. However, this approach can also be embedded within other application types.

association between the browser session and the phone's prompt. This leads to a timing vulnerability that we explore on its own.

These PBA systems make two key assumptions: 1) the nonce will not be revealed to an adversary and 2) the legitimate user will only respond to the challenge of their own authentication attempts. However, if either assumption is violated, the PBA system will fail to achieve its authentication goals. In the remainder of this section, we discuss the threat model and how an adversary may violate the PBA assumptions to gain unauthorized access.

Table 1. Attack effectiveness by PBA implementation method

PBA implementation	Responding device	Implementation defeated by	
		Malicious site?	Timing attack?
Code via SMS	Browser	Yes	Not tested
Code via Phone Call	Browser	Yes	Not tested
Code via E-mail	Browser	Yes	Not tested
App-based One-time-Code	Browser	Yes	Not tested
App-based Code Matching	Phone	Yes	Not tested
App-based "Approve" Request	Phone	Yes	Yes

3.1 Threat Model and Experiment Setup

We scope our focus to attacks on PBA systems in SSO environments. We assume that the user's phone, device, and the identity provider are not compromised. Further, we constrain the adversary such that it does not have access to the user's phone connection. We assume the adversary's goal is to defeat the PBA factor itself and either has already defeated other authentication factors (e.g., passwords) or does not need to (e.g., a single-factor PBA system).

In one scenario, which we label the "Malicious Site" scenario, the adversary interacts with the user as a malicious SSO relying party. This scenario is consistent with a phishing attack in which an adversary successfully lures a user to a phishing website that impersonates a legitimate web site that uses a specific identity provider. In our second scenario, which we label the "Timing Attack" scenario, the adversary is on path between the user's browser and the SSO identity provider and is able to see and delay/drop packets between those endpoints and to interact with the SSO provider on its own; however, it is unable to decrypt or forge packets belonging to the browser or identity provider. The Timing Attack scenario is consistent with an adversary running a malicious public WiFi network [7], that has compromised the user's residential router [27], or is naturally on-path (e.g., an ISP or nation-state adversary).

For clarity, we describe scenarios in which a user is attempting to authenticate on a client device (e.g., a desktop/laptop computer or tablet), which we refer to as "the browser," and performs the PBA step on a separate device (e.g., a phone). These actions could be done on the same device; if so, one must relax the adversary constraint against having access to the phone's connection.

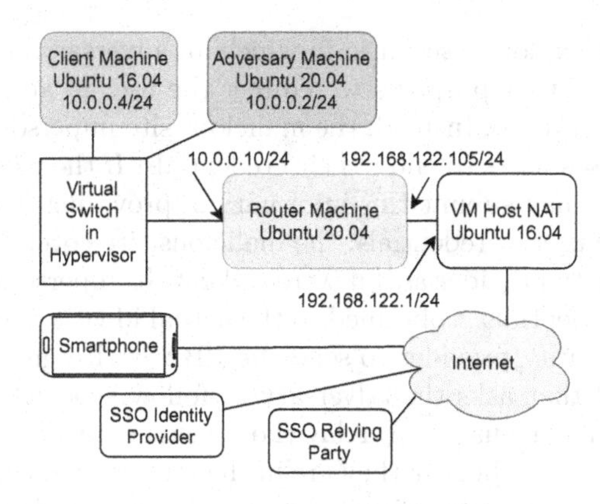

Fig. 2. Our experimental network.

In the remainder of this section, we explore these attack scenarios with different PBA implementation options. We do so using an industry-leading SSO identity provider which has 40% of the identity provider market share and is used by around 80,000 companies globally. We refer to this vendor as *AnonSSO*; we use a pseudonym for the vendor because the vendor employs current industry best practices and the vulnerabilities are inherently due to PBA options themselves, not due to the vendor's implementation. The AnonSSO vendor's approach is representative of other implementations, and there were no implementation-specific details that would prevent the results from generalizing to other implementations. Our study was approved by our Institutional Review Board (IRB) and was conducted with careful attention to ethical conduct. As we further explain in subsequent sections, our scenarios do not harm or attempt to compromise the AnonSSO system.

We perform our experiments using a set of virtual machines that are connected with the AnonSSO system via a bridged network interface. Figure 2 shows our experimental network setup for conducting PBA attacks. We host three virtual machines on a VM server. One virtual machine (top left) acts as the legitimate client, another acts as an adversary for the malicious website (top right), and a third acts as a router that is benign in the Malicious Site scenario but is adversary-controlled in the Timing Attack scenario. These three virtual machines are connected through a virtual bridge created by the physical machine's hypervisor. The router is configured with two interfaces: one to the virtual bridge and one to the hypervisor's network card via a NAT interface. The router provides connectivity to the AnonSSO authentication portals. A smartphone associated with the legitimate user connects directly to the Internet.

3.2 Impact of Malicious Relying Party Sites

Adversaries have had success in luring users into visiting malicious sites [15] and impersonating legitimate entities [32]. Previous work shown that advanced phishing toolkits [22] can mimic the site with high fidelity, which simplifies this process for adversaries.

Accordingly, we explore a scenario in which an adversary creates a malicious relying party website that purports to redirect the user to an identity provider, but actually does not do so. Instead, the malicious site impersonates the identity provider and prompts users to enter their credentials. If the user does not notice the deception, they may submit this information, providing it to the adversary. Upon receiving the user's credentials, the malicious site covertly initiates a connection to the identity provider as if it were a client. It impersonates the end-user and supplies the credentials it obtained to the actual identity provider. This process triggers the identity provider to send the PBA challenge (and transmit the nonce if necessary) that asks the adversary to follow specific instructions. The adversary uses the malicious site to relay those instructions to the real user. The user, who may incorrectly believe they are in the middle of a valid authentication attempt, may follow the instructions (e.g., by echoing the nonce to the adversary's malicious site or through a phone-based application) to respond to the PBA challenge. By doing so, the end-user effectively authorizes the adversary's authentication attempt rather than its own (either by revealing the nonce to the adversary or approving the adversary's login in the authenticator application, which violates both the first and second assumption discussed in Sect. 3). The adversary succeeds whenever the user authorizes the adversary's login.

As shown in Table 1, the PBA implementation method may require the adversary to relay PBA instructions or request user inputs via the malicious site. In other scenarios, the user may directly interact with their phone's application without requiring the adversary to issue a prompt. Our user study in Sect. 4.3 suggests that adversaries could deceive users into performing these actions.

We test the attack scenario using a legitimate client machine (top left in Fig. 2) to connect to a malicious website (the adversary machine, top right in Fig. 2). Both the client and adversary have unimpeded access to the Internet (i.e., the router machine, center of Fig. 2, forward traffic without manipulation or delay). The malicious relying party uses HTTP communication between itself and the client. It uses a custom set of login pages to mimic the login process of the identity provider while displaying user-supplied credentials and nonce values to the adversary. As mentioned above, the adversary must perform its own login attempt quickly upon receiving credentials, but this can be accomplished with an automated process (e.g., using web browser automation tools such as Selenium [30]). We omit this automation step since it has previously been explored. The nonce received by adversaries is valid for multiple minutes and remains valid during the entire attack process. Our tests confirm that an adversary can implement the scenario in a straightforward manner with few observable symptoms.

3.3 Timing Attacks on Unassociated PBA Approvals

The process of using an SSO protocol results in a specific traffic pattern involving redirection of a client from a relying party to an identity provider and back. An on-path adversary may examine traffic to determine such patterns, leveraging DNS requests to identify the servers involved with relying parties and identity providers. By recording such network traffic and browser actions, adversaries can build a database of actions for each authentication step. The adversary can later use that database when monitoring a target's traffic to time an attack.

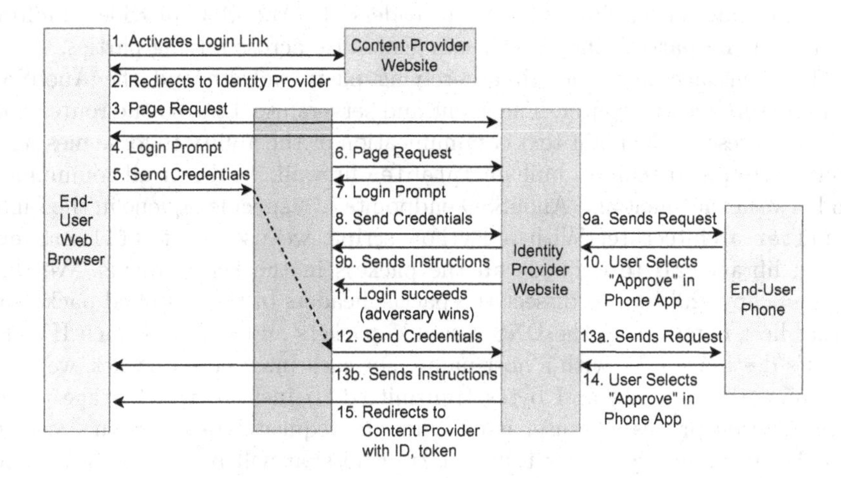

Fig. 3. An on-path adversary launching a timing attack on a PBA workflow

In the timing attack, the adversary monitors all connections from a target to a relying party. The adversary matches the target's packets to each known step in the authentication process. Once the target reaches the step that transmits login credentials, the on-path adversary then can queue the victim's packets and submit its own login request using previously-obtained credentials, as shown in Fig. 3. If the login system uses push-based authentication via a phone-based application, the adversary's attempt to log in will create a notification to the end user's phone asking for approval. Since the prompt appears at the expected time during the user own authentication attempt, the user may approve it. However, in doing so, the end user authorizes the adversary's authentication attempt instead, violating the second PBA security assumption discussed in Sect. 3.

For the timing attack to work, the PBA approval process on the phone must not be explicitly linked with the browser login session. Only the last implementation method in Table 1, the app-based "approve" request, meets this requirement. In that workflow, the "approve" button does not provide context for what session is being approved nor does it require the user to supply a unique identifier (such as a nonce or matching code). This ambiguity allows an adversary to delay and reorder interactions to gain access.

In our experiments, the adversary controls the router machine depicted in the center of Fig. 2. The adversary machine (bottom left) is not involved in this attack. The adversary pre-profiles the relying party interaction with AnonSSO and creates an annotated database with packet sizes for each event. AnonSSO prompts for a username associated with the account and ask for password in a second page. The adversary builds a transition map for the initial authentication page request, the submission of a username, and the submission of a password. For each TCP flow to AnonSSO, it tracks the number of packets and total bytes transmitted to distinguish the password submission step for pausing the legitimate transaction. Since identity providers like AnonSSO provide a uniform API for relying parties, the process can generalize across relying parties.

The client accesses a legitimate relying party website and the AnonSSO infrastructure via the router. The client and servers use TLS, so the router does not have access to the plain-text communication or the ability to forge messages. Using the router machine's built-in `iptables` firewall, we direct all communication between the client and AnonSSO endpoints to a specified queue in the Linux `netfilter` architecture. With a Python script, we use the `NetfilterQueue` Python library [21] to dequeue all the packets in the kernel queue. We then use the `Scapy` [29] tool to dissect the packet headers of the obtained packets to extract host names from the DNS response packets and associate each HTTPS packet's destination IP with a host name. For each intercepted packet, we track the cumulative packets and bytes transmitted to identify which stage of the authentication process the user is in. Once the requisite transmission occurs to send the username, we know the next transmission will be the password submission. Before reaching the password submission stage, we simply forward each packet. However, once we receive the first packet associated with password submission, we queue it and all subsequent traffic in that flow. The adversary then performs an out-of-band login, which sends a PBA approval request to the phone.

3.4 Observable Characteristics of the Attack Scenarios

The Malicious Relying Party Site and the Timing Attack scenarios have subtle symptoms. We describe these symptoms in this section. Our user study in Sect. 4 found that our participants did not have concerns about these symptoms.

For the malicious relying party scenario, the primary non-adversary controlled symptom of the attack is the lack of redirection from the relying party to an identity provider in the browser's address bar. Otherwise, an adversary can convincingly replicate the page visuals to mimic a legitimate relying party and the identity provider. An adversary may choose to continue its deception after succeeding (i.e., after the user completes the PBA process to authorize the adversary) by redirecting the user to the legitimate relying party's site. The user may notice they are not logged in and may retry the process. The user may incorrectly believe the site had an issue performing the login rather than recognizing that they had been attacked.

For the timing attack scenario, the primary non-adversary controlled symptom of the attack is that the browser will not provide confirmation of password submission and a prompt to complete the PBA process before the PBA application prompts the user for approval. Until the user proceeds with the PBA approval, the browser will appear as if it is awaiting a response from the identity provider. After the user approves the attack, the adversary can choose to drop or deliver the queued packets. If dropped, the web request will time out. If delivered, the user will receive a second PBA prompt, which if completed, will authorize the user's access.

In our own personal usage of PBAs, we occasionally receive duplicate PBA prompts for a login attempt or must retry a PBA attempt to successfully sign-in. While the causes of these scenarios are unclear, we suspect network transmission issues or server-related errors. In our user study (Sect. 4.4), we explore whether our human subjects have had similar experiences. We find that they did, and that these experiences seem to desensitize participants to such PBA attack symptoms.

4 User Study and Findings

Given the subtle symptoms of the PBA attacks, we next explore whether users notice them and whether the symptoms raise concerns. To do so, we conducted a user study to gather participants' impressions. We recruited 13 participants, conducted the study, and debriefed each participant to understand their actions. We found that the symptoms of the PBA attacks did not concern the participants.

4.1 IRB Process and Participant Recruiting

We used our organization's Institutional Review Board (IRB) to ensure appropriate protections for our human subjects. The main concern in our study was the use of distractions and ambiguity. Our goal was to measure participants' responses to phone-based security prompts on an account, without biasing the results by revealing that we were specifically monitoring their security decisions.

In our informed consent process, we indicated that our study would explore "how website design affects [the] user experience" and that the study would "measure how various design choices affect how easily and quickly a user notices that information being presented to them is important." The protocol procedures indicated that participants would use video conferencing and screen sharing software to log in to a puzzle website, complete several puzzles, and review the results. The participants received a $5 USD gift card incentive.

We recruited participants via email. Our participants were undergraduate Computer Science students, which may result in biases making them more sensitive to computing details that could reveal a security risk.

4.2 Experimental Setup

Our participants met with the researchers via video conferencing, which was necessary safety consideration during a high propagation phase of the COVID-19 pandemic. The subjects were experienced with video conferencing and, except where noted, we do not believe that the format affected the study.

The researchers used the same experimental VM infrastructure used in Sect. 3.1 using shared screen control software that enables remote control for participants. The researchers then allowed the human subjects to control that VM system via the video conferencing software. The participants were told that as part of our user interface study, the participant would need to remotely control the entire process. It was pointed out that the system may need to use smartphones, and if so, the researchers would hold the phone up to their video conferencing camera and ask the participant how to proceed.

The participants were asked to provide careful feedback about the website associated with a puzzle game. They were asked to log in to the site and were supplied with a researcher-provided account and were told that doing so would enable tracking their progress in the game. For the identity provider, we used the same vendor as in Sect. 3. The researchers ensured that participants used the researcher-provided system and credentials to avoid risk to participants.

If at any point the user expressed security concerns, we immediately ended the study to allow us to debrief the individual and address those concerns. If the participant completed the authentication process without expressing concern, we asked them to play an online game for a few minutes while commenting on any design aspects that they noticed. The requested commentary was to focus users on the website rather than the authentication process during our interview.

During the experiment, one researcher acted as the host and guided the participant through the study. The host allowed the participant to control their screen and VM through the conferencing software. Another researcher focused on performing the attack on the VM system while it was controlled by the participant. If the host needed assistance, the second researcher would provide it. Otherwise, the second researcher remained silent as if they were an observer taking notes, while actually performing the adversary actions in the experiment.

Except as noted, each interactive segment with participants ended with a short interview. We asked some questions before informing the participant of the focus on PBAs and asked others afterward. The researchers answered any questions for the participants, offered online account security advice, and then ended the session. We split the participants into groups to study both the malicious website and the packet delay attacks.

4.3 Participant Responses to the Malicious Relying Party Scenario

In the malicious website attack, we explored both SMS-based delivery of nonce values and application-based approval verification mechanisms. We omitted exploration of code delivery via email, audio phone calls, or application-based code display since they have similar user-facing characteristics as the SMS-based delivery of nonce values.

In our experiments, we created a site that was delivered over HTTP with a similar-looking URL host name (in which the period character was missing from a host name, resulting in the concatenation of a domain and host-name in the domain portion of the URL). When the host directed the participant to log in to the game site, the malicious website showed a fake variant of the Anon-SSO authentication portal. As the user entered their credentials, the adversary observed the console of the website. Once the user submitted authentication credentials, the website displayed the credentials to the adversary's console and took no subsequent action. The adversary viewed those credentials and quickly submitted a separate login attempt to the real AnonSSO authentication portal with the participant-supplied credentials. The adversary's action caused the actual AnonSSO system to send a PBA request to the host's phone.

When the PBA challenge appeared on the host's phone, the host displayed the prompt to the participant. The participant either had to choose to proceed or abort, in the case of the application-based approval process, or to enter the displayed code into the website for the SMS-based code delivery option. If the participant chose to proceed in application-based approach, the adversary's attempt was authorized. Likewise, if the participant typed in the correct code into the malicious website, it was displayed to the adversary via the website's console and the adversary could then enter the code to log in. Both of these outcomes were considered successful attacks. If the participant chose to abort the log in, the attempt was considered an unsuccessful attack.

Table 2. User study results indicating whether individuals identified attacks. One participant was disqualified due to detecting an experimental setup issue unrelated to the phone-based authenticators.

	Malicious site		Timing attack		
	SMS	App, OS context	App, No context	App, Distance	App, Screenshot
Number participants	3	3	3	2	2
Disqualified	1	0	0	0	0
Attack failed	0	0	0	0	0
Attack succeeded	2	3	3	2	2
Symptoms noted	1	1	0	0	0
No symptoms noted	1	2	3	2	2

In the second column of Table 2, we show the results of SMS-based code delivery experiments. One participant was disqualified before proceeding to the PBA test. Another did not detect the attack, but described symptoms of the attack during the post-experiment interview. The third participant did not notice the attack or any symptoms of a problem.

Our testing protocol required us to abort the user study for one of our participants before conducting the PBA attack. The disqualified participant noticed discrepancies in the site content of the fake sign-in page before reaching the stage where a password was entered and before the PBA could be tested. The participant indicated they had previously been the victim of an attack and observed an inconsistency in the animation associated with our mimicry of the vendor's site. This participant did not notice the host name's mismatch or the HTTP indicator. We thus were unable to obtain PBA data for that participant.

When we explored the application-based approval approach, none of the three participants detected the attack live, as shown in the third column of Table 2. Only one participant indicated any symptoms; the one identified was related to a mismatch in operating system on the PBA prompt. This response hinted at the value of context; however, a more sophisticated adversary would be able to observe OS details of the legitimate client and forge browser or OS headers when interacting with the identity provider, causing the results to match.

4.4 Participant Responses to Timing Attack Scenario

In this attack, the host directed the participant to log in to the game website, which started an authentication session via the vendor's authentication page. At the same time, the adversary researcher initiated a second authentication session on a separate system, supplying the same credentials. However, the adversary researcher did not submit the password credentials immediately. Instead, the adversary researcher monitored the attack script running on the router VM. The adversary researcher activated the attack script while the participant was logging in[2]. When the attack script from Sect. 3.3 observed the trigger condition, it automatically paused all packets associated with the participant's login session. Then, the adversary researcher started the login attempt for the second session. The submission of the second session's information resulted in a PBA request via the host's phone.

The host researcher then showed their phone to the participant with the PBA prompt and asked the participant what buttons should be pressed. If the victim told the host to press a button that allowed the request, the adversary's attempt was approved and the adversary could observe the success. The adversary then instructed the attack script to unpause the participant's connection and deliver all the queued packets associated with the participant's log in attempt. This resulted in a second PBA request at the host's phone, which the host then displayed to the participant and asked for instruction.

In exploring the packet pausing attack, we considered two variants: the standard authentication prompt and one enhanced with additional context. As with the lab-based study, both variants considered only the application-based "approve" request workflow in which the participant was asked to confirm whether they initiated the request or not.

In the standard authentication scenario, the user was provided with a prompt that indicated the account being signed in, the device operating system and architecture performing the log in, a rough location (country-level granularity), an indication that the log in was being performed "now," a button indicating this was correct, and one indicating it was incorrect.

As noted earlier, the only symptoms of the timing attack are that the web page where the user submits a credential is briefly delayed and the end user receives multiple requests to authenticate. As we see in the fourth column of Table 2, none of the three participants detected the attack or reported suspicious symptoms during the interview process. In fact, one of the participants indicated that receiving a second phone-based notification request "seemed pretty standard

[2] Future engineering efforts may allow the script to run continuously and to automatically identify the login session. Since our goal was to measure participant reactions, for simplicity, we manually activated it in the user study.

for [AnonSSO]." That participant proceeded through both verification prompts quickly. Another participant took longer to consider both PBA prompts, but proceeded in each case.

We next explored user behaviors when they have additional context that might alert them to something awry. In this case, the host showed the participant a false notification screen during the first authentication request with two discrepancies from a real notification: it showed that the request originated from a location that was thousands of miles away and was from an operating system that mismatched what the participant was using. As shown in the fifth column of Table 2, neither of the two participants in this scenario detected the attack or reported suspicious behavior in the interview.

In our final scenario, participants were shown a screenshot of a computer desktop in the phone-based authenticator prompt and asked if the image matched what they were trying to do. The researchers intentionally ensured that the screenshot did not match the participant's screen: the screenshot showed a different browser, a different OS, and different screen size. Further, the contents of the window did not match: one displayed a username entry page for a login to a different website whereas the study participant was viewing the password entry page for a login to the game website. Both participants chose to proceeded (as shown Table 2, column 6), despite examining the prompt for over a minute. During the post-experiment interview, both participants indicated the picture was difficult to view through the video conferencing software, so they could not clearly see the details or differences. One of the participants indicated the screenshot looked like their PC's desktop, which may have been a false assurance.

This exploration confirmed our hypothesis that the timing attack was too subtle to seem suspicious to end users and that the duplicate authentication prompt would not raise concerns for them. The user study partially refuted our hypothesis that additional context would help. The details about location and machine type provided little value for user verification. While screenshots may have been useful, the experimental setup appeared to affect the results and further study may be needed. However, the post-experiment interview indicated that even when screenshots do not match, users may still proceed anyway, as long as the image looks familiar. One participant expressed privacy concerns if accurate screenshots of the system were to appear within the PBA prompt.

4.5 Participant Feedback and Study Limitations

In our post-experiment interviews, most participants indicated that having PBAs as part of a MFA scheme increased their confidence in the security of their accounts. To them, the approach was worth the inconvenience. However, two participants indicated that it was not worthwhile.

The presence of PBAs increased some participants' confidence that they were interacting with an authentic website. One user believed that an email with a nonce serves as "proof" of security. This reaction indicates that users may be particularly vulnerable to social engineering attacks that incorporate PBAs.

User studies have inherent limitations in terms of realism, representative populations, and scale. Our use of researcher-provided credentials and a researcher observing the login may have affected realism, possibly by providing inherent assurance and by heightening user attention to the process. The video conferencing tool affected the screenshot study, but based on participant feedback and actions, it did not affect the other results. Finally, our participant pool was small, with 13 Computer Science majors, which is subject to bias. However, that bias should have increased the attack detection rate and none of our participants reported PBA attacks. This highlights real risks with PBAs in practice.

4.6 Potential Mitigations for Deployment

We recommend that deployers of PBA systems eliminate the usage of the simple application-based prompt to approve or deny a request. Instead, organizations would be more resilient against timing attacks by using the "code matching" requirement for applications since that requires the user to link the action being authorized on the phone with the device being authorized. This can entirely defeat the Packet Delay Attack.

Providing additional context about the relying party or service being authorized may allow end-users to identify mismatches. Such context was examined by our participants, but despite the presence of mismatches, they did not abort the authentication process. We recommend end-user training about how PBAs work and the symptoms of an attack.

Additional training about typo-squatting and website mimicry would also be useful to help users avoid malicious site impersonation. Because the content of a website is under the adversary's control, attackers can create convincing replicas of legitimate sites. Training users to understand URLs may help manage this risk.

Finally, we recommend that organizations consider human interaction with PBAs and minimize the number of times users must employ them. A misconfigured environment may unnecessarily prompt individuals to use phone-based authenticators, which may desensitize users and cause them not to carefully vet authentication prompts. With the sparing usage of such authentication prompts, users may review them more carefully.

5 Concluding Remarks

We explored the use of phone-based authentication systems, which are in widespread use on the Internet. Despite assurances to the contrary, we showed that these systems offer little resistance to phishing attacks. One common phone-based authenticator mechanism can also be defeated by strategic timing attacks. We explored the attack scenarios and showed that they were unnoticed by technically-inclined participants in a user study.

Acknowledgements. This material is based upon work supported by the National Science Foundation under Grant No. 1651540.

References

1. Akhawe, D., Felt, A.P.: Alice in warningland: a large-scale field study of browser security warning effectiveness. In: USENIX Security Symposium, pp. 257–272 (2013)
2. Amran, A., Zaaba, Z.F., Mahinderjit Singh, M.K.: Habituation effects in computer security warning. Inf. Secur. J.: Global Perspect. **27**(4), 192–204 (2018)
3. Anderson, B.B., Kirwan, C.B., Jenkins, J.L., Eargle, D., Howard, S., Vance, A.: How polymorphic warnings reduce habituation in the brain: insights from an FMRI study. In: ACM Conference on Human Factors in Computing Systems, pp. 2883–2892 (2015). https://doi.org/10.1145/2702123.2702322
4. Avatier: Azure active directory seamless single sign-on (2020). https://docs.microsoft.com/en-us/azure/active-directory/hybrid/how-to-connect-sso. Accessed 29 Apr 2021
5. Avatier: Which companies use multi-factor authentication with their customers? (2021). https://www.avatier.com/blog/companies-use-multi-factor-authentication-customers/. Accessed 29 Apr 2021
6. Bravo-Lillo, C., Cranor, L.F., Downs, J., Komanduri, S., Sleeper, M.: Improving computer security dialogs. In: Campos, P., Graham, N., Jorge, J., Nunes, N., Palanque, P., Winckler, M. (eds.) INTERACT 2011. LNCS, vol. 6949, pp. 18–35. Springer, Heidelberg (2011). https://doi.org/10.1007/978-3-642-23768-3_2
7. Breński, K.P.: Evil Hotspot-are public hotspots safe? Ph.D. thesis, Zakład Strukturalnych Metod Przetwarzania Wiedzy (2017)
8. Brustoloni, J.C., Villamarín-Salomón, R.: Improving security decisions with polymorphic and audited dialogs. In: Proceedings of the ACM Symposium on Usable Privacy and Security, pp. 76–85 (2007). https://doi.org/10.1145/1280680.1280691
9. Cristofaro, E.D., Du, H., Freudiger, J., Norcie, G.: Two-factor or not two-factor? A comparative usability study of two-factor authentication. CoRR abs/1309.5344 (2013). http://arxiv.org/abs/1309.5344
10. Das, S., Dingman, A., Camp, L.J.: Why Johnny doesn't use two factor a two-phase usability study of the FIDO U2F security key. In: Meiklejohn, S., Sako, K. (eds.) FC 2018. LNCS, vol. 10957, pp. 160–179. Springer, Heidelberg (2018). https://doi.org/10.1007/978-3-662-58387-6_9
11. Das, S., Wang, B., Tingle, Z., Camp, L.J.: Evaluating user perception of multi-factor authentication: a systematic review. CoRR abs/1908.05901 (2019). http://arxiv.org/abs/1908.05901
12. Dasgupta, D., Roy, A., Nag, A.: Multi-factor authentication. In: Advances in User Authentication. ISFS, pp. 185–233. Springer, Cham (2017). https://doi.org/10.1007/978-3-319-58808-7_5
13. Dhamija, R., Tygar, J.D., Hearst, M.: Why phishing works. In: Proceedings of the ACM SIGCHI Conference on Human Factors in Computing Systems, pp. 581–590 (2006). https://doi.org/10.1145/1124772.1124861
14. Doerfler, P., et al.: Evaluating login challenges as a defense against account takeover. In: The ACM World Wide Web Conference, pp. 372–382 (2019). https://doi.org/10.1145/3308558.3313481
15. Downs, J.S., Holbrook, M.B., Cranor, L.F.: Decision strategies and susceptibility to phishing. In: Proceedings of the ACM Symposium on Usable Privacy and Security, pp. 79–90 (2006). https://doi.org/10.1145/1143120.1143131
16. European Commission: Payment services (PSD 2) - directive (EU) 2015/2366 (2015). https://ec.europa.eu/info/law/payment-services-psd-2-directive-eu-2015-2366_en. Accessed 6 June 2022

17. Felt, A.P., et al.: Improving SSL warnings: comprehension and adherence. In: Proceedings of the ACM Conference on Human Factors in Computing Systems, pp. 2893–2902 (2015). https://doi.org/10.1145/2702123.2702442
18. Government of Singapore: Police advisory on scam survey leading to the misuse of singpass access to digital services (2022). https://ec.europa.eu/info/law/payment-services-psd-2-directive-eu-2015-2366_en. Accessed 6 June 2022
19. Hong, J.: The state of phishing attacks. Commun. ACM **55**(1), 74–81 (2012). https://doi.org/10.1145/2063176.2063197
20. Jover, R.P.: Security analysis of SMS as a second factor of authentication: the challenges of multifactor authentication based on SMS, including cellular security deficiencies, SS7 exploits, and sim swapping. Queue **18**(4), 37–60 (2020)
21. Kerkhoff Technologies Inc: Netfilterqueue (2021). https://github.com/kti/python-netfilterqueue. Accessed 29 Apr 2021
22. Kondracki, B., Azad, B.A., Starov, O., Nikiforakis, N.: Catching transparent phish: analyzing and detecting MITM phishing toolkits. In: Proceedings of the ACM Conference on Computer and Communications Security, pp. 36–50 (2021). https://doi.org/10.1145/3460120.3484765
23. Konoth, R.K., van der Veen, V., Bos, H.: How anywhere computing just killed your phone-based two-factor authentication. In: Grossklags, J., Preneel, B. (eds.) FC 2016. LNCS, vol. 9603, pp. 405–421. Springer, Heidelberg (2017). https://doi.org/10.1007/978-3-662-54970-4_24
24. Lee, K., Kaiser, B., Mayer, J., Narayanan, A.: An empirical study of wireless carrier authentication for SIM swaps. In: Symposium on Usable Privacy and Security, pp. 61–79 (2020)
25. Microsoft: Microsoft digital defense report (2020). https://www.microsoft.com/en-us/security/business/security-intelligence-report. Accessed 29 Apr 2021
26. Microsoft: Optimize reauthentication prompts and understand session lifetime for Azure AD multi-factor authentication (2020). https://docs.microsoft.com/en-us/azure/active-directory/authentication/concepts-azure-multi-factor-authentication-prompts-session-lifetime. Accessed 29 Apr 2021
27. Niemietz, M., Schwenk, J.: Owning your home network: router security revisited. CoRR abs/1506.04112 (2015). http://arxiv.org/abs/1506.04112
28. ReportLinker: Global multi-factor authentication (MFA) industry (2021). https://www.reportlinker.com/p03329771/Global-Multi-Factor-Authentication-MFA-Industry.html. Accessed 29 Apr 2021
29. SecDev: Scapy (2021). https://github.com/secdev. Accessed 29 Apr 2021
30. Selenium: Seleniumhq browser automation (2021). https://www.selenium.dev/. Accessed 29 Apr 2021
31. Sinigaglia, F., Carbone, R., Costa, G., Zannone, N.: A survey on multi-factor authentication for online banking in the wild. Comput. Secur. **95**, 101745 (2020)
32. Spaulding, J., Nyang, D., Mohaisen, A.: Understanding the effectiveness of typosquatting techniques. In: Proceedings of the ACM/IEEE Workshop on Hot Topics in Web Systems and Technologies (2017). https://doi.org/10.1145/3132465.3132467
33. Sunshine, J., Egelman, S., Almuhimedi, H., Atri, N., Cranor, L.F.: Crying wolf: an empirical study of SSL warning effectiveness. In: USENIX Security Symposium, pp. 399–416 (2009)
34. Zaaba, Z.F., Boon, T.K.: Examination on usability issues of security warning dialogs. Age **18**(25), 26–35 (2015)

FRACTAL: Single-Channel Multi-factor Transaction Authentication Through a Compromised Terminal

Savio Sciancalepore[1](✉), Simone Raponi[2], Daniele Caldarola[2], and Roberto Di Pietro[2]

[1] Eindhoven University of Technology (TU/e), Eindhoven, The Netherlands
s.sciancalepore@tue.nl

[2] Division of Information and Computing Technology (ICT), College of Science and Engineering (CSE), Hamad Bin Khalifa University (HBKU), Doha, Qatar
{sraponi,rdipietro}@hbku.edu.qa

Abstract. Multi-Factor Authentication (MFA) schemes currently used for verifying the authenticity of Internet banking transactions rely either on dedicated devices (namely, tokens) or on out-of-band channels—typically, the mobile cellular network. However, when both the dedicated devices and the additional channel are not available and the Primary Authentication Terminal (PAT) is compromised, MFA schemes cannot reliably guarantee transaction authenticity. The afore-mentioned situation is typical, e.g., offshore or on-board of aircraft, when only few untrusted terminals have Internet connection.

In this paper, we present FRACTAL, a new scheme providing single-channel transaction MFA through general-purpose additional authentication terminals. Moreover, the proposed solution is also resilient against a potentially-compromised PAT. FRACTAL easily scales up as per the number of multiple authentication factors, and it is extensible beyond the banking scenario, e.g., to unattended and constrained scenarios, by integrating also Internet of Things (IoT) devices as additional authentication terminals. Other than enjoying a formal verification of its security properties via *ProVerif*, FRACTAL is also supported by an extensive experimental performance assessment. Our real-world Proof-of-Concept scenarios, implemented using *Spring* micro-services, show that FRACTAL can complete a transaction in about 2 s, independently from the remote server location. The flexibility of use, the guaranteed security, and the striking performance, characterize FRACTAL as a solution with an expected high potential impact in the authentication field, for both Industry and Academia.

Keywords: Internet transactions · Network security · Cryptographic protocols

1 Introduction

The capillary diffusion of Internet services in the last decade has certified the shift of banking services from in-person to online [1], and this trend has been

C. Alcaraz et al. (Eds.): ICICS 2022, LNCS 13407, pp. 201–217, 2022.
https://doi.org/10.1007/978-3-031-15777-6_12

even magnified by the COVID19 pandemic. Nowadays, all banks offer web platforms and mobile smartphone applications allowing users with an active bank subscription to manage their funds online [2].

Despite the evident advantages, the afore-mentioned shift carries a plethora of security issues. For instance, powerful remote attackers could steal legitimate users' credentials to fully impersonate them on the web, e.g., by initiating unauthorized monetary transactions, leading to huge economic losses for both individuals and companies [3]. To offer enhanced security to their customers, many bank service providers currently implement user Multi-Factor Authentication (MFA) solutions [4]. Specifically, user MFA schemes require the client to provide multiple pieces of evidence to demonstrate to be the same physical person associated with the end-user account. Typically, user MFA occurs via either the delivery of a one-time Personal Identification Number (PIN) to a device registered by the end-user with the bank, such as a mobile cellular number, or proving to be in possession of dedicated smart card readers, released by the bank to the user [5]. Moreover, recent standards such as FIDO and FIDO2/WebAuthN provided multiple standardized mechanisms and out-of-the-box APIs to perform user authentication on several platforms [6].

In this context, mutual transaction MFA protocols focus on ensuring the authenticity of the Internet transactions. Such schemes have been widely investigated, both in the literature and in the industry/banking domain (see Sect. 6 and the survey in [4]). Nonetheless, currently-deployed solutions strongly rely on the availability of either dedicated devices (such as reader/token generators), used specifically for authentication purposes, or additional channels to the Internet connection, e.g., the cellular network, used to deliver the one-time password. When they are not available, their security relies on the security of the primary terminal adopted by the user to interact with the remote service. If such a terminal is controlled by the adversary, currently-available user MFA schemes cannot guarantee transactions mutual authentication, as they cannot reliably verify remote server's authenticity. The unavailability of the additional channel is a typical situation, e.g., on a cruise ship offshore, or during an aircraft trip, to name a few. In these scenarios, users and remote servers should establish *Strong Mutual Authentication* without a dedicated channel to the secondary devices, allowing an untrusted terminal to route messages to/from general-purpose additional authentication terminals. To the best of our knowledge, such a challenging scenario has not been addressed, yet.

Contribution. In this paper, we present *FRACTAL*, an efficient solution to enforce single-channel transaction MFA even when the main terminal is compromised. *FRACTAL* is a flexible scheme, where several user devices (e.g., general-purpose or Internet of Things (IoT) ones) can be used to demonstrate the authenticity of an online transaction, although the primary authentication terminal used to trigger the transaction could be compromised. Moreover, *FRACTAL* requires very limited effort by the user, which is required only to identify its own transaction. We discuss the security features of *FRACTAL*, and we prove its security via the verification tool *ProVerif*. Moreover, we implemented a functioning prototype of *FRACTAL* proving that, using *FRACTAL*, it is possible to successfully perform an online transaction in 2 s on average, while adding multiple devices only slightly affects its performance.

We believe that *FRACTAL* may be useful in several application scenarios—as witnessed by the *FRACTAL* supporting patent [7], that inspired this paper.

Roadmap. This paper is organized as follows. Sect. 2 introduces the scenario and the adversary model, Sect. 3 illustrates *FRACTAL*, Sect. 4 discusses the security of *FRACTAL*, Sect. 5 includes the performance evaluation of *FRACTAL*, Sect. 6 reviews the related work, and, finally, Sect. 7 tightens the conclusions.

2 Scenario and Adversary Model

2.1 Scenario

We assume an end-user, namely $\mathcal{A}$, want to access via a regular Internet connection her savings account at the bank $\mathcal{B}$. $\mathcal{A}$ and $\mathcal{B}$ are equipped with a private/public key pair, and their connection is secured via the well-known Transport Layer Security (TLS) protocol, e.g., leveraging public-key certificates. We assume that $\mathcal{A}$ uses a Primary Authentication Terminal (PAT) to interact with the bank $\mathcal{B}$. The Primary Authentication Terminal (PAT) can be either a fixed workstation or a laptop. We do not assume the presence of any particular additional interface on the PAT (e.g., biometrics).

We assume that the server of $\mathcal{B}$, i.e., the *remote server*, requires MFA to authorize any operation. To this aim, $\mathcal{A}$ has to register with $\mathcal{B}$ multiple Additional Authentication Terminals. The Additional Authentication Terminals (AATs) can be any general-purpose device in possession of $\mathcal{A}$, that she can leverage to demonstrate her identity at the authentication time. Note that $\mathcal{A}$ registers the AATs at the join with $\mathcal{B}$, and they can be added/modified through secure channels. During the registration phase, the bank B stores securely its public key and public key certificate on each AAT, to avoid any possible tampering.

Finally, we assume that the AATs could not be connected to the Internet, thus being not able to communicate directly with $\mathcal{B}$. This is a frequent situation, occurring when the end-user is in a remote location. For instance, when $\mathcal{A}$ is on a cruise ship, usually only the PAT is connected to the Internet, while any other AAT would require additional subscriptions. Another use-case would be the use of a shared on-demand terminal, that is owned by the user but rented on request.

2.2 Adversarial Model

The adversary assumed in this work, namely ADV, is in full control of the PAT. We neglect the specific tool used by the attacker to compromise the PAT, as the PAT could be either colluding with the adversary or deployed by ADV on purpose. Overall, this assumption empowers ADV, enabling him to carry out both passive and active attacks from the PAT. As a passive attacker, *ADV* is a global eavesdropper, able to detect any packet transmitted and received by the PAT. As an active attacker, *ADV* features active attacking capabilities, in line with the well-known Dolev-Yao attacker model [8]. Thus, *ADV* can inject its own messages on the channel, either by replaying eavesdropped messages or by forging new messages, impersonate either the PAT or the server, as well as perform Man In The Middle (MITM) attacks against every involved party. In addition, by having complete control of the PAT, *ADV* can tamper with the copy of the key and the public key certificate of the server stored on the PAT, e.g., by replacing them with one of his choices. Moreover, in our paper, we assume that at least one AAT is not compromised by the adversary.

One of the possible goals of the attacker is to steal money from $\mathcal{A}$, held in her account on $\mathcal{B}$. To this aim, relying on the compromised PAT, *ADV* can launch either synchronous or asynchronous MITM attacks.

In the former scenario, *ADV* waits for the end-user to perform a transaction. At that time, *ADV* launches a MITM attack and redirects any request performed by the client to a malicious server, interacting with the remote server on behalf of *A*. In the latter case, *ADV* launches the attack without waiting for actions performed by the end-user.

3 Protocol Description

3.1 Basic Protocol Flow

Figure 1 describes the initial steps required by *FRACTAL*. Note that the operations described in this section are always executed, and they do not depend on the particular scenario where the protocol is operated.

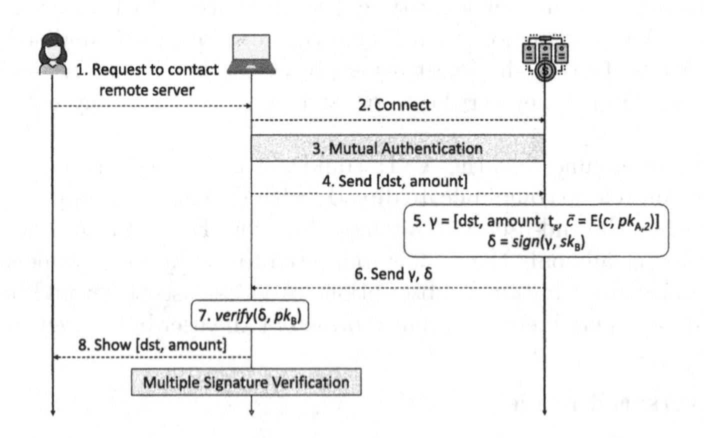

Fig. 1. Initial protocol flow of *FRACTAL*.

1. The end-user first instructs the PAT to initiate a connection with the remote server. Typically, this operation occurs through the typing of a Uniform Resource Locator (URL) in the browser.
2. The PAT sends a connection request to the public IP address of the remote server, on behalf of the end-user. Note that, if the PAT is compromised, the request can be redirected first to the IP address of the attacker, to be then re-routed to the server—e.g., the attacker carries out a MITM.
3. The PAT and the remote server carry out mutual authentication, to verify each other identities. Any mutual authentication protocol can be used here (the prototype described in Sect. 5.1 uses TLS with X.509 certificates).
4. Through the PAT, the end-user specifies the details of the desired banking transaction. This interaction typically happens through the keyboard, and involves the specification of the transaction recipient dst, and the amount to be transferred, $amount$.
5. Assume $\mathcal{A}$ registered a single AAT with the bank (N=1). Let $sk_{\mathcal{A},2}$ and $pk_{\mathcal{A},2}$ be the private/public key pair of the AAT. On the reception of the transaction details from the PAT, the remote server generates a one-time code c. The code c is encrypted using the public key of the registered AAT $pk_{\mathcal{A},2}$, generating an encrypted code $\tilde{c}$, as in Eq. 1.

$$\tilde{c} = E\left(c, pk_{\mathcal{A},2}\right), \tag{1}$$

where the operator $E\left(m, K\right)$ refers to the public-key encryption of the plaintext m using the public key K. When N AATs are registered, N encrypted codes are generated according to Eq. 1. Then, the remote server creates a public-key signature of the transaction, namely δ, as in Eq. 2.

$$\delta = sign\left(\left[dst, amount, t_s, \tilde{c}\right], sk_{\mathcal{B}}\right), \tag{2}$$

where t_s refers to the expiration time of the transactions, and $sign$ is a generic public-key signature algorithm.
6. The information about the recipient of the transaction, the amount, the timestamp, and the signature δ are delivered back to the PAT.
7. At reception time, the PAT first verifies the authenticity of the signature δ, as per Eq. 3, by using the public key of the remote server $pk_{\mathcal{B}}$.

$$verify\left(\left[dst, amount, t_s, \tilde{c}\right], pk_{\mathcal{B}}\right) \stackrel{?}{=} \gamma, \tag{3}$$

where $verify\left(\cdot\right)$ is a signature verification algorithm. Note that the PAT can verify only the authenticity of δ. Instead, it cannot verify autonomously the content of $\tilde{c}$, but it has to rely on the assistance of the AAT(s).
8. As a first verification step, the recipient and the amount involved in the transaction are showed to the end-user. Thus, the end-user can immediately realize if the intended recipient and amount match the ones showed by the PAT. However, when the PAT is compromised, this verification step is not enough to ensure transaction authenticity. Therefore, additional validation steps are performed with the assistance of the AAT(s).

The above-discussed steps are common to all the scenarios assumed in our work. The following operations, instead, depend on the scenario and capabilities of the involved terminals. We hereby consider two reference scenarios, where the AATs are equipped either with a Bluetooth channel (3.2), or with a camera (3.3), being these the most diffused interfaces in general-purposes devices.

3.2 Scenario #1

In this scenario, we assume the PAT is connected to the AATs via Bluetooth. In line with Sect. 2, we assume that the AAT(s) do not have Internet connection. Figure 2 shows the additional interactions required by *FRACTAL*, described below.

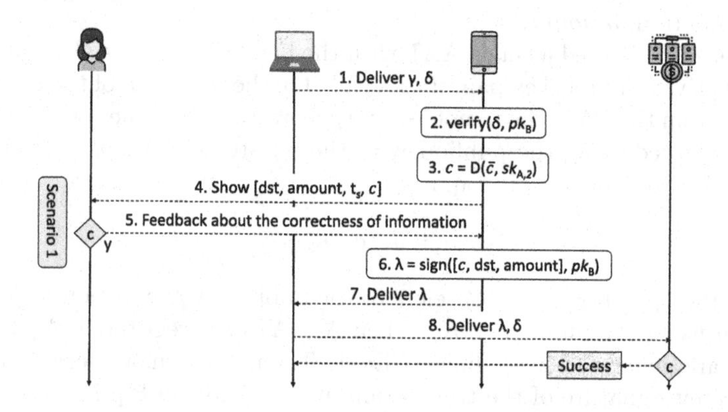

Fig. 2. Additional interactions required by *FRACTAL* in the Scenario #2.

1. Assuming the pairing between the PAT and the AAT has been already performed. The PAT delivers to the AAT all the information received from the remote server, including the transaction recipient, the amount, the validity time, the encrypted code, and the signature δ.
2. First, the AAT verifies the signature δ (2b), via the check in Eq. 3.
3. Then, using its private key $sk_{\mathcal{A},2}$, the AAT extracts c, as per Eq. 4.

$$c = D\left(\tilde{c}, sk_{\mathcal{A},2}\right), \tag{4}$$

4. The available information, i.e., the transaction recipient, amount, and validity time are visually shown to the user, e.g., via a screen, so that she can verify their consistency. If the end-user verifies the correctness of the information and the AAT supports an input method, the end-user can specify her final approval, by pressing a dedicated button. Otherwise, the end-user rejects the transaction.
5. If $\mathcal{A}$ validates the transaction, the AAT signs all the information using the public key of the remote server, generating a signature λ. The value λ is then delivered to the PAT.

6. The PAT forwards to the remote server λ and δ.
7. At reception time, the remote server checks the correctness of λ, using its private key. If the code just received matches the locally-stored one for the particular transaction, identified by its signature δ. If the correspondence is verified, the transaction is executed. Otherwise, the transaction is aborted.

3.3 Scenario #2

In this scenario, we assume the AAT(s) does not feature any means to connect with a remote PAT. Despite this is a quite restrictive assumption (usually, at least one connection mode is available), it helps modelling a scenario where the PAT and the AAT are theoretically incompatible. The only assumption on the AAT is that it includes a camera and an application that can process QR codes. Figure 3 shows the additional interactions required by *FRACTAL*, described below.

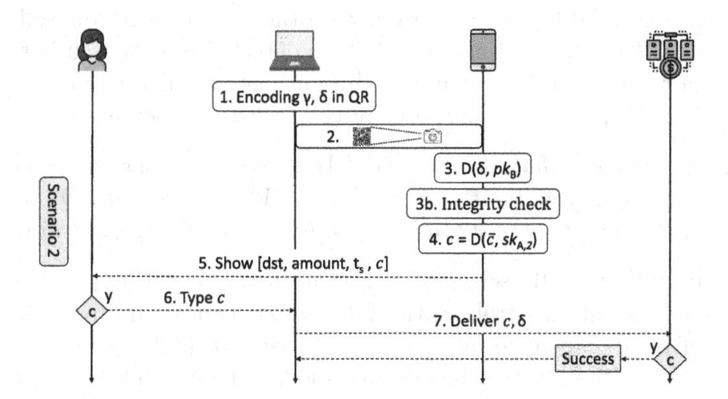

Fig. 3. Additional interactions required by *FRACTAL* in the Scenario #2.

1. The PAT creates a QR code, encoding the information received from the remote server, i.e., the transaction recipient, the amount, the time validity, $\tilde{c}$, and δ, and it shows the QR code on the screen.
2. $\mathcal{A}$ uses the camera on the AAT to acquire the QR code.
3. The processing of the QR code includes two main steps. First, the AAT validates δ (3b), through Eq. 3. Then, the AAT extracts c, as per Eq. 4 (2b).
4. The available information, i.e., the transaction recipient, amount, and validity time are visually shown to the user, e.g., via a screen, so that she can verify their consistency.
5. If the end-user verifies the correctness of the information, the end-user can proceed with the transaction by entering the one-time code c to the PAT. Otherwise, the end-user rejects the transaction.
6. In case $\mathcal{A}$ typed the code c, the PAT forwards to the remote server c and δ.
7. At reception time, the remote server checks if c matches the locally-stored one for the particular transaction, identified by its signature δ. If it matches, the transaction is executed. Otherwise, the transaction is aborted.

4 Security Considerations

4.1 Security Features

Overall, *FRACTAL* provides the following security features.

Protection Against Replay Attacks. *FRACTAL* is robust against replay attacks thanks to the support of nonces and expiration timestamps. Indeed, the signature of the transaction (δ) delivered to the PAT and the AATs includes a one-time code, c, and a timestamp, indicating the transaction expiration time. Thus, any replay after the expiration time is immediately identified and rejected. At the same time, if the adversary replays the transaction before the expiration time, being the code c one-time, the transaction is rejected, as well.

Protection Against MITM Attacks. Thanks to standard mutual authentication, *FRACTAL* protects against external MITM attacks, launched by adversaries not in control of the legitimate entities. An additional feature of *FRACTAL* is the capability to reject MITM attacks even when the PAT is compromised. Indeed, through *FRACTAL*, the end-user can always detect the mismatch between the provided information and the (supposedly authenticated) one received by the remote server. We prove such property in the following Theorem 1.

Theorem 1. *Assume that the attacker ADV compromises the PAT. Then, under the security assumptions in Sect. 2.2, ADV is prevented from misleading the user and the remote server about the authenticity of a forged transaction.*

Proof. Assume that ADV selects dst' and $amount'$ as the forged transaction recipient and amount. Controlling the PAT, ADV can obtain from the remote server an authentic signature $\delta' = sign\left[dst'||amount'||t_s||\tilde{c'}, sk_{\mathcal{B}}\right]$, where $\tilde{c'} = E\left[c', pk_{A,2}\right]$. Now, the attacker has three options. The first one is to send to the AAT the legitimate values dst and $amount$, together with the forged signature δ'. In this case, the signature verification on the AAT, reported in Eq. 3, fails. Thus, the attack would be rejected.
The second option consists in delivering to the AAT dst', $amount'$, and δ'. In this case, being δ' generated from dst' and $amount'$, the check in Eq. 3 would be successful. However, the end-user can notice that $dst \neq dst'$ and $amount \neq amount'$, thus detecting the attack and rejecting the transaction.
The third option is to report to the AAT the legitimate transaction values, i.e., $amount$, dst, and δ, but to report to the remote server the code for the legitimate transaction and the signature of the forged transaction (i.e., c, δ'). Being δ generated from both dst and $amount'$, the check in Eq. 3 would be successful. Thus, the AAT would generate the code c and it would report c to the PAT. Then, the PAT could report to the remote server δ', mimicking an approval of the end-user on the forged transaction. However, being the relationship between the one-time-code and the transaction unique, the remote server could easily verify that both $\delta \neq \delta'$ and $c \neq c'$, denying the execution of the transaction.

Note that the considerations above apply also with *asynchronous* attacks, as the AAT would pop up an unsolicited notification to the end-user. Thus, the end-user could easily realize the ongoing attack. That is, having the human in the loop is instrumental to the security of *FRACTAL*, reverting the saying that the human is the weakest link in the security chain.

In summary, the PAT has no control over the messages it routes from the remote server to the AATs. Therefore, *ADV* cannot modify such messages in a way to achieve its objectives. We formally verify this property in 4.2 via *ProVerif*.

4.2 Formal Security Analysis via ProVerif

We formally verified the security properties of *FRACTAL* using the automatic tool ProVerif [9], in line with many recent scientific contributions [10,11], and we also released the source code at [12], to allow interested readers to reproduce our results and verify our claims. The logic of ProVerif is rooted on two main assumptions. First, the cryptographic primitives used within the security protocol are inherently robust. Second, the attacker is consistent with the widely-accepted Dolev-Yao model, having the capability to read, inject, delete, and modify all the messages exchanged on the communication channel. Based on the above assumptions and user-specified security objectives, ProVerif enables the formal analysis of secrecy and authentication properties [9].

We implemented *FRACTAL* in ProVerif to verify that, even when the PAT is controlled by *ADV*, the remote server could always discriminate forged and legitimate transactions. We implemented the flow of *FRACTAL* in *Scenario #1*, and we modeled the end-user through a simple process, verifying that the values possessed by the AAT are the same typed to the PAT. We also assumed that the mutual authentication step represented in Fig. 1 has been already executed, and that the result of such a process is a session key *TLSKey*, shared between the PAT and the remote server. Finally, to model the tampering of the PAT, we leaked the session key *TLSKey* to the adversary, enabling PAT impersonation.

With reference to our security properties, ProVerif provides the output *not attacker(elem[]) is true* when the attacker does not know the value of *elem*, while the output *not attacker(elem[]) is false* is provided if the attacker knows the value of *elem*. Moreover, the output *inj-event(last_event ()) ==> inj-event(previous_event ()) is true* means that the function *last_event* is executed only when another function, namely *previous_event* is executed. Thus, as per the logic of the ProVerif tool, we defined two main events:

- *begin_EndUser(x_1,x_2)*, indicating that the End-User initiates a transaction specifying the values x_1 and x_2;
- *end_RS(x_1,x_2)*, indicating that the remote server completes a transaction with the values x_1 and x_2.

Figure 4 shows the excerpt of the output of *ProVerif*, when executed locally (recall that the source code is available at [12]).

The first query verifies that the session key *TLSKey* is known to *ADV*. As mentioned above, this condition models the tampering of the PAT. The second query verifies that the event *end_RS(x_1,x_2)*, occurring when the remote server completes a transaction with the values x_1 and x_2, happens if and only if the

```
Verification summary:
Query not attacker(TLSKey[]) is false.
Query inj – event(end_RS(x1, x2)) ==> inj – event(begin_EndUser(x1, x2)) is
true.
```

Fig. 4. Excerpt of the output provided by the *ProVerif* tool.

event *begin_EndUser(x_1,x_2)* has previously occurred. In turn, this means that the server completes a transaction with the end-user only when the end-user really initiated that transaction. Overall, the positive outcome of this query verifies that, thanks to *FRACTAL*, the remote server can always verify the end-user, even when the PAT is compromised.

5 Implementation and Performance Assessment

5.1 Implementation Details

We implemented *FRACTAL* in Java, using Spring [13]. Spring is an open-source application framework, containing a set of core features that can be used by any Java application. It also includes several extensions, that allow to build web applications on top of the Java Enterprise Edition (EE) platform. To allow type inference, conciseness, and inter-operation with mobile applications, we used the Kotlin programming language [14]. We rely on the non-relational database MongoDB to store transactions and the related details [15]. To efficiently manage cryptographic keys and X.509 certificates, we use *keytool* [16], while we used the *ZXing* library for barcode image processing [17]. Finally, we implemented the app running on the AATs using the AndroidStudio IDE [18].

We implemented the PAT and the *remote server* as standalone JAVA web applications on a dedicated machine, i.e., an Intel Core i7-3632QM, equipped with a CPU running at 2.20 GHz, 8.00 GB of RAM, and the Windows 10 Operating System (OS). Using a dedicated web application for the PAT, the user can login and insert the transaction details on a web page, and then the web app manages the interactions with the remote server. This approach allows separating the entities involved in the system, while introducing a negligible interaction delay. For the AAT, we used a *Xiaomi Mi A3* smartphone, running the Android 10 OS. Finally, our protype uses the *SHA-256* hashing algorithm and the *RSA-2048* public key signature scheme. As a reference example, in Figs. 8 and 9 (included in Annex 7), we report the screen shown to the end-user on the AAT and the PAT to validate the transaction, respectively. The web application of the PAT requires 1,342 MB of RAM, while the *apk* of the app is 3,259 KB for the *Scenario #1* and 4,142 KB for the *Scenario #2*.

5.2 Experimental Performance Assessment

For our experimental evaluation, we adopted a methodology inspired by the recent contribution in [19]. Specifically, the PAT, the AATs, and the remote server of our first scenario have been physically implemented in the same machine, thus being directly-connected to each other. However, we modeled a realistic deployment of the remote server in a random point of the World at the communication level, by introducing additional delays in the interaction between the PAT and the remote server. These additional latencies have been modeled by considering real end-to-end communication delays incurred between two real endpoints connected to the Internet. In detail, we identified ten (10) geographically distributed hosts publicly accessible through the Internet via their IP addresses, listed in Table 1. Then, we launched a set of 10,000 *ICMP Echo Requests* to IP addresses in Table 1 from an endpoint located in Doha, Qatar. Finally, we measured the Round Trip Time (RTT) values of the *ICMP Echo Requests*, as the difference between the time when the request is sent and the time when the corresponding *ICMP Echo Reply* is received. Then, the RTTs have been statistically modeled through an empirical Cumulative Distribution Function (CDF), shown in Fig. 5, and these curves have been used to model the time required to contact a specific remote server. Specifically, for each experiment, we located the remote server in one of the identified hosts, and we modeled the corresponding communication delay by extracting a random sample from the empirical CDF of this host in Fig. 5. Assuming a single AAT, the results of our investigation are reported in Fig. 6, along with the 95% confidence interval.

Table 1. Details of the hosts used for the modeling of a remote server.

ID	IP address	Nation	Location
$S1$	39.32.0.1	Pakistan	Islamabad
$S2$	8.8.8.8	USA	Mountain view
$S3$	76.74.224.13	Canada	Vancouver
$S4$	61.69.229.154	Australia	Sydney
$S5$	193.70.52.72	France	Paris
$S6$	167.71.129.73	England	London
$S7$	80.116.252.221	Italy	Rome
$S8$	202.46.34.59	China	Shenzhen
$S9$	125.30.18.121	Japan	Tokyo
$S10$	139.59.140.10	Germany	Frankfurt

Note that the location of the *remote server* has a slight impact on the latencies. The highest (average) delay is observed for $S10$, located in Germany, with a mean value of the delay of 2.262 s. Overall, we can notice that the transaction can be completed in a limited time, not impacting on the usability of the solution, while guaranteeing high level of security.

To provide further insights, we also evaluated the time to complete a transaction, increasing the AATs. As a reference, we assumed a scenario consistent with the *Scenario #1*, where the AATs are connected to the PAT via Bluetooth, with the PAT being the hub of a logical star network topology. The results are provided in Fig. 7, along with the 95% confidence interval over 100 tests.

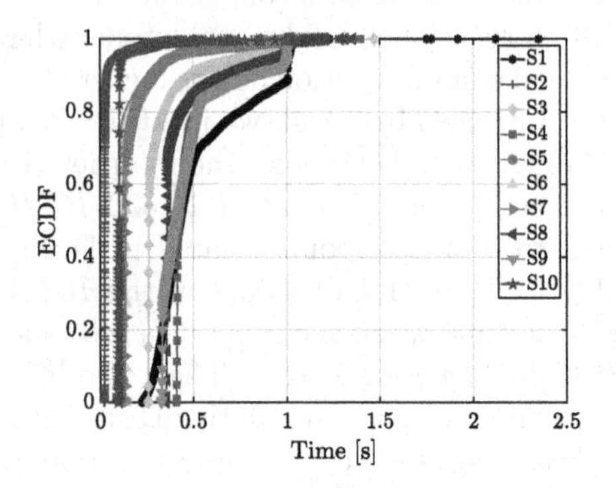

Fig. 5. CDF of the RTTs measured for 10 geographically-distributed hosts.

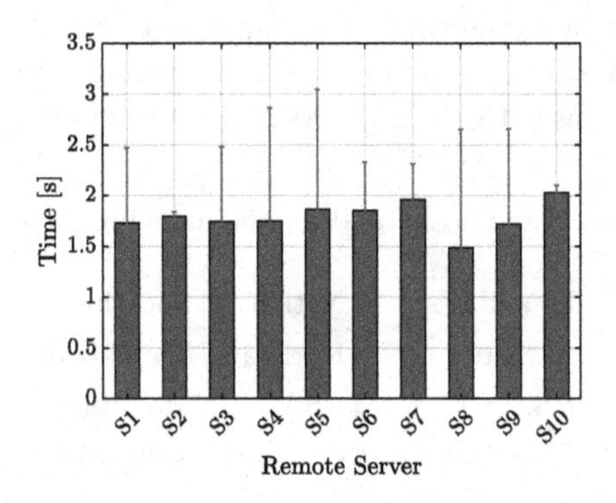

Fig. 6. Time to complete *FRACTAL*, with single AAT and single *remote server* located in mountain view (S2), neglecting the time to input the code.

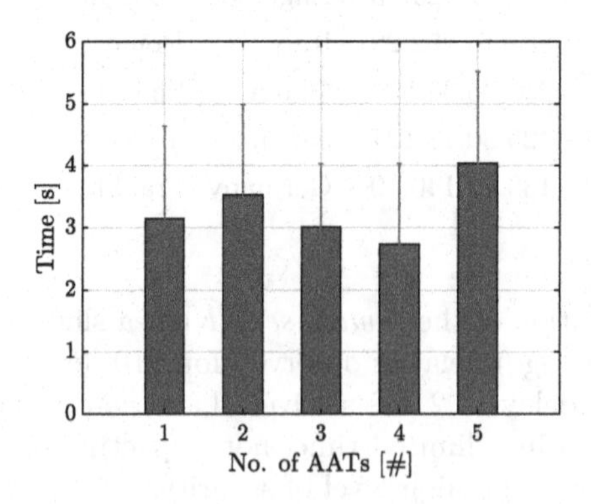

Fig. 7. Time required to complete *FRACTAL*, with multiple AATs directly connected to the PAT via Bluetooth.

As the number of AATs increases, the delay does not increase. Indeed, due to the non deterministic nature of devices interactions, when 3 and 4 AATs are connected, the average delay is less than what experienced for 1 and 2 AATs. Even in the worst case (5 AATs), the transaction can be completed in only 4.035 s, keeping the delay manageable. We remark that this computation delay does not consider the time for the user to input on the PAT.

6 Related Work and Qualitative Comparison

The majority of MFA systems in the literature focus on user/client MFA, using biometric factors to enforce client authentication. For instance, the authors in [20], proposed a two-phase authentication mechanism for federated identity management systems. Unfortunately, the reliance on biometrics prevents the use of legacy devices, not equipped with modules able to gather user biometric features. In [21], the authors addressed the problem of outsourcing the biometric features of users to untrusted servers, by designing a MFA scheme where the biometric features of users remain on their devices. Relying on biometric information, older devices are out of the equation as well. In [22], the authors introduced a zero-effort MFA system, requiring a smartphone and a smartwatch (replaceable by a smart bracelet) able to capture the gait patterns of an individual, her mid/lower body movements, and her wrist/arm movements. Given the alleged uniqueness of the combination of these patterns, only the account holder should be able to authenticate. Relying on specific sensors able to capture users' gait, the scheme is unusable in many contexts.

The authors in [23] introduced *graphical passwords* as new means for MFA. They proposed different architectures where such MFA could be employed, and they demonstrated the enhanced usability of their tool. However, their scheme relies on the delivery of SMS via out-of-band channels, not being usable when this is not available. A similar limitation can be found also in in [24]. The authors in [25] introduced a MFA technique specifically tailored to fragile communications. They contextualized the protocol in a smart grid scenario, where the user interacts with an Intelligent Electronic Device (IED) to perform the authentication. The interaction of users with online banking systems has been studied also by [26], where the authors proposed a MFA scheme without using additional devices to prove the identity of the user. However, the adversary models considered here do not assume tampering of the PAT. In a document released by the Federal Financial Institutions Examination Council [27], the authors embrace the new legal and technological changes with respect to the protection of customer information. However, this document only took into account the customer authentication, thus giving the banks' one for granted. Thus, under the assumption of a possible compromise of the PAT by an attacker, the mutual authentication feature could be easily disrupted. Note that many commercial solutions addresses two-factor and multi-factor authentication, based on the FIDO Alliance Specifications. An example is YubiKey, using *Security Keys*, i.e., hardware devices that authenticate the user after the user presses a button on the security key [28]. Requiring dedicated devices, such solutions are not applicable here. Finally, note that standard security protocols such as Extensible Authentication Protocol (EAP) cannot be used directly to solve our problem, since they all assume that the PAT is not compromised. We summarized the above discussion in Table 2. Note that all the schemes based on single-channel authentication are not effective when the the PAT is compromised. Moreover, they often require biometric interfaces on the supporting devices. Conversely, *FRACTAL* leverages a single communication channel, and it is robust also when the PAT is compromised. Besides, *FRACTAL* does not require additional dedicated interfaces on the supporting devices.

Table 2. Qualitative comparison of *FRACTAL* against competing solutions.

Ref.	Single channel	No dedicated devices as AATs	No biometric interfaces required	Robust against PAT tampering
[20]	✓	✗	✗	✓
[23]	✗	✓	✓	✓
[25]	✓	✗	✗	✓
[21]	✗	✗	✗	✓
[26]	✓	Not Applicable	Not Applicable	✗
[29]	✗	✓	✗	✓
[22]	✓	✗	✗	✓
[24]	✗	✓	✗	✓
FRACTAL	✓	✓	✓	✓

7 Conclusion

In this paper, we have presented *FRACTAL*, a flexible Multi-Factor Authentication scheme for securing banking transactions. *FRACTAL* uses a single communication channel shared between the bank servers and the Primary Authentication Terminal to provide security to the transaction, even if the Primary Authentication Terminal is compromised. The solution does not require any specialized hardware: it involves only (CoTS) Additional Authentication Terminals (AAT)—e.g. mobile phone, smart watch.

We discussed the security features of *FRACTAL*, and we proved its security using *ProVerif*. We also implemented it in a real client-server scenario, using *Spring* micro-services. Our experimental campaign demonstrated that, using a single AAT, independently from the remote server location, the transaction can be completed in about 2 s, while additional AATs can be added with just a slight impact on the end-to-end delay.

In addition to its striking security properties, we believe that our solution could also play a key role in overcoming the need for dedicated devices currently used by banks for MFA, and to reduce the need for separate out-of-band channels. Finally, the applicability of *FRACTAL* goes beyond the presented use-case, extending to other domains where additional communication channels are not available and dedicated devices are not suitable.

Acknowledgements. This work was supported by both the HBKU Technology Development Fund under contract TDF 02-0618-190005 and the NPRP-S-11-0109-180242 from the QNRF-Qatar National Research Fund. Both HBKU and QNRF are members of The Qatar Foundation. This work has been partially supported also by the INTERSCT project, Grant No. NWA.1162.18.301, funded by Netherlands Organisation for Scientific Research (NWO). The findings reported herein are solely responsibility of the authors.

Annex A

Fig. 8. Screen shown on the AAT to validate the transaction. $\mathcal{A}$ can verify that the details of the intended transaction match the ones on the screen. Then, in case of *Scenario #1*, $\mathcal{A}$ can validate the transaction by pressing *confirm*. In case of *Scenario #2*, $\mathcal{A}$ can insert the code on the PAT to verify the transaction (see Fig. 9.)

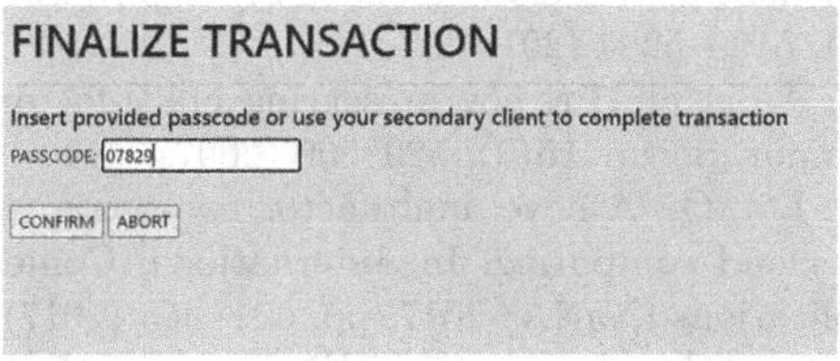

FINALIZE TRANSACTION

Insert provided passcode or use your secondary client to complete transaction

PASSCODE: 07829

CONFIRM ABORT

Fig. 9. Screen shown on the PAT to validate the transaction in case of *Scenario #2*. If the details of the transaction shown on the AAT match the intended ones, $\mathcal{A}$ can insert the code in the *passcode* field and press the *confirm* button to validate the transaction.

References

1. Chandio, F., Irani, Z., Zeki, A., et al.: Online banking information systems acceptance: an empirical examination of system characteristics and web security. Inf. Syst. Manag. **34**(1), 50–64 (2017)
2. Luo, G., et al.: Overview of intelligent online banking system based on HERCULES architecture. IEEE Access **8**, 107685–107699 (2020)
3. Carminati, M., Caron, R., Maggi, F., Epifani, I., Zanero, S.: BankSealer: a decision support system for online banking fraud analysis and investigation. Comput. Secur. **53**, 175–186 (2015)
4. Sinigaglia, F., et al.: A survey on multi-factor authentication for online banking in the wild. Comput. Secur. **95**, 101745 (2020)
5. Kiljan, S., et al.: Evaluation of transaction authentication methods for online banking. Futur. Gener. Comput. Syst. **80**, 430–447 (2018)
6. FIDO Alliance Specifications. https://fidoalliance.org/specifications. Accessed 05 Apr 2022
7. Di Pietro, R., Sciancalepore, S., Raponi, S.: Methods and systems for verifying the authenticity of a remote service. US Patent App. 16/657,088, July 2020
8. Dolev, D., Yao, A.: On the security of public key protocols. IEEE Trans. Inf. Theory **29**(2), 198–208 (1983)
9. Blanchet, B., et al.: ProVerif 2.02pl1: automatic cryptographic protocol verifier, user manual and tutorial. Technical report, September (2020)
10. Tedeschi, P., Sciancalepore, S., Eliyan, A., Di Pietro, R.: LiKe: lightweight certificateless key agreement for secure IoT communications. IEEE Internet Things J. **7**(1), 621–638 (2020)
11. Hirschi, L., Cremers, C.: Improving automated symbolic analysis of ballot secrecy for E-voting protocols: a method based on sufficient conditions. In: IEEE Euro S&P 2019, pp. 635–650 (2019)
12. CRI-LAB, Code of FRACTAL in ProVerif (2021). https://github.com/cri-lab-hbku/tdf-proverif. Accessed 05 Apr 2022
13. Spring Community. https://spring.io/why-spring. Accessed 05 Apr 2022
14. Kotlin Foundation. https://kotlinlang.org/. Accessed 05 Apr 2022
15. MongoDB Inc. https://mongodb.com. Accessed 05 Apr 2022
16. Oracle. https://tinyurl.com/y62ds856. Accessed 05 Apr 2022
17. ZXing Project. https://github.com/zxing/zxing. Accessed 05 Apr 2022
18. Jetbrains. https://developer.android.com/studio. Accessed 05 Apr 2022
19. Sciancalepore, S., et al.: On the design of a decentralized and multiauthority access control scheme in federated and cloud-assisted cyber-physical systems. IEEE Internet Things J. **5**(6), 5190–5204 (2018)
20. Bhargav-Spantzel, A., et al.: Privacy preserving multi-factor authentication with biometrics. J. Comput. Secur. **15**(5), 529–560 (2007)
21. Han, Z., Yang, L., Liu, Q.: A novel multifactor two-server authentication scheme under the mobile cloud computing. In: International Conference on Networking and Network Applications (NaNA) 2017, pp. 341–346 (2017)
22. Shrestha, B., Mohamed, M., Saxena, N.: ZEMFA: zero-effort multi-factor authentication based on multi-modal gait biometrics. In: International Conference on Privacy, Security and Trust (PST) 2019, pp. 1–10. IEEE (2019)
23. Sabzevar, A.P., Stavrou, A.: Universal multi-factor authentication using graphical passwords. In: IEEE International Conference on Signal Image Technology and Internet Based Systems 2008, pp. 625–632 (2008)

24. Mohammed, M.M., Elsadig, M.: A multi-layer of multi factors authentication model for online banking services. In: International Conference on Computing, Electrical And Electronic Engineering 2013, pp. 220–224 (2013)
25. Huang, X., et al.: Robust multi-factor authentication for fragile communications. IEEE Trans. Dependable Secure Comput. **11**(6), 568–581 (2014)
26. Boonkrong,. S.: Internet banking login with multi-factor authentication. KSII Trans. Internet Inf. Syst. **11**(1), 511–535 (2017)
27. Council, Federal Financial Institutions Examination, Authentication in an internet banking environment, FFIEC (2005)
28. Reynolds, J., et al.: A tale of two studies: the best and worst of yubikey usability. In: IEEE Symposium on Security and Privacy (SP) 2018, pp. 872–888 (2018)
29. Nagaraju, S., Parthiban, L.: Trusted framework for online banking in public cloud using multi-factor authentication and privacy protection gateway. J. Cloud Comput. **4**(1), 22 (2015)

Privacy and Anonymity

Lightweight and Practical Privacy-Preserving Image Masking in Smart Community

Zhen Liu[1], Yining Liu[1](✉), and Weizhi Meng[2]

[1] Guangxi Key Laboratory of Trusted Software, School of Computer Science and Information Security, Guilin University of Electronic Technology, Guilin 541004, China
ynliu@guet.edu.cn

[2] Department of Applied Mathematics and Computer Science, Technical University of Denmark, Lyngby, Denmark

Abstract. Advances in the Internet of Things (IoT) and telecommunications technologies (e.g., 5G) have contributed to the development of smart cities and nations (collectively referred to as smart communities). In a smart community, IoT devices can collect significant information about urban residents (e.g., a large number of images collected by cameras containing sensitive information), and such information may be shared with intermediate nodes. In real-world deployment, intermediate nodes are not completely trusted, where the information collected may be used for commercial purposes (e.g., user profiling and advertising) or malicious activities (e.g., covert surveillance). In this paper, we introduce an approach to ensure privacy-preserving image masking. Specifically, before the image is transmitted to the camera owner or the monitoring cloud platform, only sensitive areas instead of the entire image will be processed according to the camera owner's settings, which allows to significantly reduce the computational cost. Then, in order to reduce the interactions between the community data center and the IoT camera, the monitoring cloud platform performs proxy re-encryption. This allows the community data center to recover the original image without relying on the IoT device's private key. Our evaluation indicates the utility and efficiency of our approach, as compared with similar schemes.

Keywords: Smart community · Proxy re-encryption · Privacy preservation · Image masking

1 Introduction

In recent years, the smart community supported by technologies such as the Internet of Things (IoT) and 5G has become a common concept [1,13]. For

Supported in part by the Natural Science Foundation of China under Grant 62072133, in part by the Key Projects of Guangxi Natural Science Foundation under Grant 2018GXNSFDA281040, in part by the Innovation Project of Guangxi Graduate Education under Grant YCSW2022279.

C. Alcaraz et al. (Eds.): ICICS 2022, LNCS 13407, pp. 221–239, 2022.
https://doi.org/10.1007/978-3-031-15777-6_13

example, IoT cameras can collect a large number of images for real-time environmental and crowd monitoring in various applications (e.g., public safety). However, these images are very likely to contain sensitive information, such as license plates, faces, and user behaviors, which can be used for nefarious purposes (e.g., unauthorized surveillance by non-state actors). This reinforces the importance of ensuring privacy in the collected images, during transiting and storing.

Compounding the challenge is the resource-constrained nature of IoT devices (including cameras) since these devices are generally incapable of supporting heavily computational operations. Generally, images collected by IoT devices (e.g., home security cameras) are often outsourced to some monitoring cloud server for storage and subsequent processing in the data center. As the monitoring cloud center is usually considered untrustworthy or semi-trustworthy [22], there is a risk of privacy leakage if original images are transmitted without any security measures. However, conventional encryption methods [14,15] may not be practical since they have to encrypt the entire image. This prevents the utilization of these images by intermediate nodes. However, it is not essential for the intermediate nodes to have access to the raw (unencrypted) image, and only selected features are required for the application. Since IoT cameras can generally capture between 30 and 40 images per second, the computational cost associated with real-time encryption is high. Hence, this motivates our work in this paper to propose a lightweight solution.

Our solution is user-centric, in the sense that the proposed scheme identifies different types of sensitive areas that can be customized by the camera owner. For example, some camera owners may consider license plates to be sensitive while others regard a user's face as sensitive. IoT cameras can then identify specific privacy targets based on the privacy attributes set by the camera owner. Regarding the number of recognized sensitive areas, it depends on the number of sensitive factors in an image. For example, if there are three faces in an image and the sensitive area is set as a human face, then the number of recognized sensitive areas in this image is three. In addition, in a real-world IoT scenario, the attackers are often curious citizens without specialized skills, rather than professionals or intelligence agencies. Therefore, our scheme adopts a lightweight image encryption algorithm that allows cameras to process sensitive areas. Also, during the image transmission process, the intermediate nodes cannot obtain any information about the sensitive area. To reduce the interaction between the community data center and the camera owner, our proposed scheme adopts proxy re-encryption, so that the community data center can use its own private key for image recovery.

Our contributions can be summarized as follows:

- We propose a new threat model – the curious human-eye-attack hypothesis model, because in real-world IoT scenarios, potential attackers do not have professional hacking skills and are often curious citizens.

- Unlike conventional image encryption approaches, our proposed scheme allows users to customize the sensitive areas to be identified, which results in the improved flexibility.
- The encryption time of our proposed algorithm is much shorter than similar schemes, and we can simultaneously perform image recognition and encryption on IoT devices.
- Through proxy re-encryption, the community data center can recover the encrypted image without the need to interact with the encryptor, which reduces the communication overhead.

The paper structure is shown as follows. Section 2 and Sect. 3 introduce the related work and the used tools, respectively. Our proposed approach is presented in Sect. 4, followed by the evaluation setup and finding discussion in Sect. 5. Finally, the conclusion and future work are presented in Sect. 6.

2 Related Work

There is a broad range of image encryption methods in the literature, and this remains a topic of ongoing interest. For example, chaotic mapping has been utilized in several image encryption methods partly because the former is sensitive to initial circumstances, determinacy, and ergodicity [6,9]. Meng et al. [16] proposed an improved image encryption algorithm based on chaotic mapping and discrete wavelet transform domain. In this paper, the image scrambling only changes the position of each pixel, but cannot change the pixel value. Besides, chaotic mapping is often combined with other encryption methods such as DNA encryption to improve the security. Nezhad et al. [27] proposed a hybrid algorithm using the tent chaotic mapping and DNA sequencing techniques. First, the original image and the chaotic mapping are encrypted separately using DNA sequence. Then, the logical XOR operator is applied to them and the encrypted image is produced using chaotic systems. Guan et al. [7] developed a new encryption algorithm based on the IJRBP and chaotic system, which has the advantages of high encryption efficiency and the ability to resist various common attacks.

Mondal et al. [17] proposed a lightweight encryption technique for images using chaotic maps and diffusion circuits. They adopt simple bit-wise operations to reduce the computational overhead. However, Preishuber et al. [20] pointed out that the existing evaluation indicators are inadequate metrics for security analysis of chaotic mapping. In addition, chaotic mapping is computationally expensive to process a single image and is not suitable for the scene that processes real-time images. Other image encryption approaches include those based on permutation [12,19]. While permutation-based approaches are generally more secure than those based on chaotic mapping, as these approaches usually modify the entire image and lack flexibility.

In recent years, there is a trend toward designing image encryption algorithms for IoT environments. Dhall et al. [4] proposed a chaos-based, multiple-round, adaptive, and dynamic framework for image encryption with new levels of dynamism across differently functional dimensions of the entire encryption process. In another work, Rajendran et al. [21] proposed a chaotic security architecture for ensuring security of the images during transmission and storage. Although both schemes are secure and suitable for IoT environments, these algorithms also process the entire image and do not consider the user's demands.

Muhammad et al. [18] proposed a secure surveillance framework for IoT systems by integrating both video summarization and image encryption. By extracting keyframes, the approach detects key events and carries out disaster warnings. However, in the IoT environment, many situations do not meet the keyframe, such as a speeding car. Jan et al. [11] proposed an end-to-end encryption framework to reduce the response time, security overhead, and computational / communication costs. However, this method requires the participation of edge nodes, which increases the difficulty of deployment. Image encryption in the IoT environment is different from traditional image encryption. In the IoT environment, image privacy area changes with factors such as distance and users' varying demands. Also, attackers in the IoT environment tend to be curious residents. Zhang [28] proposed a new image encryption structure based on a lifting scheme. Firstly, the image is decomposed into low frequency components and high frequency components, and then chaotic mapping is used to disturb the two groups of components. Finally, a lifting scheme is used for image encryption. This scheme has a faster encryption speed and higher security.

Now, machine learning, especially deep learning, has developed rapidly. Yang et al. [25] proposed the Graph-based neural networks for Image Privacy (GIP) to infer the privacy risk of images. Yang et al. [26] proposed transferable face image privacy protection based on federated learning and ensemble models. Besides, there are many studies [2,23] aiming to protect image privacy through GAN networks. Although these methods can find a balance between image privacy and image availability, they are not suitable for real smart community environments due to their high computational complexity.

3 Preliminaries

This section aims to briefly introduce the used tools in this work, such as the ChaCha20-Poly1305 stream encryption algorithm, the bilinear map, and proxy re-encryption technology.

3.1 Yolo v5 Object Detection Algorithm

Yolo v5 is an object detection algorithm based on deep learning neural networks, and its code can be accessed via https://github.com/ultralytics/Yolov5. The number of objects in which Yolo v5 can detect is sufficient in the context of this work, and Yolo v5 is also very fast and hence can be deployed on IoT cameras.

The basic structure of Yolo v5 is mainly divided into three parts as shown in Fig. 1: BackBone, PANet, and OutPut. BackBone is mainly divided into Focus structure and CSP structure. The structure of the Focus is first adopted in Yolo v5. It mainly adds the slicing operation. The current Neck of Yolo v5 is the same as that of Yolo v4, and both adopt the structure of FPN and PAN to strengthen the capability of network feature integration. The processed image is finally output through the OutPut.

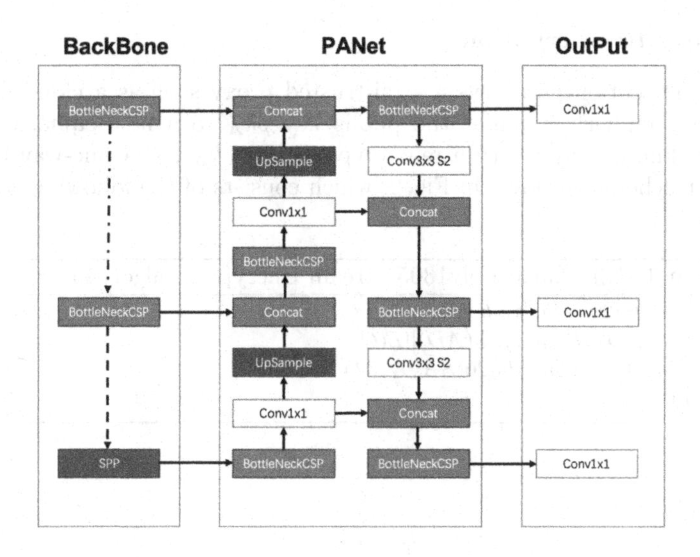

Fig. 1. The architecture of Yolo v5.

3.2 ChaCha20-Poly1305 Stream Encryption Algorithm

ChaCha20-Poly1305 stream encryption algorithm [3] was proposed by Google in 2013, which adopted the AEAD mode. Unlike block encryption algorithms (such as AES), stream encryption algorithm software can achieve higher performance and run faster on mobile devices. Thus, it is very suitable for the IoT environment.

ChaCha20-poly1305 stream encryption algorithm takes as input a 32-byte secret key SK, a 12-byte nonce N, a variable-length plaintext P, and a variable-length associate data AD, and returns a ciphertext C and a 16-byte authentication tag T. This also shows that the algorithm is very suitable for IoT environment, as shown in Algorithm 1, where the function $L : \{0,1\}^* \to \{0,1\}^{64}$ returns the length of the input as 64-bit little-endian integers, and the function $Pad : \{0,1\}^{8\kappa} \to \{0,1\}^{8(\kappa+\delta)}$ returns the κ-byte input padded with $\delta = 16 - \kappa \ mod \ 16$ zero bytes.

3.3 Bilinear Map

We say a map $e : G_1 \times G_1 \to G_T$ is a bilinear map if:

- G_1, G_T are groups of the same prime order of q.

- For all $a, b \in Z_q^*, g \in G_1$, $e(g^a, g^b) = e(g, g)^{ab}$.
- The map is non-degenerate (i.e., if $G_1 =< g >$, then $G_T =< e(g,g) >$).
- e is efficiently computable.

3.4 Proxy Re-encryption

Proxy re-encryption [5] allows a semi-trusted proxy such as a cloud server, to transform a ciphertext under the public key pk_A to a new ciphertext under another public key pk_B with a re-encryption key $rk_{A \to B}$. A one-way proxy re-encryption scheme is shown in Fig. 2, which consists of the following steps:

Algorithm 1. ChaCha20-Poly1305 Stream Encryption algorithm

1: $C \leftarrow ChaCha20 - SC\{SK, N, P\}$
2: $M \leftarrow Pad(AD)||Pad(C)||L(AD)||L(C)$
3: $T \leftarrow Poly1305 - ChaCha20(SK, N, M)$
4: **return** (C,T)

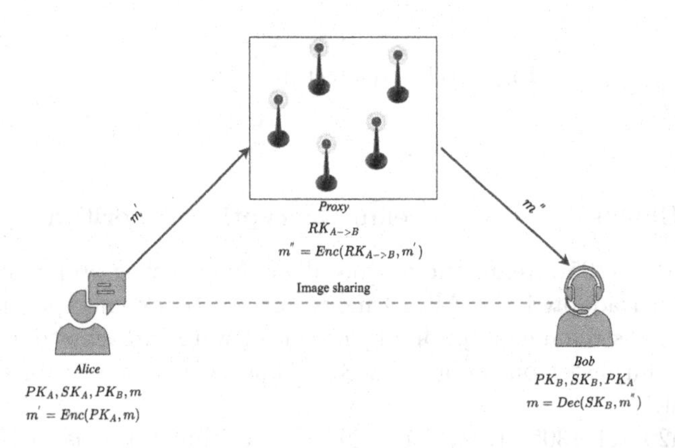

Fig. 2. Proxy re-encryption.

- Setup. Accepting a security parameter n and outputting both the master public parameters and the master secret key.
- Key Generation. Inputting an identity $id \in \{0,1\}^*$ and the master secret key, outputting secret keys sk_A, pk_A, sk_B, pk_B for users A and B, and a re-encryption key $rk_{A \to B}$ for the cloud server.
- Encryption. Inputting image m, system parameters and the public key pk_A, outputting the encrypted image m'.
- Re-encryption. Inputting image m', system parameters, and the re-encryption key $rk_{A \to B}$, outputting the re-encrypted image m''.

- Decryption. Inputting the re-encryption image m'' , system parameters, and a secret key sk_B, outputting the original image m.

However, the traditional proxy re-encryption only supports processing text, we have improved it in this work and make it suitable for image processing.

4 Our Proposed Scheme

The system model is shown in Fig. 3. There are four main parties,such as camera owner, IoT devices, monitoring cloud server and community data center. In particular, camera owner can deploy IoT devices and set sensitive factors. IoT devices are mainly responsible for real-time monitoring, dynamically identifying sensitive areas and encrypting them according to users' requirements. IoT devices and camera owner can be unified on the data owner side. The images without sensitive information will be uploaded to the monitoring cloud server for storage. To reduce the communication burden between the data owner and the community data center (the data consumer), the cloud server will perform proxy re-encryption, enabling the community data center to recover the original image using its own private key. During the process, the data owner and the community data center can complete the data communication, but the monitoring cloud server will not get any information about the sensitive area.

Threat Model. In this work, we consider a curious human-eye-attack hypothesis, because in real-world IoT scenarios, potential attackers are often curious persons who do not have professional hacking skills.

In the following parts, we mainly present our proposed scheme in detail. The notions and abbreviations are shown in Table 1.

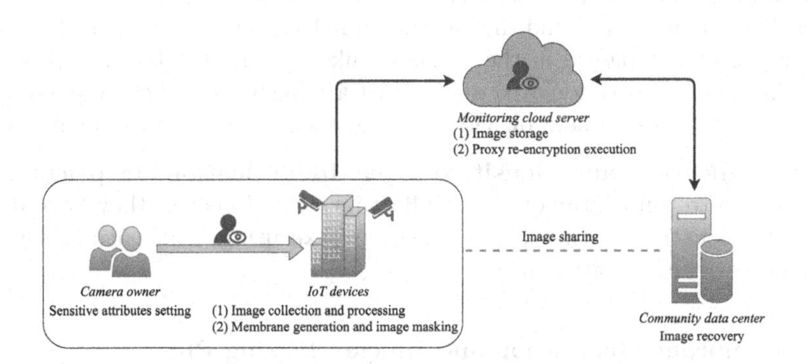

Fig. 3. System model.

Table 1. Notions and abbreviations

Notation	Description
TA	Trusted authority
ids	Identity of IoT devices
pk_D, sk_D	Data owner's private key and public key
pk_U, sk_U	Data user's private key and public key
$rk_{D \rightarrow U}$	Re-encryption key
SA	Sensetive area
SA_x, SA_y	the upper left corner's coordinates of SA
x, y	Width and length of the sensetive area
ori_{img}	Original image
en_{img}	Encrypted image
re_{img}	Re-encrypted image

4.1 Preparation and Image Pre-processing Phase

System Initialization. First, the trusted authority (TA) generates a security parameter n, a master public parameters, and a master secret key. Then according to the master secret key and the identity (ids) of IoT devices, monitoring cloud server, and community data center, TA creates two pairs of public and private keys: $\{pk_D,\ sk_D\}$ and $\{pk_U,\ sk_U\}$. Besides, it also needs to generate a re-encryption secret key $rk_{D \rightarrow U}$ according to IoT devices and the community data center's keys, and sends it to the monitoring cloud server.

The Settings of Sensitive Attributes. Camera owners can set the sensitive attributes that need to be covered according to their demands (e.g., license plates, human faces, etc.) and the location of IoT camera deployment. For example, the cameras deployed in the home are likely to use the face as the private area while the cameras deployed at the door are likely to set license plate as the sensitive area. Then camera owners can send their settings to IoT cameras.

Image Collection and Sensitive Area Identification. In practice, IoT devices have to run continuously to collect images. Moreover, they have to recognize the location of the sensitive area by executing Yolo v5 and save the coordinates of the sensitive area.

4.2 Membrane Generation and Image Masking Phase

After obtaining the size and coordinate points of sensitive areas during the image pre-processing phase, IoT devices generate a masking membrane of the same size as the sensitive area by ChaCha20-Poly1305 stream encryption algorithm according to parameters (SA_x, SA_y, x, y). ChaCha20-Poly1305 stream encryption algorithm is designed for mobile devices, which can be computed very fast by IoT devices. Besides, according to the experimental result, the time consumption for the ChaCha20-Poly1305 stream encryption algorithm is less than Yolo v5. Therefore, it is feasible to run this encryption algorithm on IoT devices. In the following subsections, the algorithm of adding membrane will be introduced in detail. The IoT device then attaches the membrane to the sensitive area of the original image to complete the encryption. Afterwards, the IoT device transmits the encrypted image, along with the size and coordinates of the sensitive area, to the camera owner and the monitoring cloud server for monitoring and storage, respectively.

Algorithm 2. Image encryption algorithm.

Input: the original image ori_{img}, four coordinate points (SA_x, SA_y, x, y) of the sensitive area.

Output: the encrypted image en_{img}.

```
1: for i = 0 → x do
2:     for j = 0 → y do
3:         mem[i][j] = ChaCha20 − Poly1305(SA[i][j])
4: for i = 0 → SA_x do
5:     for j = 0 → SA_y do
6:         if i == SA_x and j == SA_y then
7:             Find SA
8: en_img = img
9: for i = 0 → x − 1 do
10:    for j = 0 → y − 1 do
11:        en_img[SA_x + i][SA_y + j] = mem[i][j]
12: return en_img
```

- **Step 1.** IoT devices run Yolo v5 object detection algorithm to identify the sensitive area SA of the image ori_{img} and obtain four coordinate points SA_x, SA_y, x, y. (SA_x, SA_y) represents the horizontal and vertical coordinates of the upper left corner of the sensitive area. x represents the width of the sensitive area. y represents the length of the sensitive area. The sensitive area SA of the image is shown as Eq. 1.

$$SA = \begin{bmatrix} I_{SA_x,SA_y} & \cdots & I_{SA_x,SA_y+y} \\ I_{SA_x+1,SA_y} & \cdots & I_{SA_x+1,SA_y+y} \\ \vdots & \cdots & \vdots \\ I_{SA_x+t,SA_y} & \cdots & I_{SA_x+t,SA_y+y} \\ \vdots & \cdots & \vdots \\ I_{SA_x+x,SA_y} & \cdots & I_{SA_x+x,SA_y+y} \end{bmatrix} \tag{1}$$

- **Step 2.** IoT devices generate a membrane of the same size as a sensitive area and encrypt it using the ChaCha20-Poly1305 stream encryption algorithm and its public key pk_D.
- **Step 3.** IoT devices put the membrane into the sensitive area of the image to generate an encrypted image en_{img} according to Algorithm 2. Then IoT devices upload the encrypted image and four coordinate points to the monitoring cloud server.

Algorithm 3. Proxy re-encryption algorithm.

Input: the encrypted image en_{img}, the re-encryption secret key $rk_{D \to U}$.
Output: the second encrypted image re_{img}.

```
1: for i = 0 → SA_x do
2:     for j = 0 → SA_y do
3:         if i == SA_x then
4:             if j == SA_y then
5:                 find SA
6: for i = 0 → x do
7:     for j = 0 → y do
8:         en_img[SA_x + i][SA_y + j] = re − encryption(en_img[SA_x + i][SA_y + j])
9: re_img = en_img
10: return re_img
```

Algorithm 4. Image recovery

Input: the second encrypted image re_{img}, the data center's private key sk_U.
Output: the original image img.

```
1: for i = 0 → SA_x do
2:     for j = 0 → SA_y do
3:         if i == SA_x and j == SA_y then
4:             find SA
5: for i = 0 → x do
6:     for j = 0 → y do
7:         re_img[SA_x + i][SA_y + j] = ChaCha20 − Poly1305(re_img[SA_x + i][SA_y + j])
8: img = re_img
9: return img
```

4.3 Proxy Re-encryption Phase

After the monitoring cloud server receives the encrypted image with the size and coordinates of the sensitive area, it can locate the sensitive area of the encrypted image, and then perform the proxy re-encryption (see Algorithm 3) on sensitive area to encrypt en_{img} into re_{img}. After finishing re-encryption, the monitoring cloud server saves the second encrypted image re_{img} for subsequent processing and usage.

4.4 Image Recovery Phase

Camera owner can decrypt images directly. As for the community data center, it first sends the image request to the monitoring cloud server. The monitoring cloud server then searches for the requested image according to the request and sends it back to the community data center. When the data center receives the second encrypted image re_img, it executes the decryption algorithm of the ChaCha20-Poly1305 stream encryption algorithm to recover the image by using the secret key sk_U. Then it can get the original image. The algorithm is shown in Algorithm 4.

5 Evaluation and Results

In this section, we introduce the experimental setup and evaluate the performance of our proposed scheme.

5.1 Evaluation Setup

Sensitive-Area Identification Dataset. To the best of our knowledge, datasets for identifying sensitive areas are not currently available. We then integrated the commonly used datasets such as wider face dataset and Chinese city parking dataset to form a new dataset, named as the sensitive-area identification dataset. It includes three main types of sensitive areas: faces, license plates, and fingerprints.

Due to our evaluation conditions, all experiments were conducted by using Python language with Intel UHD Graphics 630 1536 MB processor, 16 GB RAM, and MacOS Catalina 10.15.7. It is worth noting that the Macbook's performance is closer to an IoT camera.

First, we trained Yolo v5 on sensitive-area identification dataset to recognize license plates, fingerprints, and faces with over 95% accuracy. The Yolo v5 with 28-layer adotps Adam optimizer and has the following settings: learning rate = 0.0001, test size = 0.2, and batch size = 64. The input image was first cropped into $3 * 640 * 640$. Then, Yolo v5 object detection algorithm was run on this computer in order to determine the location and the number of sensitive areas. After obtaining the sensitive region of the image, we performed the sensitive region encryption algorithm to obtain the encrypted image. Then the proxy re-encryption and decryption algorithm will be executed to obtain the re-encrypted image and decrypted image respectively.

As an example, we selected three typical images from the commonly used images during image processing, as given in Fig. 4. In particular, Fig. 4(a) shows the original images. Figure 4(b) describes the generated invisible membrane to cover the private location of the original image. Figure 4(c) presents the fully encrypted images, and Fig. 4(d) shows the community data center that decrypts the original image with its secret key.

5.2 Findings and Results

PSNR. The Peak Signal to Noise Ratio (PSNR) is an objective criterion for evaluating images and is used for an engineering project between the maximum signal and background noise. We use PSNR to evaluate the quality of decoded images as follows:

$$PSNR = 10 \times log_{10} \frac{(2^n - 1)^2}{MSE} \tag{2}$$

$$MSE = \frac{1}{H \times W} \sum_{i=1}^{H} \sum_{j=1}^{W} (X(i,j) - Y(i,j))^2 \tag{3}$$

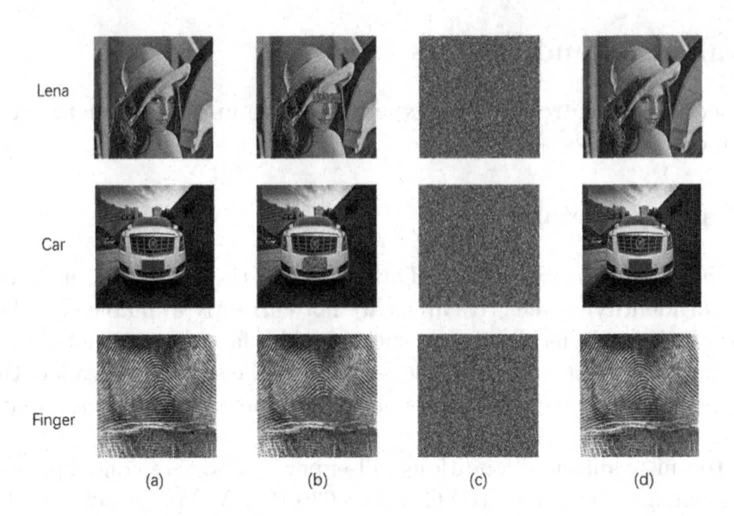

Fig. 4. Example: (a) images with 256 * 256, (b) sensitive area masking images, (c) fully encrypted images, (d) recovered images.

where (H, W) are the length and width of image I, and Y is a noise image. Generally, the larger the value of PSNR, the higher capability to recover the image. As the face recognition example, the evaluated PSNR of "Lena", "Finger" and "Car" images is ∞. We believe that the quality of the decrypted image is extremely high according to the previous work [10], which is the same as the original image quality.

Image Entropy. Image entropy is a statistical form of features, which reflects the average amount of information in the image. The one-dimensional entropy of an image represents the amount of information contained in the aggregation feature of the grayscale distribution in the image. Let P_i represent the proportion of pixels with a grayscale value of i in the image, and then define the unary grayscale entropy of a grayscale image as:

$$H = -\sum_{i=0}^{255} P_i log P_i \tag{4}$$

where P_i is the probability of a certain gray level in the image, which can be obtained from the gray level histogram. In our experiment, we calculated the image entropy of Lena, finger, and car. The image entropy of "Lena" is 7.634, "Car" image is 7.626, and "Finger" image is 7.623, which indicates that the risk of accidental information leakage is very low.

NPCR. The number of changing pixel rates (NPCR) and the unified averaged changed intensity (UACI) [24] are the two most commonly used quantities to

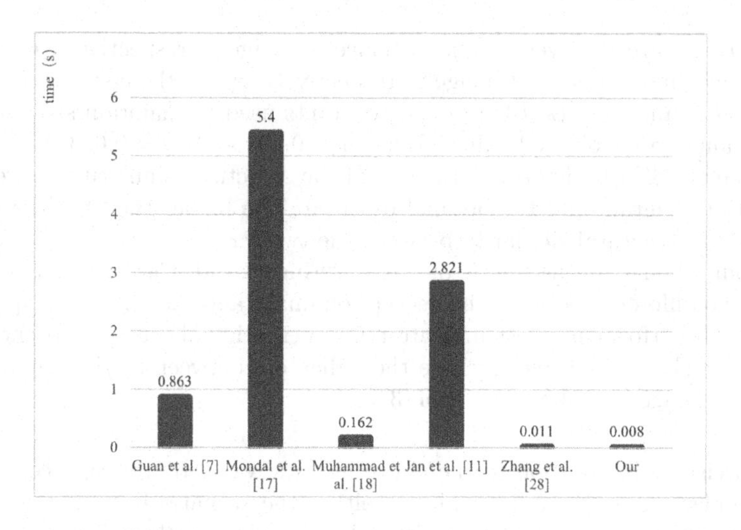

Fig. 5. Encryption time comparison with image size of 1024 * 1024.

evaluate the strength of image encryption algorithms/ciphers concerning differential attacks. NPCR represents the ratio of different grayscale values for different encrypted images at the same location, which can effectively measure the randomness between two similar images. UACI represents the average variation density between different encrypted images. UACI can estimate the intensity change between the corresponding pixels of two different shadow images. Given the two images I and J, both of which are $H * W$ in size, NPCR and UACI can be defined as follows:

$$NCPR(I,J) = \frac{1}{W * H} \sum_{i=1}^{W} \sum_{j=1}^{H} D(i,j) \tag{5}$$

$$D(i,j) = \begin{cases} 0, \; if \; I(i,j) = J(i,j) \\ 1, \; if \; I(i,j) \neq j(i,j) \end{cases} \tag{6}$$

In our experiment, we calculated the NPCR and UACI of the "Lena" image as 1 and 0.501, the NPCR and UACI of the "Car" image as 1 and 0.499, and the NPCR and UACI of the "Finger" image as 1 and 0.499. The larger the UACI value, the stronger the strength of the image encryption algorithm. Therefore, our solution can be highly robust against differential attacks.

SSIM. Structural similarity (SSMI) [8] is a measure of predicting the perceived quality between two images. Given the two images x, y and the structural similarity of the two images, the structural similarity of the two images can be calculated as follows:

$$SSIM(x,y) = \frac{(2\mu_x\mu_y + c_1)(2\sigma_{xy} + c_2)}{(\mu_x^2 + \mu_y^2 + c_1)(\sigma_x^2 + \sigma_y^2 + c_2)} \tag{7}$$

where μ_x, σ_x are the average and variance of image x, respectively. μ_y, σ_y are the average and variance of image y, respectively. σ_{xy} is the covariance of x, y. $c_1 = (k_1L)^2$ and $c_2 = (k_2L)^2$ are two constants used to maintain stability. L is the dynamic range of pixel values. Here $k_1 = 0.001$ and $k_2 = 0.003$. We can use the value of SSIM to determine the size of the structural similarity between the decryption diagram and the original diagram. The higher the SSIM value, the greater the structural similarity between the two images.

In our scheme, all the SSIMs of "Lena", "Finger" and "Car" are 1. This shows there is no difference between the decryption image and the original image structurally. All performance test indicators between original images and encrypted images are shown in Table 2, where the indicators between original images and recovered images are shown in Table 3.

Comparison. In addition to the above metrics, we also provide a comparison with similar research studies which is either the scenario is similar with ours or the computation time is very little, by considering other features such as dynamic, user-centric, and the interaction between the community data center and the image owner, as shown in Table 4. These studies can only identify one or more specific targets, and they cannot be changed dynamically according to the users' needs. In particular, the fastest encryption scheme from Zhang et al. [28] in these studies could reach 0.011 s, but it is still slower than our proposed scheme with only 0.008 s. In addition, the decryption process of these schemes requires the participation of the data owner, resulting in a high communication burden.

Table 2. Performance indicators between original images and encrypted images.

	PSNR	SSIM
Lena	∞	1
Car	∞	1
Finger	∞	1
Baboon	∞	1

Table 3. Performance indicators between original images and recovered images.

	Image entropy	NPCR	UACI
Lena	7.634	1	50.1
Car	7.626	1	49.9
Finger	7.623	1	49.9
Baboon	7.629	1	50.0

Table 4. Other functions compared with similar studies.

	Our	Guan et al. [7]	Mondal et al. [17]	Muhammad et al. [18]	Jan et al. [11]	Zhang et al. [28]
Dynamic	Yes	No	No	No	No	No
User-centric	Yes	No	No	No	No	No
Light interaction	Yes	No	No	No	No	No

5.3 Efficiency Analysis

Instead of encrypting the whole image, in our scheme, we only need to encrypt the sensitive areas selected by users. We also selected the three images as shown in Fig. 4(a) whose sizes are 1024 * 1024 pixels as an example to calculate the algorithm time for an average of one hundred masking operations.

From the experimental results, our image encryption speed is very fast, which only takes **0.033 s** for IoT devices to encrypt. This is a great improvement as compared with Yolo v5 that is 0.154 s in our experimental environment. Also, the time consumption for the cloud to run re-encryption is only 0.0479 s. In addition, the encryption time changes only slightly with varied sizes of the sensitive area. The comparison in the aspect of encryption time between our scheme and other algorithms is shown in Fig. 5. Since our scheme can dynamically process images according to the size of sensitive area, Fig. 6 shows the time consumption of image encryption and proxy re-encryption with varied sizes of the sensitive area.

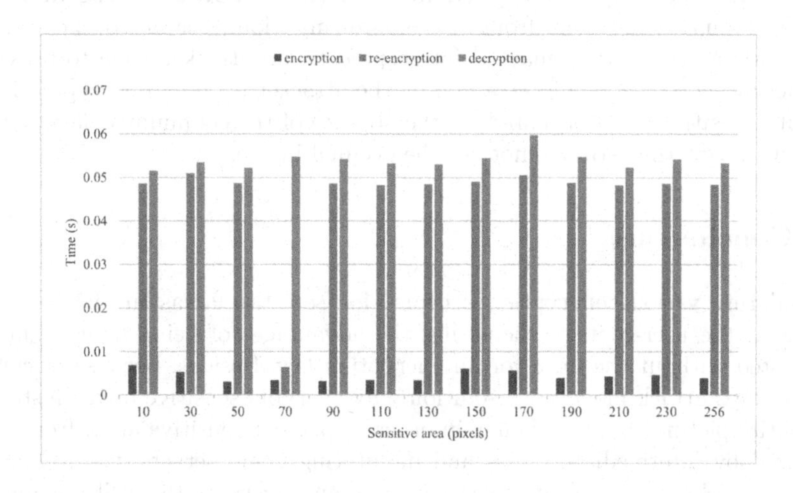

Fig. 6. Time changes with the size of the sensitive area.

5.4 Security Analysis

As illustrated in Fig. 3, our system consists of three entities: the data owner including camera owner and IoT devices, the monitoring cloud server, and the community data center. The data owner is responsible for image collection and image encryption. The encrypted images are then uploaded to the monitoring cloud server. The monitor cloud server stores encrypted images and performs proxy re-encryption. The community data center initiates a data request to the monitoring cloud server, retrieves the secondary encrypted images, and decrypts them using its own private key (i.e., obtaining the original images). In our system, both the data owner and the community data center are considered to be trusted. Therefore, we only consider the following three adversarial situations.

1) The first encrypted image is intercepted by the attacker. Images may be intercepted by attackers when they are uploaded to the monitoring cloud server by the data owner. The ChaCha20-Poly1305 stream encryption algorithm uses pseudo-random encrypted data streams as secret key. The encryption process is a byte-by-byte dissimilation of the key stream with the plaintext to obtain the ciphertext. Thus, it is not possible to obtain the original image data in polynomial time.

2) The monitoring cloud server tries to get the original image content. Monitoring cloud servers as third-party entities are considered semi-trustworthy or untrustworthy. They must not have any information about the original images. Since the images received by the monitoring cloud server are encrypted by the ChaCha20-Poly1305 stream encryption algorithm, they also cannot get any information about the original image.

3) The secondary encrypted image is intercepted by the attacker. When images are sent from the monitoring cloud server to the community data center, they may be intercepted by an attacker. Due to proxy re-encryption, the image intercepted by the attacker is still an encrypted image with sensitive areas obscured. In the absence of the community data center's private key, they still cannot get the original image.

6 Conclusions

Considering various encryption schemes for sensitive areas in the IoT environment, the user-centric scheme has the advantages of being lightweight and decrypted without the need for an encryption key. Besides, because of curious human eyes attack model, our solution only encrypts sensitive areas, instead of the entire picture. In our scheme, in order to prevent sensitive areas from being acquired by intermediate nodes and monitoring cloud servers, the IoT device performs Yolo v5 to identify sensitive areas and encrypts them. To enable the community data center to decrypt the encrypted image with its own private key, the monitoring cloud performs the proxy re-encryption algorithm and provides secondary encryption when it receives the encrypted image sent by an IoT device. The community data center receives the second encrypted image and decrypts it with its private key to obtain the original image. As the sensitive area encryption algorithm takes very little time compared to Yolo v5, it is possible to identify and encrypt sensitive areas simultaneously on IoT devices.

Also, the use of proxy re-encryption reduces interaction between the IoT device and the community data center in the system. According to the experimental results, it is proved that our proposed scheme is feasible and effective in the IoT environment compared with similar studies.

In future work, we plan to re-run our experiments on some low-performance devices, such as Raspberry Pi. We also try to deploy the program into a real IoT device. In addition, we plan to maintain and update the sensitive-area identification dataset on a regular basis.

References

1. Boccardi, F., Heath, R.W., Lozano, A., Marzetta, T.L., Popovski, P.: Five disruptive technology directions for 5G. IEEE Commun. Mag. **52**(2), 74–80 (2014). https://doi.org/10.1109/MCOM.2014.6736746
2. Chen, Z., Zhu, T., Wang, C., Ren, W., Xiong, P.: GAN-based image privacy preservation: balancing privacy and utility. In: Chen, X., Yan, H., Yan, Q., Zhang, X. (eds.) ML4CS 2020. LNCS, vol. 12486, pp. 287–296. Springer, Cham (2020). https://doi.org/10.1007/978-3-030-62223-7_24
3. De Santis, F., Schauer, A., Sigl, G.: Chacha20-poly1305 authenticated encryption for high-speed embedded IoT applications. In: Design, Automation Test in Europe Conference Exhibition (DATE 2017), pp. 692–697 (2017). https://doi.org/10.23919/DATE.2017.7927078
4. Dhall, S., Pal, S.K., Sharma, K.: A chaos-based multi-level dynamic framework for image encryption. In: Alam, M., Shakil, K.A., Khan, S. (eds.) Internet of Things (IoT), pp. 189–217. Springer, Cham (2020). https://doi.org/10.1007/978-3-030-37468-6_10
5. Green, M., Ateniese, G.: Identity-based proxy re-encryption. In: Katz, J., Yung, M. (eds.) ACNS 2007. LNCS, vol. 4521, pp. 288–306. Springer, Heidelberg (2007). https://doi.org/10.1007/978-3-540-72738-5_19
6. Gu, G., Ling, J.: A fast image encryption method by using chaotic 3D cat maps. Optik **125**(17), 4700–4705 (2014). https://doi.org/10.1016/j.ijleo.2014.05.023, https://www.sciencedirect.com/science/article/pii/S0030402614005993
7. Guan, Z., Li, J., Huang, L., Xiong, X., Liu, Y., Cai, S.: A novel and fast encryption system based on improved Josephus scrambling and chaotic mapping. Entropy **24**(3) (2022). https://doi.org/10.3390/e24030384, https://www.mdpi.com/1099-4300/24/3/384
8. Horé, A., Ziou, D.: Image quality metrics: PSNR vs. SSIM. In: 2010 20th International Conference on Pattern Recognition, pp. 2366–2369 (2010). https://doi.org/10.1109/ICPR.2010.579
9. Hua, Z., Zhou, Y., Huang, H.: Cosine-transform-based chaotic system for image encryption. Inf. Sci. **480**, 403–419 (2019). https://doi.org/10.1016/j.ins.2018.12.048, https://www.sciencedirect.com/science/article/pii/S0020025518309927
10. Huynh-Thu, Q., Ghanbari, M.: Scope of validity of PSNR in image/video quality assessment. Electron. Lett. **44**(13), 800–801 (2008)
11. Jan, M.A., Zhang, W., Usman, M., Tan, Z., Khan, F., Luo, E.: SmartEdge: an end-to-end encryption framework for an edge-enabled smart city application. J. Netw. Comput. Appl. **137**, 1–10 (2019). https://doi.org/10.1016/j.jnca.2019.02.023, https://www.sciencedirect.com/science/article/pii/S1084804519300827

12. Li, T., Zhang, D.: Hyperchaotic image encryption based on multiple bit permutation and diffusion. Entropy **23**(5), 510 (2021). https://doi.org/10.3390/e23050510, https://www.mdpi.com/1099-4300/23/5/510
13. Li, X., Lu, R., Liang, X., Shen, X., Chen, J., Lin, X.: Smart community: an internet of things application. IEEE Commun. Mag. **49**(11), 68–75 (2011). https://doi.org/10.1109/MCOM.2011.6069711
14. Liu, W., Sun, K., Zhu, C.: A fast image encryption algorithm based on chaotic map. Opt. Lasers Eng. **84**, 26–36 (2016). https://doi.org/10.1016/j.optlaseng.2016.03.019
15. Maisheri, C., Sharma, D.: Enabling indirect mutual trust for cloud storage systems. Int. J. Comput. Appl. **82**(2), 1–11 (2013). https://doi.org/10.5120/14085-0768
16. Meng, L., Yin, S., Zhao, C., Li, H., Sun, Y.: An improved image encryption algorithm based on chaotic mapping and discrete wavelet transform domain. Int. J. Netw. Secur. **22**(1), 155–160 (2020)
17. Mondal, B., Singh, J.P.: A lightweight image encryption scheme based on chaos and diffusion circuit. Multimedia Tools Appl. 1–25 (2021)
18. Muhammad, K., Hamza, R., Ahmad, J., Lloret, J., Wang, H., Baik, S.W.: Secure surveillance framework for IoT systems using probabilistic image encryption. IEEE Trans. Industr. Inf. **14**(8), 3679–3689 (2018). https://doi.org/10.1109/TII.2018.2791944
19. Naseer, Y., Shah, T., Shah, D.: A novel hybrid permutation substitution base colored image encryption scheme for multimedia data. J. Inf. Secur. Appl. **59**, 102829 (2021). https://doi.org/10.1016/j.jisa.2021.102829, https://www.sciencedirect.com/science/article/pii/S221421262100065X
20. Preishuber, M., Hütter, T., Katzenbeisser, S., Uhl, A.: Depreciating motivation and empirical security analysis of chaos-based image and video encryption. IEEE Trans. Inf. Forensics Secur. **13**(9), 2137–2150 (2018). https://doi.org/10.1109/TIFS.2018.2812080
21. Rajendran, S., Doraipandian, M.: Chaos based secure medical image transmission model for IoT- powered healthcare systems. In: IOP Conference Series: Materials Science and Engineering, vol. 1022, no. 1, p. 012106 (2021). https://doi.org/10.1088/1757-899x/1022/1/012106
22. Harichandana, B.S.S., Agarwal, V., Ghosh, S., Ramena, G., Kumar, S., Raja, B.R.K.: PrivPAS: a real time privacy-preserving ai system and applied ethics. In: 2022 IEEE 16th International Conference on Semantic Computing (ICSC), pp. 9–16 (2022). https://doi.org/10.1109/ICSC52841.2022.00010
23. Sirichotedumrong, W., Kiya, H.: A GAN-based image transformation scheme for privacy-preserving deep neural networks. In: 2020 28th European Signal Processing Conference (EUSIPCO), pp. 745–749 (2021). https://doi.org/10.23919/Eusipco47968.2020.9287532
24. Wu, Y., Noonan, J.P., Agaian, S., et al.: NPCR and UACI randomness tests for image encryption. Cyber J. Multi. J. Sci. Tech. J. Sel. Areas Telecommun. (JSAT) **1**(2), 31–38 (2011)
25. Yang, G., Cao, J., Chen, Z., Guo, J., Li, J.: Graph-based neural networks for explainable image privacy inference. Pattern Recogn. **105**, 107360 (2020). https://doi.org/10.1016/j.patcog.2020.107360, https://www.sciencedirect.com/science/article/pii/S0031320320301631
26. Yang, J., Liu, J., Han, R., Wu, J.: Transferable face image privacy protection based on federated learning and ensemble models. Complex Intell. Syst. **7**(5), 2299–2315 (2021). https://doi.org/10.1007/s40747-021-00399-6

27. Yoosefian Dezfuli Nezhad, S., Safdarian, N., Hoseini Zadeh, S.A.: New method for fingerprint images encryption using DNA sequence and chaotic tent map. Optik **224**, 165661 (2020). https://doi.org/10.1016/j.ijleo.2020.165661, https://www.sciencedirect.com/science/article/pii/S0030402620314923
28. Zhang, Y.: The fast image encryption algorithm based on lifting scheme and chaos. Inf. Sci. **520**, 177–194 (2020). https://doi.org/10.1016/j.ins.2020.02.012, https://www.sciencedirect.com/science/article/pii/S0020025520300748

Using Blockchains for Censorship-Resistant Bootstrapping in Anonymity Networks

Yang Han[1], Dawei Xu[1,2(✉)], Jiaqi Gao[2], and Liehuang Zhu[1]

[1] Beijing Institute of Technology, Beijing, China
{hanyang,liehuangz}@bit.edu.cn
[2] Changchun University, Changchun, China
xudw@ccu.edu.cn

Abstract. With Tor being a popular anonymity network, many censors and ISPs have blocked access to it. Tor relies on privately and selectively distributing IPs of circumvention proxies (i.e., bridges) to censored clients for censorship evasion. However, existing distributors are still vulnerable to blocking or compromising anonymity. This paper introduces Antiblok, a new and practical channel for bridge distribution leveraging blockchain, a globally decentralized environment. A key insight of Antiblok is that all blockchain transactions are under pseudonymous identities, allowing requesting clients to fetch bridge information while maintaining anonymity, regardless of the trustworthiness of blockchain nodes. To prevent the use of off-chain communication channels, we present an account sharing protocol based on DH key exchange. The unblockability of Antiblok depends on the economic consequences of blocking the Ethereum system. We show that Antiblok effectively thwarts client-side blocking of the distribution channel for Tor bridges, and we describe the security of our design.

Keywords: Tor · Anti-censorship · Blockchain · Covert communication

1 Introduction

In recent years, Internet surveillance has become prevalent as more of our regular activities shift online [1–3]. Public concerns over privacy have been likely at one of their peaks. Anonymous communication is a useful privacy-enhancing technology that protects the user's identity and prevents Internet activity tracking. State-of-the-art anonymity networks such as Tor [4] typically use proxies (i.e., relays) to forward their clients' traffic. Unfortunately, detecting and preventing the use of such networks is trivially achieved by enumerating ingress points. In particular, Tor's entry points (i.e., guard relays) are publicly advertised. To circumvent such blocking, Tor introduces a fraction of secret entry relays called bridges and has long focused on how to thwart Internet censorship [5–7]. However, a resourceful censor may constantly evolve new ways to block circumvention tools.

C. Alcaraz et al. (Eds.): ICICS 2022, LNCS 13407, pp. 240–260, 2022.
https://doi.org/10.1007/978-3-031-15777-6_14

Since the traffic of connections to Tor entry relays can be easily fingerprinted and blocked, Tor has employed pluggable transports (PTs). Most of the PTs (e.g., obfs4 [8]) rely on the use of bridges, but not all. For instance, meek [7] adopts a technique called domain fronting to hide Tor traffic inside the encrypted payload of an allowed HTTPS connection. Unfortunately, domain fronting has been shut down by many CDNs such as Amazon and Google [9], which makes meek much less effective. Snowflake [10] is a new way, but clients still require domain fronting to access the broker and often experience slower speeds [11]. In reality, there are about ten times more clients connecting by obfs4 than by meek or Snowflake [12]. Consequently, the major challenge facing Tor's censorship circumvention system is to bootstrap censored clients into the anonymity network. Once censors block all channels for fetching bridges (e.g., by blocking all connections to distributors), they can trivially prevent clients from using Tor to access the Internet.

Tor currently supports three channels for clients to fetch bridge IP addresses, including HTTPS, Email, and Moat [13]. The first two channels are easily blocked by censors, and moat is an effective tool built into Tor browser that uses domain fronting to stealthily transmit sensitive data via apparently innocuous HTTPS connections. However, domain fronting not only relies on the ongoing support of CDNs but does not make clients anonymous or completely hide their destination. Worse yet, some of the requesting clients are likely to be impersonated by censors who seek to block bridges by learning their IP addresses. Therefore, the purpose of Tor's bridge distribution strategy is to transmit bridge information to censored clients while mitigating the risks of bridge enumeration. To reduce the disclosure of bridges to censors, Tor limits the number of bridges returned to every censored client to three for specific time intervals and distinguishes clients by their IPs and email accounts. In addition, numerous efforts have focused on how to assign bridges to prevent malicious clients from enumerating [14–17].

In this paper, we present a new, practical, and more decentralized distribution channel for Tor bridges to mitigate the threat of blocking without degrading Tor users' anonymity. Instead of trusting centralized CDN providers, our architecture employs a blockchain system that does not require any additional trust. We call our blockchain-based bridge distribution solutions Antiblok. A major advantage of Antiblok is that it supports anonymous requests, since requesting clients are identified by blockchain addresses rather than their real identities. It is essential for privacy-conscious users who wish to hide the fact that they are attempting to access the anonymity network. In contrast to traditional anonymous techniques (e.g., ring signature [18]), blockchain technology not only incorporates anonymity but can also serve as a circumvention tool. A client within a censored area can obtain bridges by leveraging the help of a hard-to-block blockchain that acts as a bidirectional covert communication channel.

Antiblok advances the state-of-the-art by incorporating two new ideas in its design. First, to enable the stealthy transmission of bridge information, Antiblok uses a popular and widely accepted public blockchain – Ethereum. As Ethereum is a relatively mature blockchain ecosystem with a lot of free infrastructures, the barriers to extending Antiblok in Tor Browser are low. To create a covert channel, all that two parties – client and distributor – need to do are to send a transaction using an Ethereum account. More importantly, Antiblok adopts random accounts to covert transaction records between both endpoints. Second, to maintain the secrecy and confidentiality of sensitive data, we encrypt bridge information using a standard algorithm and then embed encrypted data into a particular field of transactions. Due to a large number of normal transactions in Ethereum, special transactions carrying secret messages are sufficiently concealed.

Contributions. In this effort, we explore how the unblockable Ethereum seamlessly works with Tor's censorship circumvention system. The exploration spans both theoretical and practical perspectives, and is driven by the vulnerable issue of bridge distribution blocking. We make three main contributions:

- We propose a novel bridge distributor for the Tor network. By using a public blockchain as a covert channel, clients can bypass blocking without degrading their anonymity and unlinkability.
- We design a negotiation-based account sharing mechanism that allows clients to not interact with addresses publicly related to Tor, making distinguishing special transactions from normal transactions far more difficult.
- We implemented and evaluated Antiblok on the Ethereum Rinkeby, and show that it can help censored clients to fetch information about Tor bridges.

2 Background

2.1 Tor Network

Tor is a low-latency anonymity network that consists of roughly 6,800 volunteer-run *relays* (i.e., onion routers) and two to eight million daily users [19,20]. These users rely on Tor to anonymize their Internet connections, and also to circumvent censorship and surveillance. To provide sender anonymity, Tor forwards Internet traffic through source-routed paths constructed by (typically) three relays called *circuits*, i.e., relays forming the anonymous route set are selected by the client.

Tor circuits use multiple layers (three layers by default) of encryption to hide the source and destination of communications. Encrypted traffic typically enters the network at a *guard* relay. Guards are relays that are fast and stable compared to other relays. This entry guard mechanism reduces the likelihood of the client connecting to an attacker. The second hop in a circuit is the *middle* relay, which carries traffic to the *exit* relay that acts as the egress point of Tor. Each hop in the circuit peels away one layer of encryption to reveal the next hop.

To distribute the relays used to construct the circuits, Tor introduces a set of dedicated servers called *directory authorities* (DirAuths) to maintain the network status. The IP addresses of these DirAuths are provided with the Tor software such as the Tor Browser. Every relay periodically uploads up-to-date information to all DirAuths, including its IP, keys, capabilities, etc. The DirAuths hourly vote and generate a *consensus* document listing all currently active relays in the Tor network. Then, both clients and relays download this consensus.

Censorship Circumvention of Tor. Censorship of the Tor network is possible and many governments have completely blocked the anonymity service. Tor can be censored at IP level. Censors can easily detect all Tor relays based on the IPs obtained from the publicly available consensus. Further, Tor can be censored by DPI techniques [21,22], since the characteristics of Tor traffic are fingerprintable and censors can prevent any traffic that looks like Tor communication.

Tor gets around the censors by introducing *bridges* and *pluggable transports* (PTs) [23]. Tor bridges are alternative entry relays whose IPs are not published in the consensus stored in DirAuths. With the Tor Project sending out a call for running a bridge in Nov. 2021, the number of bridges has significantly increased to about 2,600 [19]. PTs transform all Tor traffic between a client and the bridges to look similar to other innocuous traffic so that it is not identifiable. Multiple PTs have been proposed (e.g., obfs4 [8], meek [7], and Snowflake [10]), but these tools are also subject to blocking and surveillance.

Bridge Distribution. Clients that cannot directly connect to guards need to get information about the bridge to enter the anonymity network. The Tor software has a list of built-in default bridges, but they are often far more brittle compared to the advertised Tor relays. For instance, obfs4 is a recommended bridge type, but over half of the default obfs4 bridges are inoperative [24]. Unfortunately, the IPs of default bridges are hard-coded in the Tor Browser, leaving them vulnerable to censorship as well. Meek and Snowflake are two special cases, which reside on the popular cloud provider (i.e., Azure) and volunteer proxies, respectively.

Tor clients can also request a custom bridge. Similar to the DirAuth maintaining a list of available public relays, Tor introduces a *bridge authority* (BAuth) to collect running bridges in the network. BAuth sends information about bridges to *BridgeDB* [25] after testing their reachability. Subsequently, BridgeDB assigns and distributes Tor bridges based on client requests. It currently supports three manual distribution channels, visiting its website, sending it an email, and using moat to fetch from within the Tor Browser. All these approaches, however, might compromise the anonymity of clients due to identity exposure, while remaining challenging for clients located in certain heavily censored areas.

To prevent an adversary from simply enumerating bridges, BridgeDB assigns only three bridges per requested IP or email at specific time intervals, and limits email requests to addresses from Riseup or Gmail. However, such protections are ineffective against a resourceful censor, e.g., it can create large numbers of email accounts and connect from a diverse set of IP addresses. Previous works [23,26, 27] present numerous techniques for discovering Tor Bridges.

2.2 Blockchain Network

A blockchain is essentially an implementation of a distributed ledger, consisting of a continuously growing list of *transactions* that are shared among participants. All blockchain nodes employ a distributed consensus algorithm to guarantee data consistency. There are three types of current blockchains: public, consortium, and private [28]. A public blockchain allows anyone in the world to participate in the consensus process without the need for a trusted third party, which is secured by economic incentives and cryptographic verification. However, consortium and private blockchains tend to limit the read and write permissions of participants by including trusted entities and are therefore not fully decentralized.

Ethereum and Smart Contract. Ethereum [29] is the second largest blockchain platform next to Bitcoin, with a current market capitalization of over $340 billion (Feb. 14, 2022). Ethereum's cryptocurrency is known as *Ether* (ETH). Ethereum is described as a second-generation blockchain since it supports *smart contracts* – programs that run automatically when predetermined conditions are met. The Ethereum smart contract is typically written in *Solidity*, and its code is executed in a special environment called *Ethereum Virtual Machine* (EVM).

Ethereum can be viewed as a transaction-based state machine and Ethereum transactions are typically sent from and received by accounts. There are two basic types of accounts: *externally-owned accounts* (EOAs) and *contract accounts*, and both of them hold an Ether balance. The main difference between them is that contract accounts have the associated code, while EOAs do not have any code. An EOA is uniquely correlated with an address and a public/private key pair. In the creation of an EOA, a private key is randomly generated, and then a public key can be derived using the ECDSA with secp256k1 curve [30]. Finally, the last 20 bytes of the Keccak256 hash of the public key are taken as the address. Every transaction is signed by the private key of the account that initiated it.

Blockchain for Resisting Censorship. One of the most important issues with traditional centralized architecture is Internet censorship. Third parties such as central authorities can access and censor user data arbitrarily. Public blockchains (e.g., Bitcoin, Ethereum, Cardano) present a more effective means of overcoming censorship policies, as there is a financial incentive to protect stored data, as well as thousands of copies for verification and redundancy. The distribution of nodes in public blockchains is geographically diverse. For instance, Ethereum covers 70 countries [31] and Bitcoin covers 99 countries [32] (Jun. 16, 2022). We note that the underlying P2P protocol of blockchain is easy to identify, but there are many P2P applications in use now and blocking them is challenging. In addition, there are a number of websites like Etherscan that publish blockchain data, and APIs like Infura for writing data in the blockchain. Consequently, censorship resistance is seen as one of the value propositions of cryptocurrencies. The use of blockchain for censorship resistance has drawn the attention of researchers [33–35].

Covert Communication over Blockchain. Public blockchains can serve as a channel for covert communication, as special messages can be embedded in ordinary, non-secret blockchain transactions to conceal the fact that a secret message is being sent. Traditional covert channels usually lead to privacy leakage because the IP addresses or MAC addresses of transmitters and receivers are static [36]. In contrast, such identity disclosure can be prevented when employing blockchain as a steganographic channel due to its support for the pseudonymous identity. Communicating parties are identified by their blockchain addresses rather than network identities such as email or IP. Therefore, transactions remain anonymous as long as there is no link between pseudo-identity and real identity. Although Biryukov et al. [37] present a method to deanonymize Bitcoin users, a resourceful adversary can disclose the sender's IP address in only a small fraction (11%) of all transactions in the Bitcoin network.

Recently, several schemes for building covert channels over public blockchains have been proposed [38–41]. In the Ethereum system, the block interval is merely 15 s on average, which is about 40 times faster than Bitcoin and therefore more efficient. Senders use the data fields of Ethereum transactions as containers for carrying secret messages (e.g., the address field, the gas field). We note that not all fields of a transaction can be used for steganography due to the limitations of the sender's ability to control the transaction.

2.3 Public Key Encryption

We review the formal definition of Public Key Encryption (PKE). A PKE scheme $\mathcal{PKE}$ consists of a tuple of PPT algorithms $(\mathsf{KeyGen}, \mathsf{Enc}, \mathsf{Dec})$.

- **KeyGen**(1^λ): Takes as input 1^λ, and outputs a public/secret key pair $(\mathbf{pk}, \mathbf{sk})$.
- **Enc**$(\mathbf{pk}, M)$: Takes as input a public key $\mathbf{pk}$ and a message M, and outputs a ciphertext C.
- **Dec**$(\mathbf{sk}, C)$: Takes as input a secret key $\mathbf{sk}$ and a ciphertext C, and outputs either a message M, or a special rejection symbol $\perp$.

We always require $\mathcal{PKE}$ to be perfectly correct, meaning that for all key pairs $(\mathbf{pk}, \mathbf{sk})$ in the range of $\mathsf{KeyGen}(1^\lambda)$, for all messages M and for all C in the range of $\mathsf{Enc}(\mathbf{pk}, M)$ we require

$$\mathsf{Dec}(\mathbf{sk}, C) = M$$

2.4 Elliptic Curve Diffie-Hellman

The Diffie-Hellman (DH) key exchange protocol allows two parties communicating over an insecure channel to agree on a secret key. We now formally define the notion of Elliptic Curve Diffie-Hellman (ECDH) key exchange, which is a variant of the DH protocol using Elliptic Curve Cryptography (ECC). An ECDH scheme $\mathcal{ECDH}$ consists of three algorithms $(\mathsf{Setup}, \mathsf{KeyGen}, \mathsf{SharedKey})$, an elliptic curve E over $\mathbb{F}_p$, and a point $P \in E$.

- **Setup**(P): Input the system parameter P.
- **KeyGen**(P, a): On input parameter P and a natural number a. It computes $A = aP$ and outputs a tuple (A, a).
- **SharedKey**(A, b): On input a public key A and a private key b. It outputs either a shared key K_{ab} $(K_{ab} = bA)$, or a failure symbol $\perp$.

We require $\mathcal{ECDH}$ to be perfectly correct, meaning that for all corresponding elliptic curve key pairs (A, a) and (B, b) generated by KeyGen it holds

$$\mathsf{SharedKey}(A, b) = \mathsf{SharedKey}(B, a) \neq \perp$$

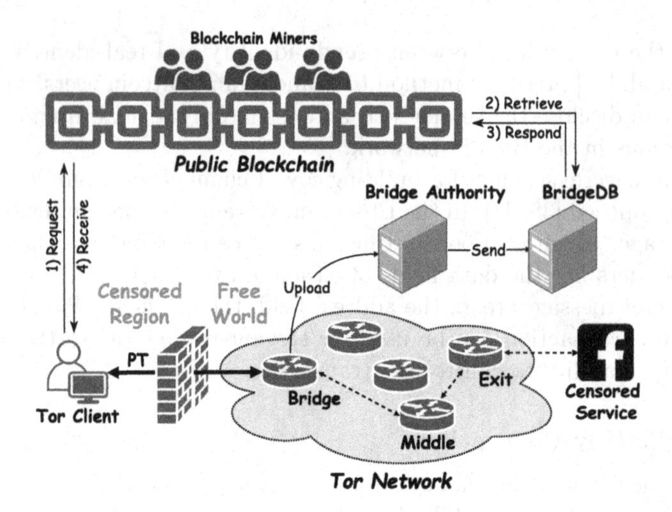

Fig. 1. Bridge distribution workflow: the architecture of Antiblok

3 Overview

3.1 System Model

Antiblok retains Tor's BAuth and BridgeDB while leveraging a public blockchain to act as an anonymous and censorship-resistant distribution channel for bridges. Figure 1 depicts the Antiblok model and the bridge distribution workflow.

- A *Tor client* is a user who cannot connect to the Tor network due to blocking and thus seeks a bridge to circumvent censorship. These clients typically need an unblockable channel since they have no means to access dedicated servers offered by the Tor Project (e.g., BridgeDB).
- A *bridge* refers to an unpublished Tor relay that facilitates entry to anonymity networks for blocked clients. Volunteers can register at the BAuth as a bridge operator and then upload their bridge information.
- The *censored service* is the destination of the client circuit and is likely to be monitored or blocked without the use of anonymity networks such as Tor.
- The *bridge authority* (BAuth) aggregates information from bridges and checks their cryptographic validity and port reachability. Next, it sends information about available bridges to BridgeDB.
- The *BridgeDB* is a collection of servers for assigning and distributing bridges to Tor clients. In addition to the 3 existing transmission channels for bridges (i.e., https, email, moat), it should provide an anonymous and hard-to-block channel to clients located in censored areas.
- The *public blockchain* offers a covert channel in the form of blockchain transactions. The bridge information used to bootstrap clients into the anonymity network is embedded in transactions recorded in the public ledger.

3.2 Threat Model

As described in Sect. 2.1, Tor currently lacks the technical means to provide an anonymous bridge distribution channel while resisting censorship. As a solution to this lack of channels, we adopt Tor's threat model [4]: a non-global adversary that can control a fraction of all relays (including bridges), and probably see the traffic at one end of a circuit but cannot observe the entire network.

More importantly, we consider a censoring adversary that attempts to block or monitor the communications between clients and blockchain nodes, and knows the public key of BridgeDB. We assume, however, that the adversary is not able to decrypt encrypted content without knowledge of the decryption key. Further, we assume that the censor is reluctant to block or significantly hinder the use of public blockchains. Even in a heavily censored area, censoring a public blockchain with a market capitalization would come at an unaffordable economic cost. Also, we assume that the censor does not have the resources to control more than 51% of the mining power in a public blockchain.

3.3 Design Goals

Antiblok seeks to provide a censorship-resistant solution using a public blockchain for bootstrapping anonymity networks, which satisfies the following properties.

- **Anonymity.** Antiblok ensures anonymity while distributing bridges to clients, i.e., neither the censor nor BridgeDB knows the real identity of users.
- **Unobservability.** Antiblok removes the direct connection to BridgeDB, rendering the censor unable to detect the client's communication with BridgeDB, even if it monitors incoming and outgoing traffic for the client.
- **Unblockability.** The censor is unable or unwilling to block communications between the client and BridgeDB, even if it can identify them.
- **Availability.** Antiblok provides a channel that is resilient to DoS attacks.
- **Minimal changes.** Antiblok can be weaved into Tor as an alternative bridge distribution channel without redesigning BAuth and BridgeDB of Tor, which allows Antiblok to be incrementally deployed on the real Tor network.

4 Antiblok Details

At a high-level, the primary aim of Antiblok is to allow Tor clients inside censored areas to successfully learn available bridges and reliably route their traffic to the destination website via three voluntarily participating relays containing a bridge, while keeping the anonymity of requesting clients. To achieve this goal, we design a request-response mechanism in the blockchain framework. As the requester, the client sends a request transaction to the blockchain. BridgeDB is the responder, which continuously retrieves the request transaction from the distributed ledger. Once BridgeDB discovers a new request, it will send a response transaction with embedded bridge information. Finally, the client gets the data from the response transaction. More concretely, Antiblok adopts a set of protocols to (1) ask clients

Algorithm 1: Client Request Procedure

```
1  Function EthereumAccountCreation(p, B):
2      ECDH.Setup(p);
3      s ← Generate_Random_Seed();
4      (ecdh_pk, ecdh_sk) ← ECDH.KeyGen(p, s);
5      sk_acct ← ECDH.SharedKey(B, ecdh_sk);
6      addr_acct ← Get_Address(sk_acct);
7      return ecdh_pk, addr_acct;
8  Function RequestTransaction(bpk, ecdh_pk, addr_acct):
9      (pk, sk) ← PKE.KeyGen(1^λ);
10     pk_enc ← PKE.Enc(bpk, pk);
11     msg ← ecdh_pk, pk_enc;
12     txn1 ← Build_Request_Transaction(x, addr_acct, msg);
13     txn1_signed ← Sign_Transaction(txn1);
14     txn1_hash ← Send_Transaction(txn1_signed);
15     txn1_receipt ← Wait_Transaction_Receipt(txn1_hash);
16     return txn1_receipt;
```

to pay for requests (to make enumeration more difficult), (2) ensure that requests are anonymous, and (3) use a hard-to-block channel to transmit Tor bridges from BridgeDB to clients, and (4) provide the encrypted transmission of data.

4.1 Client Request

Antiblok offers a Tor Browser extension that is installed on the user's computer. The user may run the extension optionally, e.g. by using the Ethereum blockchain as a transmission channel when she experiences difficulty contacting BridgeDB or wishes to keep anonymity. There are two ways to connect to Ethereum, running a node and calling an RPC service. Although the former can provide more privacy guarantees, it is complex and challenging for many users. So the extension adopts the latter mechanism for clients based on a trade-off between performance and security. Many RPC services such as Infura and QuikNode have been the primary selection for most DApp clients, since they run optimized nodes to communicate with the Ethereum blockchain. Algorithm 1 presents the pseudocode of the client request process, which considers the following two aspects:

Account Creation. Antiblok requires Tor clients to create a one-time Ethereum account for every request. The private key of this account is determined between the client and BridgeDB via ECDH. As the first step in account creation, a client defines the efficient cryptography parameter p for the ECDH algorithm (Line 2). Then, the client utilizes p and a cryptographically-random seed s to generate an elliptic curve key pair $(\mathsf{ecdh_{pk}}, \mathsf{ecdh_{sk}})$ (Lines 3–4). After obtaining BridgeDB's ECDH public key B, the shared private key $\mathsf{sk_{acct}}$ is computed (Line 5). Notably, the parameter p and the public key B are hard-coded in the Antiblok extension. For a successful creation, the client extracts the Ethereum address $\mathsf{addr_{acct}}$ from $\mathsf{sk_{acct}}$ (Line 6), which is the unique identifier of the account.

As mentioned above, Antiblok proposes an account sharing mechanism based on a key-agreement protocol. The key strength of this design is that no payment interaction is required, and the identity of the payer (i.e., the client) is anonymous due to the pseudo-identity. All communications between clients and BridgeDB occur over the blockchain transaction. A request transaction is sent from a user address ADDR to $\mathsf{addr_{acct}}$, while a response transaction is sent from $\mathsf{addr_{acct}}$ to a random smart contract address, as we will see in Sect. 4.2. The unobservability of steganography transactions is enhanced by the randomness of addresses. Note that ADDR should not be linked to other activities in order to avoid user privacy from being inferred, e.g., the source of its balance needs to be untraceable. There are many ways to quickly and easily buy ETH anonymously such as Ether ATMs and cryptocurrency exchanges [42]. On the other hand, an ECDH-based account reduces the difficulty for Tor clients to receive data. Synchronizing the Ethereum ledger to retrieve the response transaction would result in significant performance penalties for clients. Instead, Antiblok enables the client to receive bridges simply from the input field of the transaction sent from the known account.

Request Transaction. Antiblok uses Elliptic Curve Cryptography (ECC) as its PKE algorithm. ECC can offer the same level of security using a small size key as other PKE algorithms with large size keys, which makes ECC very appealing for scenarios with limited storage (such as blockchain and IoT). The most extended encryption scheme in ECC is the ECIES [43].

To defend against adversaries (e.g., censors, curious clients, etc.) that want to simply enumerate bridges (for example, for the purpose of deanonymizing users by inspecting packets sent and received by a bridge), the client needs to transfer x ETH to $\mathsf{addr_{acct}}$ before requesting bridges, which increases the economic cost of enumeration attacks [26] to some extent. Note that the minimum value of x must be able to cover fees for subsequent transactions; otherwise, the transaction will fail. However, such costs have limited impact against a determined, well-funded censor. To implement the idea of balancing the trade-off between ordinary clients and adversaries, we add a payment procedure to the request transaction.

BridgeDB's ECC public key bpk is hard-coded in the Antiblok extension, and thus easily accessible. It will be used to encrypt a 33-byte compressed public key pk and this public/private key pair $(\mathsf{pk}, \mathsf{sk})$ is generated with the secp256k1 curve implementation of ECIES (Lines 9–10). Then, a 64-byte $\mathsf{ecdh_{pk}}$ is combined with the encrypted ECC public key $\mathsf{pk_{enc}}$ to form the request message msg (Line 11). The msg is embedded in a request transaction $\mathsf{txn_1}$, and $\mathsf{txn_1}$ is signed by ADDR's blockchain private key and sent to $\mathsf{addr_{acct}}$ (Lines 12–14). Afterwards, the client is able to query whether $\mathsf{txn_1}$ was successfully included in one block by checking its transaction receipt $\mathsf{txn_{1_{receipt}}}$ (Line 15).

4.2 BridgeDB Response

BridgeDB learns about new bridges from the BAuth and assigns them to request clients based on the *bridge assignment algorithm*. Previous works [14–17] provide numerous strategies, but how to assign Tor bridges is not the focus of this effort and thus will not be discussed in depth here. Unlike clients that use RPC services, BridgeDB runs an Ethereum node to hold a copy of the up-to-date public ledger to constantly retrieve whether a new steganography transaction txn_1 carrying a request has been sent. Once the request is discovered and parsed, BridgeDB will transmit three bridges to the client through a transaction. Algorithm 2 presents the pseudocode of BridgeDB response process.

BridgeDB Retrieval. As mentioned in Sect. 4.1, BridgeDB's public keys (i.e., B and bpk) are broadly disseminated, but corresponding private keys (i.e., b and bsk) are kept secret and only BridgeDB knows them. Once a new block is added to Ethereum, BridgeDB gets all the transactions it contains and checks if any of them carry requests. For a successful retrieval, BridgeDB can take out the secret message msg from the input field of txn_1, then parse msg into two parts – $\mathsf{ecdh}_{\mathsf{pk}}$ and $\mathsf{pk}_{\mathsf{enc}}$, and finally compute the private key $\mathsf{sk}_{\mathsf{acct}}$ as well as the corresponding address $\mathsf{addr}_{\mathsf{acct}}$ (Lines 2–4). Notably, $\mathsf{addr}_{\mathsf{acct}}$ must be equal to $\mathsf{addr}_{\mathsf{R}}$ that is the output address of the transaction txn_1; otherwise, it fails on address verification (Lines 5–7). Finally, BridgeDB decodes $\mathsf{pk}_{\mathsf{enc}}$ with its private key bsk into pk that will use to encrypt the bridge information (Line 8).

Bridge Assignment. Currently, BridgeDB supports three types of distributors: https, email, and moat. After a new bridge is reported by the BAuth, BridgeDB assigns it to distributors based on an HMAC function and makes the assignment persistent. Instead of making improvements to the bridge assignment algorithm, this paper extends a distributor (i.e., Antiblok) that seeks to successfully provide clients with bridge information, while making them anonymous like Tor Browser does. Similar to the current assignment strategy based on IPs and email accounts, Antiblok assigns bridges based on Ethereum addresses (e.g., ADDR), while using the request transaction's timestamp time to limit the number of bridges returned to a client for specific time intervals (Lines 11–12). (While not yet implemented in our prototype, the blockchain-based distributor could be simulated manually. We leave such integration to the future work of Antiblok.)

Response Transaction. After assigning a subset of bridges (usually three) upon request, BridgeDB will provide the client with them. Because all transactions in a public blockchain are open and traceable, information about Tor bridges must be cryptographically protected before it is embedded in a blockchain transaction. Antiblok leverages a standard PKE algorithm to encrypt bridge information into $\mathsf{bridges}_{\mathsf{enc}}$, which will be embedded in a response transaction txn_2 (Lines 13–14). Unlike the construction of txn_1, txn_2 – a contract transaction – is signed by $\mathsf{sk}_{\mathsf{acct}}$ and sent from $\mathsf{addr}_{\mathsf{acct}}$ to a public smart contract address $\mathsf{ADDR}_{\mathsf{sc}}$ (Lines 15–16). $\mathsf{ADDR}_{\mathsf{sc}}$ is not pre-defined but is dynamically selected by BridgeDB based on the up-to-date Ethereum ledger. More concretely, BridgeDB periodically (e.g., on a weekly basis) detects the contracts that have been called most frequently during this period and then selects a function F from their list of functions. We require that the input field of a normal transaction generated by the call to F cannot be blank, so a steganography transaction can be constructed with F. Consequently, $\mathsf{bridges}_{\mathsf{enc}}$ is concealed in the data of a random smart contract, further improving the indistinguishability of txn_2, and

Algorithm 2: BridgeDB Response Procedure

```
1  Function ParseRequestData(b, bsk, msg, addr_R):
2      ecdh_pk, pk_enc ← Parse(msg);
3      sk_acct ← ECDH.SharedKey(ecdh_pk, b);
4      addr_acct ← Get_Address(sk_acct);
5      if addr_acct ≠ addr_R then
6          return ⊥;
7      end
8      pk ← PKE.Dec(bsk, pk_enc);
9      return pk;
10 Function ResponseTransaction(pk, ADDR):
11     time ← Get_Transaction_Timestamp(txn_1);
12     bridges ← Assign_Bridges(ADDR, time);
13     bridges_enc ← PKE.Enc(pk, bridges);
14     txn2 ← Build_Response_Transaction(bridges_enc);
15     txn2_signed ← Sign_Transaction(txn2);
16     txn2_hash ← Send_Transaction(txn2_signed);
17     if the account acct does not have enough ETH then
18         return ⊥;
19     end
20     txn2_receipt ← Wait_Transaction_Receipt(txn2_hash);
21     return txn2_receipt;
```

such randomness does not incur additional work for clients. If acct's balance is not sufficient to cover the transaction fee for txn_2, this request will fail because txn_2 cannot be sent (Lines 17–19); otherwise, the client can receive the encrypted bridge data from the input field of txn_2 once txn_2 has been sent, even if it is not yet included in a block.

4.3 Circuit Creation

Client Reception. Algorithm 3 presents the pseudocode of client reception process. Once a client finds a new txn_2 with $addr_{acct}$, the client extracts the ciphertext $bridges_{enc}$ from txn_2 and decrypts it with sk to obtain bridges (Line 2). Even if an adversary discovers that txn_2 is carrying a secret message, the adversary cannot learn the plaintext by decrypting it, since sk is known only to the client. Finally, the client parses bridges into $\{b_1, b_2, b_3\}$ (Line 3).

Algorithm 3: Client Reception Procedure

```
1 Function ReceiveBridges(sk, bridges_enc):
2     bridges ← PKE.Dec(sk, bridges_enc);
3     {b1, b2, b3} ← bridges;
4     return {b1, b2, b3};
```

Circuit Creation. When a censored client fetches bridges (i.e., b_1, b_2, and b_3), the client will connect to a bridge using a PT. However, for some heavily censored regions, the characteristics of Tor traffic cause it to be identified by censors even after obfuscation. There are many studies on enhancing PTs to evade censorship, but it is not the focus of this work. An illustration of a 3-hop Tor circuit is shown in Fig. 1. After successfully establishing a circuit, the client will route its traffic to censored services using the constructed circuit.

5 Security

In this section, we evaluate and analyze several important security properties of Antiblok, and provide informal proofs of their effectiveness.

Denial-of-Service (DoS) Attacks. The purpose of a DoS attack is to prevent a system or resource from serving users by flooding the target with massive traffic. Antiblok is not subject to such traditional DoS attacks [44] due to their difficulty in scaling to decentralized systems with thousands of peer nodes. By leveraging the decentralized nature of Ethereum, Antiblok can allocate bandwidth to absorb DoS attacks as they happen. But blockchain systems are not completely immune to DoS attacks. There is a body of research on blockchain DoS security, such as eclipse attacks [45], routing attacks [46], and mining-based DoS attacks [47], but most efforts to stop cryptocurrency blockchains are prohibitively costly, even for powerful censors. Additionally, a censoring adversary could disrupt Ethereum by flooding it with transactions to exhaust its resources, but the resulting economic consequences would negatively impact people within the censored area.

Security of Account Sharing. To enable a Tor client and BridgeDB to securely share a private key (i.e., $\mathsf{sk}_{\mathsf{acct}}$) without prior off-chain communication, Antiblok presents a negotiation-based protocol for account sharing that relies on Elliptic-Curve Diffie-Hellman (ECDH) key exchange. We now examine its security in the presence of both passive and active adversaries. In the *discrete logarithm attack*, a passive adversary Alice is able to eavesdrop on the communication channel and learn information (such as the parameter p, BridgeDB's ECDH public key B, and the client's ECDH public key $\mathsf{ecdh}_{\mathsf{pk}}$) without interfering with the protocol. But the security of ECDH relies on the intractability of the (computational) Elliptic Curve Diffie-Hellman Problem (ECDHP); therefore, Alice will not able to derive the private keys b and $\mathsf{ecdh}_{\mathsf{sk}}$ from B, $\mathsf{ecdh}_{\mathsf{pk}}$, p unless she solves the ECDLP. In the *man-in-the-middle attack*, an active adversary Bob runs the ECDH protocol separately with the client and BridgeDB. If they cannot distinguish each other's bits from Bob's bits, the protocol can be compromised. Antiblok will be included in the Tor software, so the client can easily verify the integrity of B. On the other hand, with the use of Ethereum as a communication channel, BridgeDB can learn $\mathsf{ecdh}_{\mathsf{pk}}$ directly from the public ledger and authenticate the computed private key with the txn_1's (default) output address, clearly preventing such attacks.

Difficulty of Enumerating Bridges. While Antiblok does not attempt to prevent long-running malicious middle relays from enumerating bridges, it increases the difficulty of leveraging excessive client requests to enumerate the IP addresses of bridges. Such enumeration allows bridges to be quickly found by an adversary, degrading the availability and security of Tor bridges. All transactions generated during the bridge distribution process need to be paid by the client. Any increase in the number of requests is therefore met with more resources provided by the corresponding request fees. That is, an adversary cannot send more requests by simply changing accounts; instead, there will be a high financial cost (i.e., paying ETH for the Ethereum transactions). For instance, on June 18, 2022, Ethereum's average transaction fee is $4.74 [48]. Under the extreme assumption that all Tor bridges are assigned to the Antiblok distributor by BridgeDB, the minimum cost for an adversary to enumerate all 2,600 bridges is roughly $8,216. Unfortunately, an adversary with sufficient funds is still able to obtain more bridges by sending a large number of transactions.

Anonymity of Requests. By monitoring communications between two entities, an adversary can collect sensitive information (e.g., identity, personal interests, location, etc.). Tor currently requires clients to directly connect to BridgeDB to request bridges, which enables distributors and potential adversaries to observe the IP addresses of clients through traffic eavesdropping. We also note that none of the existing distribution channels (i.e., https, email, moat) can preserve users' privacy. They assume that the distributor is completely trusted and authorized to know which bridges are given to a particular client, which may compromise user anonymity [14,49]. In contrast, Antiblok proposes a non-interactive distribution protocol that enables blockchain nodes to serve as a medium between clients and BridgeDB in order to provide strong anonymity guarantees to users. As long as the Ethereum address is not linked to the client's identity, our distributor cannot associate bridges with a requester's IP. Ethereum Mainnet has 2,864 nodes (Mar. 12, 2022) [50]. Although we use RPC services to limit the number of connectable nodes to around 10% of all nodes, every additional RPC node monitored by the adversary will only increase the probability of client connections being monitored by roughly 0.35%. More importantly, at least 63% of Ethereum applications use Infura API, which handles billions of requests per day. By comparison, the client sends only about 24 API requests when sending a txn_1. As a result, even though an extremely powerful censor could detect all connections between clients within the censored area and RPC Ethereum nodes, it is difficult to distinguish special requests from a large number of normal requests.

6 Evaluation

Our evaluation aims to show (1) whether Antiblok helps censored Tor clients get bridges, and (2) how much time cost will be incurred by introducing Ethereum.

6.1 Experimental Setup

BridgeDB Configuration. We manually simulate the assignment and distribution workflow. We start by applying for obfs4-type bridges on BridgeDB's website based on the IP address [25], and the response result is shown in Fig. 2.

obfs4 **188.126.94.105:8443** D734D62C13012A9B8E49F4BDCA98F355B71214DE
cert=cD8xzwtH9DWgXb21kMRulQIEUUE5pUyJqRUmB17r/zPnemkcJuHK/7v
MLAgCo0DaXlfsHg iat-mode=0
obfs4 **93.225.178.129:8443** 4F597FFA9BB6DF73C8948A7A37AFE68270553E9A
cert=y1sRF/zn37gk88jAVBebJSP4xEGKQ0QjKacFFMLv52K9b+6X1MX8J2fjm
vWcPOyuDJ2XVA iat-mode=0
obfs4 **143.47.188.119:11111** 09C926F75CFBA3748A4C813EA3B45623C8956EB3
cert=kJ0L2EBjnq+L+zuHkOljDgtov/an5Jh0OtGUolI2ScCxYfm6H/BiaNauhgrb
mVoqnutnLw iat-mode=0

Fig. 2. A result example of BridgeDB's website assigning and distributing three obfs4-type bridges for a request.

Then, we perform experiments for Antiblok. Once a request transaction txn_1 is detected, we take this example (i.e., three bridges obtained from the website) as the assignment results and transmit them to the client covertly via an Ethereum transaction txn_2. For simplicity, our experiments adopt Infura as an API to access Ethereum without really running a node as described above.

Client Configuration. The simulated client within the censored area runs on an Alibaba Cloud ECS instance created in Beijing, with CentOS and Tor Browser installed. Experiments are conducted over the live Tor network and Ethereum.

6.2 Functionality Evaluation

We note that congestion often occurs in the Ethereum ecosystem due to a large number of transactions. In the worst case, the congestion will last for many days, which certainly affects the availability of Antiblok. However, an Ethereum-based channel is still more reliable than fundamentally blocked traditional distribution channels, as blockchain systems always have the ability to self-mitigate. In order to evaluate the ability of Antiblok to bypass client-side blocking, we use Rinkeby Testnet as a communication medium between the censored client and BridgeDB. In practice, most Testnets select only a small number of nodes to validate transactions and create blocks, so their transaction frequency tends to be lower than that of the Mainnet and they are more prone to congestion.

Table 1. The average time for every process of bridge distribution workflow

Workflow	Process	Average time (s)	
Request	Create key pairs	0.012	2.217
	Send txn_1	2.205	
Retrieve	Add txn_1 to a block	15.001	16.423
	Examine the block	1.422	
Respond	Encrypt response message	0.004	1.791
	Send txn_2	1.787	
Receive	Get bridge information	1.496	1.496

The implementation details of Antiblok in Ethereum are offered in Sect. 4. In summary, BridgeDB embeds request and response messages into two Rinkeby transactions using the *web3.py* library, respectively. The average time for every process of bridge distribution workflow is shown in Table 1. While the procedure for adding the transaction txn_1 to a single block is limited by the Ethereum itself, the total time of Antiblok (i.e., roughly 22 s) is acceptable as it does not have to be real-time. For example, a censored client waits about 20 s while requesting bridge information by traditional email channels, which is comparable to Antiblok, but such channels are not able to offer anonymity. If a censor blocks all traffic to/from the Ethereum platform, we should expect that clients are not able to connect to Rinkeby nodes.

As a preparation for our experiments, BridgeDB should create two key pairs and make the public keys widely available. The first one is an ECDH key pair on the NIST256p curve. In our implementation, we use the *ecdsa* library to perform ECDH key agreement between BridgeDB and clients. Next is an ECIES key pair, which is implemented over the *eciespy* library and is used by clients to perform PKE, i.e., encrypt/decrypt the request message.

Request Transaction. The requesting client is required to perform two similar cryptographic processes, creating an ECDH key pair to compute a shared private key $\mathsf{sk}_{\mathsf{acct}}$, and an ECIES key pair used to encrypt/decrypt the response message. The process of creating two fresh key pairs takes the client about 12 ms, and deriving the private key takes about 10 ms on average. Finally, the client takes about 2.2 s to build and send a transaction carrying a request. [51] is an example of the request transaction txn_1. Afterwards, the client has to continuously query whether txn_1 has been included in one block (i.e., the finality of txn_1) to ensure that BridgeDB can retrieve a request.

Retrieval Process. To retrieve every special request transaction, BridgeDB has to constantly query the latest block included in Ethereum. We conducted group experiments according to the type of Ethereum, with a total of two groups (i.e., Testnet and Mainnet). Ten experiments were performed in each group. Overall, the average time to examine one block in Testnet is approximately 1.4 s, while the average time for Mainnet is about 2.7 s. The experimental results demonstrate that the retrieval time is much lower than the block time. Note that the time required to detect special transactions from one block is affected by the response delay of remote calls to the Ethereum API. We, therefore, recommend BridgeDB (i.e., the bridge provider) maintain an Ethereum node to reduce the time overhead of transaction retrieval.

Response Transaction. BridgeDB extracts the client's public key, which is encrypted, from the retrieved transaction, and decrypts it with BridgeDB's private key. This public key will be used to encrypt the bridge information. The time of the encryption and decryption process depends on the length of the message. In our experiments, both encryption and decryption take less than 6 ms. As an easy-to-implement design, BridgeDB can randomly create an EOA as the receiver of txn_2. However, we send txn_2 to a frequently invoked contract in order to further improve the difficulty for adversaries to detect the hidden information, and [52] is an example of txn_2. Notably, once txn_2 is sent, the client can discover it immediately without waiting for it to be added to one block.

7 Discussion

Antiblok bypasses client-side blocking of bridge distribution in Tor by employing blockchain transactions as rendezvous. Our technique is targeted at anonymously providing censored clients with bootstrapping information (i.e., bridge IPs).

Related Work. As discussed above, not only are Tor's advertised relays blocked, but secret bridges are also at risk of being blocked due to an enumeration attack. Worse yet, a resourceful censor blocks channels for distributing bridges whenever possible, fundamentally preventing users from accessing anonymity services. The early two distributors – https and email – required direct contact with BridgeDB, which had become ineffective for most censored users. Tor then presents moat [13] that relies on domain fronting supported by popular CDNs (like meek [7] does). Moat argued that censorship circumvention is achieved due to the high collateral damage of blocking the distribution channel: blocking Tor clients' requests would also result in the blocking of CDN servers. Unfortunately, moat depends on the cooperation of CDN providers, while requiring that users have to trust them (for anonymity cannot be guaranteed).

We present Antiblok, a distinctive bridge distribution channel that eliminates all reliance on CDNs. Antiblok achieves greater blocking resistance by employing Ethereum as a covert transmission channel, thus making the bridge distribution process more difficult to censor. Conceptually, Antiblok is a method of providing a distributed transmit architecture for Tor bridges.

Recabarren et al. [33] present a Bitcoin-based communication system to resist censorship, called Tithonus. The Tithonus server uses the Bitcoin blockchain and its network to provide censored clients with bootstrapping information (e.g., Tor bridge's IP). But a single transmission in Tithonus (such as a client registration message) needs to be encoded into two transactions, a preparing transaction and a redeeming transaction. Bitcoin, in comparison to Ethereum, has a slower block time and a lower embedding capacity. MoneyMorph [35] covers four blockchains (including Ethereum) as the covert transmission channel. As a case, Zcash offers the lowest cost. But blockchains with a higher market cap and more nodes would contribute to more censorship resistance. Ethereum is therefore more reliable for evading censors. MoneyMorph uses a fixed smart contract as rendezvous, which compromises the unobservability of the communication. In contrast, we introduce randomness to enhance the concealment of special transactions. In addition, even if the client runs an SPV node (aka a light node) that downloads only the header of each block, as MoneyMorph suggests, it incurs additional bandwidth overhead for a Tor client. There is also work focused on introducing blockchain technology into components of anonymity networks. For instance, SmarTor [53] replaces the DirAuths with a smart contract in order to decentralize the anonymity network and reduce trust assumptions.

Limitations. Similar to traditional (CAPTCHA- and identity-based) strategies, blockchain address-based strategy is also susceptible to enumeration. Bridge enumeration is an open problem for Tor, however, one point worth stressing is that enumeration attacks are usually enabled by an adversary with enough resources. Such adversaries would unlikely disrupt the running of blockchains. In addition, Antiblok provides some protection against automated enumeration, as learning the bridges requires paying transaction fees (i.e., ETH) in advance.

8 Conclusion

In the case of Tor, Antiblok bypasses bridge distribution blocking by reassigning the job of transmitting bridges from https, email, or moat to a public blockchain with a decentralized architecture. The key insight is that blockchain can provide users with anonymous identities, allowing censored clients to maintain anonymity while resisting censorship. We propose a covert communication scheme that employs Ethereum as a non-interactive transmission channel, and an account sharing protocol to protect the unobservability of communications.

Acknowledgements. This work was supported by the NSFC General Technology Basic Research Joint Fund (Grant No. U1836212).

References

1. Aryan, S., Aryan, H., Halderman, J.A.: Internet censorship in Iran: a first look. In: 3rd USENIX Workshop on Free and Open Communications on the Internet (FOCI 13) (2013)
2. Yadav, T.K., Sinha, A., Gosain, D., Sharma, P.K., Chakravarty, S.: Where the light gets in: analyzing web censorship mechanisms in India. In: Proceedings of the Internet Measurement Conference 2018, pp. 252–264 (2018)
3. Ramesh, R., et al.: Decentralized control: a case study of Russia. In: Network and Distributed Systems Security Symposium (2020)

4. Dingledine, R., Mathewson, N., Syverson, P.: Tor: the second-generation onion router. Technical report, Naval Research Lab Washington DC (2004)
5. Karlin, J., et al.: Decoy routing: toward unblockable internet communication. In: USENIX Workshop on Free and Open Communications on the Internet (FOCI 11) (2011)
6. Mohajeri Moghaddam, H., Li, B., Derakhshani, M., Goldberg, I.: Skypemorph: protocol obfuscation for Tor bridges. In: Proceedings of the 2012 ACM Conference on Computer and Communications Security, pp. 97–108 (2012)
7. Fifield, D., Lan, C., Hynes, R., Wegmann, P., Paxson, V.: Blocking-resistant communication through domain fronting. Proc. Priv. Enhanc. Technol. **2015**(2), 46–64 (2015). https://doi.org/10.1515/popets-2015-0009
8. Angel, Y.: obfs4 - the obfourscator. https://github.com/Yawning/obfs4/
9. Brandom, R.: Amazon web services starts blocking domain-fronting, following Google's lead (2018). https://www.theverge.com/2018/4/30/17304782/amazon-domain-fronting-google-discontinued
10. Snowflake: pluggable transport using WebRTC. https://gitlab.torproject.org/tpo/anti-censorship/pluggable-transports/snowflake/
11. Snowflake moving to stable in Tor browser 10.5 (2021). https://blog.torproject.org/snowflake-in-tor-browser-stable/
12. Users - Bridge user by transport. https://metrics.torproject.org/userstats-bridge-transport.html?transport=obfs4&transport=meek&transport=snowflake
13. BRIDGES. https://tb-manual.torproject.org/bridges/
14. Wang, Q., Lin, Z., Borisov, N., Hopper, N.: rBridge: user reputation based Tor bridge distribution with privacy preservation. In: Network and Distributed Systems Security Symposium (2013)
15. Douglas, F., Rorshach, W.P., Pan, W., Caesar, M.: Salmon: robust proxy distribution for censorship circumvention. Proc. Priv. Enhanc. Technol. **2016**(4), 4–20 (2016). https://doi.org/10.1515/popets-2016-0026
16. Zamani, M., Saia, J., Crandall, J.: TorBricks: blocking-resistant Tor bridge distribution. In: Spirakis, P., Tsigas, P. (eds.) SSS 2017. LNCS, vol. 10616, pp. 426–440. Springer, Cham (2017). https://doi.org/10.1007/978-3-319-69084-1_32
17. Nasr, M., Farhang, S., Houmansadr, A., Grossklags, J.: Enemy at the gateways: censorship-resilient proxy distribution using game theory. In: Network and Distributed Systems Security Symposium (2019)
18. Rivest, R.L., Shamir, A., Tauman, Y.: How to leak a secret. In: Boyd, C. (ed.) ASIACRYPT 2001. LNCS, vol. 2248, pp. 552–565. Springer, Heidelberg (2001). https://doi.org/10.1007/3-540-45682-1_32
19. Tor Metrics. https://metrics.torproject.org
20. Mani, A., Wilson-Brown, T., Jansen, R., Johnson, A., Sherr, M.: Understanding Tor usage with privacy-preserving measurement. In: 2018 Proceedings of the Internet Measurement Conference, pp. 175–187 (2018)
21. Choffnes, D., Gill, P., Mislove, A.: An empirical evaluation of deployed DPI middleboxes and their implications for policymakers. In: Proceedings of TPRC (2017)
22. Li, F., et al.: lib• erate,(n) a library for exposing (traffic-classification) rules and avoiding them efficiently. In: Proceedings of the 2017 Internet Measurement Conference, pp. 128–141 (2017). https://doi.org/10.1145/3131365.3131376
23. Matic, S., Troncoso, C., Caballero, J.: Dissecting Tor bridges: a security evaluation of their private and public infrastructures. In: Network and Distributed Systems Security Symposium, pp. 1–15. The Internet Society (2017)

24. Jansen, R., Vaidya, T., Sherr, M.: Point break: a study of bandwidth {Denial-of-Service} attacks against Tor. In: 28th USENIX Security Symposium (USENIX Security 19), pp. 1823–1840 (2019)
25. BridgeDB. https://bridges.torproject.org/
26. Ling, Z., Luo, J., Yu, W., Yang, M., Fu, X.: Extensive analysis and large-scale empirical evaluation of Tor bridge discovery. In: 2012 Proceedings IEEE INFOCOM, pp. 2381–2389. IEEE (2012). https://doi.org/10.1109/infcom.2012.6195627
27. Durumeric, Z., Wustrow, E., Halderman, J.A.: {ZMap}: fast internet-wide scanning and its security applications. In: 22nd USENIX Security Symposium (USENIX Security 13), pp. 605–620 (2013)
28. Zheng, Z., Xie, S., Dai, H., Chen, X., Wang, H.: An overview of blockchain technology: architecture, consensus, and future trends. In: 2017 IEEE International Congress on Big Data (BigData congress), pp. 557–564. IEEE (2017)
29. Wood, G., et al.: Ethereum: a secure decentralised generalised transaction ledger. Ethereum Proj. Yellow Pap. **151**(2014), 1–32 (2014)
30. Qu, M.: Sec 2: Recommended elliptic curve domain parameters. Certicom Res., Mississauga, ON, Canada, Technical report SEC2-Ver-0.6 (1999)
31. Ethereum Mainnet Statistics (2022). https://www.ethernodes.org/countries
32. REACHABLE BITCOIN NODES (2022). https://bitnodes.io/
33. Recabarren, R., Carbunar, B.: Tithonus: a bitcoin based censorship resilient system. arXiv preprint arXiv:1810.00279 (2018)
34. He, S., Tang, Q., Wu, C.Q., Shen, X.: Decentralizing IoT management systems using blockchain for censorship resistance. IEEE Trans. Industr. Inf. **16**(1), 715–727 (2019). https://doi.org/10.1109/tii.2019.2939797
35. Minaei, M., Moreno-Sanchez, P., Kate, A.: MoneyMorph: censorship resistant rendezvous using permissionless cryptocurrencies. Proc. Priv. Enhanc. Technol. **2020**(3), 404–424 (2020). https://doi.org/10.2478/popets-2020-0058
36. Ahsan, K.: Covert channel analysis and data hiding in TCP/IP. MA Sc. thesis, Department of Electrical and Computer Engineering, University of Toronto (2002)
37. Biryukov, A., Khovratovich, D., Pustogarov, I.: Deanonymisation of clients in bitcoin P2P network. In: Proceedings of the 2014 ACM SIGSAC Conference on Computer and Communications Security, pp. 15–29 (2014)
38. Partala, J.: Provably secure covert communication on blockchain. Cryptography **2**(3), 18 (2018). https://doi.org/10.3390/cryptography2030018
39. Gao, F., Zhu, L., Gai, K., Zhang, C., Liu, S.: Achieving a covert channel over an open blockchain network. IEEE Netw. **34**(2), 6–13 (2020)
40. Alsalami, N., Zhang, B.: Uncontrolled randomness in blockchains: covert bulletin board for illicit activity. In: 2020 IEEE/ACM 28th International Symposium on Quality of Service (IWQoS), pp. 1–10. IEEE (2020)
41. Zhang, L., Zhang, Z., Wang, W., Jin, Z., Su, Y., Chen, H.: Research on a covert communication model realized by using smart contracts in blockchain environment. IEEE Syst. J. (2021). https://doi.org/10.1109/jsyst.2021.3057333
42. Buy ethereum anonymously. https://www.cryptimi.com/buy-cryptocurrency/buy-ethereum-eth#buy-ethereum-anonymously
43. Gayoso Martínez, V., Hernández Encinas, L., Sánchez Ávila, C.: A survey of the elliptic curve integrated encryption scheme (2010)
44. Understanding denial-of-service attacks (2019). https://www.cisa.gov/uscert/ncas/tips/ST04-015
45. Heilman, E., Kendler, A., Zohar, A., Goldberg, S.: Eclipse attacks on {Bitcoin's}{peer-to-peer} network. In: 24th USENIX Security Symposium (USENIX Security 15), pp. 129–144 (2015)

46. Apostolaki, M., Zohar, A., Vanbever, L.: Hijacking bitcoin: routing attacks on cryptocurrencies. In: 2017 IEEE Symposium on Security and Privacy (S&P), pp. 375–392. IEEE (2017). https://doi.org/10.1109/sp.2017.29
47. Mirkin, M., Ji, Y., Pang, J., Klages-Mundt, A., Eyal, I., Juels, A.: BDoS: blockchain denial-of-service. In: Proceedings of the 2020 ACM SIGSAC Conference on Computer and Communications Security, pp. 601–619 (2020)
48. Average Transaction Fee Chart. https://etherscan.io/chart/avg-txfee-usd
49. Loesing, K., Murdoch, S.J., Dingledine, R.: A case study on measuring statistical data in the Tor anonymity network. In: Sion, R., et al. (eds.) FC 2010. LNCS, vol. 6054, pp. 203–215. Springer, Heidelberg (2010). https://doi.org/10.1007/978-3-642-14992-4_19
50. Ethereum Node Tracker. https://etherscan.io/nodetracker
51. An example of request transaction. https://4el5.short.gy/VMT3U0
52. An example of response transaction. https://4el5.short.gy/s9Mi4R
53. Andre, G., Alexandra, D., Samuel, K.: SmarTor: smarter tor with smart contracts: improving resilience of topology distribution in the Tor network. In: Proceedings of the 34th Annual Computer Security Applications Conference, pp. 677–691 (2018)

Repetitive, Oblivious, and Unlinkable SkNN Over Encrypted-and-Updated Data on Cloud

Meng Li[1(✉)], Mingwei Zhang[1], Jianbo Gao[1], Chhagan Lal[2], Mauro Conti[2,3], and Mamoun Alazab[4]

[1] Key Laboratory of Knowledge Engineering with Big Data (Hefei University of Technology), Ministry of Education; School of Computer Science and Information Engineering, Hefei University of Technology; Anhui Province Key Laboratory of Industry Safety and Emergency Technology; and Intelligent Interconnected Systems Laboratory of Anhui Province (Hefei University of Technology), Hefei, China
mengli@hfut.edu.cn, {mwzhang,jianbogao}@mail.hfut.edu.cn

[2] Department of Intelligent Systems, CyberSecurity Group, Delft University of Technology, Delft, The Netherlands
c.lal@tudelft.nl

[3] Department of Mathematics and HIT Center, University of Padua, Padua, Italy
conti@math.unipd.it

[4] College of Engineering, IT and Environment, Charles Darwin University, Darwin City, Australia
alazab.m@ieee.org

Abstract. Location-Based Services (LBSs) depend on a Service Provider (SP) to store data owners' geospatial data and to process data users' queries. For example, a Yelp user queries the SP to retrieve the k nearest Starbucks by submitting her/his current location. It is well-acknowledged that location privacy is vital to users and several prominent Secure k Nearest Neighbor (SkNN) query processing schemes are proposed. We observe that no prior work addresses the requirement of *repetitive query* after index update and its privacy issue, i.e., how to match a data item from the cloud repetitively in an oblivious and unlinkable manner. Meanwhile, a malicious SP may skip some data items and recommend others due to unfair competition.

In this work, we formally define the repetitive query and its privacy objectives and present an Repetitive, Oblivious, and Unlinkable SkNN scheme ROU. Specifically, we design a multi-level structure to organize locations to further improve search efficiency. Second, we integrate data item identity into the framework of existing SkNN query processing. Data owners encrypt their data item identity and location information into a secure index, and data users encrypt a customized identity range of a previously retrieved data item and location information into a token. Next, the SP uses the token to query the secure index to find the specific data item via privacy-preserving range querying. We formally prove the privacy of ROU in the random oracle model. We build a prototype based on a server to evaluate the performance with a real-world dataset.

C. Alcaraz et al. (Eds.): ICICS 2022, LNCS 13407, pp. 261–280, 2022.
https://doi.org/10.1007/978-3-031-15777-6_15

Experimental results show that ROU is efficient and practical in terms of computational cost, communication overhead, and result verification.

Keywords: Cloud computing · S*k*NN · Repetitive query · Privacy

1 Introduction

1.1 Background

Smartphones are now equipped with a Global Positioning System (GPS) module and various applications that support location-based service (LBS) [1–4]. It works by sending a data user's current location query to a Service Provider (SP). The SP matches the query with data items from data owners and retrieves corresponding results to the data user. For instance, Google Maps enable data users to find Starbucks, bars, and restaurants near their current location.

While LBSs provide practical benefits, the privacy concerns rooted in location revelation and untrusted SP [5,6] are a major hindrance towards the broad adoption of LBSs. First, the submitted locations may include data users' sensitive locations. Second, locations are tightly correlated to human activities, such as visiting a cancer hospital and meeting a friend in a hotel. Besides, there are many reports on the data leakage incidents caused by cyber attacks, hardware malfunction, or misoperation for the past decade [7,8]. Therefore, it is highly important to protect data stored on SPs. To protect data stored on SPs, Secure k Nearest Neighbor (SkNN) query processing has been proposed [9–12].

1.2 Motivations

Motivation I (Blurry Memory): Our motivation arises from real-world applications. For example, say a data user Bob submitted a S4NN query (*location X*, "*pizza shop*", 4) to the SP and obtained 4 results, namely Papa Johns, Mr. Pizza, Domino's Pizza, and Big Pizza. Among the results he went to, he was quite satisfied with the Domino's Pizza. Days later, when Bob is near the same query location and wants to dine at the previous Domino's Pizza. Unfortunately, Bob's memory of this shop somehow blurs and he submits a *repetitive query* to the SP to find the preferred pizza shop. **Motivation II (Index Update):** Following the example above, even if Bob remembers the name of the shop, the SP may happen to update the index tree as depicted in Fig. 1 such that the preferred location will not be included in the query results. In the previous query, the SP found the Domino's Pizza at di_5, i.e., data item 5, among the first 4 matched data items $\{di_2, di_4, di_5, di_6\}$. After the index update, di_5 will not be returned to Bob who submits the same S4NN query because the data items are reordered and di_5 is not among the top 4 matched data items. **Motivation III (Malicious SP):** The SP could be malicious in the sense that it has a secret agreement with some data owners to deliberately order data items, which is similar to unfair ranking where some search engines treat websites unfairly [17]. In this case, the user-preferred data item, which is not "favored" by the SP,

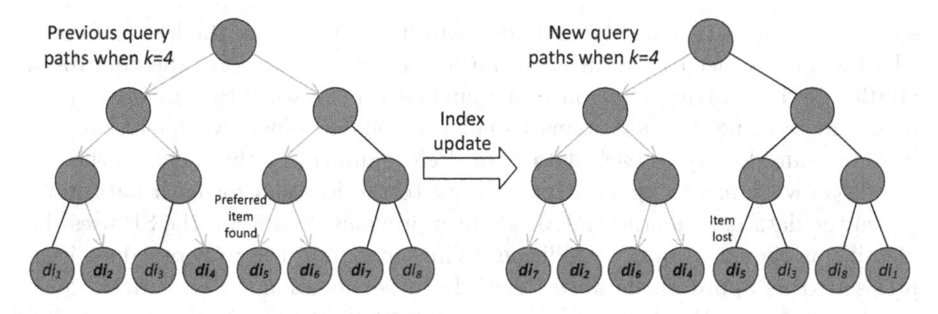

Fig. 1. Query paths before/after index update. (Data items marked in yellow are matched data items. The data item marked in red is matched and user-preferred data item. After index update, the preferred data item di_5 is moved back on the leaf level such that the search paths change. The user will not receive such an item by using the same S4NN query because the SP finds 4 matched data items before di_5.) (Color figure online)

may be put way back in the data item queue. **Motivation IV (Improving Efficiency):** There are many methods of processing locations in SkNN, such as Voronoi diagram [9], Paillier cryptosystem [10,18], and projection functions [11]. Improving search efficiency is always an ongoing goal.

Based on the first three motivations, no previous studies [9–16] have considered the requirement of repetitive query and its potential privacy problems, which lead to the following new requirements for SkNN.

- **Repetitive query after index update**: A data user queries the same data item that is returned in the previous query especially after the SP has updated the index.
- **Obliviousness**: We need to prevent the SP from knowing the data user's requirement to retrieve a previously matched data item.
- **Unlinkability**: Prevent the service provider from knowing that the specific data item has been previously matched to the data user.
- **Exclusiveness**: Prevent the SP from abandoning the preferred data item in case the SP gives priority to other data items.

1.3 Possible Solutions and Technical Challenges

A simple way of finding a preferred data item in an oblivious and unlinkable manner is to query all the locations with the same type as the preferred data item. However, this brings too many computational costs in result searching and communication overhead. Assume that the data user needs the data item di with a sequence number $*$. Intuitively, there are three approaches to finding $*$ in SP's index tree: (1) *Start from* $*$ and continue to find the other $k-1$ data items. (2) *End at* $*$ and return the obtained data items. (3) *Randomly choose* r data items before $*$ and find $k-r-1$ data items after $*$. These three approaches require special treatment on locating $*$ which makes it difficult for the SP not to notice this difference. This first one may traverse the whole index and the second one may return all the matched data items if di_* is in the last leaf node. The last one is faced with an uncertain choice of r. Therefore, the **technical challenge I** is solving the contradiction between locating the preferred data item and treating all data items equally. To enable repetitive query, we can introduce an identity to each data item. Before uploading the index to the SP, the data owner has to append the identity to the location for each data item. When the data user queries the SP about a previous data item, the SP uses the identity as the extra query condition. While matching the previous data item precisely, this approach, however, excludes other location-matched data items which may expose the data user's requirement of repetitive matching. To match other data items, we cannot use the identity directly. Therefore, the **technical challenge II** lies in the contradiction between using a preferred identity and matching other data items.

To address the above challenges, we propose ROU: a Repetitive, Oblivious, and Unlinkable query processing scheme. Specifically, we first divide the location map into a l-leveled pyramid and each level consists of a number of grids. For each level, we use the similar space encoding technique to save computational cost for both data users and the SP. The data owner (data user) encodes the data items' locations (current location) and obtains a set of leveled location codes. At the lth level, we assign an identity to each data item. The data owner computes a prefix family of the data item identity and integrates it with location codes at the l-th level. Next, the data owner inserts the integrated codes into an Indistinguishable Bloom Filter (IBF) as a secure index. The data user who is about to submit a repetitive query generates a customized identity range to compute a minimum set of prefixes. Next, the data user also integrates the prefixes with the location codes similarly and computes a query token. Finally, the SP searches the secure index by querying the token on it and returns matched data items to the data user. Our contributions are summarized as follows.

- To the best of our knowledge, we are the first to focus on the repetitive query in SkNN and we propose a repetitive, oblivious, and unlinkable query processing scheme.
- We achieve the three above-mentioned new requirements via customized identity transformation and privacy-preserving range querying. We design a multi-level structure to encode locations to accelerate the search efficiency.
- We formally define privacy and then prove it in the random oracle model. We build a prototype of ROU based on a server and a real-world dataset. Experimental results demonstrate its efficiency and practicability.

1.4 Paper Organization

The remaining of this paper is organized as follows. We discuss related work in Sect. 2. We elaborate on the system model, threat model, and design objectives in Sect. 3. In Sect. 4, we introduce the proposed space encoding. In Sect. 5, we present the ROU scheme. We formally analyze the privacy of the ROU in Sect. 6. We implement the ROU scheme and analyze its performance in Sect. 7. Lastly, we draw conclusions in Sect. 8.

2 Related Work

2.1 S*k*NN

Yao et al. [9] proposed SNN methods by asking the SP, given only an encrypted query point $E(p)$ and an encrypted database $E(D)$, to return a corresponding (encrypted) partition $E(G)$ satisfying that $E(G)$ contains SNN query answer. They name their method the secure Voronoi diagram (SVD) method that is based on special partitions over D and the Voronoi diagram of D. They partition the database D into small groups and then store the encrypted groups on the SP. Instead of returning the whole encrypted database, the SP retrieves one encrypted group for any SNN query. The SVD method does not require any new encryption schemes, but only depends on any standard encryption scheme E (e.g., RSA and AES) which means its security is the same as E.

Elmehdwi et al. [10] proposed an k-nearest neighbor search protocol based on two non-colluding semi-honest SPs that preserves both the data privacy and query privacy. They first design a basic protocol and show why it is not secure and present a fully secure kNN protocol. The basic protocol allows the data user to retrieve k records that are closest to his query by using Paillier cryptosystem [18] and secure squared Euclidean distance. The advanced protocol, however, utilizes secure bit-decomposition, secure minimum out of n numbers, secure bit-OR to avoid exposing the data access patterns in the basic protocol.

Lei et al. [11] proposed a secure and efficient query processing protocol SecEQP. They leveraged some primitive projection functions to convert the neighbor regions of a given location. Given the codes of two converted locations, the service provider computes the proximity of the two locations by judging whether the two codes are the same. This is an improvement over their previous work [14] since the two-dimensional location data is projected to high-dimensional data which expands the location space to make the converted location more secure. The data owner further embeds the codes into a similar IBFTree in order to build a secure index. The data user computes similar trapdoors by a keyed hash message authentication code. The final secure query processing is the same as [14].

2.2 Privacy-Preserving Range Querying

Li et al. [13] presented the first range query processing protocol which achieved index indistinguishability under the indistinguishability against chosen keyword attack (IND-CKA). A data owner converts each data item dt_i by prefix encoding [19] and organizes each prefix family of encoded item $F(di_i)$ into a PBTree. Then the data owner makes the PBtree privacy-preserving by a keyed hash message authentication code HMAC and Bloom filters. For each prefix pr_i, the data owner computes several hashes $\mathsf{HMAC}(K_j, pr_i)$ and inserts a randomized version $\mathsf{HMAC}(r, \mathsf{HMAC}(K_j, pr_i))$ into a Bloom filter. Each r corresponds to a node and each node relates to a prefix family, i.e., data item. Next, a data user converts a range into a minimum set of prefixes and computes several hashes

$\mathsf{HMAC}(K_j, pr_i)$ for each pr_i as a trapdoor. The service provider searches in the PBtree to find a match by using the trapdoor.

Li et al. [14] concerned processing conjunctive queries including keyword conditions and range conditions in a privacy-preserving way and presented a privacy-preserving conjunctive query processing protocol supporting adaptive security, efficient query processing, and scalable index size at the same time. Specifically, they adopt prefix encoding as in their earlier work [13] and design an indistinguishable Bloom filter (IBF), i.e., twin Bloom filter to replace the previous structure. A pseudo-random hash function H to determine a cell location $H(h_{k+1}(h_j(pr_i)) \oplus r)$, i.e., which twin cell stores '1'. Instead of building a PBTree, they construct an IBTree as the secure index.

Different from the previous works, ROU scheme can support the three new features in SkNN, namely repetitive, oblivious, and unlinkable. The novelty of ROU is in realizing the function of repetitive query by mixing customized identity range query with existing SkNN query without sacrificing privacy.

3 Problem Formulation

Before we dive into the details of ROU, we elaborate on its system model, threat model, and design objectives. Specifically, we formally define the repetitive query and its privacy objectives.

3.1 System Model

The system model, as drawn in Fig. 2, consists of a data owner O, a data user U, and SP. We define $\mathcal{DI} = \{di_1, di_2, \cdots, di_n\}$ as the set of n data items. A location loc as a pair of coordinates.

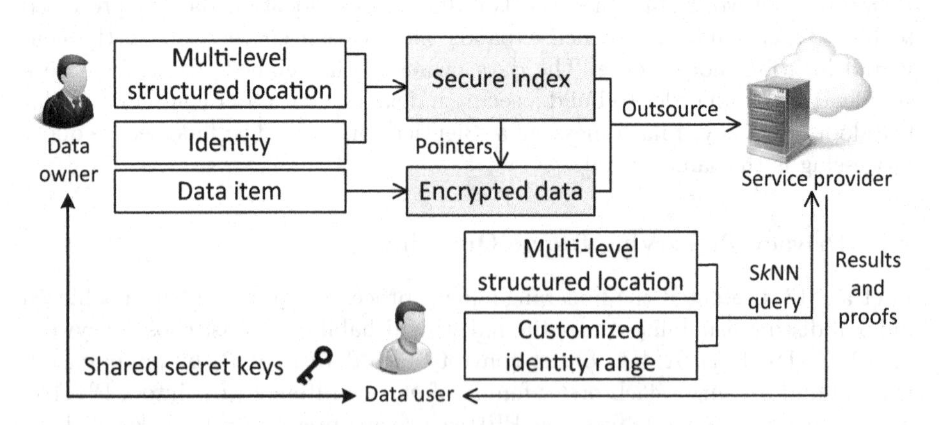

Fig. 2. ROU system model.

Data Owner: A data owner has some data items to be shared with data users. Each data item has type, location, and identity. The data owner extracts the information of each data item and calculates a secure index by using secret keys. Next, he encrypts his data item by using another secret key and a standard encryption algorithm. Each secure index has a pointer to link to the ciphertext. Finally, the data owner uploads the index and the ciphertext to the SP. The secret keys are shared with data users. We assume that ROU has only one data owner for simplicity, but also supports the multi-owner setting.

Data User: A data user generates a query token by using the type, current location, and shared secret keys, and an identity range. If the data user does not have a specific preference on a data item, the identity range is set by default. Otherwise, the identity range is computed based on the identity of the preferred data item. Next, the data user submits the query token to the SP, which retrieves corresponding results and proofs to the data user. The data user decrypts and verifies the received results. We formally define the repetitive query as follows.

Definition 1 (Repetitive Query). *A repetitive query is a single location-time predicate or a combination of location-time predicates linked by the Boolean operators [20]. Let* $Q = (pid, \mathcal{T}(t, loc, id, R))$ *be a SkNN query submitted by a data user and it is a pair of pseudo-identity pid and a query token* $\mathcal{T}$. $\mathcal{T}$ *is composed of type t, location loc, identity of a previously matched data item id, and an identity range R. Let* $Q_i = (pid_i, \mathcal{T}_i(t_i, loc_i, id_i, R_i))$ *be the ith query of data owner O. A repetitive query event, denoted by* ReQuery, *is expressed as* $(pid_i \neq pid_j) \wedge (\mathcal{T}_i \neq \mathcal{T}_j) \wedge (t_i = t_j) \wedge (loc_i = loc_j) \wedge (id_i = id_j) \wedge (R_i \neq R_j) \wedge (time.j > time.i)$ *for two queries* Q_i *and* Q_j.

SP: The SP helps the data owner to authorize the query service to a set of data users. The SP stores the secure indexes and ciphertexts uploaded from the data owners. It responds to data users' query tokens by searching over the secure indexes and returning corresponding results and proofs to data users.

3.2 Threat Model

The threats mainly arise from the behaviors of the internal entities, including the semi-honest (honest-but-curious) data owner and data users. This assumption is proposed by [21] and has been well acknowledged by existing work [11,13,14,22–24]. The SP is malicious [12,15]. Although it acts as a bridge between the data owner and data users to offer query services, it may also behave maliciously, i.e., it ignores some data items when searching the index in its database.

3.3 Design Objectives

There are four design objectives in this work: functionality, privacy, security, and efficiency.

Functionality, i.e., repetitive query after index update. ROU allows data users to query a previously matched data item even if the SP has updated the index.

Privacy. (1) Data/Index/Token Privacy. From the encrypted data item, index, and token, the adversary cannot learn any useful information about the data, data item's location, query location, and type [25–29]. (2) Obliviousness. ROU prevents the SP from knowing that the data user submitted a repetitive query. (3) Unlinkability. ROU prevents the SP from knowing that the data item referred to in the repetitive query was a retrieved data item of the data user. We define two experiments $\mathsf{PrivK}^{\mathsf{obl}}_{\mathcal{A},\Pi}$ and $\mathsf{PrivK}^{\mathsf{unl}}_{\mathcal{A},\Pi}$, based on a Probabilistic Polynomial-Time (PPT) adversary $\mathcal{A}$ and the ROU scheme $\Pi = (\mathsf{Setup}, \mathsf{Index}, \mathsf{Token}, \mathsf{Query})$, and a function S computing the minimum set of prefixes. The formal definitions are as follows.

The Adversarial Obliviousness Experiment $\mathsf{PrivK}^{\mathsf{obl}}_{\mathcal{A},\Pi}$**:**

1. $\mathcal{A}$ is given the size m of IBF and number of pseudo-random hash functions p, and outputs a pair of quintuples $q_0 = (t_0, loc_0, id'_0, R_0)$, $q_1 = (t_1, loc_1, id'_1, R_1)$ satisfying $t_0 = t_1$, $loc_0 = loc_1$, $id'_0 = 0$, $id'_1 \in \{1, n\}$, and $|\ \mathsf{S}(R_0)| = |\mathsf{S}(R_1)\ |$.
2. Secret keys are generated by using Setup, and a uniform bit $b \in \{0, 1\}$ is chosen. A query token $\mathcal{T}_b \leftarrow \mathsf{Token}(q_b)$ is computed and given to $\mathcal{A}$. We refer to $\mathcal{T}_b$ as the challenge token.
3. $\mathcal{A}$ outputs a bit b'.
4. The output of the experiment is defined to be 1, i.e., $\mathsf{PrivK}^{\mathsf{obl}}_{\mathcal{A},\Pi} = 1$ and $\mathcal{A}$ succeeds, if $b' = b$, and 0 otherwise.

The Adversarial Unlinkability Experiment $\mathsf{PrivK}^{\mathsf{unl}}_{\mathcal{A},\Pi}$**:**

1. $\mathcal{A}$ is given the size m of IBF and number of pseudo-random hash functions p, and outputs a pair of quintuples $q_0 = (t_0, loc_0, id'_0, R_0)$, $q_1 = (t_1, loc_1, id'_1, R_1)$ satisfying $t_0 = t_1$, $loc_0 = loc_1$, $di_{id'_1} \in \mathsf{Query}(q_0, \mathcal{I})$, and $|\ \mathsf{S}(R_0)\ |=|\ \mathsf{S}(R_1)\ |$, where $\mathcal{I}$ is the index tree.
2. Secret keys are generated by using Setup, and a uniform bit $b \in \{0, 1\}$ is chosen. A query token $\mathcal{T}_b \leftarrow \mathsf{Token}(q_b)$ is computed and given to $\mathcal{A}$. We refer to $\mathcal{T}_b$ as the challenge token.
3. $\mathcal{A}$ outputs a bit b'.
4. The output of the experiment is defined to be 1, i.e., $\mathsf{PrivK}^{\mathsf{obl}}_{\mathcal{A},\Pi} = 1$ and $\mathcal{A}$ succeeds, if $b' = b$, and 0 otherwise.

Definition 2 (Obliviousness). *Given a repetitive query event* ReQuery, *the SkNN scheme* Π *is oblivious if for every* $\mathcal{A}$, *it holds that* $\Pr[\mathsf{PrivK}^{\mathsf{obl}}_{\mathcal{A},\Pi} = 1] = \frac{1}{2}$. *In other words, it is trivial for* $\mathcal{A}$ *to succeed with probability* $1/2$ *by outputting a random guess. Obliviousness requires that it is impossible for any* $\mathcal{A}$ *to do better.*

Definition 3 (Unlinkability). *Given a repetitive query event* ReQuery, *the SkNN scheme* Π *is unlikable if for every* $\mathcal{A}$, *it holds that* $\Pr[\mathsf{PrivK}^{\mathsf{unl}}_{\mathcal{A},\Pi} = 1] = \frac{1}{2}$.

Security, i.e., exclusiveness. ROU prevents the SP from abandoning the preferred data item in case the SP gives priority to some locations. In other words, the data users can verify the query results that should include the preferred data item.

Efficiency. ROU should satisfy two types of efficiency requirements. (1) Low computational cost: the data owner/data user/SP spends a reasonable amount of time on computing index, token, and searching. (2) Low query latency. The data user can get the result within a reasonable amount of time. (3) Low communication overhead. It requires an acceptable amount of transmitted messages between the data owner, data users, and the SP.

4 The Proposed Space Encoding

The proposed space encoding technique is constructed on a multi-level structure to process data items. As shown in Fig. 3, there are four levels in the pyramid-like structure, i.e., l_1, l_2, l_3, and l_4. All the levels refer to the whole service area, but they are divided based on different granularity. From the second level L_2, the area is divided into more than one grid. There are 4, 16, and 64 grids in L_2, L_3, L_4, respectively. Each level encodes its grids from the number 1 prefixed with the level number such that each grid has a unique number on each level.

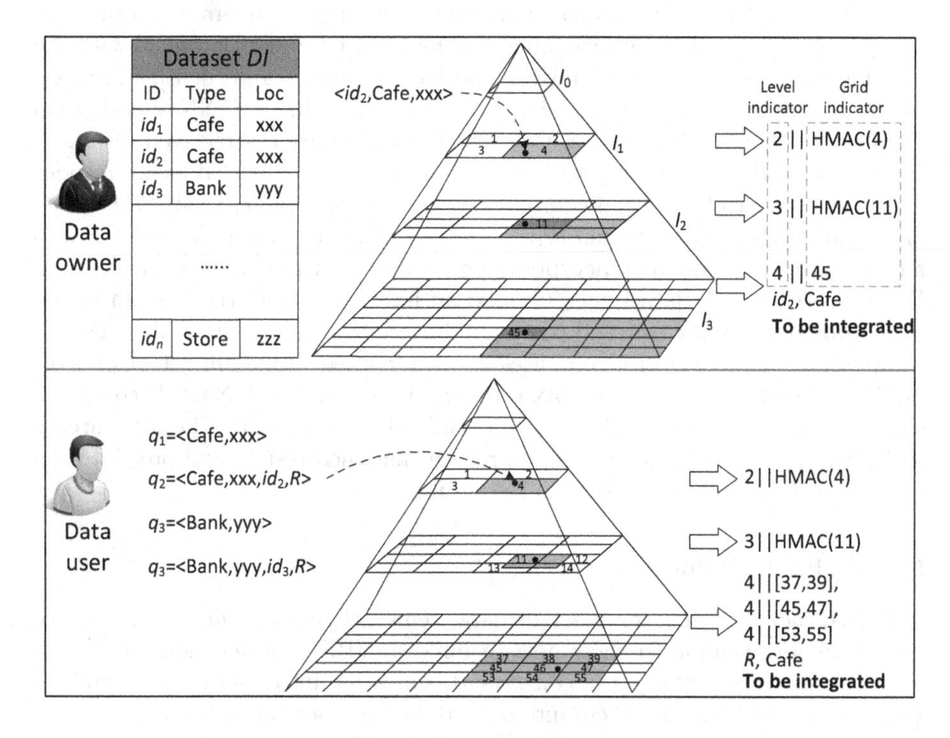

Fig. 3. Space encoding.

For the data item < id_2, Cafe, xxx>, the data owner encodes xxx from L_2 to L_4 to obtain three strings: "2||HMAC(4)", "3||HMAC(11)", and "4||HMAC(45)". Here, the real numbers before || stand for different levels and HMAC is a keyed hash message authentication code. At the last level L_4, the data owner integrates its identity id_2 with "4||HMAC(45)" as a foundation for repetitive query, which we will provide details in Sect. 5.2.

For the query q_2=<Cafe, xxxid_2, R >, the data user encodes her current location from L_1 to L_3 similarly. When reaching L_3, the data user computes a bigger grid that covers the current grid and obtain three grid number ranges, i.e., "[37,39]", "[45,47]", and "[53,55]". The size of the bigger grid is flexible and it is determined according to the data users. Next, the data user will also integrate the identity id_2 of the previously matched data item $< id_2$, Cafe, xxx> with "4|| [37, 39]", "4|| [45, 47]", and "4|| [53, 55]" similarly. The R is an identity range that covers id_2 that we will provide details in Sect. 5.3.

5 The Proposed Scheme ROU

5.1 Overview

As depicted in Fig. 4, we use the level-based space encoding to obtain location codes. We adopt the privacy-preserving range querying to generate identity prefixes for data items. Further, we integrate the repetitive query problem with the location querying problem by mixing the location codes and identity prefixes. Lastly, we leverage IBFs to build secure indexes and achieve SkNN querying via membership checking. The data users decrypt and verify the received results.

For each data item, O converts its location into a set of leveled location codes $\mathcal{LC}$ from level 2 to level l and computes a concatenated prefix family $\mathcal{PF}$ based on l, grid number, identity, and type on level l. Next, O inserts $\mathcal{LC}$ and $\mathcal{PF}$ into an IBF as a leaf node and encrypts the data item using symmetric encryption. When processing all data items, O build an index tree from the bottom to up, and submits the index tree and corresponding ciphertexts to SP. A data user U computes $\mathcal{LC}$ similarly and computes a concatenated minimum set of prefixes $\mathcal{MP}$ based on l, grid range, identity range, and type on level l. Next, U computes a query token qt based on $\mathcal{LC}$ and $\mathcal{MF}$ and submits it to SP. The SP searches the index tree by using the token and returns matched results and proofs to the U. Finally, U decrypts and verifies the results.

5.2 Index Building

A data owner O is holding a set of data items $\mathcal{DI}$. $di_i =< id_i, t_i, loc_i >$. We use di_i as an example to show how to build an IBF in a leaf node. For each di_i, O chooses a secret key K_0, converts di_i's location into a set of grid numbers $\{g_{i2}, \cdots, g_{il}\}$ and encodes them into a set of leveled location codes:

$$\begin{aligned}\mathcal{LC}_i = \{lc_{i2}, lc_{i3}, \cdots, lc_{il}\} = \{2 \mid\mid \mathsf{HMAC}_{K_0}(g_{i2}), \cdots,\\ l-1 \mid\mid \mathsf{HMAC}_{K_0}(g_{il-1}), l \mid\mid g_{il}\}.\end{aligned} \tag{1}$$

For the first $l-2$ levels, O processes di_i's location codes $\{lc_{i2}, lc_{i3}, \cdots, lc_{il-1}\}$ as follows. Given $p+1$ secret keys $K_1, K_2, \cdots, K_p, K_{p+1}$, p pseudo-random hash

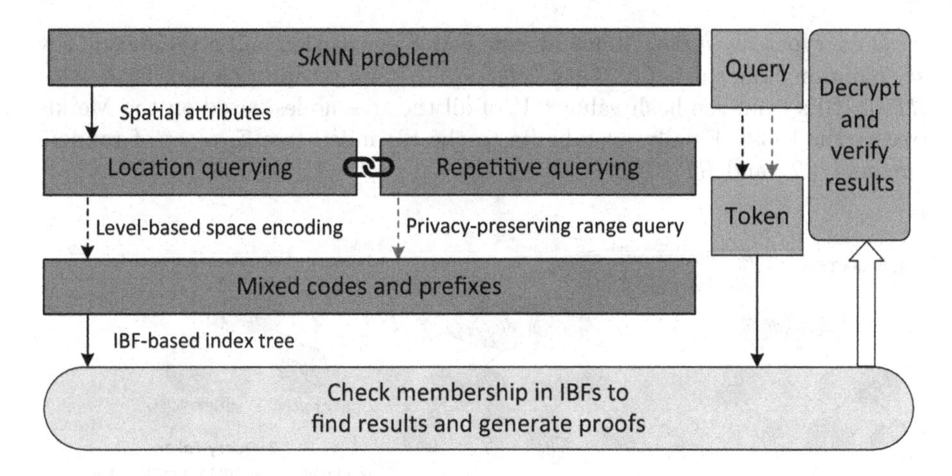

Fig. 4. ROU scheme overview.

functions $h_1, h_2, \cdots, h_p$ where $h_i = \mathsf{HMAC}_{K_i}(\cdot)$, and another hash function $H(.) = \mathsf{SHA256}(.)\%2$, O creates an indistinguishable Bloom filter IBF_O and embeds each location code lc_{iu} and a randomly chosen number r_i into IBF_i by setting for all $u \in [2, l-1]$ and $v \in [1, p]$:

$$IBF_i[H(h_{K_{p+1}}(h_v(lc_{iu})) \oplus r_i)][h_v(lc_{iu})] = 1, \tag{2}$$

$$IBF_i[1 - H(h_{K_{p+1}}(h_v(lc_{iu})) \oplus r_i)][h_v(lc_{iu})] = 0. \tag{3}$$

For the lth level, O computes a prefix family $\mathcal{PF}_{i1}$ of g_{il} by using prefix encoding [13] and a prefix family $\mathcal{PF}_{i2}$ of id_i's identity id_i. Then, O mixes $\mathcal{PF}_{i1}$ with $\mathcal{PF}_{i2}$ by concatenating their prefixes to obtain a mixed code set $\mathcal{MC}_i$. Further, O prefixes each mixed code with the level number and the type (converted into a real number). In this way, we lay a foundation for the data user to meet the requirement of repetitive query. Next, O inserts each code mc_u in $\mathcal{MC}_i$ into IBF_i by setting for all $u \in [1, |\mathcal{MC}_i|]$ and $v \in [1, p]$:

$$IBF_i[H(h_{K_{p+1}}(h_v(mc_u)) \oplus r_i)][h_v(mc_u)] = 1, \tag{4}$$

$$IBF_i[1 - H(h_{K_{p+1}}(h_v(mc_u)) \oplus r_i)][h_v(mc_u)] = 0. \tag{5}$$

When processing all data items, O obtains n IBFs and builds an index tree from the bottom to up. O sorts the n IBFs in a random order and organize them into a binary tree structure to achieve sublinear search time [11]. An index tree $\mathcal{I}$ is built as follows. Assume that IBF_1 is the father IBF of two children IBFs: IBF_2 (left child) and IBF_3 (right child), then for each $i \in [1, m]$, the value of IBF_1's ith twin is the logical OR of IBF_2's ith twin and IBF_3's ith twin.

$$\begin{aligned} &IBF_1[H(h_{K_{p+1}}(i) \oplus r_1)][i] = \\ &IBF_2[H(h_{K_{p+1}}(i) \oplus r_2)][i] \vee IBF_3[H(h_{K_{p+1}}(i) \oplus r_3)][i]. \end{aligned} \tag{6}$$

O encrypts the n data items by using AES encryption and a symmetric key sk to obtain ciphertexts $\mathcal{CT} = \{ct_1, ct_2, \cdots, ct_n\}$ and computes a root hash value RT of IBFs from the hash value HV of all the tree nodes based on the Merkle tree method [30]. Finally, O submits to the SP index tree $\mathcal{I}$, a set of random numbers, $\mathcal{CT}$, and RT.

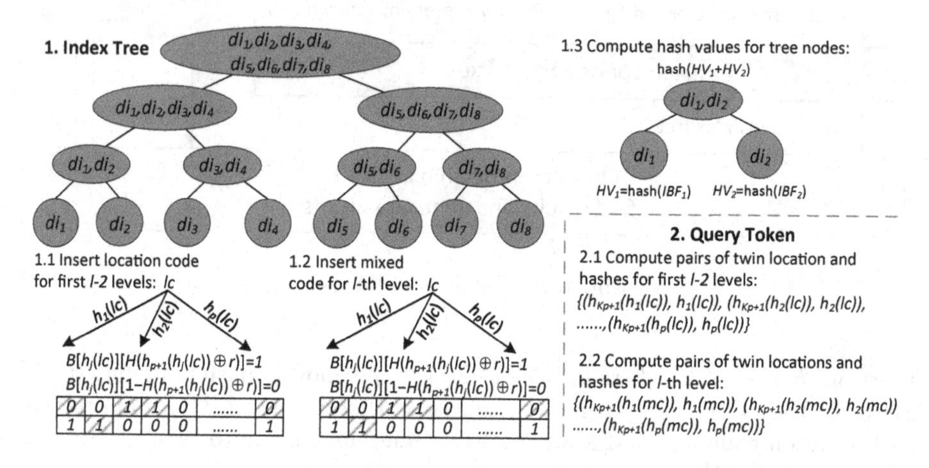

Fig. 5. Index tree and query token.

5.3 Token Generation

A data user U is standing at location loc and expecting to find the data item id_i. U converts loc into a set of leveled location codes:

$$\mathcal{LC} = \{lc_2, \cdots, lc_l\} = \{2 \mid\mid \mathsf{HMAC}_{K_0}(g_2), \cdots, \\ l-1 \mid\mid \mathsf{HMAC}_{K_0}(g_{l-1}), l \mid\mid \mathsf{Exp}(g_l)\}, \tag{7}$$

where $\mathsf{Exp}(g_l)$ expands current grid to a bigger area which consists of the nearest nine grids as shown in Fig. 3.

For each location code $lc_u, 2 \leq u \leq l-1$, U computes p hashes $h_j(lc_u), 1 \leq j \leq p$. For each $h_j(lc_u), 1 \leq j \leq p$, U computes $h_{K_{p+1}}(h_j(lc_u))$. The subtoken for lc_u is a p-pair of twin locations and hashes: $\{(h_{K_{p+1}}(h_1(lc_u)), h_1(lc_u)), \cdots, (h_{K_{p+1}}(h_p(lc_u)), h_p(lc_u))\}$. Then, O obtains a $((l-2) \times p)$-pair of twin locations and hashes. We denote the set by $\mathcal{T}_1$, i.e., the first part of the $\mathcal{T}$.

For the lth level, U computes a minimum set of prefixes $\mathcal{M}_1$ for $\mathsf{Exp}(g_{il})$ and a minimum set of prefixes $\mathcal{M}_2$ for $R_i(id_i)$. Here, we require that $R_i(id_i) =$

$$\begin{cases} [id_i, id_i+1] \vee [id_i+2, id_i+3] \vee \cdots \vee \\ [id_i + 2 \mid \mathsf{S}(1,n) \mid -2, id_i + 2 \mid \mathsf{S}(1,n) \mid -1], \; if \; id_i\%2 = 0 \\ [id_i - 1, id_i] \vee [id_i+1, id_i+2] \vee \cdots \vee \\ [id_i + 2 \mid \mathsf{S}(1,n) \mid -3, id_i + 2 \mid \mathsf{S}(1,n) \mid -2], \; otherwise \end{cases}$$

By doing so, we have that $\mid \mathcal{M}_2 \mid = \mid \mathsf{S}(id_1, id_n) \mid$. U mixes $\mathcal{M}_1$ with $\mathcal{M}_2$ by concatenating their prefixes to obtain a mixed code set $\mathcal{MC}$. Further, U prefixes each mixed code with the level number and the type. We denote the set by $\mathcal{T}_2$, i.e., the second part of the $\mathcal{T}$. Finally, U submits the query token $\mathcal{T} = (\mathcal{T}_1, \mathcal{T}_2)$ to the SP. We draw the process of tree construction and token generation in Fig. 5.

5.4 Query Processing

After receiving $\mathcal{T}$, the SP searches $\mathcal{I}$ from up to the bottom to find leaf nodes that match $\mathcal{T}$. Specifically, the SP proceeds in two stpes. (1) SP performs query processing by checking whether $IBF[H(h_{K_{p+1}}(lc)) \oplus r)][h_j(lc)] = 1$ for at least one $j \in [1, p]$ and $(H(h_{K_{p+1}}(lc)), h_j(lc))$ is one pair in $\mathcal{T}_1$. If this match continues until the leaf level, it means there is at least one data item matches the query on the first $l-2$ levels. (2) At a leaf node, the SP performs the similar query processing by using $\mathcal{T}_2$. If there is a match, the SP continues to search other matched leaf nodes. Finally, the SP returns the ciphertexts of match lead nodes and the proofs (IBFs of the branch-but-unmatched nodes) to the U.

5.5 Result Verification

O decrypts the ciphertexts and checks whether the returned data items include the preferred data item. Next, O verifies that her query does not match the IBFs in the proofs. O also recomputes the value of the root from bottom to up by using the leaf IBFs and proofs. If the computed value equals to RT, O is convinced that that results are not tampered with.

6 Privacy Analysis

6.1 Data/Index/Token Privacy

Theorem 1. *ROU is adaptive IND-CKA $(\mathcal{L}_1, \mathcal{L}_2)$-secure in the random oracle model, achieving data/index/token privacy.*

Due to the space limitation, please refer to our technical report for the detailed proofs.

6.2 Obliviousness

In the adversarial obliviousness experiment $\mathsf{PrivK}^{\mathsf{obl}}_{\mathcal{A},\Pi}$, a challenge query token $\mathcal{T}_b$ is returned to the adversary $\mathcal{A}$. Specifically, $\mathcal{T}_b$ consists of two parts $\mathcal{T}_{b1}$, $\mathcal{T}_{b2}$. Given that $loc_0 = loc_1$, we have $\mathcal{T}_{01} = \mathcal{T}_{11}$. For the second part, we require that $R(id_1')$ is a customized range satisfying $\mid \mathsf{S}(R(id_1)) \mid = \mathsf{S}([id_1, id_n])$. By doing so, we have $\mid \mathcal{T}_{02} \mid = \mid \mathcal{T}_{12} \mid$ and they are indistinguishable for using secret keys and the one-way hash functions. Therefore, $\mathcal{T}_0$ and $\mathcal{T}_1$ are indistinguishable, i.e., $\Pr[\mathsf{PrivK}^{\mathsf{obl}}_{\mathcal{A},\Pi} = 1] = \frac{1}{2}$.$\square$

6.3 Unlinkability

In the adversarial unlinkability experiment $\mathsf{PrivK}^{\mathsf{unl}}_{\mathcal{A},\Pi}$, a challenge query token $\mathcal{T}_b$ is returned to the $\mathcal{A}$. Similarly, we have $\mathcal{T}_{01} = \mathcal{T}_{11}$. Although $di_{id'_1} \in \mathsf{Query}(q_0, \mathcal{I})$, i.e., the data item $di_{id'_1}$ belongs to the previously received data items, we have also randomized $\mathcal{T}_{12}$ to make it indistinguishable from $\mathcal{T}_{02}$. Therefore, we have $\Pr[\mathsf{PrivK}^{\mathsf{unl}}_{\mathcal{A},\Pi} = 1] = \frac{1}{2}$.□

6.4 Exclusiveness

To prevent the SP from abandoning the preferred data item, we ask the data users to explicitly integrate a customized identity range in the query token. In this way, the SP can only return matched data items. Further, the SP has to generate proofs to prove that the claimed unmatched nodes do not match the query. In this way, the data users are convinced that the preferred data item is not abandoned.

7 Performance Analysis

7.1 Experiment Settings

Dataset. We use the locations of three cities, i.e., Orlando, Portland, and Atlanta, from the Yelp dataset [31]. Each location has a type and two location coordinates. After preprocessing the dataset, we obtain 10,000 data items from each of the three cities and each item is in the form of (id, t, loc).

Parameters. We vary n from 2,000 to 10,000, and k from 1 to 5. The false positive rate is set to 1%. The number of pseudo-random hash functions p is 5. According to the false positive rate equation [13], the IBF size m ranges from 1.2 to 12 KB. The lengths of secret keys, random numbers, and the symmetric key are 1024 bits, 1024 bits, and 256 bits, respectively.

Metrics. We evaluate the time of tree construction, token generation, query processing, and result verification. We evaluate the communication overhead of index tree, query token, results and proofs. We conduct each set of experiment over twenty times and compute the average time. Communication overhead is calculated by measuring the size of the transmitted messages. Since AES is applicable to the symmetric encryption of all schemes, we remove this part in comparison, but focus on the index, token, query, and verification.

Setup. We instantiate ROU on a PC server running Windows Server 2021 R2 Datacenter with a 3.7-GHz Intel(R) Core(TM) i7-8770K processor, and 32 GB RAM. We use HMAC-SHA256 as the pseudo-random function to implement the hash functions of IBF. We use AES as the symmetric encryption. We have uploaded all source codes of ROU on Github: https://github.com/UbiPLab/ROU.

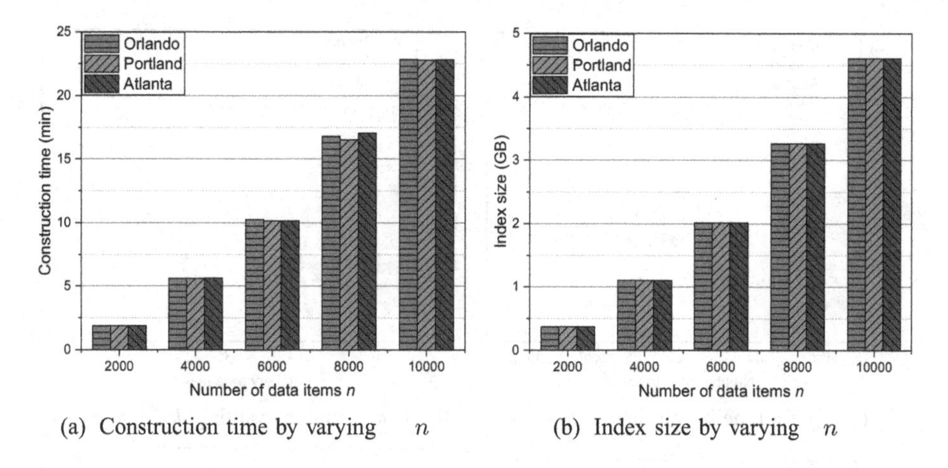

(a) Construction time by varying n (b) Index size by varying n

Fig. 6. Performance of tree construction.

7.2 Index Building

The computational cost of building an index tree as a function of n is shown in Fig. 6(a). The communication overhead of uploading the index tree to the SP as shown in Fig. 6(b). It can be observed that construction time and index size grow linearly with n. When $n = 10,000$, it costs the data owner less than 23 ms in computing an index tree of 4.6 GB.

7.3 Token Generation

We assume that the data user only want to find one specific data item that was returned in her previous query. The token size is independent of k, but not n because the size of $\mathcal{M}_2$ increases with n. There are two types of queries: ordinary query and repetitive query. Since the total number of prefixes in the two cases are the same, there will be no difference for their computational cost (35.1 ms) and communication overhead (77.1 KB) when $n = 2000$.

7.4 Query Processing

The query processing time of ROU is a function of n and k. Figure 7(a) shows that the query processing time is in the millisecond scale. When $k = 1$ and $n = 10000$, the average query processing time for Orlando is 28 ms. The difference among the three cities are caused by the different distribution of matched data items. However, Fig. 7(b) shows that when $n = 2000$, with the k increasing from 1 to 5, the query processing time does not grow much with k on each of the three lines, because we have designed a customized identity range for the data owner, which may lead to less matched results. In other words, the repetitive SkNN query does not have to return k results. In Fig. 7(c) and Fig. 7(d), the communication overhead of the SP increases with n and k because the IBF size increases with n and the number of returned nodes increases, respectively.

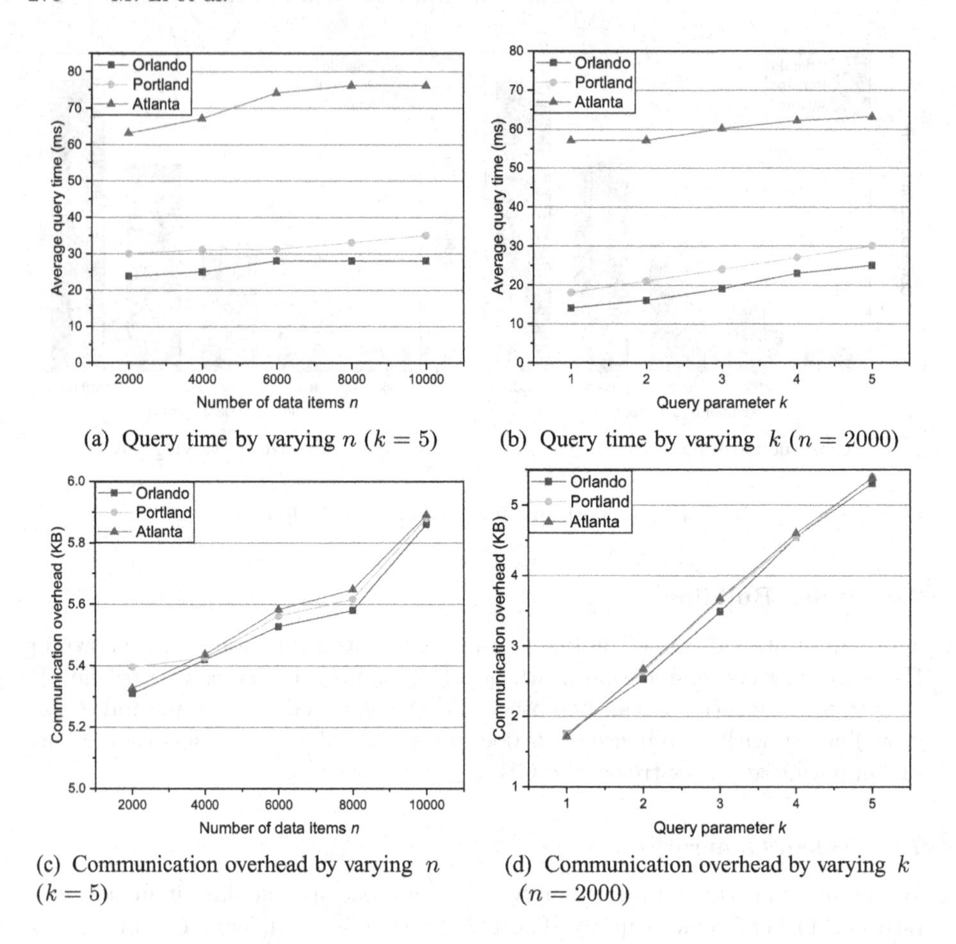

(a) Query time by varying n ($k = 5$)

(b) Query time by varying k ($n = 2000$)

(c) Communication overhead by varying n ($k = 5$)

(d) Communication overhead by varying k ($n = 2000$)

Fig. 7. Performance of query processing.

7.5 Result Verification

After the SP returns the results and proofs to the data user, the data user verifies the results by recomputing the root's hash value from the received hash values. The result verification time, as shown in Fig. 8, corresponds to the results and proofs returned by the CS. It costs the data user (in Orlando) 0.08 ms and 0.1 ms when $k = 1, n = 2000$ and $k = 1, n = 10000$, respectively. We attribute this advantage to the exclusiveness of the repetitive query.

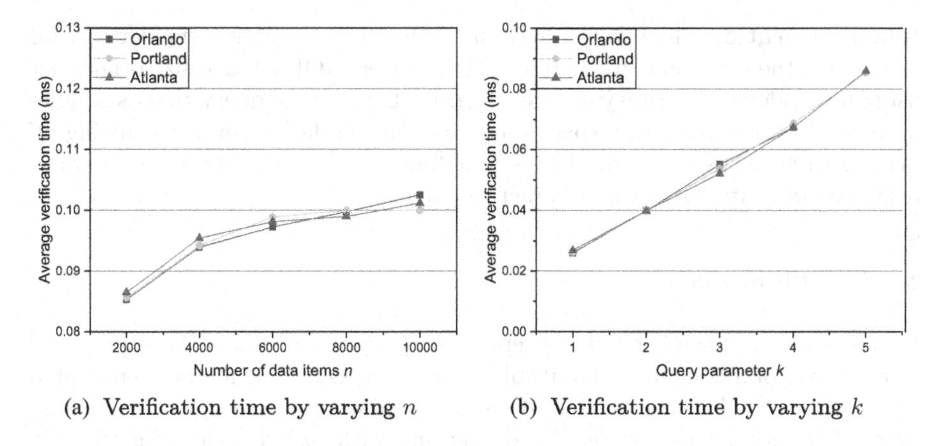

(a) Verification time by varying n (b) Verification time by varying k

Fig. 8. Performance of result verification.

Table 1. Comparison of computational costs and communication overhead.

Computational costs								
Scheme	Index building (min)		Token generation (ms)		Query processing (ms)			
					$n = 2000$		$n = 10000$	
	$n = 2000$	$n = 10000$	$n = 2000$	$n = 10000$	$k = 1$	$k = 5$	$k = 1$	$k = 5$
SecEQP [11]	1.82	22.5	33.08	512.37	16.06	28.07	17.06	30.08
ServeDB [15]	0.34	2.02	32.02	511.38	21.06	73.19	26.07	88.23
ROU	1.87	22.8	35.1	516.37	14.04	25.04	16.04	28.08
Communication overhead								
Scheme	Index building (GB)		Token generation (KB)		Query processing (KB)[1]			
					$n = 2000$		$n = 10000$	
	$n = 2000$	$n = 10000$	$n = 2000$	$n = 10000$	$k = 1$	$k = 5$	$k = 1$	$k = 5$
SecEQP [11]	0.37	4.60	41.25	43.12	n/a			
ServeDB [15]	0.31	1.86	20.14	21.18	17.99	84.56	18.35	88.49
ROU	0.37	4.60	77.10	116.19	1.75	5.31	2.03	5.86

1: messages for result verification

7.6 Comparison

We compare ROU with existing work, i.e., SecEQP [11] and ServeDB [15], which are constructed upon the same techniques. We add the type and data item identity into their schems by using privacy-preserving range query. We record the comparison results in Table 1. In **index building**, SecEQP and ServeDB also build an index tree. The cost of SecEQP is similar to ours for using multiple coordinate systems. The cost of ServeDB is lower only uses a Bloom filter as an index, thereby involving less computation time and communication overhead. In

token generation, the two comparison schemes have a slightly smaller cost for not mixing the location codes and identity prefixes. ROU's token size is large for using mix indexes. In **query processing**, ROU's average query time is smaller because the data user has a specific requirement on data item, thus cutting off many search paths when the CS is searching on the index tree. Comparison results show that ROU exhibits practical efficiency.

8 Conclusions

In this work, we have located the repetitive query in SkNN and have proposed a repetitive, oblivious, and unlinkable query processing scheme over encrypted data on cloud. The novelty of ROU is in realizing repetitive query by mixing customized privacy-preserving range querying with SkNN query. We formally define and prove the privacy of ROU. By carefully designing the index building and token generation, we achieve repetitive query in an oblivious and unlinkable manner. We implement ROU and evaluate its performance on a desktop server and a real-world dataset. The experimental results show that ROU achieves practical efficiency.

Acknowledgment. The work described in this paper is supported by National Natural Science Foundation of China (NSFC) under the grant No. 62002094 and Anhui Provincial Natural Science Foundation under the grant No. 2008085MF196. It is partially supported by EU LOCARD Project under Grant H2020-SU-SEC-2018-832735.

References

1. Liu, X., He, K., Yang, G., Susilo, W., Tonien, J., Huang, Q.: Broadcast authenticated encryption with keyword search. In: Baek, J., Ruj, S. (eds.) ACISP 2021. LNCS, vol. 13083, pp. 193–213. Springer, Cham (2021). https://doi.org/10.1007/978-3-030-90567-5_10
2. Luo, Y., Jia, X., Fu, S., Xu, M.: pRide: privacy-preserving ride matching over road networks for online ride-hailing service. IEEE Trans. Inf. Forensics Secur. (TIFS) **14**(7), 1791–1802 (2019)
3. Zhu, L., Li, M., Zhang, Z., Qin, Z.: ASAP: an anonymous smart-parking and payment scheme in vehicular networks. IEEE Trans. Dependable Secure Comput. (TDSC) **17**(4), 703–715 (2020). https://doi.org/10.1109/TDSC.2018.2850780
4. Zhu, X., Ayday, E., Vitenberg, R.: A privacy-preserving framework for outsourcing location-based services to the cloud. IEEE Trans. Dependable Secure Comput. (TDSC) **18**(1), 384–399 (2021)
5. Damodaran, A., Rial, A.: Unlinkable updatable databases and oblivious transfer with access control. In: Liu, J.K., Cui, H. (eds.) ACISP 2020. LNCS, vol. 12248, pp. 584–604. Springer, Cham (2020). https://doi.org/10.1007/978-3-030-55304-3_30
6. Li, M., Chen, Y., Zheng, S., Hu, D., Lal, C., Conti, M.: Privacy-preserving navigation supporting similar queries in vehicular networks. IEEE Trans. Dependable Secure Comput. (TDSC), **99**(2), 1–11. https://doi.org/10.1109/TDSC.2020.3017534

7. Danger within: defending cloud environments against insider threats (2018). https://www.cloudcomputing-news.net/news/2018/may/01/danger-within-defending-cloud-environments-against-insider-threats
8. 7 Most Infamous Cloud Security Breaches (2017). https://blog.storagecraft.com/7-infamous-cloud-security-breaches
9. Yao, B., Li, F., Xiao, X.: Secure nearest neighbor revisited. In: Proceeding 29th IEEE International Conference on Data Engineering (ICDE), April, pp. 733–744, Brisbane, Australia (2013)
10. Elmehdwi, Y., Samanthula, B.K., Jiang, W.: Secure k-nearest neighbor query over encrypted data in outsourced environments. In: Proceeding IEEE 30th International Conference on Data Engineering (ICDE), pp. 664–675, Chicago, USA (2014)
11. Lei, X., Liu, A. X., Li, R., Tu, G.-H.: SecEQP: a secure and efficient scheme for SkNN query problem over encrypted geodata on cloud. In: Proceeding 35th IEEE International Conference on Data Engineering (ICDE), April, pp. 662–673, Macao, China (2019)
12. Cui, N., Yang, X., Wang, B., Li, J., Wang, G.: SVkNN: efficient secure and verifiable k-nearest neighbor query on the cloud platform. In: Proceeding 36th IEEE International Conference on Data Engineering (ICDE), April, pp. 253–264, Dallas, USA (2020)
13. Li, R., Liu, A., Wang, A. L., Bruhadeshwar, B.: Fast range query processing with strong privacy protection for cloud computing. In: Proceeding 40th International Conference on Very Large Data Bases (VLDB), September, pp. 1953–1964, Hangzhou, China (2014)
14. Li, R., Liu, A.X.: Adaptively secure conjunctive query processing over encrypted data for cloud computing. In: Proceeding IEEE 33rd International Conference on Data Engineering (ICDE), April, pp. 697–708, San Diego, USA (2017)
15. Wu, S., Li, Q., Li, G., Yuan, D., Yuan, X., Wang, C.: ServeDB: secure, verifiable, and efficient range queries on outsourced database. In: Proceeding IEEE 35th International Conference on Data Engineering (ICDE), April, pp. 626–637, Macao, China (2019)
16. Chen, Y., Li, M., Zheng, S., Hu, D., Lal, C., Conti, M.: One-time, oblivious, and unlinkable query processing over encrypted data on cloud. In: Meng, W., Gollmann, D., Jensen, C.D., Zhou, J. (eds.) ICICS 2020. LNCS, vol. 12282, pp. 350–365. Springer, Cham (2020). https://doi.org/10.1007/978-3-030-61078-4_20
17. Poutinsev, F.: Unfair search engine ranking results (2021). https://honestproscons.com/unfair-search-engine-ranking-results. Honest Pros and Cons (HPC)
18. Paillier, P.: Public-key cryptosystems based on composite degree residuosity classes. In: Stern, J. (ed.) EUROCRYPT 1999. LNCS, vol. 1592, pp. 223–238. Springer, Heidelberg (1999). https://doi.org/10.1007/3-540-48910-X_16
19. Liu, A.X., Chen, F.: Collaborative enforcement of firewall policies in virtual private networks. In: Proceeding 27th ACM Symposium on Principles of Distributed Computing (PODC), August, pp. 95-104, Toronto, Canada (2008)
20. Cao, Y., Xiao, Y., Xiong, L., Bai, L., Yoshikawa, M.: Protecting spatiotemporal event privacy in continuous location-based services. IEEE Trans. Knowl. Data Eng. (TKDE) **33**(8), 3141–3154 (2021)
21. Canetti, R., Feige, U., Goldreich, O., Naor, M.: Adaptively secure multi-party computation. In: Proceeding 28th ACM Symposium on Theory of Computing (STOC), May, pp. 639–648, Philadelphia, USA (1996)

22. Boldyreva, A., Chenette, N., O'Neill, A.: Order-preserving encryption revisited: improved security analysis and alternative solutions. In: Rogaway, P. (ed.) CRYPTO 2011. LNCS, vol. 6841, pp. 578–595. Springer, Heidelberg (2011). https://doi.org/10.1007/978-3-642-22792-9_33
23. Kamara, S., Papamanthou, C., Roeder, T.: Dynamic searchable symmetric encryption. In: Proceeding 19th ACM Conference on Computer and Communications Security (CCS), October, pp. 965–976, Raleigh, USA (2012)
24. Cash, D., et al.: Dynamic searchable encryption in very-large databases: data structures and implementation. In: Proceeding 21st Annual Network and Distributed System Security Symposium (NDSS), February, pp. 1-16, San Diego, USA (2014)
25. Li, M., Chen, Y., Lal, C., Conti, M., Alazab, M., Hu, D.: Eunomia: anonymous and secure vehicular digital forensics based on blockchain. IEEE Trans. Dependable Secure Comput. (TDSC), 1 (2021). https://doi.org/10.1109/TDSC.2021.3130583
26. Li, M., Zhu, L., Zhang, Z., Lal, C., Conti, M., Alazab, M. : User-defined privacy-preserving traffic monitoring against n-by-1 jamming attack. IEEE/ACM Trans. Networking (TON), p. 1 (2022). https://doi.org/10.1109/TNET.2022.3157654
27. Li, M., Zhu, L., Zhang, Z., Lal, C., Conti, M., Alazab, M.: Anonymous and verifiable reputation system for E-commerce platforms based on blockchain. IEEE Trans. Network Serv. Manag. (TNSM) **18**(4), 4434–4449 (2021). https://doi.org/10.1109/TNSM.2021.3098439
28. Li, M., Hu, D., Lal, C., Conti, M., Zhang, Z.: Blockchain-enabled secure energy trading with verifiable fairness in industrial internet of things. IEEE Trans. Ind. Inf. (TII) **16**(10), 6564–6574 (2020). https://doi.org/10.1109/TII.2020.2974537
29. Li, M., Zhu, L., Zhang, Z., Lal, C., Conti, M., Martinelli, F.: Privacy for 5G-supported vehicular networks. IEEE Open J. Commun. Soc. (OJ-COMS), **2**, 1935–1956 (2021). https://doi.org/10.1109/OJCOMS.2021.3103445
30. Szydlo, M.: Merkle tree traversal in log space and time. In: Proceeding 10th International Conference on the Theory and Applications of Cryptographic Techniques (Eurocrypt), May, pp. 541–554, Interlaken, Switzerland (2004)
31. Yelp Open Dataset. https://www.yelp.com/dataset

Privacy-Aware Split Learning Based Energy Theft Detection for Smart Grids

Arwa Alromih[1,2](✉), John A. Clark[1], and Prosanta Gope[1]

[1] Department of Computer Science, University of Sheffield, Sheffield, UK
{asmalromih1,john.clark,p.gope}@sheffield.ac.uk

[2] Information Systems Department, King Saud University, Riyadh, Saudi Arabia

Abstract. Energy thefts are one of the critical attacks that often cause high revenue losses for utility companies around the world. Effective detection of such attacks is very important and must be implemented to comply laws and regulations that govern users' privacy. Current detection approaches rely on significant amounts of raw fine-grained smart meter data and generally do not consider privacy. On the other hand, most privacy-preserving machine learning (PPML) approaches, such as homomorphic ML and federated learning, are not well suited to the smart grid environment due to their processing complexity and communication overheads. Therefore, our contributions in this work are twofold: first, we propose an *enhanced* privacy-preserving detection model for energy thefts using the concept of Split Learning. Subsequently, since the classical Split Learning cannot be directly applied in the smart grid (SG) environment due to its communication overhead, we introduce a *new variant* of Split Learning that is more communication-efficient and suits the smart grid environment. The proposed model can ensure two advantages over the existing techniques. First, the use of Split Learning enables the training of a detection model without any need for raw data. This helps in achieving data privacy. Second, the splitting of the detection model allows the system to be more robust against honest-but-curious adversaries. Our evaluations show that the proposed detection model can ensure better privacy protection and communication efficiency, which are essential for smart grid, without compromising detection accuracy.

Keywords: Energy theft · Privacy · Split Learning · Smart grid · Communication efficiency

1 Introduction

Smart grid (SG) networks are one of the evolutionary steps toward intelligent power grids. SG added a bidirectional communication channel between different components of the electrical grid which helped in better overall facilitating of automated grid management [5]. This is done by the use of smart meters (SM) that allow the automatic collection of fine-grained data. These metering data are sent regularly to the grid and measure the energy usage and production if

C. Alcaraz et al. (Eds.): ICICS 2022, LNCS 13407, pp. 281–300, 2022.
https://doi.org/10.1007/978-3-031-15777-6_16

the household has a distributed energy resource installed in place, such as solar panels. Access to fine-grained metering data has enabled several new applications such as load management, load forecasting, demand response and billing [2].

The dependence on information and communication technologies has opened up new avenues of attack. One of the main attacks against data integrity in energy systems is energy theft. This attack involves manipulation of the fine-grained data that the smart meters send through the network. Recently, a number of approaches have been proposed for the detection of energy thefts in the smart grid. These detection approaches are mainly classified into three categories: state estimation, game theory, and machine learning (ML) techniques [5]. However, most existing detection methods access users' raw energy data without any concerns for their privacy and ignore the fact that users' private data are governed by privacy policies such as GDPR. The use of raw energy data creates new privacy vulnerabilities associated with what these data could reveal. For example, the disclosure of the real-time, fine-grained power consumption can reveal the identity of an individual or information about his/her financial, social, physical or health characteristics [16]. In particular, high power consumption may reveal that people are in the house, while low readings may indicate that the house is empty. Hence, this creates the need to develop new mechanisms to be able to build energy theft detectors without violating users' privacy.

1.1 Related Work and Motivation

In the literature, different strategies have been developed to defend against energy thefts in smart grids. However, very few have considered users' privacy in this regard, especially ML approaches [2,5]. The existing privacy-preserving energy theft detection approaches usually fall under two categories, encryption techniques [11,26], and privacy-preserving machine learning (PPML) techniques [10,16,25,27]. We first consider the encryption based approaches followed by the existing PPML techniques, and highlight their strengths and weaknesses.

The energy theft detection proposed in [26] uses a recursive filter based on state estimation to estimate the energy consumption for all users, and compare it with the true reading. If the difference is larger than a predefined threshold, then the reading is flagged as abnormal. In their work, the authors use the Number Theory Research Unit (NTRU) algorithm to encrypt users' data and preserve users' privacy. The simulation results show a detection accuracy of more than 92%, however the scheme introduces communication and computation overhead. The work presented in [11] uses the concept of harmonic to arithmetic mean (HMAM) as a detector for energy theft with fully homomorphic encrypted (FHE) data. In general, the classical FHE cannot be used with HMAM, so the authors has modified it in a way that allows HMAM to be computed. The new modified version of FHE is claimed to achieve faster computation than the original version.

The PPML approaches fall largely into two sub-categories: encryption-based ML approaches and distributed-based ML [12]. In encryption-based ML, the ML model is trained using encrypted measurements. This can be done using special-purpose encryption algorithms such as homomorphic encryption [27], functional encryption [10] or multiparty computation (MPC) [16] protocols. The authors in [10] use functional encryption (FE) to encrypt users' data and then those encrypted data are fed into a fully connected feed-forward network (FFN) model to evaluate if they are malicious or not. Functional encryption is a relatively efficient cryptosystem that allows performing computations on encrypted data without the need to decrypt it. Although FE is assumed to be efficient in terms of communication and computation, it requires an extra step where a key distribution center needs to generate and distribute keys for all participants in the system. In [16], the authors proposed a privacy-preserving theft detector by employing secret sharing to mask the fine-grained meter measurements. The use of secret sharing allows for the aggregation of data before sending them to the system operator. In addition, a convolutional-neural network (CNN) machine learning model that is based on MPC protocols is used as the detection model. The secure MPC is executed by both the SMs and the system operator in order to evaluate the CNN model. Although the results suggested an accuracy of over 90% in different CNN models, the use of the cryptographic techniques to preserve privacy introduces a high communication and computation overhead. The other category of PPML techniques are those that are based on distributed processing such as federated learning (FL) and split learning (SL). FL has been applied to detect energy thefts in [25].

Despite the aforementioned approaches, privacy-preserving energy theft detection research is still very limited and primarily relies on complicated cryptographic functions such as homomorphic encryption and MPC [5]. These methods are computationally and communicationally expensive and are not suited for smart meters which are often computationally restricted [8]. Furthermore, the use of such cryptographic techniques alongside ML introduces additional processing and communication costs and would rely entirely on the strength of the key management mechanisms [3]. The other type of PPML which is distributed-based such as FL and SL has also some weaknesses. Recent research has shown that FL is prone to privacy attacks such as membership-inference attacks and feature leakage attacks [21]. Other work in [17], showed that SL is vulnerable to reconstruction attacks and feature leakage attacks. This is especially true in an environment where aggregators and servers are considered honest-but-curious entities. An honest-but-curious entity is a type of adversary that is commonly used in the analysis of privacy properties. It is a legitimate participant of the system who will exactly follow the protocol defined but will attempt to learn all possible information from legitimately received communication [19].

1.2 Our Contribution

In this paper, we first propose an enhanced privacy-aware energy theft detection scheme which ensures users' privacy using the concept of Split Learning (SL). Although the classical SL approach has the advantage over FL in protecting users' data from reconstruction and feature leakage attacks [9], it can not be directly applied to the environment of smart grids. This is because it introduce large communication overhead. Hence, we propose a new variant of SL, called

"Three-Tier Split Learning". In this variant, aggregators are intermediate entities in the system between clients and the central server. This architecture helps reduce the communication overhead of the system. It also makes the detection approach more suitable to smart grids where aggregators and energy suppliers are considered to be honest-but-curious entities. We also consider the issue of feature leakage attacks in SL that has been studied in [17] and propose a defensive mechanism. The contributions of this work are as follows:

1. We propose an energy theft detection system which preserves the privacy of the users' data using Split Learning. The detection model combines stacked auto-encoders along with Split Learning to detect anomalies. In Split Learning, only the model updates of the split layer are sent rather than the raw data (in case of non-private detection) or the whole model updated (in case of federated learning). This is the *first work* that applies Split Learning in energy theft detection.
2. We propose a *new variant* of Split Learning, called Three Tier Split Learning, that suits the nature of the smart grid infrastructure. This enhanced version adds aggregators to the system which makes the whole ML model splits into three parts (clients, aggregators, server) rather than two (clients, server). Moreover, we introduce a means of minimising the communication overhead through aggregating the updates from the split layers.
3. We evaluate our detection model with a range of different energy theft scenarios. This is also investigated in cases where malicious clients are involved in the training phase and a possible solution is discussed.
4. We analyse the privacy of the proposed model in terms of data leakage.

In a nutshell, the major aim of our proposed scheme is to demonstrate how to achieve high accuracy detection results while preserving privacy. The remainder of this paper is organized as follows: Some preliminary knowledge is defined in Sect. 2. We introduce the system and threat models in Sect. 3. In Sect. 4, we present our proposed energy theft detection scheme. We detail our experimental setup in Sect. 5, while Sect. 6 gives the results. Section 7 concludes the paper. All important notations used throughout the paper are defined in Table 1.

2 Preliminaries

2.1 Anomaly Detection Using Auto-encoders

An auto-encoder (AE) is a special type of neural network that has mainly two parts, an encoder part and a decoder part. (See the Appendix A for more details about how a neural network works). The encoder compresses the input features to produce a latent representation that is then decoded by the decoder to reconstruct the input features [6]. Auto-encoders are trained to minimise the error between the reconstructed data and the original input where the model learns the relationships among features of the input set. After the model has converged, this reconstruction error can be used to detect anomalies. Reconstruction errors for normal packets are minimised by the auto-encoder whereas

Table 1. Notations

Symbol	Definition
CSM_i	The consumption smart meter reading of client i
PSM_i	The production smart meter reading of client i
φ	ML model function
x	Sample data point
y	Label of the sample data point x
$\hat{x}$	A modified sample data point
A_c	Activations of the client
A_a	Activations of the aggregator
ϖ	An attacker's inference model

anomaly-input data result in higher reconstruction errors. A suitable threshold is required to assess if the errors are high enough for that data to be termed anomalous [7]. Stacked autoencoders (SAEs) are constructed by stacking several AEs together. The first AE maps the input to a first latent representation. After training the first autoencoder, its decoder layer is discarded and then replaced by a second autoencoder, which has a smaller latent vector dimension. This process is repeated depending on the depth of the SAE. The depth of stacked autoencoders helps in learning more abstract features from the extracted ones [6]. Auto-encoders are best suited for anomaly detection in environments with high volume data streams such as smart grids. They can be trained to learn the representation of a single 'normal' class. Attacks (or at least anomalies) can be detected without labelling by observing the magnitude of the reconstruction error [7].

2.2 Privacy Preserving Machine Learning and Split Learning

The main idea of Privacy-Preserving Machine Learning (PPML) is to allow ML models to be trained without the need to disclose private data in its clear form [3]. Traditional privacy-preserving techniques, such as differential privacy methods and cryptographic-based techniques, were added to typical machine learning algorithms in order to make them privacy-friendly [3]. However, they either provide privacy to a certain level or increase the computation and communication costs dramatically. An alternative to these techniques is the use of decentralised ML algorithms where training is done collaboratively between the system's entities [12]. Two major methods were introduced: federated learning [14] and split learning [9]. Here, our main focus is on split learning.

Split learning (also known as split neural network) is a framework for distributed learning techniques that was developed by MIT to offer a decentralised training for a model without sharing raw data by the clients [24]. In the basic form of split learning, a neural network model W is split into two parts W_c

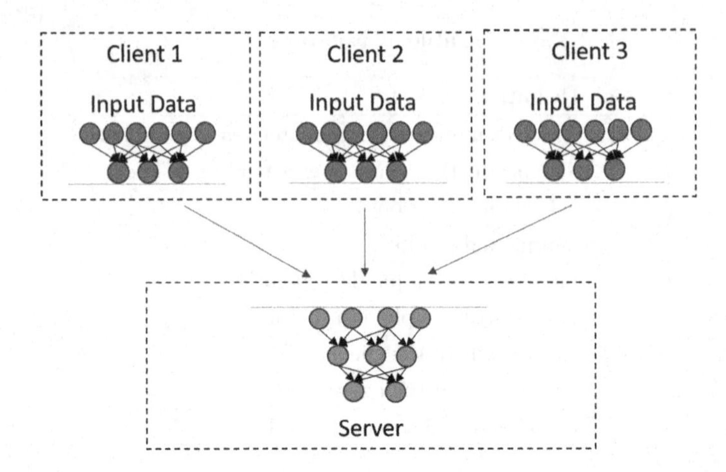

Fig. 1. Split learning setup showing distribution of layers across clients and server

and W_S as shown in figure Fig. 1. This aims to provide privacy protection for the client whilst minimising the computational load. The first part of the network, W_c, resides on the client system and the remaining part W_S resides on the server side. These parts are called client-side network and server-side network respectively. Both the clients and the server train their part of the model separately where the process starts at $t = 0$ with the client data as the input layer, and then proceeds until the split layer is reached. The output of the split layer, called activations $A_{k,t}$, is forwarded to the server to continue the training process. The server completes a full round of forward propagation to obtain the set of activations of the last layer $A_{S,t}$. The server now starts a back propagation round from the last layer up to the cut layer where the gradients at the cut layer $\nabla\ell(A_{S,t}; W_{S,t})$ are sent back to clients. At the client side, the remainder of back propagation is completed where W_c weights are updated for $t+1$. This process is continued without the need for the parties to exchange raw data until the distributed split learning network converges. The complete algorithm of split learning can be found in Appendix B. Split learning is fairly new, and has not been applied in the context of smart grids security. Our work will modify it in a way that is suitable for theft detection in this context.

3 System Model and Threat Model

3.1 System Model

As shown in Fig. 2, the system model in this paper considers a typical smart grid system with three-tier entities: clients, substations (aggregators) and the utility (server). Each user (client) is equipped with one or two smart meters. We consider two types of energy users: regular consumers and prosumers. Prosumers are those users who can produce and consume energy simultaneously. Consumers

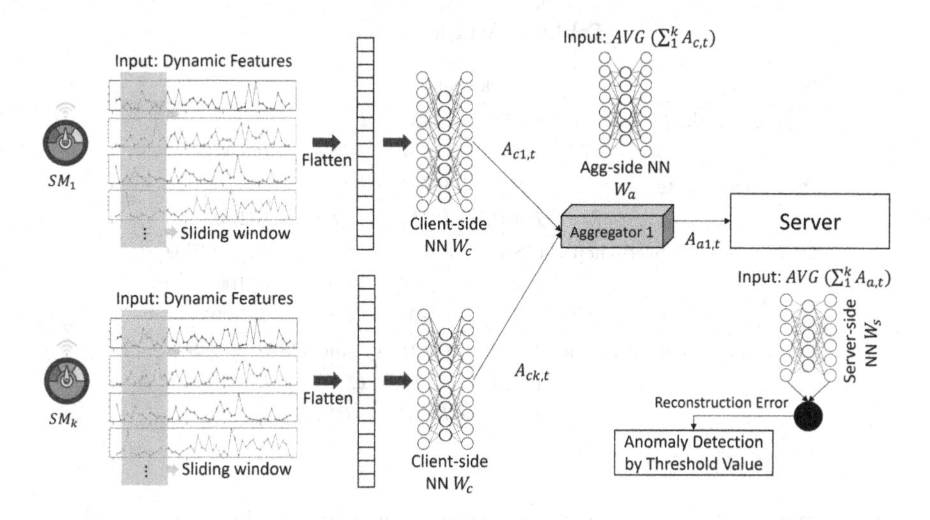

Fig. 2. System model

are equipped with one SM that is responsible for collecting energy consumption data (CSM), whereas a prosumer is additionally equipped with a smart meter for energy production data (PSM). The user's SM is responsible for collecting the energy consumption/production data, processing it and then transmitting it through the network. The smart meters send their processed data to the server via intermediate aggregators that further process the data before sending them to the server. The data sent between the three entities (smart meters, aggregators and server) are in the form of model updates, i.e. activations and gradients.

3.2 Threat Model

This work considers a practical threat model with multiple security and privacy attacks where all external and internal entities can act maliciously. The system assumes that aggregators and the server are honest but curious entities (semi-honest), i.e. they do not tamper with the system's instructions but they may try to infer information about users' behaviours. External adversaries may eavesdrop on the activations sent between the system's entities in an attempt to learn individuals' private data. We also assume that participating clients have the ability to modify and manipulate their smart meter readings and their neighbours' readings in an attempt to gain financial advantage. These attacks can be viewed as follows:

- *Energy Theft Attacks*: these attacks occur when users (both consumers and prosumers) try to modify the smart meters' readings by either decreasing a consumption smart meter CSM readings or increasing a production smart meter PSM readings. This increase/decrease can be a constant value l or a constant percentage k. We also consider *balance attacks* that were introduced

Table 2. Attack scenarios

Attack type	Attack scenario
Consumers thefts	$CSM_i = CSM_i - l$
	$CSM_i = CSM_i - (CSM_i \times k/100)$
Prosumers thefts	$PSM_i = PSM_i + l$
	$PSM_i = PSM_i + (PSM_i \times k/100)$
Consumers balanced thefts	$CSM_i = CSM_i - l$ and $CSM_j = CSM_j + l$
	$CSM_i = CSM_i - (CSM_i \times k/100)$ and $CSM_j = CSM_j + (CSM_i \times k/100)$
Prosumers balanced thefts	$PSM_i = PSM_i + l$ and $PSM_j = PSM_j - l$
	$PSM_i = PSM_i + (PSM_i \times k/100)$ and $PSM_j = PSM_j - (PSM_i \times k/100)$

in [4], where an attacker manipulates two smart meters in order to make the total net sum of the modified readings equal to the original un-manipulated data. By this, the attacker is either stealing from another client (if they have the same tariff) or collaborating with another client to steal from the grid (if they have a different tariff). In total, we consider eight different attack types of energy theft as shown in Table 2. These attacks can be viewed in four separate categories: consumer thefts, prosumer thefts, consumer balanced thefts and prosumer balanced thefts.

- *Poisoning Attacks*: these can be carried out only by internal clients that can modify the smart meter data. The goal of this attack is to try to divert the output of the complete ML model φ by using crafted data point $\hat{x}$. Let φ be ML function which maps X to Y, $\varphi(x) = y$ where y is the correct label and $\hat{X}$ be the set of poisoned data points. When the model φ is trained using $\hat{X}$, φ diverts from its normal behaviour and produces wrong outputs $\hat{Y}$. In practice, energy theft attacks discussed above (including balance attacks) are types of poisoning attacks.
- *Feature Leakage/Reconstruction Attack*: this attack compromises the privacy of the users' readings and can be launched by external or internal adversaries. The goal of the attack is to guess the values of the sensitive features of a data point given only the activations sent by the client (i.e. its model component's split layer activations). A formal definition of the attack is as follows:

Definition (Feature Leakage): Let x be an input data point with a set of features $x_1, x_2, ..x_n$ where $x = (x_1, x_2, ..x_n) \in X$, and let φ be a client model that maps X to A, $\varphi(x) = A$ where A is the set of activations from the split layer. To launch a feature leakage attack, the attacker tries to find a function ϖ that can infer x from A, $\varpi(A) = x$. The goal is to have an exact inference. However, in practice, useful inference can be approximate.

4 Proposed Theft Detection Model

Figure 2 shows how the proposed detection model works. Section 4.1 explains the "Three-Tier Split Learning" approach, which is the newly proposed variant of Split Learning and Sect. 4.2 describes the theft detection model.

4.1 Three-Tier Split Learning

Our Three-Tier Split Learning architecture follows the system design of the state-of-the-art split learning system but adds one new component which is an aggregator between the clients and the server. The newly added aggregator makes the split learning framework more applicable to the context of smart grids. We also introduce a way to calculate the intermediate updates by averaging the activations received from the clients for each client-aggregator pair before sending the results to the server. This makes the process more parallel than sequential. In our extension of split learning, the learning model W is split into 3 different parts, W_c at the client side, W_a at the aggregator and W_S at the server side. The procedure of the "Three-tier split learning" method starts as follows: each client c trains the W_c part of the network and sends the activations of the split layer to the aggregator a. Each aggregator a waits until it receives all activations from its clients and computes the average of these activations and uses it to complete a forward pass on its part of the model W_a. After completing a forward pass, the aggregator sends the activations of the last layer of its model to the server. As in the aggregator, the server waits for the activations from all aggregators and computes their average to be used as input for its part of the model W_S. After the completion of the forward pass, the server generates the gradients for the final layer and back-propagates the error to its cut layer of W_S. The gradients are then passed to the aggregators where they perform a back-propagation and send their gradients to the clients. The rest of the back-propagation is completed by clients. This process is continued until the model converges. Algorithm 1 provides the detailed instructions of the "Three-Tier Split Learning".

4.2 Energy Theft Detection Approach

The aim of this research is to explore how Split Learning can be used to train an anomaly detector to detect energy thefts without violating clients' privacy. We do so by combining the previously explained three-tier split learning method with a stacked auto-encoder. The stacked auto-encoder (SAE), as an unsupervised ML algorithm, enables us to train the detection model without the need for labeling, and the three-tier split learning provides privacy assurance as clients will not need to send their private raw data.

The architecture of our energy theft detection model is shown in Fig. 2. A stacked auto-encoder (SAE) model is split between the system's entities, one part is at the client's side, the second part is at the aggregator's side and the third part is at the server's side. The server part consists of 4 layers, while the client's and aggregator's parts can vary in depth (details can be found in

Algorithm 1. Three-Tier Split Learning Algorithm with Averaging

```
function SERVER                                          ▷ executes at round t ≥ 0
    for epoch e do
        A_t ← []
        for agg a ∈ aggregator_t do
            A_{a,t} ← AGGREGATOR(a, t)
            A_t[c] ← A_{c,t}
        end for
        A_{t.avg} ← sum(A_t)/len(S_t)
        Complete forward propagation with A_{t.avg} to get A_{S,t}
        Calculate Loss
        W_{S,t+1} ← W_{S,t} − η∇ℓ(W_{S,t}; A_{t.avg})   ▷ Back propagation part of the server
        CLIENTBACKPROP(c, t, ∇ℓ(A_{t.avg}; W_{S,t}))      ▷ k here is the last client
    end for
end function
function AGGREGATOR(a,t)                                 ▷ executes at round t ≥ 0
    for epoch e do
        A_t ← []
        for client c ∈ S_t do
            A_{c,t} ← CLIENTUPDATE(c, t)
            A_t[c] ← A_{c,t}
        end for
        A_{t.avg} ← sum(A_t)/len(S_t)
        Complete forward propagation with A_{t.avg} to get A_{a,t}
        send A_{a,t} to Server
    end for
end function
```

Table 4). Next, each client collects a set of features that includes consumption, generation and weather data. This is done at regular intervals (usually every 15–20) min. However, given the nature of timeseries data, a sliding window of multiple data points is considered as an input to the client's part of the model. This helps capture the correlation between consecutive data points. After that, the vector of features is fed to the client's part of the stacked SAE and the latent representation (client's output) is sent to the aggregator. Each aggregator uses the average of all the outputs of its clients as an input to its part of the SAE. The output of the aggregator, which is the latent representation, is sent to the server, which also uses the average of all the aggregators' outputs as the input to its part of the stacked SAE. This process is repeated until the model converges.

To detect energy thefts, the server computes a threshold which is used as a bound to detect those energy thefts. Any data point that causes a reconstruction error exceeding this predefined threshold would then be considered as anomaly. As in [6], to estimate this threshold, the server calculates the reconstruction errors of its part of the SAE for all the training dataset. Then, the threshold value is estimated by the mean and standard deviation of those reconstruction errors; it can be described as:

$$threshold = \frac{1}{d}\sum_{i=1}^{d} RE_i + \sqrt{\frac{1}{d}\sum_{i=1}^{d}(RE_i - \frac{1}{d}\sum_{i=1}^{d} RE_i)^2}$$

where RE is the reconstruction error, and d is the number of training dataset elements.

5 Experimental Setup

In this section, we give details about how we conducted our experiments, such as the dataset used, formation of the energy theft attacks, simulation environment, neural network parameters, and the evaluation metrics.

Dataset. In this work, we have used the dataset from [4]. The dataset was generated using GridLab-D, which is a power simulation tool that simulates the power flow between the power grid's entities. GridLab-D is very flexible as it allows reporting both production and consumption data that are dynamically influenced by weather data. In this dataset, there were a total of 1596 clients, 49 of which were prosumers with solar panels. Every client reported 17 different dynamic parameters every 15 min. These dynamic parameters include some electricity parameters (power consumption, power generation, voltage, current, real energy, reactive energy, reactive power and apparent power), and a set of weather-related parameters (temperature, wind speed, wind direction, pressure, humidity, solar radiation, extraterrestrial radiation, solar illumination and sky cover). The dataset also included 13 different physical features of each client's property such as: floor area, ceiling height, thermal integrity levels, number of glazing layers, glazing treatment, glass type, windows frame type, types of heating and cooling systems, and solar panel size. In all experiments, a sliding window of 16 data points (4 h) is considered as an input to the client's part of the model. This means that each sample is a vector of 285 features ([17 dynamic features * 16 data points] + 13 static features). These data samples are split into 70% for training and 30% for testing.

Energy Theft Attacks. Since all readings in our dataset are real (normal) readings, we had to modify them using mathematical functions to create malicious data points. This is widely done in energy theft detection and was first presented in [13]. These malicious points are created according to the attack scenarios presented in Sect. 3.2. In this work the value of l is chosen to be 400 which is $\frac{1}{3}$ the mean of all readings. While the value of k is set to 40 which is less than half of the reading.

Simulation Environment. The proposed Three-Tier Split Learning detection model is implemented using PyTorch [18]. PyTorch is a Python-based machine learning library that enables access to every computational node in a ML model. This allowed us to split the whole detection model into three splits.

Neural Network Parameters. In our experiments, the SAE model consists of a total of ten neural network layers. In our approach, clients had three layers, aggregators had three layers, and the server had four layers. In every experiment, the network iterated over the samples for a total of 20 epochs with a batch size of 96. The Adam optimiser is used with its default hyper-parameters as the optimization algorithm in all entities.

Performance Metrics. To evaluate the performance of the proposed model in terms of energy theft detection, we consider accuracy, recall (also known as detection rate (DR)), and precision. These basic metrics allow the calculation of other metrics using them, such as F1 or F2 scores [22].

6 Results and Discussion

In this section, we analyse the security, privacy and communication overhead of our proposed model. In order to analyse the security and privacy aspect of our system, we analyse how well our system behaves against the attacks discussed in the threat model. The following subsections gives details about the results against these threats, and provides details about our communication analysis.

6.1 Detection of Energy Thefts Attacks

Here, we evaluate how good the SAE works in our "Three-Tier Split Learning" setting in terms of energy thefts detection. For this purpose, we have also trained the same SAE in two other settings: a centralised setting and in a federated learning setting. In the centralised setting, the global model of the SAE along with all client's data are available at the server side. This setting is the basic setting where privacy is not considered. In the federated learning setting, the SAE model is trained locally at each client and a shared model is averaged at the server side. Figure 3 compares the performance of our proposed model with the centralised version and the federated learning approach in terms of accuracy, recall and precision. It is clear from the figure that the results of our proposed approach are highly comparable to the other two settings. This shows clearly that our approach achieves excellent results in detecting energy thefts while it preserves privacy compared to the centralised approach. The results are also very similar to the federated learning approach with an advantage of having lower communication overhead (discussed in Sect. 6.4). In all three settings, the training is taking place repeatedly over different batches of data and therefore the results are not linearly improving.

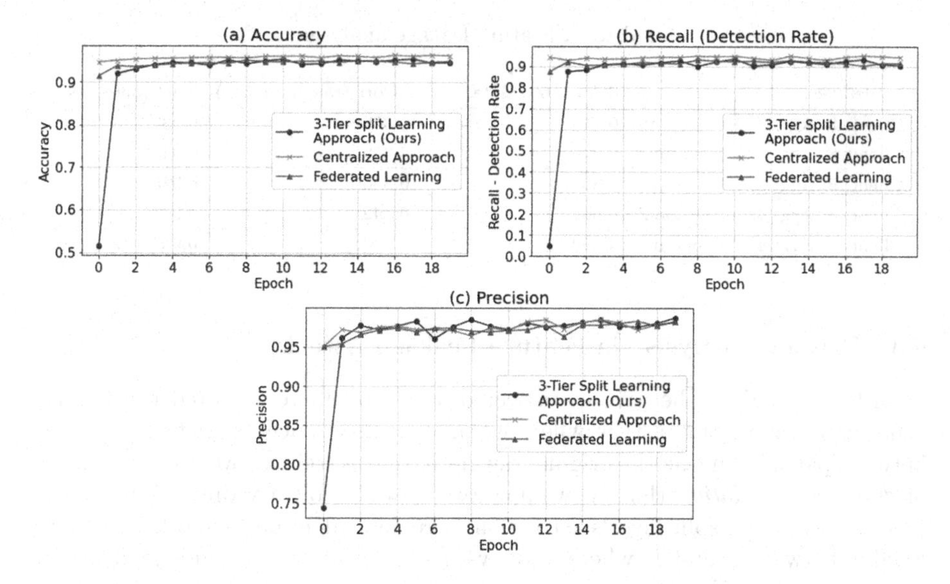

Fig. 3. Results of the detection model using three-tier split learning (proposed work), centralised detection, and federated learning

6.2 Resilience Against Poisoning Attacks

In this experiment, we have tested how well our detection approach works in the event of having poisoned training data. As explained in Sect. 3.2, poisoning attacks are attacks that can be launched whenever a collaborative ML algorithm is involved. In our approach, this is because clients are involved in training the detection model. Our results in Table 3 shows that the more poisoned data is used to train the system, the worse our detection results are. When only 20% of the data is poisoned, then the detection rate decreases up to almost 15%. Therefore, we had to find a way to overcome this. We adopt a simple solution where we randomly drop 10% of the training updates received from the clients. As can be seen in the last record of Table 3, this simple random dropping improves the detection results and make it comparably close to the normal case where no poisoned data are injected.

Table 3. Detection results with poisoned data

Percentage of poisoned training data	Accuracy	Recall (DR)	Precision
0% (no attack)	0.946	0.905	0.970
5%	0.945	0.901	0.969
10%	0.933	0.891	0.973
15%	0.870	0.760	0.974
20%	0.870	0.762	0.973
20% with random 10% dropping	0.933	0.903	0.979

Table 4. Feature leakage analysis

Model used	dCorr (raw data, A_c)	dCorr (raw data, A_a)	Theft detection
Client: 3, Agg: 3, No Dropout	0.740	0.369	0.937
Client: 3, Agg: 3, 1 Dropout	0.620	0.359	0.945
Client: 3, Agg: 3, 2 Dropout	0.557	0.366	0.935
Client: 4, Agg: 4, No Dropout	0.665	0.342	0.937
Client: 4, Agg: 4, 2 Dropout	0.490	0.340	0.934

6.3 Privacy Analysis via Feature Leakage Attack

In order to analyse the privacy aspect of our model, we explored the feature leakage/reconstruction attack where we analyse how much the activations sent between parties can leak the original raw data. In this context, we use the concept of *distance correlation* that shows how two sets of arbitrary dimension vectors (e.g. raw data and split layer's activations) are dependent on each other. It takes a value between 0 and 1, where lower values indicates greater independence of the two vectors. Note that, distance correlation is one of the best statistical measures that can show both linear and nonlinear associations which makes our evaluation more comprehensive. It is also one of the few metrics that can test dependence of two arbitrary length vectors. Distance correlation was used in [1] and [23] as part of the privacy assessment framework. Detailed information about distance correlation can be found in Appendix C.

Table 4 shows the distance correlation $dCor$ between the activations sent by the client A_c and the real raw input data. It also shows the $dCor$ between the activations sent by the aggregator A_a and the raw input data. In the first setting, both the client and the aggregator have 3 hidden layers with no dropout layers in between and as the results show, the distance correlation between the raw data and the activations sent by the clients is high (0.74). This actually suggests that it would be easy for the attacker to infer the raw data back from those activations. One possible defence was to employ dropout [15], and another one was to increase the number of hidden layers. In neural networks, Dropout is a well-known regularisation technique that is used to overcome overfitting [20]. The basic idea of dropout is to randomly deactivate neurons' activations with a probability between 0 to 1. This random dropout of activations will make it harder for the attacker to build a robust system that can infer the raw data from the activations as the attacker will be observing different activations list each time [15]. As you can see in Table 4, after adding some dropout layers and as we increase the number of hidden layers the distance correlation is decreasing and at the same time this does not affect the theft detection rate in any way. Briefly, we can say that our "Three-Tier Split Learning" approach can protect against feature leakage attacks, defined in Sect. 3.2.

Protection Against Feature Leakage Attacks: Suppose an adversary obtains the set of activations sent from the client to the aggregator A_c. He/She needs a function ϖ that can infer the original raw data x from A_c, $\varpi(A_c) = x$. However,

Table 5. Communication analysis

Method	Communication per client	Total communication
Three-tier split learning	$2S_c$	$K \times (2S_c) + L \times (2S_a)$
Classical split learning	$2S_c$	$K \times (2S_c) + K \times (2S_c)$
Federated learning	$2N$	$2KN$

our results show that the average $dCor(A_c, x)$ is less than 0.5 with only 4 layers and dropout at the client model. This implies that the probability of finding ϖ with good accuracy for any probabilistic polynomial time adversary $\mathcal{A}$, $Adv_{\mathcal{A}}^{Leak}$ is negligible, i.e., $Adv_{\mathcal{A}}^{leak}(A_c, x) \leq \epsilon$.

6.4 Communication Analysis

The objective of the proposed three-tier split learning theft detection is not only to ensure privacy and non-thefts by the smart meters, but also to ensure that the communication overhead during the whole process is minimised. In this section, we compare the communication overhead of the proposed scheme with the classical split learning approach (where aggregators are assumed to only forward clients' communications to the server without any avergaing) and with the federated learning approach (as it is the closest scheme for providing security and privacy properties to the system). To do so, we analyse the amount of data transferred by every client and the total data transferred between parties in the system. We use the following notation to mathematically measure the communication efficiencies. Notation: K= # clients, L= # aggregators, N= size of the complete model parameters (neurons), S_c= size of the split layer at the client, and S_a= size of the split layer at the aggregator. Here $K > L$ and $N \gg S_c + S_a + S_s$.

In Table 5, we can see that the communication cost, for the same neural network model, in the Three-Tier Split Learning approach is less than that of both the classical split leaning and the federated learning approach. In the three-tier split learning, every client sends the updated activations from their split layer S_c and receives the updated gradients from the aggregator with size S_c, which total $2S_c$. The same is for every aggregator which makes the total communication of one round $K \times (2S_c) + L \times (2S_a)$. In the classical split learning approach, when averaging is not implemented, the aggregator would act as a repeater, forwarding every communication between the server and the client. This will make the clients' updates and the gradients' updates be sent twice in the network. This makes the total communication in the classical split learning greater than our proposed three-tier split learning. On the other hand, clients in the federated learning approach send the full network updates to the server and the full gradients are then forwarded from the server to all clients. This makes the total communication of one round equal to $2KN$ which is significantly more than $K \times (2S_c) + L \times (2S_a)$.

Table 6. Comparison

Property	Centralised approach	Federated learning approach	Three-tier split learning (ours)
Energy theft detection	✓	✓	✓
Privacy preservation	✗	✓	✓
Resilience against poisoning attacks	✓	✓	✓
Stronger resilience against feature leakage attacks	✗	✗	✓
Higher communication efficiency	✓✗	✗	✓

✓: True. ✗: False. ✓✗: May not be true in some cases.

6.5 Summary of Comparison

In summary, our results show that the proposed approach outperforms existing ones in terms of the above properties. Our Three-Tier Split Learning approach can detect energy thefts with high recall, precision, and accuracy. It also preserves the privacy of users' data as compared to the centralised approach. In terms of resilience against poisoning attacks, all three models can detect poisoning attacks with the help of an additional procedure (for instance, in our case by adding dropout layers and randomly dropping 10% of the training data). Moreover, it is more challenging to infer features from only the split layer's activations in contrast to inferring them from the whole model updates [21]; hence our proposed model provides stronger resilience against feature leakage attacks than the federated learning approach. Furthermore, our analysis (provided in Sect. 6.4) shows that the proposed approach has higher communication efficiency than the federated approach. It should be noted that the communication efficiency of our model would also be better than the centralised approach when the feature set is larger than the split layer size, which is true in most cases. Table 6 outlines this comparison.

7 Conclusion

We have proposed a new variant of split learning, Three-Tier Split Learning, as a private collaborative machine learning algorithm to tackle the challenge of preserving users' privacy. It trains a detection model for energy thefts without the need to use raw data. It is tested on a dataset that contains malicious readings generated from various cyber-attacks including consumer thefts, prosumer theft and balance attacks. Our experiments showed that it gives a 94.6% accuracy, 90.5% recall (detection rate) and 97.0% precision. Moreover, even in the case of poisoning attacks, simply dropping 10% of the model updates can provide comparable results to those with no poisoned data. The model demonstrates good privacy preservation; the distance correlation between the updates sent from the clients and the aggregator with the raw data is low, making it difficult for attackers to infer the raw data from those updates. There is also a significant reduction in terms of communication. Thus, our proposed model ensures not only privacy-preservation, but also communication efficiency.

Appendix A: Neural Networks

Neural networks (NN), like any other network or graph, are networks that are composed of nodes *(neurons)* and edges *(weights)*. The nodes or neurons are arranged into layers starting from the input layer, followed by one or more hidden layers and finally the output layer. Each neuron is a computational unit that takes the inputs from the preceding layer and outputs the weighted sum of these inputs (plus a bias). The output of each neuron can be restricted using *activation functions.* The most used activation functions are *Rectified Linear Unit (ReLu)*, *hyperbolic Tangent function (Tanh)* and *Sigmoid.* These activation functions limit the output value within a specified range, i.e., Relu output is from 0 to +infinity, Tanh output is from −1 to 1 and Sigmoid output ranges between 0 and 1. Therefore, each neuron n in layer l calculates its output z_n^l as:

$$z_n^l = A_z^l(\sum_{j=1}^{s}(i_j^{l-1} \times w_j^l) + b_n)$$

where A is the activation function, i_j^{l-1} is the jth output from the preceding layer, w_j^l is the jth weight of that output and b_n is the bias of this neuron. There are two passes in each round (epoch) of training a neural network: a forward propagation pass and a backward propagation pass (backpropagation). In the forward pass, the input data are propagated to the input layer, then proceed to the hidden layer(s), measuring the network's predictions up to the output layer where the network outputs the prediction $\hat{y}$. This makes $\hat{y}$ equals to:

$$\hat{y} = A^L(W^L A^{L-1}(W^{L-1} .... A^2(W^2 A^1(W^1 X))...))$$

where L is the total number of layers, W^i is the weights vector of layer i and X is the input vector. This is first done using initial weights and bias (weights and bias are initialised randomly). The outputs of all neurons of the same layer are called activations. The network's error (loss) is calculated based on the output of the forward pass prediction $\hat{y}$ and the desired output y. The loss function is computed for every output of the neural network as follows: $loss = L(\hat{y}, y)$. In the backpropagation pass, the weights and biases of the network are adjusted in proportion to how much they contribute to the overall error (loss). These adjustment values are called gradients and they are sent back to along the network to update the neurons weights and bias where the updated value for each weight w will be: $w_{new} = w_{old} - \alpha(\frac{\partial loss}{\partial w})$, where α is a learning rate that controls how much we are adjusting the weights with respect the loss gradient and ∂ is the derivative of the loss in respect to the that weight w.

Appendix B: Split Learning Algorithm

This appendix provides a brief description of the three main functions in the Split Learning algorithm.

Algorithm 2. Split learning algorithm

function SERVER ▷ executes at round $t \geq 0$
  for client $c \in S_t$ **do**
    $A_{c,t} \leftarrow$ CLIENTUPDATE(c, t)
    Complete forward propagation with $A_{c,t}$ to get $A_{S,t}$
    Calculate Loss
    $W_{S,t+1} \leftarrow W_{S,t} - \eta\nabla\, l(W_{S,t}; A_{S,t})$ ▷ Back propagation part for the server
    CLIENTBACKPROP$(c, t, \nabla\ell(A_{S,t}; W_{S,t})))$
  end for
end function

function CLIENTUPDATE(c, t)
  $A_{c,t} \leftarrow \phi$
  if Client c is first client in $t = 0$ **then**
    $W_{c,t} \leftarrow randominitialize$
  else
    $W_{c,t} \leftarrow$ CLIENTBACKPROP$(W_{c-1,t-1})$
  end if
  for local epoch e **do**
    for batch $b \in B$ **do**
      Forward propagation on client part
      Concatenate the activations of cut layer to $A_{c,t}$
    end for
  end for
  send $A_{c,t}$ to Server
end function

function CLIENTBACKPROP$(c, t, \nabla\ell(A_{S,t}; W_{S,t}))$
  for batch $b \in B$ **do**
    Back propagation on client part with $\eta\nabla(A_{S,t}; W_{S,t})$
  end for
  Update model weights $W_{c,t+1}$ and send to next client
end function

Appendix C: Distance Correlation

The distance correlation of two random variables X and Y is obtained by dividing their distance covariance by the product of their distance standard deviations. This makes the distance correlation equals to

$$dCor(X, Y) = dCov^2(X, Y)/\sqrt{dVar(X)\, dVar(Y)}$$

where $dCov(X, Y)$ is the square root of the average of the product of the double-centered pairwise Euclidean distance matrices and can be calculated as $dCov^2(X,Y) := \frac{1}{n^2}\sum_{i=1}^{n}\sum_{j=1}^{n} D(x_i,x_j)D(y_i,y_j)$, where $D(x_i,x_j)$ is the Euclidean distances between the ith and jth observations. Distance correlation, in contrast to Pearson's correlation, cannot be negative, i.e. $0 \leq dCor \leq 1$.

References

1. Abuadbba, S., et al.: Can we use split learning on 1D CNN models for privacy preserving training? In: Proceedings of the 15th ACM Asia Conference on Computer and Communications Security, pp. 305–318 (2020)
2. Ahmed, M., Abid Khan, M.A., Tahir, M., Jeon, G., Fortino, G., Piccialli, F.: Energy theft detection in smart grids: taxonomy, comparative analysis, challenges, and future research directions. IEEE/CAA J. Autom. Sinica **8**(12), 1–23 (2021)
3. Al-Rubaie, M., Chang, J.M.: Privacy-preserving machine learning: threats and solutions. IEEE Secur. Priv. **17**(2), 49–58 (2019)
4. Alromih, A., Clark, J.A., Gope, P.: Electricity theft detection in the presence of prosumers using a cluster-based multi-feature detection model. In: 2021 IEEE International Conference on Communications, Control, and Computing Technologies for Smart Grids (SmartGridComm), pp. 339–345. IEEE (2021)
5. Althobaiti, A., Jindal, A., Marnerides, A.K., Roedig, U.: Energy theft in smart grids: a survey on data-driven attack strategies and detection methods. IEEE Access **9**, 159291–159312 (2021)
6. Aygun, R.C., Yavuz, A.G.: Network anomaly detection with stochastically improved autoencoder based models. In: 2017 IEEE 4th International Conference on Cyber Security and Cloud Computing (CSCloud), pp. 193–198. IEEE (2017)
7. Borghesi, A., Bartolini, A., Lombardi, M., Milano, M., Benini, L.: Anomaly detection using autoencoders in high performance computing systems. In: Proceedings of the AAAI Conference on Artificial Intelligence, vol. 33, pp. 9428–9433 (2019)
8. Gope, P., Sikdar, B.: Lightweight and privacy-friendly spatial data aggregation for secure power supply and demand management in smart grids. IEEE Trans. Inf. Forensics Secur. **14**(6), 1554–1566 (2018)
9. Gupta, O., Raskar, R.: Distributed learning of deep neural network over multiple agents. J. Netw. Comput. Appl. **116**, 1–8 (2018)
10. Ibrahem, M.I., Nabil, M., Fouda, M.M., Mahmoud, M.M., Alasmary, W., Alsolami, F.: Efficient privacy-preserving electricity theft detection with dynamic billing and load monitoring for AMI networks. IEEE Internet Things J. **8**(2), 1243–1258 (2020)
11. Ishimaki, Y., Bhattacharjee, S., Yamana, H., Das, S.K.: Towards privacy-preserving anomaly-based attack detection against data falsification in smart grid. In: 2020 IEEE International Conference on Communications, Control, and Computing Technologies for Smart Grids (SmartGridComm), pp. 1–6. IEEE (2020)
12. Jia, Q., Guo, L., Fang, Y., Wang, G.: Efficient privacy-preserving machine learning in hierarchical distributed system. IEEE Trans. Netw. Sci. Eng. **6**(4), 599–612 (2018)
13. Jokar, P., Arianpoo, N., Leung, V.C.: Electricity theft detection in AMI using customers' consumption patterns. IEEE Trans. Smart Grid **7**(1), 216–226 (2015)
14. McMahan, B., Moore, E., Ramage, D., Hampson, S., y Arcas, B.A.: Communication-efficient learning of deep networks from decentralized data. In: Proceedings of the 20th International Conference on Artificial Intelligence and Statistics, vol. 54, pp. 1273–1282. PMLR (2017)

15. Melis, L., Song, C., De Cristofaro, E., Shmatikov, V.: Exploiting unintended feature leakage in collaborative learning. In: 2019 IEEE Symposium on Security and Privacy (SP), pp. 691–706. IEEE (2019)
16. Nabil, M., Ismail, M., Mahmoud, M.M., Alasmary, W., Serpedin, E.: PPETD: privacy-preserving electricity theft detection scheme with load monitoring and billing for AMI networks. IEEE Access **7**, 96334–96348 (2019)
17. Pasquini, D., Ateniese, G., Bernaschi, M.: Unleashing the tiger: inference attacks on split learning. In: Proceedings of the 2021 ACM SIGSAC Conference on Computer and Communications Security, pp. 2113–2129, CCS 2021. Association for Computing Machinery, New York (2021)
18. Paszke, A., et al.: PyTorch: an imperative style, high-performance deep learning library. Adv. Neural Inf. Process. Syst. **32**, 8024–8035 (2019)
19. Paverd, A., Martin, A., Brown, I.: Modelling and automatically analysing privacy properties for honest-but-curious adversaries. Technical report (2014)
20. Srivastava, N., Hinton, G., Krizhevsky, A., Sutskever, I., Salakhutdinov, R.: Dropout: a simple way to prevent neural networks from overfitting. J. Mach. Learn. Res. **15**(1), 1929–1958 (2014)
21. Thapa, C., Chamikara, M.A.P., Camtepe, S.A.: Advancements of federated learning towards privacy preservation: from federated learning to split learning. In: Rehman, M.H., Gaber, M.M. (eds.) Federated Learning Systems. SCI, vol. 965, pp. 79–109. Springer, Cham (2021). https://doi.org/10.1007/978-3-030-70604-3_4
22. Tharwat, A.: Classification assessment methods. Appl. Comput. Inform. **17**(1), 168–192 (2020)
23. Turina, V., Zhang, Z., Esposito, F., Matta, I.: Combining split and federated architectures for efficiency and privacy in deep learning. In: Proceedings of the 16th International Conference on emerging Networking EXperiments and Technologies, pp. 562–563 (2020)
24. Vepakomma, P., Gupta, O., Swedish, T., Raskar, R.: Split learning for health: distributed deep learning without sharing raw patient data. arXiv preprint arXiv:1812.00564 (2018)
25. Wen, M., Xie, R., Lu, K., Wang, L., Zhang, K.: FedDetect: a novel privacy-preserving federated learning framework for energy theft detection in smart grid. IEEE Internet Things J. **9**(8), 6069–6080 (2022)
26. Wen, M., Yao, D., Li, B., Lu, R.: State estimation based energy theft detection scheme with privacy preservation in smart grid. In: 2018 IEEE International Conference on Communications (ICC), pp. 1–6. IEEE (2018)
27. Yao, D., Wen, M., Liang, X., Fu, Z., Zhang, K., Yang, B.: Energy theft detection with energy privacy preservation in the smart grid. IEEE Internet Things J. **6**(5), 7659–7669 (2019)

Attacks and Vulnerability Analysis

Query-Efficient Black-Box Adversarial Attack with Random Pattern Noises

Makoto Yuito, Kenta Suzuki, and Kazuki Yoneyama(✉)

Ibaraki University, Hitachi, Japan
kazuki.yoneyama.sec@vc.ibaraki.ac.jp

Abstract. Adversarial examples are one of the largest vulnerability of deep neural networks. An attacker can deceive the classifiers easily with the malicious inputs (called adversarial examples), which perturbations are slightly added to benign inputs. Various attack methods have been studied in both white-box and black-box settings, and some methods achieve high attack success rates even in the black-box settings; that is, the attacker is restricted to only query accesses to the target network. In this paper, we propose a simple hyperparameter-free score-based black-box ℓ_∞-adversarial attack using local uniform noises and a random search. Specifically, we construct adversarial perturbations by combining local uniform noises such as vertical-wise and horizontal-wise, and incorporate this idea into the random search method to update the perturbation sequentially. We evaluate our method in terms of attack success rates and query efficiency using models that classify common datasets CIFAR-10 and ImageNet. We show that our method achieves higher attack success rates and query efficiency than previous attack methods, especially in low-query budgets on both untargeted and targeted attack settings. We also examine attacks to adversarially trained models and discuss the effect of local uniform noises on these models. Furthermore, we show that our method achieves relatively high attack success rates and query efficiency on average against input-transformation-based defense methods, and is virtually unaffected by these defense methods.

Keywords: Black-box adversarial attacks · AI security

1 Introduction

1.1 Backgrounds

Due to recent breakthroughs in deep learning techniques, Deep Neural Networks (DNNs) have achieved state-of-the-art classification performance in various tasks. However, it has also been shown that the classification models can still be easily affected by adversarial examples [4,5,7,8,15,28,30,32] which are malicious inputs such that small perturbations are added to benign inputs in order to fool the classifiers. Adversarial attacks can cause serious security problems

Makoto Yuito—Presently, he is with IVIS, Inc.

C. Alcaraz et al. (Eds.): ICICS 2022, LNCS 13407, pp. 303–323, 2022.
https://doi.org/10.1007/978-3-031-15777-6_17

because DNNs are deployed in the real world in various applications. For example, Deng et al. [12] analyze adversarial attacks on driving models, and show that these regression models are also very vulnerable to adversarial attacks. Sharif et al. [35] show that it is possible to impersonate another individual by having the face image wear glasses, as in Adversarial Patch [6]. Therefore, in order to design robust models, it is necessary to investigate the potential risks and identify the vulnerabilities of deep learning models. Hence adversarial attacks are an important research topic.

If an adversarial example of an image x exists, attacking a classifier turns into a search problem within a small volume around a benign image x. Recently, several algorithms have been proposed to generate adversarial examples, and these methods can be classified based on several categories.

Threat Model: One of the key differences in adversarial attacks is the setting of the attacker, and there are two primary types: white-box and black-box. In the white-box setting [7,15,28,32], the attacker is assumed to have all the knowledge about the target model. The main idea of generating adversarial examples in this setting is to apply a perturbation in the direction of the gradient of the loss w.r.t. the input x. However, in reality, an attacker is likely to have access to only a limited amount of information. In the black-box setting [4,5,8,23,30], the attacker is only allowed query access to the target model. That corresponds to an attack on a web service using a pre-trained classifier (e.g., Google Cloud Vision API [2], IBM Watson Visual Recognition [3], Amazon Rekognition [1]). In this setting, the attacker needs to compute a perturbation only from the output information obtained by querying a model, which is thus more difficult setting. The main strategies for generating adversarial examples in the black-box setting are shown in Sect. 2.

Adversarial Goal: Another important difference in adversarial attacks is whether the attacker aims to misclassify the input x to a class other than the true class y (untargeted), or to misclassify the classification result to a specific target class $t(\neq y)$ (targeted). Targeted attacks, especially on classifiers with a large number of classes, are quite a difficult task.

Distance Metric: Adversarial examples are inputs with slight perturbations that are carefully crafted to cause the classifier to misclassify them. It is commonly used ℓ_p-distances between adversarial and benign examples with $p \in \{0, 2, \infty\}$.

We focus on score-based black-box adversarial attacks. Existing query-based black-box attack methods have already achieved a high attack success rate, and the main effort is now focusing on reducing the number of queries. Attacks with low queries, i.e., methods with better query efficiency, can save attackers a great deal of cost in both time and money. For example, the Google Cloud Vision API [2] limits the number of requests per minute to 1,800. High query efficiency attack methods are also effective in deceiving systems [10] that recognize the behavior of submitting many similar queries in short time as fraudulent, which is one of our motivations.

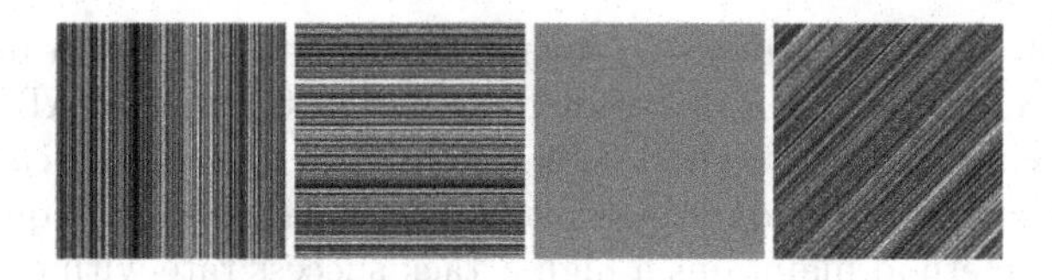

Fig. 1. Sample images of each random pattern noise (RPN) in our sampling space (from the left is vertical-wise, horizontal-wise, uniform, diagonal-wise local uniform noise).

1.2 Our Contribution

In this paper, we propose a simple but effective hyperparameter-free score-based black-box ℓ_∞-adversarial attack in computer vision. The core technique of our approach is to use susceptibility of Convolutional Neural Networks (CNNs) to noise with regional homogeneity [24,46], and specifically to construct adversarial perturbations by combining patterned noises such as vertical-wise and horizontal-wise (see Fig. 1). This idea is incorporated into an iterative random search method to sequentially update the perturbations. In a pre-specified non-orthogonal search direction, we modify the perturbation with randomly selected local uniform noises, check whether it is moving towards or away from the decision boundary using a confidence score, and repeat the perturbation update. With each update, the image moves further away from the original image and towards the decision boundary.

In Sect. 4, we conduct comparative experiments with several existing ℓ_∞-attacks using naturally and adversarially trained models and input-transformation-based defense methods.

In the experiments on the naturally trained models in Sect. 4.1, we use CIFAR-10 and ImageNet datasets to perform comparative experiments with Parsimonious, SignHunter and Square Attack. As a result, we show that our method achieves high attack success rates in both untargeted and targeted attack settings, especially in low query budgets. Specifically, in the untargeted attack on CIFAR-10, our method achieves the average query efficiency of 1.8 times while achieving a higher attack success rate than that of Square Attack. In the untargeted attack on ImageNet, our method also achieves 1.4 times higher average query efficiency than that of Square Attack.

In Sect. 4.2, we evaluate our method against several defensive models based on adversarial training that classify MNIST and CIFAR-10 datasets. In the benchmark Madry et al.'s and TRADES models on MNIST, our method achieves higher attack success rates than the other black-box methods. However, in other Clean Logit Pairing (CLP) and Logit Squeezing (LSQ) models, the results of our method are inferior to those of other black-box attacks, especially in terms of attack success rate. From this result we clarify the effect of local uniform noise in each defensive model.

In Sect. 4.3, we show attacks to several input-transformation-based defense methods that adopt the naturally trained models classifying CIFAR-10 and ImageNet as a backbone. Our method achieves an attack performance of over 90% on CIFAR-10 and over 70% on ImageNet, despite relatively small query budgets. Therefore, our method maintains a high attack success rate with or without the protection of defense methods.

Overall, our method achieves high attack performance on a wide range of target models in a hyperparameter-free manner, making it a realistic method for attackers. We also observe that our method suffers from gradient masking, and our definition of local uniform noise is highly convergent for defensive models other than gradient masking. Finally, in Sect. 4.4, we experimentally verify the effectiveness of our definition of local uniform noises and show that all of them contribute to the attack performance.

2 Related Work

There are a few different settings for adversarial attacks in the black-box setting. This section describes the differences between these settings and the main strategies. Then, we show our contribution by comparing with them.

2.1 Transfer-Based Black-Box Attacks

Most of the existing adversarial attacks assume the white-box setting, where the attacker has full access to the model architecture and the ability to perform backpropagation to obtain gradient information. On the other hand, white-box attacks can be pseudo-black-boxed by using transferability [38], called transfer-based black-box attacks. Transferability is a property that adversarial examples generated for a classifier can be used as for another same type classifiers. Papernot et al. [31] proposed a method to learn a surrogate model by querying the target model. By using the surrogate model with decision boundaries similar to the target model, they can simulate a white-box adversarial attack [15,32]. However, transfer-based attacks have some problems. First, although transfer-based attacks are theoretically possible in a decision-based setting, they often require carefully designed surrogate models, or even require many queries to extract the target model. Next, the generated adversarial examples do not always transfer well [36]. Recent studies have also proposed input transformation methods [13,25,42] to improve the transferability of adversarial examples, and showed black-box attack performances. Although such a method [25] achieves particularly high transferability, they ignore the task of extracting models and only show the attack success rates between each network architecture.

2.2 Score-Based Black-Box Attacks

In score-based black-box attacks, the attacker can obtain the predicted probabilities for each class by querying the inputs to the target model. The attacker solves an optimization problem to compute the adversarial perturbations while directly observing the output from the target model.

Gradient Estimation Based Methods. The ZOO method, proposed by Chen et al. [9], generates adversarial examples by estimating the gradient of the classifier using a coordinate-wise finite difference method. The AutoZOOM, a modified version of ZOO, was proposed by Tu et al. [39], which uses random gradient estimation and dimensionality reduction techniques to significantly improve query efficiency while maintaining attack performance. However, it still requires an enormous number of queries to the target model (13, 525 queries on average for the targeted attack on ImageNet). Hence, gradient estimation-based methods are considerably less efficient, especially for models with high-dimensional inputs.

Gradient-Free Methods. The Parsimonious Attack proposed by Moon et al. [30] solves a discrete optimization problem with local search and the greedy algorithm. On a perturbation divided into a set of n^2 square tiles, Parsimonious finds the sign of each tile by local search, and then uses the greedy algorithm to find a better solution. The SignHunter Attack proposed by Al-Dujaili et al. [4] sequentially estimates the sign of gradient in $1/2^n$ regions of the perturbation in deterministic order. Several attack methods including these [4,29,30] reduce the dimensionality of the search space of the perturbation by modifying neighboring pixels in the perturbation at once, making the computation more efficient. Andriushchenko et al. proposed the Square Attack [5], which achieved state-of-the-art attack success rates and query efficiency. Square Attack solves optimization problems by random search, which directly updates the perturbation with randomly generated square-shaped noise, as opposed to methods that invert the sign of the perturbation, such as Parsimonious and SignHunter. The DeepSearch proposed by Zhang et al. [47] generates adversarial examples close to the original images by reducing the ℓ_∞ distance of the perturbation, while using hierarchical grouping strategy like Parsimonious. However, we do not compare our method with DeepSearch because the attack success rate and query efficiency are not high (similar to those of Parsimonious) although the ℓ_∞ distance of the perturbation generated by DeepSearch is small.

On the other hand, several studies have improved query-based attacks, in which the attacker generates adversarial examples in transfer-based and query-based manner using a surrogate white-box model that is either pre-trained or trained by the attacker himself. The Subspace Attack by Guo et al. [18] uses the gradient of the surrogate model as a heuristic search direction for finite difference gradient estimation. Huang et al. proposed TREMBA [19], which learns an embedding space that can generate adversarial perturbations for a surrogate model, and significantly reduces queries compared to NES and AutoZOOM. Feng et al. [14] improved the transfer performance from the surrogate model to the target model. Their proposed $\mathcal{CG}$-Attack is robust to biases between the surrogate model and the target model by transferring partial parameters of the adversarial distribution of the surrogate model while learning the untransferred parameters based on queries to the target model. The SWITCH proposed by Ma et al. [27] continues to select loss-maximizing perturbations whenever possible when images perturbed by gradients generated from a surrogate model do not satisfy the optimization objective. Yatsura et al. proposed a meta-learning method [45] to be used in combination with random search based attacks. Their learned controller improves the attack performance by online adjustment of the parameters of the proposal distribution at each iterate during the attack. However as explained in Sect. 2.1, we do not compare our method to these methods since the attacker needs to construct a surrogate model in advance and the computational cost is high.

2.3 Defense Methods

As adversarial attacks become more prevalent, many recent studies have also focused on building defense models against them. There are several lines of research in the literature, and the defense methods are roughly consisted of two groups: input-transformation-based defense methods and adversarial training.

The input-transformation-based defense methods include denoising, input randomization, and input transformation. These methods attempt to mitigate the effects of perturbations in adversarial examples by adding image processing-like changes to an input image. Specifically, the denoising methods include low-pass filtering [34] and autoencoders [16], which attempt to remove adversarial perturbations from adversarial examples. The input randomization methods including resizing and padding [41] and the input transformation methods including JPEG Compression [17,26] attempt to mitigate the effect of adversarial perturbations.

On the other hand, adversarial training [21,28,48] aims to obtain robustness by training the model with adversarial examples, which is a more costly but more effective method than image processing defenses. In general, it is known that adversarial training defenses are more robust than other defenses in the case of MNIST and CIFAR-10. Furthermore, Madry et al. [28] show that PGD [28] is a universal first-order adversarial attack, which means that adversarial training with PGD-generated adversarial examples is resistant to many other first-order attacks. The PGD-generated adversarial examples are the basis for many adversarially trained models, including [21,28,48]. The model of Madry et al. [28] provides robust adversarial training by min-max optimization. TRADES [48] focuses on the trade-off between robust error and natural error and trains to improve both. Adversarial Logit Pairing [21] learns by matching the logit of a benign image with the corresponding logit of adversarial examples, while acquiring ancillary information such as their similarity to each other.

2.4 Differences Among Other Black-Box Methods and Our Method

We discuss more about the existing methods presented in Sect. 2.2 and clarify the differences between them and our method. First, regarding the optimization method of the perturbation, it can be observed that Parsimonious [30] has

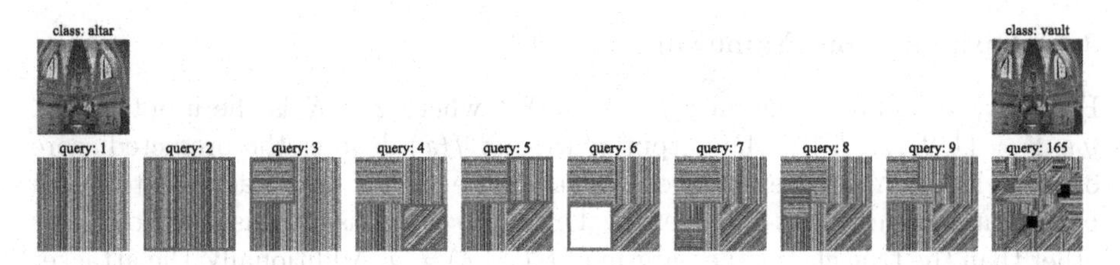

Fig. 2. An example of a sequence of adversarial perturbations on ImageNet generated by our method at each iterate. The left column shows the adversarial perturbation and the adversarial example for the first query (the attack has not yet succeeded at this point), and the right column shows those for the 165th query where the attack was successful (the class changed from *altar* to *vault*). In addition, the transition of adversarial perturbations after the first query is shown between them. The red boxes indicate the block range b determined by the SplitBlock function, i.e., the region where the noise is modified at each iterate. In the second query, we change the perturbation with a randomly picked RPN for a 1×1 region, i.e., the entire image region, and if the loss is lowered, we update the perturbation to this. In the third to sixth queries, the search is performed in 2×2 regions. After that, the perturbation update process is repeated while gradually increasing the number of segmented regions. (Color figure online)

many useless queries, partly because it uses the local search. SignHunter [4] is a deterministic search and can guarantee the attack success rate for the number of queries, but it is not very efficient. Since the convergence of the iterative random search used in Square Attack [5] is much higher than that of Parsimonious and SignHunter, an iterative random search is also used in our method.

As for the components of the perturbation, the perturbation of Parsimonious and SignHunter consist of a uniform noise in a specific segmentation range (square or rectangle shape), while the perturbation of Square Attack consists of a vertical-wise initialization and a uniform noise of a square of a certain size. On the other hand, our method places not only square-shaped but also vertical-wise, horizontal-wise and diagonal-wise uniform noise on the segmented area of squares in the image. Furthermore, while Parsimonious and Square Attack have hyperparameters that need to be tuned depending on the setting of the attack and the target model, our method does not need any hyperparameters. This feature is a great advantage in black-box attacks because it can be easily implemented in any setting.

3 Our Methods

In this section, we first recall the definitions of the threat model in the adversarial attacks and describe an optimization framework for finding adversarial perturbations against classification models. Then, we describe our black-box ℓ_∞-adversarial attack using random pattern noises and random search.

3.1 Optimization Framework

Formally, we define a classifier $f : X \to \mathbb{R}^K$ where $x \in X$ is the input image, $y \in Y = \{1, 2, \ldots, K\}$ is the output space and $f(x)$ denotes the predicted score of each class in Y. In the untargeted setting, the goal of the attacker is to find a perturbation δ such that an adversarial example $(x+\delta)$ is misclassified to classes other than the true class y, i.e., $\arg\max_{k \in Y} f_k(x+\delta) \neq y$. Additionally, the attacker also seeks to minimize ℓ_p distance, i.e.,

$$\arg\max_{k \in Y} \; f_k(x+\delta) \neq y \quad \text{s.t.} \quad \|\delta\|_p \leq \epsilon \text{ and } (x+\delta) \in X, \tag{1}$$

where $\|\cdot\|_p$ is the ℓ_p-distance norm function and ϵ is the radius of ℓ_p-ball. The task of finding a perturbation δ can be handled as a constrained optimization problem. Therefore, ℓ_p-bounded untargeted attacks aims at optimizing the following objective:

$$\min_{\delta : \|\delta\|_p \leq \epsilon} \; L(f(x+\delta), y) \tag{2}$$

where L is a loss function (typically the cross-entropy loss) and y is the true label of x. Equation 2 mostly works to minimize the score for label y. We also study the adversary in targeted setting. In the targeted setting, the attacker aims $\arg\max_{k \in Y} f_k(x+\delta) = t$ for a target label $t(\neq y)$ chosen from Y and optimizes the perturbation by minimizing the loss $L(f(x+\delta), t)$. A black-box targeted attack on a network with many output classes (large K) will be a rather difficult task.

3.2 Algorithm

In this section, we present our black-box ℓ_∞-attack. We assume that the attacker has an image $x \in X$ and a black-box classifier f. An output $f(x)$ is the predicted probabilities over K-classes w.r.t. input image x. In the untargeted setting, our goal is to find a perturbation $\delta \in \{-\epsilon, \epsilon\}^d$ such that $\arg\max \; f(x+\delta) \neq y$ under the ℓ_∞-perturbation constraint, where $\epsilon \in \mathbb{R}^+$ is the radius of ℓ_∞-ball. Our method is based on a random search [33] which is a well known iterative technique in optimization problems. If we apply this technique to the adversarial attacks, it acts as sequential updates of the perturbation. If the loss value $L(f(x+\delta^*), y)$ w.r.t. the perturbed image $(x+\delta^*)$ with the updated perturbation δ^* is lower than the prior loss value $L(f(x+\delta), y)$, this update is adopted to the current perturbation, otherwise it is discarded.

The core technique of our approach is that the perturbation is composed of noises with regional homogeneity. There are studies [24,46] showing the vulnerability of CNNs to local uniform noises. In particular, Li et al. [24] investigate how effective local homogeneous noise is for defensive models against adversarial attacks. They find that adversarial perturbations made for defensive models exhibit more homogeneous patterns than those made for naturally trained models. We therefore investigate whether local homogeneous noises can be applied

Algorithm 1 . Our Method with Random Search

Input: classifier f, original image $x \in X$, true class y, image size w, image channels c, ℓ_∞-radius ϵ, max number of iterations N

Output: adversarial perturbation $\delta \in \{-\epsilon, \epsilon\}^d$

1: $\delta \leftarrow$ initial perturbation (vertical-wise),
2: $x_{adv} \leftarrow (x + \delta).Clip(0, 1)$
3: $l \leftarrow L(f(x_{adv}), y)$, $i \leftarrow 1$
4: **if** *attack is already successful* **then**
5:     **break**
6: **end if**
7: $\mathcal{B} \leftarrow$ SplitBlock(w)
8: **while** $i < N$ **and** *attack is not successful* **do**
9:     $b \leftarrow \mathcal{B}^{(i\% len(\mathcal{B}))}$
10:     $\delta^* \leftarrow$ RPNSampling($\delta, b, w, c, \epsilon$)
11:     $x_{adv} \leftarrow (x + \delta^*).Clip(0, 1)$
12:     $l^* \leftarrow L(f(x_{adv}), y)$
13:     **if** $l^* < l$ **then**
14:         $\delta \leftarrow \delta^*$, $l \leftarrow l^*$
15:     **end if**
16:     $i \leftarrow i + 1$
17: **end while**

Algorithm 2. RPNSampling

Input: perturbation δ, block area to be modified b, image size w, image channels c, ℓ_∞-radius ϵ

Output: new updated $\delta^* \in \{-\epsilon, \epsilon\}^d$

1: $\delta^* \leftarrow \delta$
2: sample RPN uniformly $\gamma \in \{\delta_{vert}, \delta_{horiz}, \delta_{uni}, \delta_{diag}\}$
3: **for** $i = 1, \ldots, c$ **do**
4:     $\delta^*_{b,i} \leftarrow \gamma_{b,i}$
5: **end for**

Algorithm 3. SplitBlock

Input: image size w

Output: a sequence of block areas $\mathcal{B}$

1: $\mathcal{B} = \emptyset$
2: **for** $i = 1, \ldots, w$ **do**
3:     Split the whole area of image into i^2 square shaped blocks $\{b_1, b_2, \ldots, b_{i^2}\}$ with size w/i
4:     $\mathcal{B} \leftarrow \mathcal{B} \cup$ shuffled $\{b_1, b_2, \ldots, b_{i^2}\}$
5: **end for**

to generate adversarial examples (Note that, they [24] aim to generate universal adversarial perturbations, which is a deceptive perturbation for arbitrary images, and is a different objective from ours, so it is not comparable). Specifically, our method constructs perturbations with four patterned noises: vertical-wise, horizontal-wise, uniform, and diagonal-wise (henceforth, collectively referred to as random pattern noise, RPN). This represents a major difference from Square Attack [5], which updates perturbations only with uniform noise in the form of squares.

Algorithmic Scheme with Random Search. Our proposed schemes are presented in Algorithms 1, 2 and 3. First, we set a initial perturbation to the vertical-wise one. A vertical-wise initialization is a technique used in [5]. Then, we obtain the current loss by querying the perturbed image $(x+\delta)$. Since we are interested in query efficiency, the algorithm stops as soon as an adversarial perturbation is found. Therefore, the process is terminated if the attack is already successful

Table 1. Results of both untargeted and targeted attacks on Madry et al.'s naturally trained model [28] classifying CIFAR-10. We set the norm bound $\epsilon_\infty = 0.031$ and a limit of queries to 10 k.

Attack	Success rate		Avg. queries		Med. queries	
	Untargeted	Targeted	Untargeted	Targeted	Untargeted	Targeted
Parsimonious [30]	93.3%	97.3%	329	631	244	476
SignHunter [4]	88.9%	95.6%	157	370	73	311
Square Attack [5]	93.0%	96.7%	131	354	67	253
Ours	**96.4%**	**98.3%**	**72**	**242**	**28**	**132**

at the first query point (step 3 in Algorithm 1). After that, we decide the set of block areas to be modified using the SplitBlock algorithm in Algorithm 3. In a random search loop, first the algorithm picks a block area b and obtains the new perturbation δ^* updated for the area through RPNSampling in Algorithm 2. Then, an adversarial example x_{adv} is generated by adding the perturbation to the benign image. Note that, all perturbed images are clipped in the domain $[0,1]^d$. If the resulting loss corresponding to the perturbed image $(x+\delta^*)$ with the updated perturbation is lower than the current loss, the change is applied. The process is performed at most N (the maximum number of iterations) times and the attack is failure if we cannot find the adversarial perturbation until N times. Figure 2 shows a sequence of candidates of adversarial examples at each iterate generated by our method. A candidate is generated at each iterate, and the perturbation is updated if the loss at that time is lower than the previous one.

RPN Sampling. Our RPNSampling algorithm presented in Algorithm 2 returns a new perturbation δ^* updated for a given block area b to be modified. As the variation of RPNs, we focus on vertical-wise, horizontal-wise, uniform and diagonal-wise perturbations. We show the samples of each RPN in Fig. 1. In this algorithm, one of the four RPNs $\delta_{vert}, \delta_{horiz}, \delta_{uni}, \delta_{diag} \in \{-\epsilon, \epsilon\}^d$, which are randomly generated each time, is sampled as γ. The algorithm then changes the new perturbation δ^* to γ only in the region of block area b. From Fig. 2, it can be observed that one of the randomly generated RPNs is picked at each iterate and changed to that RPN only in certain regions. The effectiveness of each RPN is experimentally verified in Sect. 4.4.

Split Block. The SplitBlock algorithm shown in Algorithm 3 returns a set of elements which are block areas to be modified in the perturbation. The purpose of this function is to decide a low-dimensional space for a perturbation. In general, the input space of a deep learning classifier is very high-dimensional. Therefore, the optimization in the high-dimensional domain requires a very large number of queries and is inefficient. The optimization can be done efficiently by narrowing down the search space for solutions by making changes in some regions at a time as a group. The dimensionality reduction techniques are used in many existing methods [4,29,30], and we observe that the main difference lies in the number of region partitions.

Given image size w, the SplitBlock equally divides the perturbation into n^2 ($n \in \{1, \ldots, w\}$) square regions. Then, each divided area including its coordinates is stored in the order of the region size in the set.

While Square Attack [5] updates the perturbation by randomly selecting a square shaped region $s \times s$ of size $s(< w)$ from the image size $h \times w$, our method updates it regularly for each of the n^2 equally divided square regions. After updating all the n^2 regions, we move on to search in $(n+1)^2$ regions. This can be observed in Fig. 2. In the testing phase, we show that the non-orthogonal search direction and n^2 partitions provide a wider change area in the low query budget, which is a factor to achieve high query efficiency.

4 Experiments

In this section, we evaluate our method by comparing it with other ℓ_∞-attack methods: Parsimonious [30], SignHunter [4] and Square Attack [5]. We consider the ℓ_∞-threat model and execute attacks on both untargeted and targeted attack settings, then quantify the performance in terms of attack success rates, average queries and median queries. The attack success rate is calculated by the proportion of adversarial images which successfully fool the model. The mean and median queries are the mean and median number of queries for successful adversarial images.

In Sect. 4.1, we show results based on naturally trained models, i.e., models that are not hardened against adversarial attacks. In Sect. 4.2 and 4.3, we show results based on robust models of adversarially training and models with input-transformation-based defenses. In Sect. 4.4, we evaluate our method a little more by ablation study. Specifically, we experimentally investigate how much each of our defined RPNs contributes to the attack performance.

4.1 Experiments on Naturally Trained Models

Datasets and Target Models. We evaluate our method on CIFAR-10 [22] and ImageNet [11] datasets. CIFAR-10 is $32 \times 32 \times 3$ dimensional images having 10 classes. For CIFAR-10, we randomly choose 1,000 images from the test set for evaluation, all of which are initially correctly recognized by the target model. ImageNet has 1,000 classes. Since the size of images of ImageNet dataset is not fixed, we re-scale these images to $299 \times 299 \times 3$ (default input size of Inception-v3 model explained below). For ImageNet, we randomly choose 1,000 images belonging to 1,000 categories from ILSVRC 2012 validation set, all of which are initially correctly recognized by the target model. All images are normalized in $[0, 1]$ scale, and for all experiments, we clip the perturbed image into the input domain $[0, 1]^d$ for all algorithms by default.

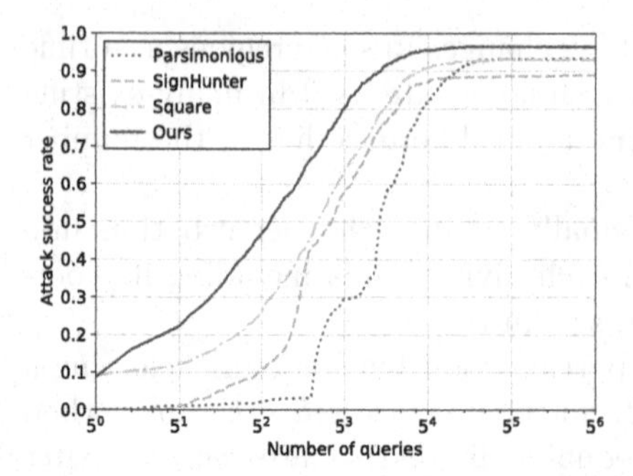

Fig. 3. Cumulative distribution of the number of queries required for untargeted attacks on CIFAR-10.

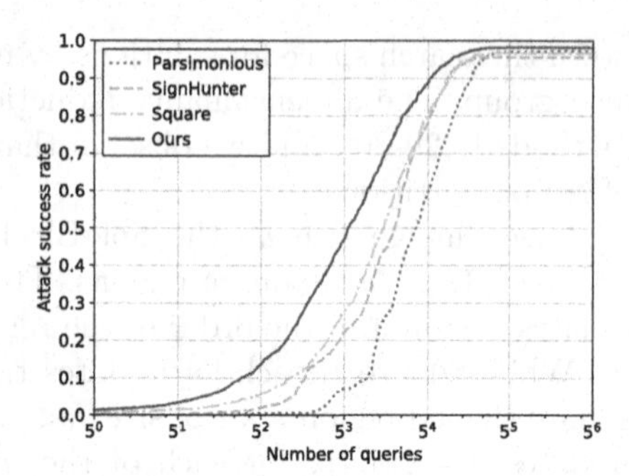

Fig. 4. Cumulative distribution of the number of queries required for targeted attacks on CIFAR-10.

Table 2. Results of both untargeted and targeted attacks on Inception-v3 classifying ImageNet. We set the norm bound $\epsilon_\infty = 0.031$ and a limit of queries to 10 k.

Attack	Success rate		Avg. queries		Med. queries	
	Untargeted	Targeted	Untargeted	Targeted	Untargeted	Targeted
Parsimonious [30]	96.1%	78.4%	1082	3495	389	2807
SignHunter [4]	94.9%	72.4%	966	3656	204	3222
Square Attack [5]	98.5%	**90.9%**	568	2592	96	1716
Ours	**98.6%**	90.2%	**416**	**2116**	**49**	**1312**

For the experiments on CIFAR-10, we use Madry et al.'s naturally trained model [28]. The model architecture and weights are available at here[1]. For the experiments on ImageNet, we use the pre-trained model provided as an application in Keras[2]. We select the Inception-v3 [37] pre-trained model in our experiments because we can see in [5] that it is robuster than some other models for ImageNet against adversarial attacks.

Method Setting. Since it is standard in the literature, we give a budget of 10 k queries per image to find an adversarial perturbation. We set the maximum ℓ_∞-perturbation of the adversarial image to $\epsilon = 0.031$ ($\approx 8/255$) on both CIFAR-10 and ImageNet. Query budgets and the maximum distortion ϵ are parameters specific to the threat model of adversarial attacks, so they are generally not considered as hyperparameters. In targeted attacks, we set the target class to $y_{target} = (y_{true} + 1)\ mod\ K$, where y_{true} is the true class, and K is the number of classes.

[1] https://github.com/MadryLab/cifar10_challenge.
[2] https://keras.io/api/applications/.

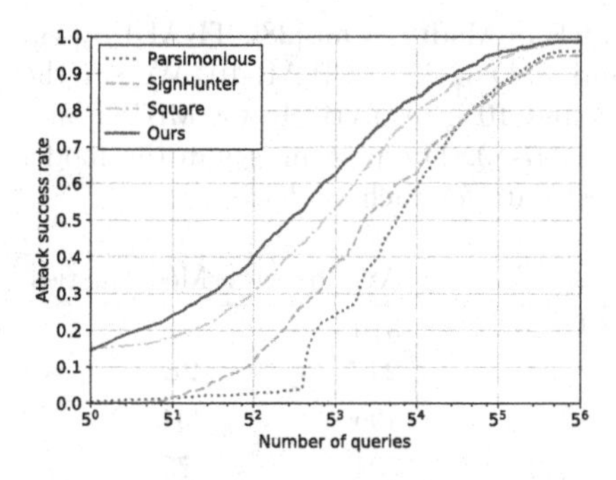

Fig. 5. Cumulative distribution of the number of queries required for untargeted attacks on ImageNet.

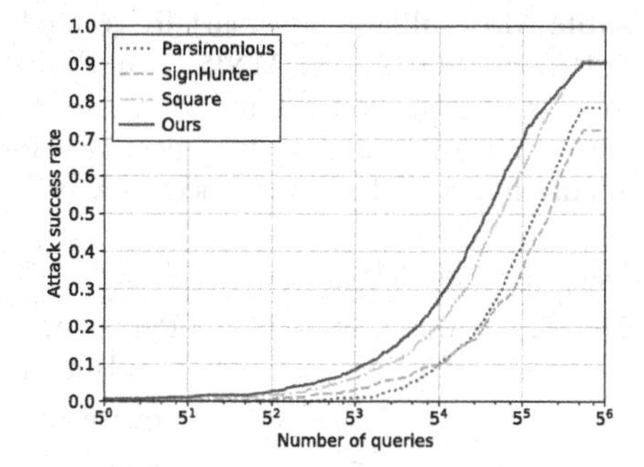

Fig. 6. Cumulative distribution of the number of queries required for targeted attacks on ImageNet.

Results on CIFAR-10. We show the results in Table 1. Our method achieves the highest attack success rates on both untargeted and targeted settings. Also at the same time, we improve the number of queries required to fool the classifiers compared to other three methods. Compared to the state-of-the-art method, Square Attack, our method achieves a higher attack success rate, 1.5 to 1.8 times higher average query efficiency, and 1.9 to 2.4 times higher median query efficiency. We also plotted the cumulative success rates in terms of the required budget in Figs. 3 and 4. Especially in low-query budgets, our method remarkably outperforms the other methods. Additionally, the success rates of Square Attack and our method at 1 query indicate the strength of the vertical-wise initialization. As hyperparameters for the comparison methods, we set $block\ size = 4$ and $batch\ size = 64$ for Parsimonious and $p = 0.05$ for Square Attack by default.

Results on ImageNet. The results are presented in Table 2, and Figs. 5 and 6. Although our method does not achieve the highest attack success rate in the targeted attack setting, it achieves higher attack success rate and query efficiency in the untargeted attack setting. As Figs. 5 and 6 show, our method achieves the highest attack success rate up to 5^5 queries in both untargeted and targeted attack settings. Additionally, we can see from Table 2 that more than half of the images are successfully attacked for the untargeted attack with 49 queries, which is about half of the median query of Square Attack. These results indicate the high query efficiency in low query budgets of our method. As hyperparameters for the comparison methods, we set $block\ size = 32$ and $batch\ size = 64$ for Parsimonious and $p = 0.05$ for Square Attack.

Table 3. Results on adversarially trained models of Madry et al. [28], TRADES [48], CLP and LSQ [21] on MNIST, and CLP and LSQ [21] on CIFAR-10. We set the norm bound ϵ_∞ and a limit of queries to 0.3 and 10 k respectively for MNIST and 0.062 ($\approx 16/255$) and 10 k respectively for CIFAR-10. The percentages in the model column indicate the natural accuracy in the test data for each model.

Dataset	Model	Attack	Success rate	Avg. queries	Med. queries
MNIST	Madry et al. [28] (99.0%)	Parsimonious	11.0%	310	58
		SignHunter	7.5%	**217**	**28**
		Square Attack	11.1%	496	204
		Ours	**11.3%**	504	73
	TRADES [48] (100.0%)	Parsimonious	7.4%	338	60
		SignHunter	5.5%	**198**	**53**
		Square Attack	**7.5%**	450	228
		Ours	**7.5%**	296	81
	CLP [21] (99.3%)	Parsimonious	87.3%	581	65
		SignHunter	24.2%	741	**6**
		Square Attack	**92.8%**	**353**	63
		Ours	80.1%	638	122
	LSQ [21] (99.1%)	Parsimonious	83.8%	418	79
		SignHunter	23.0%	852	**7**
		Square Attack	**90.1%**	**248**	68
		Ours	74.3%	666	117
CIFAR-10	CLP [21] (74.2%)	Parsimonious	99.5%	285	117
		SignHunter	**99.9%**	**109**	39
		Square	99.5%	186	41
		Ours	99.7%	178	**33**
	LSQ [21] (85.5%)	Parsimonious	77.5%	960	199
		SignHunter	83.4%	**354**	34
		Square	**85.0%**	533	29
		Ours	80.5%	627	**26**

4.2 Experiments on Adversarially Trained Models

Here we evaluate our method to robust models based on adversarial training.

Datasets and Target Models. We use some robust models classifying MNIST [44] and CIFAR-10 datasets as the same as the experiment of Square Attack [5]. MNIST is $28 \times 28 \times 1$ dimensional grayscale handwritten numeric dataset. In the experiments on MNIST, we randomly sample 1,000 images from the test set, all of which are initially correctly recognized by Madry et al.'s naturally trained model [28]. In the experiments on CIFAR-10, we use the same test data in Sect. 4.1.

In line with the experiments in [5], we use the ℓ_∞-adversarially trained models of Madry et al. [28], TRADES [48], Clean Logit Pairing (CLP) [21] and Logit Squeezing (LSQ) [21] for MNIST, and the ℓ_∞-adversarially trained models of CLP and LSQ for CIFAR-10.

Method Setting. We give a budget of 10 k queries per image to find an adversarial perturbation. We set the maximum ℓ_∞-perturbation of the adversarial image to $\epsilon = 0.3$ on MNIST, and $\epsilon = 0.062$ ($\approx 16/255$) on CIFAR-10. All experiments in this section are done in the untargeted setting.

Results on MNIST. Table 3 shows the results. In Madry et al.'s and TRADES models, SignHunter achieves better query efficiency, but has a lower attack success rate on average than the other methods. Comparing the methods with similar attack success rates, our method achieves higher attack success rates and better query efficiencies. Although our method does not achieve better performance than other methods in CLP and LSQ models, our method achieves better performance in Madry et al.'s and TRADES models, where the original robust accuracy is higher. This indicates the potential attack power of our method. As the hyperparameter for Parsimonious, we set $block\ size = 4$ and $batch\ size = 64$. As the hyperparameter for Square Attack, we set $p = 0.8$ for Madry et al.'s and TRADES models and $p = 0.3$ for CLP and LSQ models.

Results on CIFAR-10. The results are shown at the bottom of Table 3. All methods have high attack success rates overall, and there is not as large a difference in attack performance due to the shape of the uniform noise as for MNIST. In both models, our method achieves the highest median query efficiency, although not the highest average query. This suggests that the query efficiency in low query budgets of our method is high. As hyperparameters for the comparison methods, we set $block\ size = 4$ and $batch\ size = 64$ for Parsimonious and $p = 0.3$ for Square Attack.

On the Difference in Attack Success Rates in CLP and LSQ Models. It may be concluded that the difference in the attack performance of the methods in CLP and LSQ models classifying MNIST comes from the form of local uniform noise generated by each method. SignHunter considers the image as a one-dimensional vector and flips the sign in a particular segmentation range, so that a rectangular noise can be seen in the image. On the other hand, Parsimonious and Square Attack make most of the noise consist of square shaped uniform noise. The results in Table 3 show a large margin in terms of attack success rate of the attacks between these two patterns. This suggests that CLP and LSQ models are particularly vulnerable to square shaped uniform noise, which causes the large differences.

Table 4. Results on input-transformation-based defenses: Bit-Red [43], JPEG [17], FD [26], and ComDefend [20]. Each defense method adopts the backbones of Madry et al.'s naturally trained model classifying CIFAR-10 and Inception-v3 pre-trained model classifying ImageNet, respectively. We use 50 randomly selected images and set a limit of queries to 200, the norm bound ϵ_∞ to 0.031 for CIFAR-10 and 0.062 for ImageNet.

Dataset	Defense	Attack	Success rate	Avg. queries	Med. queries
CIFAR-10	Bit-Red [43] (78.0%)	Parsimonious	71.8%	61	71
		SignHunter	84.6%	32	16
		Square Attack	**92.3%**	27	14
		Ours	**92.3%**	**23**	**12**
	JPEG [17] (82.0%)	Parsimonious	48.8%	95	75
		SignHunter	73.2%	63	52
		Square Attack	85.4%	31	19
		Ours	**92.7%**	**29**	**12**
	FD [26] (86.0%)	Parsimonious	81.4%	73	70
		SignHunter	83.7%	38	28
		Square Attack	97.7%	28	8
		Ours	**100.0%**	**24**	**6**
ImageNet	Bit-Red [43] (78.0%)	Parsimonious	51.3%	86	74
		SignHunter	74.4%	60	35
		Square Attack	**84.6%**	35	22
		Ours	82.1%	**33**	**12**
	JPEG [17] (82.0%)	Parsimonious	41.7%	82	69
		SignHunter	66.7%	89	73
		Square Attack	77.1%	37	13
		Ours	**83.3%**	**36**	**8**
	FD [26] (86.0%)	Parsimonious	66.0%	55	67
		SignHunter	74.5%	35	18
		Square Attack	93.6%	27	7
		Ours	**97.9%**	**21**	**4**
	ComDefend [20] (94.0%)	Parsimonious	27.7%	80	78
		SignHunter	61.7%	72	61
		Square Attack	**74.5%**	45	25
		Ours	**74.5%**	**37**	**13**

4.3 Experiments on Input-Transformation-Based Defenses

In this section, we attack against input-transformation-based defense methods other than adversarial training.

Datasets and Target Models. Since the basic input-transformation-based defense methods are input-independent, they can be applied to various models to easily improve the defense performance against adversarial attacks. We consider four defense methods: Bit-Depth Reduction (Bit-Red) [43], JPEG Compression (JPEG) [17], Feature Distillation (FD) [26], and ComDefend [20]. All of these defense methods are input-transformation-based methods that apply a transformation to the input image to mitigate the effects of adversarial perturbation. We conduct attack experiments on models applying each defense method to Madry et al.'s naturally trained model for classifying CIFAR-10 and Inception-v3 pre-trained model for classifying ImageNet, respectively. Since ComDefend requires a separate pre-trained model for defense and is not available in CIFAR-10, we only consider ImageNet for this method. For the test data on both CIFAR-10 and ImageNet, we randomly sample 50 images from those used in Sect. 4.1, and we generate adversarial examples of these images.

Method Setting. Considering a more realistic setting, we give a budget of 200 queries, which is much less than the number of queries in the experiment in Sect. 4.1. We set the maximum ℓ_∞-perturbation of the adversarial image to $\epsilon = 0.031$ ($\approx 8/255$) on CIFAR-10, and $\epsilon = 0.062$ ($\approx 16/255$) on ImageNet. The amount of perturbation distortion on ImageNet is based on VMI-CT-FGSM [40]. All experiments in this section are done in the untargeted setting.

Results on CIFAR-10. The results are shown in upper part of Table 4. Our method outperforms the other black-box attacks against all three input-transformation-based defenses. Our method achieves an attack success rate of more than 90% for all defense methods, and when compared to the results for the case without defense methods in Sect. 4.1, it can be seen that our method is almost unaffected by these defenses. Overall, our method achieves better performance in situations where the attacker is given only a small query budget. As hyperparameters for Parsimonious, we set *block size* $= 4$ and *batch size* $= 64$.

Results on ImageNet. The results are shown in the lower part of Table 4. Our method achieves better attack performance except the attack success rate on Bit-Red. In particular, the median query of our method is about half that of Square Attack in most settings, which indicates a relatively high query efficiency of our method. The defense methods such as input transformation are very easy to apply to ImageNet with high dimensionality and are considered more realistic than adversarial training. However, such a simple defense method is not sufficient to prevent adversarial attacks. In terms of the amount of perturbation distortion in the adversarial image, these defense methods may be more robust for smaller amounts of that. As hyperparameters for Parsimonious, we set *block size* $= 32$ and *batch size* $= 64$.

4.4 Ablation Study

In this subsection, we evaluate our methodology a little more. We perform a simple ablation study to show how the individual RPNs (in Sect. 3.2) improve

Table 5. Ablation study of our method which shows how the individual RPNs (in Sect. 3.2) improve the performance. Our final method is highlighted in blue, and the results are shown below when each RPN was removed from the "All" sampling space.

Sampling space	Success rate	Avg. queries	Med. queries
All	**90.2%**	**2116**	**1312**
All – vertical-wise	88.7%	2208	1366
All – horizontal-wise	89.8%	2263	1390
All – uniform	88.8%	2237	1314
All – diagonal-wise	88.2%	2271	1468

the performance of our attack. The comparison is done for an ℓ_∞-threat model of radius $\epsilon = 0.031$. We use 1,000 test images and carry out targeted attacks against the Inception-v3 model pre-trained on ImageNet with a 10 k query budget. Results are shown in Table 5. The "All" in sampling space column means that the RPN is sampled from the all sampling space (vertical-wise, horizontal-wise, uniform and diagonal-wise), which is our final method we used in our experiments. The results when each RPN is removed from the all sampling space are shown below that. In terms of attack success rate and query efficiency, we can see that all RPNs contribute to the attack performance. In particular, when the diagonal-wise pattern is removed, the attack success rate and query efficiency are greatly degraded. In addition, based on the results, further analysis of the noise patterns will be a future challenge, assuming that the addition of new RPNs will improve the attack performance.

5 Conclusion

We proposed a query-efficient black-box attack using an iterative random search and random pattern noises. In our experiments, we show that our method achieves higher success rates than existing methods in both untargeted and targeted attacks, especially in low-query budgets. In the experiments on defensive models, we show that our method achieves high attack performance in most settings. Since our method is hyperparameter-free, it is practical and easy to apply for attackers.

References

1. Amazon Rekognition. https://aws.amazon.com/rekognition/
2. Google Cloud Vision API. https://cloud.google.com/vision/
3. IBM Watson Visual Recognition. https://www.ibm.com/cloud/watson-visual-recognition
4. Al-Dujaili, A., O'Reilly, U.M.: Sign bits are all you need for black-box attacks. In: International Conference on Learning Representations (2020). https://openreview.net/forum?id=SygW0TEFwH

5. Andriushchenko, M., Croce, F., Flammarion, N., Hein, M.: Square attack: a query-efficient black-box adversarial attack via random search. In: Vedaldi, A., Bischof, H., Brox, T., Frahm, J.-M. (eds.) ECCV 2020. LNCS, vol. 12368, pp. 484–501. Springer, Cham (2020). https://doi.org/10.1007/978-3-030-58592-1_29
6. Brown, T.B., Mané, D., Roy, A., Abadi, M., Gilmer, J.: Adversarial patch. CoRR abs/1712.09665 (2017)
7. Carlini, N., Wagner, D.: Towards evaluating the robustness of neural networks. In: IEEE Symposium on Security and Privacy 2017, pp. 39–57 (2017)
8. Chen, J., Jordan, M.I., Wainwright, M.J.: HopSkipJumpAttack: a query-efficient decision-based attack. In: IEEE Symposium on Security and Privacy 2020, pp. 1277–1294 (2020)
9. Chen, P., Zhang, H., Sharma, Y., Yi, J., Hsieh, C.: Zoo: zeroth order optimization based black-box attacks to deep neural networks without training substitute models. In: Proceedings of the AISec@CCS 2017, pp. 15–26 (2017)
10. Chen, S., Carlini, N., Wagner, D.A.: Stateful detection of black-box adversarial attacks. In: Proceedings of the SPAI 2020, pp. 30–39 (2020)
11. Deng, J., Dong, W., Socher, R., Li, L.J., Li, K., Fei-Fei, L.: ImageNet: a large-scale hierarchical image database. In: CVPR 2009 (2009)
12. Deng, Y., Zheng, J.X., Zhang, T., Chen, C., Lou, G., Kim, M.: An analysis of adversarial attacks and defenses on autonomous driving models. In: PerCom 2020, pp. 1–10 (2020)
13. Dong, Y., Pang, T., Su, H., Zhu, J.: Evading defenses to transferable adversarial examples by translation-invariant attacks. In: Proceedings of the CVPR 2019, pp. 4312–4321 (2019)
14. Feng, Y., Wu, B., Fan, Y., Li, Z., Xia, S.: Efficient black-box adversarial attack guided by the distribution of adversarial perturbations. CoRR abs/2006.08538 (2020)
15. Goodfellow, I., Shlens, J., Szegedy, C.: Explaining and harnessing adversarial examples. In: ICLR 2015 (2015)
16. Gu, S., Rigazio, L.: Towards deep neural network architectures robust to adversarial examples. In: ICLR 2015 (2015)
17. Guo, C., Rana, M., Cissé, M., van der Maaten, L.: Countering adversarial images using input transformations. In: ICLR 2018 (2018)
18. Guo, Y., Yan, Z., Zhang, C.: Subspace attack: Exploiting promising subspaces for query-efficient black-box attacks. In: Wallach, H.M., Larochelle, H., Beygelzimer, A., d'Alché-Buc, F., Fox, E.B., Garnett, R. (eds.) NeurIPS 2019, pp. 3820–3829 (2019)
19. Huang, Z., Zhang, T.: Black-box adversarial attack with transferable model-based embedding. In: ICLR 2020 (2020)
20. Jia, X., Wei, X., Cao, X., Foroosh, H.: ComDefend: an efficient image compression model to defend adversarial examples. In: Proceedings of the CVPR 2019, pp. 6084–6092 (2019)
21. Kannan, H., Kurakin, A., Goodfellow, I.J.: Adversarial logit pairing (2018)
22. Krizhevsky, A., Hinton, G.: Learning multiple layers of features from tiny images. Technical report, University of Toronto (2009)
23. Li, H., Xu, X., Zhang, X., Yang, S., Li, B.: QEBA: query-efficient boundary-based blackbox attack. In: Proceedings of the CVPR 2020, pp. 1218–1227 (2020)

24. Li, Y., Bai, S., Xie, C., Liao, Z., Shen, X., Yuille, A.: Regional homogeneity: towards learning transferable universal adversarial perturbations against defenses. In: Vedaldi, A., Bischof, H., Brox, T., Frahm, J.-M. (eds.) ECCV 2020. LNCS, vol. 12356, pp. 795–813. Springer, Cham (2020). https://doi.org/10.1007/978-3-030-58621-8_46
25. Lin, J., Song, C., He, K., Wang, L., Hopcroft, J.E.: Nesterov accelerated gradient and scale invariance for adversarial attacks. In: ICLR 2020 (2020)
26. Liu, Z., et al.: Feature distillation: DNN-oriented JPEG compression against adversarial examples. In: CVPR 2019, pp. 860–868 (2019)
27. Ma, C., Cheng, S., Chen, L., Yong, J.: Switching gradient directions for query-efficient black-box adversarial attacks. CoRR abs/2009.07191 (2020)
28. Madry, A., Makelov, A., Schmidt, L., Tsipras, D., Vladu, A.: Towards deep learning models resistant to adversarial attacks. In: ICLR 2018 (2018)
29. Meunier, L., Atif, J., Teytaud, O.: Yet another but more efficient black-box adversarial attack: tiling and evolution strategies (2019)
30. Moon, S., An, G., Song, H.O.: Parsimonious black-box adversarial attacks via efficient combinatorial optimization. In: ICML 2019, pp. 4636–4645 (2019)
31. Papernot, N., McDaniel, P.D., Goodfellow, I.J., Jha, S., Celik, Z.B., Swami, A.: Practical black-box attacks against machine learning. In: Proceedings of the AsiaCCS 2017, pp. 506–519 (2017)
32. Papernot, N., McDaniel, P.D., Jha, S., Fredrikson, M., Celik, Z.B., Swami, A.: The limitations of deep learning in adversarial settings. In: IEEE EuroS&P 2016, pp. 372–387 (2016)
33. Rastrigin, L.A.: The convergence of the random search method in the extremal control of many-parameter system. Autom. Remote Control **24**(10), 1337–1342 (1963). https://scholar.google.com/scholar?cluster=14844809834l0715230
34. Shaham, U., et al.: Defending against adversarial images using basis functions transformations. CoRR abs/1803.10840 (2018)
35. Sharif, M., Bhagavatula, S., Bauer, L., Reiter, M.K.: Accessorize to a crime: real and stealthy attacks on state-of-the-art face recognition. In: Proceedings of the ACM CCS 2016, pp. 1528–1540 (2016)
36. Su, D., Zhang, H., Chen, H., Yi, J., Chen, P., Gao, Y.: Is robustness the cost of accuracy? – A comprehensive study on the robustness of 18 deep image classification models. In: Ferrari, V., Hebert, M., Sminchisescu, C., Weiss, Y. (eds.) ECCV 2018, pp. 644–661 (2018)
37. Szegedy, C., Vanhoucke, V., Ioffe, S., Shlens, J., Wojna, Z.: Rethinking the inception architecture for computer vision. In: Proceedings of the CVPR 2016, pp. 2818–2826 (2016)
38. Szegedy, C., et al.: Intriguing properties of neural networks. In: ICLR 2014 (2014)
39. Tu, C., et al.: Autozoom: autoencoder-based zeroth order optimization method for attacking black-box neural networks. In: Proceedings of the AAAI 2019, pp. 742–749 (2019)
40. Wang, X., He, K.: Enhancing the transferability of adversarial attacks through variance tuning. In: Proceedings of the CVPR 2021, pp. 1924–1933 (2021)
41. Xie, C., Wang, J., Zhang, Z., Ren, Z., Yuille, A.L.: Mitigating adversarial effects through randomization. In: ICLR 2018 (2018)
42. Xie, C., et al.: Improving transferability of adversarial examples with input diversity. In: Proceedings of the CVPR 2019, pp. 2730–2739 (2019)
43. Xu, W., Evans, D., Qi, Y.: Feature squeezing: detecting adversarial examples in deep neural networks. In: NDSS 2018 (2018)

44. Yann, L., Corinna, C.: The MNIST database of handwritten digit (1998)
45. Yatsura, M., Metzen, J.H., Hein, M.: Meta-learning the search distribution of black-box random search based adversarial attacks. CoRR abs/2111.01714 (2021)
46. Yin, D., Lopes, R.G., Shlens, J., Cubuk, E.D., Gilmer, J.: A Fourier perspective on model robustness in computer vision. In: NeurIPS 2019, pp. 13255–13265 (2019)
47. Zhang, F., Chowdhury, S.P., Christakis, M.: DeepSearch: a simple and effective blackbox attack for deep neural networks. In: Devanbu, P., Cohen, M.B., Zimmermann, T. (eds.) ESEC/FSE 2020, pp. 800–812 (2020)
48. Zhang, H., Yu, Y., Jiao, J., Xing, E.P., Ghaoui, L.E., Jordan, M.I.: Theoretically principled trade-off between robustness and accuracy. In: ICML 2019 (2019)

Autoencoder Assist: An Efficient Profiling Attack on High-Dimensional Datasets

Qi Lei[1(✉)], Zijia Yang[1], Qin Wang[2], Yaoling Ding[3], Zhe Ma[1], and An Wang[3]

[1] Bank Card Test Center, Beijing, China
leiqiuq@outlook.com, zjyangzijia@outlook.com, ma.z@bctest.com
[2] CSIRO Data61, Sydney, Australia
[3] School of Cyberspace Science and Technology, Beijing Institute of Technology, Beijing, China
{dyl19,wangan1}@bit.edu.cn

Abstract. Deep learning (DL)-based profiled attack has been proved to be a powerful tool in side-channel analysis. However, most attacks merely focus on small datasets, in which their points of interest are well-trimmed for attacks. Countermeasures applied in embedded systems always result in high-dimensional side-channel traces, i.e., the high-dimension of each input trace. These traces inevitably require complicated designs of neural networks and large sizes of trainable parameters for exploiting the correct keys. Therefore, performing profiled attacks (directly) on high-dimensional datasets is difficult. To bridge this gap, we propose a dimension reduction tool for high-dimensional traces by combining signal-to-noise ratio (SNR) analysis and autoencoder. With the designed asymmetric undercomplete autoencoder (UAE) architecture, we extract a small group of critical features from numerous time samples. The compression rate by using our UAE method reaches 40x on synchronized datasets and 30x on desynchronized datasets. This preprocessing step facilitates the profiled attacks by extracting potential leakage features. To demonstrate its effectiveness, we evaluate our proposed method on the raw ASCAD dataset with 100,000 samples in each trace. We also derive desynchronized datasets from the raw ASCAD dataset and validate our method under random delay effect. We further propose a 2^n-structure MLP network as the attack model. By applying UAE and attack model on these traces, experimental results show all correct subkeys on synchronized datasets and desynchronized datasets are successfully revealed within hundreds of seconds. This indicates that our autoencoder can significantly facilitate DL-based profiled attacks on high-dimensional datasets.

Keywords: Side-channel analysis · Deep learning · Autoencoder

1 Introduction

Side-channel analysis (SCA) exploits the weakness of cryptographic algorithm implementations from the view of physical information such as timing [1], power

Q. Lei—Full version refers to https://eprint.iacr.org/2021/1418.

C. Alcaraz et al. (Eds.): ICICS 2022, LNCS 13407, pp. 324–341, 2022.
https://doi.org/10.1007/978-3-031-15777-6_18

consumption [2], and electromagnetic emanations [3]. When running a cryptographic algorithm, intermediate states and operations that are closely dependent on secret keys may cause leakage and be detected through SCA observation, severely threatening the security of cryptographic systems. In SCA, attack methods can be categorized into *profiled attacks* and *non-profiled attacks*. Profiled attacks indicate that a powerful adversary has access to an open clone cryptographic device for secret keys, side-channel traces, and unlimited computational power. The adversary first characterizes all types of secret information from a clone device (a.k.a., profiling phase), and then recovers secret data from the target device by utilizing their learned models (a.k.a., attack phase). Typical profiled attacks covers template attack [4] and machine learning-based attacks [5–8]. Non-profiled attacks refer to attacks where an adversary can only collect physical information from a target device. The abilities of adversaries in this type are weaker than those in profiled attacks. The attacker uses statistical analysis to derive secret information from side-channel traces. This line of attacks include simple power analysis (SPA), differential power analysis (DPA) [2], correlation power analysis (CPA) [9] and mutual information analysis (MIA) [10,11].

One of the most powerful side-channel attacks has been proven to be the deep learning (DL)-based profiled attacks. This type of attacks outperforms other types in the context of attack traces that are needed to recover keys [8] [12]. Moreover, DL-based profiled attacks have been explored on both unprotected datasets and protected datasets [8,13–18]. These datasets have accurately located *points of interest* (POIs). However, in a practical secure implementation, each trace includes tens of thousands of samples due to applied countermeasures. Directly applying neural networks on high-dimensional datasets is impractical because the design inevitably requires a very complicated network architecture and a large size of trainable parameters to learn key information. To the best of our knowledge, deep learning-based profiled attacks lack applications on high-dimensional datasets. To bridge this gap, a preprocessing of locating leakage samples and extracting features becomes a necessity.

In advance to accelerate the attacks, many solutions of preprocessing techniques for POI selection have been explored in the profiled attacks field. For instance, principal component analysis (PCA), linear discriminant analysis (LDA), and kernel discriminant analysis (KDA) have been investigated in template attacks [13,19–22]. Similarly, long short-term memory (LSTM) autoencoder in machine learning-based profiled attacks [23] has been applied on unprotected AES implementation to extract features. Additionally, convolutional layers are also considered to be a feature extraction tool [16]. However, all these solutions are applied on small datasets in that each trace contains only hundreds of samples. It is insufficient in practical scenarios where each trace consists of large sizes of samples. An efficient tool to locate and extract critical features when performing attacks is desperately necessary. We thereby propose a new approach to preprocess high-dimensional datasets.

Our approach utilizes the undercomplete autoencoder (UAE). UAE can learn a lower-dimensional representation of the data by an encoder and correspondingly decodes it into the original input space by a decoder. The structure of the encoder and decoder can be designed as fully connected layers, convolutional layers, LSTM layers, etc. Our proposed solution can outperform previous methods due to a specialized design. Specifically, our method introduces an asymmetric structure into autoencoder design to reduce the training complexity. This design differs from traditional autoencoder solutions which have symmetric autoencoder structures as in [17,23]. In our asymmetric model, the encoder and decoder have different structures. This provides benefits with flexible parameter settings and independent combinations of training models. Therefore, to accelerate DL-based attacks on high-dimensional traces, we design a preprocessing and attack model suite, which consists of signal-to-noise ratio (SNR) analysis, autoencoder, and multilayer perceptron (MLP). Our proposed model can extract key features from high-dimensional traces. To demonstrate its effectiveness, we give experiments under our model and several parallel models on the raw ASCAD datasets to compare attack results before and after the feature extraction. Experimental results indicate that our method has significantly improved the attack performance. In a nutshell, the contributions of this work are listed as follows.

- We propose an innovative trace preprocessing architecture to compress high-dimensional datasets. The preprocessing tool is composed of SNR analysis and asymmetric UAE that can extract critical trace features.
- We design a 2^n-structure MLP structure to perform profiled attack. Experimental results show that the MLP can perform successful attacks while existing attack models cannot reveal all correct keys before and after the feature extraction. Meanwhile, training a 2^n-structure MLP takes only $\frac{1}{42} \sim \frac{1}{3}$ the time compared with the state-of-the-art models.
- The preprocessing and the attack architecture are confirmed on both synchronized traces and desynchronized traces. We introduce convolutional layers in UAE to confront the desynchronization effect and further exploit the secret key. The compression rate by using UAE separately reaches **40x** on synchronized datasets and **30x** on desynchronized datasets.

Paper Structure. Section 2 introduces the raw ASCAD dataset for evaluation and provide dimension reduction techniques applied in SCA. The architectures of the asymmetric UAE and 2^n-structure MLP are presented in Sect. 3. Section 4 details the hyperparameter selection and shows experimental results on synchronized and desynchronized datasets. Section 5 concludes this work.

2 Primitives and Components

This section introduces the high dimensional traces of raw ASCAD dataset that contains a large number of samples served for profiling attacks. Then, we provide details of dimension reduction techniques (SNR and autoencoders) in SCA. Finally, we present neural networks used as attack models in profiled attacks.

2.1 Raw ASCAD Dataset

The ASCAD dataset [14] provides power traces produced by a masked AES-128 implementation on ATMega8515 microcontroller. It is used as a benchmark to evaluate machine learning techniques applied in the side-channel context. The raw ASCAD dataset consists of 60,000 traces and each trace is composed of 100,000 samples. It is split into a training set with 50,000 traces and a test set with 10,000 traces. The common used ASCAD dataset contains 700 samples with random delay $N_D = \{0, 50, 100\}$. Desynchronized subsets in ASCAD are generated via random delay technique [14]. We use the same desynchronization technique in this work to generate high-dimensional desynchronized datasets.

The leakage model in the ASCAD dataset is chosen to be S-box output, which is defined as

$$S\text{-}box(p[i] \oplus k[i]),$$

where $i \in \{1, ..., 16\}$, p *represents plaintexts while* k *is secret keys (both are measured in byte).*

2.2 Dimension Reduction Techniques in SCA

To find a possible leakage interval, SCA makes use of statistical tests (SNR, T-test) to detect leakage points and dimensional reduction skills (PCA, LDA, KDA, autoencoders) to reduce attack complexity. Our method combines SNR and autoencoder to address the high-dimensionality issue in SCA trace preprocessing. We introduce them as follows.

Signal-to-Noise Ratio Analysis. Signal-to-noise ratio (SNR) analysis has been applied to find POIs in the raw ASCAD dataset. We introduce five mainstream types of SNR defined in [14] (see Table 1). Specifically, to illustrate how we locate POIs from original traces according to SNR analysis results, we use SNRs in the third S-box (cf. Fig. 1) as an example.

The value of $snr1$ (unmasked S-box output) is low in the third S-box, which implies the first-order leakage is weak in tested traces. In contrast, $snr2$ (masked S-box output) and $snr4$ (masked S-box output in linear parts) have high correlation value that shows the real intermediates in AES traces. $snr3$ is a mask constant that leads to high SNR peaks. We can find that $snr3$ is a misleading signal when selecting the proper S-box operation period. $snr5$ shows high peaks of each S-box mask $r[i]$.

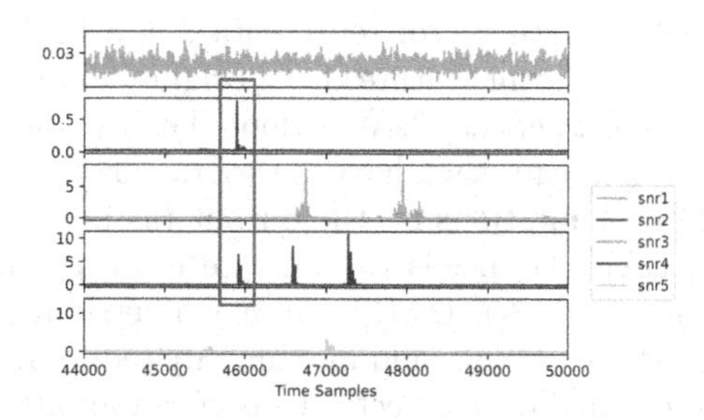

Fig. 1. SNR peaks of the third S-box

We zoom in SNR analysis result to observe the positions of SNR peaks excluding the misleading signal $snr3$. When n where $n > 1$, SNR reaches peaks at the same position (marked with a red square), we collect the highest values of each SNR type at the point $P_i (i = 1, ..., n)$. Next, we use P_{middle}, where $P_{middle} = \frac{\sum P_i}{n}$, as the middle point to trim an interval for further analysis. In the third S-box, $P_{middle} = 45900$ and the trimmed interval (the number of samples denoted as L) is $(45900 - \frac{L}{2}, 45900 + \frac{L}{2})$.

Table 1. SNR definitions

Name	Type	Target sensitive variables
*snr*1	Unmasked S-box output	S-box($p[3] \oplus k[3]$)
*snr*2	Masked S-box output	S-box($p[3] \oplus k[3]$) $\oplus r_{out}$
*snr*3	Common S-box output mask	r_{out}
*snr*4	Masked S-box output in linear parts	S-box($p[3] \oplus k[3]$) $\oplus r[3]$
*snr*5	S-box output mask in linear parts	$r[3]$

Here, we consider the POI selection from two folds. Firstly, the selected POI interval should be large enough to cover all operation information of one S-box. Secondly, a small number of POIs will help to reduce the subsequent analysis complexity. Therefore, we start attacks from a coarse sample number set $L = \{1000, 2000, 3000, 4000, 5000\}$ on the raw ASCAD dataset.

Autoencoders. The autoencoder is a type of neural network trained to represent the original information from a code. An autoencoder consists of two parts, namely *encoder* and *decoder*. An encoder $z = f(x)$ maps an input to the code while a decoder $x' = g(z)$ generates the reconstruction of original inputs. The autoencoder types that are widely adopted include undercomplete autoencoder (UAE), denoising autoencoder (DAE), and contractive autoencoder (CAE).

In this work, we focus on dimension reduction for profiling attacks. Undercomplete means the dimension of the code is lower than the input. By using smaller-length vector representation, a UAE can force the network to learn the most significant features from original data. The training goal is to minimize the loss of $L(x, g(f(x)))$. We adopt the UAE as our first choice to simplify attacks. This is because we have no extra noise on original traces (as DAE required) and no limitations on the output (as in CAE). As we noticed that implementing DAE [17] needs to generate extra corrupted data for robust training, we decide to design UAE in an asymmetric manner as a penalty to reconstruct a robust code. We would maximumly reduce hidden layers in the decoder part and use original data directly to perform an attack. In this way, we can reduce the training model complexity and shorten preprocessing time.

2.3 Neural Networks

A deep learning-based profiled attack is a task of classification that trains the neural network model for precisely predicting the correct keys. A neural network contains an *input layer*, multiple *hidden layer(s)* and an *output layer*. The input layer in SCA receives the one-dimensional side-channel traces, while the output layer provides key classification results. Model hyperparameter and optimizer hyperparameters will affect the network attack performance. Grid search optimization (GSO) [24] is widely used to select a better group of hyperparameters.

In an MLP, all layers are fully connected and each connection has a weight value (parameter) that gets updated during backpropagation. We adopt the categorical cross-entropy loss in prediction layer for MLP classification in SCA.

A convolutional neural network (CNN) is composed of three building blocks: *convolutional layers*, *pooling layers* and *fully connected layers*. Convolution layers play a key role in CNN to extract features. In the study of Zaid et al. [16], they propose to build an efficient CNN for side-channel attack. Here, we present the convolutional layer and pooling layer that are useful to detect desynchronization. They assume that the maximum random delay in the original trace is N_D. In the convolutional block for feature extraction, filter size is set to $\frac{N_D}{2}$. Pooling stride is the same as filter size ($\frac{N_D}{2}$) to preserve information related to desynchronization while also maximize the dimension reduction. It is demonstrated that smaller filter sizes help to identify local features and larger filter sizes can extract global features but inevitably cause spreading of relevant information. Besides, they recommend a small number of filters {2, 4, 8} for fast training.

It is worth noticing that convolutional neural networks can be deconstructed into the convolutional part for feature extraction and a fully connected part for classification. In CNNs, fully connected layers are similar to MLP. As convolutional structures can be applied to undercomplete autoencoder for feature extraction, in this work, only MLP is used for classification.

3 Model Design

In this section, we propose our autoencoder-assisted model that combines SNR, UAE, and MLP. The model provides efficient reduction of large-size datasets.

3.1 Model Overview

The entire model (including feature selection and classification modules) consists of three parts: *SNR*, *UAE* and *MLP*. Figure 2 shows the trace processing path. Firstly, SNR analysis is performed on original traces. We use SNR middle position P_{middle} (see Sect. 2.2) to trim traces of length L. The goal of this SNR exploration is to find a proper interval that can cover all possible leakage points and give good performance results. Then, the trimmed traces are fed into an undercomplete autoencoder. The autoencoder will compress L-sample traces by R times. SNR and UAE are the preprocessing part for feature extraction (POI selection). Each code with $\frac{L}{R}$ samples becomes the input of the MLP attack model. MLP acts as the classifier and the attacking result is presented by the key rank (guessing entropy) of correct subkeys.

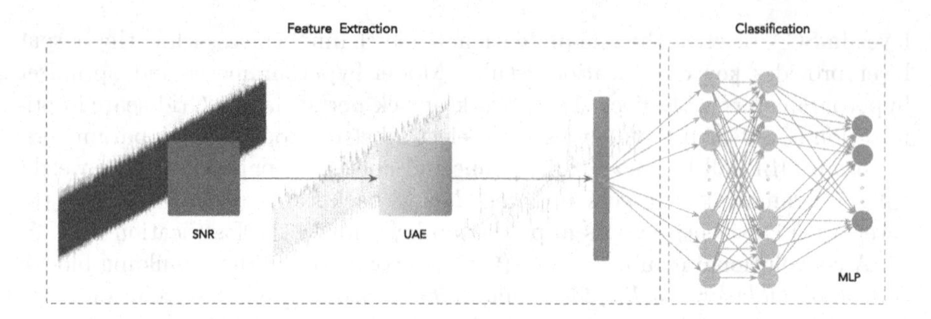

Fig. 2. Preprocessing and attack model structure

3.2 UAE Structure

Autoencoders are normally designed to be symmetric where the encoder and decoder have the same structure. SCA attack solutions utilizing autoencoders [17,23] also follow the standard symmetric structure. A heuristic thinking is that the symmetric structure can help to retrieve the original data. However, the symmetric structure increases the complexity of a UAE network. In this work, we introduce asymmetric UAE for feature extraction. Asymmetric means the structure of an encoder and a decoder are different and the encoder is more complicated than the decoder. Specifically, an encoder has one or more hidden layers. Whereas a decoder has fewer hidden layers or no hidden layer. The asymmetric structure can be viewed as a regularization that forces the network to learn more robust features without relying on a corrupted input or a regularizer. Formally, we define this asymmetric UAE model as

$$\hat{f} = I' \circ [\alpha]^{H_D} \circ C \circ [\alpha]^{H_E} \circ I, H_E > H_D,$$

where the input layer is represented as an identity function I, and the output layer I' means a reconstruction of input. The code (bottleneck layer) is denoted as C. A hidden layer is denoted as α. The number of hidden layers in encoder is H_E and decoder H_D. Here, the restriction $H_E > H_D$ shows the asymmetric feature. A hidden layer α can apply structures such as a fully connected layer, a convolutional block, or an LSTM layer. Specifically, when α represents a fully connected layer, it is formalized as

$$A \circ \lambda,$$

where A is an activation function and λ represents a linear function. When α represents a convolutional block, it is

$$\delta \circ [A \circ \gamma]^{H_{conv}},$$

where a convolutional block contains H_{conv} convolutional layers (denoted γ), an activation function (denoted A) and one pooling layer (denoted δ).

In this work, we instantiate the asymmetric UAE model for synchronized traces and desynchronized traces. We test our proposed architecture on both synchronized and desynchronized high-dimensional datasets. As for synchronized traces, the hidden layers in UAE are configured to be fully connected layers for fast training. As for desynchronized traces, convolutional layers are introduced in UAE. Similar techniques applying CNN to deal with desynchronized cases are also demonstrated in [8,16].

UAE for Synchronized Traces. An asymmetric UAE for synchronized traces is presented as

$$\hat{f}_{sync} = I' \circ C \circ [A \circ \lambda]^{H_E} \circ I,$$

where $H_D = 0$ *means no hidden layer exists.* Notably, we remove the hidden layer in the decoder to maximally simplify the structure and reduce the complexity. We illustrate this UAE structure in Fig. 3. We suppose that the input layer and the output layer have N_I nodes. Each hidden layer has N_F nodes. The bottleneck layer has N_C nodes. In this way, an N_I-dimensional trace is compressed to a N_C-dimensional code. We accordingly denote the compress ratio R from the input dimension to a code dimension as $R = \frac{N_I}{N_C}$.

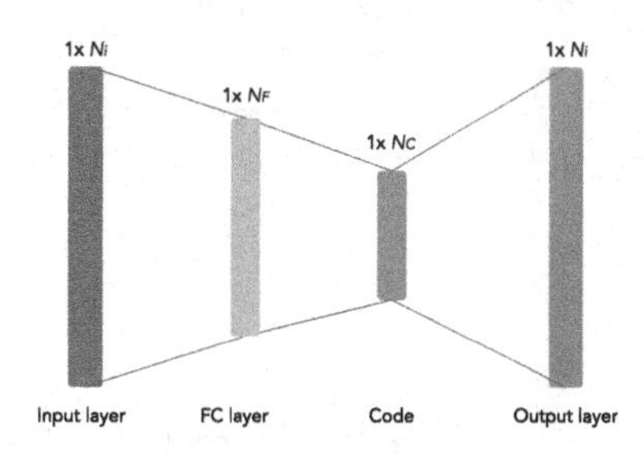

Fig. 3. Synchronized UAE

Here, we specify the trick that is used to determine the number of nodes in each fully connected layer. Assuming that $N_F^{[j]}$ is the number of neurons in the j-th fully connected layer, we define the number of nodes in j-th hidden layer is 2^n. The number of nodes in $(j-1)$-th hidden layer is twice the number of j-th layer, i.e., $N_F^{[j-1]} = 2 \times N_F^{[j]}$. We limit $N_F^{[1]} < N_I$ to ensure the number of nodes in each hidden layer to be fewer than the input layer. We denote this structure as **2^n-structure**. To set an asymmetric UAE structure, we only need to define the number of hidden layers H_E and the number of nodes $N^{[H_E]}$ in the last hidden layer. Using 2^n-structure, we perform the grid search optimization for UAE hyperparameters (see Table 4L) on the raw ASCAD dataset (synchronized traces).

UAE for Desynchronized Traces. Similarly, for desynchronized traces, a UAE is formalized as

$$\hat{f}_{desync} = I' \circ [A \circ \lambda] \circ C \circ [A \circ \lambda] \circ [\delta \circ [A \circ \gamma]] \circ I,$$

where $H_D = 1, H_E = 2$ *and* $H_{conv} = 1$. During the UAE model design, we still attempt to design the most simplified UAE. As a flatten layer at the output layer of a convolutional block is required, we add a fully connected layer in

Parameter	Values
Activation function	{sigmoid, tanh, ReLU, SeLU}
Learning Rate	0.001 (default value)
Batch size	{128,256,512}
Epochs	{10,20,30,40}
no of neurons $N^{[H_E]}$ (last hidden layer)	{64,128,256,512}
no of hidden layers H_E	{1,2,3}
Compress ratio R	{5,10,20,30,40,50}

Parameter	Values
Activation function	{sigmoid, tanh, ReLU, SeLU}
Learning Rate	One-cycle policy
Weight initialization	He uniform
Batch size	{128,256,512}
Epochs	{10,20,30,40,50,60}
Filter size K	$\{\frac{N_D}{2}, N_D, 2N_D\}$
no of filter	{2,4,8}
Compress ratio R	{5,10,20,30,40,50}

Fig. 4. Grid search optimization for UAE hyperparameters on synchronized (L) and desynchronized (R) traces

both encoder and decoder. Therefore, in our UAE, the encoder consists of a convolutional block and a flatten layer. The decoder only has a fully connected layer whose nodes number is same as which in the encoder. The structure of the asymmetric UAE for desynchronized traces is depicted in Fig. 5.

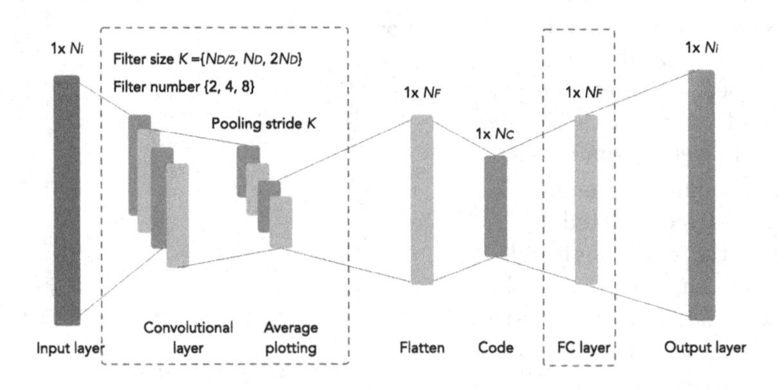

Fig. 5. UAE structure for desynchronized traces

The input layer and output layer have N_I nodes. Regarding the method of building efficient CNN architectures [16], we apply filter size $K = \{\frac{N_D}{2}, N_D, 2N_D\}$. Note that we extend the kernel size $\frac{N_D}{2}$ recommended in [16] to $\{N_D, 2N_D\}$. As smaller kernels extract local features and larger kernels extract global features, we would like to find out the kernel size influence in UAE feature extraction. The pooling stride is the same as kernel size. We perform grid search on a small number of filters from $\{2, 4, 8\}$. After average pooling, the flatten layer has N_F nodes and the decoder hidden layer has the same number of nodes. The compress ratio from input dimension to a code dimension is $R = \frac{N_I}{N_C}$. Based on CNN structure, we perform the grid search optimization for UAE hyperparameters (see Table 4R) on the desynchronized raw ASCAD dataset. *One-cycle policy* [25] is applied to update the learning rate.

3.3 MLP Structure

As for the attacking model, 2^n-structure is applied to build an MLP network. We denote the j-th hidden layer of MLP as $\mathcal{L}_j$ and the number of nodes in j-th hidden layer as $N^{[j]}$. Here, $N^{[j]} = 2^n$ and $N^{[j+1]} = 2 \times N^{[j]}$, i.e., the next hidden layer has twice the number of nodes than the previous layer. The 2^n-structure in MLP is reverse to that in UAE because a UAE tries to compress a dataset. We assume that there are H hidden layers. Then an MLP under 2^n-structure is configured by H and $N^{[1]}$. The number of nodes in the input layer is denoted as N_C (code length) while the output layer as N_O. The architecture of a 2^n-structure MLP is shown in Fig. 6. With this model, we apply hyperparameters as illustrated in Table 2 to find a simple architecture to reveal S-box subkeys with fewer traces and shorter training time.

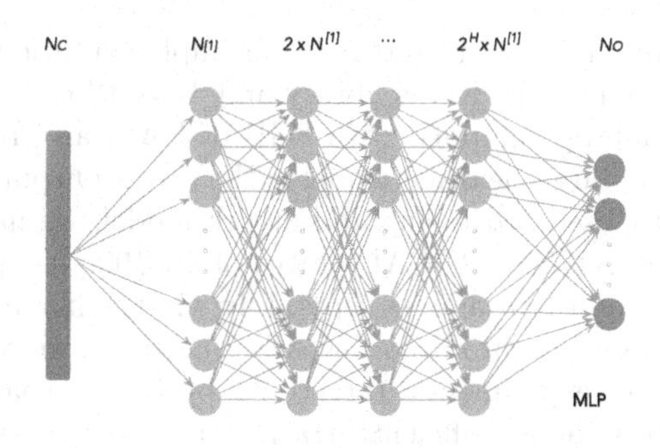

Fig. 6. 2^n-structure MLP structure

Table 2. Grid search optimization for the MLP hyperparameters

Parameter	Values
Activation function	{sigmoid, tanh, ReLU, SeLU}
Learning rate	One-cycle policy
Batch size	{50, 100, 200, 300}
Epochs	{10, 20, 25, 50, 75, 100}
no of neurons $N^{[1]}$ (first hidden layer)	{64, 128, 256}
no of hidden layers H	{2, 3, 4, 5, 6}

4 Experimental Results

We apply our preprocessing and attack model suite[1] on high-dimensional side-channel traces. The model suite is tested on synchronized and desynchronized raw ASCAD datasets. Firstly, we show experiment settings and datasets preparation for profiled attacks. Then, we grid search proper UAE and MLP hyperparameters for each dataset. For a good model evaluation, the first priority is to compare the least number of traces for guessing entropy (GE) converge to 1. We also compared our model with existing SCA attack models from the perspective of training time and trainable parameters as in most SCA studies [14,16,17].

[1] https://github.com/niki-lei/Autoencoder-Assist-An-Efficient-Profiling-Attack-on-High-dimensional-Datasets.

4.1 Experimental Configurations

For environment settings, our implementation of machine learning techniques is based on the Keras library and Tensorflow backend. We run training models on a laptop equipped with 8GB of RAM and Intel(R) Core(TM) i7-8750H CPU. The target leakage model is the S-box output S-box($p[i] \oplus k[i]$). To validate our proposed processing and attack model, we have three datasets: a synchronized dataset (raw ASCAD traces of 100,000 samples) and desynchronized datasets (50 and 100 samples separately window maximum jitter on raw ASCAD traces).

On synchronized raw ASCAD traces, we select the third S-box as the preliminary experimental target. We perform a coarse searching of applicable parameters for an efficient attack. Since there is no desynchronized dataset of raw ASCAD traces, we generate random delay traces (samples window maximum of 50 and 100) in the same way as mentioned in Sect. 2.1. Based on SNR observation, a leakage interval $[N, N+L]$ is selected for an S-box. Desynchronization parameter N_D is set to 50 and 100, respectively. Then a random integer $r < N_D$ is generated to trim each interval $[N+r, N+r+L]$ out as one item in the desynchronized dataset. In this way, we obtain the $N_D = 50$ and $N_D = 100$ desynchronized datasets.

Hyperparameters used in UAE and MLP are selected via a grid search optimization with finite sets of values. The final chosen value is obtained based on the best attack performances, namely, the least traces for the GE reaches 1. To perform a successful attack, three steps are conducted. Firstly, we propose a naïve UAE and MLP to explore a proper POI length based on SNR analysis. Then, we fix a non-optimized MLP to grid search UAE hyperparameters on the POI dataset. Finally, optimize MLP on UAE extracted features.

4.2 SNR Parameter

Since current attack models use the third S-box in ASCAD dataset as the target of evaluation, we also aim at the same S-box for model training and comparison. POI intervals are trimmed from the length set $L = \{1000, 2000, 3000, 4000, 5000\}$ according to the SNR peak central point. The trimmed dataset $\mathcal{D}^L_{S\text{-}box3}$ is labeled by $sbox(p[3] \oplus k[3])$. Each $\mathcal{D}^L_{S\text{-}box3}$ is split into three subsets: 45,000 traces as *training* set, 5,000 traces as *validation* set and 10,000 traces as *test* set for attacks. We use 2^n-structure for both asymmetric UAE and MLP. For initial setting of UAE and MLP, we use non-optimized hyperparameters. We set the initial UAE structure (as in Fig. 3) to have

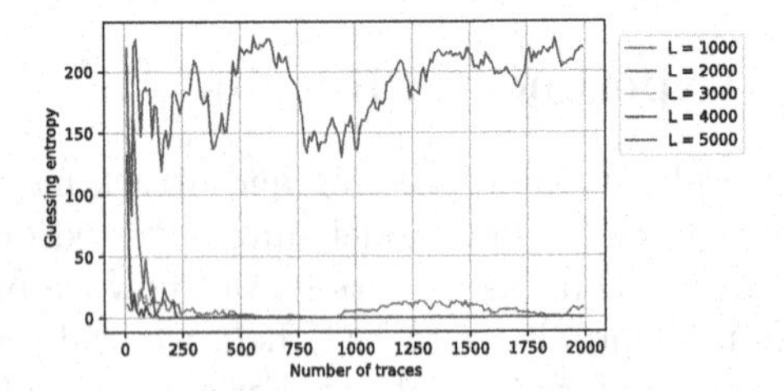

Fig. 7. Comparison of SNR intervals

2 hidden layers while the last hidden layer has $N_F^{[2]} = 256$ nodes (i.e., $N_F^{[1]} = 512 < N_I$). Compress ratio is set to 10. MLP is set to $H = 3$ and $N^{[1]} = 256$ (3 hidden layers and the first hidden layer has 256 nodes). For both UAE and MLP, activation function is ReLU. UAE is trained under 30 epochs with a batch size of 512 and learning rate of 0.001. MLP is trained under 50 epochs with a batch size of 200 and learning rate is updated under One-Cycle policy.

We use the same autoencoder and MLP attack model to compare the number of traces needed in exploiting the third byte. The attack results on the third S-box (Fig. 7) show that POI intervals $L = \{2000, 3000, 4000\}$ can help to perform a successful attack (GE = 1). The reason for the bad performance of $L = 5000$ dataset is that an autoencoder will extract a general feature and thus the useful leakage information is compressed. The interval with 3000 time samples requires the least traces. Thus, we fix the POI interval to 3000 time samples to investigate UAE and MLP architectures.

For the third S-box, we obtain the synchronized dataset $\mathcal{D}^{3000}_{S\text{-}box3}$ by trimming raw ASCAD dataset at position (44400, 47400). We also derive desynchronization datasets ($N_D = 50$ and $N_D = 100$) based on this interval. We split each desynchronized dataset into 3 subsets: 45,000 traces are used as the training set, 5,000 as validation set while 10,000 as the attack.

To test the performance of a classical attack model MLP_{best} [14] and the state-of-the-art models [16] on ASCAD dataset, we perform a profiled attack using these models on $\mathcal{D}^{3000}_{S\text{-}box3}$ directly. We also evaluate the model we proposed on $\mathcal{D}^{3000}_{S-box3}$. Experimental results show that none of these models can reveal the correct keys before UAE feature extraction.

4.3 Grid Search on UAE Hyperparameters

For synchronized and desynchronized datasets, we use corresponding UAE structures (Sect. 3.2) to optimize asymmetric UAE hyperparameters. Mean squared error is used to measure loss and Adam [26] to be optimizer.

Synchronized Traces. We apply hyperparameters in Fig. 4(L) to explore the best hyperparameters on synchronized dataset. The naive (non-optimized) MLP (in Sect. 4.2) is still used to evaluate UAE performance.

Among four activation functions, ReLU performs the best and takes less training time compared with SeLU function. Optimization is completed on a batch size of 512 and 30 training epochs with a learning rate of 0.001. The compress ratio is 40. The best UAE structure $H_E = 1$ and $N^{[H_E]} = 256$ (a UAE has one hidden layer and the last hidden layer has 256 nodes) make guessing entropy converge to 1 within 200 traces. The number of trainable parameters in this UAE is 787, 531 and it takes 104 s to reduce loss to 0.64 (Fig. 8a). Besides, we demonstrate that the number of hidden layers in the encoder has little impact on reducing loss and guessing entropy. On the contrary, the number of nodes in each layer can have a significant influence on attack results.

According to the best experiment result, we set the compress ratio to 40 in the third S-box. Equivalently, the code generated by UAE has 75 time samples. We apply this code in the next step, i.e., MLP optimization.

Desynchronized Traces. Following our UAE architecture for desynchronized traces (Fig. 5), we grid search an efficient UAE (Fig. 4(R)) to extract critical features for key revealing in the existence of random delay effect. We evaluate our proposed architecture when $N_D = 50$ and $N_D = 100$. The best hyperparameters for UAE are grid searched on $N_D = 50$ and $N_D = 100$ datasets separately. By applying the UAE architecture in Sect. 3.2, the filter size and pooling stride in convolutional layer applies not only $\frac{N_D}{2}$ in [16], we extend the larger size of $\{N_D, 2N_D\}$ for each dataset. For both datasets, optimization is completed with SeLU activation function and He uniform weight initialization. The one-cycle policy is applied to update the learning rate for the convolutional autoencoder. For desynchronized traces, the naïve MLP adds one hidden layer (i.e., $H = 4$), which is more powerful for classification but can also cause overfitting.

(i) **Random delay with $N_D = 50$.** In $N_D = 50$ desynchronized dataset, the best compress ratio is 20. The UAE with 8 filters and filter size of $2N = 100$ requires the least traces to recover the correct key on the third S-box. It takes 2,741 s to run the training on a batch size of 512 within 50 epochs. Finally, the loss of UAE reaches 0.49 (Fig. 8b). The number of trainable parameters in this UAE is 1, 588, 438 and 440 traces are needed for GE converging to 1. The best UAE hyperparameters for $N_D = 50$ dataset are listed in Table 3. It should be noted that in this experiment, the best filter size is $2N = 100$ instead of $\frac{N_D}{2}$ which uses the least number of traces. This indicates that global features extracted by a larger kernel also help distinguish the correct key candidate.

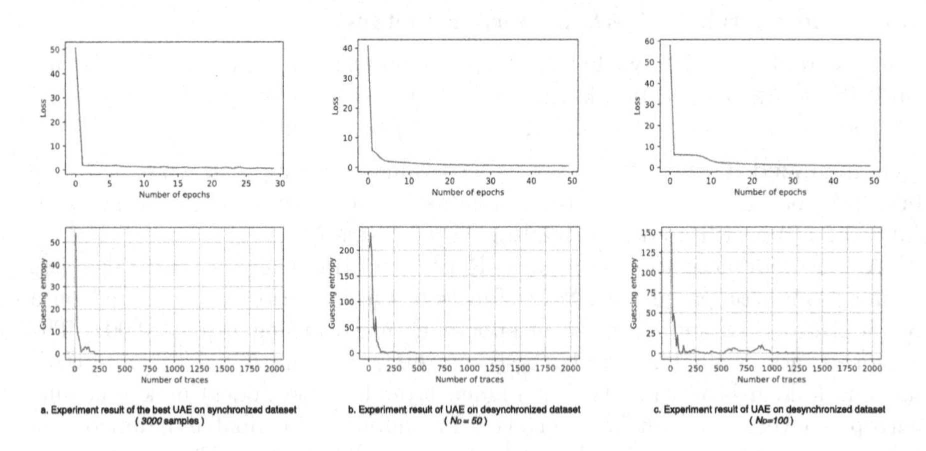

Fig. 8. UAE evaluations

(ii) **Random delay with $N_D = 100$.** For a higher random delay dataset, it needs a filter size of 50 and 4 filters to find the correct key. The $N_D = 100$ dataset is compressed by 30 times on a batch size of 512 within 50 epochs.

The training takes 2428 s to reach a loss of 0.75 (Fig. 8c). The number of trainable parameters in this UAE is 771, 544. The best hyperparameters are summarized in Table 3.
Here, we discuss the selection of different UAE filter sizes and filter numbers in profiled attacks. On $N_D = 50$ dataset, filter length of $\{\frac{N_D}{2}, N_D, 2N_D\}$ all can make GE converge to 1. While on $N_D = 100$ dataset, only $\frac{N_D}{2}$-kernel size can help exploit the correct key. Based on our experiment on 2 desynchronized datasets, we demonstrate that although $\frac{N_D}{2}$ is not always the best parameter for filter size; it is more general in datasets with different random delays. Besides, increasing the number of kernels can provide more information about trace features.

Table 3. Hyperparameters for the best UAE on desynchronized datasets ($N_D = 50$ and $N_D = 100$ traces)

Hyperparameter	Values ($N_D = 50$)	Values ($N_D = 100$)
Activation function	SeLU	SeLU
Learning rate	Once-cycle policy	Once-cycle policy
Batch size	512	512
Epochs	40	30
Filter size K	100	50
no of filters	8	4
Compress ratio	20	30

4.4 Grid Search on MLP Hyperparameters

Results on Synchronized Traces. On synchronized datasets, we present experimental results on the third byte.

Grid search optimization is conducted for MLP on parameter set in Tab. 2. We use categorical cross-entropy to evaluate guessing entropy. For fast convergence, the optimization is conducted by using Adam within 50 epochs and a batch size of 200. We also apply the one-cycle policy to update the learning rate in each epoch. ReLU is used as the activation function since it consumes less training time. Under 2^n-structure MLP, $H = 2$ and $N^{[1]} = 64$ deliver the best performance.

We compare our network performance with MLP_{best} [14] and the state-of-the-art model [16] for ASCAD dataset ($N_D = 0$) in Table 4. Experiment results are obtained from the same environment settings. We use original model parameters and hyper-parameters [14,16] for ASCAD to run experiments on our computer. The complexity of our network (in terms of trainable parameters) performs

5 times less than the MLP_{best} model. Training time is 3 times shorter than the convolutional network in [16]. Guessing entropy reaches 1 within 120 traces in our MLP network while another two models cannot reveal the correct subkey within 2000 traces. In conclusion, our 2^n-structure MLP network has a good performance while all other models fail on 3000 time sample traces (as shown in Fig. 9a). The least training time (autoencoder together with MLP) only takes around 200 s without any professional GPUs, which is proved to be satisfactorily efficient in a profiled attack.

Table 4. Comparison of performance on ASCAD 75 time samples (R = 40) (synchronized dataset)

	MLP_best [14]	State-of-the-art [16]	Our method
Complexity (trainable parameters)	227,456	4,432	46,208
Number of traces	–	–	140
Learning time (seconds)	98	160	46

Results on Desynchronized Traces. In this part, we conduct experiments based on random delay $N_D = 50$ and $N_D = 100$ traces, respectively. Same as synchronized dataset, hyperparameters in Table 2 are used for optimization.

(i) **Random delay with $N_D = 50$.** Under 2^n-structure MLP, optimization is done on $H = 3$ and $N^{[1]} = 128$ within 20 epochs and a batch size of 200. ReLU is the best activation function.
We compare our network performance with the state-of-the-art model [16] for ASCAD dataset ($N_D = 50$) in Table 5. Since the input layer only has 150-time samples, the dimension size becomes negative at the third convolutional block of existing CNNs [16]. We update the third block pooling size and pooling stride from 4 to 3 to make the attack applicable. The complexity of our network (in terms of trainable parameters) is nearly 4 times more than the state-of-the-art model. However, our model training time is over 42 times less than the network in [16]. GE reaches 1 within 170 traces in our MLP network while the state-of-the-art model requires 640 traces (Fig. 9b). Thus, our 2^n-structure MLP model is highly effective in profiled attacks.

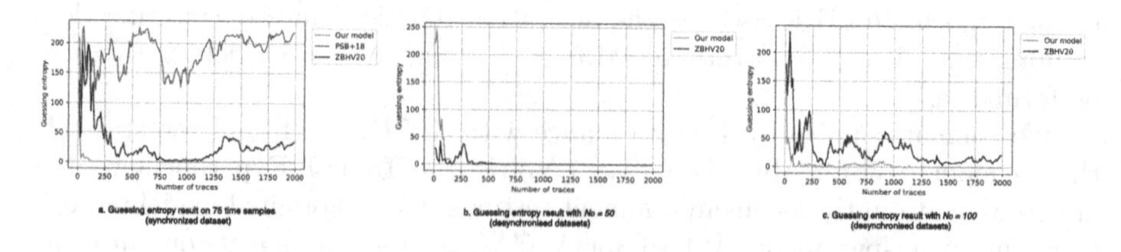

Fig. 9. Guessing entropy evaluations

(ii) **Random delay with $N_D = 100$.** Under 2^n-structure MLP, optimization is done on $H = 3$ and $N^{[1]} = 256$ MLP within 30 epochs and a batch size of 200. ReLU is the best activation function. We also compare our network performance with the state-of-the-art model [16] for ASCAD dataset ($N_D = 100$) in Table 5. Guessing entropy reaches 1 within 1150 traces in our MLP network while the other model cannot reveal the correct subkey within 2000 traces (as shown in Fig. 9c). Although the trainable parameters of our network are 7 times more than the state-of-the-art model, the training time is 4.5 times shorter than the convolutional network in [16]. On random delay $N_D = 50$ and $N_D = 100$ datasets, the short learning time (31 s and 245 s, respectively) shows an adversary can efficiently perform an attack using UAE preprocessing. This tool saves exploiting effort in terms of time-division by extracting key features from original traces.

Table 5. Comparison of performance on random delay $N_D = 50$ and $N_D = 100$ datasets

	State-of-the-art [16] ($N_D = 50$)	Our method ($N_D = 50$)	State-of-the-art [16] ($N_D = 100$)	Our method ($N_D = 100$)
Complexity (trainable parameters)	82,991	315,264	136,476	945,152
Number of traces	640	170	–	1150
Learning time (seconds)	1309	31	1120	245

5 Concluding Remarks

In this paper, we propose an efficient method to perform profiled attacks on high-dimensional datasets. We introduce the asymmetric undercomplete autoencoder to extract key features from high-dimensional leakage traces. To reduce the huge dimensionality, we combine SNR and UAE to reduce 100, 000 samples down to 3, 000 samples (by SNR) and further to 75 samples (by UAE) on the raw ASCAD dataset. We propose 2^n-structure UAE and MLP to perform the profiled attack and further investigate that our approach can work with desynchronization cases. We also equip UAE with convolutional layers to resolve the desynchronization problem as discussed in previous works [13,16]. We demonstrate that UAE is a necessary component in our approach to exploit leakage information. UAE can be a powerful preprocessing tool for an attacker, especially in a practical scenario with thousands of time samples. Although our experiments are conducted on laptops without any professional GPUs, a successful attack only takes around 200 s in total for UAE and MLP training. Experimental results indicate that our method is efficient in terms of both setup cost and training time.

References

1. Kocher, P.C.: Timing attacks on implementations of Diffie-Hellman, RSA, DSS, and other systems. In: Koblitz, N. (ed.) CRYPTO 1996. LNCS, vol. 1109, pp. 104–113. Springer, Heidelberg (1996). https://doi.org/10.1007/3-540-68697-5_9
2. Kocher, P., Jaffe, J., Jun, B.: Differential power analysis. In: Wiener, M. (ed.) CRYPTO 1999. LNCS, vol. 1666, pp. 388–397. Springer, Heidelberg (1999). https://doi.org/10.1007/3-540-48405-1_25
3. Quisquater, J.-J., Samyde, D.: ElectroMagnetic analysis (EMA): measures and counter-measures for smart cards. In: Attali, I., Jensen, T. (eds.) E-smart 2001. LNCS, vol. 2140, pp. 200–210. Springer, Heidelberg (2001). https://doi.org/10.1007/3-540-45418-7_17
4. Chari, S., Rao, J.R., Rohatgi, P.: Template attacks. In: Kaliski, B.S., Koç, K., Paar, C. (eds.) CHES 2002. LNCS, vol. 2523, pp. 13–28. Springer, Heidelberg (2003). https://doi.org/10.1007/3-540-36400-5_3
5. Hospodar, G., Gierlichs, B., De Mulder, E., Verbauwhede, I., Vandewalle, J.: Machine learning in side-channel analysis: a first study. J. Cryptogr. Eng. **1**(4), 293–302 (2011)
6. Heuser, A., Zohner, M.: Intelligent machine homicide. In: Schindler, W., Huss, S.A. (eds.) COSADE 2012. LNCS, vol. 7275, pp. 249–264. Springer, Heidelberg (2012). https://doi.org/10.1007/978-3-642-29912-4_18
7. Gilmore, R., Hanley, N., O'Neill, M.: Neural network based attack on a masked implementation of AES. In: HOST, pp. 106–111. IEEE (2015)
8. Maghrebi, H., Portigliatti, T., Prouff, E.: Breaking cryptographic implementations using deep learning techniques. In: Carlet, C., Hasan, M.A., Saraswat, V. (eds.) SPACE 2016. LNCS, vol. 10076, pp. 3–26. Springer, Cham (2016). https://doi.org/10.1007/978-3-319-49445-6_1
9. Brier, E., Clavier, C., Olivier, F.: Correlation power analysis with a leakage model. In: Joye, M., Quisquater, J.-J. (eds.) CHES 2004. LNCS, vol. 3156, pp. 16–29. Springer, Heidelberg (2004). https://doi.org/10.1007/978-3-540-28632-5_2
10. Gierlichs, B., Batina, L., Preneel, B., Verbauwhede, I.: Revisiting higher-order DPA attacks: multivariate mutual information analysis. IACR Cryptol. ePrint Arch. **2009**, 228 (2009)
11. Batina, L., Gierlichs, B., Prouff, E., Rivain, M., Standaert, F.-X., Veyrat-Charvillon, N.: Mutual information analysis: a comprehensive study. J. Cryptol. **24**(2), 269–291 (2011)
12. Lerman, L., Poussier, R., Bontempi, G., Markowitch, O., Standaert, F.-X.: Template attacks vs. machine learning revisited (and the curse of dimensionality in side-channel analysis). In: Mangard, S., Poschmann, A.Y. (eds.) COSADE 2014. LNCS, vol. 9064, pp. 20–33. Springer, Cham (2015). https://doi.org/10.1007/978-3-319-21476-4_2
13. Cagli, E., Dumas, C., Prouff, E.: Convolutional neural networks with data augmentation against jitter-based countermeasures. In: Fischer, W., Homma, N. (eds.) CHES 2017. LNCS, vol. 10529, pp. 45–68. Springer, Cham (2017). https://doi.org/10.1007/978-3-319-66787-4_3
14. Prouff, E., Strullu, R., Benadjila, R., Cagli, E., Dumas, C.: Study of deep learning techniques for side-channel analysis and introduction to ASCAD database. IACR Cryptol. ePrint Arch. **2018**, 53 (2018)
15. Carbone, M., et al.: Deep learning to evaluate secure RSA implementations. CHES **2019**(2), 132–161 (2019)

16. Zaid, G., Bossuet, L., Habrard, A., Venelli, A.: Methodology for efficient CNN architectures in profiling attacks. IACR Trans. Cryptogr. Hardw. Embed. Syst. **2020**(1), 1–36 (2020)
17. Lichao, W., Picek, S.: Remove some noise: on pre-processing of side-channel measurements with autoencoders. IACR Trans. Cryptogr. Hardw. Embed. Syst. **2020**(4), 389–415 (2020)
18. Masure, L., Dumas, C., Prouff, E.: A comprehensive study of deep learning for side-channel analysis. TCHES **2020**, 348–375 (2020)
19. Archambeau, C., Peeters, E., Standaert, F.-X., Quisquater, J.-J.: Template attacks in principal subspaces. In: Goubin, L., Matsui, M. (eds.) CHES 2006. LNCS, vol. 4249, pp. 1–14. Springer, Heidelberg (2006). https://doi.org/10.1007/11894063_1
20. Choudary, O., Kuhn, M.G.: Efficient template attacks. In: Francillon, A., Rohatgi, P. (eds.) CARDIS 2013. LNCS, vol. 8419, pp. 253–270. Springer, Cham (2014). https://doi.org/10.1007/978-3-319-08302-5_17
21. Cagli, E., Dumas, C., Prouff, E.: Enhancing dimensionality reduction methods for side-channel attacks. In: Homma, N., Medwed, M. (eds.) CARDIS 2015. LNCS, vol. 9514, pp. 15–33. Springer, Cham (2016). https://doi.org/10.1007/978-3-319-31271-2_2
22. Eisenbarth, T., Paar, C., Weghenkel, B.: Building a side channel based disassembler. Trans. Comput. Sci. **10**, 78–99 (2010)
23. Ramezanpour, K., Ampadu, P., Diehl, W.: SCAUL: power side-channel analysis with unsupervised learning. IEEE Trans. Comput. **69**(11), 1626–1638 (2020)
24. Bergstra, J., Bardenet, R., Bengio, Y., Kégl, B.: Algorithms for hyper-parameter optimization. In: Advances in Neural Information Processing Systems, vol. 24 (2011)
25. Smith, L.N.: A disciplined approach to neural network hyper-parameters: part 1 - learning rate, batch size, momentum, and weight decay. CoRR, abs/1803.09820 (2018)
26. Kingma, D.P., Ba, J.: Adam: a method for stochastic optimization. arXiv preprint arXiv:1412.6980 (2014)

TZ-IMA: Supporting Integrity Measurement for Applications with ARM TrustZone

Liantao Song, Yan Ding(✉), Pan Dong, Yong Guo, and Chuang Wang

School of Computer, National University of Defence Technology, Changsha, China
{songliantao,yanding,pandong,yguo,wangchuang}@nudt.edu.cn

Abstract. With the development of cloud computing and distributed systems, the computer system becomes increasingly complicated and open. To protect the integrity of applications, Integrity Measurement Architecture (IMA) is applied in the Linux kernel. However, traditional operating systems are complex and may contain many potential vulnerabilities. If the sensitive data used in IMA is leaked or modified, the protection mechanism will lose effectiveness. This paper proposes TZ-IMA, a security-enhanced solution to verify the integrity of applications based on ARM TrustZone technology. The system saves the encrypted reference hash value of applications and the encryption key in the normal world and in TrustZone, respectively. Before an application is executed, the integrity of application is checked by the secure world. Moreover, a vPCR module is constructed in TrustZone to protect the security of the measurement list. Based on the trusted anchor provided by TrustZone, TZ-IMA enables a challenger to prove that the attesting platform has sufficient integrity to be used. TZ-IMA is implemented on ARMv8 development board, and the evaluation results demonstrate that the overhead is only approximately 5% compared with the original system.

Keywords: Trusted computing · Remote attestation · Local appraisal · ARM TrustZone · Integrity measurement architecture

1 Introduction

The rapid growth of computer science and technology brings many challenges and security issues to open servers [11]. The complex execution environment may directly influence the integrity of applications. In recent years, with the extensive application of ARM architecture in servers fields, the administrator needs to provide a mechanism to ensure that the programs running on the server cannot be modified illegally and protect the integrity of applications on ARM platforms.

Integrity Measurement Architecture (IMA) was introduced in Linux 2.6.30 as part of the Linux integrity subsystem [30]. IMA mainly provides two mechanisms,

C. Alcaraz et al. (Eds.): ICICS 2022, LNCS 13407, pp. 342–358, 2022.
https://doi.org/10.1007/978-3-031-15777-6_19

appraisal and attestation. IMA appraisal stores the reference value of the application hash in the local system. The measurement value is compared with the reference value every time the application is loaded. IMA attestation measures and records all executable contents loaded onto the Linux system. The measurement list (MList) is protected by the trusted platform module (TPM), which is part of the Trusted Computing Group (TCG) standards. The remote challenger can validate that the programs running on the server are credible. However, traditional operating system (OS) kernel is growing larger and may contain many potential vulnerabilities. Once the sensitive data in the kernel space is leaked or modified by attackers, the protection mechanism may be bypassed. Moreover, as a coprocessor, TPM is very slow, resource limited, and must be deployed specifically on server systems, which limits the usage of IMA [1].

Fortunately, ARM TrustZone has been widely deployed on servers, mobile phones, and Internet devices [3,7,19]. Compared with TPM, TrustZone is a more flexible approach, which utilizes the CPU to construct a trusted execution environment (TEE). TrustZone-based integrity protection for system software has been widely used. SamSung proposed TIMA [20], which measures the load-time integrity of the bootloader and the kernel image, and saves them in TrustZone secure memory for further attestation. However, TIMA focused on kernel integrity rather than application integrity. Wang et al. [29] proposed a remote attestation scheme named TZ-MRAS for mobile terminal devices, but the method is not suitable for scenarios where applications frequently changed. To solve the problems mentioned above, this paper proposes an integrity measurement architecture for applications based on TrustZone technology called TZ-IMA.

Three technical challenges are observed in TZ-IMA. First, the integrity verification method should be suitable for applications frequently changed. The encryption key is the basis of system and must be carefully protected. Second, the trade-off between the integrity protection and the retrieval efficiency of the reference values should be considered. Third, conventional IMA saves the MList in the form of plaintext in the kernel space, and the attackers may obtain measuring results illegally. The confidentiality of MList which is used for remote attestation should be protected to prevent information leakage.

To address these challenges, TZ-IMA uses symmetric encryption algorithm to encrypt the reference value of applications and saves the key and the process of hash value comparison to TrustZone. The encrypted reference value of application is stored in the extended attributes of the normal world file system, which protects the reference value and realizes the rapid retrieval with the support of the file system simultaneously. Furthermore, TZ-IMA constructs a vPCR module in TrustZone and encrypts the measurement value of measuring events (ME), which guarantees both the confidentiality and integrity of MList.

In summary, our main contributions in this paper are as follows:

- An integrity verification method for applications based on TrustZone technology is proposed. The security-sensitive data and the code are isolated and will not be exposed to the OS or the user process in the normal world.

- A remote attestation scheme that enables challengers to validate the integrity of applications running on servers is proposed. TZ-IMA leverages TrustZone to serve as the trusted anchor and enhance the confidentiality of MList.
- A prototype is implemented on an ARMv8 platform. The evaluation results demonstrate that the performance overhead is only approximately 5.3%.

The remainder of this paper is organized as follows. Section 2 introduces the IMA in the Linux kernel and TrustZone structure. Section 3 introduces the threat model and design of TZ-IMA. Sections 4 and 5 describe the system in two mechanisms, local appraisal and remote attestation, respectively. Section 6 evaluates the performance overhead. Section 7 presents the related work. Section 8 concludes the paper.

2 Background

2.1 IMA

The integrity subsystem aims to protect the system's data from being modified maliciously. It follows the TCG Open Integrity Standard, which provides an integrity solution for systems by using a combination of integrity techniques such as hashing and invariance. Many functionalities are supported by IMA, such as measurement, remote attestation, local appraisal, audit, and authenticity.

Local Appraisal. IMA appraisal is the extension of Secure Boot. The goal of IMA appraisal is to detect any malicious modification of files, including remote attacks, local attacks, and even hardware attacks. It stores the reference hash value of the evaluated file in the security extended attribute "security.ima". Before the application is executed, it compares the measurement value of the file with the reference value stored in the "security.ima". If the values do not match, access to the file is denied. Furthermore, to support the application that dynamically changed, EVM is applied to protect the reference value, and the HMAC key used for the signature is stored on Trust Platform Module (TPM) for protection.

Measurement and Attestation. IMA measurement is the extension of trusted boot (or measured boot). IMA attestation records all the MEs, which triggers a measure and changes the execution status of the remote system. The MEs can be defined by IMA policy. By default, when applications are executed by the root, the kernel triggers hooks to measure the code, and creates and maintains a real-time MList which includes the path and the hash value of MEs. Platform configuration register (PCR) in TPM is used to protect the integrity of MList. A malicious code cannot delete or modify the value in the PCRs from the TPM chip. Accordingly, IMA can be used to prove the integrity of the system in a trusted boot system.

2.2 Overview of TrustZone

ARM TrustZone is a hardware isolation mechanism that is a secure extension to provide a trusted execution environment and protect sensitive data such as fingerprints, keys, and digital signatures. Figure 1 shows the architecture of TrustZone. ARMv8 processor is divided into two execution environments, Rich Execution Environment (REE) and TEE. REE is also called normal world, and TEE is called secure world or TrustZone. The processor has four privilege levels. User applications run on EL0, and the OS runs on EL1. Virtual Machine Monitor (also called Hypervisor) runs on EL2, and EL3 is reserved for low-level firmware and security monitor.

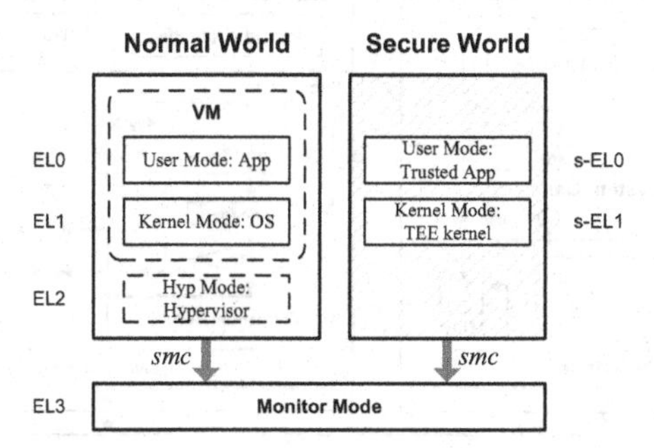

Fig. 1. Architecture of TrustZone

Compared with the normal world, the secure world has a higher privilege. The normal world cannot access the resources of the secure world, whereas the secure world can access all resources. Switching between the two worlds is carried out through the secure monitor call (*smc*). When the *smc* instruction is invoked, world switching is completed with the secure monitor. TrustZone provides an efficient method of security for the entire system by implementing hardware-based isolation in the CPU. Most TPMs can only perform preprogrammed operational and encryption algorithms currently.

3 System Design

3.1 Threat Model and Assumptions

TZ-IMA focuses on the integrity protection of applications. An attacker could launch an offline attack or remote attack, such as inducing users to run malicious programs to destroy the integrity of applications. Memory corruption vulnerabilities can be exploited to manipulate security-sensitive data in the kernel.

We assume all hardware are implemented correctly and trustworthy, and do not consider vulnerabilities in the secure world. Side channel attacks on memory and SoC or physical attacks, such as bus listening are not considered. We also assume secure boot was enabled.

3.2 TZ-IMA Framework

To protect the integrity of sensitive applications on the local system and enable the challenger to perform remote attestation, TZ-IMA is proposed. The architecture of TZ-IMA is shown in Fig. 2. Compared with the original IMA, the components of TZ-IMA are across worlds: integrity measurement hooks to intercept the user's execution in sensitive applications in the normal world and four security modules to provide integrity measurement and protection services in the secure world.

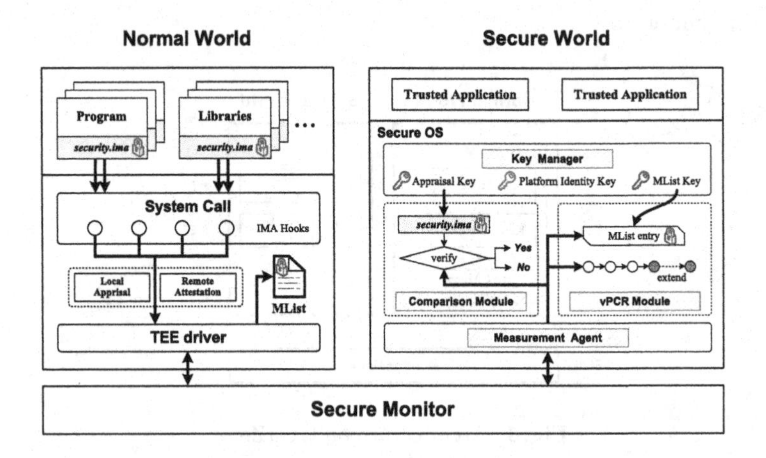

Fig. 2. Architecture of TZ-IMA

TZ-IMA reuses the hooks in the original IMA. These hooks are mainly called by system calls, such as open, execve, mmap, and finit_module. When the process is caught by hooks, the system calls the security modules in the secure world according to the policy. Four security modules are constructed in the secure world. The details are as follows.

Measurement Agent: To ensure that the applications cannot be modified maliciously, TZ-IMA constructs a measurement agent in the secure world. It is used to calculate the hash value of the application when the application is loaded into memory. When the applications newly add or change, the measurement agent is also used to generate the new hash value and the new reference value.

Comparison Module: The comparison module is used for local appraisal. When the comparison module obtains the measurement value from the measurement agent and the reference value from the normal world, it encrypts the measurement value and compares it with the reference value. If these two values do not match, access to the application is blocked.

vPCR Module: To support remote attestation, an MList is constructed in the normal world. The PCR register is virtualized to protect the integrity of the MList. Meanwhile, vPCR encrypts measurement value with MList Key to ensure the confidentiality of MList. A challenger can judge whether the remote system is trusted by validating the vPCR signature and the MList.

Key Manager: The key manager module is responsible for managing three keys used in TZ-IMA. The appraisal key and the MList key are two symmetric encryption keys and used to encrypt the reference value of the application hash and the confidentiality of MList, respectively. The platform identity key (PIK) is a private RSA key used to identify the attesting system's TrustZone. The symmetric keys are generated from the hardware unique key and chip ID, and the RSA key pair is created by a trusted central authority. These three keys are saved in the RPMB partition of eMMC.

Based on the four security modules, TZ-IMA provides two mechanisms (appraisal and attestation) to protect the integrity of applications running on the server. The details will be discussed in Sect. 4 and Sect. 5.

4 Local Appraisal Mechanism

The appraisal mechanism can detect whether the application has been altered accidentally or maliciously by appraising an application's measurement value against the reference value.

4.1 Reference Value Protection

To prevent the application from being tampered with, the reference hash value with public key signatures is always used to verify the integrity of an application. The problem of using public key signatures is that they only apply to unchanged applications [2]. However, many security-critical applications may change in the real system, which greatly limits the usage of IMA. To solve this problem, TZ-IMA uses symmetric encryption algorithm to protect the reference value, and the symmetric encryption key is called appraisal key.

The appraisal key needs to be carefully protected because it is a very sensitive information in the system. At the same time, the comparing logic also needs to be isolated from the complicated execution environment to enhance its security. Therefore, the appraisal key is saved to TrustZone and a comparison module is constructed to compare the measurement value with the reference value in the secure world. Thus, during the whole lifetime of TZ-IMA, the appraisal key used by the comparison module only stays in the secure world, and the appraisal key will never be exposed to the normal world.

Meanwhile, the reference value of a specific application must be quickly retrieved for comparison due to the complexity of the file system. Instead of maintaining a complicated data structure in the secure world, TZ-IMA stores the encrypted reference values in the files' extended attributes of the normal world file system The method not only protects the integrity of the reference value, but also realizes the rapid retrieval with the support of the file system.

4.2 Application Integrity Verification

Before an application is executed, the integrity of application needs to be verified. When the process opens and executes sensitive applications, the query is captured by the hooks, and the measurement agent in TrustZone are called to calculate the current hash value of the application. Then, the normal world OS retrieves the extended attribute value of the application and passes the encrypted reference value to the secure world. After receiving the reference value, the comparison module loads the key saved in the key manager module to decrypt the reference value. If it matches the measurement value, the integrity verification is successful, and the application can be executed. Otherwise, the authentication fails, and access is denied. The workflow of integrity verification is shown in Fig. 3.

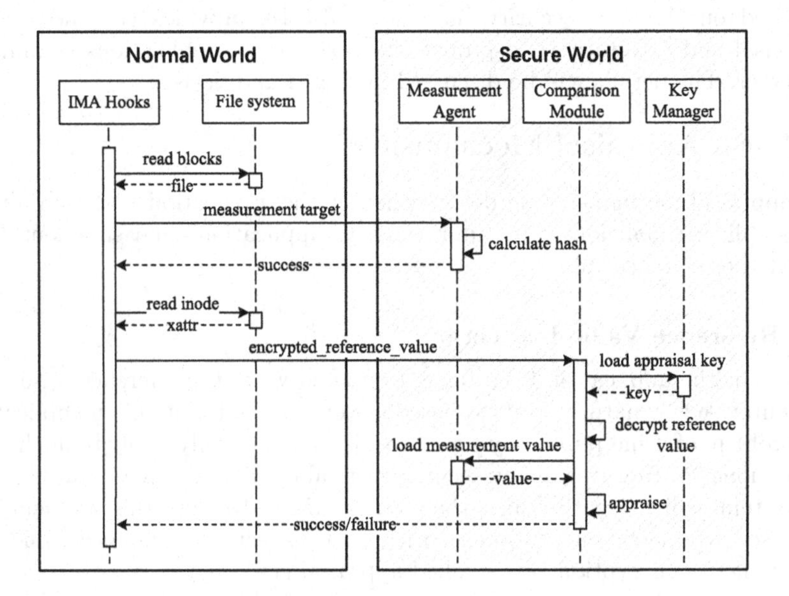

Fig. 3. Workflow of application integrity verification

4.3 Reference Value Generation and Update

When the system is deployed for the first time, a security administrator account is created and the password is protected by the UUID of the device. The security administrator calls the key manager generates an appraisal key in the secure world and calls measurement agent to calculate sensitive application hash. Then, the reference hash value is encrypted by appraisal key and the result is returned to the normal world. The normal world OS saves the encrypted reference value in the extended attribute "security.ima".

TZ-IMA is suitable for applications that often change because the appraisal key is stored in the local system. When the application or configuration file changes, the secure world verifies the identity of security administrator first based on password. Only if the authentication is passed, the measurement agent recalculates the hash value and loads the appraisal key to re-encrypt the new hash value. Finally, the encryption module returns the encrypted hash value, and the OS in the normal world stores the value to the extended attribute in the normal world. The workflow of reference value generation or update is shown in Fig. 4.

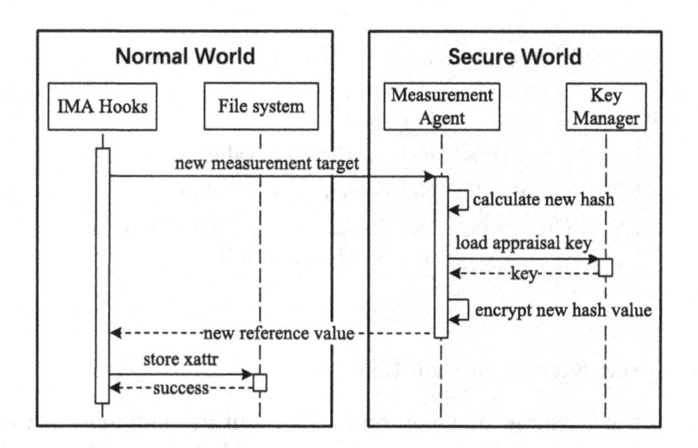

Fig. 4. Workflow of new reference value generation

5 Remote Attestation Mechanism

TZ-IMA attestation aims to provide an MList including the application execution status of the remote system and enable the tenants to detect whether sensitive applications have been modified or executed by malware. The simple MList cannot avoid being tampered by an attacker. In this section, we virtualize PCR Module and encrypt the MList to protect both the integrity and confidentiality of measuring results.

5.1 vPCR Module

Based on TrustZone provided by ARM CPU, TZ-IMA leverages pseudo trusted application to virtualize a vPCR module. vPCR is a 160-bit virtual register, which is set to all 0s when the computer is initialized. After booting, the vPCR value is updated in an "extended" manner, that is, starting from the first component, concatenating the integrity measurement value of the component m_i

with the existing vPCR value, performing a hash operation, and then storing the result back to vPCR (Eq. 1).

$$vPCR_i := Hash\,(\cdots Hash\,(Hash\,(0\,||\,m_1)\,||\,m_2)\cdots||\,m_i) \tag{1}$$

At the same time, vPCR only supports reset and expansion commands, and cannot roll back to the original value at a certain time. The interfaces provided by vPCR module is shown in Table 1. The normal world has no way to modify the value of vPCR because vPCR is saved in TrustZone and can be only called by measurement agent. Thus, vPCR Module can protect the integrity of MList.

Table 1. Interfaces provided by vPCR module.

Interfaces	Function
TEE_vPCR_Extend	Extend the hash value to vPCR
TEE_vPCR_Read	Get current vPCR value
TEE_vPCR_Quote	Use PIK_{priv} to sign the vPCR value and return the result

5.2 Encrypted Measurement List

MList records the workflow of computer system, but its confidentiality cannot be guaranteed because MList is saved in the form of plaintext in the normal world. To solve this problem, TZ-IMA generates an MList Key in the Key Manager and leverages the key to encrypt the measurement value as shown in Eq. 2. The normal world can only see the encrypted result to ensure the confidentiality of MList.

$$MList_i.value := ME.vaule\ xor\ MListKey \tag{2}$$

The workflow of attestation mechanism can be seen from Fig. 5. Before the application is executed, the measurement agent calculates the file hash and extends the measurement value to vPCR module. Then, it encrypts the measurement value with MList Key and returns the result to the normal world OS to store the encrypted file hash. The vPCR value and the MList can be used to prove the integrity status of a remote system. This manner ensures both the integrity and confidentiality of MList and saves the storage space.

5.3 Application Scenario

When a challenger initiates an integrity verification challenge, the system needs to provide an integrity report generated by a trusted entity. Figure 6 depicts the procedure of remote attestation.

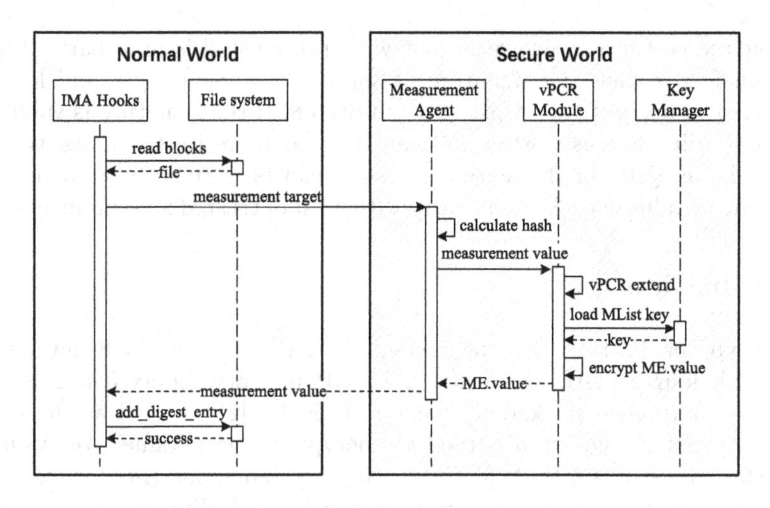

Fig. 5. Workflow of attestation mechanism

In Step 1, the challenger creates a nonpredictable 160-bit nonce and sends a challenge request message to the attesting system. The 160-bit nonce is generated to protect against a replay attack. In Step 2, the attesting system loads a protected RSA private key, PIK_{priv}, into TrustZone. Then, the secure world signs the vPCR value and the nonce provided by the challenger with PIK_{priv}. In Steps 3 and 4, the normal world OS retrieves $sig\{vPCR, nonce\}_{PIK_{priv}}$ and *MList Key* from the trusted OS and the MList from the file system, respectively. Then, the attesting system responds with a challenge response message to the challenger, including $sig\{vPCR, nonce\}_{PIK_{priv}}$ and MList in Step 5.

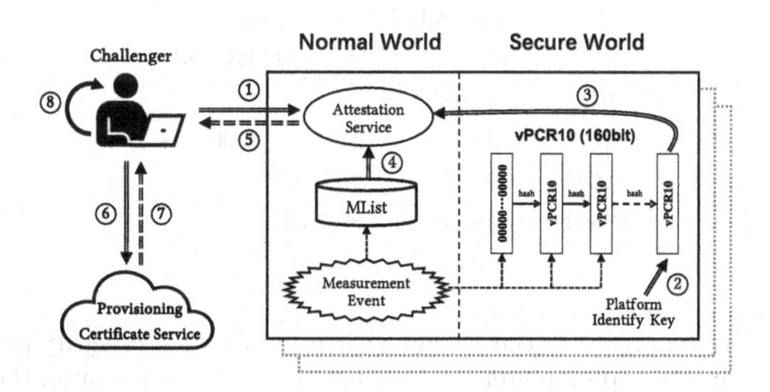

Fig. 6. Procedure of remote attestation

After the challenger receives the response, it first retrieves a trusted certificate of $certPIK_{pub}$ in Steps 6 and 7. The PIK certificate binds the verification key, PIK_{pub}, to a specific system and states that the related secret key is known only to a specific TrustZone. The validity of the certificate must be verified first by checking the certificate revocation list with the trusted issuing party. Finally, the challenger verifies the signature of $sig\{vPCR, nonce\}_{PIK_{priv}}$ and leverages *MList Key* to decrypt the MList entry in Step 8. If the signature is valid, then the vPCR value is trustworthy and can be used to verify the integrity of the MList. The integrity of the attesting system can be verified by comparing the measurement value of each MList entry with a list of trusted measurement values.

6 Evaluation

A prototype of TZ-IMA is implemented on a Phytium ARMv8 development board with four 2.6 GHz cores and a 16 GB memory. Linux 5.4.19 is modified as the normal world kernel, and OP-TEE 3.2 [21] is used as the trusted OS. HMAC-SHA1 algorithm is used to encrypt the hash value. We evaluation the performance of TZ-IMA in three parts: system boot time, system overhead on UnixBench based on macro-benchmark and LMbench based on micro-benchmark.

6.1 Boot Time

The boot time includes the time spent in initializing the kernel and userspace. Table 2 shows the experimental data of boot time on the original system without protection, the system with IMA, and the system with TZ-IMA. The effect of the IMA and TZ-IMA boot time is evaluated in the appraisal scenario and the attestation scenario.

Table 2. System boot time.

	Appraisal	Attestation	
		Measure	Extend
Origin	27.98 s		
IMA	29.27 s	29.41 s	≈90.73 s
TZ-IMA	29.46 s	29.49 s	
Overhead	+0.6%	≈−75.3%	

Original system refers to the system without application integrity protection, whose boot time is approximately 27.98 s. For the appraisal scenario, the default policy is to verify all applications owned by the root. The system verifies approximately 6000 files during boot time. Table 2 shows that IMA takes approximately 29.27 s to boot, and TZ-IMA takes approximately 29.46 s to boot. The overhead of TZ-IMA appraisal is approximately 0.6% higher than IMA appraisal, which is mainly brought by world switching.

For the attestation scenario, TZ-IMA calculates the hash value, generates the MList, and extends the measurement value to vPCR. The default policy is to measure all files read as root and all applications executed. The table above shows that booting takes approximately 29.49 s.

In comparison, the time for TPM PCR extension is much longer. A previous study showed that TPM 1.2 takes approximately 27.4957 ms to perform PCR extension per file [26]. The time spent on PCR extending during boot time can be emulated by Eq. 3:

$$T_{pcr\_extend} = t \times n_{pcr\_extend}, \tag{3}$$

where T_{pcr_extend} is the performance overhead of IMA with PCR extend, t is the duration for a single PCR extension, and n_{pcr_extend} is the times of PCR extension. While booting, IMA attestation measures approximately 3300 files, resulting in approximately 90.73 s of boot time, which is three times greater than booting without IMA. However, based on TrustZone, the overhead of TZ-IMA-attestation brought by the world switching and vPCR extension is only approximately 0.1 s, decreasing by approximately 75.3% compared with IMA appraisal.

Table 3. System performance test result of UnixBench.

No.	Test items	Origin	Appraisal	Δ(%)	Attestation	Δ(%)
1	Dhrystone 2 using register variables	2117.66	2083.78	−1.60	2115.82	−0.09
2	Double+Precision whetstone	819.73	819.32	−0.05	819.70	0.00
3	Excel throughput	747.04	727.26	−2.65	731.88	−2.03
4	File Copy 1024 bufsize 2000 maxblocks	813.83	798.08	−1.94	812.91	−0.11
5	File Copy 256 bufsize 500 maxblocks	590.01	579.46	−1.79	579.95	−1.71
6	File Copy 4096 bufsize 8000 maxblocks	1409.95	1394.68	−1.08	1401.04	−0.63
7	Pipe throughput	725.95	723.72	−0.31	725.21	−0.10
8	Pipe+based context switching	365.11	363.52	−0.44	366.18	+0.29
9	Process creation	629.37	627.38	−0.32	629.99	+0.10
10	Shell scripts (1 concurrent)	825.57	812.98	−1.53	816.28	−1.13
11	Shell scripts (8 concurrent)	1402.46	1364.08	−2.74	1385.14	−1.23
12	System call overhead	442.16	440.60	−0.35	442.39	+0.05

6.2 UnixBench

UnixBench is a tool for testing the performance of Unix systems [27]. The result of performance tests is an index value, which is obtained by comparing the test result of the test system with the test result of a baseline system. Table 3 shows the test results. The influence of TZ-IMA for most test items is less than 1% than the original system, and the performance of two items, "excel throughput" and "shell scripts", decrease by approximately 2%. The result is reasonable because the overhead is mainly caused by frequent program execution. More time is spent on hash calculation and world switching.

6.3 LMbench

LMbench [18] is a set of micro benchmarks for measuring latency and bandwidth of the most basic UNIX or Linux APIs. LMbench is used to test the system call overhead of TZ-IMA. The normalized overhead of runtime performance of the appraisal scenario and the attestation scenario compared with that of the original system is illustrated in Fig. 7.

TZ-IMA mainly influences *open/close*, *mmap*, and *process fork+execve* system calls because TZ-IMA performs security check on these system calls. TZ-IMA does not remarkably affect the performance of *syscall read*, *syscall write*, and *stat*. For most test items, the overhead of a single operation is less than 0.1 μs compared with the original system and can be neglected.

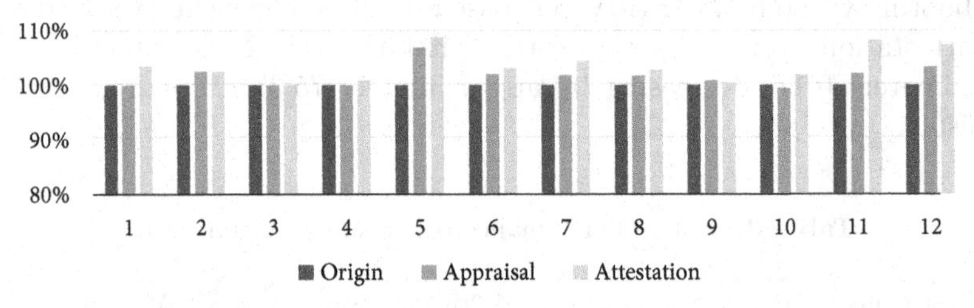

Fig. 7. Performance evaluation of LMbench (normalized syscall latency, lower is better)

7 Security Analysis

7.1 Local Appraisal

Compared with traditional IMA, TZ-IMA saves the reference hash value and the appraisal key in the normal world and the secure world, respectively. The appraisal key is saved to RPMB during offline time, and cannot be stolen or modified by attackers. To ensure the security of encryption key during runtime, TZ-IMA saves the key and transfers the reference value generation, integrity measuring, and the measurement value comparison procedure to the secure world. All appraisal key related logic is implemented in the secure world and prevents the key from being stolen or tampered by attackers during the whole lifetime of the computer. In this way, the secure extended attribute cannot be forged offline because attackers cannot get the appraisal key.

7.2 Remote Attestation

Remote attestation aims to enable challengers to observe the execution flow on the machine, which requires the platform to record the running status of the applications faithfully. The key point to the remote authentication mechanism is to prevent the MList from being forged. While the measurement value is recorded in the MList, the value is also extended to the vPCR in TrustZone to ensure the integrity of the MList. To enhance the confidentiality of MList, we also construct a MList key in the secure world to encrypt the MList entry. In TZ-IMA attestation, the vPCR extension value is stored in the secure world and cannot be modified by attackers. Although the MList in the normal world kernel space my be leaked, attackers cannot get sensitive information because the measurement results are encrypted.

7.3 Security Limitations

For both appraisal and attestation mechanism, the integrity of kernel code may be broken and the hook point check may be bypassed during the run time. This limitation can be fixed by setting the kernel code segment to read-only and depriving the kernel's ability to control certain privileged system functions [4,8]. In this manner, the check of TZ-IMA cannot be bypassed.

As assumed in the threat model, hardware-based attack or side-channel attacks are not considered in TZ-IMA. There may be bugs in trusted kernel [14,17,28]. Attackers may leverage such vulnerabilities to obtain data from TEE and even destroy the REE kernel. Defending against these attacks is not the goal of TZ-IMA. Researchers are focusing on various methods, including stronger isolation and formal verification, to solve these problems.

8 Related Work

Trusted Computing: Trusted Computing aims to make computing more secure through hardware enhancements and associated software modifications. IMA [24] was proposed in 2004, which extends TCG's trustworthiness measurement to the application layer for the first time. However, IMA can only protect the load-time integrity of applications. Davi et al. proposed DynIMA [6] to protect against run-time attack. Son et al. [26] presented batch extend and core measurement, two mechanisms to decrease the overhead of IMA. Luo et al. [16] presented Container-IMA, which enables challengers to authenticate specific containers by dividing measurement logs and introducing a container-based PCR mechanism. Bohling et al. [5] proposed a method to subvert IMA by a malicious block device.

Virtual TPM (vTPM) is a virtualization TPM solution introduced by IBM in 2006 [22], primarily for virtualized server platforms. vTPM mainly solves the problem of how to build a trust root for virtual environments and how to guarantee the security and the portability of the trust root, and provides trustworthy functions such as identity verification, integrity storage, key management, encryption, and decryption for virtual machines. However, the implementation of vTPM has the risk of leaking confidential data, and the performance overhead is relatively high. Wang et al. [32] presented SvTPM based on Intel SGX, another TEE, which prevents cloud tenants and even cloud administrators from obtaining the vTPM private key or any other sensitive data, providing strong isolation protection for vTPM on the cloud. Similarly, Raj et al. [23] leveraged ARM TrustZone in implementing fTPM, which provides security guarantees similar to TPM chips.

ARM TrustZone: TrustZone is usually applied to provide a secure execution environment in GlobalPlatform. SPROBES [8] and TZ-RKP [4] are two TrustZone-based solutions for enforcing the runtime integrity of kernel code. Santos et al. [25] proposed a TrustZone-based system called TLR, which ensures the confidentiality and the integrity of .NET mobile applications. It divides an application into secure and normal parts, and separates it from the OS and other applications. Hua et al. [12] virtualized TrustZone and leveraged TrustZone to build an isolated execution environment for each guest OS. Guan et al. proposed TrustShadow [9], a TrustZone-based system used in IoT that protects existing applications from untrusted OSs. TrustShadow makes full use of TrustZone and separates resources into different worlds. Han et al. [10] used TrustZone to monitor the integrity of IMA code in Linux kernel, while TZ-IMA focused on protecting the integrity of applications. Hua et al. [13] leveraged TrustZone in constructing an isolated execution environment for containers, which can defend against MUMA attack. Based on mandatory access control, Zhang et al. proposed iFlask [31], which protects the integrity of applications but cannot protect against offline attacks. Wang et al. [29] proposed remote attestation schemes for mobile terminals, but this method needs to modify binary programs and it is not suitable for flexible applications update in the cloud computing scenario. Ling et al. [15] also proposed a runtime process integrity verification architecture for ARM IoT nodes.

9 Conclusion

In this paper, the design and implementation of TZ-IMA, an instrumentation mechanism that supports integrity measurement for applications based on ARM TrustZone, is presented. An application integrity verification scheme and a remote attestation mechanism are proposed, which can protect the application integrity for the entire lifetime of the computing platform. The prototype and the evaluation results demonstrate that our architecture can provide better verification of application integrity, and the overhead is only approximately 5%.

Acknowledgment. This work was supported by the National Natural Science Foundation of China [grant numbers U19A2060, 62172431].

References

1. Arm trustzone. https://developer.arm.com/ip-products/security-ip/trustzone
2. An overview of the linux integrity subsystem. https://sourceforge.net/projects/linux-ima/files/linux-ima/Integrity_overview.pdf
3. Amd opteron a1100 (2016). http://www.amd.com/en-gb/products/server/opteron-a-series
4. Azab, A.M., et al.: Hypervision across worlds: real-time kernel protection from the arm trustzone secure world. In: Proceedings of the 2014 ACM SIGSAC Conference on Computer and Communications Security, pp. 90–102 (2014)
5. Bohling, F., Mueller, T., Eckel, M., Lindemann, J.: Subverting linux' integrity measurement architecture. In: Proceedings of the 15th International Conference on Availability, Reliability and Security, pp. 1–10 (2020)
6. Davi, L., Sadeghi, A.R., Winandy, M.: Dynamic integrity measurement and attestation: towards defense against return-oriented programming attacks. In: Proceedings of the 2009 ACM Workshop on Scalable Trusted Computing, pp. 49–54 (2009)
7. Foley, M.J.: Windows server on arm: it's happening. Website (2017). http://www.zdnet.com/article/windows-server-on-arm-its-happening/
8. Ge, X., Vijayakumar, H., Jaeger, T.: Sprobes: enforcing kernel code integrity on the trustzone architecture. arXiv preprint arXiv:1410.7747 (2014)
9. Guan, L., Liu, P., Xing, X., Ge, X., Zhang, S., Yu, M., Jaeger, T.: Trustshadow: Secure execution of unmodified applications with arm trustzone. In: Proceedings of the 15th Annual International Conference on Mobile Systems, Applications, and Services, pp. 488–501 (2017)
10. Han, S., Park, J.: Shadow-box v2: the practical and omnipotent sandbox for arm. Slideshow at Blackhat Asia (2018)
11. Hashizume, K., Rosado, D.G., Fernández-Medina, E., Fernandez, E.B.: An analysis of security issues for cloud computing. J. Internet Serv. Appl. **4**(1), 1–13 (2013)
12. Hua, Z., Gu, J., Xia, Y., Chen, H., Zang, B., Guan, H.: vTZ: virtualizing ARM trustzone. In: 26th USENIX Security Symposium (USENIX Security 2017), pp. 541–556 (2017)
13. Hua, Z., Yu, Y., Gu, J., Xia, Y., Chen, H., Zang, B.: TZ-container: protecting container from untrusted OS with ARM trustzone. SCIENCE CHINA Inf. Sci. **64**(9), 1–16 (2021)
14. Li, W., Xia, Y., Chen, H.: Research on ARM trustzone. GetMobile Mob. Comput. Commun. **22**(3), 17–22 (2019)
15. Ling, Z., et al.: Secure boot, trusted boot and remote attestation for ARM trustzone-based IoT nodes. J. Syst. Architect. **119**, 102240 (2021)
16. Luo, W., Shen, Q., Xia, Y., Wu, Z.: Container-IMA: a privacy-preserving integrity measurement architecture for containers. In: 22nd International Symposium on Research in Attacks, Intrusions and Defenses (RAID 2019), pp. 487–500 (2019)
17. Machiry, A., et al.: Boomerang: exploiting the semantic gap in trusted execution environments. In: NDSS (2017)
18. McVoy, L.W., Staelin, C., et al.: LMbench: portable tools for performance analysis. In: USENIX Annual Technical Conference, San Diego, CA, USA, pp. 279–294 (1996)
19. Morgan, T.P.: ARM servers: Cavium is a contender with ThunderX (2015). https://www.nextplatform.com/2015/12/09/arm-servers-cavium-is-a-contender-with-thunderx/

20. Ning, P.: Samsung Knox and enterprise mobile security. In: Proceedings of the 4th ACM Workshop on Security and Privacy in Smartphones & Mobile Devices, p. 1 (2014)
21. OP-TEE. https://github.com/OP-TEE/
22. Perez, R., Sailer, R., van Doorn, L., et al.: vTPM: virtualizing the trusted platform module. In: Proceedings of the 15th Conference on USENIX Security Symposium, pp. 305–320 (2006)
23. Raj, H., et al.: fTPM: a software-only implementation of a TPM chip. In: 25th USENIX Security Symposium (USENIX Security 2016), pp. 841–856 (2016)
24. Sailer, R., Zhang, X., Jaeger, T., Van Doorn, L.: Design and implementation of a TCG-based integrity measurement architecture. In: USENIX Security Symposium, vol. 13, pp. 223–238 (2004)
25. Santos, N., Raj, H., Saroiu, S., Wolman, A.: Using arm trustzone to build a trusted language runtime for mobile applications. In: Proceedings of the 19th International Conference on Architectural Support for Programming Languages and Operating Systems, pp. 67–80 (2014)
26. Son, J., et al.: Quantitative analysis of measurement overhead for integrity verification. In: Proceedings of the Symposium on Applied Computing, pp. 1528–1533 (2017)
27. UnixBench (2016). https://sourceforge.net/projects/unixbench5/
28. US-CERT/NIST: CVE-2015-4421 in Huawei Mate7 (2015). https://cve.mitre.org/cgi-bin/cvename.cgi?name=CVE-2015-4421
29. Wang, Z., Zhuang, Y., Yan, Z.: TZ-MRAS: a remote attestation scheme for the mobile terminal based on arm trustzone. Secur. Commun. Netw. **2020**, 1–16 (2020)
30. IMAI Wiki: https://sourceforge.net/p/linux-ima/wiki/Home/
31. Zhang, D., You, S.: iFlask: isolate flask security system from dangerous execution environment by using ARM trustzone. Futur. Gener. Comput. Syst. **109**, 531–537 (2020)
32. Zhi, W.Y.Y.: Kernel integrity measurement architecture based on TPM 2.0. Comput. Eng. **44**(3), 166–170 (2018)

FuzzBoost: Reinforcement Compiler Fuzzing

Xiaoting Li[1], Xiao Liu[2], Lingwei Chen[3], Rupesh Prajapati[4], and Dinghao Wu[4](✉)

[1] Visa Research, Palo Alto, CA, USA
xiaotili@visa.com
[2] Meta, Inc., Menlo Park, CA, USA
bamboo@fb.com
[3] Wright State University, Dayton, OH, USA
lingwei.chen@wright.edu
[4] Pennsylvania State University, University Park, PA, USA
{rxp338,duw12}@psu.edu

Abstract. Enforcing the correctness of compilers is important for the current computing systems. Fuzzing is an efficient way to find security vulnerabilities in software by repeatedly testing programs with enormous modified, or fuzzed input data. However, in the context of compilers, fuzzing is challenging because the inputs are pieces of code that are required to be both syntactically and semantically valid to pass front-end checks. Also, the fuzzed inputs are expected to be distinct enough to trigger abnormal crashes, memory leaks, or failing assertions that have not been detected before. In this paper, we formalize compiler fuzzing as a reinforcement learning problem and propose an automatic code synthesis framework called FuzzBoost to empower the input code mutations in the fuzzing process. In our learning system, we incorporate the deep Q-learning algorithm to perform multi-step code mutations in each training episode, and design a reward policy to assess the testing coverage information collected at runtime. By interacting with the system, the fuzzing agent learns to predict code mutation actions that maximizing the fuzzing rewards. We validate the effectiveness of our proposed approach and the preliminary evidence shows that our reinforcement fuzzing method can outperform the fuzzing baseline on production compilers. Our results also show that a pre-trained model can boost the fuzzing process for seed programs with similar patterns.

Keywords: Compilers · Fuzzing · Reinforcement learning

1 Introduction

Compilers are fundamental in the current computing system as they are part of the trust base of the machine. However, they contain bugs and it is non-trivial

X. Li, X. Liu and L. Chen—Work done while at PSU.

C. Alcaraz et al. (Eds.): ICICS 2022, LNCS 13407, pp. 359–375, 2022.
https://doi.org/10.1007/978-3-031-15777-6_20

to verify all the vulnerabilities due to their large codebase. For example, GCC has about 15 million lines of code [27]. Fuzzing is an effective way to find security vulnerabilities in compilers by repeatedly testing the codes with randomly modified, or fuzzed inputs [28]. It plays an important role in quality assurance, software development, and vulnerability assessment over decades [8,9,19,30]. Many existing vulnerabilities are reported by fuzzing techniques [23]. Due to the unlimited search space and limited computing resources, existing fuzzing tools explore efficient strategies in fuzzing program inputs. Especially in the scenario of compiler testing, no one can exhaustively examine the entire input space, or traverse all the possible execution paths of target compilers in practice. Therefore, a variety of strategies are designed based on fuzzing heuristics to prioritize finding interesting inputs to be fuzzed. Such fuzzing heuristics may be a random selection, or trying to maximize a specific goal, such as code coverage [15], execution timeouts, and crashes [35].

Coverage-guided testing is widely adopted by fuzzers [10,33,36], which utilizes code coverage information as the search heuristic to generate new inputs from the fuzz action of a predefined list. These exhaustive bounded searches use domain-specific heuristics and are thereby limited in applicability and scalability. Additionally, they overlook the benefit from *past experiences* in historical mutations and cannot automatically learn the common knowledge that is shared in different input seeds generated during the fuzzing boosting process. Moreover, most coverage-guided frameworks only calculate the rewards/fitness after a single mutation is taken, which yet underestimates the power of a series of mutation combinations. For instance, state-of-the-art mutation-based methods like American Fuzzing Lop (AFL) [36] add newly generated fuzzing programs after one mutation according to defined search heuristics into the seed set for the next round of fuzzing. However, for coverage-guided fuzzing, testing coverage does not increase linearly. In other words, each of these mutations may not improve the testing efficacy incrementally. They can be rejected by lexical or semantic checks in the early stage of compilation. But a trace of mutations may trigger a giant improvement as it may increase the possibility of generating more diverse input programs to enhance the code coverage of compilers.

Faced with these challenges, we formalize compiler fuzzing as a reinforcement learning problem and propose FuzzBoost to integrate the superiority of reinforcement learning to the coverage-guided fuzzing. The design of FuzzBoost is inspired by the fact that fuzzing can be modeled as a learning process with a feedback loop where the model aims to learn the mutation heuristics based on the feedback (reward) from the runtime information for evaluating the quality of current input [5]. Reinforcement learning describes the learning process by an agent interacting with the environment to learn an optimal policy by trial and error. It is usually effective for sequential decision-making problems in natural and social sciences, and engineering [3,29]. Theoretically speaking, the problem of compiler fuzzing can be seen as a problem of program synthesis, the goal of which is to cover more paths, trigger more crashes or memory leaks in compilers' execution traces while compiling new generated codes. Specifically, we model compiler fuzzing as a multi-step decision-making process where a learning task progresses with a feedback loop. The fuzzing agent initially generates new inputs with little knowledge but random heuristics. The compiler iteratively runs with the newly fuzzed input. Based on the feedback of the environment, we capture runtime information gathered from binary instrumentation techniques to evaluate the quality of input seeds according to heuristics we define in our learning cycle.

In this paper, we utilize seed programs from test suites of production compilers (GCC [11] in our research) to evaluate FuzzBoost. To demonstrate the effectiveness of our framework, we also compare it with a baseline fuzzing mechanism used in the system AFL [36], which is a widely-used fuzzing method. AFL applies mutation actions with a uniformly distributed strategy. From the results, FuzzBoost outperforms baseline random fuzzing with a higher coverage improvement on seed programs. Additionally, to better improve the efficiency of FuzzBoost on the fuzzing process, we conduct the experiments on a pre-trained model. As a result, our tool achieves a better fuzzing performance, which means that the fuzzing process can be boosted when we reuse the existing model for new seed programs in compiler fuzzing.

In summary, we make the following contributions:

- We integrate reinforcement learning to the compiler fuzzing problem and design a principled reinforcement fuzzing method to automatically generate new test seeds.
- We define reward functions to optimize the fuzzing goal and use a deep Q-learning algorithm to automatically learn a trace of high-reward mutations for given seeds which extensively leverage the knowledge in prior experiences. Our method is task-agnostic that does not rely on any other fuzzing techniques.
- We implement a prototyping tool called FuzzBoost and analyze real-world compiler fuzzing jobs. We conduct various analytical experiments and results demonstrate its testing efficacy.

2 Overview

Mutation-based fuzzing relies on generating new program inputs by mutating seed programs with heuristics. In the previous method [36], the designed fuzzer performs one-step manipulation on the provided input corpus. Then the fuzzer may select its collection of interesting fuzzed inputs after based on their performance, which is measured by capturing new crashes in the context of black-box fuzzing or capturing new path information in grey- or white-box fuzzing. However, it overlooks the potential of a trace of mutations in generating interesting fuzzed inputs, some intermediate states of which may not be good enough to attract interest or even break the compilation process due to lexical checks in early stages. Therefore, we re-model the problem as a multi-step decision-making problem that gives enough attention to these intermediate states being ignored in previous design models. Specifically, we formally model compiler fuzzing as a Markov decision-making process as described in Fig. 1.

As shown in the figure, in this multi-step decision-making process, there is an input mutation engine M, that performs a fuzzing action a, and subsequently

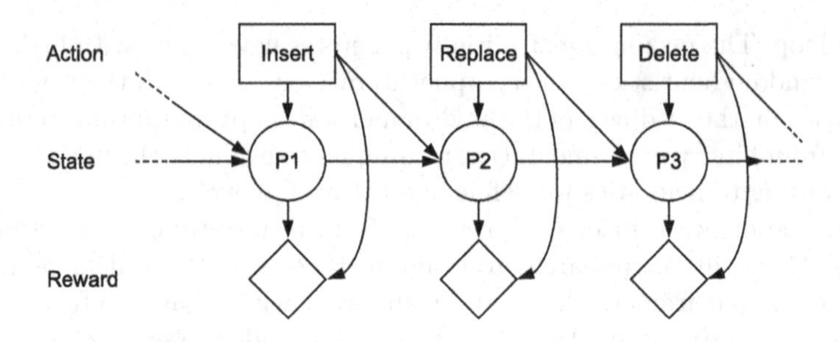

Fig. 1. Compiler fuzzing process

observes a new state s directly derived from the mutated program P_t by exercising the predicted action a on an original seed program P_{t-1}. It means that the input mutation engine predicts the program rewriting actions based on the extracted state from the seed program. After that, the engine can receive a reward r based on performed actions and system state transitions. With the given formalization, it is natural to use the Markov decision process (MDP) to model this problem. Therefore, we define the corresponding T-step finite horizon MDP as $M = (s_1, a_1, r_1, s_2, a_2, ..., s_T)$. Here s_t, a_t, r_t represent the state, action, and reward at time step $t = 1, ..., T-1$, respectively. To achieve the trace of the most effective rewrites for a seed program, we apply reinforcement learning methods [34] to deploy our formalization. Followed by prior footsteps [5,16], we use deep Q-learning algorithm [20,21] to learn the fuzzing engine.

In reinforcement learning, one episode is one complete sequence of actions that starts with an initial state configuration and ends with a terminal state. In the problem of compiler fuzzing, one episode can be formalized as generating a fuzzed program by performing one pre-designed mutation on an existing seed program (initial state), while the learning agent guides the mutation actions that aims to maximize the total reward it receives during the episode. Compared with those conventional mutation-based fuzz testing methods, we adopt the same methodology that using the coverage-guided heuristics to continuously select and generate the desired program generated from the seed set along the episode. The main difference is that, in our design, we *lazy-evaluate* the quality of the fuzzed inputs until it reaches the terminal state. To this end, our fuzzing process contains those intermediate states that might not be syntactically valid but can eventually contribute to high-quality fuzzed inputs.

Before we start the learning process, we randomly initialize a standard deep neural network. In the first episode, State 0 is represented as a program string P preprocessed from a seed program. To reduce the randomness and exhaustive space of mutation, we choose a substring of the whole input program to be our mutation target. Specifically, we extract a substring within a seed program with the window size (length of the substring) w and offset o. By observing this substring, the trained deep neural network automatically predict a mutation action to be taken in the next step. Feasible mutation actions on token-level include *insert* a token, *switch* two or more tokens, *replace* a token, or *change* the window position or size to enable another substring to observe and mutate. Once an action is decided, we run the compiler with the program after performing such a mutation and calculate the reward r of this new program with a record of the execution trace. Subsequently, it moves to the State 1 for further mutations. With the increased number of actions being taken, we deduct the reward by a discounted rate γ which is a value between 0 and 1 to enforce an expected fuzzing trace with fewer mutation actions. We iterate the mutation prediction and evaluation until a *terminal* state is achieved. During the learning process, there are four key elements in this process: *state*, *action*, *environment*, and *reward*.

2.1 State

A state S is a concrete configuration in the environment. As defined in MDP, each process has one state and when the process proceeds, the state updates. In the case of compiler fuzzing, the agent learns to interact with a given seed program. We define the state as a function regarding a given input seed program P. The interaction is performed upon the observation of selected substring within such an input, which is viewed as a series of consecutive token symbols. Formally, let Σ denote a finite set of symbols. The set of possible program inputs I in this language is defined by the Kleen closure $I := \Sigma^*$. For an input program string $P = (p_1, p_2, ..., p_n) \in I$, let

$$S(P) := \{(p_{1+i}, p_{2+i}, ..., p_{m+i}) \mid i \geq 0,\ m + i \leq n\} \tag{1}$$

denote the set of all substrings of P. We define the states of the Markov decision process to be I and I is a union set of $S(P)$. Thus, we have $P \in I$ denotes an input program and $P_0 \in S(P) \subset I$ is a substring of this input seed program. The entire state space of a seed program is $S(P)$, which is theoretically infinite since permutations in this language I can increase after mutation. In other words, the seed program can be converted to any other valid programs.

2.2 Action

Action A is the set of all possible mutation actions that the agent can perform. In most cases, actions are deterministic and should be chosen among a pre-defined list. In compiler fuzzing, we define the set of possible action A of the MDP to be pre-designed rewrite rules on the extracted substrings $S(P_o)$. The rewrite rules are designed in accord with the extracted substring and predicted type. To be specific, we categorize rewrites from two perspectives, i.e., the extracted content and the extraction window, so the agent can predict which type and on which position an action should be performed on the current input.

The rewrites of extracted content are performed on the token-level which include *insertion*, *replacement*, *re-ordering*, *deletion* and *replication*. These rewriting rules conform with the C language lexical requirements. For *insertion*, we append new tokens after the predicted index according to production rules; that is, if the last token is an operator, we randomly sample a token from the existing identifiers as its next. For *deletion*, we delete the token located at the predicted index. For *replacement*, we replace the token at the predicted index with another token randomly sampled from sets of tokens with same characteristic; e.g., if this token is a keyword of C, we select another keyword for replacement. Note, the keyword and operator token set are predefined, while identifier token set is generated by parsing the seed program. For the second type, they are designed to make a change on the extraction windows. Atomic mutations include window *left shift* and *right shift*, and window size *up* and *down* with one token length either from left or right side for each. Each of these actions do not change the input program but motivate the diversity of the extracted substring $S(P_o)$ and covers more states in the mutation space. For both types of mutations, the time step increases until the termination state is triggered on the current episode. We define a *terminate* action to early stop the mutation episode. That is, the mutation agent can proactively terminate a mutation episode while observing an extracted substring.

2.3 Environment

The environment is the world that the agent evaluates each action. The environment takes the current state and action as the input, and then outputs the reward of performing such action and calculates the next state after executing the action. In compiler fuzzing, the environment is the compiler or verifier. To observe more detailed information about the fuzzing efficacy, we develop a tool called FuzzBoost based on program execution traces. In this respect, we record dynamic traces while running the testing production compilers, i.e., GCC, on generated programs. In compiler construction, a basic block of an execution trace is defined as a straight-line code sequence with no branches except for the entry and exit points, which is considered as one of the important atomic units to measure code coverage. In our method, we capture all the unique basic blocks $B(\mathrm{T}_P)$ concerning each execution trace T_P and calculate a store of all the unique basic blocks covered by the existing test suite I' to represent our measure of interest. In our implementation of FuzzBoost, the program execution trace is collected by Pin [18], a widely-used dynamic binary instrumentation tool. Pin provides infrastructures to intercept and instrument the execution trace of a binary. During execution, Pin inserts the instrumentation code into the input program and recompile the output with a Just-In-Time (JIT) compiler. We develop a plug-in of Pin to log the executed instructions. Additionally, we develop another coverage analysis tool based on the execution trace to report all the basic blocks touched so far. It also reports whether new basic blocks are triggered by the fuzzed program and the number of new covered blocks as well. Furthermore, our environment also logs and reports abnormal crashes, memory leaks, or failing assertions of compilers with the assistance of internal errors alarms from the compiling messages.

2.4 Reward

Designing a good reward function to facilitate learning and maintaining the optimal policy is the key goal in our framework. Rewards provide evaluative

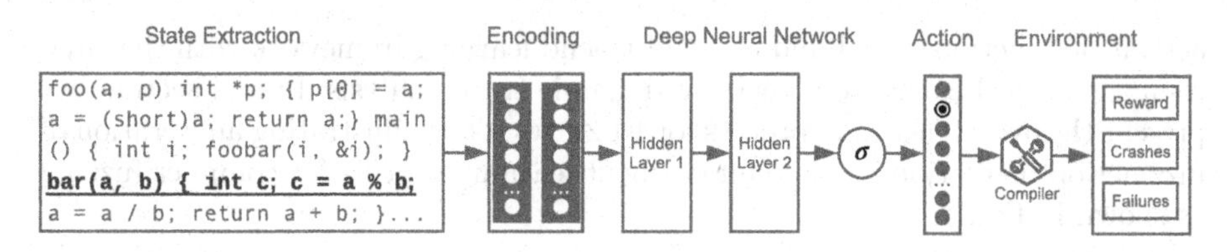

Fig. 2. Fuzz action prediction in the reinforcement learning process of compiler fuzzing

feedback to guide an RL agent to make decisions. However, rewards can be very sparse so that it is challenging for the learning problems. In the game of Go, a reward only occurs at the end of a game. In such cases, the learning process can converge slowly because of the sparse motivations. We solve this challenge by giving every mutation step a reward, so the goal of agent is to maximize the accumulated rewards until one episode terminates at step T,

$$R = \sum_{t=0}^{T} \gamma_t r_{t+1}(P), \tag{2}$$

where $\gamma_t \in (0,1)$ indicates a discount factor to gradually deduct the reward in the future. $r_{t+1}(P)$ represents the reward of generated program P at step $t+1$. In fuzz testing, the possible rewarding heuristics are program coverage, new crashes, timeout, etc. They aim at enlarging the analyzed surface in the target programs being fuzzed and digging into the program traces accordingly that are more suspicious. In compiler fuzzing, we adopt testing coverage as the reward to motivate the learning towards a vulnerability search on more areas of the compiler's code. However, unlike conventional definitions for coverage, which are usually line/function/branch coverage that require expensive computing resources to calculate, we define the reward based on the ratio of unique basic blocks covered by a certain fuzzed seed program P at step t to the entire unique basic blocks of its mutated test suites I' along the episode;

$$r(P, I') := B(\mathrm{T}_P) / \sum_{\rho \in I'} B(\mathrm{T}_\rho), \tag{3}$$

where $B(\mathrm{T}_P)$ is the number of unique basic blocks in the execution trace of a program P and $I' \subset I$ is the programs generated from this test suite. This stepwise reward r is a continuous scalar value that has a range of $(0, 1]$, where 1 is achieved when a specific execution trace covers all the basic blocks that have been tested so far by its existing fuzzed cases. The designed reward motivates the mutation steps towards the training goal: improving the compiler testing coverage by selecting a critical subsequence inside a seed program and enforcing simple mutations in a trace.

3 Designed Framework

To start a deep Q-learning process for compiler fuzzing, we propose FuzzBoost which adopts a deep neural network with two layers connected with non-linear activation functions. We build this end-to-end learning framework with the environment reward calculated based on dynamic trace analysis. In this section, we present the overall learning process for FuzzBoost by illustrating an iteration of fuzz action prediction in the reinforcement learning process for compiler fuzzing as shown in Fig. 2.

3.1 Initialization

We start with an initial input seed $P \in I$, where the choice of P is not constrained but can be any C program even not well-formed ones. We employ the GCC test suite as our sampling pool and randomly selected programs to be our seed inputs. We propose to use a neural network as the Q function to mimic the reasoning for input mutation of compiler fuzzing. This deep neural network maps states (embedding of an extracted substring from seed programs) to Q outputs for all actions A. Due to the lack of heuristics at the very beginning, the neural network is randomly initialized and reinforcely optimize the model parameters θ from the environment feedbacks, i.e., rewards, by maximizing the code mutation rewards in the episode training.

3.2 State Extraction

FuzzBoost observes a substring within a seed program to predict actions to perform. The substring is extracted from the seed program by the customized window and encoded as $S(P)$. In Sect. 2.1, we define the states of our Markov decision process to be $I = \Sigma*$. To be more specific, it is a substring P' at offset $o \in 0, ..., |P| - |P'|$ and of window size $|P'|$. To make the extracted state tractable, we define actions in Sect. 2.2 to shift and resize the window. By performing window-related actions, the fuzzing agent can see the whole program by partially observing fragments consecutively. In other words, FuzzBoost learns to select the most critical piece of code to mutate incrementally during the training process. After the sequence is extracted, we use a word embedding model to abstract the sequence into a fixed-dimensional vector for training.

3.3 Deep Q-Network

We implement the Q-learning module based on Tensorflow [1] 1.14. The deep neural network used for prediction is a forward neural network with two hidden layers connected with non-linear activation functions. The two hidden layers contain 100 and 512 hidden units respectively, and are fully connected with an input layer with 100 units (which is the max window size for input substring) and an output layer with 10 units (which is the size of action space). The goal of the training is to maximize the expected reward. Since the MDP is a finite horizon in our practical design, we adopt a discount rate $\gamma = 0.9$ to address the long-term reward. We set the learning rate $\alpha = 0.001$ to achieve our best-tuned results. We use the decayed epsilon-greedy strategy for exploration in the reinforcement learning iteration, that is, the ϵ value is set up to 1 at the very beginning and decays over time until a min value, 0.01 in our configuration, is reached. In this scenario, with the probability $1-\epsilon$, the agent selects an action $a = argmax_{a'} Q(s_t, a_t)$, which is the estimated optimum by the on-training neural network. In the meanwhile, with probability ϵ, the agent explores any other actions with a uniformly distributed choice within the action space $|A|$.

3.4 Termination

A mutation episode terminates when the agent detects a terminal state. In our design, we define three conditions that may trigger the terminal state of mutating the seed program: (1) the agent executes the "terminate" action from the neural network prediction; (2) the generated program reaches a maximum number of mutation steps; or (3) the agent generates an invalid action that triggers miscellaneous effects during the reward calculation. The first type of termination will cut the program mutation actively by FuzzBoost while the latter two are passively ended with pre-defined policies. Theoretically, the mutation trace can be generated as long as possible to achieve enough diversity. But in practice, to excessively improve the testing efficacy, we empirically set up the mutation trace length to be 20 actions to enforce our agent to learn within the shortest path. To catch the found bugs/vulnerabilities, we log and report abnormal crashes, memory leaks, or failing assertions of compilers with the assistance of internal errors alarms from the compiling messages. Moreover, in our design, all the programs that have achieved higher code coverage are kept to be the seeds and waiting for another round of fuzzing, otherwise removed from the seed pool. Therefore, the agent can still explore the entire language set even with the restricted length of learning traces during an episode. The methodology applied in our mechanism is the same as conventional coverage-guided fuzzing methods but has made mutation traces *longer* in one round (compared with 1 step in conventional fuzzing) and *predictable* by a neural network (compared with purely random in conventional fuzzing).

4 Experiments

In our research, we propose a reinforcement learning framework FuzzBoost that incrementally trains a deep neural network to predict mutation actions on a given seed program to improve the compiler testing coverage effectively. We evaluate the performance of FuzzBoost on a seed input set gathered from the GCC test suites. We randomly sample 20 C programs in the test suite as our benchmark dataset, more specifically, from the *gcc.c-torture* repository. The window size is set to be 50 to extract the substring inputs. We run FuzzBoost for four weeks to test its fuzzing efficacy and compare with the baseline random fuzzing method used in a popular tool (AFL) [36]. We also conduct an empirical analysis on starting the compiler fuzzing with a pre-trained model to investigate if it can boost our process. All measurements are performed on i7-7700T 2.90 Ghz with 12 GB of RAM.

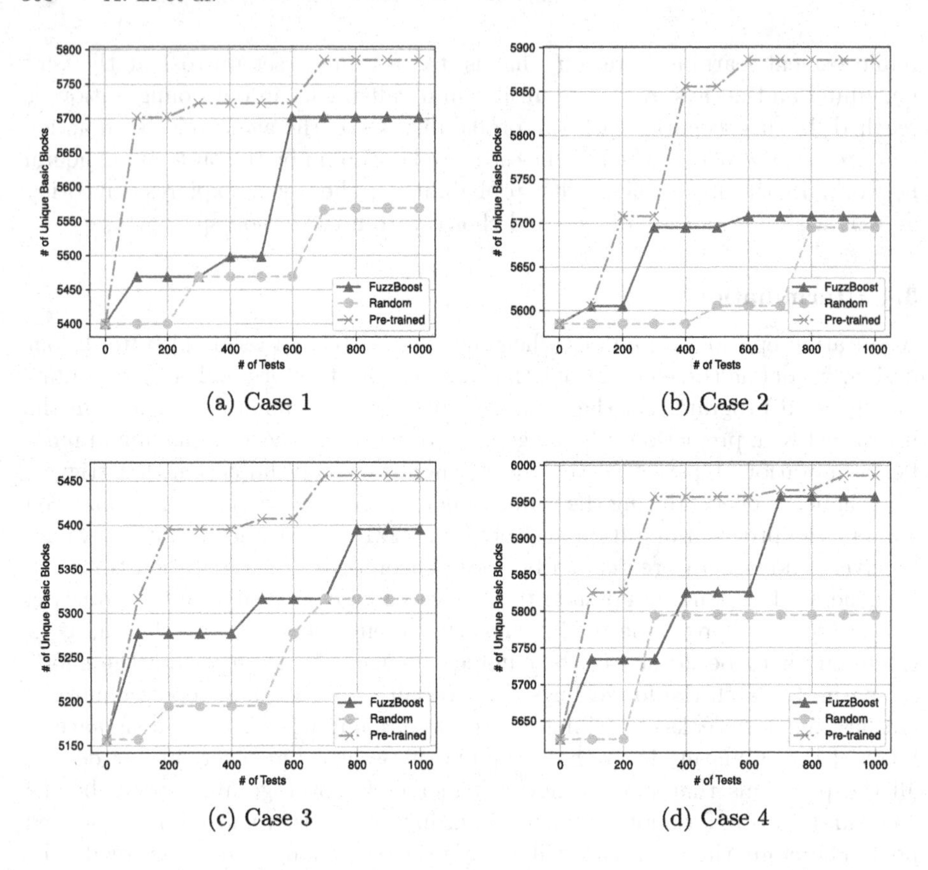

Fig. 3. Number of unique basic blocks covered by generated test suites

4.1 Fuzzing Efficacy

In our design, to improve the efficiency in this end-to-end learning process, we use an approximation of the code coverage improvement to describe the coverage information, which is the accumulated number of unique basic blocks being executed with the generated new test cases. In order to show that FuzzBoost learning algorithm learns to perform high-reward actions given a seed input observation, we compare the improved testing efficacy against a baseline with random action selection policy. The choice of the baseline method is uniformly distributed among the action space A and we terminate the actions with the same methodologies as our method described in Sect. 3.4. Random mutation is widely used in software fuzzing tools [36] which is proven to be effective while a good heuristic, such as coverage-guided, is designed.

Comparison: We perform the experiments with our method FuzzBoost and baseline method Random-based mutation strategy to fuzz each of the programs from the sampling pool. We respectively generate 1,000 new tests from seed programs for both strategies and record the accumulated number of unique basic

Table 1. Coverage improvements with different window size

Window size	50	60	70	80	90	100
Coverage improvement (%)	37.14	36.11	30.29	28.95	28.07	27.94

blocks along the execution trace. On average, our proposed method FUZZBOOST achieves higher testing coverage by 37.14% than the Random-based mutation method in terms of the number of the accumulated unique basic blocks on the seed programs. We randomly select four seed programs and illustrate the coverage improvement of comparisons between baseline method and FUZZBOOST in Fig. 3. The results in each sub-figure represent the number of unique basic blocks that different amount of test programs trigger in the compiler. We can see that FUZZBOOST gradually increases the code coverage as the model being trained to mutate programs more effectively. Our method obviously outperforms the baseline for all cases, among which the most and least improvements, 79.17% (case 1, seed1.c) and 12.24% (case 2, seed2.c) respectively, are achieved. We also observe that FUZZBOOST improves the code coverage with a faster speed than the baseline. We believe this is because our method can learn to fuzz more efficiently and generate interesting test suites with fewer mutation actions.

Window Size: Since the size of each seed program varies, and, arguably, the limited window size may restrict the diversity of mutation trace and thus put a constraint on exploring the entire seed program. As a result, the seed program cannot be thoroughly observed or mutated accordingly after one episode of fuzzing. In this part, we analyze the impact of the current framework with different window sizes on model effectiveness. We increase the initial window size $w = |P'|$ from 50 to 100 and measure the average coverage improvement to compare against the baseline strategy on seed1.c as the seeds in sample pool are generally short. Table 1 shows the experimental results. We can see the coverage improvement decreases while increasing the window size of the initially extracted substring. That is, smaller substrings are better to start with and to mutate the program than larger ones in our method. Our interpretation is that small windows narrow down the mutation space and thus reduce the action randomness, which may increase the possibility of learning a high-quality mutation trace for the model, especially when the model is highly under-trained in the beginning stage. It also indicates that our model is trained to learn better moves of small windows and accordingly select better action to improve coverage. Also, it is worth noting that the ultimate goal for fuzz testing is not the exploration of entire programs, but making control-flow changes within limited observations to boost the fuzzing process.

End State: We set up the compiler fuzzing as an end-to-end reinforcement learning framework. Unlike the problem of Go, the end state of FUZZBOOST is not deterministic in all cases. In our design, we hard-code a limit on the length of mutation traces from the computation cost point of view, but theoretically, the traces can be endless to gain enough randomness and achieve the higher

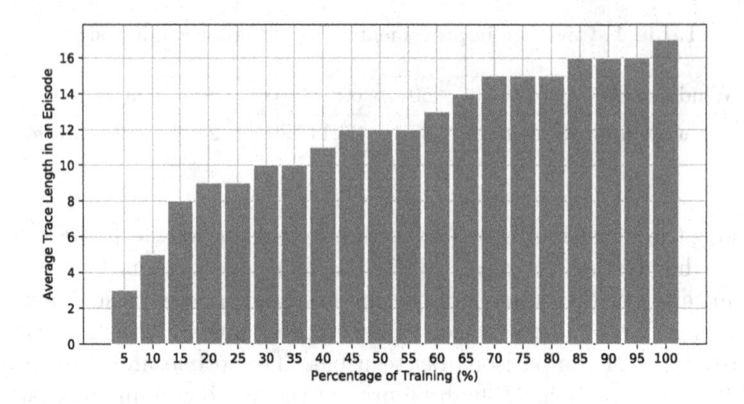

Fig. 4. Mutation length during training

reward. In the process of optimization, we provide the learning agent an action to actively terminate the episode which varies across the learning stage. Thus, to analyze how the end state evolves, we record the distribution of mutation trace lengths under different training stages. Figure 4 presents the average trace length distributions along the learning process over all training seeds. From the result, we can see that, as the training goes on, mutation trace lengths are increasing gradually. In this respect, the reward expectation of learned mutation actions are positive in a form that reinforces the model to dig more mutation opportunities in one episode to maximize the fuzzing reward.

4.2 Boosting with Pre-training

Our trained fuzzing tool learns to constantly accumulate the prior experience by training on the seed programs. This naturally lead us to the question for the sake of resource cost: given an agent which is pre-trained on seed programs $P_{train} = p_i \sim P$, can it improve testing efficiency than learning from scratch? To answer that, we use the same experimental setting as the experiments in Sect. 4.1 and reuse the seed programs from the initial 20 seed programs and craft another 9 α-equivalent programs for each seed respectively. We call a program P' is α-equivalent to program P when we only perform bound variable renaming on P. We randomly pick 80% of them to serve as P_{train} to learn an agent and the rest 20% are used for P_{test}. After pre-training on P_{train}, we save the model and use it on P_{test} to continue the trial-and-error reinforcement compiler fuzzing.

The fuzzing results under such a pre-trained model are shown in Fig. 3 and compared with the performance of FuzzBoost learned with an initially arbitrary model. The coverage improvement for the case of pre-trained model increases drastically towards the highest coverage against the one trained from scratch despite the small improvements in one of the seed programs (case 4). In addition, as the training goes on, the pre-trained model can find useful action in mutation space more quickly and generate fuzzed programs with high testing coverage.

```
1  foo (a, p)
2       int *p;
3  { p[0] = a;
4    a = (short) a;
5    return a;
6  }
7  main () {
8    int i;
9
10   foobar (i, &i);
11
12
13 }
14 foobar (a, b) {
15   int c;
16   c = a % b;
17   a = a / b;
18   return a + b;
19 }
```

Listing 1.1. Original

```
1  foo (a, p)
2       int *p;
3  { p[0] = foobar(a,p);
4    p = (short) a;
5    return a;
6  }
7  main () {
8    int i;
9    for (int a=8; a>0; a--) {
10     foobar (i, &i);
11   }
12   foobar(i, &i);
13 }
14 foobar (a, b) {
15   int c;
16   c = a % b;
17   a = c / b;
18   return a + b;
19 }
```

Listing 1.2. Mutated

4.3 Mutation Example

In this part, to demonstrate how effective FuzzBoost can achieve in program mutations for compiler fuzzing, we showcase the topmost utilized mutations in the following example. We present an original seed (on the left) and its corresponding new generations after mutations (on the right). We discuss each of these abstracted edits involved in the trace of atomic mutations. These edits help explain what is learned by the model during the reinforcement learning process. It should be noted that these mutations are not accomplished within one episode, while we just use this one example to illustrate what the most used mutations are and how they look like.

Example: By observing the results, we find: (1) the top most chosen mutation is *insertion*. Usually, the fuzzing engine tries to insert statements with keywords that do not exist in the original seed file. As shown in *line 9* to *line 11* in the *mutated file*, the fuzzing engine tries to insert a `for` statement into the seed file. By inserting these non-existing tokens, the compiler should execute the lexical analysis in a way that has not been used before; (2) the second chosen mutation is *replication* that the fuzzing engine tries to replicate statements locally as shown in *line 12* in the *mutated file.* The replication will trigger the compiler to optimize code which will improve the testing coverage; (3) the third chosen mutation is *replacement* that can replace a variable (`a`) with a function call (`foobar(a,p)`) as in *line 3* or replace a variable (`a`) with another existing variable (`p`). The replacement either makes the statement more complex to parse, causes exception handlings such as typecast, or changes the control-flow of the seed file, all of which will make the compilation different from the original paths, thereby increasing the testing coverage.

5 Discussion

It is critical to compare with related works, but we find it difficult to perform apple-to-apple comparisons. For instance, generation-based fuzzing tools, such as DeepSmith [7] and Learn&Fuzz [12], craft new programs from scratch other than mutating seed programs while our tool is built on mutation-based fuzzing that rely on seed programs to achieve the whole-program validity. Moreover, some previous methods [7,12] generate a bunch of new programs which usually get rejected at an early stage in compilation and therefore leads to a inefficient and shallow testing procedure. AFL [36] can generate new fuzzed inputs in a very fast way as it only conducts one-step random mutation on seed programs each time. However, it does not suit for compiler fuzzing because its mutation mechanism deals with random changes on inputs without considering their structure context. Compiler requires highly-structured and syntax-aware inputs, so we only compare our tool with its mutation heuristic in the paper. For NEUZZ [26], it is grey-box fuzzing that relies on the coverage analysis on target applications. But for compiler testing, the computation cost for code edge coverage is very high, and that is why we use $\#BasicBlocks$ tested as an approximation.

In this work, we do not claim our tool is better than others. Instead, we reveal our insight of leveraging the superiority of reinforcement learning for compiler fuzzing to efficiently solve a multi-step mutation-based fuzzing problem. In our mechanism, we lazy-evaluate the mutation results and consider those intermediate states in the mutation traces to explore code coverage in a deeper way. Our designed rewriting rules in mutation actions incorporate the structure context of programs, thus our fuzzed inputs can better conform with the syntax requirements of programming languages. What's more, the mutations can realize the comprehensive search in the large space to iteratively guiding the tool for the final fuzzing goals. Our experimental results and analysis comprehensively demonstrate the effectiveness of our compiler fuzzing tool.

6 Related Work

Our study is related to deep reinforcement learning and mutation-based fuzzing.

Deep Reinforcement Learning: Despite the popularity in solving the game of Go, reinforcement learning is also widely adopted as a powerful technique for program synthesis [2,5,13,16,17,22,26]. Bunel et al. [6] perform reinforcement learning on top of a supervised model with an objective that explicitly maximizes the likelihood of generating semantically correct programs. Böttinger et al. [5] use a deep Q-learning network to learn a grammar description for inputs to perform generation-based fuzzing. Researchers also propose Neurally Directed Program Search (NDPS) [31], for solving the challenging non-smooth optimization problem of finding a programmatic policy with maximal reward. Existing projects that adopt deep reinforcement learning for program synthesis focus on semantic goals toward synthesis tasks. Our target is to generate source programs that are well-formed but contain different syntactic features, which are similar to the work from Böttinger et al. [5] that aims at PDF parser fuzzing. But differently, in our design, we consider the improvement of testing coverage of compilers as the reward for reinforcement learning.

Mutation-Based Fuzzing: Mutation-based fuzzing contains two important decisions: 1) where to mutate, and 2) what new value to use for the mutation [24]. Generally mutation-based fuzzers are not aware of the expected input format or specifications, and they cannot select mutations very wisely [25]. It generates new inputs by blindly modifying the provided seeds. A well-known fuzzer that is mutation-based is called AFL [36] which randomly mutates seed inputs and incrementally add new seeds into the set with respect to defined heuristics. Several boosting techniques are proposed to improve the efficiency of mutation-based fuzzing. AFLFast [4] boosts up original AFL fuzzer by focusing on low-frequency paths that allow the fuzzer to explore more paths with limited time. Skyfire [32] applies grammar in existing seed inputs for fuzzing programs that take highly-structured inputs. Kargen and Shahmehri [14] perform mutations on the machine code instead of on a well-formed input to produce high-coverage inputs. DeepFuzz [17] utilizes an RNN-based model to generate new well-formed C programs for compiler fuzzing based on existing testsuites. In this paper, our method boosts the mutation process by using a deep neural network to predict the mutation without any training data.

7 Conclusion

In this paper, we propose FuzzBoost, a deep reinforcement learning framework to fuzz off-the-shelf compilers by generating new programs with coverage-guided dynamics. Our proposed end-to-end learning framework learns to select a trace of best mutation actions in each round towards high code coverage and performs automatically without any human supervision. It improves the testing coverage on a seed set from the GCC test suites and outperforms the baseline fuzzing agent with a random selection strategy. Moreover, we demonstrate that a pre-trained agent in our framework can generalize the strategy to new seed instances to expedite the fuzzing process, which is much faster than starting from scratch.

Acknowledgement. We gratefully acknowledge the support of NVIDIA Corporation with the donation of the Titan Xp GPU used for this research. This research was supported in part by the National Science Foundation (NSF) grant CNS-1652790 and the Office of Naval Research (ONR) grant N00014-17-1-2894.

References

1. Abadi, M., et al.: TensorFlow: a system for large-scale machine learning. In: 12th USENIX Symposium on Operating Systems Design and Implementation (OSDI 2016), pp. 265–283 (2016)

2. Becker, S., Abdelnur, H., State, R., Engel, T.: An autonomic testing framework for IPv6 configuration protocols. In: Stiller, B., De Turck, F. (eds.) AIMS 2010. LNCS, vol. 6155, pp. 65–76. Springer, Heidelberg (2010). https://doi.org/10.1007/978-3-642-13986-4_7
3. Bertsekas, D.P., Tsitsiklis, J.N.: Neuro-dynamic programming: an overview. In: Proceedings of the 34th IEEE Conference on Decision and Control, Piscataway, NJ, pp. 560–564. IEEE Publications (1995)
4. Böhme, M., Pham, V.T., Roychoudhury, A.: Coverage-based greybox fuzzing as Markov chain. IEEE Trans. Softw. Eng. **45**(5), 489–506 (2017)
5. Böttinger, K., Godefroid, P., Singh, R.: Deep reinforcement fuzzing. In: 2018 IEEE Security and Privacy Workshops (SPW), pp. 116–122. IEEE (2018)
6. Bunel, R., Hausknecht, M., Devlin, J., Singh, R., Kohli, P.: Leveraging grammar and reinforcement learning for neural program synthesis. arXiv preprint arXiv:1805.04276 (2018)
7. Cummins, C., Petoumenos, P., Murray, A., Leather, H.: Compiler fuzzing through deep learning. In: Proceedings of the 27th ACM SIGSOFT International Symposium on Software Testing and Analysis, pp. 95–105. ISSTA (2018)
8. Duran, J.W., Ntafos, S.: A report on random testing. In: ICSE, pp. 179–183 (1981)
9. Duran, J.W., Ntafos, S.C.: An evaluation of random testing. IEEE Trans. Softw. Eng. **SE-10**(4), 438–444 (1984)
10. Gan, S., et al.: CollAFL: path sensitive fuzzing. In: 2018 IEEE Symposium on Security and Privacy, pp. 679–696. IEEE (2018)
11. GCC, The GNU Compiler Collection. gcc.gnu.org (2019). http://gcc.gnu.org/
12. Godefroid, P., Peleg, H., Singh, R.: Learn&Fuzz: machine learning for input fuzzing. In: Proceedings of the 32Nd IEEE/ACM International Conference on Automated Software Engineering, ASE 2017, pp. 50–59. IEEE Press (2017)
13. Kaelbling, L.P., Littman, M.L., Moore, A.W.: Reinforcement learning: a survey. J. Artif. Intell. Res. **4**, 237–285 (1996)
14. Kargén, U., Shahmehri, N.: Turning programs against each other: high coverage fuzz-testing using binary-code mutation and dynamic slicing. In: Proceedings of the 10th Joint Meeting on Foundations of Software Engineering, pp. 782–792. ACM (2015)
15. Kifetew, F.M., Tiella, R., Tonella, P.: Combining stochastic grammars and genetic programming for coverage testing at the system level. In: Le Goues, C., Yoo, S. (eds.) SSBSE 2014. LNCS, vol. 8636, pp. 138–152. Springer, Cham (2014). https://doi.org/10.1007/978-3-319-09940-8_10
16. Li, X., Liu, X., Chen, L., Prajapati, R., Wu, D.: ALPHAPROG: reinforcement generation of valid programs for compiler fuzzing. In: Proceedings of the Thirty-Fourth Annual Conference on Innovative Applications of Artificial Intelligence (IAAI-2022) (2022)
17. Liu, X., Li, X., Prajapati, R., Wu, D.: DeepFuzz: automatic generation of syntax valid C programs for fuzz testing. In: Proceedings of the 33rd AAAI Conference on Artificial Intelligence (2019)
18. Luk, C.K.: Pin: building customized program analysis tools with dynamic instrumentation. In: Proceedings of the 2005 ACM SIGPLAN Conference on Programming Language Design and Implementation, pp. 190–200 (2005)
19. Miller, B.P., Fredriksen, L., So, B.: An empirical study of the reliability of UNIX utilities. Commun. ACM **33**(12), 32–44 (1990)
20. Mnih, V., et al.: Playing Atari with deep reinforcement learning. arXiv preprint arXiv:1312.5602 (2013)

21. Mnih, V., et al.: Human-level control through deep reinforcement learning. Nature **518**(7540), 529 (2015)
22. Rajpal, M., Blum, W., Singh, R.: Not all bytes are equal: neural byte sieve for fuzzing. arXiv preprint arXiv:1711.04596 (2017)
23. Rash, M.: A collection of vulnerabilities discovered by the AFL fuzzer (AFL-fuzz) (2019). https://github.com/mrash/afl-cve
24. Rawat, S., Jain, V., Kumar, A., Cojocar, L., Giuffrida, C., Bos, H.: VUzzer: application-aware evolutionary fuzzing. In: NDSS, vol. 17, pp. 1–14 (2017)
25. Saavedra, G.J., Rodhouse, K.N., Dunlavy, D.M., Kegelmeyer, P.W.: A review of machine learning applications in fuzzing. arXiv preprint arXiv:1906.11133 (2019)
26. She, D., Pei, K., Epstein, D., Yang, J., Ray, B., Jana, S.: NEUZZ: efficient fuzzing with neural program smoothing. In: 2019 IEEE Symposium on Security and Privacy, pp. 803–817. IEEE (2019)
27. Sun, C., Le, V., Zhang, Q., Su, Z.: Toward understanding compiler bugs in GCC and LLVM. In: ISSTA, pp. 294–305 (2016)
28. Sutton, M., Greene, A., Amini, P.: Fuzzing: Brute Force Vulnerability Discovery. Pearson Education, London (2007)
29. Sutton, R.S., Barto, A.G.: Reinforcement Learning: An Introduction. MIT Press, Cambridge (2018)
30. Takanen, A., Demott, J.D., Miller, C., Kettunen, A.: Fuzzing for Software Security Testing and Quality Assurance. Artech House, Norwood (2018)
31. Verma, A., Murali, V., Singh, R., Kohli, P., Chaudhuri, S.: Programmatically interpretable reinforcement learning. arXiv preprint arXiv:1804.02477 (2018)
32. Wang, J., Chen, B., Wei, L., Liu, Y.: Skyfire: data-driven seed generation for fuzzing. In: 2017 IEEE Symposium on Security and Privacy, pp. 579–594 (2017)
33. Wang, M., et al.: SAFL: increasing and accelerating testing coverage with symbolic execution and guided fuzzing. In: Proceedings of the 40th International Conference on Software Engineering: Companion Proceedings, pp. 61–64. ACM (2018)
34. Watkins, C.J., Dayan, P.: Q-learning. Mach. Learn. **8**(3–4), 279–292 (1992)
35. You, W., Liu, X., Ma, S., Perry, D., Zhang, X., Liang, B.: SLF: fuzzing without valid seed inputs. In: Proceedings of the 41st International Conference on Software Engineering, ICSE (2019)
36. Zalewski, M.: American fuzzy lop (2014)

Secure Boolean Masking of Gimli

Optimization and Evaluation on the Cortex-M4

Tzu-Hsien Chang[1,4], Yen-Ting Kuo[5], Jiun-Peng Chen[3,4](✉), and Bo-Yin Yang[2,4]

[1] Cybersecurity Center of Excellence Program, National Applied Research Laboratories, Taipei, Taiwan

[2] Institute of Information Science, Academia Sinica, Taipei, Taiwan
by@crypto.tw

[3] Research Center for Information Technology Innovation, Academia Sinica, Taipei, Taiwan
jpchen@ieee.org

[4] Department of Electrical Engineering, National Taiwan University, Taipei, Taiwan
r10921a19@ntu.edu.tw

[5] University of Tokyo, Tokyo, Japan

Abstract. Gimli is a highly secure permutation with high performance across a broad range of platforms. However, side-channel analysis poses a threat to the Gimli without any masking protection. To resist side-channel analysis, the current state of the art of Boolean masking in software proposes an efficient scheme of bitwise logic operations. In practice, a software implementation of masked Gimli may leak information due to pipeline registers and also due to other effects. To avoid unintentional leakage, costly overheads are required, such as more randomness and higher-order share implementation. For our implementation, we present two efficient optimal masked Gimli implementations for the ARM Cortex-M4 on the STM32F407 Discovery(a common Cortex-M4 board) and evaluate their security using TVLA. In 3-shared scenarios, our approach performs with high security with a t-statistic value bounded by a threshold of 4.5 standard deviations, which implies that leakage information cannot be detected. Furthermore, our results promise significant performance improvement for the implementation on Cortex-M processors, with a reduction of the amount of overhead for masking by 61% and 76% for 2 and 3 shared scenarios, respectively.

Keywords: Gimli · ARM Cortex-M4 · Threshold implementation · DPA

1 Introduction

In several emerging areas (e.g. sensor networks, healthcare, distributed control systems, or the Internet of Things), highly resource-constrained devices are interconnected, typically communicating wirelessly with one another, and working in

C. Alcaraz et al. (Eds.): ICICS 2022, LNCS 13407, pp. 376–393, 2022.
https://doi.org/10.1007/978-3-031-15777-6_21

concert to accomplish some task. Because the majority of current cryptographic algorithms were designed for desktop/server environments, many of these algorithms do not fit into these devices.

GIMLI [6], a high secure permutation with high performance across a broad range of platforms, is suitable for use in constrained environments [20]. Ciphers using the three operations + (add), $\lll$ (rotation) and XOR are usually called ARX ciphers, and these include SPECK [5] and ChaCha20 [7]. All of GIMLI, SPECK, and ChaCha20, can be attacked by side-channel. Side-channel analyses are physical attacks based on the exploitation of the information (typically time, power consumption, or electromagnetic radiation), which can be measured while the cryptographic algorithm is operating on the device. Differential power analysis (DPA) [21] and Correlation Power Analysis (CPA) [11] based on power consumption or electromagnetic radiation, have received significant attention since it is very powerful and does not usually require detailed knowledge of the target device to be successfully implemented.

In ARX ciphers, Addition without protection is vulnerable to DPA [21,31]. Furthermore, early work on masking for addition in software has a high-performance overhead [18]. GIMLI, replacing addition $a+b$ with a similar bitwise operation $a \oplus b \oplus ((c \wedge b) \ll 1)$ is suitable for implementing a DPA resistant cryptographic algorithm. There are different papers discussing attack and resistance of GIMLI, such as [15,16]. However, there have not been many papers in the past discussing GIMLI's resistance to SCA in optimized software implementations.

Threshold Implementation (TI) is a masking scheme based on secret sharing and multi-party computation [23–25]. TI is fairly simple to apply to a wide range of ciphers [17,18], and its implementation is not very error-prone if a known set of requirements and best practices is followed. Though early works on masking suggested using two-share TI to reduce the size of the sequential logic in hardware implementations [12,28], however, it is vulnerable to implement two-share TI for most microprocessors in practice. For example, Cortex-M3 and Cortex-M4 pipeline register leak information about Hamming distance between the current operand value and the previous one [13]. Since these problems also appear in most of the software implementation of Boolean masking, implementing a secure masking algorithm is very challenging.

Therefore, how to make Gimli have good execution efficiency in Cortex-M3 and Cortex-M4, resist Side-Channel Attacks, and evaluate the protective effect, all need to be considered.

1.1 Our Contributions

In this paper, we present an efficient and high-security level method for masked GIMLI without leakage due to pipeline registers in embedded software applications. We investigate how to implement the TI for the non-linear layer of GIMLI, finding possibilities to optimize the instruction count for ARM implementations.

First of all, we implement the GIMLI for ARM Cortex-M3 and Cortex-M4 processors and optimize it on the assembly level by using ARM features such as the flexible second operand and by minimizing the number of memory operations.

Second, the Threshold Implementation of GIMLI to resist Side-Channel Attacks, two-share and three-share masking are presented, respectively. For two-share masking, we construct masked non-linear layers of Gimli based on SecAND and SecOR [10], which do not consume entropy and could utilize the flexible second operand to reduce the cycle count. For three-share masking, we are inspired by *changing of the guard* [14], which is a generic method to generate the threshold implementation scheme.

Finally, we use the Test Vector Leakage Assessment (TVLA) method to evaluate the t-test score of the threshold implementation on Cortex-M4. These inspection methods follow the ISO/IEC 17825 [1]. To reduce leakage due to pipeline registers, we rearrange the parallel instruction of two-share TI. However, it is hard to limit the t-value to a threshold of 4.5 standard deviations. This problem could be more severe on other platforms. On the other hand, the statistical value of our three-share TI is inside the ± 4.5 [30] interval for every point in time.

As mentioned above, we qualify DPA-resistant software implementations and prove that our three-share TI is uniform without additional randomness by the reversible property of Gimli. The method of proof about uniformity can also generate a high secure three-share TI of NORX family ciphers [3] because of their similar structure. Furthermore, the three-share TI is a better choice in the strictly secured scenarios, since it can prevent information leakage due to the architecture of microprocessors. These different architectures are a trade-off between performance and security to perform the full permutation.

2 Preliminaries

2.1 GIMLI

GIMLI is a 384-bit permutation designed to achieve high security with high performance across a broad range of platforms. In this paper, we focus on the GIMLI-CIPHER, which performs Authenticated Encryption with Associated Data (AEAD).

The GIMLI Permutation. The GIMLI permutation applies a sequence of rounds to a 384-bit state. We denote by $W = \{0,1\}^{32}$ the set of bit-strings of length 32. We will refer to the elements of this set as *words*; The state is represented as a 3×4 matrix of words $W^{3\times 4}$; the rows are named x, y, z; the columns are enumerated by $0, 1, 2, 3$; the round number is denoted by r. For example, x_1^8 denotes the second 32-bit word before the execution of round 8. Finally, we use

- $a \oplus b$ to denote a bitwise exclusive or (XOR) of the values a and b,
- $a \wedge b$ for a bitwise logical and of the values a and b,
- $a \vee b$ for a bitwise logical or of the values a and b,
- $a \lll k$ for a cyclic left shift of the value a by a shift distance of k, and
- $a \ll k$ for a non-cyclic shift (i.e., a shift that is filling up with zero bits) of the value a by a shift distance of k.

Algorithm 1 describes how this state is permuted in 24 rounds: a non-linear layer starts with round 24 and ends with round 1. During each round, the state is first substituted and permuted (SP-Box). Every second round, the state is mixed linearly (alternating between a "small" or "big" swap). Finally, in every fourth round, a constant is added.

Algorithm 1: The GIMLI permutation

input : $s = (s_{i,j}) \in W^{3\times4}$
output: GIMLI$(s) = (s_{i,j}) \in W^{3\times4}$
for $r = 24$ **down to** 1 **do**
  for $j = 0$ **to** 3 **do**
    $x \leftarrow s_{0,j} \lll 24$ ▷ SP-box
    $y \leftarrow s_{1,j} \lll 9$
    $z \leftarrow s_{2,j}$
    $s_{2,j} \leftarrow x \oplus (z \ll 1) \oplus ((y \wedge z) \ll 2)$
    $s_{1,j} \leftarrow y \oplus x \oplus ((x \vee z) \ll 1)$
    $s_{0,j} \leftarrow z \oplus y \oplus ((x \wedge y) \ll 3)$
  end
  if r *mod* $4 = 0$ **then**
    $s_{0,0}, s_{0,1}, s_{0,2}, s_{0,3} \leftarrow s_{0,1}, s_{0,0}, s_{0,3}, s_{0,2}$ ▷ Small-Swap
  else if r *mod* $4 = 2$ **then**
    $s_{0,0}, s_{0,1}, s_{0,2}, s_{0,3} \leftarrow s_{0,2}, s_{0,3}, s_{0,0}, s_{0,1}$ ▷ Big-Swap
  end
  if r *mod* $4 = 0$ **then**
    $s_{0,0} \leftarrow s_{0,0} \oplus \texttt{0x9e377900} \oplus r$ ▷ Add constant
  end
end
return $(s_{i,j})$

2.2 Threshold Implementation

Threshold Implementation [25] (TI) is a special case of Boolean masking. Even in the presence of glitches, it has been proven secure against first-order differential power analysis for digital circuits. The advantages are that it does not need fresh random values after every non-linear transformation, unlike traditional masking methods.

Definition. TIs use shares with the following properties: correctness, incompleteness, and uniformity:

Correctness states that applying the sub-functions to a valid shared input must always yield a valid sharing of the correct output.
$\mathbf{z} = \bigoplus_{i=1}^{d} z_i = \bigoplus_{i=1}^{d} f_i(x, y, ...) = f(\mathbf{x}, \mathbf{y}, ...)$

Non-Completeness requires sub-functions f_i of a shared function F to be independent of at least one input share for first-order SCA resistance. That is, a function $F(x, y, ...)$ shall be split into sub-functions $f_i(x_{j \neq i}, y_{j \neq i}, ...)$. This requirement was updated in [9] to require any d sub-functions to be independent of at least one input share to achieve d-th order SCA resistance. Non-completeness ensures that the final circuit is not affected by glitches. Since glitches can only occur in sub-functions f_i, and each sub-function has insufficient knowledge to reconstruct a secret state (since it has no knowledge of at least one share x_i), no leakage can be caused by glitches.

Uniformity requires all intermediate states (shares) to be uniformly distributed. Uniformity ensures that the mean leakages are state-independent, a key requirement to thwart first-order DPA. To ensure uniformity in a circuit, it suffices to ensure uniformity for the output share of each function, as well as for the inputs of the circuit. This property is often the most difficult to achieve and most costly in terms of hardware area.

2.3 The ARM Cortex-M Processors

The ARM Cortex-M4 is a 32-bit RISC processor based on the ARMv7E-M architecture, targeting low-cost and energy-efficient microcontrollers. It is equipped with 13 general-purpose registers (GPRs, `r0-r12`), plus the link register (`lr`, which holds the return address from a subroutine), the stack pointer (`sp`), and the program counter (`pc`). The `lr` register can also be used as a GPR after its content has been saved to the stack. Besides GPRs, most existing M4s also have 32 floating-point registers ("M4f"), as is the case for the cheap, widely available, and popular STM32F407 Discovery board. Aside from performing floating-point instructions, floating point registers can be used as a cache to store frequently used constants or loop counters by using the `vmov` instruction that moves 32 bits between general-purpose and floating-point registers in exactly 1 cycle.

A very helpful feature, called "flexible second operand", can also save lots of time. In most data-processing instructions, the second operand can be a register shifted or rotated in the same instruction without causing extra latency. A shift or rotation that operates on a flexible second operand can be the arithmetic right shift (`ASR`), logical right or left shift (`LSL` and `LSR`), or rotation (`ROR`), plus rotate right extended by one bit (`RRX`). For example, the instruction `ADD r0,r1,r2,LSL #3` can calculate $r0 = r1 + (r2 \ll 3)$ in one clock cycle. As all common instructions like `EOR` and `AND` support the flexible second operand, shifts and rotates on registers can be had "for free" in many cases.

Since the functionalities we use for the implementation are also present in the ARM Cortex-M3 processor (which is missing only what's usually called DSP instructions compared to the M4), our work can be extended there without too much trouble, needing only to replace all caching in the floating point registers with stack access operations (and possibly re-optimizing).

3 Side-Channel Countermeasures

In this section, we provide three ways to avoid information leakage in GIMLI implementation. In Sect. 3.1 and Sect. 3.2, we discuss 2-share TI targeted for software implementations since they require a register stage after some operations to achieve non-completeness. This is easier to accomplish on software than highly parallel hardware implementations because each operation is stored to a register anyway. Therefore, if none of the individual terms recombines 2 shares of the same variable prior to the register write, and if each input share is independently uniform, non-completeness is always fulfilled. On the other hand, 3-share TI can be implemented on both platforms since it does not require overhead on the register areas and additional time cycles in hardware implementations.

For all the methods below, the first step is to apply random masking to all of the twelve 32-bit states. In an s-share scheme, we generate $12(s-1)$ random numbers and exclusive-or each state with $s-1$ random numbers. For example, in a 3-share scheme, state $s_{0,0}$ will xor 2 random numbers R_0 and R_1. The shared state will then be $\mathbf{s}_{0,0} = (s_{0,0} \oplus R_0 \oplus R_1, R_0, R_1)$. Though we perform these three ways on GIMLI implementation, these methods are not unique to GIMLI but are also implementable on other NORX family ciphers because of their similar structure.

3.1 2-Share with ChaCha-8 Randomness

In this method, we recall that the classical Boolean masking schemes in AND/OR gates simply put one random value in one share, and the other share is the value we want to protect xor'ed with that random value, so that xor'ing the two shares yields the sensitive intermediate and both shares are uniformly random. We use the same idea on the SP-box in GIMLI with 3 random numbers R_0, R_1, R_2, and the result is given in Eq. 1, where $\{x_0, x_1\}$, $\{x'_0, x'_1\}$ is the input, output shares for $\mathbf{x}$ respectively. $\mathbf{y}$ and $\mathbf{z}$ are the same.

$$\begin{cases} x'_0 = R_0 \oplus x_0 \oplus (z_0 \lll 1) \oplus ((y_0 \wedge z_0) \lll 2) \oplus ((y_0 \wedge z_1) \lll 2) \oplus ((y_1 \wedge z_0) \lll 2) \\ x'_1 = R_0 \oplus x_1 \oplus (z_1 \lll 1) \oplus ((y_1 \wedge z_1) \lll 2) \\ y'_0 = R_1 \oplus y_0 \oplus \quad x_0 \quad \oplus ((x_0 \vee z_0) \lll 1) \oplus ((\neg x_0 \wedge z_1) \lll 1) \oplus ((x_1 \wedge \neg z_0) \lll 1) \\ y'_1 = R_1 \oplus y_1 \oplus \quad x_1 \quad \oplus ((x_1 \wedge z_1) \lll 1) \\ z'_0 = R_2 \oplus z_0 \oplus \quad y_0 \quad \oplus ((x_0 \wedge y_0) \lll 3) \oplus ((x_0 \wedge y_1) \lll 3) \oplus ((x_1 \wedge y_0) \lll 3) \\ z'_1 = R_2 \oplus z_1 \oplus \quad y_1 \quad \oplus ((x_1 \wedge y_1) \lll 3) \end{cases} \tag{1}$$

The correctness of this equation can simply be checked by xor'ing the two shares of each variable. For example:

$$\begin{aligned} x' &= x'_0 \oplus x'_1 \\ &= (x_0 \oplus x_1) \oplus ((z_0 \oplus z_1) \lll 1) \oplus (((y_0 \wedge z_0) \oplus (y_0 \wedge z_1) \oplus (y_1 \wedge z_0) \oplus (y_1 \wedge z_1)) \lll 2) \\ &= x \oplus (z \lll 1) \oplus ((y \wedge z) \lll 2) \end{aligned}$$

If the subscript(s) of a term have 0, then we split that term into the first share, the others are combined into the second share. Because in this way, we can optimize the memory operations in assembly code more easily. $12 \times 24 = 288$ random numbers are required for the 24 rounds of GIMLI. While STM32F4 devices feature a true random number generator, it takes approximately 60 to 70 clock cycles to generate one 32-bit random integer. Hence we implement a ChaCha-8 pseudo-random number generator which reduces the clock cycle count to around 25 to 30 per random number. We choose the ChaCha-8 pseudo-random number generator because it is fast and currently considered to be 2^{256} bit security and highly unlikely to be less secure than Gimli's design security level [22], but of course, we can substitute any secure and fast stream cipher.

3.2 2-Share with Optimal Masking

In [10], a state-of-the-art masking mechanism was proposed. We can replace the non-linear operations in Eq. 1 by the operations of 2-share Threshold implementation, according to Table 1.

Table 1. Expressions for different operations.

Operation	Expression
SecAnd	$z_1 = (x_1 \wedge y1) \oplus (x_1 \oplus \neg y_2)$ $z_2 = (x_2 \wedge y1) \oplus (x_2 \oplus \neg y_2)$
SecOr	$z_1 = (x_1 \wedge y1) \oplus (x_1 \oplus \neg y_2)$ $z_2 = (x_2 \wedge y1) \oplus (x_2 \oplus \neg y_2)$

By eliminating the requirement of fresh randomness, it outperformed the classical Boolean masking schemes on software platforms. Here we utilize their result and construct a 2-share optimal masking of GIMLI.

$$\begin{cases} x'_0 = x_0 \oplus (z_0 \lll 1) \oplus (((y_0 \wedge z_0) \oplus (\neg y_1 \vee z_0)) \lll 2) \\ x'_1 = x_1 \oplus (z_1 \lll 1) \oplus (((y_0 \wedge z_1) \oplus (\neg y_1 \vee z_1)) \lll 2) \\ y'_0 = y_0 \oplus \quad x_0 \quad \oplus (((x_0 \wedge z_0) \oplus (x_0 \vee z_1)) \lll 1) \\ y'_1 = y_1 \oplus \quad x_1 \quad \oplus (((x_1 \vee z_0) \oplus (x_1 \wedge z_1)) \lll 1) \\ z'_0 = z_0 \oplus \quad y_0 \quad \oplus (((x_0 \wedge y_0) \oplus (x_0 \vee \neg y_1)) \lll 3) \\ z'_1 = z_1 \oplus \quad y_1 \quad \oplus (((x_1 \wedge y_0) \oplus (x_1 \vee \neg y_1)) \lll 3) \end{cases} \tag{2}$$

Notice that we only put the negation gate before y shares because ARM's flexible second operand only works on the second operand (the negated one) in `ORN` instructions. Since the y shares must be shifted, they need to be used as the second operands. Also, the method does require a register stage for the operations of the AND and OR gates. For the application on hardware platforms, this would be a big trade-off in terms of speed and area.

To prove that the modified function does not leak any information about any sensitive variable, we notice that in the formula of $\mathbf{x}$ and $\mathbf{y}$ shares, the only shares that contain both subscripts are only in the AND and OR gates. Since they don't leak the information about y and z, respectively, the whole calculation will not leak either since different shares are independent.

For the $\mathbf{z}$ shares, however, there is a chance that it might leak the information about y because both shares are present in both parts. We can check this very efficiently by performing a few bitwise operations on the truth tables and computing the Hamming weight. For example, a non-constant function f leaks information about function k if and only if

$$\frac{HW(k \wedge f)}{HW(f)} \neq \frac{HW(k \wedge \neg f)}{HW(\neg f)},$$

where $\mathrm{HW}(g)$ denotes the Hamming weight of the truth table of function g [10].

To confirm the $\mathbf{z}$ shares, we may then set z_0' as f and y as k and calculate the Hamming weight for all variables indexed from 0 to 15 since the formula shifts variables to the left at most 3. Therefore, we can prove that the function of 2-share optimal masking does not leak information about the sensitive state.

3.3 3-Share Threshold Implementation

To resist-first order DPA in hardware security, at least $t+1$ shares of masking is required [9] for the TI method, where t is the degree of a function and is 2 for GIMLI. We begin by constructing a threshold implementation of a 3-share GIMLI permutation.

$$\begin{cases} x_0' = x_1 \oplus (z_1 \ll 1) \oplus (((y_1 \wedge z_1) \oplus (y_1 \wedge z_2) \oplus (y_2 \wedge z_1)) \ll 2) \\ x_1' = x_2 \oplus (z_2 \ll 1) \oplus (((y_2 \wedge z_2) \oplus (y_2 \wedge z_0) \oplus (y_0 \wedge z_2)) \ll 2) \\ x_2' = x_0 \oplus (z_0 \ll 1) \oplus (((y_0 \wedge z_0) \oplus (y_0 \wedge z_1) \oplus (y_1 \wedge z_0)) \ll 2) \\ y_0' = y_1 \oplus \quad x_1 \quad \oplus (((x_1 \wedge z_1) \oplus (x_1 \wedge \neg z_2) \oplus (\neg x_2 \wedge z_1)) \ll 1) \\ y_1' = y_2 \oplus \quad x_2 \quad \oplus (((x_2 \vee z_2) \oplus (\neg x_2 \wedge z_0) \oplus (x_0 \wedge \neg z_2)) \ll 1) \\ y_2' = y_0 \oplus \quad x_0 \quad \oplus (((x_0 \wedge z_0) \oplus (x_0 \wedge z_1) \oplus (x_1 \wedge z_0)) \ll 1) \\ z_0' = z_1 \oplus \quad y_1 \quad \oplus (((x_1 \wedge y_1) \oplus (x_1 \wedge y_2) \oplus (x_2 \wedge y_1)) \ll 3) \\ z_1' = z_2 \oplus \quad y_2 \quad \oplus (((x_2 \wedge y_2) \oplus (x_2 \wedge y_0) \oplus (x_0 \wedge y_2)) \ll 3) \\ z_2' = z_0 \oplus \quad y_0 \quad \oplus (((x_0 \wedge y_0) \oplus (x_0 \wedge y_1) \oplus (x_1 \wedge y_0)) \ll 3) \end{cases} \tag{3}$$

Theorem 1. *Equation 3 constructs a threshold implementation of* GIMLI *permutation. That is, it meets the definition of* ***Correctness, Non-Completeness*** *and* ***Uniformity****.*

Proof. For **Correctness**, we need to make sure that $(\mathbf{x}', \mathbf{y}', \mathbf{z}') = (\bigoplus x_i', \bigoplus y_i', \bigoplus z_i')$. For instance, the y part can be proved: $\bigoplus_{i=0}^{2} y_i' = (x_0 \oplus x_1 \oplus x_2) \oplus (y_0 \oplus y_1 \oplus y_2) \oplus ((x_0 \oplus x_1 \oplus x_2) \vee (z_0 \oplus z_1 \oplus z_2)) \ll 1) = \mathbf{x} \oplus \mathbf{y} \oplus ((\mathbf{x} \vee \mathbf{z}) \ll 1)$. The x and z part is simple since it only contains an and gate.

For **Non-Completeness**, it can be seen that the computations of x_0', y_0', z_0' do not involve components of x_0, y_0, z_0, the ones of x_1', y_1', z_1' do not involve components of x_1, y_1, z_1 and the ones of x_2', y_2', z_2' do not involve components of x_2, y_2, z_2.

For **Uniformity**, if the mapping of Eq. 3 is an invertible mapping from (**x**, **y**, **z**) to (**x**', **y**', **z**'), it implies that if (**x**, **y**, **z**) is a uniform sharing, then (**x**', **y**', **z**') is an uniform sharing as well [8]. It is, therefore, sufficient to show that the mapping of Eq. 3 is invertible. We will do that by giving a method to compute (**x**, **y**, **z**) from (**x**', **y**', **z**').

We do this by recovering $(\mathbf{x}, \mathbf{y}, \mathbf{z})$ from 0-bit (rightmost) to 31-bit (leftmost) with the output $(\mathbf{x}', \mathbf{y}', \mathbf{z}')$. We denote the k-bit of x_0, say, by $x_{0,k}$. We rewrite the first equation by switching the output term to the right hand side:

$$x_1 = x_0' \oplus (z_1 \ll 1) \oplus (((y_1 \wedge z_1) \oplus (y_1 \wedge z_2) \oplus (y_2 \wedge z_1)) \ll 2)$$

Because the terms after x_0' have been shifted, $x_{1,0}$ can be derived from $x_{0,0}'$. Then we rewrite the equations as:

$$y_1 = y_0' \oplus x_1 \oplus (((x_1 \wedge z_1) \oplus (x_1 \wedge \neg z_2) \oplus (\neg x_2 \wedge z_1)) \ll 1)$$

$$z_1 = z_0' \oplus y_1 \oplus (((x_1 \wedge y_1) \oplus (x_1 \wedge \ \ y_2) \oplus (\ \ x_2 \wedge y_1)) \ll 3)$$

To compute $y_{1,0}$ and $z_{1,0}$, the last term is also irrelevant. Since we already knew $x_{1,0}$, we simply calculate $y_{1,0} = y_{0,0}' \oplus x_{1,0}$. And then $z_{1,0} = z_{0,0}' \oplus y_{1,0}$. We can use the same method to get the 0-bit of the other 6 shares.

Assume that $\forall a \in \{x, y, z\}, i \in \{0, 1, 2\}, n < 0, a_{i,n} = 0$, with the bit 0 to $k-1$ of all shares known, bits k of $(\mathbf{x}, \mathbf{y}, \mathbf{z})$ can be derived easily via:

$$\begin{aligned} x_{1,k} &= x_{0,k}' \oplus z_{1,k-1} \oplus ((y_{1,k-2} \wedge z_{1,k-2}) \oplus (y_{1,k-2} \wedge z_{2,k-2}) \oplus (y_{2,k-2} \wedge z_{1,k-2})) \\ y_{1,k} &= y_{0,k}' \oplus x_{1,k} \oplus ((x_{1,k-1} \wedge z_{1,k-1}) \oplus (x_{1,k-1} \wedge \neg z_{2,k-1}) \oplus (\neg x_{2,k-1} \wedge z_{1,k-1})) \\ z_{1,k} &= z_{0,k}' \oplus y_{1,k} \oplus ((x_{1,k-3} \wedge y_{1,k-3}) \oplus (x_{1,k-3} \wedge y_{2,k-3}) \oplus (x_{2,k-3} \wedge y_{1,k-3})) \end{aligned}$$

This way, we can restore $(\mathbf{x}, \mathbf{y}, \mathbf{z})$ from the output $(\mathbf{x}', \mathbf{y}', \mathbf{z}')$. Thus all possible inputs and outputs are mapped one-to-one to each other, which implies the uniformity property.

4 Implementation Details

In this section, assembly level optimizations of original and masked GIMLI for ARM Cortex-M4 processors, aimed at both high-speed and compact code-size, are presented.

4.1 Optimization on Original Gimli

We optimized the original GIMLI in two ways: (1) we first deal with the non-linear layer, namely the SP-box, of GIMLI by exploiting the "flexible second operand" feature; (2) we optimized the big/small swap steps by minimizing the amount of memory operations.

SP-box: In order to exploit the flexible second operand feature, we slightly modify the SP-box function. Instead of calculating the rotations of y-states with a ROR instruction, we skip this and fix the missing rotations by rotating registers when they are AND-ed or XOR-ed, using the flexible second operand. Without using the flexible second operand, each round needs 15 operations. By not rotating the y-states beforehand and some rearrangement of the instructions, we can reduce that to 10 instructions, which is 33% smaller.

Notice that with 12 state words loaded into registers, we have only 2 other registers (here denoted a_0, a_1) available as scratch space.

Algorithm 2: Optimization on Original GIMLI

input : States before the SP-box (x_0, y_0, z_0)
output: States after the SP-box (x_0, y_0, z_0)

1 ROR x_0, #8
2 AND a_0, z_0, y_0, ROR, #23
3 EOR a_0, x_0, a_0, LSL, #2
4 ORR a_1, x_0, z_0
5 EOR a_1, x_0, a_1, LSL, #1
6 AND x_0, x_0, y_0, ROR, #23
7 EOR x_0, z_0, x_0, LSL, #3
8 EOR x_0, x_0, y_0, ROR, #23
9 EOR z_0, a_1, z_0, LSL, #1
10 EOR y_0, a_2, y_0, ROR, #23

Swap: To avoid the penalty of using slow memory operations, we want to minimize save and load instructions. Since the small and big swaps operate alternatively, the states will return to their former places after 2 small and 2 big swaps. With a total of 6 small and big swaps, this means that all states will return to their original places after 24 rounds of GIMLI permutation.

If we can keep track of where the states are before and after the swap, we can simply continue to the next round with this order of registers without actually swapping anything. For example, let `r1-r12` be the content of the 12 state words. The first non-linear layer is performed on (`r1, r4, r7`), (`r2, r5, r8`), (`r3, r6, r9`), and (`r4, r8, r12`). After the first small swap, the next non-linear layer should be performed on (`r2, r4, r7`), (`r1, r5, r8`), (`r4, r6, r9`), and (`r3, r8, r12`). In this way, the linear layer can be simply omitted.

For software implementation, we modify the source code given in [19] by changing the C implementation of GIMLI permutation to assembly code because that is the bottleneck of the performance. It is also where our three methods differ. And the C implementation acts as a baseline to our optimization that we compare to in the next section.

4.2 Implementation Details of Masked Gimli

Based on the idea shown in Sects. 3.1 and 3.2, we develop Algorithm 3 (SecSP1), a one round 2-share GIMLI SP-box refreshed with given random numbers, and Algorithm 4 (SecOptSP1), one round optimal masking of 2-share GIMLI. s^0 and s^1 are the 2 shares of one column in the state and R_i are random numbers generated by `ChaCha-8`.

Notice that the optimization on the original GIMLI can also be applied here as well: every shift of the **y**s can be delayed and effectively performed using the flexible second operand. We can actually count the number of operations used in these expressions of a 2-share SP-box: $2(\texttt{ROR}) + 24(\texttt{EOR}) + 12(\texttt{AND/ORR}) = 38$, which is about 4 times that of the original version. But because of the overhead of random number generation, the actual cost of Algorithm 3 is much higher.

Algorithm 3: 2-share SP-box (`SecSP1`)

---

input : $s^0 = (x_0, y_0, z_0), s^1 = (x_1, y_1, z_1), R_0, R_1, R_2 \in \{0,1\}^{32}$
output: $(x'_0 \oplus x'_1, y'_0 \oplus y'_1, z'_0 \oplus z'_1) = SP(x_0 \oplus x_1, y_0 \oplus y_1, z_0 \oplus z_1)$
1 $(x_0, x_1) \leftarrow (x_0 \lll 24, x_1 \lll 24)$
2 $(s_0, s_1, s_2, s_3) \leftarrow (x_0 \wedge z_0, x_0 \wedge \neg z_1, x_1 \wedge \neg z_0, x_1 \vee z_1)$
3 $(t_0, t_1) \leftarrow (R_0 \oplus s_0 \oplus s_1 \oplus s_2, R_0 \oplus s_3)$
4 $(u_0, u_1) \leftarrow (x_0 \oplus (t_0 \ll 1), x_1 \oplus (t_1 \ll 1))$
5 $(y'_0, y'_1) \leftarrow (u_0 \oplus (y_0 \lll 9), u_1 \oplus (u_1 \lll 9))$
6 $(s_0, s_1, s_2, s_3) \leftarrow (z_0 \wedge (y_0 \lll 9), z_0 \wedge (y_1 \lll 9), z_1 \wedge (y_0 \lll 9), z_1 \wedge (y_0 \lll 9))$
7 $(t_0, t_1) \leftarrow (R_1 \oplus s_0 \oplus s_1 \oplus s_2, R_1 \oplus s_3)$
8 $(u_0, u_1) \leftarrow (x_0 \oplus (t_0 \ll 2), x_1 \oplus (t_1 \ll 2))$
9 $(z'_0, z'_1) \leftarrow (u_0 \oplus (z_0 \ll 1), u_1 \oplus (z_1 \ll 1))$
10 $(s_0, s_1, s_2, s_3) \leftarrow (x_0 \wedge (y_0 \lll 9), x_0 \wedge (y_1 \lll 9), x_1 \wedge (y_0 \lll 9), x_1 \wedge (y_0 \lll 9))$
11 $(t_0, t_1) \leftarrow (R_2 \oplus s_0 \oplus s_1 \oplus s_2, R_2 \oplus s_3)$
12 $(u_0, u_1) \leftarrow (z_0 \oplus (t_0 \ll 3), z_1 \oplus (t_1 \ll 3))$
13 $(x'_0, x'_1) \leftarrow (u_0 \oplus (y_0 \lll 9), u_1 \oplus (y_1 \lll 9))$

We can count the operations on these expressions of the optimal 2-share SP-box as well: $2(\texttt{ROR}) + 18(\texttt{EOR}) + 12(\texttt{AND/ORR}) = 32$, which is about 3 times the non-shared version without additional memory manipulation because of the increased variables.

For further optimization, we notice that the linear layers (swap) happen every two rounds, it means that during the consecutive rounds where no swap happens in-between, the inputs to non-linear layers are the same three 32-bit states. Therefore, we can apply SP-box twice to the states without loading and saving to further reduce the number of memory instructions. Algorithm 5 shows the process of this idea applied to 2-share optimal masking, where $\mathbf{s}_{i,j}$ represents the 2 shares of 3 states (s_i, s_{4+j}, s_{8+j}) and `SP2` is one round SP-box applied twice in a row.

Algorithm 4: Optimal 2-share SP-box (SecOptSP1)

input : $s^0 = (x_0, y_0, z_0), s^1 = (x_1, y_1, z_1)$
output: $(x'_0 \oplus x'_1, y'_0 \oplus y'_1, z'_0 \oplus z'_1) = SP(x_0 \oplus x_1, y_0 \oplus y_1, z_0 \oplus z_1)$
1 $(x_0, x_1) \leftarrow (x_0 \lll 24, x_1 \lll 24)$
2 $(s_0, s_1, s_2, s_3) \leftarrow (x_0 \wedge z_0, x_0 \vee z_1, x_1 \vee z_0, x_1 \wedge z_1)$
3 $(t_0, t_1) \leftarrow (s_0 \oplus s_1, s_2 \oplus s_3)$
4 $(u_0, u_1) \leftarrow (x_0 \oplus (t_0 \ll 1), x_1 \oplus (t_1 \ll 1))$
5 $(y'_0, y'_1) \leftarrow (u_0 \oplus (y_0 \lll 9), u_1 \oplus (u_1 \lll 9))$
6 $(s_0, s_1, s_2, s_3) \leftarrow (z_0 \wedge (y_0 \lll 9), z_0 \vee \neg(y_1 \lll 9), z_1 \wedge (y_0 \lll 9), z_1 \vee \neg(y_0 \lll 9))$
7 $(t_0, t_1) \leftarrow (s_0 \oplus s_1, s_2 \oplus s_3)$
8 $(u_0, u_1) \leftarrow (x_0 \oplus (t_0 \ll 2), x_1 \oplus (t_1 \ll 2))$
9 $(z'_0, z'_1) \leftarrow (u_0 \oplus (z_0 \ll 1), u_1 \oplus (z_1 \ll 1))$
10 $(s_0, s_1, s_2, s_3) \leftarrow (x_0 \wedge (y_0 \lll 9), x_0 \vee \neg(y_1 \lll 9), x_1 \wedge (y_0 \lll 9), x_1 \vee \neg(y_0 \lll 9))$
11 $(t_0, t_1) \leftarrow (s_0 \oplus s_1, s_2 \oplus s_3)$
12 $(u_0, u_1) \leftarrow (z_0 \oplus (t_0 \ll 3), z_1 \oplus (t_1 \ll 3))$
13 $(x'_0, x'_1) \leftarrow (u_0 \oplus (y_0 \lll 9), u_1 \oplus (y_1 \lll 9))$

Algorithm 5: 24 rounds of GIMLI permutation

input : $\mathbf{s}^0 = (s^0_{i,j}), \mathbf{s}^1 = (s^1_{i,j}) \in W^{3\times4}$
output: $\text{GIMLI}(\mathbf{s}^0 \oplus \mathbf{s}^1) = (s^0_{i,j}, s^1_{i,j}) \in W^{3\times4,2}$
$\mathbf{s}_{0,0}, \mathbf{s}_{1,1}, \mathbf{s}_{2,2}, \mathbf{s}_{3,3} \leftarrow \texttt{SP1}(\mathbf{s}_{0,0}), \texttt{SP1}(\mathbf{s}_{1,1}), \texttt{SP1}(\mathbf{s}_{2,2}), \texttt{SP1}(\mathbf{s}_{3,3})$
$s^0_{1,0} \leftarrow s^0_{1,0} \oplus \texttt{0x9e377900} \oplus 24$
$\mathbf{s}_{1,0}, \mathbf{s}_{0,1}, \mathbf{s}_{3,2}, \mathbf{s}_{2,3} \leftarrow \texttt{SP2}(\mathbf{s}_{1,0}), \texttt{SP2}(\mathbf{s}_{0,1}), \texttt{SP2}(\mathbf{s}_{3,2}), \texttt{SP2}(\mathbf{s}_{2,3})$
$\mathbf{s}_{3,0}, \mathbf{s}_{2,1}, \mathbf{s}_{1,2}, \mathbf{s}_{0,3} \leftarrow \texttt{SP2}(\mathbf{s}_{3,0}), \texttt{SP2}(\mathbf{s}_{2,1}), \texttt{SP2}(\mathbf{s}_{1,2}), \texttt{SP2}(\mathbf{s}_{0,3})$
$s^0_{2,0} \leftarrow s^0_{2,0} \oplus \texttt{0x9e377900} \oplus 20$
$\mathbf{s}_{2,0}, \mathbf{s}_{3,1}, \mathbf{s}_{0,2}, \mathbf{s}_{1,3} \leftarrow \texttt{SP2}(\mathbf{s}_{2,0}), \texttt{SP2}(\mathbf{s}_{3,1}), \texttt{SP2}(\mathbf{s}_{0,2}), \texttt{SP2}(\mathbf{s}_{1,3})$
$\mathbf{s}_{0,0}, \mathbf{s}_{1,1}, \mathbf{s}_{2,2}, \mathbf{s}_{3,3} \leftarrow \texttt{SP2}(\mathbf{s}_{0,0}), \texttt{SP2}(\mathbf{s}_{1,1}), \texttt{SP2}(\mathbf{s}_{2,2}), \texttt{SP2}(\mathbf{s}_{3,3})$
$s^0_{1,0} \leftarrow s^0_{1,0} \oplus \texttt{0x9e377900} \oplus 16$
$\mathbf{s}_{1,0}, \mathbf{s}_{0,1}, \mathbf{s}_{3,2}, \mathbf{s}_{2,3} \leftarrow \texttt{SP2}(\mathbf{s}_{1,0}), \texttt{SP2}(\mathbf{s}_{0,1}), \texttt{SP2}(\mathbf{s}_{3,2}), \texttt{SP2}(\mathbf{s}_{2,3})$
$\mathbf{s}_{3,0}, \mathbf{s}_{2,1}, \mathbf{s}_{1,2}, \mathbf{s}_{0,3} \leftarrow \texttt{SP2}(\mathbf{s}_{3,0}), \texttt{SP2}(\mathbf{s}_{2,1}), \texttt{SP2}(\mathbf{s}_{1,2}), \texttt{SP2}(\mathbf{s}_{0,3})$
$s^0_{2,0} \leftarrow s^0_{2,0} \oplus \texttt{0x9e377900} \oplus 12$
$\mathbf{s}_{2,0}, \mathbf{s}_{3,1}, \mathbf{s}_{0,2}, \mathbf{s}_{1,3} \leftarrow \texttt{SP2}(\mathbf{s}_{2,0}), \texttt{SP2}(\mathbf{s}_{3,1}), \texttt{SP2}(\mathbf{s}_{0,2}), \texttt{SP2}(\mathbf{s}_{1,3})$
$\mathbf{s}_{0,0}, \mathbf{s}_{1,1}, \mathbf{s}_{2,2}, \mathbf{s}_{3,3} \leftarrow \texttt{SP2}(\mathbf{s}_{0,0}), \texttt{SP2}(\mathbf{s}_{1,1}), \texttt{SP2}(\mathbf{s}_{2,2}), \texttt{SP2}(\mathbf{s}_{3,3})$
$s^0_{1,0} \leftarrow s^0_{1,0} \oplus \texttt{0x9e377900} \oplus 8$
$\mathbf{s}_{1,0}, \mathbf{s}_{0,1}, \mathbf{s}_{3,2}, \mathbf{s}_{2,3} \leftarrow \texttt{SP2}(\mathbf{s}_{1,0}), \texttt{SP2}(\mathbf{s}_{0,1}), \texttt{SP2}(\mathbf{s}_{3,2}), \texttt{SP2}(\mathbf{s}_{2,3})$
$\mathbf{s}_{3,0}, \mathbf{s}_{2,1}, \mathbf{s}_{1,2}, \mathbf{s}_{0,3} \leftarrow \texttt{SP2}(\mathbf{s}_{3,0}), \texttt{SP2}(\mathbf{s}_{2,1}), \texttt{SP2}(\mathbf{s}_{1,2}), \texttt{SP2}(\mathbf{s}_{0,3})$
$s^0_{2,0} \leftarrow s^0_{2,0} \oplus \texttt{0x9e377900} \oplus 4$
$\mathbf{s}_{2,0}, \mathbf{s}_{3,1}, \mathbf{s}_{0,2}, \mathbf{s}_{1,3} \leftarrow \texttt{SP2}(\mathbf{s}_{2,0}), \texttt{SP2}(\mathbf{s}_{3,1}), \texttt{SP2}(\mathbf{s}_{0,2}), \texttt{SP2}(\mathbf{s}_{1,3})$
$\mathbf{s}_{0,0}, \mathbf{s}_{1,1}, \mathbf{s}_{2,2}, \mathbf{s}_{3,3} \leftarrow \texttt{SP1}(\mathbf{s}_{0,0}), \texttt{SP1}(\mathbf{s}_{1,1}), \texttt{SP1}(\mathbf{s}_{2,2}), \texttt{SP1}(\mathbf{s}_{3,3})$
return $(\mathbf{s}^0, \mathbf{s}^1)$

The details for a 3-share threshold implementation are much the same in Algorithm 6. The input is now 3 shared states ($s^0 = (x_0, y_0, z_0), s^1 = (x_1, y_1, z_1), s^2 = (x_2, y_2, z_2)$) and the output is $SP(x_0 \oplus x_1 \oplus x_2, y_0 \oplus y_1 \oplus y_2, z_0 \oplus$

Algorithm 6: 3-share SP-box

input : $s^0 = (x_0, y_0, z_0), s^1 = (x_1, y_1, z_1), s^2 = (x_2, y_2, z_2)$
output: $(x_0' \oplus x_1' \oplus x_2', y_0' \oplus y_1' \oplus y_2', z_0' \oplus z_1' \oplus z_2') = SP(x_0 \oplus x_1 \oplus x_2, y_0 \oplus y_1 \oplus y_2, z_0 \oplus z_1 \oplus z_2)$

1 $(x_0, x_1, x_2) \leftarrow (x_0 \lll 24, x_1 \lll 24, x_2 \lll 24)$
2 $y_1' \leftarrow ((y_2 \lll 9) \oplus x_2 \oplus (((x_2 \vee z_2) \oplus (\neg x_2 \wedge z_0) \oplus (x_0 \wedge \neg z_2)) \ll 1))$
3 $z_1' \leftarrow (z_2 \oplus (y_2 \lll 9) \oplus (((x_2 \wedge (y_2 \lll 9)) \oplus (x_2 \wedge (y_0 \lll 9)) \oplus (x_0 \wedge (y_2 \lll 9))) \ll 3))$
4 $y_2' \leftarrow ((y_0 \lll 9) \oplus x_0 \oplus (((x_0 \wedge z_0) \oplus (x_0 \wedge z_1) \oplus (x_1 \wedge z_0)) \ll 1))$
5 $z_2' \leftarrow (z_0 \oplus (y_0 \lll 9) \oplus (((x_0 \wedge (y_0 \lll 9)) \oplus (x_0 \wedge (y_1 \lll 9)) \oplus (x_1 \wedge (y_0 \lll 9))) \ll 3))$
6 $x_2' \leftarrow (x_0 \oplus (z_0 \ll 1) \oplus ((((y_0 \lll 9) \wedge z_0) \oplus ((y_0 \lll 9) \wedge z_1) \oplus ((y_1 \lll 9) \wedge z_0)) \ll 2))$
7 $x_1' \leftarrow (x_2 \oplus (z_2 \ll 1) \oplus ((((y_2 \lll 9) \wedge z_2) \oplus ((y_2 \lll 9) \wedge z_0) \oplus ((y_0 \lll 9) \wedge z_2)) \ll 2))$
8 $y_0' \leftarrow ((y_1 \lll 9) \oplus x_1 \oplus (((x_1 \wedge z_1) \oplus (x_1 \wedge \neg z_2) \oplus (\neg x_2 \wedge z_1)) \ll 1))$
9 $z_0' \leftarrow (z_1 \oplus (y_1 \lll 9) \oplus (((x_1 \wedge (y_1 \lll 9)) \oplus (x_1 \wedge (y_2 \lll 9)) \oplus (x_2 \wedge (y_1 \lll 9))) \ll 3))$
10 $x_0' \leftarrow (x_1 \oplus (z_1 \ll 1) \oplus ((((y_1 \lll 9) \wedge z_1) \oplus ((y_1 \lll 9) \wedge z_2) \oplus ((y_2 \lll 9) \wedge z_1)) \ll 2))$

$z_1 \oplus z_2)$, where the SP box computes the result of Eq. 3. Then, we follow the same procedure as in Algorithm 5 to construct the 24 rounds of 3-share GIMLI.

The total number of operations for the 3-share threshold implementation is 3(`ROR`) + 36(`EOR`) + 27(`AND/ORR`) = 66, which is about 7 times the non-shared version.

5 Experiments and Results

The software was cross-compiled using the GNU Compiler Collection for ARM Embedded Processors version 9.2.1 with the options `-mthumb -mcpu=cortex-m4` and tested on a STM32F407 discovery board. The length was measured by the number of assembly code instructions, while the cycle count was measured using the internal clock cycle counter. Note that the reported cycles include the overheads for calling/returning from the considered functions, while all input data was assumed to be already word aligned.

5.1 Comparison of the Implementations

Table 2 provides a comparative overview of the implementation results. The cycles are counted from the beginning of AEAD encryption of 1024 bytes associated data and 1024 bytes plaintext, while the lengths are only counted by the assembly code for 24 rounds of GIMLI permutation. In Table 2, the **original** method is pure C implementation [19] and the **non-shared** method represents the assembly implementation in Sect. 4.1.

Our GIMLI implementation is much quicker than the C implementation. Even with 3-share protection, our result is comparable with the original unprotected C implementation. The growth in cycles as the number of shares increases is meeting our prediction as well. The cycle count of the 2-share with optimal masking implementation is about 4 times that of non-masked, and that of the 3-share TI implementation is about 7 times the ones of non-shared. These reference assembly codes can be found online[1]. On the other hand, according to Table 2, a 2-share implementation with ChaCha-8 is slower than a 3-share Threshold implementation. If a 2-share Threshold implementation cannot meet the required security level, a 3-share Threshold implementation is a better choice compared with the 2-share implementation with ChaCha-8.

We also tried implementing only by M3 instructions, and the results are shown in Table 3. The main difference between M3 and M4 is the memory manipulation instructions. Without the floating-point registers as the temporary memory, the cost of store and load instructions increases as the number of shares increases.

Table 4 shows the performance of masked algorithms compared with the baseline unmasked ones. The numbers are how much slower the masked version was than the unmasked version. Our methods have a significant improvement on both scenarios compared with [29], even accounting for the Cortex-M4 to Cortex-M3 difference.

Table 2. The results for GIMLI-AEAD (1024 message bytes and 1024 ad bytes) under benchmark clock (24 MHz).

Methods	Cycles	Speed (cycles/byte)	Length
Original	1037161	506.4	–
Non-shared	151551	74.0	987
2-share (chaha-8 randomness)	2213676	1080.9	336 (not unrolled)
2-share (optimal masking)	615383	300.5	4247
3-share (TI)	1159939	566.4	8378

5.2 Leakage Detection of Side-Channel Analysis

We adopt the test vector leakage assessment (TVLA) methodology to perform leakage detection. All the experiments here are based on ChipWhisperer-Lite Two-Part Version [26]. The program ChipWhisperer Capture [27] retrieves power samples from the control board, storing power traces and input data.

To complete the DPA test at Security Level 4 of ISO/IEC 17825 [1], we focus on the first permutation and capture two sets of $l = 100000$ power traces corresponding to the selected plaintexts and randomly plaintexts and compute the Welch's t-test to identify the differentiating features between the trace sets.

Figure 1 shows the T-test of three different versions of our implementation. Each picture can be separated into three parts by the black lines: 1. loading key and plaintext and applying the random mask to the state; 2. the 24-round GIMLI permutation; and 3. recovering the state and return. To reduce leakage introduced by pipeline registers, we rearranged the parallel instructions in 2-share TI [13]. Figure 1c shows the t-test results where the parallel instructions

[1] https://github.com/kuruwa2/pqm4/tree/master/gimli24v1-aead.

Table 3. The results for GIMLI-AEAD (1024 message bytes and 1024 ad bytes) on M3 under benchmark clock (24 MHz).

Methods	Cycles	Speed (cycles/byte)	Length
Non-shared	151551	74.0	987
2-share (chaha-8 randomness)	2286875	1116.6	335 (not unrolled)
2-share (optimal masking)	615826	300.7	4250
3-share (TI)	1247448	609.1	8276

Table 4. The amount of overhead for masking.

Methods	2 shares	3 shares
Reference [29]	10.50	31.41
Ours	4.06	7.65

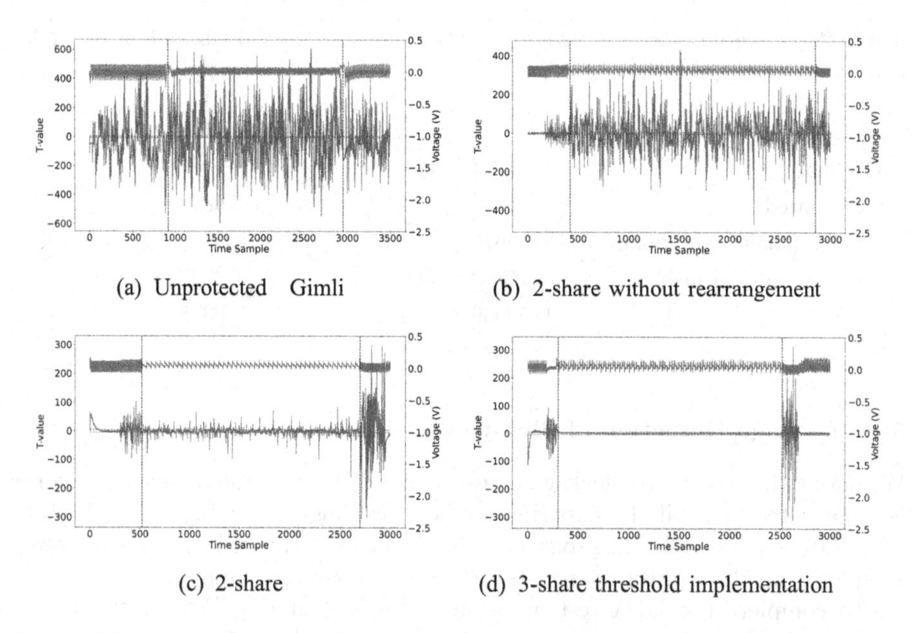

(a) Unprotected Gimli (b) 2-share without rearrangement

(c) 2-share (d) 3-share threshold implementation

Fig. 1. (a) T-test of unprotected GIMLI (b) T-test of 2-share with optimal masking GIMLI without rearrangement (c) T-test of 2-share with optimal masking GIMLI (d) T-test of 3-share threshold implementation GIMLI

are rearranged. However, 2-share TI still has some unintentional information leakage. We suspect that there are some unexpected buffers producing unintentional leakage in Cortex-M3 and -M4 [4]. On the other hand, We can see that the t-statistic value of the 3-share masked GIMLI permutation is inside the ± 4.5 [30] interval, corresponding to 99.999% confidence that a difference shown is not due

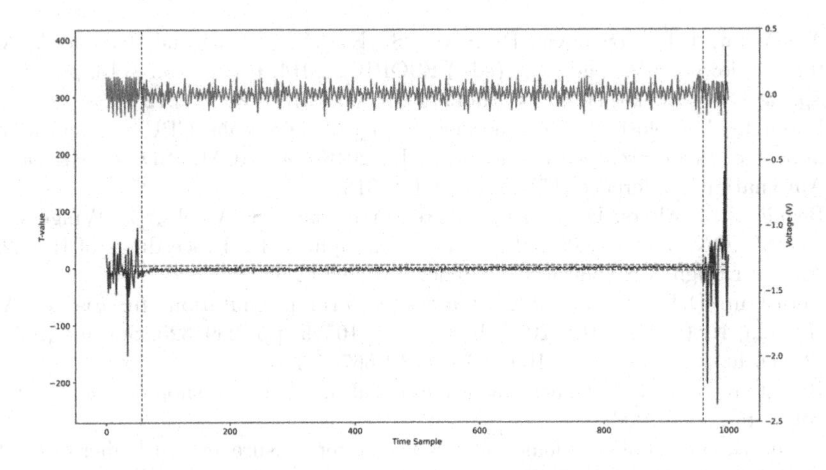

Fig. 2. T-test of 3-share threshold implementation GIMLI with l = 1000000 power traces

to random chance. To detect the leakage of 3-share masked GIMLI permutation, we capture with $l = 1{,}000{,}000$ power traces and use TVLA to perform a leakage assessment. Figure 2 shows the T-test of the first 12 round of 3-share Threshold implementation with $l = 1{,}000{,}000$ power traces. We can see that the t-statistic value is inside the ± 4.5 interval. In addition, since the first and third parts contain the public value, such as the ciphertext, it is normal that the t-statistic value exceeds ± 4.5 [2].

6 Conclusion

Our results significantly improve the performance of GIMLI implementations on Cortex-M3 and -M4 processors, especially NORX ciphers such as GIMLI are popular for their simplicity in preventing timing attacks. Compared to the original C implementation, the instruction count is reduced by 85%. In addition to this, the overhead of masking is reduced by 61% and 76% for 2-shared and 3-shared, respectively. Even if we rearranged the parallel instructions, the currently widely used 2-shared, GIMLI still exists for unintentional information leakage. Finally, we have completed a 3-share Threshold Implementation and passed the safety inspection of ISO/IEC 17825 Level 4.

References

1. ISO/IEC 17825:2016 information technology - security techniques - testing methods for the mitigation of non-invasive attack classes against cryptographic modules. Standard, International Organization for Standardization, Geneva, CH (2016)
2. Abdulrahman, A., Chen, J.P., Chen, Y.J., Hwang, V., Kannwischer, M.J., Yang, B.Y.: Multi-moduli NTTS for saber on cortex-m3 and cortex-m4. Cryptology ePrint Archive, Report 2021/995 (2021). https://ia.cr/2021/995

3. Aumasson, J.-P., Jovanovic, P., Neves, S.: NORX: parallel and scalable AEAD. In: Kutyłowski, M., Vaidya, J. (eds.) ESORICS 2014. LNCS, vol. 8713, pp. 19–36. Springer, Cham (2014). https://doi.org/10.1007/978-3-319-11212-1_2
4. Barenghi, A., Pelosi, G.: Side-channel security of superscalar CPUs: evaluating the impact of micro-architectural features. In: 2018 55th ACM/ESDA/IEEE Design Automation Conference (DAC), pp. 1–6 (2018)
5. Beaulieu, R., Shors, D., Smith, J., Treatman-Clark, S., Weeks, B., Wingers, L.: The SIMON and SPECK lightweight block ciphers. In: Proceedings of the 52nd Annual Design Automation Conference, pp. 1–6 (2015)
6. Bernstein, D.J., et al.: GIMLI: a cross-platform permutation. In: Fischer, W., Homma, N. (eds.) CHES 2017. LNCS, vol. 10529, pp. 299–320. Springer, Cham (2017). https://doi.org/10.1007/978-3-319-66787-4_15
7. Bernstein, D.J., et al.: Chacha, a variant of salsa20. In: Workshop Record of SASC, vol. 8, pp. 3–5 (2008)
8. Bilgin, B.: Threshold implementations as countermeasure against higher-order differential power analysis (2015)
9. Bilgin, B., Gierlichs, B., Nikova, S., Nikov, V., Rijmen, V.: Higher-order threshold implementations. In: Sarkar, P., Iwata, T. (eds.) ASIACRYPT 2014. LNCS, vol. 8874, pp. 326–343. Springer, Heidelberg (2014). https://doi.org/10.1007/978-3-662-45608-8_18
10. Biryukov, A., Dinu, D., Le Corre, Y., Udovenko, A.: Optimal first-order boolean masking for embedded IoT devices. In: Eisenbarth, T., Teglia, Y. (eds.) CARDIS 2017. LNCS, vol. 10728, pp. 22–41. Springer, Cham (2018). https://doi.org/10.1007/978-3-319-75208-2_2
11. Brier, E., Clavier, C., Olivier, F.: Correlation power analysis with a leakage model. In: Joye, M., Quisquater, J.-J. (eds.) CHES 2004. LNCS, vol. 3156, pp. 16–29. Springer, Heidelberg (2004). https://doi.org/10.1007/978-3-540-28632-5_2
12. Chen, C., Farmani, M., Eisenbarth, T.: A tale of two shares: why two-share threshold implementation seems worthwhile—and why it is not. In: Cheon, J.H., Takagi, T. (eds.) ASIACRYPT 2016. LNCS, vol. 10031, pp. 819–843. Springer, Heidelberg (2016). https://doi.org/10.1007/978-3-662-53887-6_30
13. Corre, Y.L., Großschädl, J., Dinu, D.: Micro-architectural power simulator for leakage assessment of cryptographic software on arm cortex-m3 processors. Cryptology ePrint Archive, Report 2017/1253 (2017). https://ia.cr/2017/1253
14. Daemen, J.: Changing of the guards: a simple and efficient method for achieving uniformity in threshold sharing. In: Fischer, W., Homma, N. (eds.) CHES 2017. LNCS, vol. 10529, pp. 137–153. Springer, Cham (2017). https://doi.org/10.1007/978-3-319-66787-4_7
15. Gruber, M., et al.: DOMREP - an orthogonal countermeasure for arbitrary order side-channel and fault attack protection. IEEE Trans. Inf. Forensics Secur. **16**, 4321–4335 (2021)
16. Gruber, M., Probst, M., Tempelmeier, M.: Statistical ineffective fault analysis of GIMLI. In: 2020 IEEE International Symposium on Hardware Oriented Security and Trust (HOST), pp. 252–261 (2020)
17. Gupta, N., Jati, A., Chattopadhyay, A., Sanadhya, S.K., Chang, D.: Threshold implementations of GIFT: a trade-off analysis. Cryptology ePrint Archive, Report 2017/1040 (2017). http://eprint.iacr.org/2017/1040
18. Jungk, B., Petri, R., Stöttinger, M.: Efficient side-channel protections of ARX ciphers. Cryptology ePrint Archive, Report 2018/693 (2018). https://eprint.iacr.org/2018/693

19. Kannwischer, M.J.: m4-crypto-eng-assignments (2020). https://github.com/mkannwischer/m4-crypto-eng-assignments/tree/master/gimli24v1-aead
20. Khan, S., Lee, W.K., Hwang, S.O.: A flexible Gimli hardware implementation in FPGA and its application to RFID authentication protocols. IEEE Access **9**, 105327–105340 (2021)
21. Kocher, P., Jaffe, J., Jun, B.: Differential power analysis. In: Wiener, M. (ed.) CRYPTO 1999. LNCS, vol. 1666, pp. 388–397. Springer, Heidelberg (1999). https://doi.org/10.1007/3-540-48405-1_25
22. Miyashita, S., Ito, R., Miyaji, A.: PNB-focused differential cryptanalysis of ChaCha stream cipher. Cryptology ePrint Archive, Report 2021/1537 (2021). https://ia.cr/2021/1537
23. Nikova, S., Rechberger, C., Rijmen, V.: Threshold implementations against side-channel attacks and glitches. In: Ning, P., Qing, S., Li, N. (eds.) ICICS 2006. LNCS, vol. 4307, pp. 529–545. Springer, Heidelberg (2006). https://doi.org/10.1007/11935308_38
24. Nikova, S., Rijmen, V., Schläffer, M.: Secure hardware implementation of non-linear functions in the presence of glitches. In: Lee, P.J., Cheon, J.H. (eds.) ICISC 2008. LNCS, vol. 5461, pp. 218–234. Springer, Heidelberg (2009). https://doi.org/10.1007/978-3-642-00730-9_14
25. Nikova, S., Rijmen, V., Schläffer, M.: Secure hardware implementation of nonlinear functions in the presence of glitches. J. Cryptol. **24**(2), 292–321 (2011)
26. O'Flynn, C.: Chipwhisperer-lite (cw1173) two-part version (2016)
27. O'Flynn, C.: ChipWhisperer - the complete open-source toolchain for side-channel power analysis and glitching attacks (2018)
28. Reparaz, O., Bilgin, B., Nikova, S., Gierlichs, B., Verbauwhede, I.: Consolidating masking schemes. In: Gennaro, R., Robshaw, M. (eds.) CRYPTO 2015. LNCS, vol. 9215, pp. 764–783. Springer, Heidelberg (2015). https://doi.org/10.1007/978-3-662-47989-6_37
29. Weatherley, R.: Performance of masked algorithms. In: Lightweight Cryptography Primitives Documentation (2020). https://rweather.github.io/lightweight-crypto/performance_masking.html
30. Whitnall, C., Oswald, E.: A critical analysis of ISO 17825 ('Testing methods for the mitigation of non-invasive attack classes against cryptographic modules'). In: Galbraith, S.D., Moriai, S. (eds.) ASIACRYPT 2019. LNCS, vol. 11923, pp. 256–284. Springer, Cham (2019). https://doi.org/10.1007/978-3-030-34618-8_9
31. Yan, Y., Oswald, E., Vivek, S.: An analytic attack against ARX addition exploiting standard side-channel leakage. Cryptology ePrint Archive, Paper 2020/1455 (2020). https://eprint.iacr.org/2020/1455. https://eprint.iacr.org/2020/1455

DeepC2: AI-Powered Covert Command and Control on OSNs

Zhi Wang[1,2], Chaoge Liu[1,2](✉), Xiang Cui[3](✉), Jie Yin[1], Jiaxi Liu[1,2], Di Wu[4], and Qixu Liu[1,2]

[1] Institute of Information Engineering, Chinese Academy of Sciences, Beijing, China
{wangzhi,liuchaoge,yinjie,liujiaxi,liuqixu}@iie.ac.cn
[2] School of Cyber Security, University of Chinese Academy of Sciences, Beijing, China
[3] Cyberspace Institute of Advanced Technology, Guangzhou University, Guangzhou, China
cuixiang@gzhu.edu.cn
[4] Huawei Technologies Co., Ltd., Shenzhen, China
wudi94@huawei.com

Abstract. Command and control (C&C) is important in an attack. It transfers commands from the attacker to the malware in the compromised hosts. Currently, some attackers use online social networks (OSNs) in C&C tasks. There are two main problems in the C&C on OSNs. First, the process for the malware to find the attacker is reversible. If the malware sample is analyzed by the defender, the attacker would be exposed before publishing the commands. Second, the commands in plain or encrypted form are regarded as abnormal contents by OSNs, which would raise anomalies and trigger restrictions on the attacker. The defender can limit the attacker once it is exposed. In this work, we propose DeepC2, an AI-powered C&C on OSNs, to solve these problems. For the reversible hard-coding, the malware finds the attacker using a neural network model. The attacker's avatars are converted into a batch of feature vectors, and the defender cannot recover the avatars in advance using the model and the feature vectors. To solve the abnormal contents on OSNs, hash collision and text data augmentation are used to embed commands into normal contents. The experiment on Twitter shows that command-embedded tweets can be generated efficiently. The malware can find the attacker covertly on OSNs. Security analysis shows it is hard to recover the attacker's identifiers in advance.

Keywords: Online social networks · Command and control · Covert communication · Neural networks

1 Introduction

Command and control (C&C) plays an essential role in an attack. It is widely used in Advanced Persistent Threat (APT), ransomware, or botnet scenarios.

C. Alcaraz et al. (Eds.): ICICS 2022, LNCS 13407, pp. 394–414, 2022.
https://doi.org/10.1007/978-3-031-15777-6_22

In a C&C system, an attacker needs to send commands to the compromised hosts via a C&C channel [2]. The hosts can be common computing devices such as PCs, servers, routers, and cameras, which have some vulnerabilities and can be infected by malware. They try to get and execute the commands from the attacker and carry out attack tasks such as DDoS, spam, crypto-mining, and data exfiltration. The major feature of a C&C system is that it has a one-to-many C&C channel, which receives commands from the attacker and forwards them to the compromised hosts. The process for the malware getting the commands is called addressing. C&C channel is a vital component in a C&C system. The attacker needs to keep the C&C channel robust and block-resistant to maintain the communication with the malware.

In recent years, the attackers have begun to utilize online web services [35], such as online social networks (OSNs), cloud drives, and online clipboards, to build the C&C channel. For example, Hammertoss (APT-29) [8] used Twitter and GitHub to publish commands and hide communication traces. HeroRat [27] used Telegram for C&C communication on Android devices. Turla [6] utilized Gmail to receive commands and exfiltrate information to the operators.

OSNs have some features to build a good C&C channel. It is nearly impossible for OSNs to go offline, and users can access OSNs anytime with a networked device. Then, visiting OSNs is allowed by most anti-virus software, and it ensures the availability of the commands. As many people use the OSNs, the attacker's accounts can hide among ordinary users. Also, it is easy to limit the accounts but not easy to shut down the OSNs. With the help of dynamic addressing, the malware can obtain commands from multiple accounts.

However, there are two main problems with building C&C channels on OSNs. First, the attacker's identifiers are reversible and predictable, which will cause the C&C channel to shut down before use. To help the malware addressing, the attacker's identifiers, i.e., ids, links, tokens, and DGAs (Domain Generation Algorithms), have to be hard-coded into the malware. Once the malware is analyzed by defenders, the reversible hard-coding will expose the C&C channel, and the attacker's accounts can be calculated in advance. Second, the commands published on OSNs are abnormal and will also expose the attacker's accounts. In most cases, commands are published in plain, encoded, or encrypted forms. They are regarded as abnormal contents on OSNs. They will expose C&C activities and raise anomalies, triggering restrictions on the attacker's accounts and interrupting the C&C activities. If the OSNs block the accounts, it is difficult for the malware to retrieve new commands.

In this paper, from the attacker's perspective, we use AI technology to overcome the above two problems and propose an AI-powered OSN C&C channel called DeepC2. The main idea of DeepC2 is as follows. To overcome the first problem, a neural network model is used for addressing. The malware finds the attacker's accounts through the feature vectors, which are extracted from the attacker's avatars by a neural network model. As the neural network models are poorly explainable [16], defenders cannot calculate and predict the avatars and accounts through the model and vectors. To solve the second problem and eliminate the abnormal content, we propose embedding the commands into contextual and readable content (we take Twitter and tweets as examples). To achieve this, the attacker uses data augmentation to generate numerous tweets and uses hash collision to get the command-embedded tweets. In the addressing process, Twitter Trends are used as the rendezvous point. The attacker posts tweets to a trending topic, and the malware finds the attacker under the topic. After addressing, the commands can be parsed from the attacker's tweets.

The contributions of this paper are summarized as follows:

- We propose a novel covert command and control scenario on OSNs.
- We introduce neural networks to solve the problem of reversible hard-coding in C&C addressing. By using feature vectors and a model, it is easy for the malware to find the attacker while hard for defenders to locate the attacker in advance.
- We propose a method for embedding commands into natural semantic tweets to avoid anomalies caused by abnormal contents on OSNs.
- We present experiments on Twitter to demonstrate the feasibility of the proposed methods and analyze their performance and security.

Ethical Considerations. The combination of AI and network attacks is an upward trend. We cannot stop the evolution of cyberattacks, but we should draw attention to the defenses in advance. This work aims not to inspire malware authors to write more efficient malware but to motivate security researchers and vendors to find solutions for an emerging threat. To this end, we intend to provide this work to build a possible scenario to help prevent this kind of attack in advance.

The remainder of this paper is structured as follows. Section 2 describes relevant backgrounds and related work. Section 3 presents the methodology for building the covert C&C channel. Detailed implementations are demonstrated in Sect. 4. Section 5 is the evaluations on the experiments. Section 6 discusses possible countermeasures. Conclusions are summarized in Sect. 7.

2 Background and Related Work

In this section, we present the background and related work of DeepC2.

2.1 Command and Control on OSNs

In the cases of addressing using OSN platforms, the defenders should find and limit the attacker's accounts in advance. By reverse-engineering a malware sample, the defenders will know the addressing process in detail [22]. If reversible methods like DGAs or IDs are used, the attacker's accounts can be calculated in advance. The defenders can limit the accounts, making the C&C channel unusable. The defenders will also know the attacker's accounts and commands by running a sample. However, when they get the commands this way, the malware in the wild also gets the commands, which is a failure from the defense perspective [28]. Therefore, the key issue in such an attack is to design a block-resistant C&C channel that even the defenders know the detailed information about the channel, it is hard to get the attacker's identifiers and limit the C&C **in advance**.

Some works build C&C channels on OSNs. Stegobot [20] uses the images shared by OSN users to build the C&C channel. The social network is regarded as a peer-to-peer network to connect the malware and the attacker. Information is hiding in images using steganography. However, the attacker's account can be obtained through reverse-engineering. Sebastian et al. [26] proposed to build a covert C&C channel on Twitter. The commands are encrypted tweets with a keyword, for example, #walmart AZEF, where #walmart is the keyword, and AZEF is the command cipher. However, this method also has the problem of abnormal contents on OSNs. Kwak et al. [15] proposed a video steganography-based C&C channel on Telegram, which can transfer large-sized secret files. Pantic et al. [21] proposed an anomaly-resistant C&C on Twitter. They used tweet-length as a command character and encoded each symbol in commands into numbers from 1 to 140. They collected tweets at different lengths from Twitter. When publishing a command, they chose tweets at specified lengths and posted them. The malware can get the commands by calculating the lengths of the tweets. However, this method has a low capacity and does not solve the reversible attacker accounts.

2.2 Easy Data Augmentation

Data augmentation is a technique to solve the insufficiency of training data. By applying data augmentation, researchers can enlarge the existing dataset to meet the needs of training works and promote the normalized performances of neural network models. In this work, the attacker needs to generate numerous tweets for hash collisions. Wei et al. [33] proposed Easy Data Augmentation (EDA) techniques. They used Synonym Replacement (SR), Random Insertion (RI), Random Swap (RS), and Random Deletion (RD) to generate sentences with similar meanings to the given sentences. Examples are shown in Table 1 in Appendix A.

The augmented sentences may not be grammatically and syntactically correct and may vary in meaning. However, due to differences in language, culture, and education, there are many grammatically incorrect tweets on Twitter. The Internet is diverse and inclusive. The attacker should ensure that the tweets have semantics but do not need them to be "correct".

2.3 AI-Powered Attacks

This work provides a new scenario on the malicious use of AI. The combination of AI and network attacks is an upward trend. For covert communication, Rigaki et al. [25] proposed using GAN to mimic Facebook chat traffic to make C&C communication undetectable. StegoNet [18] and EvilModel [31] hide malware in the neural network models to deliver malicious payloads covertly. The model parameters are replaced by the malicious payloads. DeepLocker [14] can carry out targeted attacks stealthily. DeepLocker trains the target attributes into an AI model and uses the model's outputs as a symmetric key to encrypt the malicious payload. Target detection is conducted by the AI model. When the input attributes match the target attributes, the secret key will be derived from the model to decrypt the payload and launch attacks on the target. MalGAN [13] generates adversarial malware that can bypass machine learning-based detection models. A generative network is trained to minimize the malicious probabilities of the generated adversarial examples predicted by the detector. More evasion methods [1] [30] were proposed after MalGAN.

3 Methodology

This section introduces methodologies for building a covert C&C channel on OSNs.

3.1 Threat Model

In this work, the C&C channel is built from the attackers' perspective. This work has three prominent roles: attackers, OSNs, and defenders. **Attackers.** We consider adversaries to be attackers capable of neural network and artificial intelligence technics and have the ability to use various system vulnerabilities to get into a system. **OSNs.** OSNs have the ability to limit the abnormal content and accounts based on their term of services. They can also actively detect autonomous and abnormal behaviors and limit the specious accounts according to their regulations. **Defenders.** We consider defenders to be third-party unrelated to attackers and OSNs. Defenders have access to the vectors from the prepared pictures and the structure, weights, implementation, and other detailed information of the neural network model. Defenders also have the ability to reverse engineer the malware sample to obtain its detailed implementation.

3.2 Approach Overview

Overall Workflow. We take Twitter as the OSN platform to demonstrate the method. The main workflow of DeepC2 contains four steps and is shown in Fig. 1.

(1) The attacker trains a neural network model with some pictures that may be selected as Twitter account avatars in subsequent steps. Then, the attacker extracts the feature vectors of these pictures using the trained model (see Fig. 2), embeds the model and the feature vectors into the malware, and publishes the malware to the wild.
(2) The attacker visits Twitter Trends, selects a topic according to the predefined rules, and then generates command-embedded tweets based on the selected trending topic.
(3) The attacker selects a picture as its Twitter account avatar and publishes the commands-embedded tweets using the account in the selected topic.

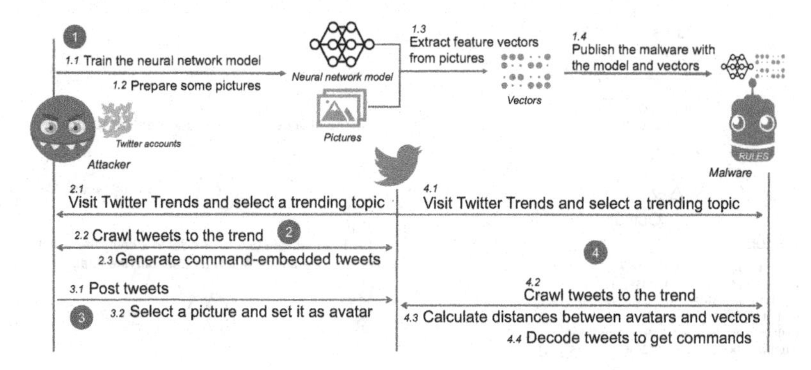

Fig. 1. Overall workflow of DeepC2

(4) The compromised hosts infected with the attacker's malware visit Twitter Trends periodically, select a topic synchronously according to the pre-defined rules, and then crawl the tweets and tweeters' avatars in the selected topic to find the attacker's account. The compromised hosts calculate the distances between the crawled avatars and the built-in feature vectors. If a distance is below a threshold, it is considered that the attacker's account is found. The commands can be obtained from the tweet.

The workflow mainly contains two key parts: dynamic addressing and command embedding. The dynamic addressing guides the compromised host to find the attacker and get the command successfully. Rather than reversible methods, it uses a neural network model, picture feature vectors, and Twitter topics to avoid the attacker's identifiers being exposed and predicted. The command embedding uses hash collision and EDA to embed commands into natural semantic tweets, thus avoiding anomalies caused by abnormal contents on OSNs. Next, we will describe these two parts in detail.

3.3 Dynamic Addressing

The dynamic addressing includes three main elements: the neural network model, avatars & feature vectors, and Twitter trending topics. In the following section, we will describe the elements from three aspects: What (is the element), How (to use), and Why (is the element).

Neural Network Model. *What.* The attacker needs a neural network model to extract feature vectors, and the malware needs the model to identify the attacker's accounts. Due to the limited resources in the host devices, the attacker cannot use the big-sized pre-trained models like VGGs, AlexNets, and Inception. Therefore, the attacker needs to build and train a model itself.

How. The model is used differently for attackers and malware (see Fig. 3). For the attacker, the model is used to extract feature vectors from avatars. The attacker

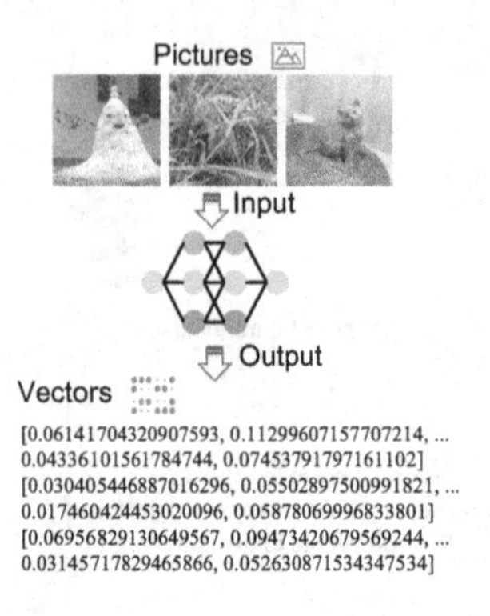

Fig. 2. Extract feature vectors

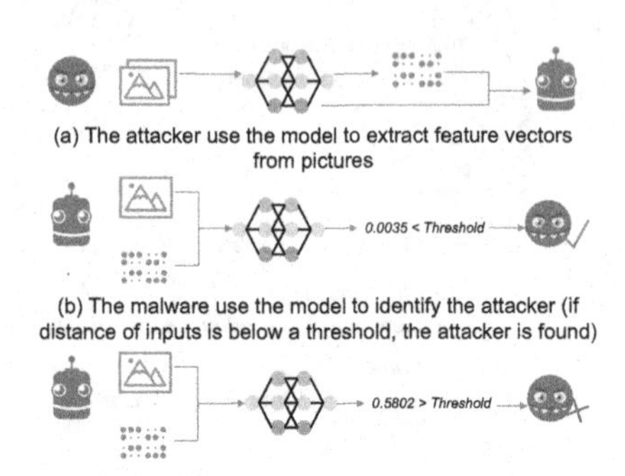

(a) The attacker use the model to extract feature vectors from pictures
0.0035 < Threshold
(b) The malware use the model to identify the attacker (if distance of inputs is below a threshold, the attacker is found)
0.5802 > Threshold

Fig. 3. Neural network model use

feeds the model with a batch of pictures, and the model outputs a batch of vectors that represent the pictures. The feature vectors and the model are published with the malware. For the malware, the model calculates the distances between avatars from Twitter users and the vectors to identify attackers. A selected vector and a crawled avatar are fed into the model, and then the model outputs the distance of the inputs.

Why. Using neural network models has the following advantages: 1) It is not easy to reverse the neural network model. Convolution is a lossy process. Combined with some intentionally introduced losses, it is hard to calculate the attacker's identifiers in advance. 2) The neural network models are fault-tolerance that similar inputs will generate similar outputs. 3) The neural network trained with gradient descent has good generalization ability [5]. It can help the malware identify attackers accurately and does not mistakenly identify someone else as the attacker.

Converting an image to a vector is similar to image-hashing [4]. In image-hashing, similar inputs have similar output hashes. However, image-hashing is not suitable for this work. There are two types of image-hashing methods. The non-neural network-based image-hashing methods are reversible, and defenders can build images that produce similar vectors according to the given hashes. For neural network-based methods [34], the learning tasks are more complex than DeepC2. The cryptographic hash algorithms are also unsuitable because they are sensitive to changes. As pictures uploaded to OSNs are compressed or resized, avatars are different from the original images, which will cause hashes to change due to the avalanche effect [32]. Therefore, the neural network model is suitable for this task.

Avatars and Vectors. *What.* The feature vectors are the abstract expression of the attacker's Twitter avatars. They are generated by the model and represent the attacker's avatars. A feature vector sample is shown in Fig. 2. Each vector contains a group of floating-point numbers, and the amount is determined by the model.

How. As stated above, the attacker gets the feature vectors from the model. The malware puts a feature vector and an avatar in the model and gets the distance between the inputs. The model will convert the avatar into another vector and calculate the distance. To prevent replay and enhance security, it is recommended that each avatar and vector be used only once. The attacker will change the current account and avatar when a command is published, and the malware will also delete the used vectors. The malware can get updates to the vectors and model and exploits from the C&C server. The malware can carry at least one vector in design when published. Due to various situations, the malware may not be able to run on time. To ensure that the compromised hosts can go online as expected, it is suggested that the malware is published with more vectors.

Why. The reason for using the feature vectors are as follows: 1) They are the natural output of the model and are easy to get for both the attacker and the malware. 2) It's difficult to reverse the vector generation process to get another picture that can produce a similar vector. 3) The vectors are distributed in a continuous interval (see Sect. 5.5), and each position has a large value space, which ensures the security of the C&C channel.

Twitter Trends. *What.* Twitter Trends contains hot topics that have had many discussions in the past 24 h. Usually, each topic contains 1 to 3 keyword(s). Twitter Trends is updated every 5 min.

How. The attacker defines a set of rules to select the trending topics, and the malware selects the topics synchronously with the attacker. In this work, we use Twitter API to get the trending topics. Twitter Trends API returns the top 50 topics in a chosen area specified by a location ID (WOEID, where on Earth ID). There are detailed tweet volumes if the volume exceeds 10 K over the past 24 h. In the experiments, we obtained trends from Johannesburg, South Africa and selected the last topic above 10 K discussions from the returned trends. The test servers in different regions fetched the same topics with the WOEID. Twitter API can be abused by attackers. However, attackers have more choices in real scenarios. They can utilize third parties that provide Twitter content queries, including tweets, trends, and user profiles. Alternatively, they can also write their implementations that use raw HTTP requests to obtain the content.

Why. Using Twitter Trends have the following advantages: 1) Twitter Trends provides a rendezvous point for the malware to find attackers among Twitter users; 2) Twitter Trends changes with the tweet volume under different topics and is updated every 5 min, which is difficult to predict; 3) Since normal users also discuss different topics, attackers can hide among them. Therefore, we use Twitter Trends for DeepC2.

3.4 Command Embedding

The attacker uses hash collision and easy data augmentation (EDA) to generate commands-embedded tweets. In this work, we take publishing an IP address as an example to illustrate the process of publishing commands. If IP addresses can be published, other messages like domains, shortening-URL IDs, or online-clipboard IDs can also be published in the same way.

(1) **Hash Collision.** To convey an IP address to the malware through tweets, the attacker splits the IP address into two parts first, as shown in Fig. 4. Each part is expressed in hexadecimal. For each tweet, the attacker calculates its hash and compares whether the first 16 bits of the hash are identical to one IP part. If two parts of an IP address collide, a successful hash collision occurs. The attacker posts the collided tweets in order. When addressing, the malware can get the IP address by calculating the hashes of tweets posted by the attacker and concatenating the first 2 bytes of hashes. In this way, 16 bits can be conveyed in one tweet.

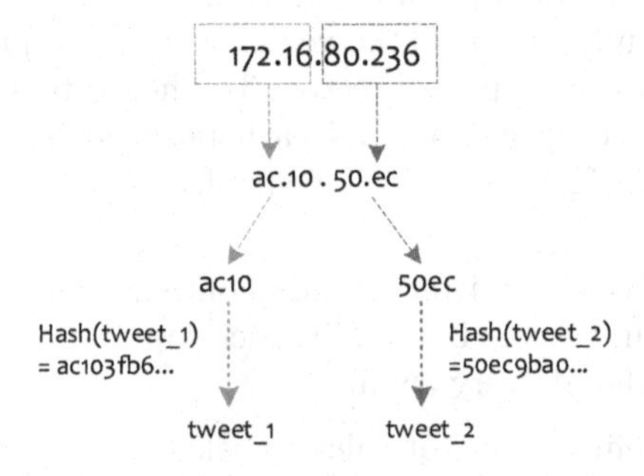

Fig. 4. Hash collision

(2) **Tweets Generation.** To perform a successful collision, the attacker needs numerous tweets. The new tweets are generated using EDA. After selecting a topic, the attacker crawls the trending tweets to generate more sentences. We crawled 1 K tweets for each selected trend in the experiments. Before generating new sentences using EDA, we cleaned the tweets first. As there are word deletions and swaps during augmentation, if a tweet is too short, the generated sentences may not contain the trending words. Thus, we filtered out tweets with less than ten words. Additionally, there were retweeted tweets that did not contain the trending words, so we filtered them out and retained only the original tweets. Then, we removed unnecessary chars like emojis, links, tabs, and line breaks in each tweet. Duplicate tweets were removed at last. Normally there were 400 to 900 tweets left. We used EDA to generate 50 sentences for each remaining tweet. It will get us 20 K to 45 K new sentences. It is still insufficient for a hash collision. We converted all sentences to the upper case and added punctuation (".", "..", "...", "!", "!!" and "!!!") at the end of each sentence. It resulted in 140 K to more than 300 K sentences in total, which greatly increased the success rate for a hash collision (see Sect. 5.2).

It's not recommended to convey a whole IP address in a tweet because it needs too many tweets to perform a successful collision. Two 16 bits will reduce the calculation greatly. Also, it is not deterministic for a successful hash collision. If a collision fails, the attacker can crawl more tweets or add more noise to the sentences. The attacker needs to post the two final tweets in order so that the malware can correctly recover the IP address.

There may be different situations where the compromised hosts cannot go online as expected, and the defenders can put on a saved avatar and post tweets with fake commands. In case of this happening, authentication like a digital signature with asymmetric key pairs is recommended to ensure a secure communication.

4 Implementation

In this section, we demonstrate the proposed convert C&C channel is feasible by presenting a proof-of-concept experiment on Twitter.

4.1 Siamese Neural Network

(1) **Architecture.** The Siamese Neural Network (SNN) [3] is effective in measuring the similarity between two inputs. The two inputs accepted by SNN will feed into two identical neural networks to generate two outputs. Like "Siamese" twins sharing the same organs, the identical neural networks share the same architecture and weights. The similarity between two inputs can be measured by calculating the distance between two outputs. We use Euclidean distance in this work. Figure 5 shows the architecture of the SNN. In this work, the two identical neural networks are CNNs [17]. It contains four convolutional layers and three fully connected layers. It accepts a 3-channel 128-pixel image as the input and generates 128 outputs to make up a feature vector.
The contrastive loss function [12] is used during the training. For two image inputs of the CNNs, Y is a binary label assigned to the pair, where $Y = 0$ represents the images being similar, and $Y = 1$ means that the images are different. G_1 and G_2 are two vectors generated by identical CNNs. Let $D_w = \|G_1 - G_2\|$ be the Euclidean distance between the vectors, w be the weights of the network, and $m > 0$ be a margin (radius around G). The loss function is:

$$L = (1 - Y)\frac{1}{2}(D_w)^2 + Y\frac{1}{2}(\max(0, m - D_w))^2$$

(2) **Training.** The model was implemented with Python 3.6 and PyTorch 1.5. To train the model, we crawled avatars of different sizes from 115,887 Twitter users and randomly selected 19,137 sets of avatars to build the dataset. Twitter provides 4 different sizes of avatars: 48×48, 73×73, 200×200 and 400×400. We randomly chose avatars of size 400×400 to make up input

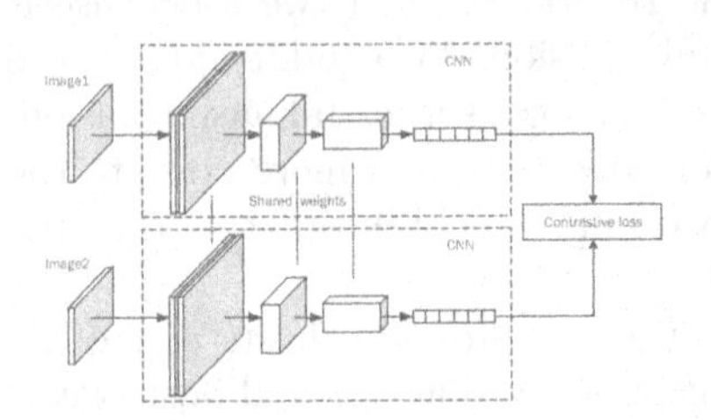

Fig. 5. Architecture of Siamese neural network

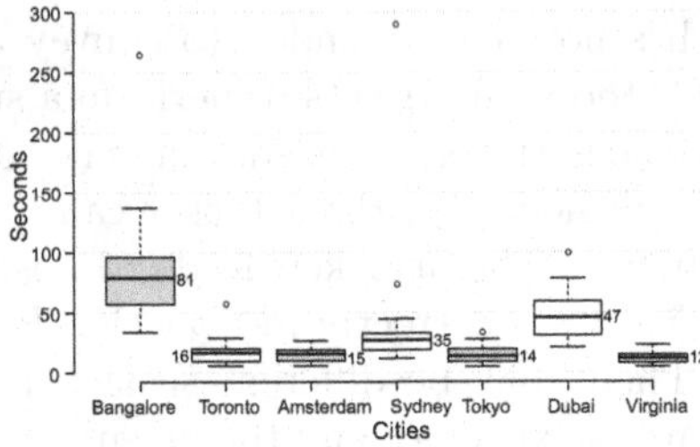

Fig. 6. Time cost for finding attacker

pairs with label 1. Due to the lack of original pictures of the avatars, we used avatars with sizes of 200×200 and 400×400 from the same user to make up input pairs with label 0. The ratio of input pairs marked as 0 and 1 is 1:2. Based on a preliminary experiment (Appendix B), the threshold for Euclidean distance was set to 0.02.

(3) **Performance.** To test the performance, we conducted the training process several times. The model converged rapidly during training. After 10–20 epochs, 100% accuracy on the test set was obtained. The size of a trained model is 2.42 MB. We used avatars from all 115,887 users to make up the validation set, for a total of 463,544 pairs (115,887 pairs with label 0 and 347,657 pairs with label 1, 1:3 in ratio). Evaluations show that the model reached an accuracy of more than 99.999%, with only 2–4 mislabeled pairs. Different from traditional machine learning works, we need to avoid hijacking the attacker's accounts, which means mislabeling from *not the same* to *the same* (false positive) is forbidden, while some mislabeling from *the same* to *not the same* (false negative) is allowed. The original labels of the mislabeled pairs were all 0, which means no avatar collision occurred with the trained models. It ensured the security of the attacker's accounts.

4.2 Experiments on Twitter

(1) **Environments.** To simulate the compromised hosts worldwide, we used 7 Ubuntu 18.04 x64 virtual servers with 1 GB ROM and 1 vCPU located in Bangalore, Toronto, Amsterdam, Sydney, Tokyo, Dubai, and Virginia. The code for the attacker was run on another virtual server with the same configuration in San Francisco. Both codes for the malware and attacker were implemented with Python 3.6.

(2) **Commands and Avatars.** We prepared 40 photos taken with mobile phones as avatars for the attacker's accounts. The photos were cut to 400×400 and converted into vectors by a trained model. The malware was published with the model and the vectors. The malware and attacker selected a trending topic once an hour in this experiment. Then, the attacker generated and posted the tweets, and the malware crawled related tweets 5 min later. In this experiment, the time was logged in a file when the attacker

completed a hash collision, the malware crawled a batch of tweets, and the malware started and finished the comparisons. The original commands and the recovered commands were also logged in a file. Afterward, we used the logs to compare the post time and the retrieval time and determine the correctness of the recovered commands.

(3) **Results.** We sent 47 commands using the 40 avatars. Due to frequent visits to Twitter trends, the selected topics are sometimes the same as the previous ones. Although it does not matter in real scenarios, we chose to wait for the next trending topic to evaluate the success rate of hash collisions more objectively. All commands in the experiments were received and parsed correctly by the seven hosts. During the tests, the attacker completed the tweet collection, tweets generation, and hash calculation in 13.8 s on average and reached a success rate of 90.28% for hash collisions. After selecting a trending topic, the malware attempted to crawl 1 K tweets and usually obtained 800–900 non-repeated tweets (only original tweets were saved for retweeted tweets). The malware needed to crawl the avatars of the tweeters and calculate the distances to identify the attacker. Due to the different network and device conditions, the time this process required varied. The time costs for the malware to find the attacker are shown in Fig. 6. It takes 5 s to 4.45 min to find the attacker after crawling the tweets. During the experiments, some of our tweets received several "likes" from Twitter users. It shows the sentences generated by EDA did not cause anomalies and were acceptable. After the malware got the IPs, the attacker deleted the tweets.

5 Evaluation

In this section, we evaluate the performance of different parts in DeepC2. **Environment:** The evaluation was performed on an Ubuntu 18.04 x64 virtual server with 1 GB ROM and 1 vCPU, and the code was implemented with Python 3.6.

5.1 Tweets Generation

To test the efficiency of tweet generation for the attacker, we selected 79 trending topics from 4 randomly selected English-speaking areas around the world (San Francisco, London, Sydney, and Johannesburg). One-thousand tweets were crawled for each topic. Additionally, we cleaned the crawled tweets using the method in Sect. 3.4 and generated 50 new sentences using EDA for each remaining tweet. The trending topics may contain one or more words. With random deletion and random swap adopted in EDA, keywords in the topics may be deleted, or position changed in the newly generated sentences. The malware cannot find the attacker's accounts if the attacker posts sentences without exact keywords. Therefore, the number of sentences with accurate keywords and the quantity of all generated sentences were also recorded.

In the 79 selected topics, 55 contained only one word, and 24 contained more than one word. With the percentage of words in each sentence to be changed set to 0.1, 89.54% of the newly generated sentences contained accurate keywords for the 55 single-word topics, and 77.55% contained accurate keywords for the 24 multi-word topics. The time cost is linearly related to the number of the new sentences, as shown in Fig. 7. As mentioned in Sect. 3.4, EDA obtains 20 K to 45 K sentences in this experiment. According to the test, generating the sentences costs 3 to 10 s. It is acceptable for the attacker to prepare sentences for a hash collision.

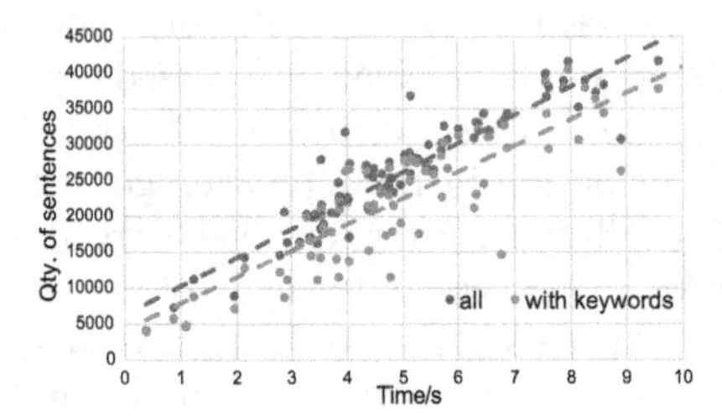

Fig. 7. Efficiency of tweets generation

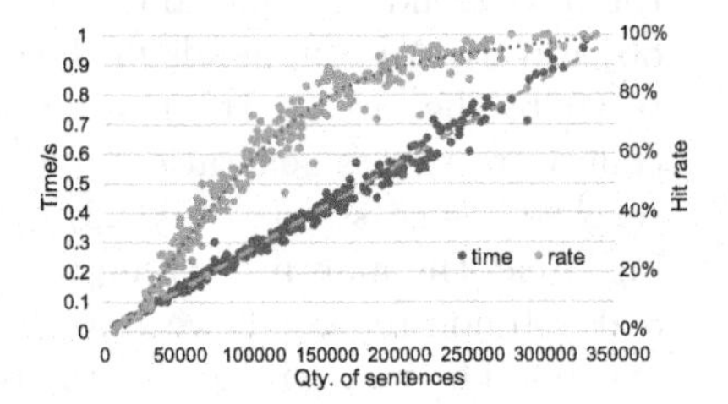

Fig. 8. Time costs and hit rate of hash collisions

5.2 Hash Collision

We used the sentences generated above to test the efficiency of hash collisions. To prepare different numbers of sentences, we followed the method in Sect. 3.4, converted cases, and added punctuation at the end of the sentences. We got four batches of new sentences incrementally for each topic by adding two conversions once. We also collected 100 IP addresses as commands from a threat report [10]. We call a batch of sentences "hit" an IP if the batch succeeds in the hash collision. We used these new sentences and hashlib in Python 3.6.9 to calculate SHA-256 hashes on the virtual server with a single thread and recorded the time costs and hit rate of hash collisions with different quantities of sentences.

As shown in Fig. 8, it took less than 1 s to calculate the hashes. In theory, 65,536 (2^{16}) sentences will hit an IP, which is ideal, as a hash collision is probabilistic. The experiment showed there should be at least 200 K sentences to obtain a 90% hit rate and more than 330 K for a nearly 100% hit rate. As mentioned in Sect. 3.4, there are usually 140 K to more than 300 K sentences for the hash collision, and it will result in a hit rate above 75%. During the experiments on Twitter, the attacker obtained an average of 219,335 sentences for hash collision and reached a hit rate of 90.28%, which was also acceptable for practical purposes. Moreover, the attacker can crawl more trending tweets and generate more sentences in real scenarios.

5.3 Avatar Recognition

To test the efficiency of avatar recognition by the malware, we used the 40 vectors above and 1,000 crawled avatars of size 400 × 400 to calculate the distances on the virtual server. The average time cost of extracting features from 1 K avatars and calculating 1 K distances was 11.92 s. It is also acceptable for the malware in such hardware conditions. In real scenarios, this process may take longer as the malware should crawl the avatars first, which varies due to different network conditions. Compared with the experiments on Twitter (Fig. 6), crawling the avatars is the most time-consuming process during the addressing.

5.4 Crawling Tweets

In this experiment, the malware crawl 1 K tweets 5 min after the selection of the trending topic. In real scenarios, attackers can customize the waiting time, crawling volume, and frequency. In this part, we'll show how the attacker determines the appropriate parameters.

We used the method in Sect. 3.3 to collect the trending topics. Then, we used the attacker's account to post tweets that contained the keywords. The malware started to find the attacker's account using the keywords after waiting for 5, 10, 20, 30, 45, 60, 90, 120, 150, and 180 min. The malware recorded how many tweets were crawled to find the attacker. We collected 56 groups of data. Figure 9 shows the relation between the crawled tweet volume and waiting time. After waiting 5 min, the malware found the attacker within 1 K tweets in all cases. After waiting for 1 h, in 88% of cases, the malware found the attacker within 1 K tweets and 98% within 3 K tweets. After waiting for 3 h, the malware could still find the attacker within 1 K tweets in 68% of cases and within 3K tweets in 89% of cases. As the waiting time is 5 min in the experiments on Twitter, it is appropriate to crawl 1,000 tweets.

The tweets may be more frequently updated if the attackers choose topics from larger cities such as New York and Los Angeles, and it may require the malware to crawl more tweets with the same waiting time. Additionally, if attackers choose top-ranked topics from the trending list, the malware also needs to crawl more tweets with the same waiting time. Moreover, it is also different if attackers choose to publish commands at midnight in the selected city. The parameters should be customized with different needs when applied in real scenarios.

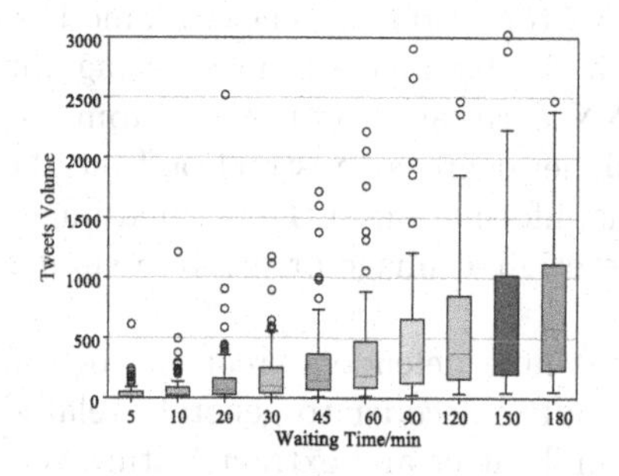

Fig. 9. Crawling volume and frequency

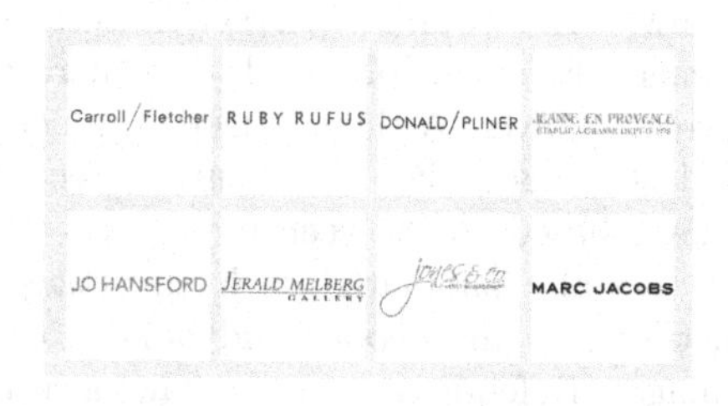

Fig. 10. A group of avatars that have distances below 0.02

5.5 Security Analysis

In this part, we discuss the security risks from the perspective of defenders.

Save and Reuse Avatars. Although it is difficult to guess the avatars used by the attacker, the defender can monitor the behaviors of the compromised hosts to identify the attacker's accounts. The defender can reuse the attacker's avatars when the next appointed time arrives. They can also select a topic and post tweets that contain fake commands. This scenario will not work for the hosts always online because each avatar is used only once. However, hosts that go online after being offline and missing a command will recognize the defender's accounts as attackers and get an incorrect command. Therefore, authentication is recommended to ensure secure C&C communication, as stated in Sect. 3.4.

Collide an Avatar. Defenders can try to collide an avatar. It sounds feasible but is hard practically. We analyzed the composition of the vectors. The 40 vectors in Sect. 4.2 contain 5,120 numbers. The numbers follow a normal distribution and constitute a continuous interval from −0.350 to 0.264. Each vector value is taken from the interval, which is ample space and hard to enumerate or collide. It ensures the security of the attacker's avatars and vectors.

However, we still attempted a collision for avatars. Using a trained model, we made more than 0.6 billion calculations on the distances between 115,887 pairs of crawled avatars. 2,050 avatar pairs have a distance below 0.02 (0.00031%), of which 81 pairs are below 0.01 (0.000012%). By analyzing these pictures, we found they share similar styles in that they all have a large solid color background, especially a white background (mainly logos) (see Fig. 10). As avatars are prepared by attackers, they can avoid this type of picture. They can use colorful pictures taken by their cameras instead of pictures from the Internet.

Train a GAN. Defenders may train a GAN with saved avatars to generate similar images. Considering the computational costs, it is not feasible. As the avatars can be animals, plants, arts, etc., the training target is too divergent to be capable with GAN. Additionally, training a GAN needs numerous data, and the attacker's avatars are insufficient for building a training set.

Train a Decoder. Defenders have access to vectors and neural network models, so they can attempt to recover and derive a similar image from cheating the malware. CNN makes protection possible. CNN learns abstract features from raw images. Each convolution layer generates a higher degree of abstraction from the previous layer. As layers deepen, much of the information in the original image is lost. This makes it difficult to recover the original image or derive a similar image based on the vectors.

We also simulated such an attack. We assume defenders treat the neural network as an encoder and build a corresponding decoder to generate related images. Defenders can also crawl avatars from Twitter and extract feature vectors using the model. The avatars and vectors make up the training data for the decoder. We trained such a decoder to generate numerous images from vectors and calculated the distance between the original image and the generated image. Due to the losses introduced by CNN and image conversion, the lowest distance we got is 0.0504, larger than the threshold. As avatars retrieved by the malware are not in the size of 128×128, more conversion and compression will be introduced to the images. It's also challenging to attack the C&C in this way.

Attack the Model. Defenders can attack the neural network model to let the malware make incorrect decisions on attacker's accounts. There are some works on neural network Trojan attacks [19], which make this attack possible. As the target of this attack is a neural network model, it may affect some compromised hosts but does not influence the other hosts. Other unaffected hosts can still make correct decisions on the attacker's accounts.

Generate Adversarial Samples. As the model and feature vectors are known to defenders, it is a white-box non-targeted adversarial attack in this scenario [23]. Defenders can generate an adversarial sample to fool the model. Adversarial attacks aim at misclassifying the original target. Although CNN has 128 outputs, they don't represent 128 classes. Each output is a value in the feature vector. A slight perturbation of the value will result in a distance higher than the threshold. Therefore, it's not applicable to attack the C&C in this way.

6 Possible Countermeasures

There are some ways to enhance the security of DeepC2, and we discuss them in Appendix C. In this section, we discuss the possible countermeasures.

Behavior Analysis. Traditional malware detection methods such as behavior analysis and traffic analysis can be applied to detect the malware [11]. There are periodic behaviors of the malware. They need to visit Twitter Trends periodically. After selecting a trending topic, they need to crawl tweets and avatars to find attackers. This series of operations can make up a behavioral pattern. In addition, the periodic net flow is also a noticeable feature.

Collaboration. In this scenario, it is recommended that security analysts share the malware samples to the communities and the related OSNs once they appear so that every party can contribute to the mitigating works. OSNs can detect attackers in real-time by running the samples and actively monitoring activities related to the malware and attackers. They can calculate the distances between the uploaded avatars and vectors and block the attackers as soon as the corresponding avatars are detected. This may need a large-scale calculation but is an effective way to mitigate this attack. Meanwhile, OSNs can also help to trace the attackers behind the accounts. Therefore, we believe the cooperation between OSNs and security communities is essential to mitigate this attack.

Improvement on OSNs. There are many ways to utilize OSNs, so OSNs should take measures to avoid abuse. The attackers should maintain some Twitter accounts. The accounts can be stolen from ordinary users, registered in bulk using automated programs [24], or brought from underground markets [9]. Therefore, we suggest OSNs apply more complex human-machine verification during the registration and manage the misbehaved social bots under the terms of services (ToS). Cracking down on underground account transactions is also necessary. While working on this work, we found some websites selling Twitter accounts in bulk. We cannot predict how they got the accounts and how the buyers use the accounts. Since it violates Twitter ToS [29], related parties should limit illegal account transactions. We have reported it to Twitter.

As AI can be used to launch cyberattacks, security vendors should also consider the malicious use of AI so that the attacks can be detected when they are applied in real scenarios in the future.

7 Conclusion

This paper discussed a novel covert command and control scenario, DeepC2, on OSNs by introducing AI technologies. By utilizing the poor explainability of neural network models, the addressing process can be concealed in AI models rather than exposed as reversible hard-coding. For issuing commands covertly, we use easy data augmentation and hash collision to generate contextual and readable command-embedded tweets to avoid abnormal content on OSNs. We conduct experiments on Twitter to show the feasibility and efficiency. Furthermore, we analyze the security of the avatars. We also discussed possible countermeasures to mitigate this kind of attack.

AI is also capable of attacks. With the popularity of AI, AI-powered attacks will emerge and bring new challenges to cybersecurity. Cyberattacks and defense are interdependent. We believe countermeasures against AI attacks will be applied in future computer systems, and protection for computer systems will be more intelligent. We hope the proposed scenario will contribute to future protection efforts.

Acknowledgements. This work is supported by the National Natural Science Foundation of China (No. 61902396), the Youth Innovation Promotion Association CAS (No. 2019163), the Strategic Priority Research Program of Chinese Academy of Sciences (No. XDC02040100), the Key Laboratory of Network Assessment Technology at Chinese Academy of Sciences, and Beijing Key Laboratory of Network Security and Protection Technology.

A Easy Data Augmentation

Table 1 is the example of EDA with an original sentence from Twitter. Bold words represent parts that have changed from the original sentence. The newly generated sentences are not grammatically correct. Because there are many grammatically incorrect sentences on the Internet, sentences generated using EDA can also be accepted by Internet users and confused with normal content.

Table 1. Sentences generated by EDA

Operation	Sentence
None	Our TAXII server is going to be taking a short nap at 11am ET today for an update.
SR	Our TAXII server is *endure* to be taking a short nap at 11am ET today for an update.
RI	Our TAXII server is going to be taking a short nap at 11am *cat sleep* ET today for an update.
RS	Our *short* server is going to be taking a *TAXII* nap at 11am ET today for an update.
RD	Our server is to be taking a short nap at 11am ET today for an update

SR: synonym replacement. RI: random insertion.
RS: random swap. RD: random deletion.

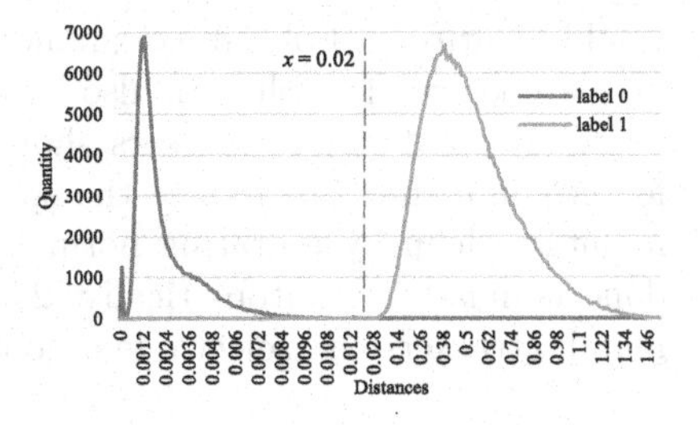

Fig. 11. Threshold for distance

B Threshold for Distance

A threshold is needed to determine whether two avatars share the same source. We use a trained model to calculate the distances on the validation set, which contains 115,887 pairs with label 0 and 347,657 pairs with label 1. We record the distances of every comparison, sort them by value and label, and count their frequencies to learn the boundary between the "same" avatars and different avatars. As shown in Fig. 11, the distances of all pairs with label 1 and only four pairs with label 0 are larger than 0.02, and the remaining pairs with label 0 are less than 0.02. It shows that 0.02 is a proper threshold for the determination. In real scenarios, attackers can choose a threshold less than 0.02, as the undistributed avatars and distances are within the authority of attackers.

C Enhancement

As proof of concept, the parameters in this work are conservative. There are ways to enhance the security of DeepC2.

In the model's design, the vectors can be longer than 128, making analysis and collisions for avatars even more difficult. The threshold of distances can also be lower than 0.02, as the undistributed avatars and the distances are within the authority of attackers. They can balance efficiency and accuracy according to the needs. Additionally, more losses can be introduced during the processing of avatars, like compression, deformation, format conversion, etc., making it harder to recover the avatars.

For addressing, the attacker can select more topics. Attackers can publish commands on the topics, and the malware can choose one randomly to find attackers. Attackers can also use other fields in OSNs to convey customized content. For instance, attackers could comment on a tweet, and the malware would identify and obtain commands from attackers' profiles. Other platforms, like Weibo and Tumblr, can also be utilized.

As stated before, attackers should maintain some accounts to publish different commands. To reduce the specious behaviors of accounts, attackers can maintain them by imitating normal users or social bots [7]. This work can be done manually or automatically [24]. When attackers need to publish a command, attackers can select one account and maintain other accounts as usual.

References

1. Anderson, H.S., Kharkar, A., Filar, B., Evans, D., Roth, P.: Learning to evade static PE machine learning malware models via reinforcement learning. CoRR abs/1801.08917 (2018). http://arxiv.org/abs/1801.08917
2. Bailey, M., Cooke, E., Jahanian, F., Xu, Y., Karir, M.: A survey of botnet technology and defenses. In: 2009 Cybersecurity Applications Technology Conference for Homeland Security, pp. 299–304 (2009)
3. Bromley, J., et al.: Signature verification using a "Siamese" time delay neural network. Int. J. Pattern Recogn. Artif. Intell. **7**(4), 669–688 (1993). https://doi.org/10.1142/S0218001493000339
4. Buchner, J.: ImageHash-PyPi (2020). https://pypi.org/project/ImageHash/
5. Chatterjee, S., Zielinski, P.: On the generalization mystery in deep learning. CoRR abs/2203.10036 (2022). https://doi.org/10.48550/arXiv.2203.10036
6. Faou, M.: From Agent.BTZ to ComRAT v4: a ten-year journey. Technical report, ESET, May 2020
7. Ferrara, E., Varol, O., Davis, C., Menczer, F., Flammini, A.: The rise of social bots. Commun. ACM **59**(7), 96–104 (2016)
8. FireEye: Uncovering a malware backdoor that uses twitter. Technical report, FireEye (2015)
9. Google: Google search (2021). https://www.google.com/search?q=buy+twitter+accounts
10. Group-IB: Lazarus arisen: architecture, techniques and attribution. Technical report, Group-IB (2017)
11. Gu, G., Perdisci, R., Zhang, J., Lee, W.: BotMiner: clustering analysis of network traffic for protocol- and structure-independent botnet detection. In: Proceedings of the 17th USENIX Security Symposium, pp. 139–154. USENIX Association (2008)
12. Hadsell, R., Chopra, S., LeCun, Y.: Dimensionality reduction by learning an invariant mapping. In: 2006 IEEE Computer Society Conference on Computer Vision and Pattern Recognition (CVPR 2006), vol. 2, pp. 1735–1742 (2006)
13. Hu, W., Tan, Y.: Generating adversarial malware examples for black-box attacks based on GAN. CoRR abs/1702.05983 (2017). http://arxiv.org/abs/1702.05983
14. Kirat, D., Jang, J., Stoecklin, M.P.: Deeplocker - concealing targeted attacks with AI locksmithing. Technical report, IBM Research (2018)
15. Kwak, M., Cho, Y.: A novel video steganography-based botnet communication model in telegram SNS messenger. Symmetry **13**(1), 84 (2021). https://doi.org/10.3390/sym13010084
16. Lecue, F., et al.: Explainable AI: foundations, industrial applications, practical challenges, and lessons learned, February 2020. https://xaitutorial2020.github.io/
17. LeCun, Y., et al.: Backpropagation applied to handwritten zip code recognition. Neural Comput. **1**(4), 541–551 (1989). https://doi.org/10.1162/neco.1989.1.4.541

18. Liu, T., Liu, Z., Liu, Q., Wen, W., Xu, W., Li, M.: StegoNet: turn deep neural network into a stegomalware. In: Annual Computer Security Applications Conference, ACSAC 2020, New York, NY, USA, pp. 928–938. Association for Computing Machinery (2020). https://doi.org/10.1145/3427228.3427268
19. Liu, Y., et al.: Trojaning attack on neural networks. In: 25th Annual Network and Distributed System Security Symposium, NDSS 2018 (2018)
20. Nagaraja, S., Houmansadr, A., Piyawongwisal, P., Singh, V., Agarwal, P., Borisov, N.: Stegobot: a covert social network botnet. In: Filler, T., Pevný, T., Craver, S., Ker, A. (eds.) IH 2011. LNCS, vol. 6958, pp. 299–313. Springer, Heidelberg (2011). https://doi.org/10.1007/978-3-642-24178-9_21
21. Pantic, N., Husain, M.I.: Covert botnet command and control using twitter. In: Proceedings of the 31st Annual Computer Security Applications Conference, ACSAC 2015, pp. 171–180. ACM (2015). https://doi.org/10.1145/2818000.2818047
22. Plohmann, D., Yakdan, K., Klatt, M., Bader, J., Gerhards-Padilla, E.: A comprehensive measurement study of domain generating malware. In: 25th USENIX Security Symposium, Austin, TX, pp. 263–278. USENIX Association, August 2016
23. Qiu, S., Liu, Q., Zhou, S., Wu, C.: Review of artificial intelligence adversarial attack and defense technologies. Appl. Sci. **9**(5), 909 (2019)
24. Quora: How can I create bulk twitter accounts automatically? (2020). https://www.quora.com/How-can-I-create-bulk-Twitter-accounts-automatically
25. Rigaki, M., Garcia, S.: Bringing a GAN to a knife-fight: adapting malware communication to avoid detection. In: 2018 IEEE Security and Privacy Workshops, SP Workshops 2018, San Francisco, CA, USA, pp. 70–75. IEEE Computer Society (2018). https://doi.org/10.1109/SPW.2018.00019
26. Sebastian, S., Ayyappan, S., Vinod, P.: Framework for design of graybot in social network. In: 2014 International Conference on Advances in Computing, Communications and Informatics (ICACCI), pp. 2331–2336. IEEE (2014)
27. Stefanko, L.: New telegram-abusing android rat discovered in the wild, June 2018. https://www.welivesecurity.com/2018/06/18/new-telegram-abusing-android-rat/
28. Taniguchi, T., Griffioen, H., Doerr, C.: Analysis and takeover of the bitcoin-coordinated pony malware. In: Proceedings of the 2021 ACM Asia Conference on Computer and Communications Security, pp. 916–930. ACM (2021)
29. Twitter: Twitter terms of service (2020). https://twitter.com/en/tos
30. Wang, J., Liu, Q., Wu, D., Dong, Y., Cui, X.: Crafting adversarial example to bypass flow-&ML-based botnet detector via RL. In: RAID 2021: 24th International Symposium on Research in Attacks, Intrusions and Defenses, San Sebastian, Spain, 6–8 October 2021, pp. 193–204. ACM (2021). https://doi.org/10.1145/3471621.3471841
31. Wang, Z., Liu, C., Cui, X.: EvilModel: hiding malware inside of neural network models. In: IEEE Symposium on Computers and Communications, ISCC 2021, Athens, Greece, 5–8 September 2021, pp. 1–7. IEEE (2021). https://doi.org/10.1109/ISCC53001.2021.9631425
32. Webster, A.F., Tavares, S.E.: On the design of S-Boxes. In: Williams, H.C. (ed.) CRYPTO 1985. LNCS, vol. 218, pp. 523–534. Springer, Heidelberg (1986). https://doi.org/10.1007/3-540-39799-X_41
33. Wei, J.W., Zou, K.: EDA: easy data augmentation techniques for boosting performance on text classification tasks. In: Proceedings of the 2019 Conference on Empirical Methods in Natural Language Processing and the 9th International Joint Conference on Natural Language Processing, EMNLP-IJCNLP 2019, Hong Kong, China, pp. 6381–6387 (2019). https://doi.org/10.18653/v1/D19-1670

34. Xia, R., Pan, Y., Lai, H., Liu, C., Yan, S.: Supervised hashing for image retrieval via image representation learning. In: Proceedings of the Twenty-Eighth AAAI Conference on Artificial Intelligence, pp. 2156–2162. AAAI Press (2014)
35. Yin, J., Lv, H., Zhang, F., Tian, Z., Cui, X.: Study on advanced botnet based on publicly available resources. In: Naccache, D., et al. (eds.) ICICS 2018. LNCS, vol. 11149, pp. 57–74. Springer, Cham (2018). https://doi.org/10.1007/978-3-030-01950-1_4

Artificial Intelligence for Detection

ODDITY: An Ensemble Framework Leverages Contrastive Representation Learning for Superior Anomaly Detection

Hongyi Peng[1(✉)], Vinay Sachidananda[1], Teng Joon Lim[2], Rajendra Patil[1], Mingchang Liu[1], Sivaanandh Muneeswaran[1], and Mohan Gurusamy[1]

[1] National University of Singapore, Singapore, Singapore
hongyi_peng@u.nus.edu, {comvs,dcsrsp,dcslium, e0503509,gmohan}@nus.edu.sg
[2] University of Sydney, Sydney, Australia
tj.lim@sydney.edu.au

Abstract. Ensemble approaches are promising for anomaly detection due to the heterogeneity of network traffic. However, existing ensemble approaches lack applicability and efficiency. We propose ODDITY, a new end-to-end data-driven ensemble framework. ODDITY use Diverse Autoencoders trained on a pre-clustered subset with contrastive representation learning to encourage base-leaners to give distinct predictions. Then, ODDITY combines the extracted features with a supervised gradient boosting meta-learner. Experiments using benchmarking and real-world network traffic datasets demonstrate that ODDITY is superior in terms of efficiency and precision. ODDITY averages 0.8350 AUPRC on benchmarking datasets (10% better than traditional machine learning algorithms and 6% better than the state-of-the-art semi-supervised ensemble method). ODDITY also outperforms the state-of-the-art on real-world datasets regarding better detection accuracy and speed. Moreover, ODDITY is more resilient to evasion attacks and has a promising potential for unsupervised anomaly detection.

Keywords: Anomaly detection · Ensemble methods · Semi-supervised settings · Intrusion detection · Auto encoder

1 Introduction

Digital equipment's prevalence threatens network security. Network administrators must contend with continually changing threat landscapes, increasing attack intensity, and complexity. As the first line of defense, the timely detection of anomalies in network traffic such as DDoS attacks, brute force attacks, botnet communications, and network/port scans, has become a significant focus in both academia and industry.

Increased uses of machine learning in anomaly detection reveals that detection of abnormal network traffic can be more challenging than binary classification. Lack of annotated data, especially anomalies, limits supervised methods'

C. Alcaraz et al. (Eds.): ICICS 2022, LNCS 13407, pp. 417–437, 2022.
https://doi.org/10.1007/978-3-031-15777-6_23

predictive capabilities. Unsupervised methods detect anomalies by measuring if a data point deviates significantly from others in terms of distance or density. Single model unsupervised method may not be sufficient due to the variety of anomaly sources. For example, a classifier that can detect a rapid surge in network traffic may not detect malware, intrusion, or misuse-related network anomalies. Ensemble learning has been proposed to combine multiple models by balancing diversity and accuracy of each individual model.

The state-of-the-art ensemble methods take a semi-supervised stacking approach. To promote diversity, semi-supervised methods employ a set of heterogeneous unsupervised representation learners as base-learners and then train a meta-learner in a supervised manner to integrate all the base-learners [3,25,35].

1.1 Research Questions

While semi-supervised ensemble approaches produce compelling results, they have several limitations. First of all, **the quality of representations learned is not guaranteed.** In previous publications [24,25,35], representation quality relied entirely on base-learner selection. Each base-learner in the framework work as an off-the-shelf black box. Thus, the mapping from raw data to representations is not learnable and depends on base-learner's hidden internal process. This problem limits generalization because deployment requires reselecting base-learners using cross validation.

Secondly, **the diversity among base-learners is not guaranteed.** There is no comprehensive strategy for selecting heterogeneous base-learners, nor are there any methodical procedures to ensure the diversity among them. As seen in Fig. 1, more recent work introduces a feature selection procedure to promote diversity [35]. However, this method relies on expensive investigations on representation generated by each base-learner and impairs the efficiency of the whole framework. There are ensemble frameworks that are based on homogeneous base-learners, such as autoencoders [10,26,30]. However, promoting diversity among homogeneous base-learners is difficult, as all individual learners use the same learning mechanism and are trained with the same dataset. As a result, they are highly correlated and produce sub-optimal performances.

The objective of addressing the gaps between extracting diverse representations and constructing an efficient semi-supervised ensemble framework leads us to the following research questions: (1)***RQ-1:*** *How to train base-learners to extract diverse and meaningful representations from raw data?* (2)***RQ-2:*** *How to achieve good trade-off between diversity and accuracy of among those base-learners?* (3)***RQ-3:*** *How to fully utilize the representations learned by base-learners and combine them?*

1.2 Overview

To address the aforementioned questions, we propose a novel and end-to-end framework `ODDITY` for anomaly detection. `ODDITY`'s advantage is due to a combi-

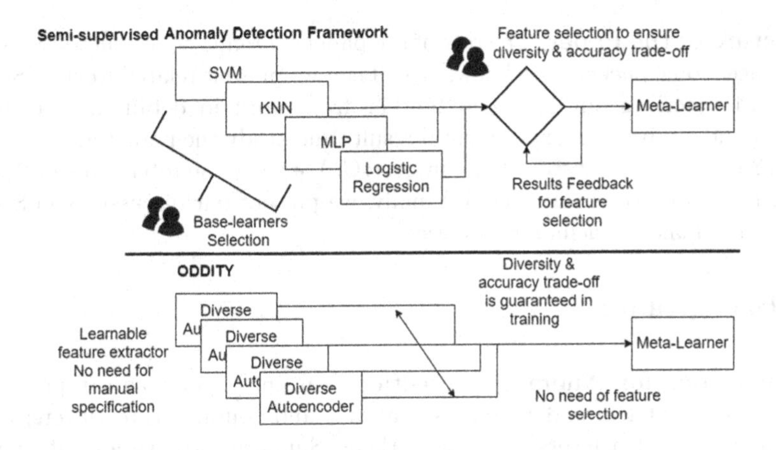

Fig. 1. ODDITY: a more generalized data-driven and end-to-end framework.

nation of design choices, not just one. We summarized its benefit and how each component helps as follows:

(1) **Base-learners: Learn to be different** ODDITY introduces an novel base-learner called *Diverse Autoencoders (DA)*. As a set of homogeneous base-learners, DAs learn from the dataset to extract diverse and complementary features. Each DA encoder is trained using contrastive representation learning on one of the pre-clustered subsets of the dataset. In addition to the diversity imposed by the partitioning of datasets, diverse autoencoders are further finetuned to produce distinct predictions. To the best of our knowledge, this is the first time that contrastive representation learning and diversity loss have been combined with the usage of an autoencoder for feature extraction.
(2) **Meta-learner: Efficient Gradient Ensemble** Regarding the ensemble methods, ODDITY extends the stacking approach. It augments the original feature space with features extracted by the encoder of diverse autoencoders to enrich the data representation. Then, It adopts a gradient boosting meta-learner to fully utilize the augmented features. ODDITY does not require empirical selection of base-learners or expensive feature-selection to promote diversity, making it a generalized and end-to-end anomaly detection framework.
(3) **Superior performance and efficiency** We evaluate ODDITY on multiple datasets and show that ODDITY outperforms reference methods in terms of both accuracy and efficiency. We also observe that the stacking-based ensemble framework of ODDITY can be easily adapted to an unsupervised setting.
(4) **Superior resilience against attack** We investigate how black-box evasion attacks impair the performance of supervised and semi-supervised anomaly detection techniques by deceiving the classifier. We observe that, due to the semi-supervised nature and the disagreement among base-leaners, ODDITY is more resilient than other techniques against evasion attacks.

Structure of the Paper. The rest of the paper is organized as follows. Section 2 introduces some background knowledge. Then we present related work in Sect. 3. Section 4 describes our proposed ODDITY framework in detail. In Sect. 5, we present and analyze the experimental results and study the potential of applying ODDITY in an unsupervised setting in Sect. C. We show the robustness of ODDITY against evasion attacks in Sect. 6. Finally, we present our discussions in Sect. 7, and Sect. 8 makes concluding remarks.

2 Background

Autoencoder for Anomaly Detection. An autoencoder is a type of neural network that learns data representations, and contains three layers: input, output, and hidden layers connecting them. Suppose we have an autoencoder with input dimension D and hidden dimension H, the encoder aims to learn the mapping $\mathsf{E} : \mathbb{R}^D \mapsto \mathbb{R}^H$. The decoder tries to learn the mapping $\mathsf{D} : \mathbb{R}^H \mapsto \mathbb{R}^D$. Considering an unlabeled dataset $\mathcal{U}$ with N samples, and each sample is a vector $\boldsymbol{x}_i$ with dimension D. Then the reconstruction error is defined as: $L = \sum_{i=1}^{N} \|\boldsymbol{x}_i - \mathsf{D}(\mathsf{E}(\boldsymbol{x}_i))\|^2$. The hidden layers learn to extract abstract representations by minimizing the reconstruction error during training. In the prediction phase, the autoencoders tend to perform poor reconstruction for abnormal points. The decision rule is straightforward-if the reconstruction error of the data point is larger than the threshold, it will be classified as anomalies.

Ensemble Methods. Ensemble learning tries to combine multiple learners to improve performance [37]. These learners are known as base-learners, suggesting a base-learning algorithm. Homogeneous approaches use a single base-learning algorithm. Building base-learners with different algorithms is called a heterogeneous approach. There are three main categories of methods in ensemble learning: boosting [15], bagging [7] and stacking [34]. An effective ensemble requires base-learners that are accurate and diversified [20]. It is intuitive to believe that each base-learner should be different; otherwise, combining identical individual learners would result in no performance improvement. Simultaneously, the individual learner's performance can not be very poor; otherwise, their combination will exacerbate the added error. Creating an ensemble of heterogeneous base-learners looks promising in encouraging diversity because heterogeneous base-learners are based on distinct learning algorithms and hypotheses. However, there is no systematic approach of either choosing diverse base-learners or ensuring a beneficial trade-off between accuracy and diversity.

3 Related Work

This section reviews the related works on semi-supervised frameworks and generating an ensemble from autoencoders.

Semi-supervised Outlier Ensemble. Micenková et al. have proposed a semi-supervised stacking-based framework BORE to leverage the strength of both supervised and unsupervised methods [24,25]. Aggarwal and Sathe [2,3] extend previous works by using heterogeneous base-learners finetuning the meta-learner with few labels. XGBOD [35] relates with our method the most. XGBOD improve the stacking-based anomaly detection framework by proposing three new feature selection methods and chooses XGBoost [11] as the final supervised meta-learner. However semi-supervised ensemble methods suffer from the following limitations: (1) They employs recurring experiments or heuristics to find more applicable heterogeneous base-learners, limiting their generalizability and scalability. (2) They adopts expensive feature selections to promote diversity among base-learner.

Ensemble of Autoencoders. Introducing diversity to homogeneous autoencoders is challenging. Many studies have been done to migrate these two issues. RandNet, developed by Chen et al. [10] targets unsupervised outlier detection. The diversity of autoencoders in RandNet is brought by bagging and randomly blocking some connections between the input layer and the hidden layer. Sarvari et al. [30] propose the Boost-based Autoencoders Ensemble (BAE). BAE is built sequentially as the training data sampled of the next autoencoder depending on the reconstruction errors obtained from the previous one (i.e. the larger the error, the less probably a data point is sampled in the next iteration). Both BAE and RandNet are unsupervised methods and they combine all autoencoders by averaging. However, sequentially training each autoencoder in BAE on an iterative basis is time-consuming, and naive averaging is less effective. Those methods differ from ours in (1) They are targeting unsupervised anomaly detection, which may not fully leverage the advantage of unsupervised learning in representation learning and the advantage of supervised learning in terms of performance. (2) Compared to introducing diversity by randomization, our approach adopts learn-to-be-diverse approach. (3) Our methodology train multiple diverse autoencoders in parallel, avoiding time-consuming sequential ensemble methods like [30].

4 Proposed Approach: ODDITY

This section elaborates on our approach. First, we introduce diverse autoencoders with experiments to demonstrate their capability of extracting diverse representations. Then we outline the whole framework of ODDITY.

4.1 Diverse Autoencoders

This section presents Diverse Autoencoders (DAs) that leverage contrastive representation learning to extract collections of diverse features. We denote an ensemble of M DAs as $\mathcal{E} = \{(\mathsf{E}_j, \mathsf{D}_j)\}^M$

As shown in Fig. 2, training DAs is a two-step process. The initial step is to train the encoder to learn the mapping from raw data to feature representation by contrasting positive pairs against negative pairs. For an unlabeled dataset $\mathcal{U} = \{\boldsymbol{x}_i\}^N$, we first cluster the dataset into M clusters $C_j = \{C_1, C_2 ... C_M\}$. Each encoder E_j is dedicated to learn the underlying representation of data

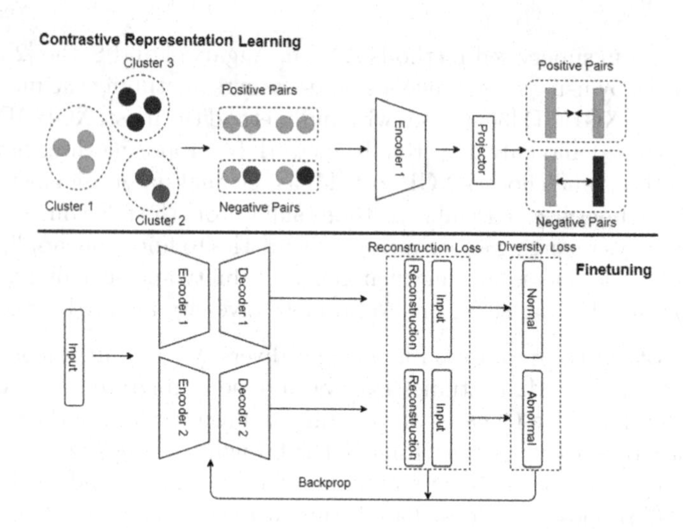

Fig. 2. Training of diverse autoencoders: contrastive representation learning and fine-tuning.

points within a single clxxuster C_j. This pre-clustering strategy avoids providing the same training data to each DA. A good representation of C_j should map data points map data points within the cluster closely together, whereas data points from other cluster should be projected farther apart. data points in the cluster C_j. Thus the positive pair can constructed using (x_a, x_b) where $x_{a,b} \in C_j$. On the opposite side, the negative pair is defiend as (x_g, x_h) where $x_{g,h} \in \mathcal{U} \setminus C_j$. The loss function of contrastive representation learning of encoder E_j is :

$$L_{contrastive}(\mathcal{U}, \mathsf{E}_j) = \text{dis}(p(\mathsf{E}_j(x_a)), p(\mathsf{E}_j(x_b))) - \text{dis}(p(\mathsf{E}_j(x_g)), p(\mathsf{E}_j(x_h))) \quad (1)$$

where $\text{dis}(a, b)$ calculates the L^p distance. As suggested *SimCLR*, we add one addition non-linear transformation $p(x)$ between the representation and the contrastive loss to improve the quality of learned representation [23]. Minimizing the contrastive loss function for all M encoders on teaches each encoder to extract the compact representation from the corresponding cluster.

In addition to the diversity imposed by the partitioning of datasets, we also introduce prediction-level diversity in the second stage. Complete M DAs in $\mathcal{E}$ is trained using a new loss function:

$$L(\mathcal{U}, \mathcal{E}, c) = \sum_{i=1}^{N} \sum_{j=1}^{M} \|\boldsymbol{x}_i - \mathsf{D}_j(\mathsf{E}_j(\boldsymbol{x}_i))\|^2 - c\, L_{div}(\mathcal{U}, \mathcal{E}) \quad (2)$$

For each DA, the first term in Eq. 2 is the regular reconstruction loss [5]. As introduced in Sect. 2, minimizing the reconstruction error improves the accuracy of the detection accuracy of autoencoders. The second term is the diversity loss where c is called the diversity factor. Maximizing this term should promote diversity by encouraging DAs to give different predictions. One way to design the diversity loss function is by averaging the pair-wise similarity of all pairs of DAs. However, computing the pair-wise similarity of M DAs takes C_m^2 comparisons (C_m^2 is the combination number), which significantly slows down the process. Thus, instead of computing pair-wise similarity, we measure the dissimilarity between DAs' reconstructions and the average reconstruction. Although there is no widely acknowledged official definition of diversity at the moment [37], we choose four typical and common measurements of dissimilarity: variance, KL-divergence, cosine similarity, and entropy:

$$L_{div}(\mathcal{U}, \mathcal{E}) = \begin{cases} \frac{1}{MN} \sum_{i=1}^{N} \sum_{j=1}^{M} ||\mathsf{D_j}(\mathsf{E_j}(x_i)) - \mathcal{C}||^2 & \text{Var} \\ \frac{1}{MN} \sum_{i=1}^{N} \sum_{j=1}^{M} KL(p_j(x_i), \frac{1}{M} \sum_{j=1}^{M} p_j(x_i)) & \text{KL-Div} \\ \frac{1}{MN} \sum_{i=1}^{N} \sum_{j=1}^{M} \frac{<\mathsf{D_j}(\mathsf{E_j}(x_i)), \mathcal{C}>}{||\mathsf{D_j}(\mathsf{E_j}(x_i))|| \, ||\mathcal{C}||} & \text{Cos} \\ \frac{1}{MN} \sum_{i=1}^{N} \sum_{j=1}^{M} -p_j(x_i) \frac{1}{M} \sum_{j=1}^{M} \log(p_j(x_i)) & \text{Entropy} \end{cases} \quad (3)$$

where $\mathcal{C}$ is the average reconstruction of M DAs $\mathcal{C} = \frac{1}{M} \sum_{j=1}^{M} \mathsf{D_j}(\mathsf{E_j}(x_i))$ and $p_j(x_i) = \phi(\mathsf{D_j}(\mathsf{E_j}(x_i)))$, and ϕ is the softmax function that normalizes the output of diverse autoencoders. The diversity loss term is another mechanism that trains DAs to produce diverse and complementary features. Minimizing the loss function with non-zero c rewards a trade-off between reconstruction accuracy and the diversity among DA, which benefits the subsequent ensemble.

Better Trade-off Leads to Better Ensemble. To investigate whether the two-stage training of DAs helps in achieving a better trade-off between diversity and accuracy. We compare diverse DA with RandNet [10] and BAE [30], we train DAs (Diverse AE), normal autoencoders (AE), RandNet and BAE that contains two base-learners (i.e., M=2) on the digits dataset [13] with the same architecture. Figure 3 demonstrates how the reconstruction and diversity losses vary with the number of epochs.

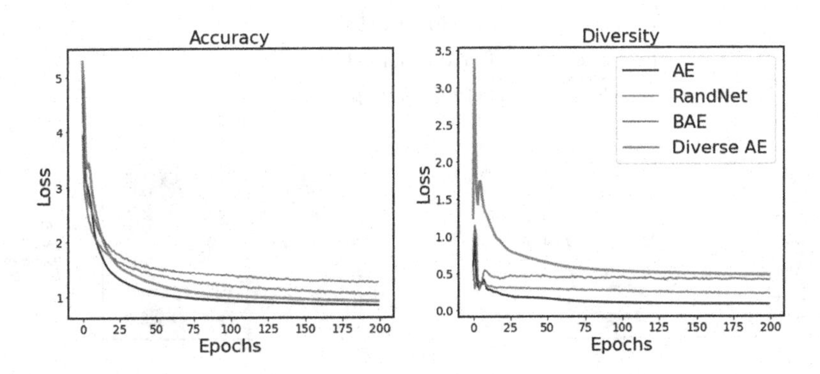

Fig. 3. Losses of different autoencoders ensemble. Comparing with other ensemble methods of autoencoders, Diverse autoencoders converge at the largest diversity while preserving small loss.

For normal autoencoders, with the decrease of the reconstruction loss, the diversity loss drops to near zero, implying that features extracted by two ordinary autoencoders are almost the same. In comparison, DAs provide significantly more diversity while achieving almost identical reconstruction loss. In other words, DAs can fulfill the fundamental requirements for base-learners, that is base-learners should be accurate and diverse. When compared to randomization-based systems such as RandNET and BAE, DAs deliver a higher level of accuracy while preserving a considerably higher level of diversity.

To demonstrate the effect of improved trade-off on ensemble results, we train two DAs with variance based diversity loss on a subset of a real network intrusion datasets UNSWNB-15 [27]. We evaluate each DA's detection accuracy against various types of malicious attacks. Then, by averaging their prediction scores (i.e. reconstruction loss), we generate the simplest ensemble of these DAs. We evaluate the ensemble's detection accuracy. The findings are summarized in Table 1.

Due to the training scheme of DA that promotes diversity, each DA is only capable of detecting a subset of attacks. The types of attacks that DA1 can detect and the attacks that DA2 can detect are complimentary. Consequently, combining DA1 and DA2 with the simplest averaging results in perfect accuracy.

Table 1. Ensemble results of simple ensemble of DAs

	DA 1	DA 2	AVG(DA1, DA2)
Reconnaissance	0.5380	0.4665	1
Backdoor	0.0960	0.9125	1
DoS	0.2812	0.7461	1
Exploits	0.7087	0.3074	1
Analysis	0.0886	0.9143	1
Fuzzers	0.6195	0.3883	1
Worms	0.8636	0.1818	1
Shellcode	0.5105	0.4894	1
Generic	0.1226	0.9728	1

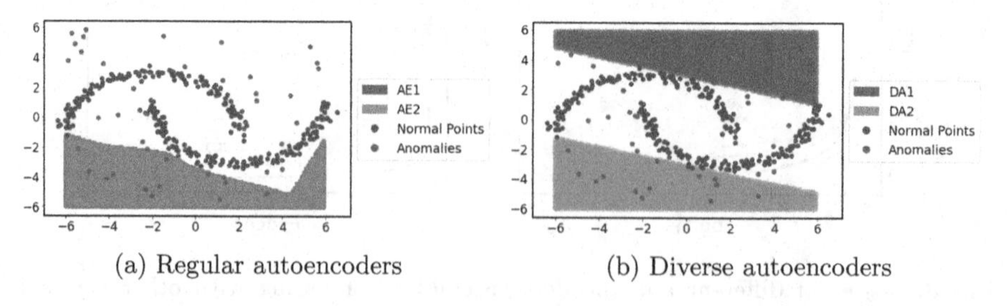

(a) Regular autoencoders (b) Diverse autoencoders

Fig. 4. Decision regions of regular autoencoders and diverse autoencoders

We create a synthesized dataset to demonstrate why diverse autoencoders enhances anomaly detection. As illustrated in Fig. 4, green moon-like points denote normal data, whereas red dots denote anomalies obtained by sampling from a uniform distribution. On the synthesized dataset, we train two regular autoencoders and two diverse autoencoders using the same $2 \rightarrow 1 \rightarrow 2$ architecture. We color the decision zone of each autoencoder into orange and blue so that all points in the decision region are predicted to be anomalies. As can be seen, the decision areas of the two regular autoencoders are largely overlapping, implying that without the diversity term, the two regular autoencoders provide identical predictions, and combining them results in minimal improvements. However, the two diverse autoencoders provide complementary predictions, and combining them can boost the performance.

4.2 Gradient Boosting Ensembles of Diverse Autoencoder

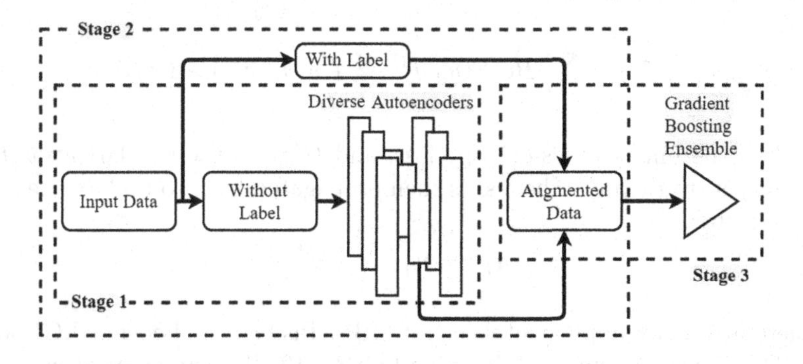

Fig. 5. Framework of ODDITY

While combining varied autoencoders improves performance, an efficient ensemble method and an efficient supervised meta-learner are required to fully utilize the power of semi-supervised learning. As depicted in Fig. 5, we demonstrate the overall framework of ODDITY. ODDITY is a stacking-based three-stage framework that uses gradient boosting ensembles (GBE) to leverage both the raw data and the representations learnt by DAs. Let $\mathcal{L}$ be a labeled dataset with N entries $\mathcal{L} = \{(\boldsymbol{x}_i, y_i)\}^N$ where $\boldsymbol{x}_i$ is a D dimensional column vector and $y_i \in \{0, 1\}$ are the corresponding label. Then the original feature space can be denoted as $X \in \mathbb{R}^{N \times D}$. We also define the unlabeled dataset $\mathcal{U}$ to be $\mathcal{U} = \{\boldsymbol{x}_i\}^{N_0}$ This unlabeled dataset can be generated by ignoring the labels or just taking out those points belonging to normal activities. ODDITY contain three stages:

1. After training multiple DAs on $\mathcal{U}$, the encoding of DAs can be viewed as a process that extracts highly informative features from X and stores these features in their hidden units.

2. ODDITY utilizes a set of compact and diverse features $F \in \mathbb{R}^{N \times MH}$ extracted by DAs. M is the number of DAs and H is the hidden dimension of DA. Then ODDITY stacks newly extracted feature F with original features X to form an augmented feature space:

$$X^* = [X, F] \in \mathbb{R}^{N \times (D+MH)} \tag{4}$$

3. This augmented dataset $\mathcal{L}^* = \{(\boldsymbol{x}_i^*, y_i)\}$ is deemed to contain richer representation and can improve the performance of supervised classifiers. Trained on L^*, a gradient boosting ensembles classifier that gives prediction $\hat{y}_i$ can be represented as:

$$\hat{y}_i = \sum_{1}^{K} w_k f_k(\boldsymbol{x}_i^*), \quad f_k \in \mathcal{F} \tag{5}$$

Where f_k is a weak classifier and w_k is the importance factor. Gradient ensembles classifiers add a new weak classifier f_t in t'th iteration and minimize the loss function:

$$\mathcal{L}^{(t)} = \sum_{1}^{N} BCE(y_i, \hat{y_i}^{(t-1)} + w_t f_t(\boldsymbol{x}_i^*)) + \Omega(f_t) \tag{6}$$

BCE is the binary cross-entropy loss, and $\Omega(f_t)$ is the regularization terms to prevent overfitting. The loss is then propagated back to update the importance factor.

$$w_t^{(t)} = w_t^{(t-1)} - \eta \frac{\partial \mathcal{L}}{\partial w_t} \tag{7}$$

There are many implementations of GBE. Instead of choosing XGBoost in XGBOD [35], we implement GBE using LGBM [19]. To demonstrate empirically that GBE is the best meta-learner for ODDITY, we first train various supervised classifiers on the raw data. Then, we replace GBE in ODDITY with such supervised classifiers to assess their efficiency in utilizing the diverse representation derived by DAs via AUPRC improvement. We train five diverse autoencoders on the benchmarking dataset (introduced in Sect. 5.1) and all of them have the same architectures: $D \to D/4 \to 4 \to 4/D \to D$, where D is the input dimension of the dataset. The results is provided in Table 2. Although all supervised meta-learners benefit from the diverse representations extracted by DAs, GBE in ODDITY improves the most and achieves the best performance.

Table 2. Comparison between GBE and other supervised meta-learner

	LR		SVM		MLP		GBE	
	without DA	with DA	without DA	with DA	without DA	with DA	without DA	with DA (ODDITY)
Letter	0.2639	0.3036	0.3606	0.4211	0.4297	0.469	0.5357	0.7848
Optdigits	1	1	1	1	0.9947	0.9917	0.9985	1
Pendigits	0.9658	0.9293	0.9547	0.9598	0.9642	0.948	0.9808	0.9971
Satellite	0.8538	0.8675	0.8560	0.8606	0.8807	0.9367	0.9549	0.9668
Mnist	0.9241	0.9269	0.9259	0.9348	0.9775	0.9836	0.9764	0.9814
Speech	0.1965	0.24	0.1784	0.1996	0.1695	0.2215	0.14	0.28
Average	0.7006	0.7123	0.7126	0.7293	0.7360	0.7584	0.7644	**0.8350 (+ 9%)**

5 Performance Evaluation

On numerous datasets, we compare our approach with conventional methods, gradient boosting ensemble techniques, and the state-of-the-art semi-supervised framework. The comparison results are provided in 5.3.

5.1 Benchmarking Datasets

To evaluate the performance and applicability of ODDITY, we use several multi-dimensional anomaly detection datasets that are commonly used for benchmarking [2,4,22,24,29,31]. We also add three real network traffic datasets, namely, KDD99, UNSW-NB15 [27] and IDS-2018 [1]. All of these three datasets aim to measure the efficiency of network intrusion detection systems (NIDS). Two significant distinctions exist between these network incursion datasets and benchmarking datasets. To begin, while benchmarking datasets are typically more challenging, network intrusion datasets can accurately reflect the real challenges associated with installing anomaly detection algorithms in a security system. As a result, it urges us to evaluate a variety of criteria when evaluating the effectiveness of anomaly detection algorithms, including precision, false-positive rate, and run-time efficiency. Second, network intrusion datasets are significantly larger, which complicates the training and testing of anomaly detection systems. Table 5 in Appendix A gives more details about each dataset.

5.2 Experiments

We randomly split every dataset so that 60% of the dataset is used for training and the remaining 40% is for testing. We select several baselines from five different categories for a comprehensive comparison:

- **Conventional supervised classifiers** including logistic regression, SVM [14], multi-layer perceptron (MLP) [6], kNN [21]. All of these methods are implemented using Sklearn [28].
- **Gradient Boosting Ensemble Methods**, GBE. We apply GBE directly to the dataset without feature augmentation but keeping the same architecture and the same hyperparameters as the meta-learner one in ODDITY.

- **Randomization-based autoencoders with ODDITY framework**, RandNet [10] and BAE [30]. Since both of them are targeting unsupervised methods, to make the performance comparable, we replace the diverse autoencoder in ODDITY with them and use the same architecture of autoencoders. this comparison aims to demonstrates whether adding diverse autoencoders can outperfors existing randomization-based algorithms by achieving higher accuracy. Also, we implement GBAE, where normal autoencoders replace the diverse autoencoders in ODDITY for ablation studies.
- **Semi-supervised Framework**, here, we choose XGBOD for the comparison. Comparing the performance of ODDITY with XGBOD can illustrate whether ODDITY can provide better predictive capability while solving the drawbacks of XGBOD. The unsupervised heterogenous base-learners of XGBOD in our experiments are kNN, Isolation Foreset, SVM, and Principal Component Analysis.
- ODDITY **Family**, all four variants of ODDITY that adopts variance-based loss function, KL-divergence-based loss function, cos-based loss function and entropy-based loss function respectively.

We implement ODDITY using Pytorch and LGBM. *Since there is no public source code for RandNet and BAE, we also implement them using Pytorch based on the original paper.* For every dataset, we set the number of diverse autoencoders M to be five, and all of them have the same architectures: $D \rightarrow D/4 \rightarrow 4 \rightarrow 4/D \rightarrow D$, where D is the input dimension of the dataset. Since the bottleneck hidden dimension H is four, the dimension of augmented features space will be $X^* = D + MH = D + 20$. That is to say, ODDITY will generate 20 new features for each data record. We adopts SGD optimizer with 0.01 learning rate to train DA. As suggested in [12], due to the imbalanced nature of anomaly detection datasets, we deem AUPRC the proper metric to measure accuracy. AURPC is defined as: AUPRC $= \sum_{i=1}^{N}(RC_i - RC_{i-1})PR_i$ where PR_i is the precision at the i th threshold and RC_i is the recall at the i th threshold.

We first evaluate the AUPRC score of our approach and baselines on benchmarking datasets. Then, we make a more comprehensive comparison between ODDITY and the state-of-the-art semi-supervised ensemble methods XGBOD on real network intrusion datasets. All experiments run on a computer with an Intel Xeon CPU, an Nvidia P100 GPU, and a 12 GB RAM.

5.3 Results

Benchmarking Datasets. Experimental results provided in Table 3 is an average of 10 independent trials. ODDITY achieves compelling results. Among all the variants of ODDITY, ODDITY-var achieves the best performance. In summary, ODDITY has significant advantages over traditional supervised classifiers. Notably, on the Letter dataset, where traditional methods perform poorly, ***ODDITY-var improves the AUPRC by more than 100% in comparison with SVM.*** Although GBE, XGBOD, and GBAE can achieve satisfying results on most of the datasets, ODDITY demonstrates uniformly superior predictive capabilities.

ODDITY-var has the best AUPRC scores in 5 out of 6 datasets, and the average AUPRC of ODDITY-var among all the datasets is 0.8350, which is 4% higher than the second-best, 6% better than the state of the art XGBOD and ***10% better than conventional supervised classifier.*** The advantages of ODDITY over LGBM suggest the importance of unsupervised representation learning, and the advantages of ODDITY over GBAE illustrate that the accuracy and diversity trade-off achieved by diverse autoencoders can further improve the performance. We also observe that ODDITY is overall superior and more generalized than XGBOD. A better trade-off between accuracy achieved by ODDITY also leads to better ensemble results compared with randomization-based techniques RandNet+GBE and BAE+GBE.

Table 3. Benchmark of ODDITY and other supervised & semi-supervised anomaly detection methods

Algorithm	Letter	Optdigits	Pendigits	Satellite	Mnist	Speech	Unweighted
	AUPRC	AUPRC	AUPRC	AUPRC	AUPRC	AUPRC	Average
Conventional Supervised Classifiers							
Logistic regression	0.2639	**1**	0.9658	0.8538	0.9241	0.1965	0.7006
SVM	0.3606	**1**	0.9547	0.8560	0.9259	0.1784	0.7126
MLP	0.4297	0.9947	0.9642	0.8807	0.9775	0.1695	0.7360
kNN	0.3462	**1**	0.9812	0.9239	0.9243	0.1209	0.7161
Gradient Boosting Ensemble Methods							
GBE	0.5357	0.9985	0.9808	0.9549	0.9764	0.1400	0.7644
Randomization-based Autoencoders with ODDITY framework							
GBAE	0.5815	0.9740	0.9860	0.9599	0.9829	0.1644	0.7748
RandNet+ GBE	0.5773	0.9784	0.9810	0.9587	0.9767	0.1488	0.7705
BAE+GBE	0.6481	0.9784	0.9806	0.9583	**0.9832**	0.1593	0.7847
Semi-supervised Framework							
XGBOD	0.6264	0.9967	0.9887	0.9385	0.9734	0.1747	0.7831
ODDITY Family							
ODDITY-KL	0.5125	0.9679	0.9886	0.9595	0.9814	0.1324	0.7570
ODDITY-cos	0.5804	0.9712	0.9768	0.9582	0.9779	0.1481	0.7697
ODDITY-Entropy	0.6889	0.9723	0.9836	0.9605	0.9789	0.1996	0.7957
ODDITY-var	**0.7848**	1	**0.9971**	**0.9668**	0.9814	**0.2815**	**0.8350**

We compare the feature importance map between ODDITY and GBE. We observe that the added features extracted by DAs play a vital role in the classification and result in improved performance. The result is provided in Fig. 7 and discussed in Appendix B.

However, we also notice that all techniques can not achieve decent results on the speech dataset. The 400 dimensions of the speech dataset might cause these failures. This is because traditional methods, especially kNN and SVM, are not suitable for handling such high-dimensional vectors. Moreover, the 20 new features generated by ODDITY are only a small portion compared with the input and can not contribute a lot to the final results.

It is natural to extend ODDITY to unsupervised learning by substituting an unsupervised classifier for the final supervised meta-learner. We compare multiple unsupervised meta-learners to determine which one is the greatest fit. The best results are then compared to those obtained using other unsupervised techniques. Appendix C contains the results and discusses them. In summary, ODDITY with an appropriate unsupervised meta-learner outperforms all other approaches even in the unsupervised setting.

Network Intrusion Datasets. Given that the ODDITY-var outperforms the baseline methods in benchmarking, we train it using the same architecture on real-world network intrusion detection datasets to provide a more extensive and comprehensive comparison with the XGBOD.

As indicated in Sect. 5.1, in addition to AUPRC, we examine precision, recall, f1-score, accuracy, false-positive rate, and runtime speed when evaluating the effectiveness and scalability of anomaly detection systems. Table 2 summarizes the outcome. While both techniques perform satisfactorily on network intrusion datasets, ODDITY still has significant advantages. ODDITY has a higher AUPRC for the KDD and UNSW-NB15 datasets, which results in increased precision, recall, and f1-score. Moreover, ODDITY has lower false-positive rates (FPR), which is a critical criterion for an anomaly detection system (3% lower for UNSW-NB15 and 0.3% lower for KDD). ODDITY's also has significant advantage interms of speed and runtime efficiency. On average, ***ODDITY accelerates training by 6 times and prediction by more than 50 times.*** On the UNSW-NB15 and KDD datasets, training ODDITY is 5.7 times and 6.69 times faster than XGBOD, respectively. Additionally, ODDITY occurs 52 times and 36.25 times during the prediction process. Even though both approaches produce excellent results on the IDS-2018 datasets, ODDITY still has a much better runtime efficiency.

Table 4. Comparison of ODDITY and XGBOD in real network intrusion datasets

	AUPRC	Precision	Recall	F1	Accuracy	FPR	Train time	Test time
UNSW-NB15								
XGBOD	99.73	95.37	96.00	96.83	99.20	5.87	3 min 16 s	24 s
ODDITY	**99.85**	**97.81**	**98.33**	**96.9**	**99.24**	**2.64**	**34 s**	**460 ms**
KDD								
XGBOD	99.95	99.74	99.56	99.59	99.97	0.85	6 min	43.5s
ODDITY	**99.99**	**99.82**	**1**	**99.91**	**99.99**	**0.56**	**53.8 s**	**1.2 s**
IDS-2018								
XGBOD	1	1	1	1	1	0	5 min 28 s	34 s
ODDITY	1	1	1	1	1	0	**27.6 s**	**234 ms**

6 Robustness Against Evasion Attacks

As machine learning-based anomaly detection systems became increasingly common, studies on their security and robustness in the presence of hostile adversaries increased. Numerous recent research has demonstrated that machine learning models' output can be arbitrarily adjusted with unnoticeable changes to the input [9,33]. Anomaly detection systems face unprecedented threats-malicious behavior and records attempt to evade detection by introducing a minor perturbation. Thus, it is critical for an ideal anomaly detection system to be resistant to evasion attacks.

6.1 Threat Model

For an anomaly detection engine h and some anomalous data points $(x, y = 1)$, the aim of evasion attacks is to fool the detection engine to misclassify the data points as normal points by introducing small or imperceptible changes to the input feature. $h(x') = 0$ where$x' = x + \delta$ and $\delta < \rho$ where ρ is a very small value.

These evasion attacks on anomaly detection systems might be classified according to the adversary's capabilities. While black-box attacks merely require queries to the target model, white-box attacks require the adversary to have full knowledge of the target model. For anomaly detection, the black-box threat model is more applicable. An attacker may query the decision engine to obtain matched anomaly scores from the detection engine. Every query to the model costs time and money and increases the risk of detection. From the perspective of defenders, anomaly detection systems are robust against evasion attacks when the rate of successfully fooling the decision engine is low even when the budget of queries B is large.

Algorithm 1: SimBA in pseudocode

```
Data: h, x, y, ϵ
Result: δ
δ = 0;
p = p_h(y|x);
B = 0;
Q is an arbitrary set of orthonormal basis of input space ;
while B < B_max do
    Randomly pick a vector q ∈ Q without replacement;
    Randomly pick a direction a ∈ {ϵ, −ϵ};
    p′ = p_h(y|x + δ + aq);
    if p' ¡ p then
        δ = δ + aq;
        p = p′;
    end
    B + +
end
```

6.2 Attack Algorithms

We launch simulated evasion attacking using the strategy introduced in SimBA (Simple black-box attacks) [17]. SimBA is a simple yet powerful strategy that exploits the confidence scores to construct adversarial data points. It's proven to achieve state-of-the-art success rate with an unprecedented low number of black-box queries. SimBA strategy can be described using the pseudocode 1 below: for any direction q and some step size ϵ, one of $x + \epsilon q$ or $x - \epsilon q$ is likely to decrease the anomaly scores $p_h(y|x)$. We therefore repeatedly pick random directions q and either add or subtract them.

6.3 Experimental Results

We launch attacks to the supervised methods and semi-supervised methods used in Sect. 5.3. We purposefully chose datasets for this experiment that most anomaly detection approaches archive decent results on, particularly the optdigits, the pendigits, the satellite, and the mnist dataset so that it will be more challenging for evasion attacks to bypass the decision engine. The orthonormal basis Q is generated from the $O(N)$ Haar distribution in the input feature space. We set the step size $\epsilon = 0.001$ and we increase the budget from 5 to 30, which means $||\delta||^2 \leq \sqrt{B * \epsilon} = \sqrt{0.003} = 0.05$. Since every input feature is scaled between 0 to 1, the SimBA attacks will only introduce a very small perturbation to the input.

Figure 6 illustrates how the success rate increases as the number of inquiries increases. Both supervised anomaly detection techniques (i.e., logistic regression, SVM, KNN, MLP) and ensemble-based methods (LGBM, XGBOD) are vulnerable to SimBA attacks. Within 30 inquiries, attackers can successfully deceive those decision engines. The sole exception is KNN on the Optdigits dataset, which maintains a near-zero success rate throughout the procedure. However, ODDITY is significantly more resilient than the others. Even when the budget for queries increases, ODDITY maintains a low success rate.

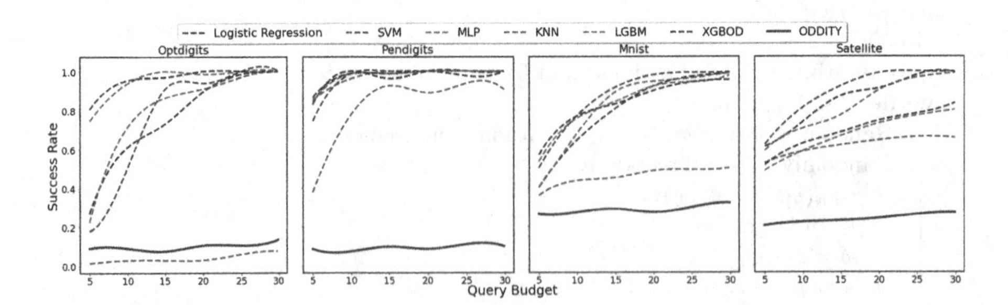

Fig. 6. Generally, success rate increase with query budget. Among all other methods, ODDITY is more robust against SimBA attacks by restricting the success rate to be low when we increase the query budget.

The reasons behind ODDITY's resiliency are (1) Diverse autoencoders in ODDITY that augment the input features, which act as redundancy and error checking. The additional features are retrieved from the original input and contain information about it. When you augment the input with certain perturbations, the decision engine can correct the input depending on the augmented feature. (2) The attacker is also unaware of the mechanism of various autoencoders, which implies the attacker is dealing with two Black Boxes. Thus, attacker cannot forecast the output of DAs when x is perturbed which increases the difficulty. (3) It's more difficult to deceive all DAs simultaneously. Because we encourage variety and disagreement among DAs, developing an attack mechanism that tricks all DAs will become increasingly difficult.

7 Discussion

ODDITY outperforms other supervised and semi-supervised techniques in terms of accuracy, efficiency and resistance to evasion attacks.

We elucidate why ODDITY achieves superior performance from three distinct perspectives: First of all, the variance-bias trade-off is a commonly used theoretical framework to understand anomaly detection techniques. Small variances and bias result in superior performance and robustness. ODDITY reduces model-centric bias by effectively combining representation learned by multiple diverse autoencoders. Besides, ODDITY, as a staking-based ensemble learning framework, introduces DAs to ensure each base-learner is independent and less correlated. Combining these base-learners yields better results since complementarity is more critical than pure accuracy. The mechanism in DAs that promotes diversity also relates to the fact that anomalies occur for a variety of reasons in real network traffic. They frequently exhibit very distinct patterns. Every DA is trained to be an expert at identifying specific types of attacks that make full use of each DA's predictive capability.

However, in unsupervised settings, the advantage of ODDITY framework is more limited since there are no labels to guide the internal feature pruning. The unsupervised meta-learners aggregate new features in a somewhat blind way. However, We still observe that the superiority of MCD + ODDITY compared with other unsupervised methods demonstrates the compelling potential for exploring ensemble methods in an unsupervised setting.

8 Conclusion and Future Works

We proposed a novel semi-supervised ensemble framework for anomaly detection, namely, ODDITY, extending the idea of combining unsupervised representation learning and supervised ensemble learning. In ODDITY, we propose DAs that leverage contrastive representation learning to extract diverse features and enrich the input representation. DAs makes ODDITY an end-to-end and data-driven framework, meaning deploying ODDITY does not need to select a set of heterogenous base-learners based on experience and expensive experiments. Experiments show that ODDITY is accurate, robust, and efficient. The future directions can be on investigating to empirically determine the performance with different hyper-parameters and visualize what DAs learned to provide interoperability.

A Experimental Datasets

Details of each dataset used in the experiment are provided in Table 5. Due to the size of the KDD99, UNSW-NB15, and IDS-2018 datasets, we randomly choose a portion of them.

Table 5. Summary of datasets

Datasets	Features	Normal	Anomaly
Letter [29]	32	1600	100
Optdigits [2]	64	5216	150
Pendigits [31]	16	6870	156
Satellite [22]	36	6435	2036
Mnist [4]	100	7603	700
Speech [24]	400	3686	61
KDD99	41	70458	2371
UNSW-NB15 [27]	49	50801	6313
IDS-2018 [1]	78	52238	19491

B Feature Importance map

The feature importance map in Fig. 7 reveals the importance of features for the Letter dataset and initially has 32 features (column 1–32), and DAs in ODDITY extract 20 more features (column 33–52). As shown in Fig. 7, the final classifier LGBM in ODDITY assign high importance factors on features (column 51, column 54, column 48, etc.) results in improved performance.

C ODDITY in Unsupervised setting

Replace the final supervised meta-learner with an unsupervised classifier to extend ODDITY to unsupervised learning. We incorporate ODDITY with three unsupervised classifiers, namely HBOS [16], Isolation Forest [22] and MCD [18]. Above mentioned methods are implemented using PyOD [36]. By choosing AUROC as the metrics, the hyperparameters and architecture of ODDITY remain the same as in Sect. 5.2. Table 6 summarizes the experimental results of averaging of ten trials. After utilizing the diverse features extracted by DA, the ROC of kNN improves by 0.3 %, the ROC of IF improves by 1.3 %, and the ROC of MCD improves by 8%. Since MCD + MCD outperforms others, we further compare the performance of MCD +ODDITY with other commonly used unsupervised anomaly detection techniques, including kNN, IF, PCA [32], and LOF [8]. ODDITY shows compelling potential in unsupervised anomaly detection by outperforming all other methods (Table 7).

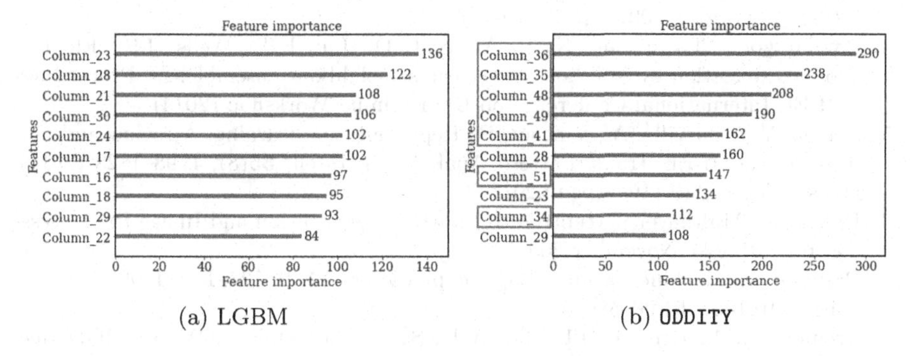

(a) LGBM (b) ODDITY

Fig. 7. Feature Importance of LGBM and ODDITY on Letter dataset

Table 6. ODDITY in unsupervised learning

Dataset	HBOS AUROC		Isolation Forest AUROC		MCD AUROC	
	without ODDITY	with ODDITY	without ODDITY	with ODDITY	without ODDITY	with ODDITY
Letter	**0.6223**	0.6080	0.6614	**0.6913**	0.6287	**0.7802**
Optdigits	**0.8184**	0.7820	0.6028	**0.6463**	0.3803	**0.4323**
Pendigits	0.9223	**0.9332**	**0.9548**	0.9202	0.8423	**0.9548**
Satellite	**0.7519**	0.7074	0.6750	**0.6835**	**0.7977**	0.7726
Mnist	0.6082	**0.6324**	0.8001	**0.8197**	0.8262	**0.8694**
Speech	0.4327	**0.4572**	0.4895	**0.5003**	0.4830	**0.4840**
KDD99	**0.4943**	0.4914	0.4903	**0.4911**	0.4914	**0.4965**
UNSWNB15	0.8451	**0.9465**	0.8125	**0.8377**	0.8219	**0.8407**
Average	0.6869	**0.6947**	0.6858	**0.6988**	0.6589	**0.7038**

Table 7. Comparison of unsupervised anomaly detection methods

Algorithm	Letter AUROC	Optdigits AUROC	Pendigits AUROC	Satellite AUROC	Mnist AUROC	Speech AUROC	KDD99 AUROC	UNSWNB15 AUROC	Unweighted Average
HBOS	0.6223	0.8184	0.9223	0.7519	0.6082	0.4327	0.4943	0.8451	0.6869
Isolation Forest (IF)	0.6614	0.6028	0.9548	0.6750	0.8001	0.4895	0.4903	0.8125	0.6858
PCA	0.5860	0.5329	0.9451	0.5918	0.8513	0.4601	0.5009	0.8724	0.6675
LOF	0.8504	0.5681	0.3863	0.5842	0.6913	0.4549	0.5076	0.5435	0.5733
MCD	0.6287	0.3803	0.8423	0.7977	0.8262	0.4830	0.4914	0.8219	0.6589
MCD + ODDITY	0.7802	0.4323	0.9548	0.7726	0.8694	0.4840	0.4965	0.8407	**0.7038**

References

1. Cse-cic-ids2018 datasets. https://www.unb.ca/cic/datasets/ids-2018.html. Accessed 23 June 2021
2. Aggarwal, C.C., Sathe, S.: Theoretical foundations and algorithms for outlier ensembles? ACM SIGKDD Explor. Newsl. **17**(1), 24–47 (2015)
3. Aggarwal, C.C., Sathe, S.: Outlier Ensembles, pp. 1–34. Springer, Cham (2017). https://doi.org/10.1007/978-3-319-54765-7
4. Bandaragoda, T.R., Ting, K.M., Albrecht, D., Liu, F.T., Wells, J.R.: Efficient anomaly detection by isolation using nearest neighbour ensemble. In: Proceedings of IEEE International Conference on Data Mining Workshop (2014)
5. Bengio, Y., Courville, A., Vincent, P.: Representation learning: A review and new perspectives. IEEE Trans. Pattern Anal. Mach. Intell. **35**(8), 1798–1828 (2013). https://doi.org/10.1109/tpami.2013.50
6. Bow, S.T.: Multilayer perceptron. In: Pattern Recognition and Image Preprocessing, pp. 201–224, November 2002
7. Breiman, L.: Machine learning. Bagging predictors **24**(2), 123–140 (1996). https://doi.org/10.1007/BF00058655
8. Breunig, M.M., Kriegel, H.P., Ng, R.T., Sander, J.: LOF. ACM SIGMOD Rec. **29**(2), 93–104 (2000). https://doi.org/10.1145/335191.335388
9. Carlini, N., Wagner, D.: Adversarial examples are not easily detected: bypassing ten detection methods. In: Proceedings of the 10th ACM Workshop on Artificial Intelligence and Security, pp. 3–14 (2017)
10. Chen, J., Sathe, S., Aggarwal, C., Turaga, D.: Outlier detection with autoencoder ensembles. In: Proceedings of the 2017 SIAM International Conference on Data Mining, pp. 90–98, September 2017. https://doi.org/10.1137/1.9781611974973.11
11. Chen, T., Guestrin, C.: Xgboost: a scalable tree boosting system. In: Proceedings of the 22nd ACM SIGKDD International Conference on Knowledge Discovery and Data Mining, pp. 785–794 (2016)
12. Davis, J., Goadrich, M.: The relationship between precision-recall and ROC curves. In: Proceedings of the 23rd International Conference on Machine Learning, pp. 233–240 (2006)
13. Dua, D., Graff, C.: UCI Machine Learning Repository (2017)
14. Erfani, S.M., Rajasegarar, S., Karunasekera, S., Leckie, C.: High-dimensional and large-scale anomaly detection using a linear one-class SVM with deep learning. Pattern Recogn. **58**, 121–134 (2016). https://doi.org/10.1016/j.patcog.2016.03.028
15. Freund, Y., Schapire, R.E.: A decision-theoretic generalization of on-line learning and an application to boosting. J. Comput. Syst. Sci. **55**(1), 119–139 (1997). https://doi.org/10.1006/jcss.1997.1504
16. Goldstein, M., Dengel, A.: Histogram-based outlier score (HBOS): a fast unsupervised anomaly detection algorithm. KI-2012 Poster 59–63 (2012)
17. Guo, C., Gardner, J., You, Y., Wilson, A.G., Weinberger, K.: Simple black-box adversarial attacks. In: International Conference on Machine Learning, pp. 2484–2493 (2019)
18. Hardin, J., Rocke, D.M.: Outlier detection in the multiple cluster setting using the minimum covariance determinant estimator. Comput. Stat. Data Anal. **44**(4), 625–638 (2004)
19. Ke, G., et al.: Lightgbm: a highly efficient gradient boosting decision tree. Adv. Neural Inf. Process. Syst. **30** (2017)

20. Krogh, A., Vedelsby, J.: Neural network ensembles, cross validation, and active learning. Adv. Neural. Inf. Process. Syst. **7**, 231–238 (1994)
21. Liao, Y., Vemuri, V.: Use of k-nearest neighbor classifier for intrusion detection. Comput. Secur. **21**(5), 439–448 (2002)
22. Liu, F.T., Ting, K.M., Zhou, Z.H.: Isolation forest. In: 2008 Eighth IEEE International Conference on Data Mining (2008). https://doi.org/10.1109/icdm.2008.17
23. Liu, X., et al.: Self-supervised learning: generative or contrastive. IEEE Trans. Knowl. Data Eng. (2021)
24. Micenková, B., McWilliams, B., Assent, I.: Learning outlier ensembles: the best of both worlds-supervised and unsupervised. In: Proceedings of the ACM SIGKDD 2014 Workshop on Outlier Detection and Description under Data Diversity (ODD2). New York, pp. 51–54. Citeseer (2014)
25. Micenková, B., McWilliams, B., Assent, I.: Learning representations for outlier detection on a budget. arXiv preprint arXiv:1507.08104 (2015)
26. Mirsky, Y., Doitshman, T., Elovici, Y., Shabtai, A.: Kitsune: an ensemble of autoencoders for online network intrusion detection. arXiv:1802.09089 (2018)
27. Moustafa, N., Slay, J.: UNSW-NB15: a comprehensive data set for network intrusion detection systems (UNSW-NB15 network data set). In: 2015 Military Communications and Information Systems Conference (MilCIS) (2015)
28. Pedregosa, F., et al.: Scikit-learn: machine learning in python. J. Mach. Learn. Res. **12**, 2825–2830 (2011)
29. Rayana, S., Akoglu, L.: Less is more: Building selective anomaly ensembles with application to event detection in temporal graphs. In: Proceedings of the 2015 SIAM International Conference on Data Mining (2015)
30. Sarvari, H., Domeniconi, C., Prenkaj, B., Stilo, G.: Unsupervised boosting-based autoencoder ensembles for outlier detection. In: PAKDD 2021. LNCS (LNAI), vol. 12712, pp. 91–103. Springer, Cham (2021). https://doi.org/10.1007/978-3-030-75762-5_8, https://arxiv.org/pdf/1910.09754v1.pdf
31. Sathe, S., Aggarwal, C.: Lodes: Local density meets spectral outlier detection. In: Proceedings of the 2016 SIAM International Conference on Data Mining (2016)
32. Shyu, M.L., Chen, S.C., Sarinnapakorn, K., Chang, L.: A novel anomaly detection scheme based on principal component classifier. Miami Univ. Dept. of Electrical and Computer Engineering, Technical report (2003)
33. Szegedy, C., et al.: Intriguing properties of neural networks. arXiv:1312.6199 (2013)
34. Wolpert, D.H.: Stacked generalization. Neural Netw. **5**, 241–259 (1992)
35. Zhao, Y., Hryniewicki, M.K.: XGBOD: improving supervised outlier detection with unsupervised representation learning. In: 2018 International Joint Conference on Neural Networks (2018)
36. Zhao, Y., Nasrullah, Z., Li, Z.: PyOD: a python toolbox for scalable outlier detection. J. Mach. Learn. Res. **20**(96), 1–7 (2019)
37. Zhou, Z.H.: Ensemble Methods: Foundations and Algorithms. CRC Press, Boca Raton (2012)

Deep Learning Based Webshell Detection Coping with Long Text and Lexical Ambiguity

Tongjian An(✉), Xuefei Shui, and Hongkui Gao

DAS-Security Co., Ltd., 310051 HangZhou, China
pacino.an@dbappsecurity.com.cn

Abstract. Webshell is a web page used by hackers for communicating with the web server and network intrusion. To detect webshell, deep learning methods are proposed to automatically extract features and mine semantics from PHP script. Although deep learning methods show a promising perspective, challenges still exist including the challenge of text selection when coping with long PHP script, the challenge of coping with lexical ambiguity in programming language and the challenge of decline of generalization ability to unseen PHPs if training samples are not treated reasonably. To resolve these challenges, we propose a two-stage deep learning webshell detection method. In stage one, a TextRank based sentence-level text selection model is proposed to preserve code semantic via extracting high-value code lines. While in stage two, after the tokenization of selected code lines, a CodeBert based token embedding model is utilized to resolve lexical ambiguity and generate token representation vectors. Down-stream task specified classifier is further utilized to detect webshell and to fine-tune the token embedding model. In training procedure, we split training/validation data set more reasonably to prevent data leakage. Besides, for the additional man labour of disposing false alarm in practical application, an extra large benign PHP data set is utilized for more solid validation. Experiments demonstrate the effectiveness of our method.

Keywords: Webshell detection · Deep learning · Text selection · Lexical ambiguity · CodeBert · TextRank

1 Introduction

Webshell is a piece of web scripting language program written for providing remote access and code execution to server functions [1]. Hackers may use webshells to do one or more of the following harmful tasks: 1) stealing information, 2) tampering databases, 3) modifying the home page of a website, 4) uploading malware, 5) intruding other machines in internal network, 6) escalating hacker's privileges to cause more serious damages [2]. In addition to the sever consequences caused by webshell, a research of Microsoft Detection and Response Team (DART) announced a steady increasing number of these attacks. From August 2020 to January 2021, DART registered an monthly average of 140,000 encounters of webshell threats on servers, almost double the 77,000 monthly average in 2019 [3]. Thus the detection of webshell is crucial in cyber-security more than ever.

C. Alcaraz et al. (Eds.): ICICS 2022, LNCS 13407, pp. 438–457, 2022.
https://doi.org/10.1007/978-3-031-15777-6_24

The webshell detection methods can be classified to 3 main categories by the features they are using and by how the features are generated: traditional/heuristic method, machine learning method and deep learning method.

Traditional/heuristic detection methods are usually based on string/signature matching algorithm and some statistically information of webshell like number of function calls, information entropy, etc. Tu et al. [5] proposed a detection method based on the optimal threshold values to identify suspicious webshell files that contain malicious codes from web applications by matching functions and signatures with webshell database. The list of selected suspicious webshell files are further checked by web administrators. Wang et al. [6] proposed an detection algorithm based on information entropy of PHP special strings, using a threshold of normal file information entropy to detect whether or not a PHP file contains webshell. Croix et al. [7] combined five heuristic features: fuzzy hashes, signature, dangerous routines, information entropy and obfuscation degree to detect PHP webshells. And the weight of each feature is determined by a genetic algorithm. Although traditional/heuristic methods can efficiently detect existing webshell, they are also subject to lack of generalization ability, i.e. fail to detect unseen webshell of which the signatures or patterns are not in the database.

Machine learning detection methods can automatically discover the relationships between features and webshell/benign PHPs by learning from large and varied data. Guo et al. [8] proposed a naive-bayes classifier based PHP webshell detection model combined with opcode sequence which generated by the Vulcan logic dumper (VLD) compilation. Method based on occurrence value of sensitive function call is studied in [9] and experiments are conducted using different machine learning classifiers including support vector machine (SVM), neural network (NN), decision tree (DT), naive-bayes (NB). Huang et al. [10] proposed ensemble learning method of multiple detectors using statistical features of high risk functions and variables. Experiments results showed that ensemble learning method yields a lower false positive rate. Although machine learning methods preform better than traditional methods, they have the disadvantages of using handcrafted features which need to be delicately designed by specialists [11, 13] and of lack PHP code semantic awareness because they commonly use statistically engineered features to train a model.

Deep learning detection methods have been proposed to mitigate defections of machine learning methods. Deep learning algorithms such as Convolution Neural Network (CNN) and Long Short-Term Memory (LSTM) have the abilities of extracting features automatically from numerous data. Lv et al. [11] proposed a CNN based webshell detection algorithm and achieved a higher detection accuracy than machine learning method. CNN is also combined with reinforcement learning approach for automatic feature selection in [12]. LSTM was exploited in work [13] combined with a down-sampling token filter method. The LSTM model is fed with a sequence of word vectors that generated from PHP script. Li et al. [4] proposed a word2vec method to vectorize words from each line in PHP script, then GRU and attention mechanism are utilized to further mine code semantic.

Although these proposed deep learning methods show a promising prospect in webshell detection task, they are subject to challenges of coping with long text and lexical ambiguity. Besides, in model training procedure, there is also a challenge of decline of model generalization ability if samples are not treated reasonably. We will illustrate these challenges and corresponding solutions in following sections. In this paper, the meaning of term 'PHP' is same with term 'PHP script' unless stated otherwise.

2 Challenges in Deep Learning Based Webshell Detection

In this section, three existed challenges in webshell detection, i.e. long text challenge, lexical ambiguity challenge and model generalization challenge will be illustrated in subsection sequence. In each subsection, we will also introduce the intuition of method utilized to solve the corresponding challenge. Finally the contributions of our paper will be extracted.

2.1 Long Text Challenge

In the scope of natural language processing (NLP), it is challenging when deep learning models process long text input. Long input leads to an expensive model computation and memory space cost. This is because the higher complexity in nowadays models such as LSTM [33], Transformer [34] and Bert [14]. Take self-attention module in Transformer for example, the computational and space complexity are both $O(nL^2)$, where L is the length of input and n is the size of hidden vector. It is obviously a drastically resource increasing for longer input length in training and inference procedure, thus make deep learning model even impossible to use in practical if long text is not well handled. To mitigate this challenge in NLP scope, some researchers utilized an "upper-stream" text selection method such as text truncation from head [21] and sentence selection [22, 32] before they fed text into deep learning model.

Long text challenge also exists in webshell detection task. We collected over 100 k PHP samples from GitHub high-ranking projects and our production systems, finding that the medians of token count of benign and webshell samples are 1 k and 3 k, respectively (token here means PHP code identifier, operator or keywords, etc.). Detailed statistics are listed in Appendix 1. In former research, some text selection methods have been utilized such as token filter [11, 13] and length fixed text truncation [4]. However, token filter methods cut the whole text into inconsecutive fragments. So down-stream deep learning model can hardly be exploited for further mining code semantic because neighboring context is not sufficiently preserved. Text truncation methods preserve a part of consecutive text and throw away the remainder, so they face the problems of overlooking important information and holding reluctant information when coming to long PHP text. To deal with long text, a more semantic friendly text selection method is necessary in webshell detection task.

In text analysis perspective, token filter methods belong to a micro lexical-level granularity which might ignore the semantic in neighboring context, whereas text truncation method belongs to macro document-level granularity which might lose the focus of high-value information of the text. Thus a medium sentence-level granularity text selection method should be researched for both preserving consecutive context and removing reluctant information from long PHP text.

2.2 Lexical Ambiguity Challenge

In the scope of NLP, lexical ambiguity is the potential for multiple interpretations of a word or phrase. Lexical ambiguity implies it difficult or impossible to understand the word meaning without some additional contextual information. This challenge also exists in unnatural language such as programming language [24]. For instance, two 'status' tokens exist in the following PHP code fragment:

```
if (! $curl['status']){$status = false;}
```

The first 'status' token in '$curl['status']' represents a key or a field member, while the second 'status' token in '$status = false' represents a local variable which has a totally different code semantic meaning with the first 'status' token. Besides, there might exist multiple '['status']' references and multiple '$status' assignments in this PHP script. For a more concise semantic comprehension, different representation vectors of these 'status' tokens should be generated by token embedding method. However, commonly used embedding method such as word2vec [4, 11, 13, 23] or one-hot encoding method generate identical vector for identical token, thus fail to deal with lexical ambiguity.

Lexical ambiguity in NLP scope can be addressed by Bert-like model architectures [14, 18, 25]. These architectures are mainly based on Transformer encoder which contains a series of self-attention modules. Since attention scores are mutually calculated between tokens in each location, the output embedding vector is not forced to invariant for identical token which is not the case in Word2vec model architecture. By combining a pretrained model with fine-tuning on down-stream tasks, these Bert-like models dominant in many NLP application. Besides, Bert-like models have also been successfully exploited in cyber-security scope including malware detection [18], cyber-security name entity recognition [19], reverse engineering [20], etc. In webshell detection area, there are also some bert-like projects pushed to GitHub [40, 41]. However, these methods have not been systematically researched and not been evaluated on a large evaluation data set appropriately. So in webshell detection task, the possibility of resolving lexical ambiguity by bert-like model should be studied.

Besides these two challenges in terms of data processing and model architecture, challenge also exist in model generalization aspect. Sample data in webshell detection model training should be carefully treated to mitigate model generalization challenge. As illustrated in next subsection, this challenge may be aggravated by data leakage problem which originate from a complete randomly training/validation sample split procedure.

2.3 Model Generalization Challenge

Training data sets used in many webshell detection literature are often consist of benign samples downloaded from GitHub public projects. After a de-duplication process, all benign samples are mix together with webshell samples into a whole data set, from which training and validation data set are randomly chosen [4, 8, 26, 27]. However, this seemingly reasonable sample split process will cause data leakage to an extent in the validation data set of benign sample. As a consequence, model with relatively lower generalization ability will be selected in training procedure.

The main reason lies in that benign PHP scripts in the same GitHub project may have bunch of project-specific code expressions such as same class names, same function names and same programming styles. And a randomly chosen data split method will split a part of one benign project in training data set and the other part of the exact same project in validation data set. This randomly split method may cause a high similarity between training and validation data set.

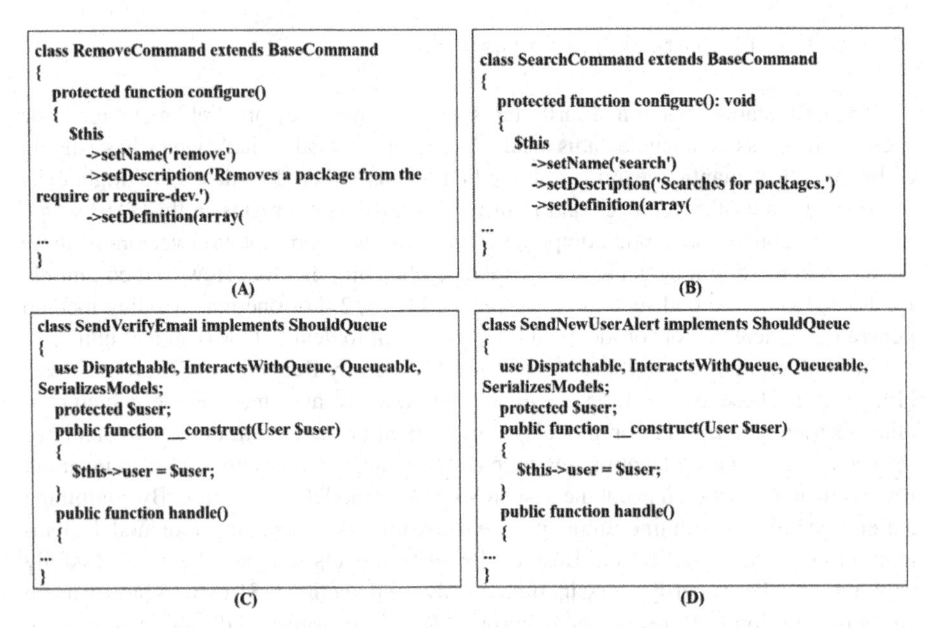

Fig. 1. 4 pieces of PHP scripts from two high ranking GitHub PHP project.

Take these 4 PHP scripts in Fig. 1 for instance, script (A) and (B) both belong to a high ranking (26 k + stars) GitHub PHP project [16], and script (C) and (D) to another high ranking (16 k + stars) GitHub PHP project [17]. It is shown that project-specific classnames and code styles do exist between (A) and (B), which is the same situation between (C) and (D). In a randomly data split method, if (A) and (C) are chosen to training data set while (B) and (D) to validation data set, then via the performance on this validation data set what will be selected is a webshell detection model that identify some project-specific text content as important features such as class name "BaseCommand" in both (A) and (B) and "use Dispatchable" expression in both (C) and (D). In other words, instead of mining complex semantic features from code logic, a model that easily remember these semantic irrelevant and project-specific features will perform well in validation set and thus be selected as a suitable model to detect those unseen PHPs. This obviously will cause a somewhat decline of generalization ability. So it is necessary to amend the data set split method to a moderate level of randomness. The selected model should mine code semantic in depth and learn from non-project-specific information.

2.4 Our Contributions

To mitigate these challenges in webshell detection task, we present a new method with following contributions:

Contribution 1: propose a TextRank based sentence-level text selection model that preserve relatively consecutive code semantic and filter out reluctant information. This contribution aims to address long text challenge in webshell detection. While text selection is always accomplished by statistically based token filter or by simple text truncation in the past, our method use a delicate model to handle it.
Contribution 2: propose a more concise code token vector representation model using CodeBert framework by taking lexical ambiguity into consideration. As far as we known, Bert-like framework has never been systematically studied in webshell detection in former works.
Contribution 3: use a training/validation data set split method with randomness in project granularity to address data leakage problem and promote model generalization ability. This data leakage problem has always been overlooked in former literature, while our paper raise it and address it.

Besides these 3 contributions, we use an extra larger benign PHP data set (86266 PHPs) to adequately validate False Positive Rate (FPR). FPR is a more crucial performance indicator even comparing with False Negative Rate (FNR) in practical applications, since webshell alarms need further checked and handled by human labor in Intrusion Detection Systems (IDS). Higher FPR is, higher labor cost the system needs. And this data set is the largest among webshell detection literature to the best of our knowledge.

The remainder of our work is organized as follows: method framework and details are illustrated in Sect. 3. Section 4 includes experiments setup data set description and result comparisons. Finally, the conclusion is draw in Sect. 5.

3 Our Method

In this section, we will firstly present an overview of our method architecture including sub-model connections, model inputs/outputs and data flows. Then we will illustrate specific implement details of each model and how these models are used to cope with long text and lexical ambiguity. In the end of this section, all implements are combine together for a further clarification of our method in a more intuitive way.

3.1 Method Overview

Overview of our method architecture is shown in Fig. 2. It is a two-stage deep learning method consists of three sub-models: a text selection model, a token embedding model and a down-stream classifier. Model functions and inputs/outputs are briefly listed below.

Text selection model aims at extracting high-value information from source code and cutting off reluctant information on sentence-level granularity. The input of text

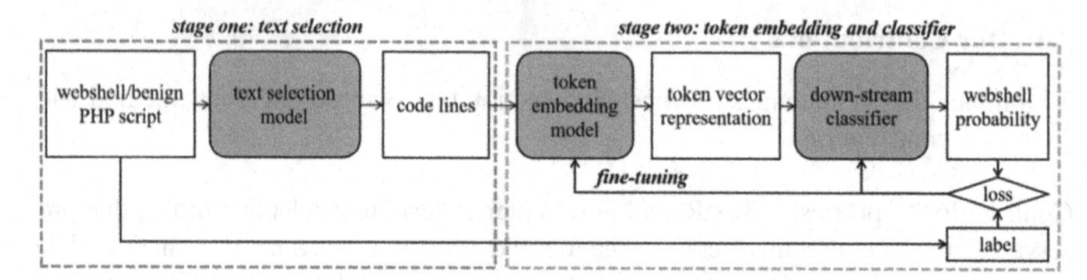

Fig. 2. Overview of our method architecture.

selection model is PHP script, the output is crucial semantic information which in our case is a number of complete code lines.

Token embedding model receives code tokens of the selected code lines as input and outputs vector representation of every single token and also an overall vector representation of the whole selected lines. For identical code token in various code contexts, different vector representations are generated by taking lexical ambiguity into consideration. To capture semantic information in a more extensive way, as usual way of NLP training, we adopt an embedding model which is already pretrained in advance with task irrelevant code. Then we fine-tuned this model with task relevant webshell/benign PHPs.

Then the token vectors are fed to a down-stream classifier to further mine code semantic. A many of models can be used as classifier including deep learning and machine learning models. These classifiers output the probability of a PHP script belongs to webshell. Model loss is calculated and utilized to train the classifier and to fine-tune the embedding model.

3.2 Text Selection Model

In our method, TextRank model is adopted to extract high-value text from PHP script. TextRank [28], a variation of PageRank [29], is a weighted graph-based model proposed for automatic extraction-based summarization. TextRank produces a graph nodes ranking without the need of labeled data. Ranking score calculation is mainly according to the similarity between nodes as formula (1) below. Nodes with higher ranking score represent better summarization of text.

$$WS(V_1) = (1-d) + d * \sum\nolimits_{V_j \in in(V_i)} \frac{W_{ji}}{\sum\nolimits_{V_k \in out(V_j)} W_{jk}} WS(V_j) \tag{1}$$

In formula (1), d is the damping coefficient; V_i and V_j represent graph node; W_{ji} represents the weight of the edge (similarity) between V_i and V_j; $in(V_i)$ is set of node that point to the node V_i while $out(V_j)$ is set of node pointed by node V_j; $WS(V_i)$ is the ranking score of node V_i and is calculated in an iterative way. This calculation method allows the use of TextRank algorithm with different languages ranging from natural language to programming language given a suitable definition of nodes and similarity between them.

In our implement, graph nodes of TextRank are PHP code lines which separated by End-of-Line (EOF) character. The similarity between a node pair is calculated by token

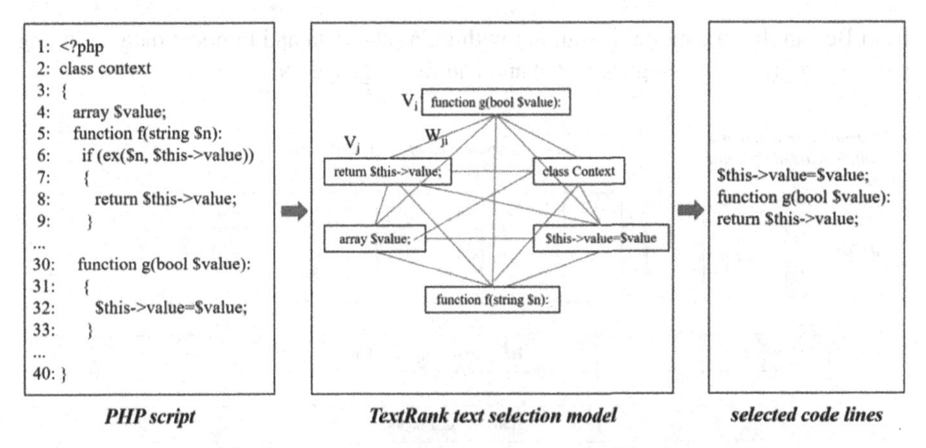
1: <?php
2: class context
3: {
4: array $value;
5: function f(string $n):
6: if (ex($n, $this->value))
7: {
8: return $this->value;
9: }
...
30: function g(bool $value):
31: {
32: $this->value=$value;
33: }
...
40: }
V_i function g(bool $value):
V_j
W_ji
return $this->value;
class Context
array $value;
$this->value=$value
function f(string $n):
$this->value=$value;
function g(bool $value):
return $this->value;
PHP script
TextRank text selection model
selected code lines

Fig. 3. TextRank based text selection model schema

co-occurrence similarity mentioned in [28]. Then the ranking score of each code line is calculated by iterations of formula (1). Important code lines are selected in a descending order of ranking score, starting with the highest score and ending when the count of selected tokens has achieved a threshold which in our work is 512 or when reaching the end of PHP script. Since TextRank is an unsupervised algorithm, the classifying loss of stage two will not be back propagated to text selection model in stage one. It means that the parameters of TextRank remain identical in training and inference phase.

Note that similarity calculation is a plug-in of TextRank model. This means co-occurrence similarity is an instance and can be replaced by similarity based on AST [42] or sentence embedding. We choose co-occurrence similarity in our implementation as an instance because this method needs no other package or software dependency and the calculation overhead is relatively low. This means co-occurrence similarity can be deployed in wide-ranging environments. If the deploy environment can fulfill the computational resource and dependency, other similarity calculation method can also be utilized.

The model schema with an intuitive example is shown in Fig. 3. Code line 8, 30 and 32 are selected from PHP script. Note that code lines contain only brace character ('{' or '}') or only string '< ?php' are filtered out from nodes in TextRank because these lines retain little semantic information.

3.3 Token Embedding Model

To cope with lexical ambiguity challenge in token embedding procedure, we adopt CodeBert which is essentially a Bert framework pretrained language model proposed in programming language scope [15]. Pretrained language model is trained in advance by a large amount of data via a self-supervised way and is later fine-tuned by a moderate amount of data in downstream tasks. As a Bert counterpart in programming language scope, CodeBert is also based on stacked Transformer encoder and is pretrained with codes of six programming languages including PHP, java, python, etc. CodeBert differs from Bert in that the model is training with unimodal data and bimodal data including pairs of programming language (PL) and natural language (NL).

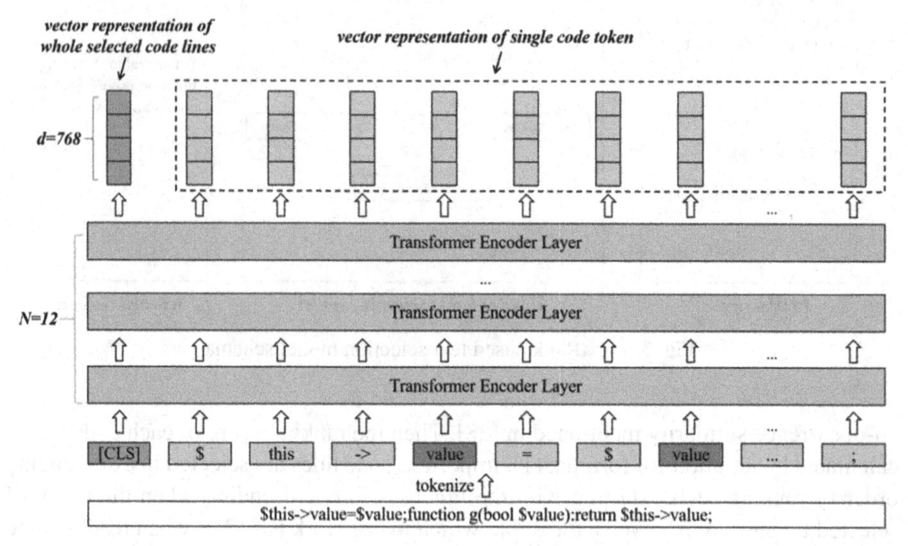

Fig. 4. CodeBert architecture.

The schema of CodeBert with 12 Transformer encoder layers and 768-dimension embedding vector is shown as Fig. 4. Attention mechanism in Transformer calculates an attention score between every token pair, no matter how long distance these pairs have. Thus identical tokens, for instance 2 'value' tokens in Fig. 4, pay attention to a different part of context and has a distinctive vector representation after training. Comparing to word2vec model in which identical token always have identical representation, CodeBert take lexical ambiguity into consideration by dynamically calculating token vector representation with context.

In our work, we use the official open source pretrained model parameters of CodeBert as the initial parameters of our token embedding model [30]. Then the parameters are fine-tuned by collected webshell/benign PHP scripts in training phase. We tokenize output code lines of text selection model into code tokens which are further fed into CodeBert to generate token vector representations. Among multiple layers of CodeBert, we use the output vector of last Transformers layer as representations of code tokens. Representation vector of whole selected code lines is noted by output vector in [CLS] location. Since [CLS] vector contains synthetic semantic information of PHP script, it can be directly used to detect webshell by combining with a light machine learning classifier such as softmax. However, down-stream deep learning classifier can also be exploited on token vectors to mine more semantic information.

3.4 Down-Stream Classifier

For further mining code semantics, we use TextCNN model as a down-stream webshell/benign classifier. TextCNN has been utilized in cyber-security scope such as code analysis and vulnerability detection [35, 36]. CNN is a neural network that includes convolutional computations originally applied to the image domain. It was transformed by researchers in [31] to TextCNN for text field application. TextCNN performs corresponding convolution and pooling operations on the input text vector using various size of sliding windows to capture both local features and global features. The model structure of TextCNN (with an intuitive example of 5 code token vectors and the vector length is 4) is shown in Fig. 5.

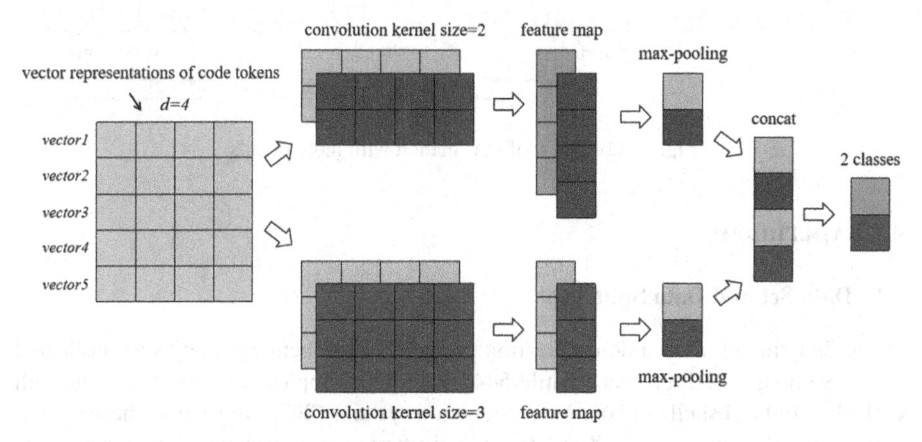

Fig. 5. TextCNN architecture.

Note that convolution kernel has length in two dimensions, one of which always aligns with the dimension of code token representation vector while the other is the kernel size. So a slide of convolution kernel on input is more like a n-gram feature extraction method where n is equal with the kernel size. Max-pooling operation are taken on the feature maps which are outputted by the convolution layer. Then a concat layer gathers all features in one hidden vector. Finally a fully connected layer transforms this hidden vector to output vector. The length of the output vector is equal to 2 which means 2 classes (webshell and benign).

Other down-stream classifier such as GRU, LSTM and Softmax can also be utilized and easily replace TextCNN in method architecture. A comparison study is performed in Sect. 4 between TextCNN, GRU, LSTM and Softmax which uses [CLS] of CodeBert as an input.

After place all model implements on Fig. 2, a more detailed architecture of our webshell detection method is shown in Fig. 6.

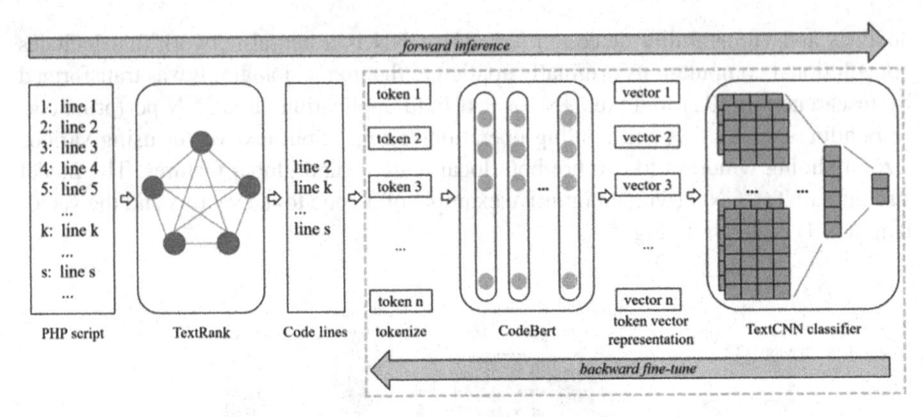

Fig. 6. Overview of our method with more details.

4 Experiment

4.1 Data Set and Data Split Type

In our experiment, after a de-duplication process, 21285 benign samples are collected from 38 GitHub PHP projects, while 5444 webshell samples are collected from both GitHub (2769 webshell) and our own production systems (2675 webshell). These benign and webshell samples are noted as data set D. We first randomly chose 12 benign projects (including 2037 benign PHPs) and 544 webshells from D to form an evaluation data set. Then for the split of training/validation data set from the rest of D (noted as D*), we use two types of data split for showing the existence of data leakage.

Split Type 1: randomly shuffle samples in D*. Then cut 80% of D* for training and 20% remainder for validation.
Split Type 2: for benign project, randomly shuffle in project granularity. Then cut 80% of benign projects for training and the 20% remainder projects for validation. While for webshell sample, the split data set remains the same with Split Type 1.

Split Type 1 and 2 are shown in Fig. 7 for more clearance. Green color part is for training data, gray color part is for validation data, while orange color part is for evaluation data. Note that evaluation data is identical for both Split Type 1 and 2.

Besides of benign samples mentioned above, we further collected 86266 benign samples from 862 GitHub projects to adequately evaluate the false positive rate indicator which is a major concern in practical webshell detection application. Webshell and benign sample sources are listed in Table 6 and Table 7 in Appendix 2.

4.2 Evaluation Criteria

We use 4 indicators to evaluate our method as formula (2). TP (true positive) and TN(true negative) are the number of webshell and benign samples that the model outputs correctly

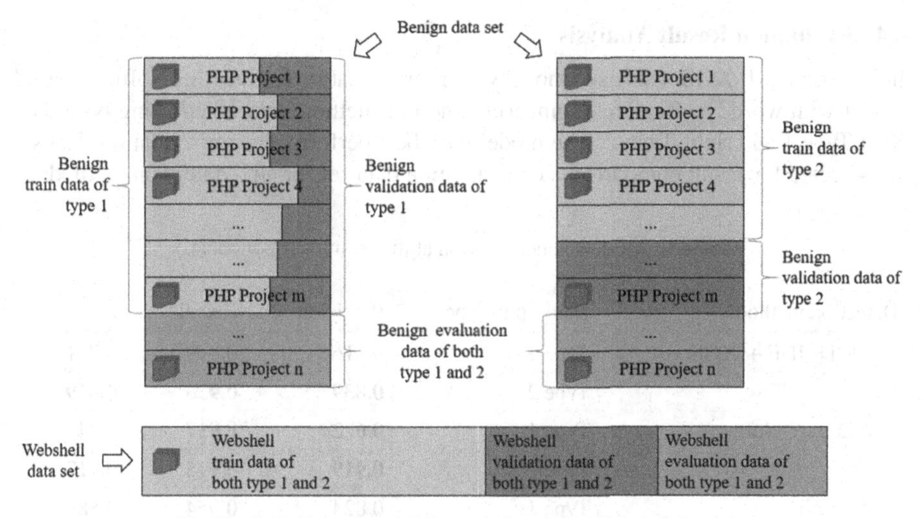

Fig. 7. Split Type 1 and 2 for training and validation data set.

while FP(false positive) and FN(false negative) are the number of webshell and benign samples that the model outputs incorrectly:

$$
\begin{aligned}
&Precision = TP/(TP + FP)\\
&Recall = TP/(TP + FN)\\
&F1 = (2 \times Precision \times Recall)/(Precision + Recall)\\
&FPR = FP/(FP + TN)
\end{aligned}
\tag{2}
$$

4.3 Model Parameters and Experiment Setup

In text selection model, the TextRank damping coefficient is set to 0.85, and the number of iterations is 200. In CodeBert model, the number of Transformer encoder layers is 12, the dimension of the hidden layer is 768 and the number of attention heads is 12. The number of convolution kernels of the TextCNN model is set to 256, and the size of the convolution kernel is set to 3, 4, 5, and 6 respectively. The learning rate is set to 0.00001 and the dropout is 0.1.

We adopt 3 series of experiments to validate the effectiveness of our proposed method.

Experiment 1: we train 3 models using both Split Type 1 and Split Type 2 for the demonstration of data leakage in validation data set and model generalization ability affected by this leakage.
Experiment 2: we compare our method with 3 available webshell detection software and 3 deep learning based webshell detection method in [4, 11, 13] to show the effectiveness of our method in an end to end way.
Experiment 3: we deploy an ablation study to further verify the effectiveness of our sub-models by changing one sub-model while keeping other sub-models fixed.

4.4 Evaluation Result Analysis

In Experiment 1, XGBoost based model with n-gram feature engineering, softmax based model with word2vec feature engineering and our method are trained using both data Split Type 1 and Split Type 2. The model with best performance on validation data set are selected for result comparisons on evaluation data set for all 3 detection methods.

Table 1. Models generalization ability results of 3 models.

Detection method	Data Split Type	Precision	Recall	F1
2-g + TF-IDF + XGBoost	Type 1	0.616	0.899	0.731
	Type 2	**0.839**	**0.923**	**0.879**
Word2vec + Softmax	Type 1	0.622	**0.917**	0.741
	Type 2	**0.919**	0.915	**0.917**
Our method	Type 1	0.824	0.954	0.884
	Type 2	**0.991**	**0.994**	**0.993**

In can be seen from Table 1 that methods trained by data Split Type 2 both perform better than Split Type 1 on evaluation data set which is unseen in training phase. Though the recall of Word2vec with softmax in Split Type 1 slightly surpasses Split Type 2, the precision decrease largely. This means the decline of generalization ability actually exists when model is trained by GitHub benign projects via data Split Type 1 which based on randomly shuffle in script granularity. Split Type 2 with project granularity shuffle should be used for avoiding the data leakage in validation set and for enhancing the ability of generalization to unseen PHPs. So in other experiments, we will adopt only data Split Type 2 and drop Split Type 1.

Note that because the webshell projects on GitHub is just collections of webshells written by different attackers from everywhere. Unlike benign projects, the webshells in these projects dose not suffer from data leakage because they were not written in same code style or using same identifier name. So random shuffle in webshell script granularity should not decrease the model generalization ability. To prove this hypothesis, we adopt an addition ablation experiment. This experiment shuffle benign samples in project granularity as Split Type 2 and compare the performance between two webshell shuffle granularities, i.e., script granularity and project granularity. And the detection methods remain same with Experiment 1. The result of this experiment support out hypothesis, that is random shuffle in webshell script granularity will not cause a data leakage problem. Detailed result is shown in Appendix 3 for not interrupting the smooth of our paper.

In Experiment 2, we introduce 3 similar deep learning based webshell detection methods which also contain a text selection model, a token embedding model and a down-stream classifier. Besides, 3 frequently used webshell detect software are also tested. FPR indicator of these 6 approaches and our method are further evaluate on 86266 benign PHPs. The results are shown in Table 2. Note that the Webshell Killer software has 2 detection mode to choose, one mode is for higher recall and the other mode is for higher precision.

Table 2. Results of 7 webshell detection approaches.

Detection method	Text selection model	Token embedding model	Down-stream classifier	Precision	Recall	F1	FPR of 86266 Benign PHPs	Recall of ob-webshells
D-shield [37]	/	/	/	0.984	0.915	0.949	0.004	0.915
Webshell Killer[1] [38]	/	/	/	0.985	0.864	0.921	0.004	0.855
Webshell Killer[2]	/	/	/	**1**	0.763	0.865	**0.001**	0.778
Cloud Walker [39]	/	/	/	0.986	0.776	0.868	0.003	0.752
Method 1 [4]	Truncation	Word2vec + Attention	GRU	0.978	0.923	0.950	0.006	0.915
Method 2 [11]	High frequency token	BoW	CNN	0.924	0.897	0.910	0.020	0.889
Method 3 [13]	Down sampling	Word2vec	LSTM	0.980	0.987	0.984	0.005	0.957
Our method	TextRank	CodeBert	TextCNN	0.991	**0.994**	**0.993**	0.002	**0.974**

Our method surpasses all detection approaches in recall and F1 score while Webshell Killer software in mode 2 achieves the highest precision, however, at the expense of a lowest recall. Since these detection software are commonly based on signatures and rule-based models, they achieve on average higher precision and a lower recall than the set of deep learning based models. Comparing with other deep learning models, our method performs better on all 4 indicators. The lower FPR (0.002) means that our method need a lower manual labour in disposing of the false alarm in practical application, with a percentage of 33% (0.002/0.006), 10% (0.002/0.02) and 40% (0.002/0.005) of deep learning method 1, 2 and 3, respectively. Our evaluation data set also contains 117 ob-webshells (short for obfuscated webshells), our method also performs better than other approaches in the recall indicator of these webshells.

In Experiment 3, we firstly fix the token embedding model and down-stream classifier while choosing the text selection sub-model from truncation, high frequency token reservation and TextRank. Secondly, we fix the text selection sub-model and down-stream classifier while choosing the token embedding model from CodeBert, Word2vec and One-hot. The aim of these two ablation studies is to verify the effectiveness of our proposed TextRank and CodeBert sub-models in webshell detection. The results are shown in Table 3 and Table 4.

It is shown in Table 3 that when token embedding model is fixed to CodeBert and down-stream classifier is fixed to GRU, TextCNN, LSTM or Softmax, TextRank text selection model performs better than truncation method and high frequency token method. Besides, there is only a slightly performance promotion of TextCNN comparing to GRU or LSTM when TextRank and CodeBert are utilized, implying text selection and token embedding model play a more important role in webshell detection task. Similar results can be seen in Table 4 that CodeBert achieves a higher F1 score and a lower FPR

Table 3. Ablation study results of text selection model.

Text selection model	Token embedding model	Down-stream classifier	Precision	Recall	F1	FPR of 86266 Benign PHPs
Truncation	CodeBert	Softmax	0.950	0.941	0.946	0.013
High frequency token	CodeBert	Softmax	0.922	0.895	0.909	0.020
TextRank	CodeBert	Softmax	**0.981**	**0.974**	**0.978**	**0.005**
Truncation	CodeBert	LSTM	0.965	0.950	0.957	0.009
High frequency token	CodeBert	LSTM	0.934	0.915	0.925	0.017
TextRank	CodeBert	LSTM	**0.991**	**0.993**	**0.992**	**0.002**
Truncation	CodeBert	GRU	0.981	0.958	0.969	0.005
High frequency token	CodeBert	GRU	0.927	0.910	0.918	0.019
TextRank	CodeBert	GRU	**0.991**	**0.993**	**0.992**	**0.002**
Truncation	CodeBert	TextCNN	0.956	0.969	0.963	0.012
High frequency token	CodeBert	TextCNN	0.938	0.914	0.926	0.016
TextRank	CodeBert	TextCNN	**0.991**	**0.994**	**0.993**	**0.002**

Table 4. Ablation study results of token embedding model.

Text selection model	Token embedding model	Down-stream classifier	Precision	Recall	F1	FPR of 86266 Benign PHPs
TextRank	One-hot	TextCNN	0.937	0.895	0.915	0.016
TextRank	Word2vec	TextCNN	0.960	0.969	0.964	0.011
TextRank	CodeBert	TextCNN	**0.991**	**0.994**	**0.993**	**0.002**
Truncation	One-hot	TextCNN	0.901	0.820	0.859	0.024
Truncation	Word2vec	TextCNN	0.934	0.941	0.938	0.018
Truncation	CodeBert	TextCNN	**0.956**	**0.969**	**0.963**	**0.012**
High frequency token	One-hot	TextCNN	0.870	0.776	0.820	0.031
High frequency token	Word2vec	TextCNN	0.887	0.884	0.886	0.030

(continued)

Table 4. (*continued*)

Text selection model	Token embedding model	Down-stream classifier	Precision	Recall	F1	FPR of 86266 Benign PHPs
High frequency token	CodeBert	TextCNN	**0.938**	**0.914**	**0.926**	**0.016**

than Word2vec and One-hot when text selection model and down-stream classifier are fixed. Thus the effectiveness of our method is verified via an end to end experiment 2 and an ablation experiment 3.

5 Conclusion

In this work, a two-stage model for webshell detection is proposed to mitigate the challenge of coping with long text in deep learning detection method and the challenge of programming language lexical ambiguity. A TextRank based sentence-level text selection method is utilized to extract high-value information from PHPs and a CodeBert based token embedding method is utilized to generate token vectors. TextCNN is used as a down-stream classifier to detect webshell and to fine-tune the embedding model. We prove the existence of data leakage of training if benign data is collected from GitHub project but not reasonably split. To avoid the decline of generalization ability caused by this data leakage, we propose a more reasonable data split method which shuffle samples in project granularity. We collect 26729 PHP samples to train and evaluate our method. Besides, an extra large benign data set (86266 samples) is also used to test false positive indicator which is a crucial concern in practical application. Evaluation results show our method improve the performance of webshell detection.

Appendix

Appendix 1. Statistical of PHP Samples (Table 5)

Table 5. Statistics of our collected PHP samples.

Label	Sample count	Count of	Mean	Std	Min	25%	50%	75%	Max
Benign	107551	Line	167	553	1	34	71	150	8.6e04
		Token	3574	2.0e04	10	406	1006	2718	3.0e06
Webshell	5444	Line	692	1328	1	7	49	1014	1.8e04
		Token	2.9e04	6.7e04	11	344	3414	3.6e04	1.1e06

Appendix 2. Examples of sample source

Table 6. Webshell sample sources from GitHub.

URLs of webshell sample
https://github.com/ysrc/webshell-sample
https://github.com/xl7dev/WebShell
https://github.com/tanjiti/webshellSample
https://github.com/webshellpub/awsome-webshell
https://github.com/DeEpinGh0st/PHP-bypass-collection/
https://github.com/tdifg/WebShell
https://github.com/malwares/WebShell
https://github.com/lhlsec/webshell
https://github.com/oneoneplus/webshell
https://github.com/vnhacker1337/Webshell
https://github.com/backlion/webshell

Table 7. Part of benign sample sources from GitHub.

URLs of benign sample
https://github.com/laravel/laravel
https://github.com/symfony/symfony
https://github.com/composer/composer
https://github.com/DesignPatternsPHP/DesignPatternsPHP
https://github.com/Seldaek/monolog
https://github.com/nextcloud/server
https://github.com/bcit-ci/CodeIgniter
https://github.com/PHPMailer/PHPMailer
https://github.com/monicahq/monica
https://github.com/nikic/PHP-Parser

Appendix 3. Experiment result of two webshell shuffle granularities (Table 8)

Table 8. Models generalization ability results of two webshell shuffle granularities.

Detection method	Webshell shuffle granularity	Precision	Recall	F1
2-g + TF−IDF + XGBoost	Script granularity	0.836	0.921	0.877
	Project granularity	0.839	0.923	0.879
Word2vec + Softmax	Script granularity	0.921	0.915	0.918
	Project granularity	0.919	0.915	0.917
Our method	Script granularity	0.989	0.994	0.992
	Project granularity	0.991	0.994	0.993

References

1. Kim, J., Yoo, D.H., Jang, H., Jeong, K.: WebSHArk 1.0: a benchmark collection for malicious web shell detection. J. Inf. Process. Syst. **11**(2), 229–238 (2015)
2. Hannousse, A., Yahiouche, S.: Handling webshell attacks: a systematic mapping and survey. Comput. Secur. **108**, 102366 (2021)
3. Web shell attacks continue to rise. https://www.microsoft.com/security/blog/2021/02/11/web-shell-attacks-continue-to-rise/. Accessed 10 Feb 2022
4. Li, T., Ren, C., Fu, Y., et al.: Webshell detection based on the word attention mechanism. IEEE Access **7**, 185140–185147 (2019)
5. Tu, T.D., Guang, C., Xiaojun, et al.: Webshell detection techniques in web applications. In: Fifth International Conference on Computing, Communications and Networking Technologies (ICCCNT), pp. 1–7. IEEE (2014)

6. Wang, C., Yang, H., Zhao, Z., et al.: The research and improvement in the detection of PHP variable webshell based on information entropy. J. Comput. **28**, 62–68 (2016)
7. Croix, A., Debatty, T., Mees, W.: Training a multi-criteria decision system and application to the detection of PHP webshells. In: 2019 International Conference on Military Communications and Information Systems (ICMCIS), pp. 1–8. IEEE (2019)
8. Guo, Y., Marco-Gisbert, H., Keir, P.: Mitigating webshell attacks through machine learning techniques. Future Internet **12**(1), 12 (2020)
9. Kurniawan, A., Abbas, B.S., Trisetyarso, A., et al.: Classification of web backdoor malware based on function call execution of static analysis. ICIC Express Lett. **13**(6), 445–452 (2019)
10. Huang, W., et al.: Enhancing the feature profiles of web shells by analyzing the performance of multiple detectors. In: Peterson, G., Shenoi, S. (eds.) DigitalForensics 2020. IFIP Advances in Information and Communication Technology, vol. 589, pp. 57–72. Springer, Cham (2020). https://doi.org/10.1007/978-3-030-56223-6_4
11. Lv, Z.-H., Yan, H.-B., Mei, R.: Automatic and accurate detection of webshell based on convolutional neural network. In: Yun, X., et al. (eds.) CNCERT 2018. CCIS, vol. 970, pp. 73–85. Springer, Singapore (2019). https://doi.org/10.1007/978-981-13-6621-5_6
12. Wu, Y., et al.: Improving convolutional neural network-based webshell detection through reinforcement learning. In: Gao, D., Li, Qi., Guan, X., Liao, X. (eds.) ICICS 2021. LNCS, vol. 12918, pp. 368–383. Springer, Cham (2021). https://doi.org/10.1007/978-3-030-86890-1_21
13. Qi, L., Kong, R., Lu, Y., et al.: An end-to-end detection method for webshell with deep learning. In: 2018 Eighth International Conference on Instrumentation & Measurement, Computer, Communication and Control (IMCCC), pp. 660–665. IEEE (2018)
14. Devlin, J., Chang, M.W., Lee, K., et al.: Bert: pre-training of deep bidirectional transformers for language understanding. arXiv preprint arXiv:1810.04805 (2018)
15. Feng, Z., Guo, D., Tang, D., et al.: Codebert: a pre-trained model for programming and natural languages. arXiv preprint arXiv:2002.08155 (2020)
16. https://github.com/composer/composer. Accessed 10 Feb 2022
17. https://github.com/monicahq/monica. Accessed 10 Feb 2022
18. Oak, R., Du, M., Yan, D., et al.: Malware detection on highly imbalanced data through sequence modeling. In: Proceedings of the 12th ACM Workshop on Artificial Intelligence and Security, pp. 37–48, November
19. Hou, J., Li, X., Yao, H., et al.: Bert-based Chinese relation extraction for public security. IEEE Access **8**, 132367–132375 (2020)
20. Li, X., Qu, Y., Yin, H.: Palmtree: learning an assembly language model for instruction embedding. In: Proceedings of the 2021 ACM SIGSAC Conference on Computer and Communications Security, pp. 3236–3251 (2021)
21. Akbik, A., Bergmann, T., Blythe, D., et al.: FLAIR: an easy-to-use framework for state-of-the-art NLP. In: Proceedings of the 2019 Conference of the North American Chapter of the Association for Computational Linguistics (Demonstrations), pp. 54–59 (2019)
22. Ding, M., Zhou, C., Yang, H., et al.: Cogltx: applying bert to long texts. Adv. Neural. Inf. Process. Syst. **33**, 12792–12804 (2020)
23. Yong, B., et al.: Ensemble machine learning approaches for webshell detection in Internet of things environments. Trans. Emerg. Telecommun. Technol. **33**(6), e4085 (2020)
24. Delorey, D. P., Knutson, C. D., Davies, M.: Mining programming language vocabularies from source code. In: PPIG, p. 12 (2009)
25. Liu, Y., Ott, M., Goyal, N., Du, J., et al.: A robustly optimized bert pretraining approach. arXiv preprint arXiv:1907.11692 (2019)
26. Zhu, T., Weng, Z., Fu, L., et al.: A web shell detection method based on multiview feature fusion. Appl. Sci. **10**(18), 6274 (2020)

27. Ai, Z., Luktarhan, N., Zhao, Y., et al.: WS-LSMR: malicious webshell detection algorithm based on ensemble learning. IEEE Access **8**, 75785–75797 (2020)
28. Mihalcea, R., Tarau, P.: Textrank: Bringing order into text. In: Proceedings of the 2004 Conference on Empirical Methods in Natural Language Processing, pp. 404–411 (2004)
29. Page, L., Brin, S., Motwani, R., et al.: The PageRank citation ranking: bringing order to the web. Stanford InfoLab (1999)
30. https://github.com/microsoft/CodeBERT. Accessed 17 Feb 2022
31. Kim, Y.: Convolutional neural networks for sentence classification. In: Proceedings of the 2014 Conference on Empirical Methods in Natural Language Processing (EMNLP), pp. 1746–1751 (2014)
32. Min, S., Zhong, V., Socher, R., et al.: Efficient and robust question answering from minimal context over documents. In: Proceedings of the 56th Annual Meeting of the Association for Computational Linguistics, pp. 1725–1735 (2018)
33. Hochreiter, S., Schmidhuber, J.: Long short-term memory. Neural Comput. **9**(8), 1735–1780 (1997)
34. Vaswani, A., Shazeer, N., Parmar, N., et al.: Attention is all you need. Adv. Neural Inf. Process. Syst. **30** (2017)
35. Lee, Y.J., Choi, S. H., Kim, C., et al.: Learning binary code with deep learning to detect software weakness. In: KSII the 9th International Conference on Internet (ICONI) 2017 Symposium (2017)
36. Lu, S., Guo, D., Ren, S., et al.: Codexglue: A machine learning benchmark dataset for code understanding and generation. arXiv preprint arXiv:2102.04664 (2021)
37. http://www.d99net.net/. Accessed 24 Mar 2022
38. https://edr.sangfor.com.cn/api/download/WebShellKillerTool.zip. Accessed 24 Mar 2022
39. https://github.com/chaitin/cloudwalker. Accessed 24 Mar 2022
40. https://github.com/lyccol/CodeBERT-based-webshell-detection. Accessed 09 Jun 2022
41. https://github.com/5wimming/bert-webshell. Accessed 09 Jun 2022
42. Backes, M., Rieck, K., Skoruppa, M., et al.: Efficient and flexible discovery of PHP application vulnerabilities. In: 2017 IEEE European Symposium on Security and Privacy (EuroS&P), pp. 334–349 (2017)

SimCGE: Simple Contrastive Learning of Graph Embeddings for Cross-Version Binary Code Similarity Detection

Fengliang Xia[1], Guixing Wu[2(✉)], Guochao Zhao[1], and Xiangyu Li[1]

[1] University of Science and Technology of China, Hefei, China
{xflmail,iyao,xylil}@mail.ustc.edu.cn
[2] Suzhou Institute For Advanced Research, University of Science and Technology of China, Suzhou, China
gxwu@ustc.edu.cn

Abstract. Binary code similarity detection (BCSD) has many applications in computer security, whose task is to detect the similarity of two binary functions without having access to the source code. Recently deep learning methods have shown better efficiency, accuracy, and potential in BCSD. Most of them reduce losses by the Siamese network, and they ignore some shortcomings of the Siamese network. In this paper, we introduce the idea of contrastive learning into graph neural networks and experimentally demonstrate that the way of training graph models by contrastive learning is significantly better than Siamese. In addition, we found that Principal Neighbourhood Aggregation for Graph Nets (PNA) has the best ability to extract structural information of control flow graph (CFG) among various graph neural networks.

Keywords: Binary code similarity detection · contrastive learning · graph neural network

1 Introduction

The same source code with different compilers, architectures, and optimization levels will generate different binary codes. BCSD determines the similarity of two binary functions without having access to the source code. It has many applications in computer security, such as software plagiarism detection, malware analysis, virus detection. A single bug at the source code may propagate to hundreds or more devices with different hardware architectures and software platforms. Cross-version binary code similarity detection is one of the fundamental tasks in computer security. In face of increasingly complex computer security threats, security practitioners are increasingly demanding the ability to detect cross-version binary code similarity.

Compared to traditional graph search algorithms, recently deep learning approaches have shown better efficiency, accuracy, and potential. Gemini [1]

This work was supported by the Natural Science Foundation of Jiangsu Province, China (BK20141209).

C. Alcaraz et al. (Eds.): ICICS 2022, LNCS 13407, pp. 458–471, 2022.
https://doi.org/10.1007/978-3-031-15777-6_25

introduces deep learning methods to the problem of binary code similarity detection. It manually extracts basic features in the form of attribute control flow graphs (ACFG), generates graph embeddings by Structure2Vec [2], and finally reduce losses using its first proposed Siamese network. Siamese network is currently a widely used training architecture in binary code similarity detection [3–5,11]. Figure 1 is a schematic representation of the Siamese network, which computes cosine similarity for each pair of graph embeddings. In contrast to Gemini manual feature extraction, Semantic-Aware [3] extracts the features of each basic block in the CFG by BERT [7]. Similarly, it uses Siamese networks on the graph embedding model to reduces losses and uses cosine distances to calculate graph similarity.

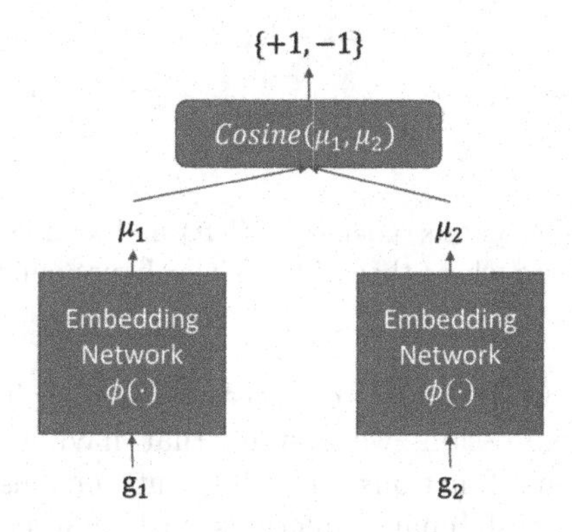

Fig. 1. Siamese structure diagram: it calculates a cosine similarity for each pair of graph embeddings

Although deep learning-based binary code similarity detection has achieved good results, there are still some areas for improvement.

Firstly, Siamese neural networks are prone to overfitting, and it is problematic to mark the similarity of function pairs with the same source as 1 and those with different sources as −1. For example, there are two functions in the OpenSSL dataset libcrypto file, as is shown in Fig. 2, from different source codes but they are very similar. At this point, it would be inappropriate to set the similarity of this function pair to either −1 or 1.

Secondly, Corso et al. (2020) has mathematically demonstrated that graph neural networks with a single aggregator (e.g., Structure2vec [2], MPNN [6]) cannot capture different types of messages during message passing, which would result in insufficient extraction of information about the whole graph representation.

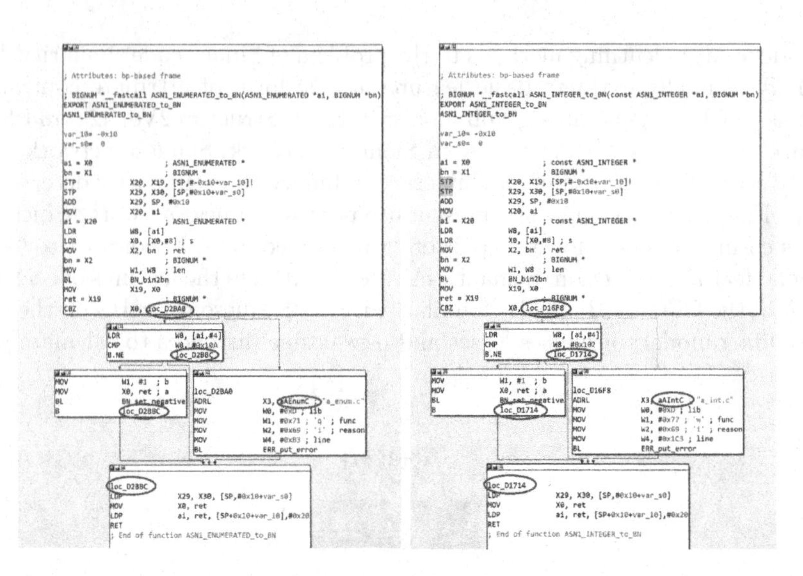

Fig. 2. CFGs of functions ASN1_INTEGER_to_BN (left) and ASN1_ENUMERATED_to_BN (right): the parts circled in red are where they differ (Color figure online)

To solve the first, we introduce contrastive learning into graph neural networks. For ease of expression, we assume that having x_i, x_i^+, x_i^-. (x_i, x_i^+) denotes a pair of binary functions with different versions of the same source. (x_i, x_i^-) denotes a pair of binary functions with different source codes. In a high-dimensional space, the essence of training graph neural networks is to push embeddings of x_i and x_i^+ closer together and to break up embeddings of x_i and x_i^-. In Siamese, it pushes x_i^+ to x_i according to cosine similarity of which the value is equal to 1, breaks up x_i and x_i^- according to cosine similarity of which the value is equal to -1. In contrastive learning, pushing x_i^+ to x_i and breaking up x_i and x_i^- are done simultaneously, pushing x_i^+ to x_i considering x_i^-, breaking up x_i and x_i^- considering x_i^+. In theory, contrastive learning is superior to Siamese.

To solve the second, we extract structural information of the CFG by the PNA, which uses multiple aggregators and scalers to pass different types of messages between nodes.

Our contributions are as follows:

1. We propose a novel framework for extracting CFG embeddings.
2. In the CFG representation, we extract CFG embeddings by PNA.
3. We introduce contrastive learning into the graph representation.
4. We conducted experiments on three datasets, and the results show that our proposed approach achieves better performance than previously.

2 Related Work

2.1 BCSD

Cross-version binary code similarity detection plays a critical role in computer security by analyzing two binary files $B1 = \{x_{11}, x_{12}, ..., x_{1n}\}$, $B2 = \{x_{21}, x_{22}, ..., x_{2n}\}$ that derive from the same project but are compiled in different versions. There are two tasks related to this:

1. Function matching: for each binary function x_{1i} in $B1$, if exist, find its match x_{2j} in the other binary $B2$.
2. Similarity score: for each pair of binary functions x_{1i} and x_{2j}, compute a similarity score ranging from -1 to 1, indicating how likely they are similar to each other. (1 being identical, -1 being completely different)

Traditional methods compute graph similarity using graph matching algorithms, which are slow and inefficient. Recent deep learning methods are widely used in BCSD. (Xu et al. 2017) proposes a GNN-based model called Gemini but requires manual extraction of features from the basic blocks of CFG. (Zuo et al. 2018) use an NLP model on this task [12]. They treat a token as a word and a block as a sentence and use LSTM to encode the semantic vector of the sentence. To obtain ground truth block pairs, they modified the compiler to add a primary block-specific annotator for annotating each generated assembly block with a unique ID. The approach has an obvious disadvantage; it requires expert experience and domain knowledge for the supervised process. Semantic-Aware [3] proposed a method for extracting basic block features using the Bert model of the NLP domain, treating instructions in basic blocks as words and basic blocks as sentences, which achieved a considerable improvement. However, it still did not break the limits of the Siamese framework.

2.2 BERT

Bert [7] is the most influential pre-training model in the NLP field, which uses the structure of a transformer to learn the meaning of each word and the contextual relationships using an attention mechanism. It has two pre-training tasks. The masked language model (MLM) task masks out words in a sentence and allows the model to predict these masked words, allowing the model to learn language-related knowledge. The second task is the next sentence prediction (NSP) task, which is a classification task that allows the model to distinguish between two sentences to learn about the relationship between them. The pre-trained Bert model was fine-tuned for most downstream tasks with excellent results.

2.3 Graph Neural Network

Following Scarselli et al. (2008) [13], who proposed learning node representations and graph representations, graph neural networks have developed a wide variety of graph models. For example, the graph convolutional network GCN [14] update node embeddings by convolutional layers; GraphSAGE [15] uses an aggregation function to merge nodes and their neighbors, and the graph attention network GAT [8] receive information from neighboring nodes by attention mechanism. MPNN [6] designs a holistic framework for graph representation learning with a message passing phase and a readout phase. The message passing phase runs several steps to receive information from neighboring nodes. The readout phase computes the embedding of the entire graph. In addition to MPNN, the graph network GN [16] and the non-local neural network NLNN [17] are also holistic frameworks for graph learning. PNA [9] is a recent study of graph models, mathematically demonstrating the need for multiple aggregators, which is a combination of multiple aggregators with a novel architecture combining degree scalers.

2.4 Contrastive Learning

Contrastive learning assumes a set of pairwise examples $D = \{(x_i, x_i^+)\}_{i=1}^m$, where x_i and x_i^+ are semantically related. It aims to learn valid representations together with pushing away non-neighbors by drawing semantically close to them (Hadsell et al., 2006). The key to contrast learning is constructing positive example pairs (x_i, x_i^+). In visual embeddings [20], an efficient solution is to perform two random transformations (e.g., cropping, flipping, warping and rotating) of the same image as (x_i, x_i^+). In sentence embedding [19], a straightforward solution is to use the randomness of dropout to have the same sentence pass through the neural network twice, producing different vectors to form (x_i, x_i^+). In contrastive learning, there are two metrics, **Alignment** and **Uniformity** (Wang and Isola (2020)) [18] used to measure the quality of representations. Align is used to calculate the distance between pairs of positive examples, given distribution of pairs of positive examples p_{pos}, and between embeddings of paired examples (assuming all representations have been regularized) to calculate the aligned average distance, ℓ_{align} as shown in Eq. (1):

$$\ell_{align} \triangleq \mathop{\mathbb{E}}_{(x,x^+)\sim p_{pos}} \| f(x) - f(x^+) \|^2 \tag{1}$$

On the other hand, Uniformity measures whether the distribution of all representations is uniform as shown in Eq. (2), where p_{data} denotes the data distribution. Optimization of these two metrics is the aim of contrast learning: embeddings of positive example pairs should be kept close, and embeddings of random examples should be scattered over the hypersphere.

$$\ell_{uniform} \triangleq log \mathop{\mathbb{E}}_{x,y \overset{i.i.d.}{\sim} p_{data}} e^{-2\|f(x)-f(y)\|^2} \tag{2}$$

3 Methods

3.1 Overall Structure

The input data for our model is CFG of the binary code decompiled by IDA Pro [21]. The overall architecture is shown in Fig. 3. In the semantic extraction part, we use a BERT model with two pre-training tasks to extract the features of the basic blocks. In the structure extraction part, we use a PNA with multiple aggregators to extract the features and structural embeddings of the graph. For the training architecture, we follow the contrastive framework [20], and use an in-batch negative cross-entropy objective [22].

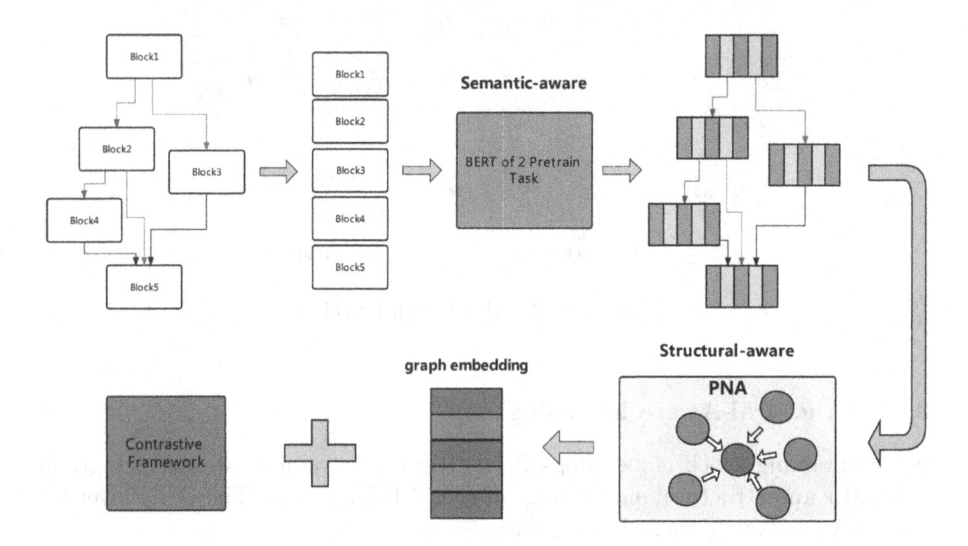

Fig. 3. The overview of our model structure consists of three parts: the semantic-aware part, the structure-aware part, and the contrastive framework.

3.2 Semantic-Aware Modeling

In this paper, we pre-train Bert using two classical tasks for BERT pre-training (MLM and NSP). The corpus is a collection of basic blocks of each version of the CFG and treats each basic block as a sentence and two basic blocks with edges as consecutive sentences, as is shown in Fig. 4. Our vocabulary consists of about ninety microcode instructions. Because our vocabulary is small, we embed each token as a 128-dimensional vector instead of 768, and accordingly, we shrink our hidden layer to a 128*8-dimensional, 12-layer transformer structure. In the MLM task, we masked 15% of the tokens for training. In the NSP task, we took two blocks of each edge as consecutive sentences and an arbitrary edge in the block set for the starting block of each edge to training as a negative example. Finally, we take the average of all tokens in the last layer of Bert as the embedding of the blocks.

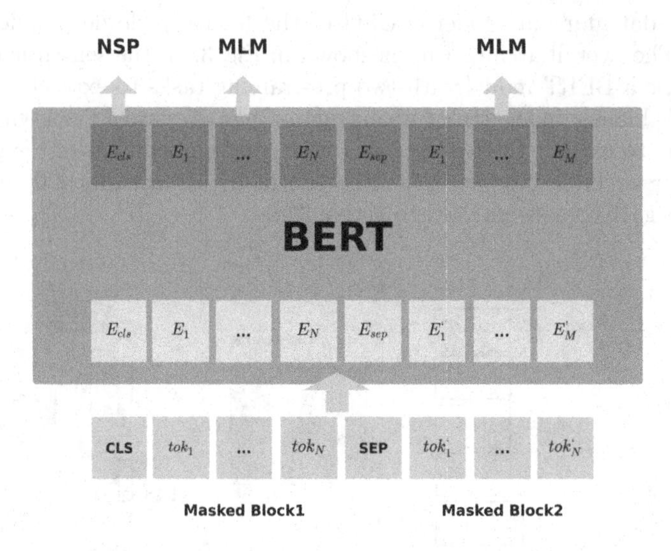

Fig. 4. NSP and MLM in BERT

3.3 Structural-Aware Modeling

After obtaining block embeddings from BERT pre-training, we extract graph semantics and structural embeddings of each CFG by PNA. The GNN layer for the PNA messaging phase is as follows:

$$X_i^{t+1} = U(X_i^t, \bigoplus_{(j,i)\epsilon E} M(X_i^t, E_{j\rightarrow i}, X_j^t)) \quad (3)$$

where $E_{j\rightarrow i}$ is a feature of the edge (j,i) (if exists), M and U are neural networks. U decreases the size of the connection message (in the space $\mathbb{R}^{13F}$) back to $\mathbb{R}^F$, where F is the dimension of the hidden features in the network. The process of M is shown in Fig. 5, where each node captures different types of messages from neighboring nodes with four aggregators and then goes through three scalers to obtain 12 different representations of the original node. We use the output of the PNA as a feature of the CFG node and then use *sum* to obtain an embedding of the whole graph.

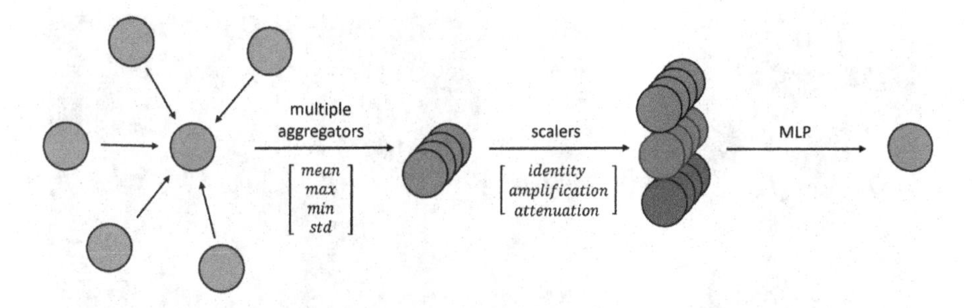

Fig. 5. Structural-aware diagram for the principal neighbourhood aggregation or PNA.

3.4 Contrastive Framework: Simple Contrastive Learning of Graph Embeddings (SimCGE)

After obtaining the graph embeddings, instead of using Siamese, we use the contrastive learning framework and take a cross-entropy objective with in-batch negatives: let g_i and g_i^+ be the representations of x_i and x_i^+ with N pairs of (x_i, x_i^+) mini-batch training with a loss function of :

$$\ell = -log \frac{e^{sim(g_i, g_i^+)/T}}{\sum_{j=1}^{N-1} e^{sim(g_i, g_j^+)/T}} \tag{4}$$

where T is a hyperparameter called the temperature coefficient and $sim(g_1, g_2)$ is the cosine similarity $\frac{\mathbf{g}_1^\top \mathbf{g}_2}{\|\mathbf{g}_1\| \cdot \|\mathbf{g}_2\|}$. Figure 6 is a diagram of the in-batch negative example, which treats all non-positive examples within the same batch as negative examples. After computing the similarity, it is similar to a classification problem, with the first column in one class and the others in another. The key to contrastive learning is to obtain the corresponding x_i and x_i^+. In BCSD, each version of the binary function x_i is naturally the x_i^+ of the corresponding function in the other version, which fits perfectly with the idea of contrastive learning. Compared to the Siamese approach of pulling x_i^+, pushing x_i^- separately, it is more reasonable to contrastive learning pulling x_i^+ considering x_i^- and pushing x_i^- considering x_i^+ simultaneously.

4 Experiment

4.1 Dataset

We train the model on gnu_debug [10] and evaluate it on three publicly available datasets: OpenSSL.1.0.1f, OpenSSL.1.0.1u, and busybox.1.21.stable, with the same source code compiled into different versions of the binary code and then decompiled into CFG using IDA Pro. In a high-dimensional space,our goal is to push embeddings of x_i and x_i^+ closer together and to break up embeddings of x_i and x_i^-. Our versions of the binary code include cross-platform (x86 &

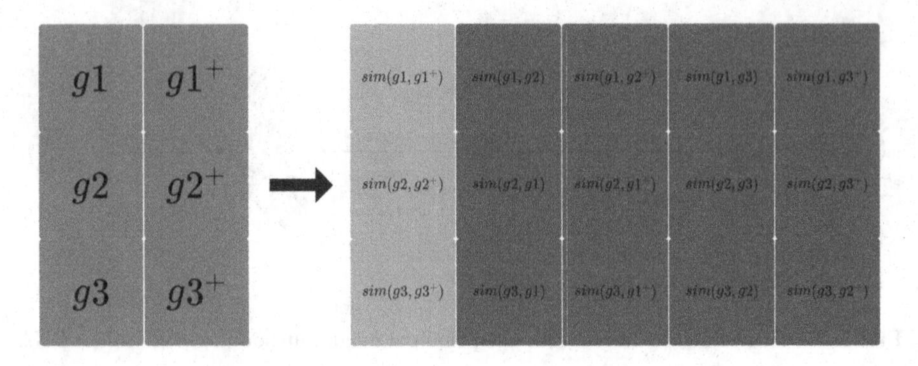

Fig. 6. In-batch diagram, which treats all non-positive examples within the same batch as negative examples

arm), cross-compiler (gcc-8.2.0 & clang-7.0), cross-optimisation level (O3 & O2), and cross-bit (64 & 32). We have done extensive experiments on these different comparison versions.

4.2 Evaluation Metrics

This task of BCSD is similar to the recommendation system, thus we evaluate the model by MRR and *Recall@K*. MRR is used for evaluating ranking tasks, which uses the multiplicative inverse of the rank of the first correct answer. *Recall@K* means whether the ranking of the true pair is in the top K scores of the highest. Given two binary files $B1 = x_{11}, x_{12}, ..., x_{1n}$ and $B2 = x_{21}, x_{22}, ..., x_{2m}$, we assume that there are T pairs of matching binary functions $(x_{11}, x_{21}), (x_{12}, x_{22})...$ (x_{1T}, x_{2T}), the rest are unmatched. For any function x_{1i} in $B1$, the BCSD can rank the functions in $B2$ based on similarity to x_{1i}, and we denote $rank_{x_{1i}}$ as the position of the correct matching function x_{2i} for x_{1i} among all ranked functions. The formulas for MRR and *Recall@K* are represented as Eqs. (5) and (7) respectively, $hit@K\,(x_{1i})$ indicates whether x_{2i} is in the top K functions that are most similar to x_{1i} as defined in Eq. (6).

$$MRR = \frac{1}{T}\sum_{i=1}^{T}\frac{1}{\text{rank}_{x_{1i}}} \tag{5}$$

$$hit@K\,(x_{1i}) = \begin{cases} 1, \; rank_{x_{1i}} \leq K \\ 0, \;\; \text{otherwise} \end{cases} \tag{6}$$

$$\text{Recall}\ @K(B1, B2) = \frac{\sum_{i=1}^{T} hit@K\,(x_{1i})}{T} \tag{7}$$

4.3 Compared Methods

As our model is cross-version specific, and Gemini as well as Semantic aware is cross-architecture specific (it is included in cross-version), we did a two-part comparison, taking the cross-architecture part out to compare with Gemini and Semantic aware, and then we focused on comparing Siamese with SimCGE in the cross-version part.

Our Model. It consists of BERT (2 tasks) + PNA + SimCGE containing semantic extraction, structural information extraction, contrastive learning components.

Cross-Architecture Comparison

Table 1 shows the results of the cross-architecture comparison.

Gemini uses Structure2vec to compute graph embeddings of CFGs, where each block is an 8-dimensional manually selected feature, trained by using the Siamese architecture.

Semantic-Aware is BERT (4 tasks) + MPNN + 11-Resnet layers + Siamese, which contains semantic-aware modeling, structure-aware modeling, and sequential-aware modeling.

Table 1. Cross-architecture comparison table

Model	gcc-64-O2 x86 vs arm		gcc-64-O3 $\times$ 86 vs arm	
	MRR	Recall@1	MRR	Recall@1
Gemini	0.6069	0.5491	0.543	0.476
Semantic-Aware	0.7922	0.7421	0.6855	0.6114
Our model	0.8317	0.7555	0.8296	0.7536

Cross-Version Comparison

Table 2 shows the performance of our model on each dataset. To facilitate comparison with other models, we take the average value as the overall performance of the model performance, as is shown in Table 3.

BERT(2 tasks) uses the BERT of both the MLM and NSP pre-training tasks and then sums the sum of all nodes as structural information.

GAT + Siamese BERT (2 tasks) + GAT + Siamese, GAT [8] is using the attention mechanism as a messaging mechanism.

GAT + SimCGE BERT (2 tasks) + GAT + SimCGE

MPNN + Siamese BERT (2 task) + MPNN + Siamese, MPNN uses GRUs to pass messages between nodes.

MPNN + SimCGE BERT (2 task) + MPNN + SimCGE

PNA + Siamese BERT (2 task) + PNA + Siamese

Table 2. Experimental results for each version data of our model

Dataset	Version	Recall@1	Recall@5	MRR
OpenSSL.1.0.1u	gcc_arm_64_O3 vs gcc_x86_64_O3	0.75703	0.93131	0.83144
OpenSSL.1.0.1u	gcc_arm_64_O3 vs clang_arm_64_O3	0.55877	0.78497	0.65607
OpenSSL.1.0.1u	gcc_arm_64_O3 vs gcc_arm_64_O2	0.80728	0.94616	0.86772
OpenSSL.1.0.1u	gcc_arm_64_O3 vs gcc_arm_32_O3	0.69462	0.89381	0.77973
OpenSSL.1.0.1u	gcc_arm_64_O3 vs clang_x86_32_O1	0.53361	0.77896	0.64047
OpenSSL.1.0.1f	gcc_arm_64_O3 vs gcc_x86_64_O3	0.75883	0.9286	0.8328
OpenSSL.1.0.1f	gcc_arm_64_O3 vs clang_arm_64_O3	0.56281	0.78025	0.65678
OpenSSL.1.0.1f	gcc_arm_64_O3 vs gcc_arm_64_O2	0.81377	0.94409	0.87149
OpenSSL.1.0.1f	gcc_arm_64_O3 vs gcc_arm_32_O3	0.6954	0.89189	0.78052
OpenSSL.1.0.1f	gcc_arm_64_O3 vs clang_x86_32_O1	0.52085	0.7699	0.63134
busybox.1.21.stable	gcc_arm_64_O3 vs gcc_x86_64_O3	0.90159	0.97956	0.93494
busybox.1.21.stable	gcc_arm_64_O3 vs clang_arm_64_O3	0.70905	0.86583	0.77902
busybox.1.21.stable	gcc_arm_64_O3 vs gcc_arm_64_O2	0.85548	0.93318	0.89027
busybox.1.21.stable	gcc_arm_64_O3 vs gcc_arm_32_O3	0.85334	0.94377	0.89383
busybox.1.21.stable	gcc_arm_64_O3 vs clang_x86_32_O1	0.57447	0.76841	0.65927
MEAN		0.70646	0.87605	0.78038

Table 3. Experimental results of various models cross versions

Model	Recall@1	Recall@5	MRR
BERT2	0.60736	0.78133	0.68288
gat + siamese	0.62234	0.8035	0.70111
gat + SimCGE	0.64254	0.82011	0.71954
mpnn + siamese	0.66772	0.84142	0.74274
mpnn + SimCGE	0.6883	0.85724	0.76148
pna + siamese	0.68132	0.85434	0.75631
our model	0.70646	0.87605	0.78038

4.4 Training

We trained our BERT using hugging-face, BERT with max_length = 128, lr = 0.00001, bsz = 896, and spent 40 h training our BERT on 4 A100 s. We used pytorch lightning and pytorch_geometric to train the graph neural network with bsz = 256, roughly 5,000 steps per graph neural network, lr = 0.0001, and the Adam for optimizer.

4.5 Results

In cross-architecture comparison, our model improves significantly more at the O3 optimization level. We can see from all three models that the task at the

O3 level is relatively more difficult. But in our model, the difference between O3 and O2 is small, we speculate that this is because the Siamese-trained model embedding is not uniformly distributed, resulting in insufficient robustness of the model.

In the cross-version comparison, we compared various graph neural networks in Siamese and SimCGE, and from Table 3, we can see that PNA has the best performance, followed by MPNN, and the worst by GAT. Among the various GNNs, SimCGE outperforms Siamese. We have specifically analyzed the reasons why SimCGE outperforms Siamese, as is shown in Fig. 7, where the uniform of SimCGE is much better than Siamese. SimCGE's trained model allows for a more uniform distribution of CFG embeddings, although at the expense of a bit of alignment, overall, it is easier to find x_i^+ for each x_i.

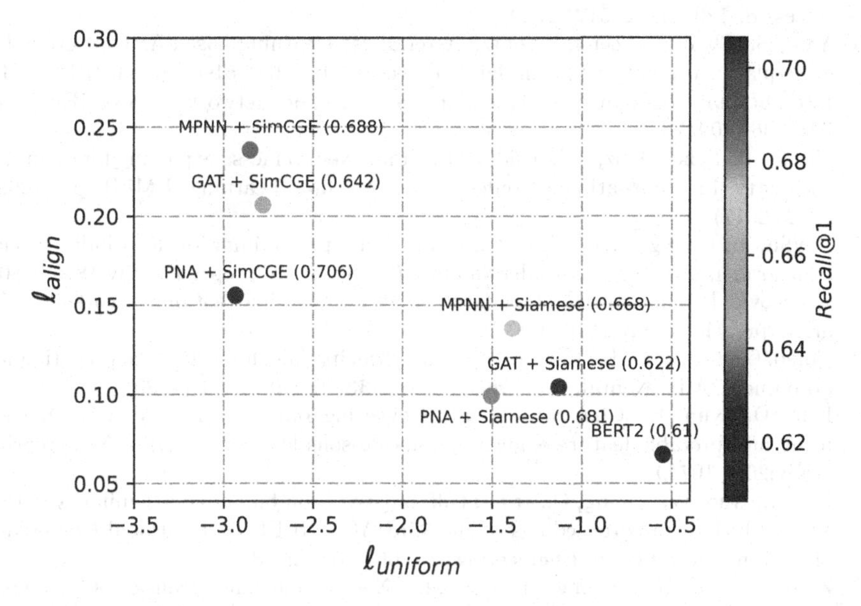

Fig. 7. Comparison of align and uniform of each model: colors of points and numbers in brackets are the performance of each model on *Recall*@1, both align and uniform are the smaller the better. These two indicators are explained in Eqs. (1) and (2) respectively.

5 Conclusion

In this paper, we propose a novel training framework for binary code similarity detection, which consists of three parts: semantic extraction, structural information extraction, and contrastive learning. For the semantic extraction part, we use BERT with two pre-training tasks; for the structural information extraction part, we use PNA and contrastive learning instead of traditional Siamese architecture. The experiments were carried out with three datasets and showed that contrastive learning significantly outperformed Siamese, with PNA best extracting structural information from control flow graphs.

References

1. Xu, X., Liu, C., Feng, Q., et al.: Neural network-based graph embedding for cross-platform binary code similarity detection. In: Proceedings of The ACM SIGSAC Conference on Computer and Communications Security, vol. 2017, pp. 363–376 (2017)
2. Dai, H., Dai, B., Song, L.: Discriminative embeddings of latent variable models for structured data. In: International Conference on Machine Learning. PMLR, pp. 2702–2711 (2016)
3. Yu, Z., Cao, R., Tang, Q., et al.: Order matters: semantic-aware neural networks for binary code similarity detection. In: Proceedings of the AAAI Conference on Artificial Intelligence, vol. 34, no. 01, pp. 1145–1152 (2020)
4. Guo, H., Huang, S., Huang, C., et al.: A lightweight cross-version binary code similarity detection based on similarity and correlation coefficient features. IEEE Access **8**, 120501–120512 (2020)
5. Yang, S., Cheng, L., Zeng, Y., et al.: Asteria: deep learning-based AST-encoding for cross-platform binary code similarity detection. In: 2021 51st Annual IEEE/IFIP International Conference on Dependable Systems and Networks (DSN). IEEE, pp. 224–236 (2021)
6. Gilmer, J., Schoenholz, S.S., Riley, P.F., et al.: Neural message passing for quantum chemistry. In: International Conference on Machine Learning. PMLR, pp. 1263–1272 (2017)
7. Devlin, J., Chang, M.W., Lee, K., et al.: Bert: pre-training of deep bidirectional transformers for language understanding (2018). arXiv preprint arXiv:1810.04805
8. Veličković, P., Cucurull, G., Casanova, A., et al.: Graph attention networks (2017). arXiv preprint arXiv:1710.10903
9. Corso, G., Cavalleri, L., Beaini, D., et al.: Principal neighbourhood aggregation for graph nets. Adv. Neural. Inf. Process. Syst. **33**, 13260–13271 (2020)
10. Kim, D., Kim, E., Cha, S.K., et al.: Revisiting binary code similarity analysis using interpretable feature engineering and lessons learned (2020). arXiv preprint arXiv:2011.10749
11. Liu, B., Huo, W., Zhang, C., et al.: Diff: cross-version binary code similarity detection with dnn. In: Proceedings of the 33rd ACM/IEEE International Conference on Automated Software Engineering, pp. 667–678 (2018)
12. Zuo, F., Li, X., Young, P., et al.: Neural machine translation inspired binary code similarity comparison beyond function pairs (2018). arXiv preprint arXiv:1808.04706
13. Scarselli, F., Gori, M., Tsoi, A.C., et al.: The graph neural network model. IEEE Trans. Neural Netw. **20**(1), 61–80 (2008)
14. Kipf, T.N., Welling, M.: Semi-supervised classification with graph convolutional networks (2016). arXiv preprint arXiv:1609.02907
15. Hamilton, W., Ying, Z., Leskovec, J.: Inductive representation learning on large graphs. Adv. Neural Inf. Process. Syst. **30** (2017)
16. Battaglia, P.W., Hamrick, J.B., Bapst, V., et al.: Relational inductive biases, deep learning, and graph networks (2018). arXiv preprint arXiv:1806.01261
17. Wang, X., Girshick, R., Gupta, A., et al.: Non-local neural networks. In: Proceedings of the IEEE Conference on Computer Vision and Pattern Recognition, pp. 7794–7803 (2018)
18. Wang, T., Isola, P.: Understanding contrastive representation learning through alignment and uniformity on the hypersphere. In: International Conference on Machine Learning. PMLR, pp. 9929–9939 (2020)

19. Gao, T., Yao, X., Chen, D.: Simcse: simple contrastive learning of sentence embeddings (2021). arXiv preprint arXiv:2104.08821
20. Chen, T., Kornblith, S., Norouzi, M., et al.: A simple framework for contrastive learning of visual representations. In: International Conference on Machine Learning. PMLR, pp. 1597–1607 (2020)
21. IDA Pro Homepage. https://www.hex-rays.com/ida-pro/
22. Henderson, M., Al-Rfou, R., Strope, B., et al.: Efficient natural language response suggestion for smart reply (2017). arXiv preprint arXiv:1705.00652

FN2: Fake News DetectioN Based on Textual and Contextual Features

Mouna Rabhi[1(✉)], Spiridon Bakiras[2], and Roberto Di Pietro[1]

[1] Division of Information and Computing Technology, College of Science and Engineering, Hamad Bin Khalifa University, Doha, Qatar
{mora33056,rdipietro}@hbku.edu.qa

[2] Singapore Institute of Technology, Singapore, Singapore
spiridon.bakiras@singaporetech.edu.sg

Abstract. Fake news is a serious concern that has received a lot of attention lately due to its harmful impact on society. In order to limit the spread of fake news, researchers have proposed automated ways to identify fake news articles using artificial intelligence and neural network models. However, existing methods do not achieve a high level of accuracy, which hinders their efficacy in real life. To this end, we introduce FN2 (Fake News detectioN): a novel neural-network based framework that combines both textual and contextual features of the news articles. Among the many unique features of FN2, it utilizes a set of explicit contextual features that are easy to collect and already available in the raw user metadata. To evaluate the accuracy of our classification model, we collected a real dataset from a fact-checking website, comprising over 16 thousand politics-related news articles. Our experimental results show that FN2 improves the accuracy by at least 13%, compared to current state-of-the-art approaches. Moreover, it achieves better classification results than the existing models. Finally, preliminary results also show that FN2 provides a quite good generalization—outperforming competitors—also when applied to a qualitatively different data-set (entertainment news). The novelty of the approach, the staggering quantitative results, its versatility, as well as the discussed open research issues, have a high potential to open up novel research directions in the field.

Keywords: Fake news · Fake news detection · Online social media · Neural networks · Textual features · Contextual features

1 Introduction

Online social networks (OSNs), like Facebook and Twitter, enjoy an enormous popularity among end-users. Indeed, the number of OSN subscribers has been increasing exponentially and reached 3.6 billion in 2020 [24]. OSNs offer a convenient way for users to create, access, and share information, making news consumption easy and available to everyone. Consequently, the majority of users

C. Alcaraz et al. (Eds.): ICICS 2022, LNCS 13407, pp. 472–491, 2022.
https://doi.org/10.1007/978-3-031-15777-6_26

today do not get their news from traditional sources, such as newspapers or TV news programs. Instead, they use OSNs as their first source of information. According to a study done by the PEW research center, more than two-thirds (68%) of adults are getting their news from social networks [5]. However, information can be shared on OSNs without any check on its trustworthiness or accuracy, which has led to a significant increase in fake news circulation.

Fake news intentionally present misleading information to the end-users [29] and it is a basic tool to fuel propaganda [13]. The topic has gained a lot of attention recently, due to its harmful impact on several aspects of our society. Fake news is considered a threat to numerous fields, including journalism, public health, democracy, and the economy. For a recent survey on how fake news propagate, the reader is pointed to [18]. Fake news harmful effects on society and individuals have been recorded extensively [12]. For example, during the 2016 U.S. elections, various false rumors about the candidates were widely spread on OSN platforms [32]— with a significant impact on the election results. Another case that demonstrates the dramatic effect of fake news is the loss of $130 billion in the U.S. stock market after the spread of false rumors stating that president Obama got injured in an explosion [19].

The aforementioned examples emphasize the importance of identifying fake news, with the objective to stop their propagation [18] or, at the very least, to label them with a low trustworthiness score. Fake news detection is usually regarded as a classification problem, i.e., it aims to distinguish between "True" and "Fake" news [15]. However, even though the problem formulation is quite simple, the problem in itself is very challenging for several reasons. First, fake news is intentionally made to misinform the reader; therefore, it is not easy for the average OSN user to detect it as such. Second, social media data is massive, noisy, and consists of a multitude of sources.

To overcome these challenges, the research community has built solutions, based on artificial intelligence (AI) techniques, that aim to automate the identification of fake news articles. Most existing approaches focus on the writing style [17], that is, they capture the semantics and linguistic features of the news articles. For instance, the authors in [1,14,27] apply natural language processing techniques and neural networks to identify fake news, using just the news text itself. Nevertheless, in addition to the news text, contextual information about the news article can be useful as well [22]. In fact, previous studies have shown the effectiveness of user profile features and social context in detecting fake news [23]. However, very few studies to date address fake news classification by combining textual and contextual data. For example, Zhang, Dong, and Yu [30] propose a framework that uses news text, news creators, and subjects to identify fake news. Similarly, Wang [28] introduces a model that uses the news text and additional metadata about the news article to classify news. However, the identification accuracy of existing models is rather low, which limits their effectiveness.

To overcome the cited limitations, our work introduces a novel fake news identification framework called *FN2*. In order to solve the underlying bi-label inference problem, FN2 trains a model that is able to predict the correct label of a news article, using both the news text and its metadata. Despite receiving huge attention from the leading research communities, the necessity of an efficient detection model that achieves good accuracy and works in real-case scenarios still exists. Existing contributions make use of explicit and implicit contextual data that are not always easy to collect in real-case scenarios. To overcome these issues, we propose FN2, a framework that only utilizes explicit metadata which are already available. That is, we do not consider implicit profile features that are not directly available in the users' raw data (such as personality, location, profile image, political bias, etc.). Such information requires much more effort to collect and can be easily manipulated. FN2 exploits a convolutional neural network (CNN) to extract textual features from the news text, and a feed-forward neural network (FNN) to extract contextual features. Our experimental results on a large news dataset demonstrate that FN2 improves the classification accuracy significantly compared to the state-of-the-art models. The main contributions of this work are summarized as follows:

- We provide a framework for fake news classification that exploits both news text and news metadata.
- We demonstrate the correlation between the news text and the metadata related to the news article.
- We demonstrate the effectiveness of contextual features in classifying fake news, and we show that by combining textual and contextual features, we can outperform existing state-of-the-art models based on textual features only.
- We demonstrate the generality of the proposed model.

The rest of this paper is organized as follows. Section 2 provides an overview of previous work on fake news detection. Section 3 describes in detail the FN2 framework and Sect. 4 presents the details of the experimental setting used in FN2. Section 5 details the results of our experimental evaluation. Finally, Sect. 7 concludes our work and discusses a few directions for future work.

2 Related Work

Fake news detection is typically achieved by training neural network models on a variety of features from the news articles. In particular, the vast majority of existing approaches rely on either textual or contextual features to do the training [3]. In the following subsections, we review the most relevant models from each category.

2.1 Fake News Detection Based on Textual Features

Fake news detection based on textual features targets the lexical and semantic properties of the news article to classify its credibility [31]. For instance,

Rashkin et al. [20] applied linguistic analysis to detect fake news. They first compared the languages of real and fake news to extract the linguistic characteristics of fake news articles. The authors looked into the news articles from many different perspectives, such as the existence of solid and weak subjectivity, the degree of dramatization (using a lexicon from Wiktionary), and hedges. They discovered that false news statements tend to use more first and second person pronouns compared to trustworthy news. Moreover, they showed that fake news tend to use more exaggerating words compared to trustworthy news. Next, the authors used the linguistic features to train a long short-term memory (LSTM) model to detect the reliability of the news statement. The LSTM model takes as input a sequence of words from the news text and outputs its reliability. The authors reported an accuracy of 56% on a test dataset of 1076 news statements collected from politifact.com.

Ahmed, Traore, and Saad [1] proposed a fake news detection approach that leverages n-grams and machine learning algorithms. Specifically, the authors utilized the term frequency and inverse document frequency (TF-IDF) for feature extraction, and employed six different machine learning models for classification on an equally distributed (true vs. fake news) dataset of 2000 news articles. The highest accuracy score was obtained with the combination of unigrams and linear SVM, on a test dataset of 400 news articles. Furthermore, the authors observed that the accuracy of the proposed algorithm decreases when using larger n-grams. However, the dataset used in the cited study is rather small.

Nasir, Khan, and Varlamis [14] employed a hybrid CNN-RNN model to detect fake news. First, the CNN model is used to extract the local features from the text input. The output of the CNN is then passed to the recurrent neural network (RNN), which learns the long-term dependencies of the local features. The authors utilized two datasets to test the efficiency of their proposed method: (i) the FA-KES dataset that contains 804 news articles about the Syrian war, where 426 articles are labeled true and 378 are labeled fake; and, (ii) the ISOT dataset that consists of 45,000 articles equally distributed between fake and true. The ISOT dataset is used for training, while the FA-KES dataset is used for testing. This approach ensures the generalization of the proposed method. The authors reported an accuracy of $60 \pm 0.7\%$ on the test dataset.

2.2 Fake News Detection Based on Contextual Features

Granik and Mesyura [8] proposed a simple approach for fake news detection, using a Naive Bayes classifier. First, they showed that there exist common properties between fake news statements and spam email, e.g., they often use a limited set of words and have a lot of grammatical mistakes. Then, they used the writing style of the news article, along with some metadata, to detect its reliability. The authors tested their methods on a dataset of 2282 Facebook posts that shared news articles. In addition to the posts, they also collected some information about the post, such as the number of likes, shares, and comments. The authors reported an accuracy of 74% using this simple classifier. Nevertheless, the dataset used in the cited work is quite small and unbalanced (it contains more true news than fake news). Furthermore, they did not pre-process the textual data, e.g., they did not remove stop words or use stemming.

Wang [28] incorporated metadata information about the news article in his fact checking model. He employed a convolutional neural network on a 12.8 k statement dataset collected from politifact.com. The model takes two inputs: the news text and the metadata related to the news, such as author, affiliation, source, etc. The text input is transformed into vector representation using word embedding. The output of the embedding layer is then passed to the CNN to extract the features. The metadata is processed similarly, using a different embedding layer, and leverages a CNN and a bidirectional LSTM (Bi-LSTM) network to extract its feature representation. The author has shown that the additional information related to the news article improves the fake news identification accuracy.

Zhang, Dong, and Yu [30] proposed a framework to score the credibility of a news article, its author, and its subject. They introduced a new hybrid feature learning unit to learn the explicit and implicit features from news text, creators, and subjects. Then, they used a deep diffusive neural network to fuse the features together and to predict the credibility. The proposed model is validated using a collected dataset from politifact.com. In particular, they collected 14,055 articles that were later organized as a network, where news articles, authors, and subjects represent the nodes, and the links represent the relation between them. The authors reported an accuracy of 62% in detecting fake news articles. In their work, they only used the subject and the authors as contextual information. However, other information about the news article could also be useful to improve performance.

Yang et al. [29] introduced TI-CNN, a model that combines both images and text to identify fake news. First, the model learns the latent features from both textual and image information, using CNN. Then, the explicit and latent features are fused to form a new set of features that is used to detect fake news. The model is trained with a scrapped dataset—containing 20,015 news articles—that focuses on the U.S. presidential election only.

Kaliyar et al. [10] introduced EchoFakeD, a model that uses news content and the social context of the news article to detect its trustworthiness. The authors focus more on user-based engagements and the context-related group of people (echo-chamber) sharing the same opinions.

Do et al. [4] propose a Fake news detection model that considers the news content and the social context. The model comprises three main components: feature learning, classifier, and mean-field. The feature learning component takes multiple inputs. Each input is transformed into high-level features that are concatenated to obtain a shared representation of the inputs. The shared representation is then passed to a classifier. This classifier consists of several fully connected layers followed by a softmax classifier to produce the class-specific probabilities. Finally, these probabilities are passed through the mean-field components. These components aim to smooth the class probability values by leveraging the correlation between the news articles.

Freire et al. [6] introduce Crowd Signals. This approach combines opinions from a high number of users in order to indicate whether a piece of news is fake or not. This approach has a limitation: it depends on explicit user opinion—which is not always guaranteed to be available—to classify the analyzed news. The authors have proposed a solution to overcome this limitation by using implicit user opinions to detect fake news. The implicit opinions are inferred from the behavior of users concerning the dissemination of the news analyzed.

Shu et al. [22] have introduced TriFN, a framework that models a tri-relationship between publishers, news articles, and users to detect fake news. TriFN extracts features using both user-news interactions and publisher-news relations. Different classifiers were tested with the resulted features. The model performances were evaluated using the FakeNewsNet dataset. The results show that the social context could effectively be exploited to improve false news detection.

3 Methodology

This work is rooted on the intuition that user profile information and metadata can be very valuable in improving the accuracy of fake news identification. To this end, our main contribution is a multi-modal embedding framework that incorporates news text and contextual information to detect fake news. A schematic representation of the proposed model is shown in Fig. 1. The model consists of two objects: text news and contextual data. We assume that we have M labeled news documents $D = \{N_i, l_i\}_{i=1}^{M}$, where N_i denotes news article i and $l_i \in \{0, 1\}$ represent the label related to that article. Here, '0' represents true news and '1' represents fake news. N_i can be further decomposed as $N_i = \{T_i, C_i\}$, where T_i denotes the textual representation extracted from the news text, while C_i denotes the contextual data, such as news source, speaker, speaker's reliability, etc. In the following sections, we first describe how to extract the representation for a news document N_i, by learning the features $\tilde{T}_i$, and $\tilde{C}_i$. We then show how to concatenate the learned features together to classify the news.

3.1 Textual Features Representation

Before detecting the linguistic patterns from the news text, we perform a preliminary data-cleaning process, removing redundancy, filtering out useless words, and converting them into feature vectors T_i that are used during feature learning.

Textual Data Preprocessing. Prior to learning the news article's text vector representation, we remove the stop words from the text. Stop words are the most commonly used words in a language, such as "a", "the", "is", "so", etc. Stop words do not add any valuable information and are, therefore, considered as noise when learning features. For the same reasons, punctuation is also regarded as noise and it is removed from the news text.

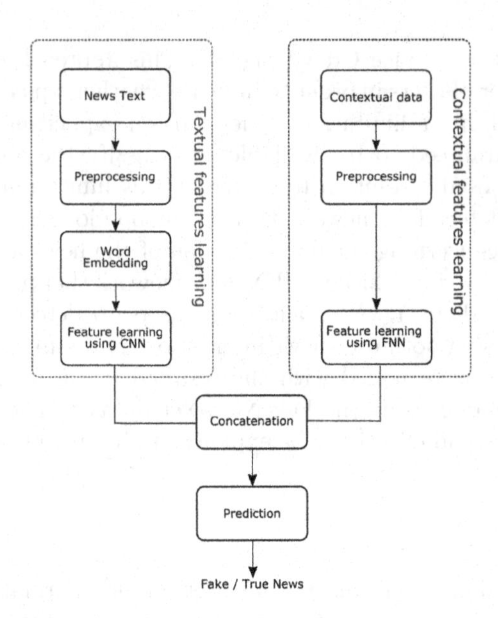

Fig. 1. Illustration of FN2's schematic diagram

Word Embedding. After eliminating noise, the news texts need to be converted into vectors of numerical values before being fed into the neural network algorithm. To this goal, word embedding is widely used in literature. Word embedding embeds the original words into vectors with low dimensions. Existing models are either static or contextual. Static embedding models such as *Word2Vec* and *Glove* do not capture the semantic meaning of the words, i.e., they assume that a word has the same meaning under different contexts. Consequently, they generate the same word vector under different contexts. For example, the word "orange" will have the exact vector representation in "orange juice" as in "Orange S.A.". That is why polysemous words make static word embedding a hard classification problem [2].

To better represent the words, contextual word embedding models are usually adopted. Such models generate the representation of each word based on the other words appearing in the sentence. In this work, we utilize BERT (Bidirectional Encoder Representation from Transformers) [26], a bidirectional, pretrained word embedding model based on multi-layer transformers, an attention mechanism that is capable of learning the contextual meaning of the words in a sentence. The two key factors that make BERT outperform other embedding methods are: (i) Mask Language Model (MLM); and, (ii) Next Sentence Prediction (NSP). MLM randomly masks a percentage of words from the input. The masked input sentence is then fed into the model to predict the masked words, based on the context of the surrounding non-masked words. With this process, BERT is capable of understanding the context more accurately. As per NSP, it is needed to identify the relationship between sentences. It learns to predict whether two input sentences are subsequent or they have just a random relationship. In our work, we leveraged the $BERT_{base}$ model. BERT takes as an input a sequence of, at most, 512 tokens. Therefore, we only consider statements with length less than 512 words.

Textual Features Learning. We use a convolutional neural network to extract textual features from the news text [11]. Our CNN model is composed of a one-dimensional convolutional layer followed by a max-pooling layer. The convolutional layer takes the vector representation of a news text T_i and learns its latent features. To learn the latent features, a convolution operation is performed between a filter with a window size k and a sequence of k consecutive words. This process produces one feature. More specifically, let $w_{i,j}$ be the jth word in the news article i. The vector representation of the news text T_i can be written as $T_i = \{w_{i,1}, w_{i,2}, \ldots, w_{i,n}\}$, where n is the maximum length of the news article. News articles with length less than n are padded with zeros. Let W represent the convolutional filter used to extract the features. The convolution operation is performed between a filter of length k and every sequence of k consecutive words $T_{i,j:j+k} = \{w_{i,j+1}, w_{i,j+2}, \ldots, w_{i,j+k}\}$. This produces a feature vector $\tilde{T}_{i,j}$ as presented in the following equation:

$$\tilde{T}_{i,j} = f(W * T_{i,j:j+k} + b) \tag{1}$$

where $b \in \mathbf{R}$ is the bias and $f(\circ)$ represents a non-linear activation function—in our case, the rectified linear unit (ReLU). Once the filter iterates over all possible windows of words, we obtain a feature map $\tilde{T}_i$ as shown in Eq. (2).

$$\tilde{T}_i = \{\tilde{T}_{i,1}, \tilde{T}_{i,2}, \ldots, \tilde{T}_{i,n-k+1}\} \tag{2}$$

The convolution layer output is then passed through a max-pooling layer. The max-pooling layer reduces over-fitting and decreases the dimensionality of the features without affecting the network's performance [25]. The max-pooling layer retains the maximum value over a spatial window, i.e., it only keeps the important features.

3.2 Contextual Features Representation.

The news article contextual features C_i describe the metadata information related to the news document and the user-based representation. A representative list of the contextual features used in this work is listed below:

- **Source:** This indicates where the news document was published, e.g., Tweet, Facebook Post, Interview, etc.
- **Tags:** This identifies the main topic stated in the news document.
- **Author's name:** This indicates the name of the article's author.

- **Author's credibility:** This metric reflects the reliability of the underlying author. Authors with low credibility are more likely to spread fake news compared to credible authors. We define user credibility as follows:

$$credibility = 1 - \frac{\#of\ fake\ news}{\#of\ total\ news\ published}$$

Contextual Features Preprocessing. As most contextual features are textual, we first split the features into words. Then, we create the vocabulary for each contextual feature. Finally, we represent each word of the vocabulary with an integer, in order to create a vector representation that is fed into a neural network for learning.

Contextual Features Learning. We use a two-layer deep feed-forward neural network to perform feature learning. The input layer takes as input the contextual feature vector $C_i = \{S_i, G_i, A_i, C_a\}$, where S_i is the source of the news document i, G_i is the tag associated to the news document i, A_i represents the author of the news document i, while C_a indicates the author's (A_i) credibility. The contextual features pass through the hidden layers in order to learn the dependencies among them. The output is the contextual learned features $\tilde{C_i}$ that can be represented as follows:

$$\begin{aligned} \tilde{C_i} &= \sigma'(h_1 * F + e) \\ h_1 &= \sigma(C_i * D + d) \end{aligned} \tag{3}$$

where C and F represent, respectively, the weights of the hidden layer and the output layer. Vectors d and e are the respective bias vectors, and functions σ and σ' denote the activation functions associated to the different layers.

3.3 Multi-modal Concatenation

As shown in Fig. 2, the two learned feature vectors $\tilde{T_i}$ and $\tilde{C_i}$, obtained from the two proposed modalities, are combined together using a simple concatenation technique in order to obtain the final multimodal representation of the news document $\tilde{N_i} = \{\tilde{T_i}, \tilde{C_i}\}$. The fuse of the two feature representations is done with a simple concatenation technique, i.e., the two feature vectors are stacked one after another to form $\tilde{N_i}$. The resulting representation, $\tilde{N_i}$, goes through a fully connected neural network and then to a sigmoid dense layer to classify the news document.

4 Experiments

In this section, we describe the dataset that we collected for our experiments, and also list the experimental settings we used to evaluate FN2. Then, we briefly discuss the baselines to which FN2 is compared against.

4.1 Dataset

While there are some public datasets on fake news detection that contain metadata, e.g., LIAR, we chose to create a new dataset from Politifact to contain more up-to-date statements that cover more contemporary writing styles. Politifact is a fact-checking website, where reporters and editors analyze the credibility of statements made by U.S. politicians. Politifact collects news statements from different sources, e.g., online social media, news articles, conferences, etc., rate their accuracy, and then publish the statement and the evaluation report. The evaluation score is organized over six qualitative levels, indicating the truthfulness of the statements and it ranges from "True" for accurate statements to "Pants on fire" for ludicrous statements. In addition to the news text, Politifact provides additional contextual data about the news documents, such as the main topics covered in the article, the date of publication, and other information related to the author, including the author's name, a short bio, the author's political party, and the history of the author's fact-checked publications. We scraped the Politifact website to collect news statements and related metadata published between 2009 and 2021. We collected 16,172 news documents, out of which 51.3% were labeled as "True" and 48.7% as "False". The statistical properties of the entire dataset are summarized in Table 1.

Table 1. Statistical properties of the collected dataset

Property	Value
# of true statements	8299
# of fake statements	7873
# of authors	4238
# of sources	63

To evaluate the performance of our framework, we adopted standard performance metrics, namely: accuracy, precision, recall, and F1 score [9].

4.2 Experimental Settings

As mentioned previously, the Politifact website uses six credibility labels: "True", "Mostly True", "Half True", "Barely True", "False", and "Pants on fire". Nevertheless, our objective in this work is to perform binary classification, so we grouped labels {"True", "Mostly True", "Half True"} to represent true news, and labels {"Barely True", "False", "Pants on fire"} to represent fake news. The proposed framework is depicted in Fig. 2.

The textual embedding is performed using the pre-trained $BERT_{large}$ model, available in [7]. For the rest of the implementation, we used Keras. We implemented a CNN consisting of a one-dimensional (1D) convolutional layer followed by a max-pooling layer for textual feature extraction. The convolutional layer

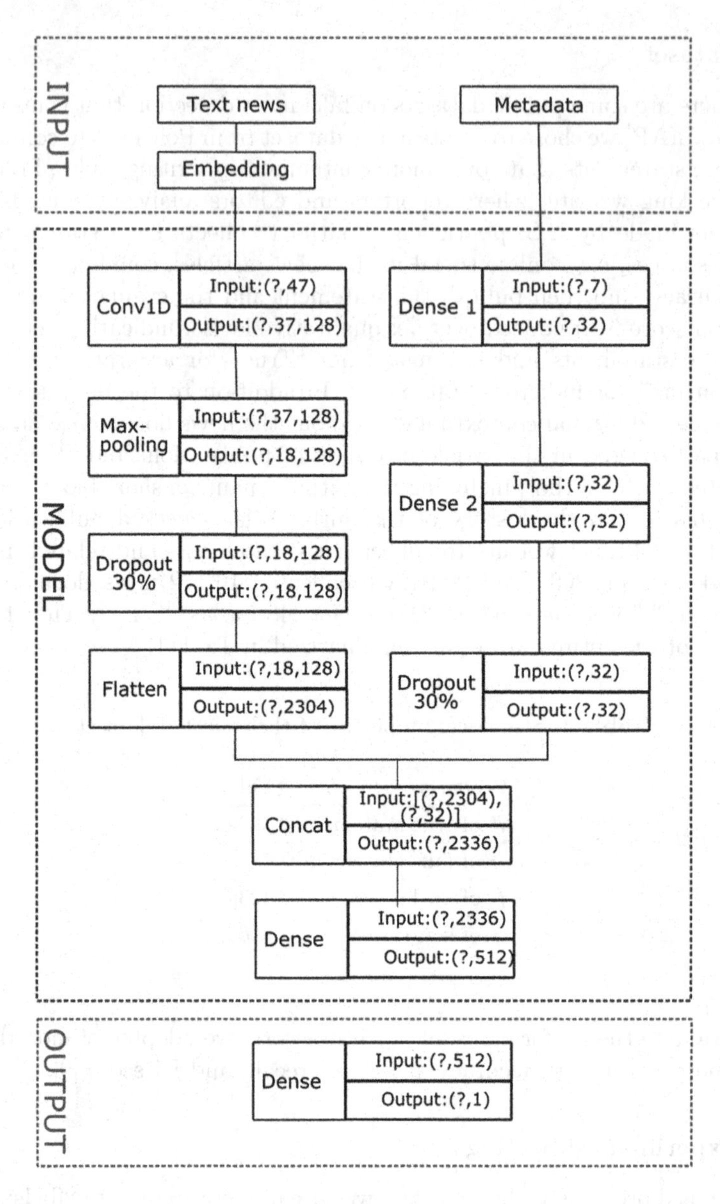

Fig. 2. Illustration of FN2's architecture

has 128 filters of size 10 and uses ReLU as the activation function. The resulting feature vectors are then fed into a max-pooling 1D layer with a window size of 2 to downsample the dimensionality of the features and to reduce the number of parameters. This approach aims at decreasing the computational cost without affecting the network's efficiency. Furthermore, to avoid over-fitting, we added a dropout layer with a rate of 0.3. Finally, we used a flatten layer to transform the pooled feature map into a one-dimensional vector.

To learn the metadata features, we implemented two dense layers with 32 hidden nodes each. We then added a dropout layer with a rate of 0.3 to avoid over-fitting. Finally, the features learned from both CNN and FNN are concatenated and passed through a dense layer with 512 hidden nodes, and then to a sigmoid dense layer to obtain news classification. The model is trained for 20 epochs, using a binary cross-entropy, Adam optimizer with a batch size of 64.

4.3 Competing Approaches

For comparison, we selected the state-of-the-art fake news detection models from two different categories: (i) textual models that employ neural networks; and (ii) contextual models that use similar contextual features. Specifically, we compare the performance of our model against the following four competitors:

- **LIWC** [16] : The Linguistic Inquiry and Word Count (LIWC) model is based on large lexicons of word categories that represent psycholinguistic processes (e.g., positive emotions, perceptual processes), summary categories (e.g., words per sentence), as well as part of speech categories (e.g., articles, verbs). More specifically, LIWC reads an input text and counts the percentage of words that reflect different emotions, thinking styles, social concerns, and even parts of speech. The LIWC analysis can be considered as somewhat similar to word embedding, yet, each dimension in an LIWC vector has a clear label. For fake news classification, LIWC has been employed to extract meaningful features from the news articles, which are then fed into an ML algorithm to learn the classification. In this work, we use LIWC features with Random Forest for fake news classification [22].
- **LSTM** [20]**:** This LSTM model classifies the reliability of news articles from their text only. The text news is first converted into 100-dimensional vectors using a 100-dim embedding GLOVE. The embedded vectors are then passed to the LSTM layer. The LSTM layer is composed of 300 hidden units and is followed by a fully connected dense layer that performs the classification.
- **Hybrid CNN-RNN** [14]**:** This model consists of two main components that extract features from the text news: a convolutional neural network and a recurrent neural network. The CNN extracts local features from the text news, while the RNN learns the long-term dependencies of the local features. First, the news text is transformed into vectors using word embedding. The embedding matrix is then fed into a one-dimensional convolutional layer, followed by a max-pooling layer. The local features extracted by the CNN layer are then provided to an LSTM layer that outputs the long-term dependencies

of the local features. Finally, the feature vectors are classified using a fully connected dense layer.

- **Hybrid CNN** [28]**:** This model captures the representation of the news text using CNN, and the representation of the metadata using CNN and Bi-LSTM. It then leverages the learned representations from the two different sources to identify fake news. To feed one source, the news text is first transformed into a vector representation and fed into a CNN to extract textual features. To feed the other source, the metadata is encoded and then passed into a CNN layer to capture the dependencies between the different features. Max-pooling is performed on the output of the CNN layer, followed by a Bi-LSTM layer. The feature representations from the news text and the metadata are then concatenated and fed into a dense layer to classify the news article.

Note that all the aforementioned solutions were evaluated on different datasets. Therefore, to ensure a fair comparison with FN2, we implemented all four models and trained them on our own dataset described in Sect. 4.1.

5 Results and Analysis

In this section, we present the results of our experimental evaluation that aims to investigate the accuracy of the proposed model. Our main goal is to answer the following questions:

- **Q1:** Did FN2 improve fake news detection by combining news article with metadata?
- **Q2:** How effective are metadata in detecting fake news?

5.1 Fake News Detection Performance

To answer **Q1**, we compare our model against the four competing approaches previously listed. We split the dataset into training (80%) and testing (20%), and repeated the process three times. The average results of the comparison are presented in Table 2.

Table 2. Performance comparison for all models using the collected dataset

	Accuracy	Precision	Recall	F1 score
LIWC [16]	0.601	0.613	0.5624	0.589
LSTM [20]	0.572	0.589	0.427	0.493
CNN-RNN [14]	0.646	0.653	0.574	0.611
Hybrid CNN [28]	0.568	0.564	0.523	0.542
FN2	**0.765**	**0.801**	**0.710**	**0.753**

From these results, we can make the following observations:

- For textual-based models, we observe that the performance of the CNN-RNN model is better than the LSTM model. This indicates that the CNN-RNN model can better capture the lexical and semantic properties of the news articles.
- FN2 achieved the best performance among all the other models. The accuracy of FN2 surpasses both the textual-based models and hybrid models by a margin of at least 13%.
- FN2 performs better than the models using textual features. This indicates that the extracted metadata features contain complementary information that helps boost fake news identification. However, this is not valid in the Hybrid CNN model, i.e., the accuracy of the textual-based models is higher than hybrid CNN. This could be explained by the fact that hybrid CNN was mainly designed for 6-class classification and that the parameters were tuned to increase the accuracy of multi-class classification.
- FN2 outperforms competitors' models in terms of all the evaluation metrics. For example, FN2 achieves an improvement of 13%, 15%, 14% compared to the CNN-RNN model in terms of accuracy, precision, and recall, respectively. The performance gap is even larger for the other models.

To illustrate the importance of metadata in detecting fake news and to answer **Q2**, we investigate the performance of two variants of the FN2 model:

- **FN2/Co:** This is a variant of the FN2 framework that does not use the metadata. The model encodes the text news and then feeds it into a convolutional neural network for feature extraction. Finally, the features are passed to a sigmoid dense layer to classify the news article.
- **FN2/T:** This is another variant of the FN2 framework that uses only the metadata to perform the classification. The processed metadata is fed into a feedforward neural network to learn the contextual features, and the resulting features are then passed to a sigmoid dense layer to classify the news article.

Results, reported in Fig. 3, allow us to provide the following observations:

- When we remove the contextual information, the performance of FN2/Co decreases dramatically. In particular, both accuracy and F1 score drop by 17% and 16%, respectively. This suggests that the contextual information brings important information to the fake news identification model.
- A similar observation can be made for FN2/T, i.e., when we eliminate the textual information from the classification model. One interesting remark is that FN2/T performs slightly better than FN2/Co, suggesting that news metadata are at least as valuable as textual data.

Therefore, our conclusion is that the textual and the contextual data of news articles are very valuable, and they contribute roughly equally to the fake news classification problem. We can, thus, confirm that they contain complementary information. However, one of the main challenges a model can face when using contextual features to detect fake news is cold-start issues. In this work, one of the contextual features, i.e., author's credibility, can suffer from the cold-start

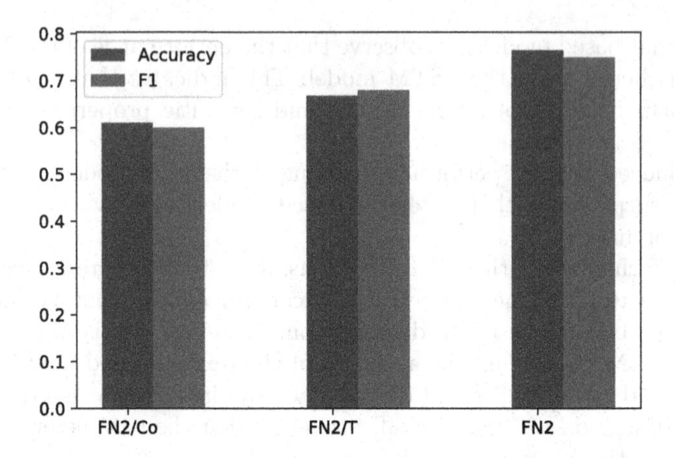

Fig. 3. Impact of removing textual or contextual information on FN2

issue. When a new author joins the network, we have little information about their credibility, which may affect the prediction results. Therefore, we study the impact of a cold start on FN2. For an author that has joined recently, we do not have any history of their reliability. Hence, we set it to zero. To evaluate the impact of cold-start on FN2, we randomly set the author's credibility to zero for both training and testing data, i.e., we randomly set 10% to 100% of the author's credibility to zero. The results are illustrated in Fig. 4. We observe that the cold start issue has slightly reduced the accuracy of FN2 to reach 71.85% when 30% of the author's credibility in the dataset is set to zero. Even when we do not have any information about any author, FN2 still outperforms its competitors and achieves an accuracy of 69.1%. This small impact of the cold-start problem can be explained by the fact that the model uses textual and contextual features for detection. In other words, combining the contextual features with the news content features has significantly reduced the effect of the cold start problem.

6 Model Evaluation on FakeNewsNet

Although many models are proposed for fake news classification, the issue of model generalization remains still an unresolved challenge. In this context, we aim to demonstrate the generality of the FN2 framework. More specifically, Fig. 5 depicts the training and validation values experienced by our model for accuracy and loss, as a function of the number of epochs. We notice that, with increasing number of epochs, the model's validation accuracy increases while the loss decreases significantly. We also observe that, as the training loss decreases steadily, the validation loss decreases as well in a similar pattern. These results show that the features learned by FN2 capture the semantics provided by data, and that the model does not experience either under- or over-fitting.

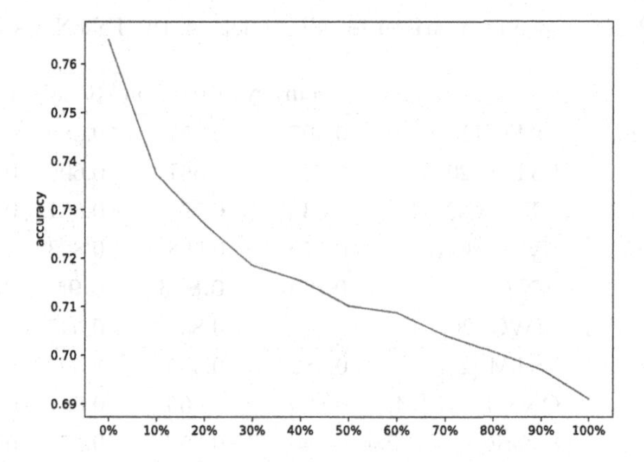

Fig. 4. Impact of the cold start problem in terms of the authors' credibility feature.

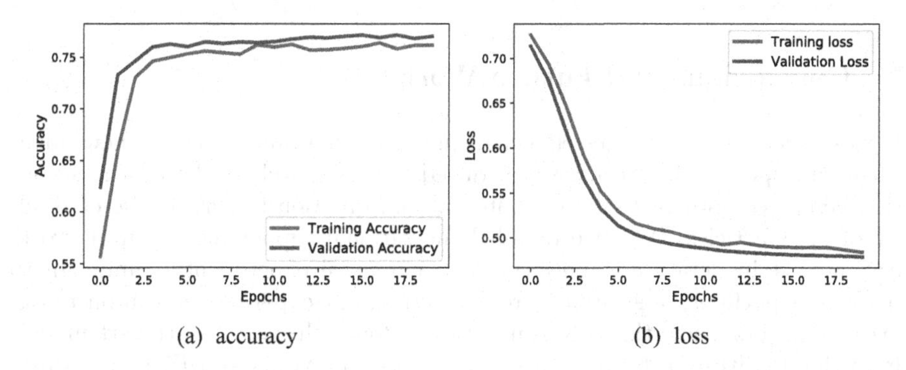

(a) accuracy (b) loss

Fig. 5. Training and validation accuracy and loss graphs of FN2 on the Poltifact dataset

To provide further support to the above formulated hypothesis about the excellent qualities enjoyed by our model, we test FN2 on a fake news multi-dimensional benchmark dataset, called FakeNewsNet [21]. The dataset contains political and entertainment news collected from two different fact-checking platforms: Politifact and GossipCop. Table 3 depicts the results of our evaluation. It shows that our model is able to generalize on unseen data, and outperforms the other models on all the evaluation metrics. Moreover, the results indicate that FN2 performs very well in non-political fields as well. Indeed, FN2 was trained with political statements from Politifact; however, when asked to classify entertainment news from GossipCop, FN2 was able to identify fake news with an accuracy of 87.4%, outperforming its competitors by at least 9.7%. These results demonstrate that the features learned by FN2 capture the semantics of the data, and that the model is not experiencing any under- or over-fitting. To summarize, our findings show that the FN2 model can be applied to various domains and across different datasets.

Table 3. Performance comparison for all models on the FakeNewsNet dataset

		Accuracy	Precision	Recall	F1
Politifact	LIWC [20]	0.807	0.751	0.880	0.811
	LSTM [20]	0.737	0.667	0.695	0.680
	CNN-RNN [14]	0.632	0.59	0.861	0.704
	Hybrid CNN [28]	0.579	0.608	0.823	0.7
	FN2	**0.875**	**0.866**	**0.95**	**0.904**
GossipCop	LIWC [20]	0.776	0.857	0.547	0.668
	LSTM [20]	0.582	0.666	0.593	0.627
	CNN-RNN [14]	0.624	0.666	0.581	0.623
	Hybrid CNN [28]	0.646	0.652	0.473	0.548
	FN2	**0.874**	**0.866**	**0.855**	**0.862**

7 Conclusions and Future Work

In this paper, we have addressed a challenging and relevant topic: fake news identification. Specifically, we have introduced FN2, a novel ML-based framework that combines both textual and contextual information to classify the credibility of news articles. We have detailed the rational supporting our framework, described it in detail, and performed an extensive experimental campaign to test our hypothesis. The results are staggering: our experiments, run on recognized data sets, show that FN2 outperforms four other state-of-the-art models from the literature in terms of accuracy, precision, recall, and F1 score. Moreover, we also demonstrated the correlation between fake news and contextual information, and showed that using contextual features improves the fake news classification accuracy by at least 13%. Further, our model has also been shown to enjoy a quite good generalization. Indeed, while being trained on politics-oriented news, it also outperforms competing models on a qualitative different data set (entertainment news).

Several interesting future directions can be further investigated. First, an extension of the FN2 model from binary classification to multi-classification will be considered. Second, it would be interesting to perform an analytical study on the quality of the results when converting multi-class labels to binary class labels—hence helping finding out the best binary split for the dataset that produces the best discrimination among the Politifact sub-labels. Third, we plan to examine whether certain social features that relate to the news article, such as the number of likes, number of shares, etc., can further enhance the accuracy of our model. Another interesting direction would aim at overcoming BERT's intrinsic limitation on the number of word tokens (512), limiting our FN2 model to only news statements with less than 512 words. Therefore, we plan to extend FN2 to consider longer news articles. Finally, due to the lack of publicly available state-of-the-art fake news detection models, FN2 was only compared to four models. However, in the future, we plan to consider and re-implement more baseline models for comparison, such as TriFN.

Acknowledgments. The authors would like to thank the reviewers that, with their comments, helped to improve the quality of the paper, and Dr. Chuan Yue for shepherding this contribution.

This work was partially supported by NPRP-S-11-0109-180242, from the QNRF-Qatar National Research Fund, a member of The Qatar Foundation. The information and views set out in this publication are those of the authors and do not necessarily reflect the official opinion of the QNRF.

References

1. Ahmed, H., Traore, I., Saad, S.: Detection of Online Fake News Using N-Gram Analysis and Machine Learning Techniques. In: Traore, I., Woungang, I., Awad, A. (eds.) ISDDC 2017. LNCS, vol. 10618, pp. 127–138. Springer, Cham (2017). https://doi.org/10.1007/978-3-319-69155-8_9
2. Cai, L., Song, Y., Liu, T., Zhang, K.: A hybrid BERT model that incorporates label semantics via adjustive attention for multi-label text classification. IEEE Access **8**, 152183–152192 (2020)
3. Collins, B., Hoang, D.T., Nguyen, N.T., Hwang, D.: Fake News Types and Detection Models on Social Media A State-of-the-Art Survey. In: Sitek, P., Pietranik, M., Krótkiewicz, M., Srinilta, C. (eds.) ACIIDS 2020. CCIS, vol. 1178, pp. 562–573. Springer, Singapore (2020). https://doi.org/10.1007/978-981-15-3380-8_49
4. Do, T.H., Berneman, M., Patro, J., Bekoulis, G., Deligiannis, N.: Context-aware deep Markov random fields for fake news detection. IEEE Access **9**, 130042–130054 (2021)
5. Elisa Shearer, K.E.M.: News use across social media platforms 2018, August 2020
6. Freire, P.M.S., da Silva, F.R.M., Goldschmidt, R.R.: Fake news detection based on explicit and implicit signals of a hybrid crowd: an approach inspired in meta-learning. Expert Syst. Appl. **183**, 115414 (2021)
7. Google-Research: Google-research/bert: Tensorflow code and pre-trained models for bert (2019). https://github.com/google-research/bert
8. Granik, M., Mesyura, V.: Fake news detection using Naive Bayes classifier. In: 2017 IEEE First Ukraine Conference on Electrical and Computer Engineering (UKRCON), pp. 900–903. IEEE (2017)
9. Junker, M., Hoch, R., Dengel, A.: On the evaluation of document analysis components by recall, precision, and accuracy. In: Proceedings of the Fifth International Conference on Document Analysis and Recognition. ICDAR 1999 (Cat. No.PR00318), pp. 713–716 (1999). https://doi.org/10.1109/ICDAR.1999.791887
10. Kaliyar, R.K., Goswami, A., Narang, P.: EchoFakeD: improving fake news detection in social media with an efficient deep neural network. Neural Comput. Appl. **33**(14), 8597–8613 (2021). https://doi.org/10.1007/s00521-020-05611-1
11. Kaliyar, R.K., Goswami, A., Narang, P., Sinha, S.: FNDNet-a deep convolutional neural network for fake news detection. Cogn. Syst. Res. **61**, 32–44 (2020)
12. Kaliyar, R.K., Kumar, P., Kumar, M., Narkhede, M., Namboodiri, S., Mishra, S.: DeepNet: an efficient neural network for fake news detection using news-user engagements. In: 2020 5th International Conference on Computing, Communication and Security (ICCCS), pp. 1–6. IEEE (2020)

13. Martino, G.D.S., Cresci, S., Barrón-Cedeño, A., Yu, S., Di Pietro, R., Nakov, P.: A survey on computational propaganda detection. In: Bessiere, C. (ed.) Proceedings of the Twenty-Ninth International Joint Conference on Artificial Intelligence, IJCAI 2020, pp. 4826–4832 (2020). ijcai.org
14. Nasir, J.A., Khan, O.S., Varlamis, I.: Fake news detection: a hybrid CNN-RNN based deep learning approach. Int. J. Inf. Manag. Data Insights **1**(1), 100007 (2021)
15. Pan, J.Z., Pavlova, S., Li, C., Li, N., Li, Y., Liu, J.: Content Based Fake News Detection Using Knowledge Graphs. In: Vrandečić, D., Bontcheva, K., Suárez-Figueroa, M.C., Presutti, V., Celino, I., Sabou, M., Kaffee, L.-A., Simperl, E. (eds.) ISWC 2018. LNCS, vol. 11136, pp. 669–683. Springer, Cham (2018). https://doi.org/10.1007/978-3-030-00671-6_39
16. Pennebaker, J.W., Francis, M.E., Booth, R.J.: Linguistic inquiry and word count: Liwc 2001. Mahway: Lawrence Erlbaum Associates **71**(2001), 2001 (2001)
17. Potthast, M., Kiesel, J., Reinartz, K., Bevendorff, J., Stein, B.: A stylometric inquiry into hyperpartisan and fake news. arXiv preprint arXiv:1702.05638 (2017)
18. Raponi, S., Khalifa, Z., Oligeri, G., Di Pietro, R.: Fake news propagation: a review of epidemic models, datasets, and insights. ACM Trans. Web (2022). https://doi.org/10.1145/3522756
19. Rapoza, K.: Can 'fake news' impact the stock market? December 2020. https://www.forbes.com/sites/kenrapoza/2017/02/26/can-fake-news-impact-the-stock-market/?sh=f625ee52fac0
20. Rashkin, H., Choi, E., Jang, J.Y., Volkova, S., Choi, Y.: Truth of varying shades: analyzing language in fake news and political fact-checking. In: Proceedings of the 2017 Conference on Empirical Methods in Natural Language Processing, pp. 2931–2937 (2017)
21. Shu, K., Mahudeswaran, D., Wang, S., Lee, D., Liu, H.: Fakenewsnet: a data repository with news content, social context, and spatiotemporal information for studying fake news on social media. Big Data **8**(3), 171–188 (2020)
22. Shu, K., Wang, S., Liu, H.: Beyond news contents: the role of social context for fake news detection. In: Proceedings of the Twelfth ACM International Conference on Web Search and Data Mining, pp. 312–320 (2019)
23. Shu, K., Zhou, X., Wang, S., Zafarani, R., Liu, H.: The role of user profiles for fake news detection. In: Proceedings of the 2019 IEEE/ACM International Conference on Advances in Social Networks Analysis and Mining, pp. 436–439 (2019)
24. Tankovska, H.: Number of social network users worldwide from 2017 to 2025, January 2021. https://www.statista.com/statistics/278414/number-of-worldwide-social-network-users/
25. Umer, M., Imtiaz, Z., Ullah, S., Mehmood, A., Choi, G.S., On, B.W.: Fake news stance detection using deep learning architecture (CNN-LSTM). IEEE Access **8**, 156695–156706 (2020)
26. Vaswani, A., et al.: Attention is all you need. Adv. Neural Inf. Process. Syst. 5998–6008 (2017)
27. Verma, A., Mittal, V., Dawn, S.: FIND: fake information and news detections using deep learning. In: 2019 Twelfth International Conference on Contemporary Computing (IC3), pp. 1–7. IEEE (2019)
28. Wang, W.Y.: Liar, liar pants on fire: a new benchmark dataset for fake news detection. In: Proceedings of the 55th Annual Meeting of the Association for Computational Linguistics (Volume 2: Short Papers), pp. 422–426 (2017)
29. Yang, Y., Zheng, L., Zhang, J., Cui, Q., Li, Z., Yu, P.S.: TI-CNN: convolutional neural networks for fake news detection. arXiv preprint arXiv:1806.00749 (2018)

30. Zhang, J., Dong, B., Yu, P.S.: Deep diffusive neural network based fake news detection from heterogeneous social networks. In: 2019 IEEE International Conference on Big Data (Big Data), pp. 1259–1266. IEEE (2019)
31. Zhou, X., Wu, J., Zafarani, R.: SAFE: similarity-aware multi-modal fake news detection. In: Lauw, H.W., Wong, R.C.-W., Ntoulas, A., Lim, E.-P., Ng, S.-K., Pan, S.J. (eds.) PAKDD 2020. LNCS (LNAI), vol. 12085, pp. 354–367. Springer, Cham (2020). https://doi.org/10.1007/978-3-030-47436-2_27
32. Zhou, X., Zafarani, R.: A survey of fake news: fundamental theories, detection methods, and opportunities. ACM Comput. Surv. (CSUR) **53**(5), 1–40 (2020)

Malware Detection with Limited Supervised Information via Contrastive Learning on API Call Sequences

Mohan Gao[1], Peng Wu[1,2](✉), and Li Pan[1,2]

[1] School of Electronic Information and Electrical Engineering, Shanghai Jiao Tong University, Shanghai, China
[2] Shanghai Key Laboratory of Integrated Administration Technologies for Information Security, Shanghai Jiao Tong University, Shanghai, China
catking@sjtu.edu.cn

Abstract. Malware is a software capable of causing damage to computer systems. Conventional malware detection methods either require feature engineering to extract specific features or require a large amount of labeled data to train an end-to-end deep learning model. Both feature engineering and labelling are laborious. In this paper, we propose a semi-supervised contrastive learning malware detection method based on API call sequences with limited label information, called SCLMD. Specifically, a heterogeneous graph is constructed from API behavior to express the rich relationships among labeled and unlabeled software. After extracting the structural and sequential features of software by two encoders, we adopt the cross-view contrastive learning to obtain the shared and consistent feature of software. A hybrid positive selection strategy is designed to select positive pairs for contrastive learning by the guidance of the limited label information. Experimental results on two real world datasets show that the SCLMD outperforms the baseline methods, especially when the supervised information is limited.

Keywords: Malware detection · Contrastive learning · Heterogeneous graph neural network

1 Introduction

Malware is any software designed to cause damage or occupy the resources of the target computer by means of executable codes, scripts, etc. Huge economic losses in the world caused by massive malware pose a considerable challenge for malware mitigation. Malware detection is the process of discovering and classifying malware based on its characteristics. According to whether the malicious code is executed in the analysis, malware detection techniques can be divided into static detection techniques and dynamic detection techniques [2]. Static techniques extract binary code, string, byte sequence, file name, etc. from the content of the software code as the signatures of the software. The detection can be completed by matching the signatures of the sample with the feature library

C. Alcaraz et al. (Eds.): ICICS 2022, LNCS 13407, pp. 492–507, 2022.
https://doi.org/10.1007/978-3-031-15777-6_27

of malware [13]. Static detection technology is simple and fast, but malware can camouflage itself by means of encryption, encapsulation, and packing to avoid detection [21,22,26]. Dynamic detection techniques extract dynamic features for detection by monitoring and recording the running state and behavior of software [8], which can effectively combat the camouflage of malware. Application Programming Interface (API) call is a typical dynamic feature. If malware wants to achieve its damage goals, it must call some APIs according to a certain pattern, which is difficult to disguise. Therefore, the malware detection based on API call is more robust and effective.

The traditional API call-based malware detection methods mainly focus on machine learning model based on feature engineering. Experts exploit feature engineering to extract the API call features that are important for malware detection first, and then adopt machine learning model on the extracted features. Extracting effective features manually are laborious and time consuming for a large number of new types of malware. With the development of deep learning technology, many recent works have proposed end-to-end methods that use deep learning models to automatically extract features from raw software behavior data to detect and classify malware. Deep learning-based methods require a large amount of labeled data to train models. However, in practice, the number of labeled malware is limited. Labeling malware requires a lot of domain knowledge and it is expensive to manually construct large-scale labeled dataset. Therefore, how to make full use of a large amount of unlabeled data for malware detection is an urgent problem to be solved.

We propose a Semi-supervised Contrastive Learning Malware Detection method (SCLMD) based on API call sequences with limited supervised information. Although a large number of software may not have labels, they have rich relationships via their API call sequences. Their relationships not only help to perform unsupervised learning, but also make the unlabeled software obtain supervised information from the limited number of labeled software. Therefore, in order to make full use of the large amount of unlabeled software, we first construct a heterogeneous network from API call sequences to express the complex relationships between software and APIs. Then, a heterogeneous network representation learning method is adopted to extract the software relationship features from the heterogeneous network. However, when constructing the heterogeneous network, the sequential information of the API call sequence is lost. Therefore, the features extracted from the heterogeneous network ignore the important API sequential features. On the other hand, the API call sequence itself has rich sequence information, so we adopt models to extract the sequential features from the API call sequence directly. Heterogeneous network-based features and sequence-based features come from different views of software, respectively. In order to fully utilize the unlabeled software data to extract shared and consistent features of software, we propose a contrastive learning framework for software feature extraction. Contrastive learning is a self-supervised learning method that does not require any label information. To better guide malware detection with limited label information, we propose a hybrid positive selection strategy for contrastive learning to introduce the limited label information into the contrastive learning framework. Experiments on Alibaba Cloud Malware Detection (ACMD) Dataset show that the performance of the proposed SCLMD is better than that of other malware detection methods based on API call sequences, especially when the ratio of labeled samples is relatively little.

Our main contributions are as follows:

1. For the first time, we use a contrastive learning framework in malware detection to extract shared and consistent features from structural view and sequential view of software;
2. We propose a hybrid positive selection strategy to guide the contrastive learning by selecting positive pairs based on the limited label information;
3. Experiments show that our proposed method can effectively detect and classify malware. Especially when there are limited labels, our method has a large performance improvement compared to other methods.

2 Related Work

2.1 Malware Detection

Malware detection methods can be divided into static detection methods and dynamic detection methods according to whether the features used for detection are generated during software runtime.

Static detection methods detect and/or classify malware by extracting binary sequences from the software as its signature codes and matching the codes in malware signature database [13]. However, static detection methods can only detect malware with known malicious signatures, and is difficult to detect emerging malware with unknown signatures. In addition, malware will also disguise its signature by means of encryption, encapsulation, and packing to avoid detection, which brings great challenges to methods based on signature matching [21,22,26].

The dynamic detection methods use the features generated during the software running process, especially the API call sequence [1,8,17]. In order to achieve the illegal goals, malware must call APIs in a specific pattern, and researchers can find the behavioral intent of any software by analyzing its API call sequence patterns. Therefore, the dynamic detection method can effectively resist the obfuscation and camouflage technology of malware. According to the sequence nature of API calls, many methods use sequence learning methods, such as Recurrent Neural Networks (RNN), to extract features from API call sequences [16,29,34]. But training sequence learning methods usually require large amounts of labels, as the sequence learning method itself cannot make good use of the relationship between unlabeled data. Thus they may perform poorly when labels are limited.

2.2 Graph Neural Network

Graph-structured data can reflect the complex relationships between entities, which can not only contribute to unsupervised learning, but also help unlabeled samples obtain supervised information from limited labels. Graph Neural Networks (GNN), such as GCN [15], GAT [30], can extract features from the graph-structured data effectively even when the labels are sparse. The heterogeneity of nodes and edges in heterogeneous graphs brings additional challenges for feature extraction, and many heterogeneous GNNs are also proposed to learn their embedding, such as HAN [6] and HGT [14]. The GNNs have been successfully applied in many areas, such as medicine research [12] and recommendation systems [28]. There are also many GNN based malware detection methods. MatchGNet [32] propose a Graph Matching Network model to learn the graph representation based on the invariant graph modeling of the program's execution behaviors, which can detect malware with less false positives. GDroid [7] maps apps and APIs into a large heterogeneous graph by building the "App-API" and "API-API" edges based on the invocation relationship and the API usage patterns, converting the malware detection problem into a node classification task. These methods all extract the relationships between entities by constructing heterogeneous networks, but they may ignore the sequential information of API call sequences.

2.3 Contrastive Learning

Contrastive learning is a self-supervised learning method that learns the shared consistent features of samples by contrasting their different views [4,9,10,19]. It also has been applied to graph neural networks to improve their representation learning ability. DGI [31] builds local patches and global summary as positive pairs, and utilizes Infomax [18] theory to contrast. Along this line, GMI [24] is proposed to contrast between center node and its local patch from node features and topological structure. GCC [25] focuses on pretraining with contrasting universally local structures from any two graphs. In heterogeneous domain, DMGI [23] conducts contrastive learning between original network and corrupted network on each single view, and designs a consensus regularization to guide the fusion of different meta-paths. HeCo [33] propose a novel heterogeneous graph neural network with co-contrastive learning, choose network schema and metapath structure as two views to collaboratively supervise each other. However, how to adopt contrastive learning to enhance the performance of specific application tasks, such as malware detection, still requires a lot of exploration.

3 Problem Formulation

Application programming interfaces (APIs) are interfaces of the operating system called by any software with a specific order to achieve its functionalities. API call sequence directly reflects software's inherent behavior features which is difficult to disguise and deceive, so malware can be effectively detected based on the API call sequence.

Given a set $< \mathcal{S}, \mathcal{A}, \mathcal{Y} >$, where $\mathcal{A} = \{a_1, \ldots, a_q\}$ denotes the set of q APIs, $\mathcal{S} = \{s_1, \ldots, s_p\}$ denotes the set of p software where a software $s_i \in \mathcal{S}$ is represented by an API call sequence, i.e., $s_i = \{s_{i1}, \ldots, s_{iT} | s_{it} \in \mathcal{A}\}$, $\mathcal{Y} = \{y_0, \ldots, y_c\}$ denotes the set of labels with one type of good software and c types of malware, the purpose of Malware detection with limited supervised information is to classify the software in $\mathcal{S}$ when the number of labeled samples r is much smaller than the number of total samples p, i.e., $r \ll p$.

4 The Proposed Model: SCLMD

In this section, we propose SCLMD, a new semi-supervised contrastive Learning malware detection method based on API call sequences. The overall architecture is shown as Fig. 1.

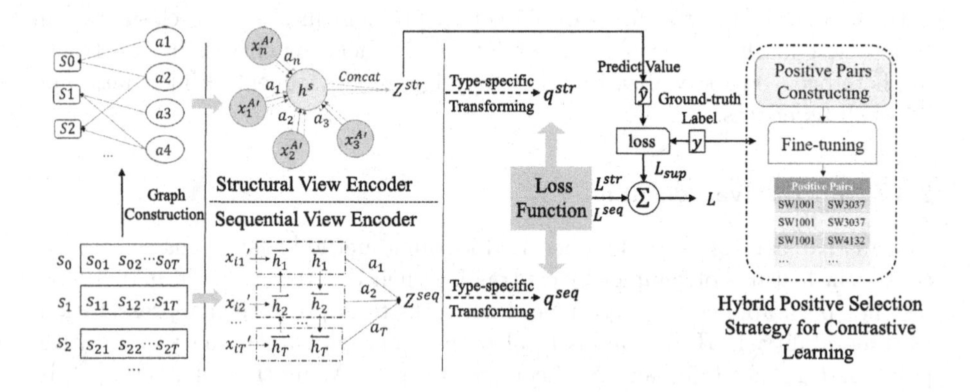

Fig. 1. The overall architecture of SCLMD.

When supervised information is limited, in order to make full use of a large amount of unlabeled software for training, we first construct a heterogeneous network based on software API call sequences to express the complex relationships between software and APIs. The relationships not only enable all software learn features from each other, but also make unlabeled software obtain supervised information from the limited number of labeled software. A heterogeneous network based on API call sequence is defined as $G = \{\mathcal{A}, \mathcal{S}, \mathcal{C}\}$ where $\mathcal{C}$ is the set of edges. An edge $c \in \mathcal{C}$ from an API $a \in \mathcal{A}$ to a software $s \in \mathcal{S}$ represents that the software s calls the API a during its runtime. Since different API may show different importance and GAT can learns the importance of different neighbors through the attention mechanism to obtain node representation, a GAT-based structural view encoder is proposed to learn the structural feature of software on the heterogeneous network G.

Although the structural features embed rich relationship between software and APIs, they ignore important sequential properties of API sequence. In order to utilize the rich sequential properties, a sequential view encoder is designed based on Bi-GRU with attention mechanism to extract the sequential feature of the software. Bi-GRU combines forward and reverse hidden states of GRU to effectively capture the semantic associations between API sequences, the reset gate and update gate in Bi-GRU can alleviate the problems of gradient disappearance and explosion with a reasonable computational cost.

The structural feature and sequential feature of a software are complementary with each other. A semi-supervised contrastive learning framework is proposed to extract the consistent features shared by two types of features. In order to make the extracted consistent feature better for malware detection, a hybrid positive selection strategy for contrastive learning is proposed to construct the set of positive pairs for contrastive learning based on the TF-IDF similarity and the provided limited supervised information.

4.1 Structural View Encoder

The structural view encoder is designed based on GAT [30] to learn features from the heterogeneous network. Let $X^A \in \mathbb{R}^{q\times d^A}$ and $X^S \in \mathbb{R}^{p\times d^S}$ be the initial feature matrix of API and software respectively, where d^A and d^S are their feature dimensions respectively, and each row of them, i.e., x_i^A and x_j^S, is the initial feature. If no initial feature matrix is provided, one-hot feature matrix will be adopted. Due to the heterogeneity of nodes, the features of different types of nodes are first projected into the same feature space by two type-specific transformation matrices $W^A \in \mathbb{R}^{d\times d^A}$ and $W^S \in \mathbb{R}^{d\times d^S}$.

$$\begin{aligned} x_i^{A'} &= W^A \cdot x_i^A \\ x_j^{S'} &= W^S \cdot x_j^S \end{aligned} \tag{1}$$

where $x_i^{A'} \in \mathbb{R}^d$ and $x_j^{S'} \in \mathbb{R}^d$ are the projected features and d is the dimension of the projected features.

Since APIs of each software may play different roles and show different importance, here we adopt a multi-head node-level attention mechanism to perform structural feature extraction of software.

$$h_j^h = ||_{h=1}^{H} \sigma(\sum_{i\in N_j} a_{ij} \cdot x_i^{A'}) \tag{2}$$

$$a_{ij} = \frac{e^{e_{ij}}}{\sum_{k\in N_j} e^{e_{kj}}} \tag{3}$$

$$e_{ij} = LeakyReLu\left(a^T \left[x_j^{S'} \middle|\middle| x_i^{A'}\right]\right) \tag{4}$$

where $a \in \mathbb{R}^{2d}$ is an attention vector, a_{ij} is the attention weights of API i for software j, N_j is the neighbor set of j, $||$ denotes the concatenate operation, H

is the number of heads, and $h_j^h \in \mathbb{R}^{Hd}$ is the concatenated multi-head embedding for software j. Finally, h_j^h is passed through a linear layer to obtain the representation of software j from the structural view:

$$z_j^{str} = W \cdot h_j^h + b \tag{5}$$

where $W \in \mathbb{R}^{d\times Hd}$ is the weight matrix, and $b \in \mathbb{R}^d$ is the bias vector. $z_j^{str} \in \mathbb{R}^d$ is the representation of software j from the structural view.

4.2 Sequential View Encoder

Given the API call sequence $\{s_{it} | t \in [1, T]\}$ of a software s_i, we first project each API s_{it} into the latent space through a transformation matrix $W_e \in \mathbb{R}^{d\times d^A}$.

$$x_{it}^{A'} = W_e x_{it}^A \tag{6}$$

A Bidirectional GRU (Bi-GRU) [5] is adopted to get the hidden states of all APIs in a API sequence by summarizing information from both directions of the sequence.

$$\begin{aligned} \overrightarrow{h_{it}} &= \overrightarrow{GRU}\left(x_{it}^{A'}\right), t \in [1, T] \\ \overleftarrow{h_{it}} &= \overleftarrow{GRU}\left(x_{it}^{A'}\right), t \in [T, 1] \end{aligned} \tag{7}$$

where $\overrightarrow{h_{it}} \in \mathbb{R}^{d/2}$ is the forward hidden state and $\overleftarrow{h_{it}} \in \mathbb{R}^{d/2}$ is the backward hidden state. The hidden state of any API is obtained by concatenating $\overrightarrow{h_{it}} \in \mathbb{R}^{d/2}$ and $\overleftarrow{h_{it}} \in \mathbb{R}^{d/2}$ i.e., $h_{it} = \left[\overrightarrow{h_{it}}, \overleftarrow{h_{it}}\right]$, $h_{it} \in \mathbb{R}^d$.

Not all APIs in a sequence contribute equally to the representation of the software's behavior intention. An attention mechanism is adopted to learn the importance of each API and aggregate the weighted hidden states of APIs to form the representation of the software from the sequential view.

$$z_i^{seq} = \sum_t a_{it} \cdot h_{it} \tag{8}$$

$$a_{it} = \frac{e^{u_{it}{}^T \cdot u_w}}{\sum_t e^{u_{it}{}^T \cdot u_w}} \tag{9}$$

$$u_{it} = tanh\left(W_\omega h_{it} + b_\omega\right) \tag{10}$$

where W_ω and b_ω are the weight matrix and bias vector of a one-layer MLP, $u_{it} \in \mathbb{R}^d$ is a projected hidden representation of h_{it}, $u_w \in \mathbb{R}^d$ is an attention vector, a_{it} is the attention weights of h_{it}, and $z_i^{seq} \in \mathbb{R}^d$ is the representation of software i from the sequential view.

4.3 Hybrid Positive Selection Strategy for Contrastive Learning

In order to fully utilize the unlabeled data to extract shared and consistent features of software from structural and sequential views, a contrastive learning framework is adopted for software feature learning. Contrastive learning is a self-supervised learning method that learn representation by contrasting positive pairs against negative pairs. The samples of positive pair should come from different views, and usually are the two augmented views of the same sample. However, when the size of training set is not large enough, the number of positive pairs may be too small to train model well. Meanwhile, the traditional positive pair construction method cannot take advantage of the supervised information to guide the feature learning for malware detection. Thus we propose a hybrid positive selection strategy for contrastive learning, which first construct positive pairs based on TF-IDF similarity, and then fine-tune the positive pairs based on limited label information.

TF-IDF (term frequency-inverse document frequency) is a numerical statistic that is intended to reflect how important a word is to a document in a collection or corpus. By regarding the software as document and the API as word, then the TF-IDF value of each API can measure how important the API is to a software. As a result, the higher the similarity of the TF-IDF vectors of two software is, the more similar the two software are in behavior. To this end, for each software, we calculate the cosine similarity of the TF-IDF vectors between it and any other software, and select top k ones with the largest cosine similarity to form the k positive pairs with it. In this way, the positive pairs set is initially constructed.

When some supervised information is provided, the positive pairs set can be further fine-tuned to better guide the feature learning. The fine-tuning is carried out based on the intuition that if two software have the same label, they are positive pairs of each other and should be added into positive set if they are not in it, while if they have different labels, they are not positive samples of each other and should be removed from positive set if they are in it. After the fine-tuning, the set of positive pairs of software i is denoted as P_i. And we naturally treat remaining nodes that are not positive samples in the same training batch with i as negative samples, denoted as N_i.

4.4 Loss

After getting the z_i^{str} and z_i^{seq} for software i from two views, we feed them into two Multi-Layer Perceptrons(MLP) with one hidden layer separately to map them into the same space where contrastive loss is calculated:

$$\begin{aligned} q_i^{str} &= MLP_{str}(z_i^{str}) \\ q_j^{seq} &= MLP_{seq}(z_i^{seq}) \end{aligned} \tag{11}$$

with the positive pairs set P_i and negative sample set N_i, the contrastive loss under structural view is defined as:

$$L^{str} = \frac{-1}{|V|} \sum_{i \in V} log \frac{\sum_{j \in P_i} e^{sim(q_i^{str}, q_j^{seq})/\tau}}{\sum_{k \epsilon (P_i \cup N_i)} e^{sim(q_i^{str}, q_k^{seq})/\tau}} \tag{12}$$

where $sim(u, v)$ denotes the cosine similarity between two vectors u and v, and τ denotes a temperature parameter. The loss considers multiple positive pairs, which is different from traditional infoNCE loss [19,35], that usually only focuses on one positive pair in the numerator of Eq. (12). For two software in a pair, the target embedding is from the structural view (q_i^{str}) and the embedding of positive and negative samples are from the sequential view (q_i^{seq}). In this way, we realize the cross-view self-supervision.

The contrastive loss L^{seq} under sequential view is similar as L^{str}, but differently, the target embedding is from the sequential view while the embedding of positive and negative pairs is from the structural view:

$$L^{seq} = \frac{-1}{|V|} \sum_{i \in V} log \frac{\sum_{j \in P_i} e^{sim(q_i^{seq}, q_j^{str})/\tau}}{\sum_{k \in (P_i \cup N_i)} e^{sim(q_i^{seq}, q_k^{str})/\tau}} \tag{13}$$

To take full advantage of supervised information, a semi-supervised framework is adopted by combing the above contrastive loss with a cross-entropy based supervised loss.

$$L_{sup} = -\sum_{i \in S_l} y_i \log \hat{y}_i \tag{14}$$

$$\hat{y} = softmax(W_{sup} \cdot z^{str} + b_{sup}) \tag{15}$$

where $S_l \subset S$ is the subset of software with labels, $\hat{y}$ is the predicted label, W_{sup} and b_{sup} are the weight matrix and bias of linear classifier. The supervised loss is applied to structural embedding rather than sequential embedding, because the unlabeled samples can acquire the supervised information from the limited number of labeled samples indirectly through the edges of the constructed heterogeneous graph.

The overall loss function is given as follows:

$$L = \lambda \cdot L^{str} + (1 - \lambda) \cdot L^{seq} + L_{sup} \tag{16}$$

where λ is a coefficient to balance the effect of two views. We can optimize the proposed model via back propagation and learn the embedding of software. In the end, the structural representation of software z^{str} is adopted to perform downstream tasks, because the graph structure helps unsupervised learning and enables the unlabeled samples obtain supervised information from the labeled ones.

5 Experiments

5.1 Datasets

We employ the following two real world malware datasets. The dataset statistics is in Table 1.

ACMD[1]: The Alibaba Cloud Malware Detection dataset contains API instruction information from 4978 benign Windows software and 8909 Windows

[1] https://tianchi.aliyun.com/competition/entrance/231694/information?lang=en-us.

Table 1. Summary statistics of the datasets.

Dataset	#software	#API	#classes
ACMD	13887	295	8
ACSAC	10079	304	6

malware. There are 7 types of malware, including infected virus, Trojan Horse program, mining program, DDoS Trojan, extortion virus, etc.

ACSAC[2]: A subset of the Alibaba Cloud's 3rd Annual Security Algorithm Challenge dataset, which contains API instruction information from 5,000 benign software and 5,079 malware, including infectious viruses, Trojans, DDOS Trojan, ransomware, etc.

We spliced the APIs called during the software execution process into an API sequence according to time order.

5.2 Baselines

The malware detection based on API call sequence is a typical sequence classification problem. Thus two sequence learning methods, i.e., BiLSTM [11] and BiGRU [5], are adopted as the comparison baselines. In both methods, the representations of software are obtained by aggregating the hidden states of all APIs in the sequence with an attention mechanism. On the other hand, the malware detection may be beneficial from the rich relationships between software expressed by the heterogeneous graph constructed above. Thus two heterogeneous graph learning methods, i.e., R-GCN [27] and HGT [14], are adopted as comparison baselines by learning the representations of software on the constructed heterogeneous graph. Besides above general baselines, some specifically designed API call sequences based methods are adopted. MaMaDroid [20] extracts features for malware detection by building a Markov Chain based behavioral model from the API call sequences. GDroid [7] maps software and APIs into a large heterogeneous graph, converting the original problem into a node classification task. LGMal [3] extracts local and global features for malware detection by combining the stacked convolutional neural network and graph convolutional networks.

It is worth noting that in both Windows and Android, API call sequences have similar functions and characteristics, the malware detection methods based on API call sequence all identify malware by mining the special pattern of the sequence. These similarities make the methods to work in both systems, so we choose two typical state of arts Android malware detection methods, MaMadroid and GDroid, as comparison algorithms in our experiments. All methods are trained in a supervised manner by inputting the representation of software into a linear classifier.

[2] https://tianchi.aliyun.com/competition/entrance/231668/information?lang=en-us.

Table 2. The model comparison on two datasets ACMD and ACSAC with different training ratio. The best performance is highlighted in boldface.

Datasets	Metrics	Training ratio	R-GCN	HGT	BiLSTM	BiGRU	MaMa Droid	GDroid	LGMal	SCLMD
ACMD	Macro-F1	70%	0.5607	0.5750	0.5466	0.5718	0.5699	0.5773	0.5921	**0.6009**
		20%	0.5403	0.5634	0.5310	0.5551	0.5517	0.5621	0.5704	**0.5741**
		10%	0.5310	0.5032	0.5116	0.4997	0.5480	0.4728	0.4652	**0.5543**
		5%	0.5203	0.3823	0.4520	0.4122	0.5248	0.3729	0.4032	**0.5292**
		2%	0.4527	0.3454	0.4089	0.4007	0.4804	0.3320	0.3564	**0.5220**
	Micro-F1	70%	0.7679	0.8092	0.8112	0.8184	0.8262	0.7749	0.8319	**0.8450**
		20%	0.7452	0.7686	0.7558	0.7764	0.8113	0.7654	0.7922	**0.8201**
		10%	0.7305	0.7060	0.7258	0.7301	**0.8041**	0.7425	0.7301	0.8008
		5%	0.6642	0.6402	0.6813	0.6433	**0.7873**	0.6954	0.7033	0.7722
		2%	0.6485	0.6145	0.6285	0.6187	0.7648	0.6087	0.5951	**0.7685**
ACSAC	Macro-F1	70%	0.6973	0.6931	0.7701	0.7792	0.7560	0.7849	0.7734	**0.7883**
		20%	0.6884	0.6368	0.7109	0.7401	0.7221	0.6921	0.6848	**0.7412**
		10%	0.6778	0.6119	0.7016	0.7126	0.7180	0.6636	0.6432	**0.7354**
		5%	0.6523	0.5431	0.6556	0.6664	0.7002	0.6028	0.5907	**0.7135**
		2%	0.6526	0.4197	0.5914	0.6014	0.6303	0.3955	0.4215	**0.6729**
	Micro-F1	70%	0.8473	0.8710	0.8943	0.8921	0.9071	**0.9084**	0.8928	0.9061
		20%	0.8402	0.8158	0.8303	0.8498	0.8976	0.8824	0.8629	**0.8997**
		10%	0.8291	0.8028	0.8005	0.8190	**0.8879**	0.8510	0.8356	0.8602
		5%	0.8056	0.7549	0.7720	0.7855	**0.8753**	0.7455	0.7012	0.8450
		2%	0.7810	0.6810	0.7109	0.7340	0.8236	0.6232	0.6000	**0.8322**

5.3 Comparative Results

For all methods, the learned embeddings of software are used to train a linear classifier. To compare the different methods when different amount of labels are provided, we randomly select 70%, 20%, 10%, 5% and 2% of each dataset as training set, and choose 10% as validation set. The rest of dataset is set as test set. We report the test performance of each method which has the best performance on the validation set. The Macro-F1 and Micro-F1 of each method is shown in Table 2. It can be seen that Macro-F1 is much smaller than Micro-F1 on both datasets. This is because both datasets are unbalanced. More than one third of ACMD and half of ACSAC is benign samples and the rest contains the other several classes. Micro-F1 can easily get high value on unbalanced dataset when the largest class is classified accurately, even though some small classes are classified poorly. Macro-F1 is a more strict metric for unbalanced dataset, as Macro-F1 takes the average value of F1 score of all classes as the result. Thus Macro-F1 is more strict and important in our experiments. On both datasets, the Macro-F1 of SCLMD outperforms those of other methods on all training ratios. The Micro-F1 of SCLMD is also better than most of methods, except slightly worse than MaMaDroid and GDroid on some cases. This may be because MaMaDroid and GDroid prefer to classify the software as benign ones, which makes them better in Micro-F1 sometimes, but worse than SCLMD in Macro-F1 all the time. On both datasets, the performance of all methods declines as long as the training ratio gets small. However the decline degree of SCLMD is smaller than other methods. For example, for SCLMD, the Macro-F1 of 2% training ratio is 86.87% of that of 70% training ratio, while for R-GCN, HGT, BiLSTM, BiGRU, MaMaDroid GDroid and LGMal, the ratio is 80.73%, 60.07%, 74.81% , 70.08%, 84.74%, 57.5% and 60% respectively. This demonstrate that the contrastive learning framework can take full use of unlabeled data to extract consistent feature of structural and sequential view, which help the method work well when the supervised information is limited.

5.4 Ablation Study

To investigate the effect of contrastive learning framework and hybrid positive selection strategy in the model, we designed Three variants of SCLMD. $SCLMD_{str}$: The representation of the software from structural view is adopted alone to feed into a linear classifier. $SCLMD_{seq}$: The representation of the software from sequential view is adopted alone to feed into a linear classifier. $SCLMD_1$: The fine-tuning based on labels in hybrid positive selection strategy is removed. The positive pairs are selected only based on the TF-IDF similarities.

Table 3. The comparison results of SCLMD and its variants on two datasets ACMD and ACSAC with different training ratio. The best performance is highlighted in boldface.

Datasets	Metrics	Training ratio	$SCLMD_{str}$	$SCLMD_{seq}$	$SCLMD_1$	SCLMD
ACMD	Macro-F1	70%	0.5757	0.5931	0.5932	**0.6009**
		20%	0.5445	0.5467	0.5507	**0.5741**
		10%	0.5377	0.5266	0.5467	**0.5543**
		5%	0.5122	0.4771	0.5206	**0.5292**
		2%	0.4025	0.3772	0.5014	**0.5220**
	Micro-F1	70%	0.7898	0.8221	0.8133	**0.8450**
		20%	0.7710	0.7809	0.7928	**0.8201**
		10%	0.7363	0.7412	0.7476	**0.8008**
		5%	0.7131	0.7051	0.7345	**0.7722**
		2%	0.6606	0.6407	0.7022	**0.7685**
ACSAC	Macro-F1	70%	0.7185	0.7407	0.7775	**0.7883**
		20%	0.6686	0.7069	0.7333	**0.7412**
		10%	0.6066	0.6671	0.7249	**0.7354**
		5%	0.5872	0.5282	0.6346	**0.7135**
		2%	0.4638	0.3814	0.6089	**0.6729**
	Micro-F1	70%	0.8727	0.8803	0.8837	**0.9061**
		20%	0.8288	0.8456	0.8706	**0.8997**
		10%	0.7490	0.8063	0.8286	**0.8602**
		5%	0.7638	0.7539	0.7936	**0.8450**
		2%	0.7071	0.6580	0.7526	**0.8322**

The comparative results of each variant is shown in Table 3. It can be seen that SCLMD$_{seq}$ performs better than SCLMD$_{str}$ when the ratio of labeled samples is larger than 20%, while SCLMD$_{str}$ outperforms SCLMD$_{seq}$ when the ratio is less than 10%. This shows that when the supervised information is large, the sequential view encoder can effectively extract sequential information to achieve better performance. On the other hand, when the supervised information is limited, the structural view encoder achieve better performance as it can use software's rich relationship to help perform unsupervised learning, and enable unlabeled samples obtain supervised information from the limited supervised information. That both SCLMD$_{seq}$ and SCLMD$_{str}$ are worse than SCLMD demonstrates the importance of contrastive learning framework in extracting consistent features from sequential and structural views. SCLMD$_1$ is also worse than SCLMD, which illustrates the proposed hybrid positive selection strategy can effectively use supervised information in contrastive learning framework.

5.5 Parameter Sensitivity

In this section, we investigate the sensitivity of two parameters, i.e., the number threshold of positive pairs k and the balance coefficient of views λ.

The results with different k is shown in Fig. 2. On both datasets, with the increase of k, the performance goes up first and then declines for nearly all training ratios. The optimum points of both datasets are located around 16 to 32. The Balance coefficient of views λ can control the importance of two views. The results with different λ is shown in Fig. 3. It is shown that when the training ratio is large, the model with a small value of λ has better performance. Because at this time, the sequential view loss is more important, and the sequential view encoder can benefit from more label information. When the training ratio is small, a larger value of λ is more favorable. Because at this time, the structural view loss is more important, and the structural view encoder can take advantage of the relationship between samples to achieve better performance with sparse labels.

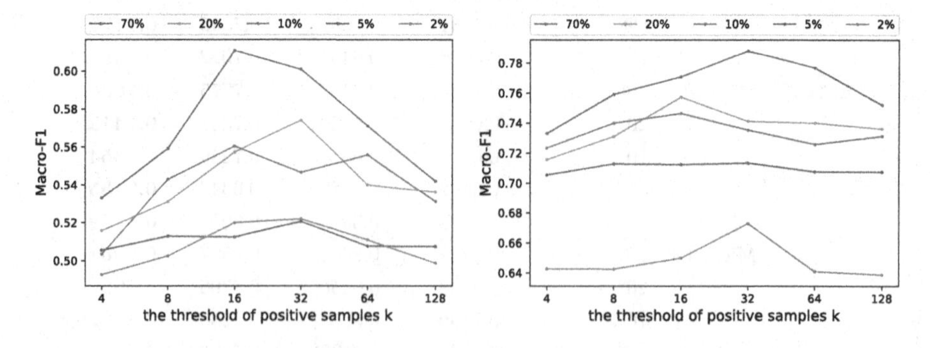

Fig. 2. The comparisons of SCLMD with different threshold of positive pairs k. (ACMD on left and ACSAC on right)

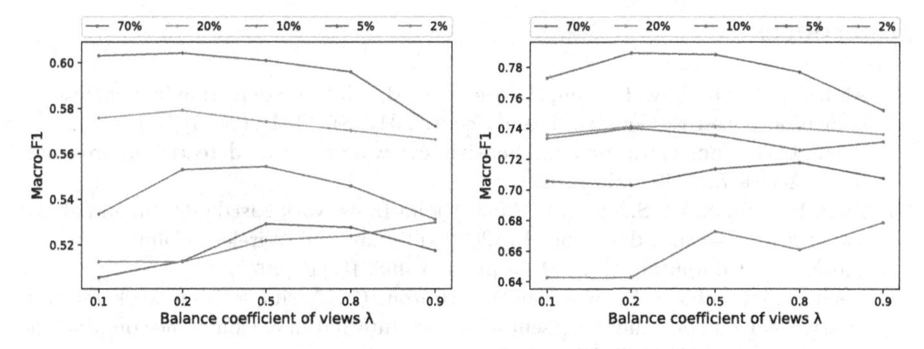

Fig. 3. The comparisons of SCLMD with different Balance coefficient of views λ. (ACMD on left and ACSAC on right)

6 Conclusion

We propose SCLMD which is a semi-supervised contrastive learning malware detection method based on API call sequences. SCLMD uses two view encoders to capture the structural and sequential features of software's API calling behavior. The two views are mutually supervised to learn the consistent representations of software by means of cross-view contrastive learning. Hybrid positive selection strategy is designed to introduce supervised information in contrastive learning. Extensive experiments on two real-world datasets verify the effectiveness of SCLMD, especially when the supervised information is limited.

Acknowledgments. This work is supported by National Natural Science Foundation of China (No. 62002219, 62172278), Shanghai Sailing Program (No. 19YF1424700), Startup Fund for Youngman Research at SJTU (SFYR at SJTU).

Appendix

A Implementation Details

For the proposed SCLMD, we use Glorot initialization and Adam optimizer. Under the condition of training ratio 70%, We manually adjust and set the learning rate to 0.01, the temperature parameter τ is set to 0.5. The number of attention heads H is set to 3. The number k of positive pairs for each sample is set to 32. The balance coefficient λ is set as 0.5. The maximum length of the API call sequence is set to 6000.

For all methods, we set the input dimension as 128, hidden dimension as 60 and representation dimension as 64. The source code of SCLMD are publicly available on Github[3].

[3] https://github.com/Noctilux-M/SCLMD.

References

1. Ahmadi, M., Sami, A., Rahimi, H., Yadegari, B.: Malware detection by behavioural sequential patterns. Comput. Fraud Secur. **2013**(8), 11–19 (2013)
2. Aslan, Ö.A., Samet, R.: A comprehensive review on malware detection approaches. IEEE Access **8**, 6249–6271 (2020)
3. Chai, Y., Qiu, J., Su, S., et al.: LGMal: A joint framework based on local and global features for malware detection. In: 2020 International Wireless Communications and Mobile Computing (IWCMC), pp. 463–468. IEEE (2020)
4. Chen, T., Kornblith, S., Norouzi, M., Hinton, G.: A simple framework for contrastive learning of visual representations. In: International Conference on Machine Learning, pp. 1597–1607. PMLR (2020)
5. Cho, K., et al.: Learning phrase representations using rnn encoder-decoder for statistical machine translation. arXiv preprint arXiv:1406.1078 (2014)
6. Dong, Y., Ziniu, H., Wang, K., Sun, Y., Tang, J.: Heterogeneous network representation learning. In: IJCAI, vol. 20, pp. 4861–4867 (2020)
7. Gao, H., Cheng, S., Zhang, W.: GDroid: android malware detection and classification with graph convolutional network. Comput. Secur. **106**, 102264 (2021)
8. Gavriluţ, D., Cimpoeşu, M., Anton, D., Ciortuz, D.: Malware detection using machine learning. In: 2009 International Multiconference on Computer Science and Information Technology, pp. 735–741. IEEE (2009)
9. Hassani, K., Khasahmadi, A.H.: Contrastive multi-view representation learning on graphs. In: International Conference on Machine Learning, pp. 4116–4126. PMLR (2020)
10. He, K., Fan, H., Wu, Y., Xie, S., Girshick, R.: Momentum contrast for unsupervised visual representation learning. In: Proceedings of the IEEE/CVF Conference on Computer Vision and Pattern Recognition, pp. 9729–9738 (2020)
11. Hochreiter, S., Schmidhuber, J.: Long short-term memory. Neural Comput. **9**(8), 1735–1780 (1997)
12. Hosseini, A., Chen, T., Wu, W., Sun, Y., Sarrafzadeh, M.: Heteromed: heterogeneous information network for medical diagnosis. In: Proceedings of the 27th ACM International Conference on Information and Knowledge Management, pp. 763–772 (2018)
13. Hu, G., Venugopal, D.: A malware signature extraction and detection method applied to mobile networks. In: 2007 IEEE International Performance, Computing, and Communications Conference, pp. 19–26. IEEE (2007)
14. Ziniu, H., Dong, Y., Wang, K., Sun, Y.: Heterogeneous graph transformer. In: Proceedings of The Web Conference, vol. 2020, pp. 2704–2710 (2020)
15. Kipf, T.N., Welling, M.: Semi-supervised classification with graph convolutional networks. arXiv preprint arXiv:1609.02907 (2016)
16. Kwon, I., Im, E.G.: Extracting the representative API call patterns of malware families using recurrent neural network. In: Proceedings of the International Conference on Research in Adaptive and Convergent Systems, pp. 202–207 (2017)
17. Lansheng, H., Kunlun, G.: Behavior detection of malware based on combination of API function and its parameters. Appl. Res. Comput. **30**(11), 3407–3410 (2011)
18. Linsker, R.: Self-organization in a perceptual network. Computer **21**(3), 105–117 (1988)
19. Liu, X., et al.: Generative or contrastive. IEEE Trans. Knowl. Data Eng. Self-supervised learn. (2021)

20. Mariconti, E., Onwuzurike, L., Andriotis, P., De Cristofaro, E., Ross, G., Stringhini, G.: Mamadroid: detecting android malware by building Markov chains of behavioral models. arXiv preprint arXiv:1612.04433 (2016)
21. Murad, K., Shirazi, S.N.-H., Zikria, Y.B., Ikram, N.: Evading virus detection using code obfuscation. In: Kim, T., Lee, Y., Kang, B.-H., Ślęzak, D. (eds.) FGIT 2010. LNCS, vol. 6485, pp. 394–401. Springer, Heidelberg (2010). https://doi.org/10.1007/978-3-642-17569-5_39
22. O'Kane, P., Sezer, S., McLaughlin, K.: Obfuscation: the hidden malware. IEEE Secur. Priv. **9**(5), 41–47 (2011)
23. Park, C., Kim, D., Han, J., Hwanjo, Yu.: Unsupervised attributed multiplex network embedding. In: Proceedings of the AAAI Conference on Artificial Intelligence, vol. 34, pp. 5371–5378 (2020)
24. Peng, Z., Huang, W., Luo, M., Qinghua Zheng, Yu., Rong, T.X., Huang, J.: Graph representation learning via graphical mutual information maximization. In: Proceedings of The Web Conference, vol. 2020, pp. 259–270 (2020)
25. Qiu, J., et al.: Gcc: graph contrastive coding for graph neural network pre-training. In: Proceedings of the 26th ACM SIGKDD International Conference on Knowledge Discovery & Data Mining, pp. 1150–1160 (2020)
26. Roundy, K.A., Miller, B.P.: Binary-code obfuscations in prevalent packer tools. ACM Comput. Surv. (CSUR) **46**(1), 1–32 (2013)
27. Schlichtkrull, M., Kipf, T.N., Bloem, P., van den Berg, R., Titov, I., Welling, M.: Modeling relational data with graph convolutional networks. In: Gangemi, A., et al. (eds.) ESWC 2018. LNCS, vol. 10843, pp. 593–607. Springer, Cham (2018). https://doi.org/10.1007/978-3-319-93417-4_38
28. Shi, C., Li, Y., Zhang, J., Sun, Y., Philip, S.Y.: A survey of heterogeneous information network analysis. IEEE Trans. Knowl. Data Eng. **29**(1), 17–37 (2016)
29. Torres, J.F., Hadjout, D., Sebaa, A., Martínez-Álvarez, F., Troncoso, A.: Deep learning for time series forecasting: a survey. Big Data **9**(1), 3–21 (2021)
30. Veličković, P., GCucurull, G., Casanova, A., Romero, A., Lio, P., Bengio, Y.: Graph attention networks. arXiv preprint arXiv:1710.10903 (2017)
31. Velickovic, P., Fedus, W., Hamilton, W.L., Liò, P., Bengio, Y., Hjelm, R.D.: Deep graph infomax. ICLR (Poster) **2**(3), 4 (2019)
32. Wang, S., Philip, S.Y.: Heterogeneous graph matching networks: application to unknown malware detection. In: 2019 IEEE International Conference on Big Data (Big Data), pp. 5401–5408. IEEE (2019)
33. Wang, X., Liu, N., Han, H., Shi, C.: Self-supervised heterogeneous graph neural network with co-contrastive learning. In: Proceedings of the 27th ACM SIGKDD Conference on Knowledge Discovery & Data Mining, pp. 1726–1736 (2021)
34. Yazi, A.F., Çatak, F.Ö., Gül, E.: Classification of methamorphic malware with deep learning (LSTM). In: 2019 27th Signal Processing and Communications Applications Conference (SIU), pp. 1–4. IEEE (2019)
35. Young, T., Hazarika, D., Poria, S., Cambria, E.: Recent trends in deep learning based natural language processing. IEEE Comput. Intell. Mag. **13**(3), 55–75 (2018)

Semi-supervised Context Discovery for Peer-Based Anomaly Detection in Multi-layer Networks

Bo Dong, Yuhang Wu(✉), Micheal Yeh, Yusan Lin, Yuzhong Chen, Hao Yang, Fei Wang, Wanxin Bai, Krupa Brahmkstri, Zhang Yimin, Chinna Kummitha, and Verma Abhisar

Visa, 900 Metro Center Blvd, Foster City, CA 94404, USA
{bdong,yuhawu,miyeh,yusalin,yuzchen,haoyang,feiwang,wabai, krbrahmk,yimzhang,chinnaiiitb,abverma}@visa.com

Abstract. User-related cyber security attacks could cause tremendous losses to any organization. Detecting such threat can be formulated as anomaly detection problem in multilayer networks where each layer of the multilayer networks contain different contextual information regarding the users. While there have been many works proposed for peer-based anomaly detection, there has been little endeavor in discover the appropriate context (peers) for anomaly detection in multilayer networks. In this paper, we propose a context discovery method, which integrates the relations provided by each individual network layer and detects the anomalous nodes in networks based on the optimized peers of nodes with (or without) limited feedback from cybersecurity experts. The proposed system addresses the frequently encountered challenges when conducting anomaly detection, i.e., feedback sparsity, and the newly emerged challenge associated with multilayer networks, i.e., finding peers of each node based on conflicting information provided by individual layers. The proposed system is capable of capturing the anomalies in multilayer networks and outperforms the widely used peer-based anomaly detection algorithms on both synthetic and real-world sensor network and cybersecurity datasets.

Keywords: Anomaly detection · Multi-layer network · Cybersecurity

1 Introduction

The COVID-19 pandemic has changed the businesses' operating model unprecedentedly. Many organizations allow employees to work from home, remote access to on-premises servers. Unfortunately, cyber-crime has thrived amid the chaos [6]. Such changes create massive cyber security challenges, yet organizations

Bo Dong—This work was conducted when Bo was an internship student in Visa Research.

C. Alcaraz et al. (Eds.): ICICS 2022, LNCS 13407, pp. 508–524, 2022.
https://doi.org/10.1007/978-3-031-15777-6_28

strive for finding solutions to prevent network incidents. When a malicious attack taken place, a hacker may first take over an account of a benign user, login the account and start searching for any off-guarded devices. The hacker will take the devices offline, injecting malicious software, and demand payment to restore their functionality. Detecting cyber attacks in the network from both user and device perspective can prevent organization from tremendous losses.

To combat diverse attack patterns in account takeover, it is critical to deploy a model to detect anomalies in user's and device's behavior. A good and flexible profile describes the "normal" pattern of such entities is crucial for this type of approach. However, due to the unpredictable nature of humans and the diverse type of devices, an one size fits all model to describe the normal behavior of user/device may not work. The anomaly detection system should understand the context of an user based on its interaction with different resources in their daily jobs, and model the device's normality based on its location and communication patterns.

One simple but efficient method to identify the right context of a given entity is to determine its peers, or peer-groups [22]. By comparing an entity's behavior with its peers (e.g., nearest neighbors), the detection system can easily identify anomaly entity behavior since it is "stands out of the crowd". Compared to the methods only considering entity's own history, because peer-based anomaly detection factorized out the common behavior of entities inside a peer-group, it can significantly reduce false alarms that caused by systematic change (e.g., change caused by outside environment) and expose more idiosyncratic risk caused by each entity. The success of the peer-group based method highly depends on the accuracy of the context detection, in another word, how to select the right peers for a specific anomaly detection problem. We propose a method that is able to discover a suitable context to describe user's/device's normal behavior in a multi-layer network so that the peer-group algorithm yields better accuracy in the complex network scenario. To take into account the interaction between users/devices, we model the account take over detection problem as node anomaly detection in the multi-layer network. On the user side, a layer in the network represents a bipartite graph - one type of a nodes can be users, another type of nodes correspond to a specific network resource, for instance, websites, cloud applications, server, etc. On the device side, a layer correspond to a property of interaction between devices, it can be physical distance between devices, but can also be the connectivity strength between two devices.

While anomaly detection has been extensively studied in many domains by leveraging the connectivity information between entities or nodes in a network [4], these approaches have mainly focused on the topological perspective of networks, and most of the approaches have been assuming the input networks to consist of only a single layer [1,5,10,21]. On top of that there has been little effort done in graph-based contextual anomaly detection even if networks are suitable for discovering contextual anomalies [13]. However, given a node, determining meaningful *peers* from multiple layers is not trivial. Compared with deriving peers from a single-layer graph, where one can directly apply kNN or community detection algorithms, obtaining a consensus of the peer information generated from the multiple layers is not straightforward. Secondly, in most real-world anomaly detection applications, the availability of high quality human labels are often sparse. Formulating the learning problem as a fully-supervised problem is unrealistic. To address the above issues, we design a multi-layer-graph-based contextual anomaly detection system that leverages the peer-group analysis to detect anomalies in both unsupervised and semi-supervised fashion.

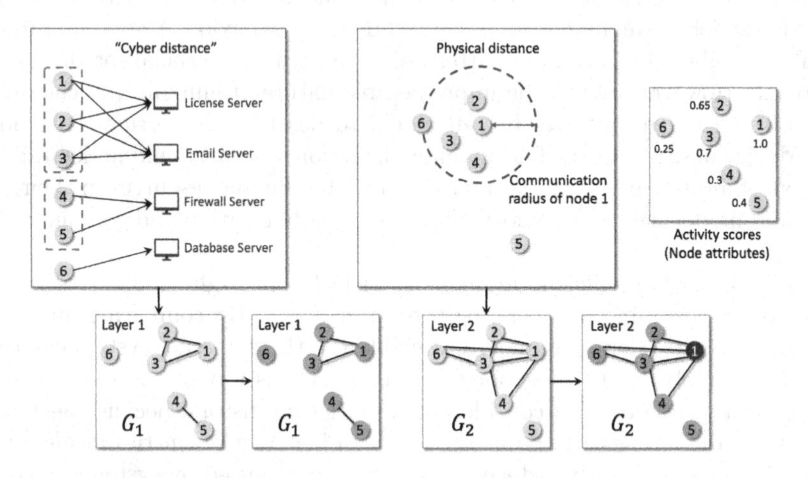

Fig. 1. Toy example to show the peer-groups in a multi-layer network setting. Upper left and middle panels illustrate nodes behavior in cyber and physical space, which respectively lead to the creation of network layers 1 (G_1) and 2 (G_2). The upper right panel shows the predefined activity scores as the node attributes. Lower panels show the topology of G_1 and G_2 and how anomalies are detected via peer-groups within each layer. Nodes are color-coded: grey denotes unlabeled nodes; green denotes predicted negatives (normal); and purple denotes predicted positives (abnormal). (Color Figure Online)

Let us start by walking through the toy example as show in Fig. 1 first. The network consists of 6 nodes, each denotes an user able to access an online server and communicate with another user physically onsite via a mobile radio within limited range. There are two distance measures: nodes visit similar sets of servers are regarded as close in terms "cyber distance", while nodes able to talk to each other within the radio communication radius are close in physical distance. Then for the same set of nodes, different distance measures render different dimensions of proximity. In this work, we formulate these dimensions by different network layers. In network layer 1 (G_1), a pair of nodes visited the same set of servers are connected due to their proximity in cyber distance, while in layer 2 (G_2), a pair of nodes able to establish a radio connection are linked due to the proximity in physical space. Obviously, each layer owns different connection topology, although they share the same set of nodes. This leads to a different 1st-hop peers at a different layer for the same node, e.g., node 1's peers are nodes 2 and 3 in layer 1 but are nodes 2, 3, 4, and 6 in layer 2.

Given the two layers and their graph topology with predefined node attributes (such as the activity scores estimated via domain knowledge), one can compute each node's anomaly score based on how much its attribute deviates from its peers. For node 1, even though it has a high attribute value of 1.0, since it has high attribute peers (nodes 2 and 3), it is regarded as normal within G_1. But in G_2, it is far away from the peer-group attribute average (nodes 2, 3, 4 and 6) and thus regarded as abnormal.

To resolve the conflict and exploit the information from both layers, the proposed approach merges G_1 and G_2 by (i) encouraging the clustering of nodes, (ii) boosting the alignment between node attributes and topology, and formulate a new merged graph G_{merge} with a weighted adjacency matrix, as shown in Fig. 2. In this way, each node in G_{merge} obtains a new set of peers carrying aggregated contextual information across the layers, and correspondingly, increase the contrast between anomaly nodes and its peers.

Besides the unsupervised approach, consider the scenario where feedback from investigators are accessible. We then leverage the feedback labels to conduct semi-supervised learning. As shown in the bottom-left panel in Fig. 2, domain knowledge based expert feedback to a single node (node 4) is presented, so that the system can take a human label into account when optimizing its parameters (the system is learned to gave more weights to the Layer 1 since the system tends treat node 5 as node 4's peer instead of node 1's or node 3's), which leads to the final results aligning better with the ground truths than any single layer based predictions.

To the best of the authors' knowledge, this is the first anomaly detection framework that explores the optimal peer-group discovery problem in a multilayer network setting. The design of our system can be applicable to different security-related applications, including intrusion detection in service networks, malfunction detection sensor networks, etc. We demonstrate the effectiveness of our system with both synthetic and real-world datasets. Through evaluations, we show that our system outperforms comparing methods when detecting anomalies in both unsupervised and semi-supervised settings.

2 Related Work

In this section, we present a summary of the literature related to this work, including contextual anomaly detection, peer analysis, graph-based anomaly detection and multilayer network analysis.

Contextual Anomaly and Peer Analysis. Contextual anomalies are determined when data points are compared against meta-information associated with these data points [14]. When detecting anomalies based on the specific context (e.g., time, location, gender, etc.), because the relation between similar data points is considered together, higher detection accuracy can be achieved with a relatively low false alarm rate [8]. Among the approaches of the contextual anomaly, peer-group analysis is one of the important approaches. It was originally proposed by Bolton *et al.* [2] for monitoring the behavior of accounts in

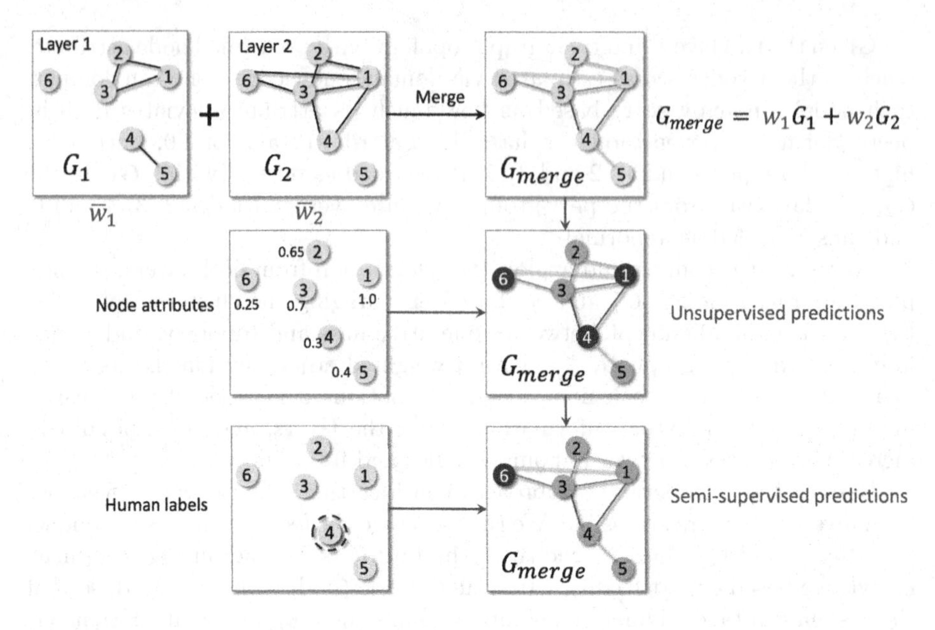

Fig. 2. Toy example of weighted multi-layer network merge and anomaly detection. The left panels show the original network layers G_1 and G_2, the node attributes, and the human labels, where the topology of G_1 and G_2 are predefined or constructed via certain distance measures. The panels on the right show the network merge process with trainable weights $\bar{w}_1$ and $\bar{w}_2$ (w_1 and w_2 are the normalized correspondences) and the scenarios of both the unsupervised and semi-supervised settings in detecting anomalies in our system. Nodes are color-coded: grey denotes unlabeled nodes; green denotes predicted negatives, and purple denotes predicted positives. A human labeled node is marked by dashed circles. Black edges show the unweighted connections within each layer, while colored edges indicate their different weights in the merged network. (Color Figure Online)

temporal sequences for credit card fraud detection. The method was first used to choose accounts whose profile are the most similar to a target account to construct the context for the target point, namely, the "peer-groups", then, the behavior of the peer-group is summarized at each subsequent time point. The anomaly of a target account is determined by comparing its behavior with the summary of the peer accounts (context). Until now, the peer-group analysis has been used in multiple contextual anomaly detection applications, such as finance [16], insider threat detection [11], sensor networks [14], etc. However, there is limited works on how to select the peer-group accordingly.

Graph-Based Anomaly Detection. Graphs naturally represent the inter-dependencies by edges between related objects, which provide rich contextual information of each data point [1]. Multiple static and dynamic graph-based anomaly detection approaches have been proposed on attributed/plain graphs. In the most recent advancement of this topic, Liu *et al.* [19] separated graph features of the attributed networks into context and content features. The context feature is used to define the community of a node, and the content feature is used to compare with the node's community Ding *et al.* [9] provided an autoencoder-based solution, which used graph convolutional network (GCN) to encode the interactions between different information modalities, using the reconstruction errors of nodes from both structural and attribute perspectives to identify anomalies.

Multilayer Graph in Network Analysis. Network data encapsulates multiple aspects of identity interactions, e.g., social networks contain interactions among both friends and coworkers [7], proteins in human body interact differently in different human organs and tissues [25]. The different types of relations can be represented by multiple layers in graphs. How to integrate the information across layers, thus became an emerging research topic. Recent literature mainly focuses on learning the embedding of the multilayer networks [15], while anomaly detection in the multilayer network has yet been thoroughly studied. It is still not clear how one can integrate the multilayered information to identify the anomaly across different layers. Further discussion regarding the contextual anomaly detection in multilayer anomaly detection remains an open problem.

To the best of our knowledge, our work is the first to address the contextual anomaly detection problem in multilayer graphs, and the first attempt to address the context discovery problem in this direction.

3 Method

3.1 Peer-Based Anomaly Detection

Given a set of nodes $\mathcal{V} = \{v_1, ...v_N\}$ in a graph G, their scalar attributes $\mathcal{A} = \{a_1, ...a_N\}$, as well as the distance between each pair of nodes $\mathcal{E}[i, j]$. The peers of a node $v_i \in \mathcal{V}$ are the set of n_p nodes $\mathcal{V}_i' \subset \mathcal{V}$ close to v_i in a given single layer graph $G = (\mathcal{V}, \mathcal{E})$. Usually, peers in the group $\mathcal{V}_i'$ are selected based on the K-nearest neighbors of $\mathcal{V}$ [2,11,14,16] determined by the distances. Compared to the anomaly detection methods only considering v's own feature a, because peer-based anomaly detection factorized out the common behavior of entities inside a peer-group $\mathcal{V}_i'$, it can significantly reduce false alarms that caused by systematic change (e.g., change caused by outside environment) and expose more idiosyncratic risk caused by each entity.

A peer-group anomaly score of node i can be defined as a weighted Z-score:

$$s_i = \frac{|a_i - \mu_i|}{\sigma_i} \tag{1}$$

where the weighted mean of peer-group $\mathcal{V}_i'$, denoted as μ_i, is used to represent the common behaviors of entities inside of the peer-group. The standard deviation of peer-groups σ_i, denoted as σ_i are computed by:

$$\mu_i = \frac{\sum_{v_j \in \mathcal{V}_i'} \mathcal{E}[i,j] a_j}{\sum_{v_j \in \mathcal{V}_i'} \mathcal{E}[i,j]} \tag{2}$$

$$\sigma_i = \sqrt{\frac{\sum_{v_j \in \mathcal{V}_i'} \mathcal{E}[i,j](a_j - \mu_i)^2}{\sum_{v_j \in \mathcal{V}_i'} \mathcal{E}[i,j]}} \tag{3}$$

Node v_i will have a high anomaly score when its attribute differs greatly when comparing to its peers.

3.2 Anomaly Detection with Multi-layer Graph

The peer-based anomaly provides a straightforward way to model the context of entity without knowing any prior knowledge about the historical pattern of the data, and it has been widely deployed in finance and security industry. However, when the graph contains multiple layers, each layer may have a separate definition of $\mathcal{E}[i,j]$. For example, anyone can interact with friends, colleagues, family members in the same week, which compose three separate layers in a social network, now let's consider the problem of identifying the peers of this person based on the three layers. We suspect no one can give out a good definition without looking into the specific anomaly detection problem the user trying to solve. A good definition of a peer-group is really depends on the context of the problem.

The objective of the proposed method is to learn the context by generate a merged graph from different graph layers so that the new graph will provide optimal nearest neighbors (peers) for a specific peer-based anomaly detection problem. We begin by defining the specific type of graph in interest, a *multilayer graph.*

Definition 1. *A multilayer graph* $G = (\mathcal{V}, \mathbf{E}, \mathbf{A})$ *consists of a set of attributed nodes* $\mathcal{V} = \{v_1, ...v_N\}$ *with attribute* $\mathcal{A} = \{a_1, ...a_N\}$*, and multiple set of edges* $\mathbf{E} = \{\mathcal{E}_1, ..., \mathcal{E}_M\}$ *where each* $\mathcal{E}_i \in \mathbf{E}$ *stores the edge information for its corresponding layer. We use* $G_m = (\mathcal{V}, \mathcal{E}_m)$ *to denote the* m*-th layer graph and* a_i *to denote the attribute (scalar) associated with* v_i*.*

Each layer of a multilayer graph typically models one type of relationship among the nodes. We assume that there exists a merged graph that reveals each nodes' peers.

Definition 2. *A merged graph* $G_{merge} = (\mathcal{V}, \mathcal{E}_{merge})$ *of a multilayer graph* G *with* m *layers is a single layer graph that 1) is generated by merging the layers in* G*, i.e.,* $G_{merge} = merge(G_1, ..., G_M)$ *and 2) contains peer relationship information.*

We describe the proposed multilayer-graph-based anomaly detection system in a feedforward fashion. As depicted in Fig. 3, the proposed system has three major components: layer merger, anomaly detection, and optimization. The layer merger module combines the multilayer graph into a single layer graph based on the learned weights $\bar{w}_m$'s using weighted-sum merge function as described in Definition 2. The anomaly detection module computes the anomaly score of each node using its attribute and peer information given by the merged graph (see Eq. 1). The optimization module is responsible for refining the weights used in the merge function. We provide a training method that could be run either fully unsupervised or semi-supervised. When the training method is run in unsupervised mode, the weights are learned by optimizing both the Embedding Clustering EC loss (see Eq. 4) and alignment loss (see Eq. 7). When the training method is allowed to ask for human feedback, the weights are refined with the ranking loss (see Eq. 8) using the limited labels provided by the user.

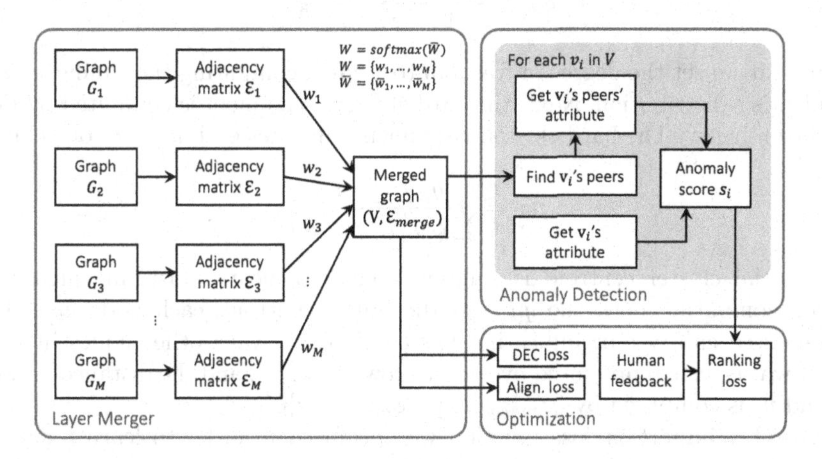

Fig. 3. Multilayer-network-based anomaly detection system, $w_1, w_2, ...w_M$ are the learnable parameters.

The specific merge function we studied in this work is the *weighted-sum merge function.* The weighted-sum merge function merges layers by computing the weighted sum of the adjacency matrix of each layer (i.e., $\mathcal{E}_{merge} = w_1\mathcal{E}_1 + ... + w_M\mathcal{E}_M$) where $w_m = \frac{e^{\bar{w}_m}}{\sum_n e^{\bar{w}_n}}$ and $\bar{w}_m$ of each $\mathcal{E}_m$ are learnable parameters.

3.3 Weight Optimization

To learn the weights of layers with limited node labels, it is important to view the problem from a top-down perspective instead of look into individual node separately. Here we cast the problem as a *consensus clustering problem.* The objective of the consensus clustering problem is to cluster nodes of a given graph G_{merge} into K peer-groups. Since $\mathcal{E}_{merge}$ contains the integrated relation between nodes, we extract row i of $\mathcal{E}_{merge}$, denoted as x_i, as the relational embedding of node i. The cluster operation can be conducted on the relational embedding of all nodes in G_{merge}. Specifically, we optimize the EC loss function [23,24]. The intuition of the loss function is to reduce the intra-cluster distances and keep the nodes inside of the original peer-groups.

$$\mathcal{L}_{\texttt{EC}} = \sum_i \sum_k p_{ik} \texttt{log} \frac{p_{ik}}{q_{ik}} \tag{4}$$

where the q_{ik} and p_{ik} are the soft and hard cluster assignments. The soft cluster assignment is defined by the similarity between a given node i's relational embedding (i.e., x_i) and cluster centroid k (i.e., c_k) measured with Student's t-distribution [23] as shown in Eq. 5. The soft cluster assignment indicates the "degree" of a node belong to each cluster:

$$q_{ik} = \frac{\sum_{k'} 1 + \|x_i - c_{k'}\|^2}{1 + \|x_i - c_k\|^2} \tag{5}$$

Note that we set the degree of freedom to 1 when computing the similarity with Student's t-distribution. Next, the hard cluster assignment is computed by Eq. 6 as shown below. The hard cluster assignment allocates each node to one cluster:

$$p_{ik} = \frac{q_{ik}^2 / \sum_{i'} q_{i'k}}{\sum_{k'} q_{ik'}^2 / \sum_{i'} q_{i'k'}} \tag{6}$$

The initial cluster centroid is computed using k-means clustering algorithm. When computing both the q_{ik} and the initial centroid, each node $v_i \in V$ is represented as a vector indicating v_i's connectivity with other nodes in V. In other words, v_i is represented by the i-th row of $\mathcal{E}_{merge}$, and the distance between v_i and v_j is computed by $\|\mathcal{E}_{merge}[i,:] - \mathcal{E}_{merge}[j,:]\|^2$.

Another important loss function we optimized in order to learn a suitable merge function is the *alignment loss* which attempts to align the node attribute with the clustering of nodes. It is important to put nodes that have similar attribute values into the same cluster to reduce potential false alarms. The alignment loss of a pair of nodes (e.g., v_i and v_j) is computed by Eq. 7.

$$\mathcal{L}_{\texttt{align}} = \sum_i \sum_j \texttt{similarity}(a_i, a_j) \texttt{log} \sum_k q_{ik} q_{jk} \tag{7}$$

where similarity() is a function that outputs the similarity between v_i's attribute a_i and v_j's attribute a_j. Any function that returns non-negative similarity can be used. In our implementation, we simply compute the similarity by converting the difference to similarity, i.e., $a_{\Delta\texttt{max}} - \texttt{abs}(a_i - a_j)$ where $a_{\Delta\texttt{max}}$ is the maximum possible difference between any pairs of nodes' attribute. Both q_{il} and q_{jl} are computed with Eq. 5.

In the case where a small set of labeled anomalous nodes is given, which is the scenario where the bottom row of Fig. 2 is considered, we use the ranking loss as shown in Eq. 8 to use such information:

$$\mathcal{L}_{\texttt{rank}} = \texttt{max}(V_0 - V_1, 0) \tag{8}$$

where vector $V_0 = \{s_i | i \in l^-\}$ is the mini-batch samples from the labeled normal nodes while vector $V_1 = \{s_i | i \in l^+\}$ is the mini-batch samples from the labeled anomalous nodes, both containing anomaly scores defined in Eq. 5. This loss function would enforce the few anomaly nodes rank higher than all the normal nodes so that the information in the few labeled data can be largely utilized.

In summary, the overall loss for our system is shown in Eq. 9.

$$\mathcal{L} = \alpha \mathcal{L}_{\texttt{EC}} + \beta \mathcal{L}_{\texttt{align}} + \gamma \mathcal{L}_{\texttt{rank}} \tag{9}$$

where α, β, and γ are hyper-parameters to trade-off the contribution of different terms.

4 Experiment

In this section, we first discuss the datasets and setup of the experiments, then analyze the results accordingly. We leverage a synthetic dataset, a real-world sensor network dataset, and a real-world cybersecurity dataset to conduct the evaluation. The statistics of each dataset is summarized in Table 1. The dataset we used have the following properties:

- **R1. Ability to be formatted as a graph:** The pair-wise distances between different nodes are measurable, and the edges are undirected and weighted.
- **R2. Multiple layers in a graph:** There are multiple ways to split a single graph into multiple layers, such as by the types of the edges, by the timestamps of the edges, as introduced in [17].

Table 1. Dataset summary

Name	#layers	#nodes
Synthetic	6	500
Sensor network	2	54
Cybersecurity	3	1,926

- **R3. Anomalies can be derived from the node attribute:** The data contains meaningful node attributes, e.g., access frequency of malicious accounts inside of organizations, records of malfunctioning sensors inside of sensor networks, etc.
- **R4. (Optional) Availability of anomaly feedback:** For the semi-supervised setting, the dataset should also contain feedback of the anomalies, i.e., ground truths of the anomalies.

4.1 Experimental Settings

The settings of our system are as follows: The weight for the merging module is initialized with random number. We use the initial merging weight to obtain the merged graph G_{merge}. With the initial merged graph G_{merge}, the initial clustering centroids are computed with k-means clustering algorithm. The weight $\bar{W}$ is trained for n_{iter} iterations. In the case where the algorithm is allowed to ask for user's feedback, the algorithm would display the top h and bottom h nodes (based on each node's current anomaly score estimated using current G_{merge}) to the user and ask the user to annotate the $2h$ provided nodes. Note, the value h is a hyper-parameter which should be set based on how much label the user is willing to label. The algorithm does not have to ask for the user's feedback every iteration; instead, the algorithm could ask for the user's feedback every few iteration. Once again, the number of iteration between user's feedback should be set based on how much effort is expected from the user, in our experiment we set h to be 10% of total nodes. After obtained the feedbacks from users, $\bar{W}$ is updated based on the overall loss mentioned in Eq. 9, and G_{merge} is regenerated with the updated $\bar{W}$. After multiple iterations, the learned $\bar{W}$ is returned to the user as the selected context for this anomaly detection problem. We implemented the system in PyTorch. We trained the model with a batch size of 128, and a total of 100 epochs. The algorithm takes less than an hour to estimate the weights in each of the datasets. Learning rate of the system is 0.1. We ran all the experiments on NVIDIA Tesla P100 GPUs. All of the performance metrics are in area under curve (AUC).

4.2 Performance on Synthetic Data and Ablation Study

For the synthetic dataset, we simulate a six-layer graph that is constructed based on the scenario of 500 users accessing resources within an organization. The users are nodes, and if two users access the same resource, there exists an undirected weighted edge between them (R1 and R2). The weight of the edge is sampled from the range [0, 1].

We generate 6 layers for our multilayer graph with 3 good layers being relevant to the anomaly detection task, where the anomaly can be identified by comparing its attribute with its peers attribute, while the other 3 bad layers contain random attributes. The 500 users form 5 clusters when consider only the 3 relevant layers. This provides us with the expectation that the final learned weights for the good layers should be higher than the bad layers. For each user node, we determine his/her peers by finding the 50 closest user nodes to him/her. For each cluster of the 5 clusters, we randomly assign a mean and standard deviation to form a Gaussian distribution, which further assigns the attribute for the users in that cluster (R3). In order to inject anomaly users, we vary a subset (5%) of users to have his/her attribute to be 3 standard deviation away based on the Gaussian distribution of his/her cluster. We also generate feedback labels for the nodes (R4), where based on injected anomalies.

Table 2. AUC-ROC of single-layer baseline performance on synthetic data.

Graph	AUC
Ground truth	1.0000
Average	0.4370
Layer 0	0.4500
Layer 1	0.5438
Layer 2	0.5063
Layer 3	0.4527
Layer 4	0.4969
Layer 5	0.4917

Table 3. AUC-ROC of the proposed system on synthetic data with hyperparameter analysis.

α \ β	0.0	0.1	1.0	10.0
0.0	-	0.9992	0.9972	0.9983
0.1	0.7817	0.9984	0.9985	0.9978
1.0	0.9990	0.8908	0.9981	0.9997
10.0	0.7964	0.7972	0.8538	0.9979

(a) Unsupervised, $\gamma = 0$

α \ β	0.0	0.1	1.0	10.0
0.0	-	0.9991	0.9981	0.9988
0.1	0.9992	0.9993	0.9989	0.9983
1.0	0.8485	0.9043	0.9993	0.9988
10.0	0.7953	0.9976	0.9992	0.9655

(b) Supervised: $\gamma = 0.1$

α \ β	0.0	0.1	1.0	10.0
0.0	-	0.9990	0.9981	0.9987
0.1	0.7917	0.9989	0.9991	0.9985
1.0	0.9976	0.9362	0.9979	0.9991
10.0	0.9984	0.8066	0.8648	0.9992

(c) Supervised: $\gamma = 1$

α \ β	0.0	0.1	1.0	10.0
0.0	-	0.9205	0.9982	0.9992
0.1	0.5216	0.5589	0.9982	0.9995
1.0	0.5012	0.8497	0.9995	0.9987
10.0	0.9012	0.9987	0.8339	0.9992

(d) Supervised: $\gamma = 10$

For the experiments on the synthetic dataset, we conducted sensitivity analysis in both unsupervised and supervised settings. We compared our system's performances with the traditional peer-grouping baseline which used a single graph layer for peer selection [2,11,14,16] in Table 2. The on-duty layer can be any of the individual graph layer or the averaged layer of all the individual layers. For comparison, we ran three sets of experiments: 1) baseline methods, 2) unsupervised while changing α and β, and 3) supervised while changing α, β, and γ. The performance of the baseline and our system are shown as AUC in Table 2 and Table 3, respectively. As can be seen in Table 3, our system, in general, outperforms the baseline in both unsupervised and supervised settings.

4.3 Performance on Sensor Network Data

We choose the widely-used Intel Lab data as one of the real-world datasets [20]. The Intel Lab data stems from the 54 sensors' activity logs in the Intel Berkeley Research lab, from February 28th to April 5th, 2004. During this time many of the sensors where malfunctioned and finally went offline. Our objective is to identify individual sensor's failure due to its own malfunctioning instead of the environment reasons.

The dataset includes the coordinates of each sensor inside of the lab, their sensor readings of temperature, humidity, light and voltage recorded every 31 s and the probability of communication success between every two sensors. We observe large amount of sensors failed after March 25th, which makes the result less meaningful. So we exploited the data before March 18th for training and the data between March 18th and March 25th for testing.

Let each of the sensors be a node in the graph, where this graph contains two layers that describe the relations between the nodes. One weighted by the euclidean distance between them (based on the coordinate information), the other weighted by the probability of communication success between them. The proposed system will learn to weight these two layers to generate the optimal context G_{merge} for anomaly detection. For the initial anomaly scores, we use the voltage readings from each sensor. We take the readings of temperature and humidity as the ground-truth for anomaly: for an individual sensor, if any time stamps without any recorded data (voltage assigned to 0), has a temperature reading above 80 Celsius or humidity reading below 10% is considered malfunction of the sensor (ground-truth anomaly). Since we observe a significant difference between each feature's distribution between day and night, we further split the dataset into a day scenario and a night scenario, where the cut-off time is 7:00 pm/am in each day. Because our objective is to predict individual malfunction sensor, we only marked out anomaly sensors as the ground-truth if majority of sensors operate normally in any timestamp.

For the experiments on the Intel lab dataset, we predicted the voltage anomaly analyses overtime for each individual sensors. We compared the performance of our system with the following baselines in Table 4:

- **peer-grouping on coordinate layer (PG-coord):** As described in [2,11, 14,16], the anomalies are only determined by considering the peer-groups generated in a single relation (the coordinate layer) in the graph.
- **peer-grouping on communication layer (PG-comm):** The anomalies are only determined by considering the peer-groups generated in the communication layer in the graph.
- **peer-grouping on both layers (PG-both):** The anomalies are determined by considering the peer-groups generated by the averaged adjacency matrix of coordinate and communication layer in the graph.

The experimental results show that our method outperforms the traditional peer-grouping method in both day and night settings, which indicates it has the capability to find the right peer-groups for anomaly detection.

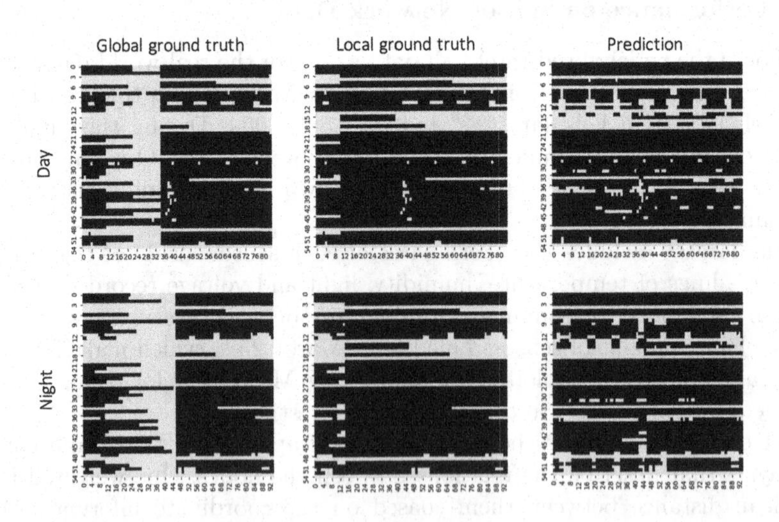

Fig. 4. Sensor voltage anomalies: global (suspicious environment problem), contextual (individual problem), and our predictions overtime. In each plot, Y axis indicates the ID of sensor, X axis represents the timestamps. Our objective is to identify the contextual anomaly with the proposed method, the detected anomaly regions are highlighted.

Table 4. AUC-ROC of anomaly detection on sensor network.

	PG-coord	PG-comm	PG-both	Ours
Night	87.9	83.0	87.0	**89.1**
Day	89.7	82.7	86.6	**90.0**

Our proposed system aims to detect contextual anomalies (individual sensor malfunction compared to its peer-group) instead of global anomalies (majority sensor failures due to environment change). Hence we leverage peer-group analysis to generate anomaly scores. To ensure our method truly detects the contextual anomalies, we visualize the sensors' voltage global and local anomaly ground truth over time, along with our prediction results in Fig. 4. In the figures, the x-axis are time, and the y-axis are sensor IDs. The lighter colors represent higher anomaly scores. As can be seen in the subplots of global ground truths, the lighter-colored vertical lines represent malfunction throughout all the sensors in that particular time, meaning they are global malfunctions due to environmental change. However, our model is designed to detect individual sensor's anomaly. As shown in the prediction plots, our predictions resemble the local anomalies shown in local ground truths well.

4.4 Performance on Intrusion Detection Data

The cybersecurity dataset is collected from real-world system logs with the goal of detecting potential intrusion threats in a large corporation. All personal identifiable data is encrypted to avoid any leaking of personal information[1]. The dataset contains three graph layers: (i) wiki access layer, (ii) cloud application access layer, (iii) user server access layer. The wiki system is a centralized database containing domain knowledge about various company products. A bad actor may visit a very large number of wiki pages, gather confidential information, and ex-filtrate these packed information at a later time. The cloud applications record user activities when they log in remotely. The employee accounts may get hacked resulting in excessively long time of a specific application usage, or much more diverse application usage compared to his/her peer's normal behavior. The user server access data provides extra information to determine the peer-groups of the user appearing in the wiki and cloud application dataset. The dataset contains five threats that be classified as "need more investigation" by the security analytics team. In the experiment, we first transform the three bipartite graphs (user and the corresponding resource usage) into homogeneous graph, and the weight of the graph indicates the number of common resources (e.g., shared wiki pages visited) they used between two user nodes in a two month aggregated basis. We used the total number of wiki pages visited in a two month window as the attribute of each user node, and train the system to

[1] Due to the proprietary nature, we are not revealing the actual statistics of the dataset.

learn the optimal weight of the three layers to create the context for anomaly detection. Besides AUC-ROC, we also pay attention to the relative rank of the anomaly in all the data-points (measured by NDCG and mAP) since it strongly correlated with the workload of investigators. An ideal anomaly detection system should pop the true alarms as high as possible so that reviewers no need spend hours, even days, to go through hundreds of false alarms before a correct hit. We compared the performance of our system with the following baselines in Table 5:

- **Local outlier factor (LOF):** LOF [3] finds anomalous data points by measuring the local deviation of a given data point with respect to its neighbours based on $\mathcal{A}$.
- **Histogram-based outlier score (HBOS):** HBOS [12] is an efficient and unsupervised method to detect outliers by assuming feature independence [12], the input of the HBOS is $\mathcal{A}$ too.
- **Isolation forest (iForest):** iForest [18] introduces a different method that explicitly isolates anomalies using binary trees, it directly targets anomalies without the process of normal instance profiling, the iForest can only take $\mathcal{A}$ as the input.
- **Single layer peer-grouping on wiki, Application, Server layer (PG-Wiki, PG-App, PG-Server):** We run the traditional peer-grouping algorithm, the peers are retrieved based on $\mathcal{E}$ on the corresponding layers, and the anomaly score is computed based on the peer-group method introduced in [2,11,14,16].

From the results, we could see LOF performs the worst, which means the local deviation on $\mathcal{A}$ is not strong enough to differentiate the anomaly users and normal users. It is mainly because anomaly users may form its own clusters and the anomaly score generated by LOF may not be able to identify these clusters as anomaly clusters. HBOS and iForest perform competitively well, however, PG-based algorithm can achieve even better results by considering context-related information. For example, the wiki and application access data records users daily routine behavior, users who share similar wiki or application access patterns are more likely in the same team, so they may have similar frequency in terms of visiting wiki pages. We observe if each layer is assigned with the same weight (PG-Avg.) provide worst result than using any of the single layer pattern alone, which is expected since the merged layer may not able to provide the good context for anomaly detection. Because the proposed method merge the three context in an adaptive manner based on Eq. 9, the proposed method performs the best.

Table 5. Intrusion detection performance the system is evaluated with AUC-ROC, NGCG (Normalized Discounted Cumulative Gain), and mAP (mean Average Precision).

Method	AUC-ROC	NDCG	mAP
LOF	97.1	29.3	4.7
HBOS	98.3	47.3	9.9
iForest	98.5	41.9	12.7
PG-Wiki	98.4	48.9	20.4
PG-App	98.5	47.5	18.5
PG-Server	98.6	39.2	12.0
PG-Avg.	97.7	45.2	15.0
Ours	**98.9**	**61.3**	**28.9**

5 Conclusion

In this work, we propose a multilayer-graph-based anomaly detection system where the anomalies is defined by considering the context (peer-groups) generated by combining individual layers. The proposed system can be trained in both unsupervised and semi-supervised fashion. We evaluate our system on both synthetic and real-world datasets. Our system is shown to capture the contextual anomalies in graphs, and outperform the compared methods.

References

1. Akoglu, L., Tong, H., Koutra, D.: Graph based anomaly detection and description: a survey. Data Min. Knowl. Disc. **29**(3), 626–688 (2014). https://doi.org/10.1007/s10618-014-0365-y
2. Bolton, R.J., Hand, D.J., et al.: Unsupervised profiling methods for fraud detection. Credit Scoring Credit Control VII, 235–255 (2001)
3. Breunig, M.M., Kriegel, H., Ng, R.T., Sander, J.: LOF: identifying density-based local outliers. In: Proceedings of the 2000 ACM SIGMOD International Conference on Management of Data, 16–18 May 2000, Dallas, Texas, USA, pp. 93–104 (2000)
4. Chandola, V., Banerjee, A., Kumar, V.: Anomaly detection: a survey. ACM Comput. Surv. **41**(3), 15:1–15:58 (2009)
5. Chen, Z., Hendrix, W., Samatova, N.F.: Community-based anomaly detection in evolutionary networks. J. Intell. Inf. Syst. **39**(1), 59–85 (2012). https://doi.org/10.1007/s10844-011-0183-2
6. Deloitte: Impact of COVID-19 in cybersecurity (2021)
7. Dickison, M.E., Magnani, M., Rossi, L.: Multilayer Social Networks. Cambridge University Press, Cambridge (2016)
8. Dimopoulos, G., Barlet-Ros, P., Dovrolis, C., Leontiadis, I.: Detecting network performance anomalies with contextual anomaly detection. In: 2017 IEEE International Workshop on Measurement and Networking (M&N), pp. 1–6. IEEE (2017)

9. Ding, K., Li, J., Bhanushali, R., Liu, H.: Deep anomaly detection on attributed networks. In: SDM, pp. 594–602 (2019)
10. Eberle, W., Holder, L.B.: Anomaly detection in data represented as graphs. Intell. Data Anal. **11**(6), 663–689 (2007)
11. Eldardiry, H., et al.: Multi-source fusion for anomaly detection: using across-domain and across-time peer-group consistency checks. J. Wirel. Mob. Networks Ubiquitous Comput. Dependable Appl. **5**(2), 39–58 (2014)
12. Goldstein, M., Dengel, A.: Histogram-based outlier score (hbos): a fast unsupervised anomaly detection algorithm. KI-2012: Poster Demo Track **9**, 59–63 (2012)
13. Hayes, M.A., Capretz, M.A.M.: Contextual anomaly detection in big sensor data. In: 2014 IEEE International Congress on Big Data, Anchorage, AK, USA, 27 June–2 July 2014, pp. 64–71. IEEE Computer Society (2014)
14. Hayes, M.A., Capretz, M.A.M.: Contextual anomaly detection framework for big sensor data. J. Big Data **2**(1), 1–22 (2015). https://doi.org/10.1186/s40537-014-0011-y
15. Interdonato, R., Magnani, M., Perna, D., Tagarelli, A., Vega, D.: Multilayer network simplification: approaches, models and methods. Comput. Sci. Rev. **36**, 100246 (2020)
16. Kim, Y., Sohn, S.Y.: Stock fraud detection using peer group analysis. Expert Syst. Appl. **39**(10), 8986–8992 (2012)
17. Kivelä, M., Arenas, A., Barthelemy, M., Gleeson, J.P., Moreno, Y., Porter, M.A.: Multilayer networks. J. Complex Networks **2**(3), 203–271 (2014)
18. Liu, F.T., Ting, K.M., Zhou, Z.: Isolation forest. In: Proceedings of the 8th IEEE International Conference on Data Mining (ICDM 2008), 15–19 December 2008, Pisa, Italy, pp. 413–422. IEEE Computer Society (2008)
19. Liu, N., Huang, X., Hu, X.: Accelerated local anomaly detection via resolving attributed networks. In: IJCAI, pp. 2337–2343 (2017)
20. Madden, S., et al.: Intel lab data. Web page, Intel (2004)
21. Noble, C.C., Cook, D.J.: Graph-based anomaly detection. In: Getoor, L., Senator, T.E., Domingos, P.M., Faloutsos, C. (eds.) Proceedings of the Ninth ACM SIGKDD International Conference on Knowledge Discovery and Data Mining, Washington, DC, USA, 24–27 August 2003, pp. 631–636. ACM (2003)
22. Thiprungsri, S., Vasarhelyi, M.A.: Cluster analysis for anomaly detection in accounting data: an audit approach. Int. J. Digital Account. Res. **11** (2011)
23. Xie, J., Girshick, R., Farhadi, A.: Unsupervised deep embedding for clustering analysis. In: International Conference on Machine Learning, pp. 478–487 (2016)
24. Zhang, H., Basu, S., Davidson, I.: Deep constrained clustering - algorithms and advances. Arxiv Preprint 1901.10061 (2019)
25. Zitnik, M., Leskovec, J.: Predicting multicellular function through multi-layer tissue networks. Bioinformatics **33**(14), i190–i198 (2017)

Peekaboo: Hide and Seek with Malware Through Lightweight Multi-feature Based Lenient Hybrid Approach

Mingchang Liu(✉), Vinay Sachidananda, Hongyi Peng, Rajendra Patil, Sivaanandh Muneeswaran, and Mohan Gurusamy

National University of Singapore, Singapore, Singapore
{dcslium,comvs,rspatil,gmohan}@nus.edu.sg, dcshongp@nus.edu, e0503509@u.nus.edu

Abstract. In this paper, we propose – Peekaboo – a multiple feature-based lenient hybrid analysis for malware detection and classification. Our solution uses application programming interface (API) calls and operational codes (opcodes) extracted dynamically and statically as the behavioral features, and uses Recurrent Neural Network (RNN) to model both static and dynamic malicious behaviors. Peekaboo carries out dynamic analysis for a subset of samples, and static analysis for all samples in a large corpus, leading to lenient hybrid analysis. Peekaboo novelty lies in reducing the computational overhead of dynamic analysis but also utilizes multiple features to improve the model performance, making it lightweight and suitable for real-world deployment for malware detection and classification at a large scale.

We have conducted multiple sets of experiments by training and evaluating Peekaboo on a large dataset, our results show a 99.67% binary classification (benign vs. malicious) accuracy and 96.30% multi-class classification (classifies samples into malware classes) accuracy with a FPR as low as 0.45%. In comparison with our baseline model, Peekaboo enables us to increase the accuracy for binary classification by more than 1% and 5% in the multi-class setting. In addition, we tested Peekaboo on unseen malware classes, and it improved the accuracy by almost 4% compared to our baseline.

Keywords: Hybrid malware analysis · Malware detection · Malware classification · Neural networks · Machine learning

1 Introduction

The number of new malware and its variants has accelerated in recent years. In the first quarter of 2019, McAfee recorded more than 60 million new malware. Malware writers use code obfuscation, encryption, and polymorphism to create new malware versions from old ones [17]. These sorts of malware are a key source of dangerous zero-day exploits, and they constitute a big cybersecurity problem. Malware now causes the most economic impact of any cyber threat [19].

C. Alcaraz et al. (Eds.): ICICS 2022, LNCS 13407, pp. 525–545, 2022.
https://doi.org/10.1007/978-3-031-15777-6_29

The signature-based detection accuracy of the known malware is high with very few false positives. The signature-based detection methods cannot detect the new variants of the existing malware so the state-of-the-art has been focusing on behavior-based detection through static and dynamic features to capture unknown variants of existing malware [3]. However, they have limitations. For instance, most of them use a single type of feature from malware samples and ignore the rest of the useful information. Some of them execute dynamic analysis for a large number of samples, which significantly increases the computational overhead. In this work, we show our efforts to overcome these limitations.

1.1 Problem Statement and Research Challenges

Most of the modern behavior-based approaches rely on a single behavioral feature of the malware extracted from either static or dynamic analysis. As a single feature cannot cover all aspects of the malware, it potentially leads to lower accuracy and a higher false positive rate (FPR). Given the strengths and limitations of both static and dynamic analysis, they are better used in conjunction to achieve a good malware detection performance [33–36]. Therefore, hybrid analysis is becoming increasingly popular.

In hybrid analysis, analyzing a large number of samples efficiently can be challenging since, it requires each sample to undergo both static and dynamic analysis [12,13,35,36]. Static analysis can be executed efficiently since it does not require huge computational capacity. However, the total runtime of dynamic analysis for a large number of samples may increase significantly. In this case, the possible solutions are to 1) use powerful host machines and 2) intentionally shorten the runtime [28]. However, the first solution may not be always possible due to practical constraints whereas the second solution may cause premature termination of the analysis before the malicious behaviors are captured.

Thus, we need a method that can not only utilize both static and dynamic features to further improve the performance of malware detection and classification but also reduce the computational overhead when running dynamic analysis in real-world deployment. The deployment of hybrid analysis in real-world situations for large-scale malware detection leads to the following research questions which we address in this paper:

- **RQ-1**: how to integrate the information obtained from different sources for malware detection and classification.
- **RQ-2**: how to address the imbalanced feature extraction costs in hybrid analysis due to the inefficiency of dynamic analysis in practical scenarios without risking earlier termination of the dynamic analysis.
- **RQ-3**: how to make hybrid analysis more efficient for a large corpus of samples without needing to significantly increase the computational capacity.

Appendix A gives some background about the different types of malware analysis, some terminologies we will be using throughout the paper, and the machine learning (ML) methods we use in the work.

1.2 Approach Overview

In this paper, we present Peekaboo, a novel behavior-based model that uses hybrid analysis. It consists of two major components: data extraction and partial feature integration. In data extraction, a subset of the samples is first selected at random and executed in a virtual environment to collect API calls. Second, the entire corpus of samples undergoes static analysis to collect the opcodes. Unlike the traditional hybrid analysis [12,13,35,36], only a small subset of samples is selected by Peekaboo to overcome the inefficiency of dynamic analysis when the sample size is large. We propose a few-shot learning (FSL) method to mitigate the issue of API call dataset being small. In partial feature integration, we train two RNNs to transform the API calls and opcodes into numeric features and apply a partial multi-view learning method to integrate them into a single feature set, which is then used to train a final ML model. By harnessing a huge amount of data from static analysis and a relatively small amount of data from dynamic analysis, Peekaboo not only improves the performance of the model but also makes it more efficient, practical, and lightweight.

Intuitively, Peekaboo can be thought of as having two models, one as the primary model (e.g., the opcode model) and another as a secondary model (e.g., API call model). *To the best of our knowledge, this is the first research attempt that proposes to utilize such an idea to address the real-world challenge that dynamic analysis is not suitable for analyzing a large corpus of samples in malware detection systems under a multi-feature and lenient hybrid analysis setting.*

1.3 Results Overview

We have conducted multiple sets of performance evaluation and benchmarking experiments. We evaluated Peekaboo on a large dataset consisting of 34,000 samples including a public dataset, and 12,000 recent malicious samples collected from VirusTotal[1]. Peekaboo can achieve an accuracy of 99.67% with a FPR of 0.45% for detecting if a given software is malicious. In a multi-class classification setting where Peekaboo attempts to identify the correct malware class for a given software, it can achieve an accuracy of 96.30% with a FPR as low as 0.45%. Peekaboo helps improve the performance of the model trained on a single feature by at least 1% and 5% in accuracy in both binary and multi-class classification respectively compared to our baseline. The FPR can also be significantly reduced. In addition, we evaluated Peekaboo on a set of samples from previously unseen malware classes, and it can successfully detect all of them.

1.4 Research Contributions

The goal of this research is to develop an approach to make use of multiple features from lenient hybrid analysis for malware detection and classification more efficient and practical. We make the following contributions in this work:

[1] VirutTotal: https://www.virustotal.com/.

- We propose a multi-feature extraction mechanism based on lenient hybrid analysis for a large corpus of benign and malicious samples. It improves the performance of malware detection and classification.
- We propose a FSL method for model training on API calls in malware detection and classification tasks based on a recently proposed FSL approach in NLP. We show the FSL method for API call is effective in making the model training converge faster and improving the performance.
- We propose, by using a partial multi-view learning method, a feature integration process that integrates the features collected using both static and dynamic analysis from a large corpus of benign and malicious samples into a single model for malware detection and classification.

Structure of the Paper. Section 2 reviews related works that have been proposed for malware detection and classification. Section 3 discusses Peekaboo in details. Section 4 presents performance evaluation. Results, analysis and discussion are presented in Sect. 5. Finally, Sect. 6 concludes the paper.

2 Related Work

We briefly discuss the related works that have been conducted by using advanced statistical and ML models for malware detection and classification.

2.1 Single Feature-Based Approaches

Some of the existing works have only used a single feature extracted from the malware samples. Common features that are used in current state-of-the-art detection and classification systems include API calls, opcodes, binary code, etc.

Observed sequences recorded from system calls are used as input data to train a hidden Markov model (HMM) for each malware family [3]. Pranamulia et al. [4] proposed a similar approach to use profile hidden Markov model (PHMM) to analyze the sequences of system calls.Duarte-Garcia, et al. [14] proposed to classify malware by using the API call sequences generated from sandbox in a semi-supervised manner.

Makandar et al. [7] proposed to use a support vector machine (SVM) to identify the malware after extracting the features from the malware image. Kim et al. [6] proposed to use generative adversarial networks (GAN) and autoencoders to capture the variants of existing malware with the aim to detect zero-day attacks. Ye et al. [2] proposed DeepAM, a malware detection framework that takes the Windows API calls extracted statically from the malware executable as inputs. DeepSign [1] and DeepOrigin [5] capture the variants of existing malware by generating and classifying signatures of malware that is invariant to code changes.

API calls collected in dynamic analysis have widely experimented for malware detection. Pascanu et al. [15] proposed to use the language model RNN to analyze the sequences of API calls made by the malware during runtime. Athiwaratkun et al. [16] also proposed a similar idea to use an RNN with LSTM and GRU architecture to analyze the system calls made by the malware at a character level. A recent work proposed by Rabadi and Teo [24] uses the function name as well as the arguments of the API calls collected during dynamic analysis. Neurlux [25] feeds the entire analysis report generated by Cuckoo Sandbox into a language model for malware detection and classification.

Zolotukhin et al. [8] proposed a method based on opcode sequences for malware classification. Yewale and Singh [10] proposed a malware detection method that uses the opcode frequency. A similar idea was proposed by Manavi and Hamzeh [9] to convert opcode frequencies into an image for training machine learning algorithms.

2.2 Multiple Feature-Based Approaches

To achieve better performance, many researchers have also constructed multiple feature-based data from both static and dynamic analyses. Shijo and Salim [35] proposed a hybrid analysis approach by integrating printable string information (PSI) and API call sequences extracted from static and dynamic analyses respectively. Islam et al. [36] proposed to use function length frequency and PSI as static features and API calls as dynamic features for malware classification.

More recently, Zhang et al. [12] proposed multiple feature-based approach for the malware clustering algorithm. Multiple features are extracted by using static and dynamic analysis tools for base clustering. Next, they rely on forward step-wise selection to select the clustering ensemble and combine the clustering results via a mixture model. Zhang et al. [13] developed a feature extraction method such that the opcode-based and API calls-based features are extracted and fed into a CNN and back-propagation neural network respectively to obtain a high-level representation of the original features. The two sets of high-level features are then combined and used to train a multi-class classifier.

3 Peekaboo: Our Proposed Approach

Peekaboo uses API call and opcode sequences as input data to our malware detection and classification model. More specifically, API call sequences are dynamically extracted by executing malware samples in a virtual environment, while opcode sequences are statically extracted by disassembling the binaries. Next, two individual models are trained to transform the API calls and opcodes into numeric features. Then the two feature sets are integrated into one for the final model training. Figure 1 shows the workflow of Peekaboo.

3.1 Lenient Hybrid Analysis

To make the hybrid analysis more efficient for a large number of samples, we propose the idea of lenient hybrid analysis. Lenient hybrid analysis can be seen as a speedy version of the traditional hybrid analysis (i.e., both static and dynamic

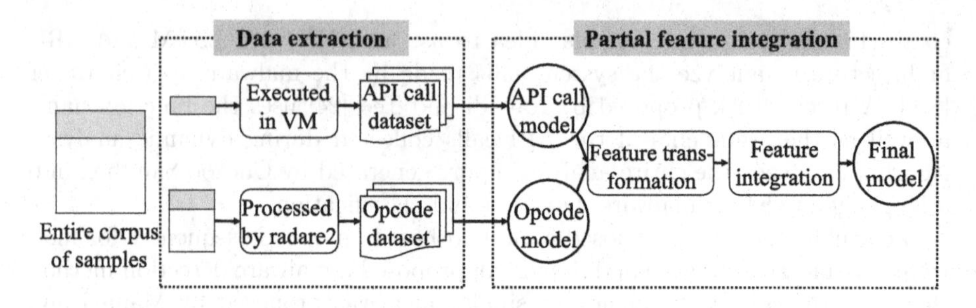

Fig. 1. The workflow of Peekaboo.

analyses are executed for every sample) when the number of samples to analyze is large. The idea behind lenient hybrid analysis is to select a subset of samples from this large corpus based on the computational resources available and execute dynamic analysis only for samples in the subset. Meanwhile, the entire corpus undergoes static analysis. Then we can treat static analysis as the primary resource of information and dynamic analysis as the secondary one. By integrating the extra information the dynamic analysis provides into the information we obtain from static analysis, the whole system can perform better malware detection and classification. *Peekaboo is the first approach that enables us to do so, it makes hybrid analysis more efficient, practical, and lightweight at a large scale in real-world deployment.*

3.2 Extraction of API Calls and Opcodes

To trace API calls dynamically, malware samples are executed in a virtual environment; to extract opcodes, the malware binaries are analyzed and disassembled using a disassembly tool. The data extraction is performed for both malware and benign samples. We first select a random and small subset of size n from the entire corpus of samples of size N. Here we set $n \ll N$. For each sample in the subset, we perform dynamic analysis to extract API calls. Then for all the samples in the entire corpus, we perform static analysis to extract opcodes. Let a_i be the API call sequence and o_i the opcode sequence for example i. For n examples in the selected subset, we have API call data $\{a_1, a_2, a_3, ..., a_n\}$. For a total number of N examples in the entire corpus, we have opcode data $\{o_1, o_2, o_3, ..., o_N\}$.

A graphical representation of the API calls and opcodes datasets is shown in Fig. 2. The length of the sequences is drawn to the same in the figure for easy visualization, the actual length of the API call and opcode sequences can be quite different.

3.3 Partial Feature Integration

We purposely set $n \ll N$ so there would be a large number of samples that are missing the API call data. In fact, there is a whole block of API call data missing

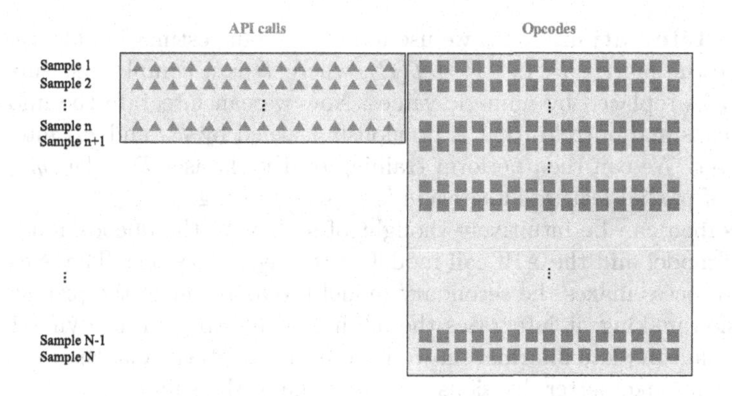

Fig. 2. A graphical representation of the API call and opcode datasets.

(shown in Fig. 2). Partial multi-view learning problems are commonly encountered in many domains such as medical diagnosis where the ML model needs to handle incomplete multi-modality data. In Peekaboo, we use the ScoreComp proposed by Yuan et al. [22] to tackle the problem of block-wise missing data. Each data source is treated independently, where a base classifier is learned such that the data source is converted into prediction scores. Then the block-wise missing data problem is transformed to a missing value imputation problem.

Feature Transformation. We formally describe the problem in the following by instantiating ScoreComp in the context of using API calls and opcodes for malware detection and classification. Consider two labeled datasets $\mathcal{D}_A = \{a_i, y_i\}_{i=1}^n$ and $\mathcal{D}_O = \{o_i, y_i\}_{i=1}^N$ where $\mathcal{D}_A$ denotes the API call dataset and $\mathcal{D}_O$ the opcode dataset. Since $n \ll N$, we must first perform FSL for the API call model. Here, we apply a data-centric FSL algorithm $\mathcal{A}$ to $\mathcal{D}_A$ to obtain $\mathcal{D}_A^{FSL} = \mathcal{A}(\mathcal{D}_A)$. The details about algorithm $\mathcal{A}$ are given in the following Sect. 3.4.

First, we perform classifications by choosing a learning algorithm $\mathcal{L}$, and corresponding loss functions on $\mathcal{D}_A$ and $\mathcal{D}_O$ individually:

$$F_A = \mathcal{L}(\mathcal{D}_A^{FSL}), \quad F_O = \mathcal{L}(\mathcal{D}_O),$$

where F is a function that maps an API call or opcode sequence to a vector of prediction scores, and $\mathcal{D}_A^{FSL} = \mathcal{A}(\mathcal{D}_A)$.

Note that after the models are trained, we obtain two functions F_A and F_O. For dataset $\mathcal{D}_A$ and $\mathcal{D}_O$, we can then construct two matrices P_A and P_O. We then use these two matrices to form an integrated matrix:

$$P_{i,:} = \begin{cases} [F_A(a_i)^T \mid F_O(o_i)^T], & \text{if sample } i \text{ is in both } \mathcal{D}_A \text{ and } \mathcal{D}_O \\ [\mathbf{NaN}^T \mid F_O(o_i)], & \text{otherwise} \end{cases},$$

where we set **NaN** to be a column vector of NaN's, and the resulting matrix P is a matrix with block-wise NaN's.

Feature Integration. Next, we use a missing value estimation algorithm $\mathcal{I}$ to complete the matrix P. Let $\tilde{P} = \mathcal{I}(P)$, where $\tilde{P}$ is a complete feature matrix with NaN's replaced by numeric values. Now we can integrate the information of API calls and opcodes that is previously learned by F_A and F_O into a single feature set. We can then perform training on the dataset $\mathcal{D} = \{p_i, y_i\}_{i=1}^{N}$. The process of partial feature integration is depicted in Fig. 3.

Peekaboo can be intuitively thought of as having the opcode model as the primary model and the API call model as the secondary one. The feature integration process makes the secondary model the assistant of the primary model in decision making: it integrates the additional information of dynamic analysis into the complete information we have from the static analysis so the whole system can make better decisions than any one of the sub-models.

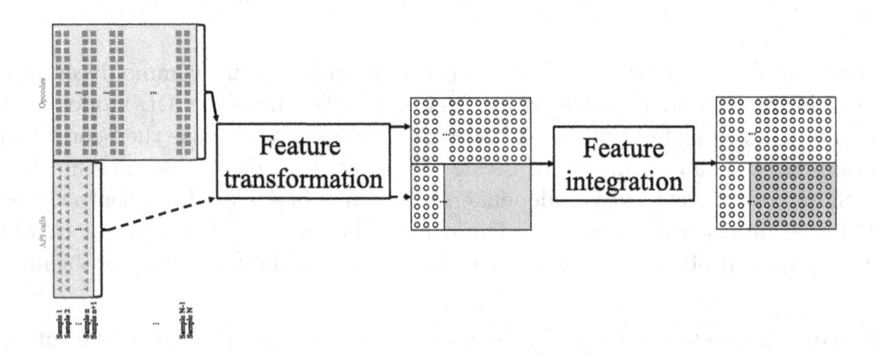

Fig. 3. The process of partial feature integration.

3.4 Few-shot Learning (FSL) for API Call Model

To transform the features to numeric features for partial feature integration, we must first perform individual model training. Since the size of the API call dataset $\mathcal{D}_A$ is intentionally made small for more efficient dynamic analysis, we need to perform FSL ensure the model trained over this dataset can generalize.

The API call sequences can be thought of as sequential data since they monitor the sequence of function calls the software makes to interact with the system resource and the flow of the program. Naturally, it resembles text classification in NLP. Therefore, some FSL methods in text classification may be applied in API call sequence classification. We modify a recently proposed data-centric FSL technique named EDA by Wei et al. [21] for the API call sequences.

EDA is a data-centric FSL approach that augments existing text data by adding randomness. Wei et al. [21] proposed a few methods in their paper to generate new text data, namely, Synonym Replacement, Random Insertion, Random Swap, and Random Deletion. Synonym Replacement and Random Insertion involve operations of synonyms of the words in the corpus. Since we do not have or define such synonyms of API calls, we only used the remaining two methods,

Algorithm 1: Few-shot learning for API call sequences

```
Input: D_A = {a_i, y_i}_{i=1}^n
Result: Dataset for few-shot learning D_A^{FSL}
Initialization: number of iterations r, probability p, empty hash tables D_{tmp1},
  D_{tmp2};
split D_A into D_A^{Train}, D_A^{Val} and D_A^{Test};
for iter = 1, 2, ..., r do
  for i = 1, 2, ..., n^{Train} do
    lb ← y_i;
    convert a_i to an array of API function names while preserving the
      order: lst ← a_i;
    make two copies lst_1 and lst_2 of lst;
    randomly draw two indices j, k of lst;
    c_1 ← lst_1[j], c_2 ← lst_1[k];
    lst_1[j] ← c_2, lst_1[k] ← c_1;
    D_{tmp1}[lst_1] = lb;
    randomly draw an index m of lst;
    c ← lst_2[m];
    sample u ~ Unif[0, 1];
    if u ≤ p then
      lst_2.remove(c);
    end
    D_{tmp2}[lst_2] = lb
  end
end
D_A^{FSL} ← D_A^{Train} ∪ D_A^{Val} ∪ D_A^{Test} ∪ D_{tmp1} ∪ D_{tmp2}
```

i.e., Random Swap and Random Deletion for $\mathcal{D}_A$. The FSL algorithm for API calls is outlined in Algorithm 1.

The sequences in $\mathcal{D}_A^{FSL}$ may no longer represent the exact behaviors we observed during execution. However, the purpose of doing so is not to create new API call sequences that fully represent the sample's exact behaviors or can be successfully carried out in practice. The FSL algorithm introduces more randomness so by performing $F_A = \mathcal{L}(\mathcal{D}_A^{FSL})$, we make F_A more robust to overfitting. The randomness is carefully controlled by the hyper-parameters r and p so that a reasonable amount of randomness is injected but does not overwhelm the original behaviors of the sample.

3.5 Achieving Real-Time Detection

In the above, we have described Peekaboo in a fixed-dataset fashion where a set of opcode and API call sequences are collected prior to any malware detection. However, in real-world scenarios, we often have to stream in the incoming samples and determine if they are malicious efficiently. Peekaboo can be naturally extended for real-world deployment in such scenarios. In real-world scenarios, each incoming sample will only have the opcode extracted statically together with the additional information provided by API calls for real-time detection by Peekaboo as described above. Peekaboo can then maintain a database of API calls by extracting the API calls dynamically from randomly selected samples from the malware database to update the API call database. This update can be set periodically and independent of the real-time detection process so that we can still use the updated API call information for malware detection in Peekaboo without affecting the efficiency of the real-time malware detection.

4 Performance Evaluation

We first consider the binary classification case, where we train a binary classifier to identify a given instance as malicious or benign. Next, we extend our experiment to the multi-class classification setting where we separate the benign software and malware and classify the malware into the correct classes in the meanwhile.

4.1 Datasets

In this work, we used public datasets. The API call sequence (APIMDS) dataset was collected and published with open access by Ki et al. [18]. They obtained the malware samples from the malware dataset provided by VirusTotal. Nowadays, most of the machines are running more recent versions of Windows operating systems, so we set up a Windows 10 environment and selected a small set of samples to collect the API calls. The opcode sequence dataset is obtained by disassembling the software binaries using radare2[2]. The opcodes extraction was done using radare2's Python plug-in r2pipe[3].

The malware samples open-sourced by VirusTotal were first seen in their database in late 2019. In total, we gathered 34,000 benign and malicious samples. Among all the samples collected, over 80% were collected after late 2019. The benign software was downloaded from Softpedia[4]; they were the most popular downloads by the time they were collected.

Similarly, we obtained a very recent 12,000 malicious samples collected from VirusTotal and performed the dynamic and static analysis.

4.2 Data Preparation

The benign software receives a single label "benign". The malware receives a label "malicious" and an additional label of the malware class. We do not use any detection labels provided by anti-virus vendors since different vendors have their own way of labeling malware class. To consistently assign labels for the malware, we use a state-of-the-art malware labeling tool called AVClass2[5] by

[2] Radare2 version 3.9.0: https://www.radare.org/n/radare2.html.
[3] R2pipe version 4.0.0: https://github.com/radareorg/radare2-r2pipe.
[4] Softpedia: https://www.softpedia.com/.
[5] AVClass2 source code: https://github.com/malicialab/avclass.

Sebastián et al. [20]. Table 1 summarizes the distribution of benign as well as different malware classes of the entire corpus of samples.

Among 34,000 samples, we randomly selected 4,500 samples for more efficient dynamic analysis in practice. This number is chosen based on our computational capacity. In practice, this number can be adjusted based on the computational resources available and the sample size.

In all the experiments, we split the entire dataset into three mutually exclusive sets, namely, training (80%), validation (10%), and test (10%) sets. The training and validation sets are used to train and fine-tune the models. The test set is held out and not revealed to the model during the entire training and validation phase. This setting simulates the situations where the model is trained on known and detected malware samples and then used to detect unknown or undetected malware. After the training and validation are complete, we then evaluate the model on the hold-out test set. The results presented in the following are obtained by evaluating the model on the hold-out test set.

4.3 Experiments

Baseline. In Peekaboo, the individual API call model can be seen as an assistant of the opcode model. Hence, we would expect the final model to perform at least as well as the opcode model. Given this intuition, in the following experiments and evaluations, we use the performance of the opcode model as a baseline to assess how much Peekaboo helps us improve the performance.

Our experiments begin with individual model training for feature transformation. Then we perform the feature integration. Finally, we train a model based on an integrated complete feature set.

Tokenization of the API Calls and Opcodes. The sequences of API calls and opcodes are treated as sequences of text. Each API call or opcode is considered to be a word, and each sequence is considered to be a sentence. Text sequences are used as input data for training in the NLP task. However, it is very difficult for any model to process text sequences. A common approach for training a model using text sequences in NLP is that the text sequences are first tokenized and then converted to numeric sequences. We perform similar tokenization and conversion procedures for API call and opcode sequences. Conceptually, the tokenization splits on white space, and the characters between two white spaces are then regarded as a single token. In our setting, each API call or opcode will be naturally regarded as a token.

Model Architecture. We choose RNN to model the API call and opcode sequences. Among different types of RNN architectures, long short-term memory (LSTM) and gated recurrent unit (GRU) has been proved to be very effective in various NLP tasks and well-known for modeling long-range dependencies within sequential data. Hence, we use both LSTM and GRU in our RNN architecture.

Table 1. Distribution of benign and malware classes.

Malware class	Percentage
Benign	29.43%
Worm	16.31%
Downloader	14.37%
Grayware	13.79%
Virus	11.55%
Backdoor	7.54%
Ransomware	2.83%
Rogueware	2.53%
Spyware	1.65%

Table 2. Distribution of unseen malware classes.

Malware class	Percentage
Bot	35.75%
Rootkit	23.46%
Clicker	19.55%
Keylogger	11.73%
Hoax:smshoax	4.47%
Hoax	3.35%
Dialer	1.68%

The first layer of the RNN is the embedding layer, followed by a one dimensional convolutional layer with ReLU activation. Next, a one dimensional max-pooling layer is used to reduce the size of the network. The LSTM and GRU layer are then employed and followed by a fully-connected layer which flattens the output. The final output layer is a fully-connected layer with softmax activation. Dropout and regularization are applied to reduce over-fitting.

Model Training for Feature Transformation. The opcode dataset is large so we proceed as per the normal training procedure. We perform Algorithm 1 for the API call dataset and follow the hyper-parameters suggested in the original paper of EDA [21].

After the models F_A and F_O are trained, we transform the API call and opcode sequences into prediction score matrices. Then, the features are integrated as discussed in Sect. 3.3. Finally, we train another model based on the integrated feature set.

We extend our experiments to the multi-class case by repeating the above process and changing the target output of the model to the multi-class labels.

Feature Integration. The feature integration step can be accomplished by using a missing value estimation algorithm. Similar to [22], we also used several popular algorithms for comparison and benchmarking purposes.

KNNimpute Algorithm. K-nearest neighbor (KNN) is a non-parametric model that can make a prediction for an unseen example by finding similar training examples. The KNNimpute algorithm [23] works in a similar way: the missing values of an example are approximated by other values of the k training examples that are the closest to it based on other available features.

Singular Value Decomposition. Singular value decomposition (SVD) is a popular approach for matrix completion. SVD for matrix completion can be solved by an algorithm in the form of an expectation and maximization (EM) algorithm. The original matrix with missing values is decomposed by SVD, and the data matrix is reconstructed from the SVD to approximate the missing values. These two steps are repeated until convergence.

Expectation and Maximization Algorithm. Expectation and maximization (EM) algorithm consists of two steps, i.e., E step and M step. The EM algorithm is an optimization algorithm that maximizes the log-likelihood of the expected complete data in an iterative manner under the (conditional) distribution of unobserved data points.

4.4 Evaluation on Unseen Malware Classes

The final model was also evaluated on a small set of malware samples that are of previously unseen malware classes. This simulates a more generic situation where the model is trained on samples of known malware classes and tries to capture the malware from unknown classes. These samples come from malware classes the multi-class classifier has not seen, so it will not classify them correctly. We only evaluated this set of samples from unseen malware classes in binary classification settings (benign vs. malicious).

This dataset consists of only malware samples; in total there are 179 malicious samples. Again, the malware class labels are assigned by AVClass2. Table 2 shows the class distribution. This additional dataset follows the same data extraction and partial feature integration of Peekaboo as depicted in Fig. 1.

5 Results and Discussion

Table 3 shows the performance of the API call model, opcode model, and the final model with the best performance for both binary and multi-class classifications. We extend the definition of FPR to a multi-class setting, where a false positive is a benign sample classified into any other class by the model.

Table 4 shows the performance of the final model we constructed using the integrated feature set. We experimented with various ML algorithms for benchmarking purposes. The ML models we used include logistic regression, k-nearest neighbors, linear support vector machine (SVM), SVM with radial basis function (RBF) kernel, Decision Tree, Random Forest, AdaBoost, Gaussian Naíve Bayes, and quadratic discriminant analysis (QDA). We report accuracy and FPR in the table.

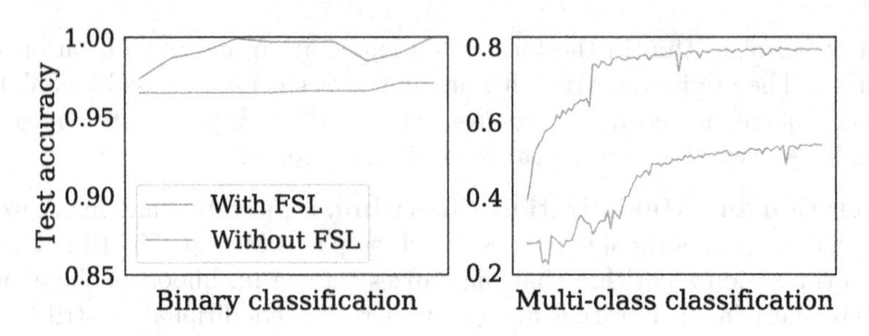

Fig. 4. Test accuracy for both binary and multi-class API call model as number of epochs increases.

Table 3. Performance of binary and multi-class classification models.

		Accuracy	False positive	True positive	False negative	True negative
API call model F_A	Binary	100.00%	0.00%	100.00%	0.00%	100.00%
	Multi-class	80.00%	0.00%	–	–	–
Opcode model F_O	Binary	98.50%	1.56%	98.52%	1.48%	98.44%
	Multi-class	90.80%	2.34%	–	–	–
Final model	Binary	99.67%	0.45%	99.71%	0.29%	99.55%
	Multi-class	96.30%	0.45%	–	–	–

Table 4. Performance of the final model using different ML model for feature integration. (Logistic Regression (LR), k-Nearest Neighbors (KNN), Linear SVM (LSVM), Decision Tree (DT), quadratic discriminant analysis (QDA))

		Binary			Multi- class		
		KNN	SVD	EM	KNN	SVD	EM
LR	Accuracy	99.53%	99.40%	99.53%	96.03%	96.17%	96.23%
	FPR	0.56%	0.45%	0.45%	0.56%	0.45%	0.45%
KNN	Accuracy	99.63%	99.67%	99.67%	95.57%	95.57%	95.57%
	FPR	0.45%	0.45%	0.45%	0.45%	0.45%	0.34%
LSVM	Accuracy	99.37%	99.37%	99.53%	96.13%	95.63%	96.03%
	FPR	0.45%	0.45%	0.45%	0.68%	0.68%	0.68%
RBF	Accuracy	99.53%	99.50%	99.57%	96.30%	96.17%	96.17%
	FPR	0.56%	0.56%	0.56%	0.45%	0.45%	0.45%
DT	Accuracy	99.60%	99.67%	99.67%	89.17%	88.47%	89.37%
	FPR	0.56%	0.56%	0.56%	0.45%	0.68%	0.68%
Random	Accuracy	99.60%	99.57%	99.63%	93.03%	91.60%	94.17%
	FPR	0.45%	0.45%	0.45%	2.14%	0.45%	0.68%
AdaBoost	Accuracy	99.53%	99.63%	99.63%	67.77%	51.20%	78.03%
	FPR	0.56%	0.56%	0.56%	67.76%	90.08%	41.94%
Naíve	Accuracy	99.40%	99.37%	99.37%	95.33%	95.37%	95.47%
	FPR	0.45%	0.45%	0.45%	2.14%	1.13%	1.58%
QDA	Accuracy	99.33%	99.37%	99.47%	95.37%	94.07%	95.33%
	FPR	0.67%	0.45%	0.45%	1.01%	0.79%	1.35%

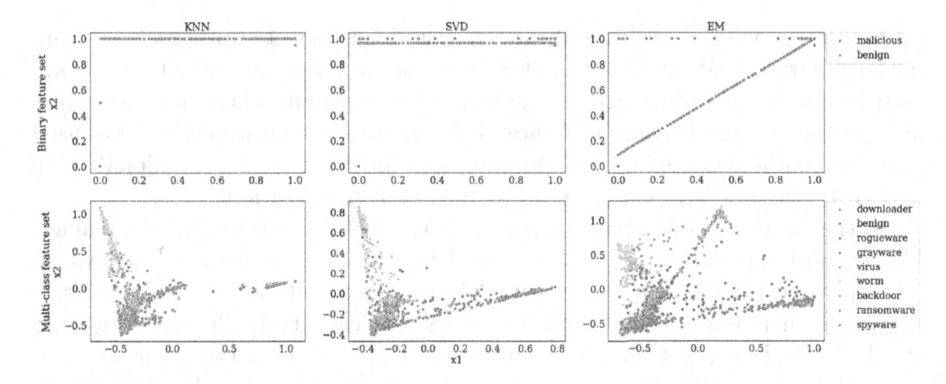

Fig. 5. The visualization of the integrated feature sets. The dimension was reduced by PCA to two in the multi-class case.

5.1 Analysis and Discussion

Model Training for Feature Transformation. The performances of individual models trained on the API call and opcode datasets are essential for feature integration since they directly influence the quality of the integrated features. The API call model can do very well for binary classification. Although this means the model can do well on a small subset of samples, it can make the feature integration assign more precise values to the missing ones. The API call model only has an 80% accuracy for multi-class classification. This is expected because the subset is relatively small, and it is difficult for a model trained on such small dataset to do well in this case.

The opcode model achieves a 98.5% accuracy for binary classification but with a high FPR of 1.56%. In the multi-class classification task, the opcode model performs much better than the API call model with a 90.8% accuracy, but it still a high FPR of 2.34%. The opcode model can perform correct classifications at a satisfactory level, but it has a high FPR. The FPR can be lowered after the feature integration.

Algorithm 1 for API calls is effective to improve the model convergence and performance. The test accuracy of both binary and multi-class API call models is shown in Fig. 4. We evaluated the model on the same test set with and without FSL. In the binary classification setting, the model trained with FSL converged within the first few epochs whereas the model converged slowly without it using the same learning rate. In particular, the API call model without FSL under multi-class settings can only achieve an accuracy slightly above 50%; this can lead to poor feature integration. FSL can improve the accuracy by almost 30% for the API call model in this case.

Feature Integration. We selected 4,500 samples for dynamic analysis so they have a complete set of features; the remaining samples are missing the output from the API call model. These missing outputs are to be filled in by the data imputation algorithms.

Figure 5 shows scatter plots of the integrated features. The integrated feature set for binary classification is of 2-dimension because two individual models are used for feature transformation. The feature set for multi-class classification is of higher-dimension because the dataset has multiple malware classes. We used principle component analysis (PCA) to project the data to a 2-dimensional space defined by the first two principle components for visualization.

In the binary classification setting, x_2 is the feature that has missing values. Ideally, both x_1 and x_2 need to be small for the benign class. In the feature set integrated by KNNimpute, x_1 of all the data points from the benign class is below 0.6 whereas x_2 of some data points are very close to 1. The second feature of all of the data points in the malware class is close to 1 but some of them have relatively small values for x_1. A similar pattern is observed in the feature set integrated by the SVD. Further analysis shows the KNNimpute algorithm performs much better than the SVD. Since x_2 is to be filled in, we say we have a bad assignment to x_2 when the algorithm assigns a large value to it for a benign sample or a small value for a malware sample. Out of all assignments, the bad ones are only 1.37% for KNNimpute whereas the bad assignments account for 33.52% of the total for the SVD.

In the feature set integrated by the EM algorithm for binary classification, most of the data points in the benign class have relatively small values for both x_1 and x_2. The EM algorithm performs the best for feature integration and only has 0.46% bad assignments. The correct assignment of the values can help lower the high FPR of the individual models.

In the multi-class classification setting, all of the algorithms perform similarly by plotting the first two principle components of the feature set. The EM algorithm performs much better at separating the benign class from the rest of the malware classes, which helps us lower the FPR. The results of the final models confirmed this observation since the lowest false positive is achieved by using the feature set integrated by the EM algorithm.

Final Model Performance. As shown in Table 4, all models can perform well at binary classification regardless of the feature set. For the multi-class classification, AdaBoost performs very poorly with extremely lower accuracy and a high FPR. The rest of the ML models perform quite well, achieving an overall accuracy of around 95% with a low FPR.

Out of all the configurations presented in Table 4, we select two of them as our final models. For binary classification, the following configuration performs the best: EM algorithm for feature integration and KNN for classification. We do not choose the configuration of the SVD for feature integration and KNN for classification because the analysis in Sect. 5.1 shows the feature set integrated by the SVD is less reliable than the one by EM algorithm.

For multi-class classification, we select this configuration as our final model: KNNimpute for feature integration and SVM with RBF kernel for classification.

By using the final models, we increased the accuracy for binary classification and lower the FPR as well as false negative rate by more than 1% compared to the baseline binary opcode model. The improvements are even larger for the multi-class classification: we improved the accuracy by more than 5% and lower the FPR by almost 2%.

False Positives. There are 4 false positives in the test set for both binary and multi-class classifications by the final model. None of these samples exists in API call dataset, which means they all had missing features before feature integration.

In the binary classification setting, all 4 benign samples were first incorrectly classified by the individual opcode model (i.e., a large value for x_1), which leads to the EM algorithm incorrectly assigning large values for x_2. Hence, both x_1 and x_2 being large leads to the final incorrect classification.

A similar pattern is observed in the multi-class classification: these 4 benign samples were incorrectly classified by the individual opcode model into other malware classes, and the missing features were filled in based on the incorrect prediction scores. The output from the opcode model could significantly impact the feature integration. Therefore, having a well-performing opcode model F_O is essential in Peekaboo.

5.2 Comparison with Related Works

We have listed other recent related works that use API calls, opcode sequences or both as features for malware detection for performance comparison. The results are summarized in Table 5, which shows the binary classification results. The evaluation metrics are accuracy, precision, recall, and F1-score. Peekaboo outperforms other state-of-the-art in all the evaluation metrics by a relatively huge margin.

Table 5. Comparative analysis

	Accuracy	Precision	Recall	F1-Score
Peekaboo	99.67%	99.81%	99.71%	99.76%
Ye et al. [2]	98.20%	98.60%	97.80%	98.20%
Manavi et al. [9]	93.20%	95.60%	91.00%	93.20%
Zhang et al. [13]	95.10%	92.40%	79.80%	85.60%
Masbo et al. [11]	94.00%	94.00%	94.00%	94.00%
Vinayakumar et al. [37]	96.30%	96.30%	96.20%	96.20%
Venkatraman et al. [38]	96.30%	91.80%	91.50%	91.60%

5.3 Evaluation on Unseen Malware Classes

We use the trained opcode model as the baseline: for these small sets of malware samples, the trained opcode model can only detect 96.09% of them, giving a false negative rate of 3.91%. We can see from Fig. 6 that some data points have relatively small values for feature x_1 since the individual opcode model classified them incorrectly. By contrast, the feature x_2 is large for all data points, meaning 1) the individual trained API call model performs well to correctly transform the features into prediction scores, and 2) more importantly the EM algorithm can correctly assign the values for those samples that are missing API call sequences.

The trained KNN classifier can correctly detect all the malware samples using the feature set integrated by the EM algorithm. The performance of the final model is significantly better than that of the individual opcode model which we use as a baseline. Even though Peekaboo is trained on the samples of known malware classes, it can still detect the samples of unseen malware classes.

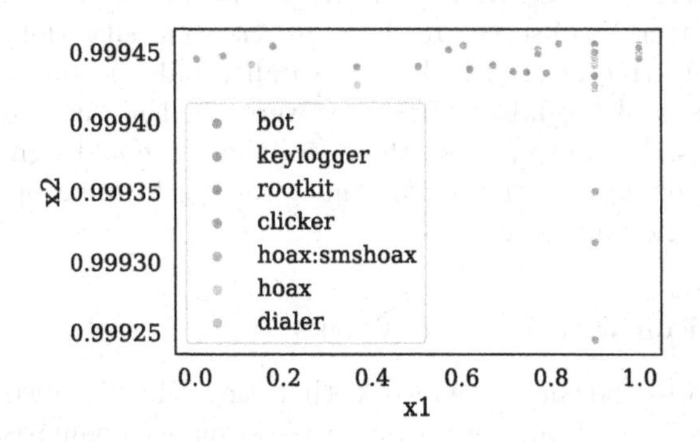

Fig. 6. The visualization of the integrated feature sets of the malware samples from previously unseen classes.

6 Conclusion and Future Work

In this paper, we present a novel approach that uses both static and dynamic analysis for constructing a better-performing model in a multi-feature setting. This approach reduces the burden of dynamic analysis, a resource-consuming task, by only utilizing a small portion of the samples. By using a data imputation algorithm to integrate the features, we can use the information we obtain from dynamic analysis to improve the model performance. The accuracy and FPR can be significantly improved after the feature integration. We demonstrated outstanding performance through various experiments, benchmarking, and analysis on a large-scale dataset. We only considered two features, namely, API calls and opcodes, for malware detection and classification in this work, but Peekaboo can be generalized to work with more than two features.

In this work, we only consider the malware samples that are not packed or obfuscated. Code obfuscation and packing have non-negligible effects on opcode extraction. In future works, we will consider de-obfuscation/unpacking of the code and it's effects on Peekaboo.

Acknowledgment. This research is supported by the National Research Foundation, Prime Minister's Office, Singapore under its Corporate Laboratory@University Scheme, National University of Singapore, and Singapore Telecommunications Ltd.

A Background

Static Analysis refers to the analysis of a binary file without executing it.

Dynamic Analysis refers to the analysis of a binary file by executing it in a controlled and well-monitored environment e.g., a virtual machine or sandbox.

Terminologies. In the malware analysis context, a feature often means a type of data extracted from the samples that can characterize the maliciousness. The use of this term in malware analysis is different from that in the usual machine learning setting where features represent the attribute of the observations. The term "multiple features" here refers to multiple kinds of features in the malware analysis setting. For example, Peekaboo uses API calls and opcodes as features.

Few-Shot Learning is a learning strategy that can improve the model generalization ability when the sample size is small. FSL is essential to Peekaboo when training the model on the API call dataset since we only select a small portion of the entire corpus of samples for dynamic analysis.

Multi-view Learning is a learning strategy that deals with data consisting of different views. A view can be a set of features obtained from one domain. In our setting, one view is the API calls collected during dynamic analysis and the other is the opcodes from static analysis. Multi-view learning aims to integrate the data for model training or use custom learning strategies to teach learners to consume data from different views to perform well on a common task. Partial multi-view learning is a task that specializes in handling missing views.

References

1. David, O., Netanyahu, N.S.: DeepSign: deep learning for automatic malware signature generation and classification. In: International Joint Conference on Neural Networks (IJCNN), vol. 2015, pp. 1–8 (2015)
2. Ye, Y., Chen, L., Hou, S., Hardy, W., Li, X.: DeepAM: a heterogeneous deep learning framework for intelligent malware detection. Knowl. Inf. Syst. **54**(2), 265–285 (2017). https://doi.org/10.1007/s10115-017-1058-9
3. Imran, M., Afzal, M.T., Qadir, M.A.: Using hidden Markov model for dynamic malware analysis: first impressions. In: 2015 12th International Conference on Fuzzy Systems and Knowledge Discovery (FSKD), pp. 816–821 (2015)
4. Pranamulia, R., Asnar, Y.D., Perdana, R.S.: Profile hidden Markov model for malware classification: usage of system call sequence for malware classification. In: International Conference on Data and Software Engineering (ICoDSE), vol. 2017, pp. 1–5 (2017)

5. Cordonsky, I., Rosenberg, I., Sicard, G., David, E.: DeepOrigin: end-to-end deep learning for detection of new malware families. In: International Joint Conference on Neural Networks (IJCNN), vol. 2018, pp. 1–7 (2018)
6. Kim, J., Bu, S., Cho, S.: Zero-day malware detection using transferred generative adversarial networks based on deep autoencoders. Inf. Sci. **460**, 460–461 (2018)
7. Kancherla, K., Mukkamala, S.: Image visualization based malware detection. In: 2013 IEEE Symposium on Computational Intelligence in Cyber Security (CICS), pp. 40–44 (2013)
8. Zolotukhin, M., Hämäläinen, T.: Detection of zero-day malware based on the analysis of opcode sequences. In: 2014 IEEE 11th Consumer Communications and Networking Conference (CCNC), pp. 386–391 (2014)
9. Manavi, F., Hamzeh, A.: A new method for malware detection using opcode visualization. In: Artificial Intelligence and Signal Processing Conference (AISP), vol. 2017, pp. 96–102 (2017)
10. Yewale, A., Singh, M.: Malware detection based on opcode frequency. In: International Conference on Advanced Communication Control and Computing Technologies (ICACCCT), vol. 2016, pp. 646–649 (2016)
11. Masabo, E., Kaawaase, K.S., Sansa-Otim, J., Ngubiri, J., Hanyurwimfura, D.: Improvement of malware classification using hybrid feature engineering. SN Comput. Sci. **1**, 17:1–17:14 (2020)
12. Zhang, Y., Rong, C., Huang, Q., Wu, Y., Yang, Z., Jiang, J.: Based on multi-features and clustering ensemble method for automatic malware categorization. In: IEEE Trustcom/BigDataSE/ICESS, vol. 2017, pp. 73–82 (2017)
13. Zhang, J., Qin, Z., Yin, H.B., Ou, L., Zhang, K.: A feature-hybrid malware variants detection using CNN based opcode embedding and BPNN based API embedding. Comput. Secur. **84**, 376–392 (2019)
14. Duarte-Garcia, H.L., et al.: A semi-supervised learning methodology for malware categorization using weighted word embeddings. In: 2019 IEEE European Symposium on Security and Privacy Workshops, pp. 238–246 (2019)
15. Pascanu, R., Stokes, J.W., Sanossian, H., Marinescu, M., Thomas, A.: Malware classification with recurrent networks. In: 2015 IEEE International Conference on Acoustics, Speech and Signal Processing (ICASSP), pp. 1916–1920 (2015)
16. Athiwaratkun, B., Stokes, J.W.: Malware classification with LSTM and GRU language models and a character-level CNN. In: 2017 IEEE International Conference on Acoustics, Speech and Signal Processing (ICASSP) (2017)
17. Elhadi, A.A., Maarof, M.A., Barry, B.I., Hentabli, H.: Enhancing the detection of metamorphic malware using call graphs. Comput. Secur. **46**, 62–78 (2014)
18. Ki, Y., Kim, E., Kim, H.K.: A novel approach to detect malware based on API call sequence analysis. Int. J. Distrib. Sens. Networks **11**, 659101 (2015)
19. The cost of cybercrime. (2019). https://www.accenture.com/_acnmedia/PDF-96/Accenture-2019-Cost-of-Cybercrime-Study-Final.pdf#zoom=50
20. Sebastián, S., Caballero, J.: AVclass2: massive malware tag extraction from AV labels. In: Annual Computer Security Applications Conference (2020)
21. Wei, J., Zou, K.: EDA: easy data augmentation techniques for boosting performance on text classification tasks. ArXiv, abs/1901.11196 (2019)
22. Yuan, L., Wang, Y., Thompson, P., Narayan, V., Ye, J.: Multi-source learning for joint analysis of incomplete multi-modality neuroimaging data. In: International Conference on Knowledge Discovery & Data Mining, pp. 1149–1157 (2012)
23. Troyanskaya, O., et al.: Missing value estimation methods for DNA microarrays. Bioinformatics **17**(6), 520–525 (2001)

24. Rabadi, D., Teo, S.: Advanced windows methods on malware detection and classification. In: Annual Computer Security Applications Conference (2020)
25. Jindal, C., Salls, C., Aghakhani, H., Long, K., Kruegel, C., Vigna, G.: Neurlux: dynamic malware analysis without feature engineering. In: Proceedings of the 35th Annual Computer Security Applications Conference (2019)
26. Subedi, K.P., Budhathoki, D.R., Dasgupta, D.: Forensic analysis of ransomware families using static and dynamic analysis. In: IEEE Security and Privacy Workshops (SPW), vol. 2018, pp. 180–185 (2018)
27. Aghakhani, H., et al.: When malware is packin' heat. limits of machine learning classifiers based on static analysis features. In: NDSS (2020)
28. Kumar, N., Mukhopadhyay, S., Gupta, M., Handa, A., Shukla, S.: Malware classification using early stage behavioral analysis. In: 2019 14th Asia Joint Conference on Information Security (AsiaJCIS), pp. 16–23
29. Kang, B., Kim, T., Kwon, H., Choi, Y., Im, E.: Malware classification method via binary content comparison. In: RACS (2012)
30. Shalaginov, A., Banin, S., Dehghantanha, A., Franke, K.: Machine learning aided static malware analysis: a survey and tutorial. ArXiv, abs/1808.01201 (2018)
31. Egele, M., Scholte, T., Kirda, E., Krügel, C.: A survey on automated dynamic malware-analysis techniques and tools. ACM Comput. Surv. **44**, 6:1–6:42 (2008)
32. Or-Meir, O., Nissim, N., Elovici, Y., Rokach, L.: Dynamic malware analysis in the modern era—A state of the art survey. ACM Comput. Surv. (CSUR) **52**, 1–48 (2019)
33. Sihwail, R., Omar, K., Ariffin, K.A.: A survey on malware analysis techniques: static, dynamic, p. 8. hybrid and memory analysis, Int. J. Adv. Sci. Eng. Inf. Technol. **8**(4-2), 1662–1671 (2018)
34. Gandotra, E., Bansal, D., Sofat, S.: Malware analysis and classification: a survey. J. Inf. Secur. **5**, 56–64 (2014)
35. Shijo, P.V., Salim, A.: Integrated static and dynamic analysis for malware detection. Procedia Comput. Sci. **46**, 804–811 (2015)
36. Islam, M., Tian, R., Batten, L., Versteeg, S.: Classification of malware based on integrated static and dynamic features. J. Network Comput. Appl. **36**(2), 646–656 (2013)
37. Vinayakumar, R., Alazab, M., Soman, K.P., Poornachandran, P., Venkatraman, S.: Robust intelligent malware detection using deep learning. IEEE Access **7**, 46717–46738 (2019)
38. Venkatraman, S., Alazab, M., Vinayakumar, R.: A hybrid deep learning image-based analysis for effective malware detection. J. Inf. Secur. Appl. **47**, 377–389 (2019)

TapTree: Process-Tree Based Host Behavior Modeling and Threat Detection Framework via Sequential Pattern Mining

Mohammad Mamun(✉) and Scott Buffett

National Research Council Canada, Fredericton, NB, Canada
{Mohammad.Mamun,Scott.Buffett}@nrc-cnrc.gc.ca

Abstract. Host behaviour modelling is widely deployed in today's corporate environments to aid in the detection and analysis of cyber attacks. Audit logs containing system-level events are frequently used for behavior modeling as they can provide detailed insight into cyber-threat occurrences. However, mapping low-level system events in audit logs to high-level behaviors has been a major challenge in identifying host contextual behavior for the purpose of detecting potential cyber threats. Relying on domain expert knowledge may limit its practical implementation. This paper presents TapTree, an automated process-tree based technique to extract host behavior by compiling system events' semantic information. After extracting behaviors as system generated process trees, TapTree integrates event semantics as a representation of behaviors. To further reduce pattern matching workloads for the analyst, TapTree aggregates semantically equivalent patterns and optimizes representative behaviors. In our evaluation against a recent benchmark audit log dataset (DARPA OpTC), TapTree employs tree pattern queries and sequential pattern mining techniques to deduce the semantics of connected system events, achieving high accuracy for behavior abstraction and then Advanced Persistent Threat (APT) attack detection. Moreover, we illustrate how to update the baseline model gradually online, allowing it to adapt to new log patterns over time.

Keywords: Process tree · Behavioral anomaly detection · Sequential pattern mining · APT detection

1 Introduction

Since modern information systems have become critical and essential components of contemporary businesses and organisations, insider threat detection is becoming a rapidly growing topic of study in the cybersecurity domain. An emerging cyberattack, known as APT, poses a huge threat to these information systems, first by breaching hosts inside a target system and then stealthily infiltrating additional hosts through the internal network to steal sensitive information. Since attackers often sabotage legitimate services executing on endpoints,

C. Alcaraz et al. (Eds.): ICICS 2022, LNCS 13407, pp. 546–565, 2022.
https://doi.org/10.1007/978-3-031-15777-6_30

it is critical to detect malicious behaviour on endpoint computers promptly and efficiently following a breach, prior to major harm being caused. Several recent studies demonstrate that malicious behaviour can be detected by leveraging patterns of benign behavior against other, seemingly benign actions that, when combined, signal something potentially more destructive [1–4].

Unfortunately, the volume of log events produced by a typical host is huge. For instance, a single desktop computer, let alone servers in large enterprise network, can generate over a million events each day [5]. Processing massive amounts of audit log events and filtering out irrelevant system events in order to recognize representative host behavior requires a tedious manual effort [6]. Existing solutions to this problem include techniques such as tag propagation [7] and graph matching [1,8,9], that mostly rely on domain expert knowledge or on a knowledge store of expert-defined rules [10]. To address this issue, our objective is to develop an efficient method for extracting representative behavior (i.e. [10,11]) for cyber analyst investigation. More precisely, we automate the extraction of host behavior using procedural task analysis on system log events and then aggregate semantically related tasks to construct baseline host behavior. Due to the fact that recurring or similar tasks have been aggregated together, TapTree can significantly reduce the number of events to analyze.

Existing anomaly detection approaches convert user operations into sequences to analyze sequential relationship between log entries, and then employ sequence processing techniques, such as deep learning [2,3,12–14], natural language processing [15], to learn from previous events and predict the next event. These methods at the log-entry level model user behaviour and indicate discrepancies as anomalies. However, this approach is oblivious to other relationships. For example, a user's daily activity is *relatively* regular over time in terms of the logical relationship among periods [1]. Moreover, event logs may also be generated concurrently by many threads, aliases, or tasks [2]. If this relationship in the log is disregarded, prediction methods based on continuous logs may suffer a loss of reliability.

We construct a baseline behavior model and assess its ability to correctly detect malicious behaviour on a recently released APT attack dataset (DARPA OpTC public dataset [16]). Evaluation results on 14 randomly selected hosts from the OpTC dataset show that TapTree recognizes targeted host behavior with accuracy over 99% with false positive rate (FPR) of less than 0.8% when using tree pattern queries, for a given candidate partial match threshold. We show that this threshold can then be adjusted to cast a wider net and achieve a perfect 100% recall on malicious behaviour detection, while still keeping the FPR relatively low at 2.9%. For the sequential pattern-based analysis, we show that FPR can be further reduced to below 0.1%, while maintaining high accuracy (>99.9%) and recall (67%), whereas recall can be improved to 86% while still maintaining FPR <1%. Moreover, we quantify the proportion of process trees that are reduced in number as a result of similar pattern aggregation. Our results demonstrate that TapTree can reduce the number of process-trees from raw audit logs by 98% percent after aggregation and thus substantially reduce the analysis overhead associated with abnormal behavior investigation. Our major contributions are summarized:

- We present TapTree, a process tree-based host behavior modeling. TapTree automatically encapsulates host contextual behaviors from raw audit log events using system generated process-tree. To our knowledge, this is the first approach to host behavioral abstraction that utilizes system process trees to aggregate semantically equivalent patterns.
- To reduce analysis overhead for the analyst and enable efficient detection, TapTree considers noise reduction, optimized tree growth such as forward pruning and aggregation of similar behaviors.
- As part of validation of proposed behavioral abstraction model, we conduct a systematic evaluation by abstracting benign behavior in a given context e.g. APT attack against Darpa-OpTC dataset. We propose the use of sub-tree pattern and sequential pattern queries to detect discriminatory behaviour and identify insider threats automatically. Experimental results using Darpa-OpTC data demonstrate that TapTree's baseline behavior is effective against both benign and malicious behaviors. In terms of mining speed, TapTree outperforms baseline generation model without aggregation by two orders of magnitude.

1.1 Analyzing the Problem

Two characteristics must be satisfied for behavioral model to be deemed effective: (i) behavioral distinctiveness to accurately represent the host behavior and (ii) behavioral consistency to identify deviant behaviors [17]. We study ways to satisfy these requirements while decreasing or eliminating the bulk of false positive occurrences.

To address this issue, our approach is to construct heterogeneous temporal process trees from homogeneous system events representing the target behavior and then use the forest (group of trees) to build the host model. Note that the number of system events might be enormous, highly interconnected, and noisy. For instance, the installation of a single package during an APT campaign may create over 50 thousand system log events [10]. In addition, the number of log entries containing information about suspicious/malicious activity is quite small. Rather than using all trees, we select the most discriminating patterns in the forest by grouping similar/redundant actions to aid analysts' analysis (e.g. if a behavior is a subset of another, it gets merged). Such representative behavior-specific forest (with modest number of trees) is simpler to interpret, faster to match, and easier to maintain.

A representative pattern should be frequent in the intended representative behavior and rare in deviant behavior. For example, a system administrator often logs into servers and does certain tasks that a human resources professional performs seldom. A sample benign/malicious tree pattern from Darpa-OpTC is presented in Fig. 2 containing the number of edges occurrences in the pattern.

Detection algorithms are primarily based on two pattern matching techniques: 1) a tree search method for pattern matching (Sect. 3.3) and 2) a sequential pattern mining classification algorithm (Sect. 3.3) on the sequence generated from the Temporal tree set as discussed in Sect. 3.2.

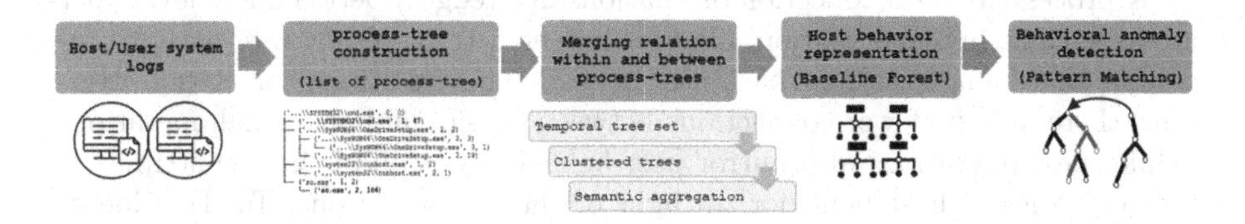

Fig. 1. TapTree pipeline for behavioral anomaly detection

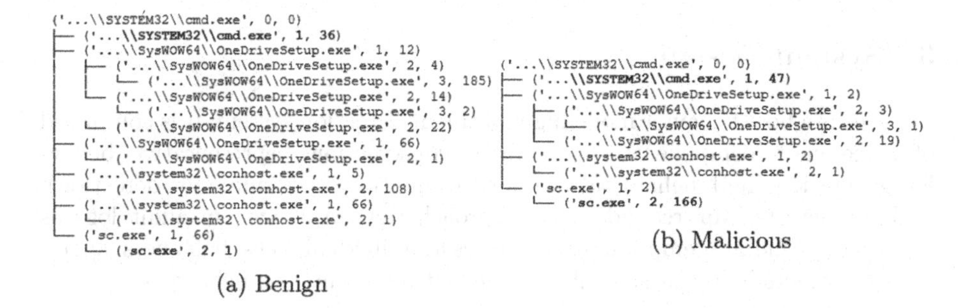

Fig. 2. Sample task trees (process name, depth, occurrence-edge)

2 Overview

2.1 Assumption

TapTree is designed to work with complex event data structures that are both hierarchical and sequential in nature (filiation-relationship of events). We presume that behaviors are audited at the kernel level and their activities are logged in system-call audit logs. The integrity and security of the underlying audit log monitoring platform (SIEM security) are beyond the scope of this study and are thus considered to be a component of the trusted computing base.

Rather than focusing on a single user's sessions [10] or a single user's one day data [18], we target multiple users' whole dataset (7 days data) for behavior modeling and validation. Results show that our approach is broadly applicable across hosts, days/sessions behavior.

A naive way to obtain the semantic representation of a host behavior is to gather up individual tasks (collection of process-trees) derived from its component events. This approach, however, may overlook the *relative weights of relationships* between events (edges in the tree) and *noisy relations* in the representative behavior.

A process-tree is a collection of relationships (edges) between low-level operations such as process-creation, file-opening, etc., triggered by user activity. We assume that benign tasks (typical user behavior) have a strong correlation. However, there may be fewer connections between benign and abnormal operations. While these operations also mirror user behavior, not all of them contribute to the semantics of host behavior. In light of these observations, TapTree identifies the relationships that are frequent within a task (process-tree) and across tasks/behaviors. This approach provides a higher discriminative weight to the relationships in the process-tree that are less prevalent.

3 System Design

TapTree is a host behavior modeling and threat detection system. It is comprised of three main components: process-tree construction, representative behavior generation, and behavioral stability evaluation (e.g. anomaly detection). Figure 1 depicts TapTree's detailed approach that takes system audit logs as input data, generate temporal process-trees as individual tasks/behaviors, aggregates/abstracts behavior semantics to output representative behaviors.

3.1 Process-Tree Construction

Hierarchical structure of process tree derived from an audit log reflects causal relationships between running processes of a computer system. Besides providing a holistic view of the system process life cycle, this property of the system process tree offers valuable contextual information about an event's proximity continually evolving over time [19]. The first component of TapTree is a process-chain based heuristic technique for mapping relationships between log entries that reflect hosts' behavior across several streams, such as file operation, authentication, flow, etc., into a task-process-tree.

TapTree primarily considers three types of relationships for generating process-trees: 1) the filiation relationship that forms a hierarchy across all running operating system processes, 2) sequential relationship between traces and 3) the logical relationship among tasks.

A task-process-tree (see Fig. 2) is a temporal semantic behavior tree represented by a tuple $T := (V, E, R)$:

- V is a set of nodes where $v \in V$ represents a path to the program (e.g. `\\System32\\conhost.exe`) that initiates an event (e.g. `Process-creation`)
- $E \subset V \times V \times R$ is a set of directed edges where $e = (u, v, r) \in E$ denotes a chronologically ordered relationship between executing programs.

- R is a set of possible occurrences between nodes V, where $r \in R$ is a positive integer. Therefore, each $e \in E$ is assigned a weight $w(u, v) : \mathbb{R}^+$ that implies the frequency/occurrence of the two program $(u \rightarrow v)$ invoking each other.

The process-tree data associated with a target behavior/task is used to construct a program-path tree, as these behaviors are typically executed by a single thread. This work focuses on the program-paths since our empirical findings show that using program-path instead of raw events is quite effective at abstracting host behaviors. Additionally, it offers significant computational benefits over more complex provenance/knowledge graph models.

3.2 Fusion of Host Behaviors

A behavior instance, such as a process-tree in our case, consists of a series of events connected semantically. Program-path identifies the path to the program initiated by an event. A fine-grained associations between these behavior instances can provide high-level abstraction for generating effective behavioral model.

TapTree consolidates behavior instances using two widely used approaches for behavioral abstraction— path-based approach [2,3,20] by splitting up process-trees into paths and contextual-representation based approach [10] by extracting sub-tree as an instance of a behavior. The following sections cover TapTree's approach to behavior consolidation.

Temporal Tree Set Generation. A temporal tree set is a collection of unique task process-trees where the trees with the same number of elements and relations/edges are merged. Because the trees are weighted, the weight of an edge between two or more similar trees is equal to the sum of the weights of its edges. As the trees are weighted, the edge's weight is equal to the maximum of its edges' weights.

Formally, a temporal tree set $F \subseteq \mathbf{T}$ of task trees is a set of $(n \geq 0)$ disjoint weighted directed trees such that,

- For all $P = (V_P, E_P, R_P), Q = (V_Q, E_Q, R_Q) \in F$, $V_P \neq V_Q$, $E_P \neq E_Q$, $R_P \neq R_Q$,
- For all $r \in R$ where $V_P = V_Q$ and $E_P = E_Q$, $p \in P$ and $q \in Q$ $r = Max(\mathbf{w}_p(e_p), w_q(e_q))$.

Clustering Trees. This method consolidates the relationships within a tree in order to avoid repeating patterns. Duplicate relations/edges at the leaf level of the tree are merged in this stage.

A clustering of the leaves of the tree T can be defined by cutting a subset of edges $C \subseteq E$. One method for achieving this is to solve the *max-diameter min-cut partitioning problem* [21]. We define a partition level $\{L_1, L_2, \ldots, L_N\}$ of L to be an admissible clustering if it can be obtained by removing some edge set C from E and assigning leaves of each of the resulting connected components to a set L_i(where $N \leq |C|+1$).

Let $T = (V, E, R)$ be a directed tree containing two edges $e_1 = (u_1, v_1, r_1)$ and $e_2 = (u_2, v_2, r_2)$ with $u_1 \neq v_1, u_2 \neq v_2, u_1 = u_2, v_1 = v_2$ where $\{v_1, v_2\}$ are leaf nodes. Merging e_1 and e_2 results a new tree $T' = (V', E', R')$, where $V' = (V \backslash \{u_2, v_2\})$, $E' = (E \backslash \{e_2\})$, $r' \in R' = r$, or $Max(\mathbf{w}(e_1), w(e_2))$ if $r_1 \neq r_2$.

Table 1. TapTree baseline model generation and matching efficiency

Method	#Trees	Baseline generation (in s)	Pattern matching (in ms)
Temporal tree set	3501	135.51	–
Clustered trees	2372	163.25	65.3992
Semantic aggregation	36	4301.83	47.0608

Semantic Aggregation. After redundant instance aggregation, we deduce the semantics of a behavior instances naturally by combining trees derived from clustered trees. Identifying a pattern, whether it is a new one to aggregate or a previously discovered one, can help in averting instances of repetitive behavior.

Recall, a naive way to obtaining the semantic representation of a behaviour instance is to consider all the trees derived from the events. However, this approach may work only if the baseline semantics of behavior (temporal tree set) is decently small or it does not need updating over time. In practice, this technique is not efficient from the view point of detection (matching) for a large enterprise system where thousands of flow of events need to be examined in a certain period. Additionally, this assumption is frequently incorrect due to the way tree relations are weighted differently to represent the semantics of behaviour and the effect of noisy events.

Induced Subtree: Given a tree pair (T_1, T_2) where $T_1 := (V_1, E_1, R_1), T_2 := (V_2, E_2, R_2)$, we say T_2 is an induced subtree of T_1 denoted by $T_2 \preceq T_1$, if and only if,

1. $V_2 \subseteq V_1$ and $E_2 \subseteq E_1$,
2. Filiation relationships in T_2 must be preserved in T_1. That is, parent-child relations for all $e = (u, v) \in T_2$ is identical to that of T_1,
3. The left-to-right ordering of siblings in T_2 must be a subordering of the associated nodes in T_1.

Growing Baseline Trees: Using consecutive growth options (forward, backward, and inward) as described in [8] for searching a given behavior pattern against baseline patterns ensures a complete and non-repetitive search in the pattern space. In this manner, behavior pattern trees are iteratively merged if they are not *induced subtrees* of the baseline trees in order to construct a baseline behavior model.

Let $T_a = (V_a, E_a, R_a)$ and $T_b = (V_b, E_b, R_b)$ be two directed trees. A merging of two trees $(T_a \cup T_b)$ results a new tree $T_{ab} = (V_{ab}, E_{ab})$ such that $E_{ab} = E_a \bigcup E_b$ and $V_{ab} = V_a \bigcup V_b$ that satisfies $V_{a1} \in V_b$ or $V_{b1} \in V_a$ where V_{a1} and V_{b1} are the roots of T_a and T_b respectively.

Table 1 outlines a comparative analysis of the aforementioned methods for behavior consolidation in relation to baseline construction. We present the volume of behavior patterns and the execution time (in seconds) required to generate the pattern in each phase of the baseline generation model in the number of trees and baseline generation column of the table.

3.3 Behavioral Anomaly Detection

Behavioral anomaly detection (BAD) is expected to effectively resolve a variety of security issues by detecting deviations from a host's normal behavioral patterns. BAD enables the monitoring of applications for malicious behavior (e.g. intrusion, compromise detection), thereby improving protection against *Zero Day* attacks. Host behavior abstraction model discussed in the previous section can be used for behavioral anomaly detection such as APT attacks. Given the behavior representation for any host or server we can utilize (1) unsupervised model/ one-class classifier (tree pattern matching) or (2) supervised model/ binary classifier (sequential pattern matching) to identify evidence of anomalous behavioral events.

The tree pattern matching algorithm compares a sequence of operations to a baseline model in order to determine whether a task is abnormal. The tree search method allows for a trade-off between recall and the false positive rate in detection. Note that an exhaustive search of the tree will always return the prototype that is closest to the input vector. However, alternative search methods can be used to determine a task that is a close match to the baseline but not necessarily an exact match.

Sequential pattern-based analysis works specifically on a set of event *traces* (i.e. sequences) extracted from the task trees, and identifies common temporal patterns that reside in those sequences. This can be used to establish a model of baseline activity, against which new activity can be measured to determine the likelihood that the new activity appears as expected and is not malicious, or to construct a classification model on labeled data in the case that sufficient samples of malicious activity can be obtained.

In the following, we discuss tree pattern queries and sequential pattern queries in detail that were used to evaluate TapTree's efficacy in identifying behavioral abnormalities.

Tree Pattern Queries. We conduct a systematic study of tree matching algorithms that determine the likelihood of a pattern occurring by performing a recursive comparison on each node of the tree. When a mismatch is detected, the comparison procedure is terminated.

Typically, queries on trees are executed using one of two classic graph traversal strategies: breadth-first search (BFS) or depth-first search (DFS). We use a modified DFS graph-querying algorithm for tree pattern queries. DFS can expand one intermediate result at a time, starting from the first variable in the pattern and continuing to the next ones until the whole pattern is matched. DFS can expand a single intermediate result at a time, beginning with the first variable in the pattern and progressing through the remaining variables until the entire pattern is matched.

Let (T_i, P) be the baseline trees (target host behavior model) T and a pattern tree P pair where children of all nodes are labelled and ordered. P matches at node t if there is a $1-1$ mapping from nodes of P to T such that: 1) root of P, $R_P \leftrightarrow t$ and 2) if $\exists(i \in P \leftrightarrow j \in T)$, all the children follows. Let λ_v be the path from R_P to v. v matches T at node $u \in T$ if λ_v matches T at u.

Exact Match: In this method, the pattern tree P must be matched *exactly* with any of the trees in the baseline patterns with respect to node label, inheritance, and order relationship. An exact match of a pattern P into a baseline tree T is a mapping $\mathcal{F}_{exact} : P \rightarrow T$ for each nodes of P that satisfies:

- For each $u \in P$, $label(u) = label(\mathcal{F}(u))$
- If $\exists u_i \rightarrow u_j \in P$ then $\mathcal{F}(u_i)$ is a parent of $\mathcal{F}(u_j) \in T$. If $u_i \Rightarrow u_j \in P$, $\mathcal{F}(u_j)$ is a descendant of $\mathcal{F}(u_i) \in T$
- For any edge $e : u_i \Rightarrow u_j \in P$ where $label(u_i) = label(\mathcal{F}(u_i))$ and $label(u_j) = label(\mathcal{F}(u_j))$, $e(c) \leq \mathcal{F}(e(c))$ where c is the frequency count of the relation such as $u_i \Rightarrow u_j$
- For any $(u_i, u_j) \in P$ if u_i is to the right of u_j, $\mathcal{F}(u_i)$ is to the right of $\mathcal{F}(u_j)$.

Partial Match: In this method, the pattern tree P must be matched partially with any of the trees in the baseline model such that the root element and all elements connected directly and indirectly to the root are matched with respect to node label, inheritance and order relationship to the baseline tree. A partial match pattern P into a baseline tree T is a mapping $\mathcal{F}_{partial} : P \rightarrow T$ that returns R that satisfies:

- Let $\exists R$
- If $\exists u_i \rightarrow u_j \in P$ then $\mathcal{F}(u_i)$ is a parent of $\mathcal{F}(u_j) \in T$. If $u_i \Rightarrow u_j \in P$, $\mathcal{F}(u_j)$ is a descendant of $\mathcal{F}(u_i) \in T$
- $\exists$ edge $e : u_i \Rightarrow u_j \in P$ where $label(u_i) = label(\mathcal{F}(u_i))$ and $label(u_j) = label(\mathcal{F}(u_j))$, $e(c) \leq \mathcal{F}(e(c))$ where c is the frequency count of the relation such as $u_i \Rightarrow u_j$ implies $e \in R$
- $R \subseteq P$

Scoring Matched Patterns: While *exact matches* do not require a threshold for detection, *partial matches* require the computation of a score in order to determine if they are anomalous. Following pattern matching, we establish a threshold for detecting malicious task trees. That is, pattern task-process-trees with a score greater than the threshold are deemed abnormal.

The percentage of items that match is used to calculate the score for a partial match. We consider the same weight or variable weight based on the depth of the element in the tree for the scoring calculation. Our intuition here is to prioritize the matches that are deeper in the tree. For a given pattern T, let R be the partial match tree for the pattern, T be the baseline tree, ω be the weight and δ represent the threshold. Partial match for the same weight is determined by:

- $k = \sum_{i=1}^{|R|} \omega_i (= 1)$ and $l = \sum_{i=1}^{|T|} \omega_i (= 1)$
- $x = k/l$
- If $x \geq \delta$ then *Match* Else *Not Match*

A partial match with variable weight calculates the pattern matching score based on the item's depth in a tree. Partial match for the variable weight is determined by:

- $k = \sum_{i=1}^{|R|} \omega_i (= depth(R_i))$ and $l = \sum_{i=1}^{|T|} \omega_i (= depth(R_i)$
- $x = k/l$
- If $x \geq \delta$ then *Match* Else *Not Match.*

Sequential Pattern Analysis. We propose the use of sequential pattern analysis on the set of task trees as a means for exploiting the temporal nature of the data. Specifically, a set of *traces* of activity is extracted from each task tree, where each trace is a sequence of actions performed as part of that task. Given a set of such traces, sequential pattern analysis, using such techniques as sequential pattern mining and classification, can be conducted to identify common patterns of interest, which can then be used to help determine the likelihood of a task containing malicious activity.

Sequential pattern mining (SPM) [22,23] is a collection of techniques that focus on the identification of frequently occurring patterns of items (i.e., objects, events, etc.), where ordering of these items is preserved. Let I be a set of *items*, and S be a set of *input sequences*, where each $s \in S$ consists of an ordered list of *itemsets*, or sets of items from I, also known as *transactions*. A sequence $\langle a_1 a_2 \ldots a_n \rangle$ is said to be *contained* in another sequence $\langle b_1 b_2 \ldots b_m \rangle$ if there exist integers $i_1, i_2, \ldots, i_n$ with $i_1 < i_2 < \ldots < i_n$ such that $a_1 \subseteq b_{i_1}, a_2 \subseteq b_{i_2}, \ldots, a_n \subseteq b_{i_n}$. A sequence $s \in S$ *supports* a sequence s' if s' is contained in s. The support $sup(s')$ for a sequence s' given a set S of input sequences is the percentage of sequences in S that support s', and is equal to $sup(s') = |\{s \in S | s \text{ supports } s'\}| / |S|$. A sequence s' is deemed a *sequential pattern* if $sup(s')$ is greater than some pre-specified minimum support. Such a pattern with a total cardinality of its itemsets summing to n is referred to as an *n-sequence* or *n-pattern*. A sequential pattern s' is a *maximal sequential pattern* in a set S' of sequential patterns if $\forall s'' \in S'$ where $s'' \neq s'$, s'' does not contain s'. The general goal of sequential pattern mining is then to identify the set S'

that contains all (and only those) sequences that are deemed sequential patterns according to the above. In some cases, the set consisting of only maximal sequential patterns is preferred.

Given a supervised learning model in which input sequences are assigned and labeled according to two or more classes, sequence classification [24–26] techniques can be used to attempt to classify new sequences, by using frequent sequential patterns as features in the classification. In addition to the above SPM model, consider the addition of a set of class labels and a labeling function $\ell : S \rightarrow L$ that labels the input set. S is thus a set of *examples*, where each example $s \in S$ can be represented by a set of features from the set SP' of frequent sequential patterns. Selected features should exhibit each of the following properties:

- High frequency
- Significantly higher representation in one class than the other(s)
- No redundancy.

Given these identified features, standard machine learning based classification methods such as SVM or Naïve Bayes can be used to build a classification model and label new instances accordingly by considering the feature patterns that they do and do not contain.

Sequential pattern analysis can be used either 1) to obtain a baseline model for normal activity, against which new activity can be measured in order to identify potential anomalies, or 2) to train a supervised classifier on labeled data containing both baseline and malicious activity in order to classify new instances as benign or malicious. In the former case, sequential pattern mining can be used to identify frequent patterns that occur typically in the baseline activity, which can then be used to measure common volumes of noise, i.e. activity that does not conform to identified regular patterns, within the baseline data in order to ascertain a tolerable level. Noise from new activity can then be measured against these tolerance levels, and any excessively noisy activity can then be identified as potentially anomalous. For the latter, a classifier can be trained to detect malicious traces, as outlined above, if the required labeled data exists.

4 Experiment and Evaluation

Our experiment is conducted on a workstation equipped with an Intel Core i7-10750H processor running at 2.6 GHz, 128 GB of RAM, and six cores. For data extraction, process tree construction, training, and tuning model we use Spark 3.0.2 and Python libraries.

4.1 Experiment Dataset

We choose OpTC dataset as it enables significant study in the area of process trees. Process trees encapsulate sequential data in log events that are semantically related but chronologically distorted. Enterprise operating systems are, by definition, process-oriented. Each process can be traced back to the process that launched it; each state change can be traced back to the process that caused it to occur.

We evaluate TapTree mainly on OpTC dataset: a benign dataset from 5 hosts, a malicious dataset from 16 hosts. To perform in-depth analysis and mining of log entries within a day/host, we separate each log entry into seven key characteristics (object, action, processID, ParentProcessID, image path, time, and host) and constructed process-trees from the raw events for the benign and malicious hosts. This way, we were able to generate 91,242 process trees from the benign dataset, but only 39 from the malicious dataset.

We consider user-specific artifacts because the same user may approach a task differently each time it is completed, or because various users may offer similar actions for a given task. Therefore, for benign activity dataset, we target all activities conducted by five targeted hosts over a seven-day period.

Malicious activities in Darpa-OPTC dataset has been generated mainly from three APT activities on windows systems: Remote Code Execution and Shell Code Injection aka Beacon (Cobalt Strike), Remote Code Execution and Lateral Movement aka Powershell Empire, Credential Harvesting aka Customized Mimikatz. APT activities come with an extremely small percentage of any dataset (less than 0.001% in Darpa-OpTC dataset). For example, Darpa-OpTC's red team targets just 27 hosts out of 1000 networked hosts for launching a series of APT attacks while engaging in benign activities. We capture malicious events from all the hosts over

Prior to matching, the evaluation dataset is filtered to remove trees that do not meet specified thresholds. For example, the smallest threshold tested used a minimum of 3 nodes and a minimum depth of 2, resulting in 44608 benign trees and 27 malicious trees, whereas the strictest threshold tested used a minimum of 5 nodes and a minimum depth of 3, resulting in 13,146 benign trees and 16 malicious trees.

4.2 Evaluation

We evaluate the effectiveness of TapTree for behavior abstraction from three aspects. (1) Comparative study on the effectiveness of tree and sequence based data mining for user behavior footprint (2) How does query accuracy vary when pattern size in queries changes? (3) How does the amount of training data affect query accuracy? (4) The performance penalty associated with various threshold and pattern queries. Evaluations are conducted on each of the tree pattern queries and sequential pattern analysis approaches using the experiment dataset from DARPA OpTC dataset.

Tests are carried out at several thresholds such as the percentage of partial match, the minimum number of nodes and depth (tree pattern queries approach), and minimum likelihood of trace maliciousness needed for classification (sequential pattern analysis approach). Performance is measured using k-fold ($k = 10$) cross-validation for the larger set of benign instances, and leave-one-out cross-validation for the smaller set containing malicious instances.

To analyze TapTree's accuracy in abstracting host behavior and compare the efficacy of behavior models (see Sect. 3.2), we use five users' data (chosen at random from 1000 windows hosts) to construct a baseline host model, and any user except those five for pattern matching. Table 1 shows the average run-time required to match a tree. While the time required to generate baseline trees from semantic aggregation is longer than that required to generate clustered trees or a Temporal Tree set (4301.83 s versus 163.25 and 135.51 s, respectively), the result is a significantly smaller number of behavioral trees for matching (36 trees compared to 3501 trees). This reduction in the tree size helps improve tree matching time. That is, compared to Clustered Tree, the average time required to match a new behaviour tree with semantic aggregation can be lowered by 39%. Although there is a cost associated with the baseline generation time associated with semantic aggregation of trees, the baseline tree generation process only needs to be performed once, whereas matching trees is a recurrent activity.

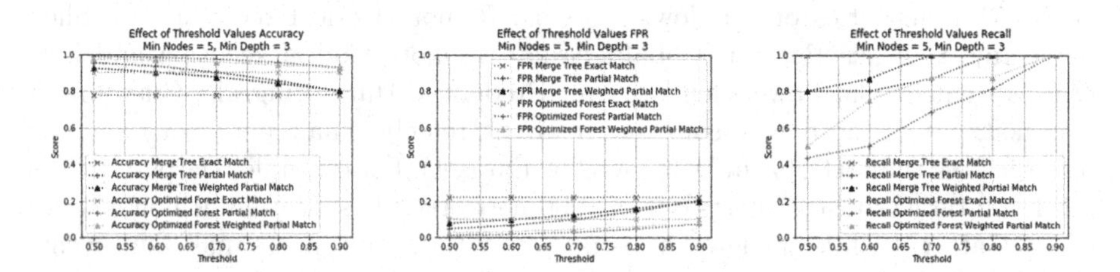

Fig. 3. Performance of TapTree on different baseline method: Semantic aggregation vs clustered Trees.

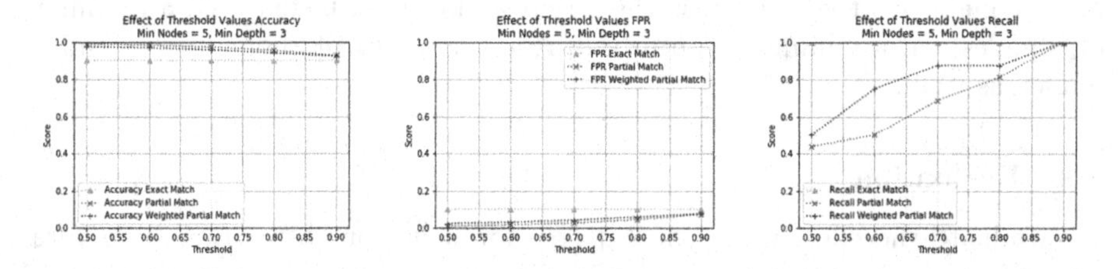

Fig. 4. Performance of TapTree on different threshold (baseline method: Semantic aggregation)

Tree Pattern Matching Methods: TapTree methods were found to perform extremely well when compared with existing methods, particularly when leveraging various threshold scores for partial matches as well as conducting proper tree filtering based on minimum node and depth values. Table 2a and 2b depict results of TapTree methods (Semantic aggregation, Clustered Tress and Sequence

Table 2. TapTree performance compared to other methods on OpTC dataset [16]. TapTree data is in the *tree* format (#Minimum depth = 2, #Minimum nodes = 3)

Method	Accuracy	FP Rate
TapTree (Semantic aggregation)	0.9913	0.008
TapTree (clustered trees)	0.9739	0.026
TapTree (sequence mining)	0.9901	0.001
DeepTaskAPT(Trace) [2]	0.9854	0.011
DeepTaskAPT(Task) [2]	0..9641	0.006
DeepLog [3]	0.8354	0.161
Random Forest	0.9052	0.083

(a) partial match threshold = 0.5

Method	Recall
TapTree (Semantic aggregation)	1.0 (FPR $\approx$ 0.029)
TapTree (clustered trees)	1.0 (FPR $\approx$ 0.067)
TapTree (sequence mining)	0.8571
DeepTaskAPT (Trace) [2]	0.7587
DeepTaskAPT (Task) [2]	0.8299
DeepLog [3]	0.7202
Random Forest	0.6784

(b) partial match threshold = 0.9

Mining) against existing approaches (DeepTaskAPT (Trace) [2], DeepTaskAPT (Task) [2], DeepLog [3] and Random Forest). When using weighted partial matching filtering based on five *minimum nodes* and three *minimum depth* for trees, Semantic aggregation achieved high accuracy (0.9913) over all existing approaches, while posting a false positive rate just slightly higher than DeepTaskAPT (Task). Increasing the partial match percentage threshold score from 0.5 to 0.9 allows both TapTree methods to achieve a recall score of 1.0, which means that they were able to capture all malicious tasks, while only increasing false positive rates to 0.029 and 0.067, respectively. While the Sequence Mining TapTree method posted accuracy and recall scores slightly lower than semantic aggregation, it produced the lowest false positive rate of all methods, and outperformed all existing methods for each of the accuracy, fp-rate and recall metrics. Performance of sequential pattern-based analysis is examined in detail in the next section.

Table 3 illustrates the performance of tree pattern queries algorithms discussed in Sect. 3.3 based on semantic aggregation baseline model (see Sect. 3). Except in cases when perfect matches are required, matching threshold scores (discussed in section Tree pattern queries) has a significant impact on TapTree's anomaly detection performance. The use of partial matches enables finer

Table 3. TapTree performance with different threshold scores and tree pattern queries algorithms (baseline model: semantic aggregation, Min Node = 5, Min Depth = 3)

Method	Threshold	Accuracy	Recall	FP rate
Exact match	–	0.900547	1.0	0.099574
Partial match (same weight)	0.9	0.926911	1.0	0.073178
	0.7	0.973028	0.6875	0.026624
	0.5	0.989439	0.4375	0.009889
Partial match (variable weight)	0.9	0.923948	1.0	0.076145
	0.7	0.957301	0.875	0.042599
	0.5	0.975232	0.5	0.02419

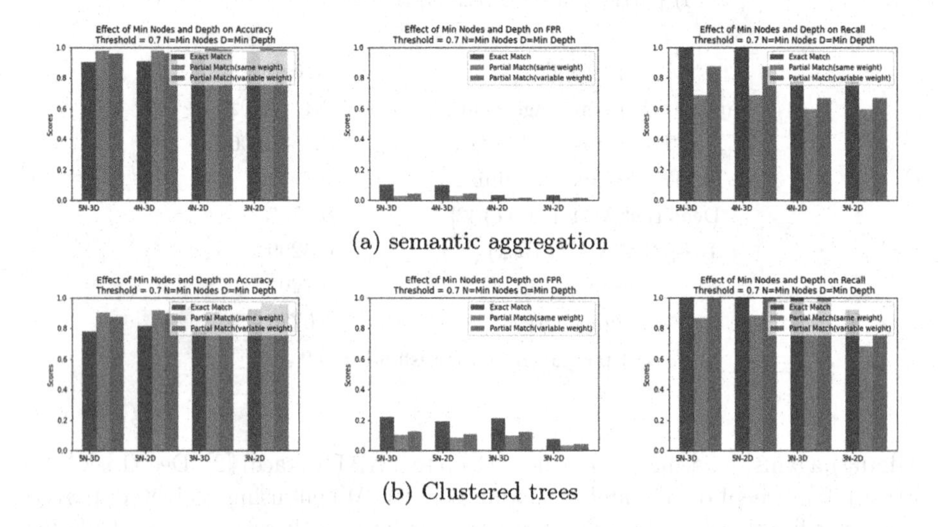

(a) semantic aggregation

(b) Clustered trees

Fig. 5. TapTree performance when jointly tuning minimum number of nodes and tree depth (N-D)

tuning of TapTree's performance to optimize results for recall, accuracy, and false positive rate (FPR). A higher threshold score value ensures improved recall performance, whereas a lower score improves accuracy and FPR.

Figure 3 depicts the comparative performance on semantic aggregation vs clustered tree with respect to accuracy and recall. As the semantic aggregation baseline method achieves the best performance, Fig. 4 describes the impact of partial match thresholds with a given minimum number of nodes and tree depth. In Fig. 5, we jointly tune *minimum number of nodes* (N) and *tree depth* (D). TapTree clearly achieves the best performance when N-D is 3-2 (accuracy/FPR) and 5-3 (recall) for a given threshold of 0.7.

Sequential Pattern-Based Methods: Performance was also measured for a sequential pattern-based classifier tasked with discerning malicious activity

Table 4. Results of sequential pattern-based malicious behaviour detection, at various classification thresholds

Threshold	TPR	TNR	Precision	Accuracy	FPR
0.1	1	0	0.00037	0.00037	1
0.2	1	0	0.00037	0.00037	1
0.3	0.92063	0.54612	0.00074	0.54633	0.45388
0.4	0.86243	0.99012	0.03080	0.99021	0.00988
0.5	0.83069	0.99717	0.09653	0.99724	0.00283
0.6	0.67196	0.99936	0.27632	0.99938	0.00064
0.7	0.42857	0.99997	0.81375	0.99990	0.00003
0.8	0.30159	1	1	0.99988	0
0.9	0	1	–	0.99978	0

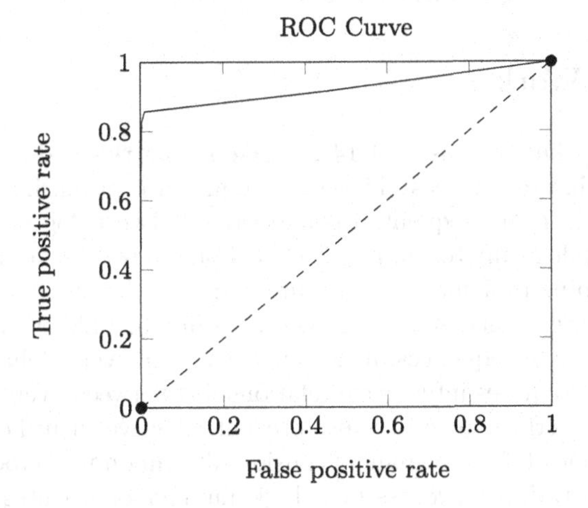

Fig. 6. Receiver operating characteristics (ROC) curve to demonstrate performance of sequential pattern-based classifier at various threshold values

from benign, trained on labeled benign and malicious data extracted from the task trees, using techniques introduced in Sect. 3.3. In all, the training set consisted of 5,15,888 benign traces and 189 malicious cases, with k-fold ($k = 10$) cross-validation employed for the benign cases, and leave-one-out cross-validation employed for the malicious cases. Testing was conducted at various thresholds, where for a threshold of x, a test case was classified as malicious if and only if the classifier's deemed likelihood of maliciousness was greater than or equal to x.

Table 4 shows performance at each threshold, in terms of true positive rate TPR (i.e. recall, the percentage of malicious cases correctly classified as such), true negative rate TNR (i.e. the percentage of benign cases correctly classified as such), precision, accuracy and false positive rate FPR. Figure 6 depicts the receiver operating characteristics (ROC) curve, plotting the true positive rate against the false negative rate.

Results show that the method is highly effective at correctly classifying both benign and malicious traces, as evidenced by the high true positive and negative rates, particularly at the 0.4 and 0.5 thresholds, as well as an ROC curve that lies well to the top and left of the dashed line in the graph representing random guess.

Precision is low at most levels due to the highly unbalanced nature of the data, meaning that even high accuracy levels can result in a large number of false positives. For example, at the 0.5 threshold, benign cases are correctly classified almost 98% of the time, however this amounts to 1460 false positives compared to 156 true positives at that threshold level. However, at higher threshold levels the precision performs remarkably well, reaching 81% at 0.7 with only 16 false positives (FPR = 0.0003) while still identifying 43% of malicious cases, and 100% precision at 0.8 while still identifying 30% of malicious cases.

5 Related Work

Sequence approaches such as [2–4,14,27] take log entries and concatenate them chronologically into sequences. These techniques are primarily concerned with capturing temporal and sequential connections between log entries, and often make use of deep learning techniques such as Long Short-Term Memory (LSTM) or machine learning tool such as signature kernel, to learn from previous events and forecast future events. Although deep learning, like LSTM, may recall long-term dependencies in sequences, it does not compare every behavior of the user explicitly [1], and ignores interactive relationships between events or hosts [1,2]. This can hinder performance possibly prevent effective identification of APTs. Additionally, some of them demand a considerable amount of labeled (malicious) data during the training process or a high number of features for model creation that might not be available in real-world deployment. In [1,2], the authors address some of these issues through alternative methods such as finding logical relationships between user tasks prior to applying deep learning [2], and utilizing a graph that depicts a user's interaction with hosts [1,28].

Meanwhile, a recent approach known as *SK-Tree* [18] uses streaming trees to represent computer processes, and presents a malware detection algorithm leveraging a machine learning tool (signature kernel [29]) for time series data, with promising results. While the SK-Tree study focused on one day of data from a single user (0201), we attempt to expand our reach and leverage more of the dataset to include data captured from multiple users over multiple days.

6 Limitations

This section discusses some of the inherent limitations of the design choice, as well as the ramifications and potential extensions of this work.

TapTree's design relies on process-tree to abstract host activities. Therefore, it may not be effective at detecting attacks that do not result in spawning new processes in the operating system. For example, attacks such as buffer overflows, which do not involve the creation of a new process, are not protected by TapTree. Baseline model may require periodic retraining due to semantic shifts in user/host behaviour and addition of previously unseen behaviour patterns. An analyst can identify new host behaviours over the course of time, sanitise them, and decide carefully whether to include the new behaviours in the baseline model for re-training. Our empirical experiment shows that TapTree can recognise an unknown tree pattern in milliseconds, while retraining with fewer new patterns takes seconds.

The lack of attack dataset on which to train malicious behavior classifiers prompts further investigation into sequential pattern-based methods for developing baseline models. Also worthwhile of further study is a comparison of our methods to relative graph-based schemes where scalability is an issue, as well as investigation into more efficient partial matching, a robust baseline model, and the exploration of new tree matching algorithms.

Despite the algorithms' poor worst-case time complexity for constructing the initial baseline tree, which is quadratic in the number of inputs, experimental evaluations show that they perform exceptionally well in terms of average run time, especially for pattern matching.

7 Conclusion

We present a detailed study on the effectiveness of performing sequential pattern matching for anomaly detection. To facilitate this, we present TapTree, a task-process-tree based model for APT detection on system log data that represents host log data in such a way that facilitates detection of malicious behaviour. Two distinct approaches for this detection are explored. The first attempts to match new data to existing baseline behaviour, represented as *temporal trees*, in an attempt to identify anomalies that could signify attack behaviour. The second extracts sequential behaviour called *traces from the trees* for existing baseline and malicious behaviour samples, and uses sequential pattern mining to identify critical patterns for use in a malicious behaviour classification model.

As for detection performance evaluated using the DARPA OpTC dataset, we demonstrate that one particular TapTree tree matching algorithm, semantic aggregation, achieved high accuracy over all existing approaches, while both semantic aggregation and Clustered Trees were found to achieve a perfect recall by adjusting tree matching thresholds, while still maintaining low false positive rates. The sequential pattern-based TapTree method, on the other hand, posted the lowest false positive rate of all methods, and outperformed all existing methods for each of the accuracy, fp-rate and recall metrics.

Acknowledgement. We would like to thank the Communications Security Establishment Canada team, especially Dr. Benoit Hamelin for supporting the project and providing the materials needed for this work. A special thanks to Kevin Shi from the University of Windsor for all the support during his co-op term with NRC.

References

1. Liu, F., Wen, Y., Zhang D., Jiang, X., Xing, X., Meng, D.: Log2vec: a heterogeneous graph embedding based approach for detecting cyber threats within enterprise. In: Proceedings of the 2019 ACM SIGSAC Conference on Computer and Communications Security, pp. 1777–1794 (2019)
2. Mamun, M., Shi, K.: DeepTaskAPT: insider apt detection using task-tree based deep learning. arXiv preprint arXiv:2108.13989 (2021)
3. Du, M., Li, F., Zheng, G., Srikumar, V.: DeepLog: anomaly detection and diagnosis from system logs through deep learning. In: Proceedings of the 2017 ACM SIGSAC Conference on Computer and Communications Security, pp. 1285–1298 (2017)
4. Tatam, M., Shanmugam, B., Azam, S., Kannoorpatti, K.: A review of threat modelling approaches for apt-style attacks. Heliyon **7**(1), e05969 (2021)
5. Lee, K.H., Zhang, X., Xu, D.: LogGC: garbage collecting audit log. In: Proceedings of the 2013 ACM SIGSAC Conference on Computer & Communications Security, pp. 1005–1016 (2013)
6. Liu, Y., et al.: Towards a timely causality analysis for enterprise security. In: NDSS (2018)
7. Hossain, M.N., et al.: SLEUTH: real-time attack scenario reconstruction from cots audit data. In: The 26th USENIX Security Symposium, pp. 487–504 (2017)
8. Zong, B., et al.: Behavior query discovery in system-generated temporal graphs. arXiv preprint arXiv:1511.05911 (2015)
9. Han, X., Pasquier, T., Bates, A., Mickens, J., Seltzer, M.: UNICORN: runtime provenance-based detector for advanced persistent threats. arXiv preprint arXiv:2001.01525 (2020)
10. Zeng, J., Chua, Z.L., Chen, Y., Ji, K., Liang, Z., Mao, J.: WATSON: abstracting behaviors from audit logs via aggregation of contextual semantics. In: Proceedings of the 28th Annual Network and Distributed System Security Symposium, NDSS (2021)
11. Mamun, M., Lu, R., Gaudet, M.: Tell them from me: an encrypted application profiler. In: Liu, J.K., Huang, X. (eds.) NSS 2019. LNCS, vol. 11928, pp. 456–471. Springer, Cham (2019). https://doi.org/10.1007/978-3-030-36938-5_28
12. Zhang, K., Xu, J., Min, M.R., Jiang, G., Pelechrinis, K., Zhang, H.: Automated it system failure prediction: a deep learning approach. In: 2016 IEEE International Conference on Big Data (Big Data), pp. 1291–1300. IEEE (2016)
13. Zheng, P., Yuan, S., Wu, X., Li, J., Lu, A.: One-class adversarial nets for fraud detection. In: Proceedings of the AAAI Conference on Artificial Intelligence, vol. 33, no. 01, pp. 1286–1293 (2019)
14. Liu, X., et al.: LogNADS: network anomaly detection scheme based on semantic representation. Future Generation Computer Systems **124**, 390–405 (2021)
15. Nammous, M.K., Saeed, K.: Natural language processing: speaker, language, and gender identification with LSTM. In: Chaki, R., Cortesi, A., Saeed, K., Chaki, N. (eds.) Advanced Computing and Systems for Security. AISC, vol. 883, pp. 143–156. Springer, Singapore (2019). https://doi.org/10.1007/978-981-13-3702-4_9

16. Weir, C., Arantes, R., Hannon, H., Kulseng, M.: Operationally transparent cyber (OpTC) (2021)
17. Mazzawi, H., et al.: Anomaly detection in large databases using behavioral patterning. In: 2017 IEEE 33rd International Conference on Data Engineering (ICDE), pp. 1140–1149. IEEE (2017)
18. Cochrane, T., Foster, P., Chhabra, V., Lemercier, M., Salvi, C., Lyons, T.: SK-tree: a systematic malware detection algorithm on streaming trees via the signature kernel. arXiv preprint arXiv:2102.07904 (2021)
19. Kent, A.D.: Comprehensive, multi-source cyber-security events data set. Technical report, Los Alamos National Lab. (LANL), Los Alamos, NM, USA (2015)
20. Wang, Q., et al.: You are what you do: hunting stealthy malware via data provenance analysis. In: NDSS (2020)
21. Balaban, M., Moshiri, N., Mai, U., Jia, X., Mirarab, S.: TreeCluster: clustering biological sequences using phylogenetic trees. PLoS One **14**(8), e0221068 (2019)
22. Agrawal, R., Srikant, R.: Mining sequential patterns. In: Proceedings of the eleventh international conference on data engineering, pp. 3–14. IEEE (1995)
23. Mooney, C.H., Roddick, J.F.: Sequential pattern mining-approaches and algorithms. ACM Comput. Surv. (CSUR) **45**(2), 1–39 (2013)
24. Lesh, N., Zaki, M.J., Ogihara, M.: Mining features for sequence classification. In: Proceedings of the Fifth ACM SIGKDD International Conference on Knowledge Discovery and Data Mining, pp. 342–346 (1999)
25. Lesh, N., Zaki, M.J., Oglhara, M.: Scalable feature mining for sequential data. IEEE Intell. Syst. Appl. **15**(2), 48–56 (2000)
26. Xing, Z., Pei, J., Keogh, E.: A brief survey on sequence classification. ACM SIGKDD Explor. Newsl. **12**(1), 40–48 (2010)
27. Shen, Y., Mariconti, E., Vervier, P.A., Stringhini, G.: Tiresias: Predicting security events through deep learning. In: Proceedings of the 2018 ACM SIGSAC Conference on Computer and Communications Security, pp. 592–605 (2018)
28. Li, Z., Cheng, X., Sun, L., Zhang, J., Chen, B.: A hierarchical approach for advanced persistent threat detection with attention-based graph neural networks. Secur. Commun. Netw. **2021**, Article ID 9961342 (2021). https://doi.org/10.1155/2021/9961342.
29. Király, F.J., Oberhauser, H.: Kernels for sequentially ordered data. J. Mach. Learn. Res. **20**(31), 1–45 (2019)

Network Security and Forensics

Dependency-Based Link Prediction for Learning Microsegmentation Policy

Steven Noel(✉) and Vipin Swarup(✉)

The MITRE Corporation, McLean, VA 22102, USA
{snoel,swarup}@mitre.org

Abstract. This paper describes a novel approach for predicting future links in cyber networks and applying the predictions to learn optimal microsegmentation policy rules. While link prediction has been applied for anomaly detection in computer networks, ours is the first application of link prediction for formulating network access policy. Link prediction adds an element of adaptivity for building baseline policy models, by predicting near-term requirements for network access. For predicting new links, those observed by at least one member of a node group are predicted to occur for all other members. This is a novel departure from the usual approach to link prediction, which is based on node affinity rather than shared dependencies. In our experiments with real enterprise network data, our approach significantly outperforms traditional link prediction, in which we apply established formulas for node similarity when comparing affinity-based versus dependency-based edge induction. For robustness to variation in future network behavior, we tune link prediction models by applying a low-pass signal filter to the prediction-quality curve and adaptively blend argmax and center of mass to optimize the prediction sensitivity parameter.

Keywords: Microsegmentation · Network access policy · Link prediction

1 Introduction

The increased complexity of enterprise networks and inadequacies of legacy security controls has led to the development of "zero trust" principles [1, 2]. Zero trust assumes that attackers could already be operating inside of a network. An enterprise must continually assess risks to its network assets and deploy protections to mitigate those risks, including requiring authorization for each request for access to network resources.

A key technology for implementing zero trust is network *microsegmentation* [3]. Traditionally, network segmentation has been oriented on "north-south" traffic, i.e., client-server interactions across a security perimeter. But in today's complex enterprise networks, security perimeters are ineffective since most traffic flows "east-west" (server to server), requiring more granularity in network access controls.

One data-driven approach to the formulation of microsegmentation policy is to cluster nodes based on similarities in their observed traffic links, and to define access rules at the cluster level. However, clustering has the potential to induce false positive links,

C. Alcaraz et al. (Eds.): ICICS 2022, LNCS 13407, pp. 569–588, 2022.
https://doi.org/10.1007/978-3-031-15777-6_31

increasing exposure while providing no benefit to the organization. There is an established area of research known as graph *link prediction*. To date, link prediction has not been applied for the formulation of network access policy. Also, the semantics of links for other application domains (such as social networks) significantly differ from those for cyber networks.

To address the challenges of data-driven formulation of microsegmentation rules, we adapt formulas from the field of link prediction for measuring the similarity of nodes in terms of their neighborhood graph topology. Recognizing the inherent differences in the semantics (meaning) of links for cyber networks, we introduce dependency-based rather than affinity-based semantics. That is, once we have formed a group of sufficiently similar nodes, the links to *other nodes* observed for at least one member of the group are predicted to occur for all other members (shared dependencies). This differs from traditional link prediction, which predicts that similar nodes will link with *each another* (shared affinity). In our experiments with real enterprise network data, our dependency-based link prediction significantly outperforms traditional affinity-based prediction, by orders of magnitude in many cases.

Link prediction and clustering algorithms have a tunable similarity threshold for associating nodes. Tuning this threshold is largely unexplored in the literature. In our experiments, link prediction quality curves tend to vary erratically, so we apply low-pass signal filtering to extract the main signal. We then estimate the optimal tuning value through an adaptive blend of argmax and center of gravity.

The observed and predicted new links can be mapped to corresponding policy rules and enforced within the microsegmentation architecture of a network. They can also be applied as a baseline policy for resiliency optimization algorithms. Given threat and defense scenarios (assumed or actual), such algorithms can find an optimal balance between access to mission-critical resources and the effort required by an adversary to reach attack goals. Our approach is deployed as a component of MITRE's Adaptive Resiliency Experimentation System (ARES), a full-stack solution that combines off-the-shelf components for data collection and security enforcement with AI-powered technology for optimizing the resiliency of a network.

Here is summary of the key contributions of this work:

1. This is the first known application of link prediction for formulating network access policy.
2. We adapt link prediction to adhere to the semantics of cyber networks (dependencies rather than affinities between nodes).
3. We automatically tune link prediction sensitivity, applying signal filtering and other techniques to address erratic fluctuations in observed performance curves.
4. Our predicted links provide input to algorithms for optimizing network resiliency to given cyberattack scenarios.

The next section describes previous work that is most relevant to our approach. Section 3 describes our approach and its role within a larger capability stack. Section 4 describes results applying our approach to enterprise network data. Section 5 summarizes this work, conveys our conclusions, and suggests directions for future work.

2 Previous Work

There is an emerging trend of data-driven models for automation and intelligence in cybersecurity [4, 5]. Still, applications of data science for security have predominantly focused on detection and response rather than prevention. Thus, automated generation of policy rules remains a largely unexplored problem. A notable exception is the application of data mining for the analysis and management of firewall policy rules [6], although that predates the development of microsegmentation architecture. For the generation of microsegmentation rules, preliminary work has proposed the application of clustering algorithms [7, 8]. Our experimental results indicate that clustering yields low precision in predicting links, which increases the attack surface within a network with no benefit of providing needed availability to network resources.

Our work is the first known application of link prediction to the problem of learning network access policy rules. In the absence of threat sources, these rules can be deployed in full on a network to be defended. These rules can also form a baseline for optimizing network access policy for maximum resilience in the face of threats (assumed or actual) [9]. In such cases, a subset of baseline rules can be deployed, with certain allow rules removed based on optimization tradeoffs. An evaluation framework has been proposed to assess microsegmentation for network security [10].

Link prediction itself is an active area of research [11–14], which includes approaches based on similarity indices, probabilistic methods, and machine learning. While research addressing the broader problem of evolving graphs (e.g., through graph representation learning) is still sparse [15], an approach has been described for distributed link prediction in dynamic graph streams [16]. Current state-of-the-art models for link prediction use heuristic learning of graph structure features; a multi-scale extension of such models shows improved performance [17].

Link prediction has been extended to handle multiplex (layered) networks [18], which has potential application to learning security policy across multiple network layers. Link prediction has been extended to higher-order structures involving more than single links, i.e., cliques on three nodes [19]. Our dependency-based semantics can predict higher-order link structures of arbitrary cardinality, based on the amount of shared incoming or outgoing links for grouped sets of nodes. Within link prediction, the problem of automatically selecting a value for a node similarity threshold (defining a set of sufficiently similar nodes for link induction) has remained unexplored [20]. We solve this by analyzing the prediction-quality as a function of similarity threshold, applying a low-pass signal filter for robustness to variation in future network behavior.

Link prediction for cybersecurity has largely focused on anomaly detection, e.g., simulated evolution to combine heuristics for differentiating normal network activity from anomalous events [21], anomaly detection via statistical techniques based a random dot product graph model [22], tensor factorization for learning patterns of normal activity from user authentication logs [23], and detecting malicious authentication events via unsupervised graph learning with a logistic regression link predictor [24].

In the commercial sector, microsegmentation policies have been automatically adapted through workload labeling [25]. Microsegmentation policy has been generated through clustering based on application implementation information [26]. There are applications of link prediction for dynamic routing [27], anomaly detection [28, 29], insider threat protection [30], and forecasting for cyber situational understanding [31].

3 Approach

3.1 Overview

The implementation of microsegmentation is a complex process. It requires ubiquitous log collection, comprehensive threat intelligence, network behavioral analytics, deployment orchestration, and advanced situational understanding. MITRE's Adaptive Resiliency Experimentation System (ARES) provides a full-stack software solution that optimizes resiliency for operational networks through zero-trust architecture via microsegmentation as well as authentication system configuration, service redundancy, and cyber deception. Figure 1 shows our approach to link prediction for microsegmentation policy as a component of ARES.

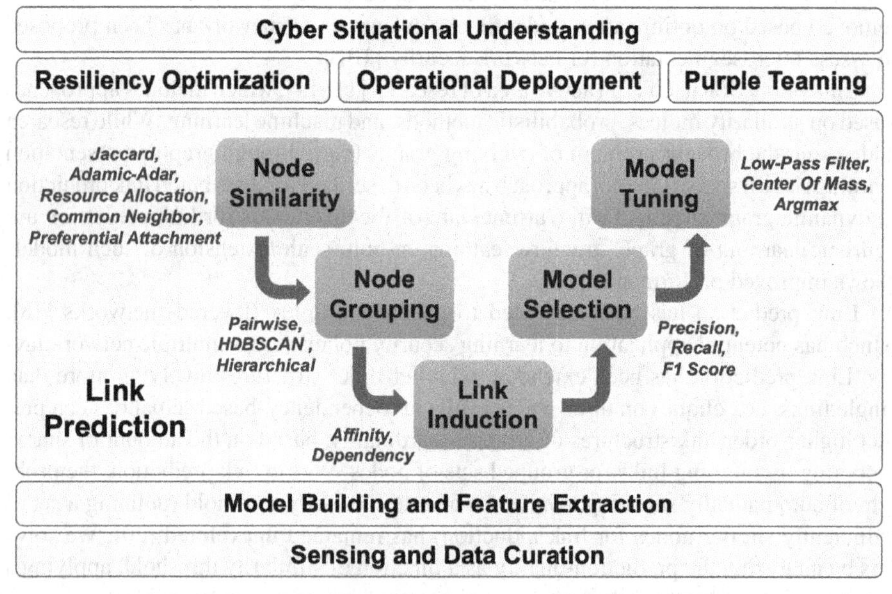

Fig. 1. Link prediction functionality as a component of MITRE's ARES software stack for optimizing the resiliency of an operational network.

In ARES, flow log data is collected from sensors that span the network and then curated, i.e., analyzed, merged, and persisted as a coherent and searchable whole. From the collected flow data, ARES builds a graph model of observed network traffic, and then extracts features for each network host (graph node) based on topological properties of the traffic graph. ARES performs link prediction through a process of (1) measuring host similarities based on traffic-graph features, (2) forming groups of nodes deemed sufficiently similar, and (3) inducing predicted links based on selected semantics (affinity or dependency-based). It then measures the performance of each link prediction algorithmic combination, and then choose and tune the best performing one.

In the absence of hypothesized or actual threats, the predicted links can be merged with the observed ones as a graph model, mapped to policy rules, and deployed in the network to enforce policy. Or if there is some security situation of concern, the links graph model can be used as input to a baseline policy for resiliency optimization, e.g., genetic algorithms to find a policy with the best balance between access to mission-critical resources and effort required by an adversary to reach attack goals [9], which can be deployed on the network. For deployed microsegmentation policy, adversarial attacks can be carried out, leveraging tools such as Cobalt Strike [32] and CALDERA [33]. Cyber situational understanding is maintained by exporting key results from the ARES stack into the CyGraph tool [34–36].

3.2 Node Similarities and Grouping

Different graph models, features, and similarity/distance measures are possible for clustering and link prediction tasks. A variety of graph models (and corresponding implementations) are employed in ARES for link prediction and other tasks. The most general form is a labeled, attributed, directed graph. Labels are needed as unique identifiers for nodes and edges, e.g., for mapping to policy rules. Relevant attributes for nodes or edges include mission criticality, attack vulnerability condition, statistics for traffic volume and temporal aspects, etc. In this paper, for simplicity and consistency with previous work in link prediction, we employ a simple undirected graph model, with the node similarity measures described below.

Early work in link prediction [37] includes the *Jaccard coefficient* as a measure of similarity for a pair of nodes u and v:

$$|\Gamma(u) \cap \Gamma(v)| / |\Gamma(u) \cup \Gamma(v)| \tag{1}$$

Here, $\Gamma(u)$ denotes the set of neighbors of u and $|\cdot|$ denotes set cardinality. The Jaccard coefficient measures the number of common links (to other nodes) for a node pair, i.e., the likelihood of them having common features. That early work also considers the *Adamic-Adar index*:

$$\sum_{w} 1/\log|\Gamma(w)|, \text{ for } w \in \Gamma(u) \cap \Gamma(v) \tag{2}$$

This refines the simple counting formula of the Jaccard coefficient by weighting rarer features more heavily. That early work also considers the *preferential attachment score*:

$$|\Gamma(u)||\Gamma(v)| \tag{3}$$

Preferential attachment is the idea that nodes adjacent to many other nodes are likely to themselves become attached, according to a model of network growth.

More recently [38], a *resource allocation index* has been proposed:

$$\sum_{w} 1/|\Gamma(w)|, \text{ for } w \in \Gamma(u) \cap \Gamma(v) \tag{4}$$

Under an assumption that for unlinked nodes, their common neighbors play the role of resource transmitters, the resource allocation index is a measure of the total resources transmitted between nodes u and v. Resource allocation index has a similar form to Adamic-Adar index. Their differences are insignificant if the degree of common neighbor w is small and are great if the degree of w is large. Thus, resource allocation index provides better performance for networks of higher average degree.

Even more recently [39], the Common Neighbor and Centrality based Parameterized Algorithm (CCPA) has been proposed for link prediction:

$$\alpha \cdot |\Gamma(u) \cap \Gamma(v)| + (1 - \alpha) \cdot \frac{N}{d_{uv}} \tag{5}$$

Here, N is the total number of nodes in the graph, d_{uv} is the graph distance between u and v. Given that closeness centrality (average shortest graph distance between two nodes) is N/d_{uv}, then CCPA is seen as a tradeoff between closeness centrality and the number of shared neighbors (the denominator of Jaccard coefficient), with $\alpha \in [0, 1]$ as the tradeoff parameter. In our experiments, we use the default value of $\alpha = 0.8$.

As shown in Fig. 1, after similarity measures have been computed for pairs of nodes in the observed link graph (the *Node Similarity* phase of ARES processing), those measures are applied to form groups of nodes deemed to be sufficiently related. In the *Node Grouping* phase, for the node similarities computed via Eqs. (1) through (5), a threshold $t \in [0, 1]$ is applied. If the similarity measure s_{uv} for a pair of nodes meets or exceeds the threshold ($s_{uv} \geq t$), then the node pair constitutes a group. Predicted links are then induced for these groups (pairs) of sufficiently similar nodes via Link Induction (as described in the next section).

In the Node Grouping phase, ARES also applies clustering algorithms as an alternative way of forming groups of sufficiently similar nodes. As for the pairwise groups selected through link prediction similarity formulas, the groups formed through clustering algorithms are passed as input to the Link Induction phase of ARES processing. The clustering algorithms learn a partitioning of the node set in which each node lies within a single cluster only. On the other hand, the groupings learned by link prediction similarities are pairwise (2 nodes) only, with each node appearing in many groups (not a partitioning of the node set as for clustering). Also, clustering yields higher-cardinality sets of nodes (beyond just 2 nodes), based on the relative distances between points and the nature and configuration of the clustering algorithm.

One form of clustering that ARES applies in Node Grouping is Hierarchical Density-Based Spatial Clustering of Applications with Noise (HDBSCAN) [40]. This is a non-parametric algorithm that clusters points in high-density regions, marking those in low-density regions as outliers. HDBSCAN has the advantage of being relatively insensitive to configuration parameters, i.e., being largely parameter free. Our experiments use (inverse) Jaccard coefficient as the distance measure for HDBSCAN.

During the Node Grouping phase, ARES also applies agglomerative hierarchical clustering [41]. Our experiments use average linkage, as intermediate between single linkage and complete linkage. For hierarchical clustering, we experimented with a variety of binary-based distance measurements. We report results using Rogers-Tanimoto distance [42], which performs slightly better than the others in our experiments.

3.3 Link Induction

In the *Link Induction* phase (part of Fig. 1), ARES processes the groups of nodes learned from each combination of node similarity function and grouping algorithm. For each group of nodes for a given algorithmic combination, it induces a set of new (not previously observed) links that are predicted to occur in the future. The resulting set of predicted links are then assessed by comparing them with links that are observed in a subsequent period.

Traditionally, link prediction has been applied to application domains in which the graph nodes represent entities of the same type, e.g., users in a social network. In such domains, graph edges (links) represent affinity relationships. For computer networks, links represent connectivity between hosts (nodes) of different types (e.g., from a client to a server). For such networks, we postulate that dependency (rather than affinity) is the more correct criterion for link prediction.

Our method is novel in that it also considers patterns of network connections that are consistent with dependency relationships. That is, rather than predicting that two similarly behaving nodes will themselves link together (show affinity with one another), it predicts that two such nodes will link with other nodes in the same way (have common dependencies). In terms of overall prediction quality, our experiments with real cyber network data show that dependency-based link prediction significantly outperforms traditional (affinity-based) link prediction.

This is shown in Fig. 2, for a small illustrative example. In this example, *Node B* connects with 4 other nodes (*C-F*), while *Node A* connects with only 3 of those other nodes (*C-E*). Applying the Jaccard coefficient as a similarity measure, the similarity of *Node A* and *Node B* is ¾, which meets our similarity threshold in this example. We thus consider *Node A* and *Node B* to be a sufficiently similar pair (the Node Grouping phase).

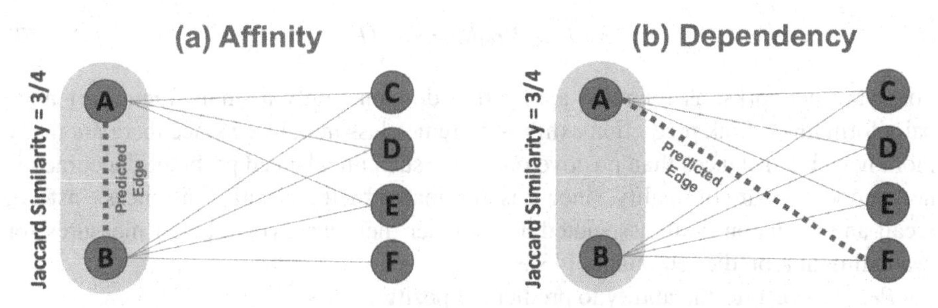

Fig. 2. Traditional (affinity based) versus ARES (dependency based) predicted links.

Now, for the Link Induction phase, we consider two kinds of semantics – traditional link prediction and link prediction using the semantics we apply for network policy formulation. In our illustrative example, Fig. 2(a) shows how traditional (affinity based) link prediction predicts that *Node A* and *Node B* will themselves become linked. Then in Fig. 2(a), our alternate link prediction semantics (dependency based) predicts that *Node A* will become linked with *Node F*. This is because *Node F* is the set of nodes with which Node B is linked (but with which *Node A* has not yet linked).

In this example, as for all the link prediction algorithms in the Node Grouping phase, the sufficiently similar group is a pair of nodes. Each node is potentially paired with all other nodes, i.e., node group membership is overlapping. However, the clustering algorithms are designed to learn an optimal partition of the nodes, so that node group membership is non overlapping.

Regardless, the Link Induction phase treats all learned node groups in the same way. That is, each learned group resulting from a given algorithmic combination in the previous phases is considered separately. For each such group, Link Induction induces a corresponding set of predicted links (using either affinity-based or dependency-based semantics). This introduces an additional layer of algorithmic combinations, each producing its own set of predicted links. Each resulting sets of predicted links is then assessed during the Model Selection phase, as described in the next section.

3.4 Model Selection and Tuning

In the *Model Selection* phase (in Fig. 1), ARES assesses the quality of the links predicted by each combination of algorithms employed in the previous phases. For this, it compares the links predicted by a given combination against links that are indeed observed during the model selection period (after the training period).

More formally, let us denote the potential future links that are predicted to be observed as *Predicted*. Links that are observed during the Model Selection period are denoted *Actual*. The set of true positive links *TP* are then those that were predicted and indeed observed:

$$TP = Predicted \cap Actual \tag{6}$$

False positive links are those that were predicted but were not observed:

$$FP = Predicted - TP \tag{7}$$

For cyber networks, as for most application domains, only a small fraction of node pairs form links. Link prediction exhibits extreme class imbalance since there are many more negative instances than positive ones. For such unbalanced problems, accuracy is not a good measure of quality, since it is dominated by the negative instances. Instead, recall and precision (with associated areas under their curves) are better measures for evaluating link prediction quality [43].

Recall measures the ability to predict the positive class:

$$Recall = |TP|/|Actual| = |TP|/|TP + FN| \tag{8}$$

Recall is particularly relevant when false negatives are more of a concern (relative to false positives). For network policy optimization, false negatives translate to denying edges that are indeed needed for operations. This does not introduce additional security risk, but it impacts the organizational mission.

Precision is the accuracy of the positive class:

$$Precision = |TP|/|Predicted| = |TP|/|TP + FP| \tag{9}$$

Precision is particularly relevant when false positives are more of a concern (relative to false negatives). For network policy optimization, false positives translate to allowing edges that are not needed operationally. This introduces additional security risk, with no mission availability reward.

F1 score F_1 is the harmonic mean of recall and precision:

$$F_1 = (2 \cdot Recall \cdot Precision)/(Recall + Precision) \tag{10}$$

The harmonic mean is suited to such fractional measures as recall and precision. Because of the product term in the numerator (as opposed to a sum for arithmetic mean), the harmonic mean punishes extreme values of recall or precision. For example, the arithmetic average for $Recall = 1$ and $Precision = 0$ is 0.5, but the F1 score is zero.

For operational cyber security, a weighted F1 score could be applied, with recall weighed more heavily in low-risk situations, and precision weighed more heavily in high-risk ones. In fact, a mission-versus-risk weight is a primary operational user input to the ARES solution, also used for user preference in trading off mission availability versus thwarting the adversary for optimizing resiliency in the face of a given cyberattack scenario.

In the Model Selection phase, ARES measures the link prediction performance of each algorithmic combination carried out in the previous phases. In all cases (except one), each algorithmic combination has a threshold parameter that determines the how similar hosts must be to group them together. The performance measures (recall, precision, and F1) all vary as a function of that grouping sensitivity threshold. ARES uses the area under each of these curves as an accumulated performance measure. By default (and for the results reported here), the default performance measure is area under the F1 score curve. Model Selection then chooses the best performing algorithm for operational deployment. The union of the observed and predicted links becomes the learned policy graph, representing the links (network connection) that are to be allowed (with deny by default).

Once ARES has selected a link prediction model (combination of algorithms in the Node Similarity, Node Grouping, and Link Induction phases), it needs to tune the sensitivity threshold for the selected Node Grouping algorithm. The performance curves for operational networks often fluctuate abruptly as a function of sensitivity threshold. To address this, ARES applies a low-pass filter to the performance curves, treating them as a noisy signal. We report results using a 4th-order low-pass Butterworth filter B_4 [44] applied to the F1 score curve $F_1(t)$:

$$\widehat{F_1}(t) = B_4[F_1(t)] \tag{11}$$

Here, $\widehat{F_1}(t)$ is the resulting lowpass-filtered F1 score signal (as a function of similarity threshold t). The Butterworth filter is maximally flat in its passband and has quick roll-off around the cutoff frequency, so that the desired frequencies are best selected. This class of filter does exhibit ringing in its response to a step in the input signal, but that improves with a lower cutoff frequency.

For a low-pass filter, the cutoff frequency is the frequency at which the filter begins to attenuate higher-frequency signal components. For the digital Butterworth filter implementation we employ, the cutoff is expressed in terms of the critical frequency at which the gain drops to $1/\sqrt{2}$ (-3 decibels below) that of the passband, with frequency normalized to unity at the Nyquist frequency (one-half of the sampling rate). In our experiments, the filtered performance curve exhibits noticeable ringing for the higher cutoff frequencies, and the filtered signal follows more closely the erratic shape of the original unfiltered signal. ARES applies a cutoff frequency of 27 units, so that the filtered signal is more of a general trend, with less ringing.

For assigning the operationally deployed value of similarity threshold, the most obvious solution is to apply the argmax function, yielding t_{opt}^{argmax}:

$$t_{opt}^{argmax} = \text{argmax}\widehat{F_1}(t) \tag{12}$$

Since it considers only a single point along the performance curve, argmax can be interpreted as the greediest (least risk averse) estimate. A more conservative (risk averse) estimate is to calculate the threshold t_{opt}^{cog} as the performance curve's center of mass:

$$t_{opt}^{cog} = \frac{1}{M} \sum_{i=1}^{n} i \cdot \widehat{F_1}(t_i) \tag{13}$$

Here, $M = \sum_{i=1}^{n} \widehat{F_1}(t_i)$ is the total mass (area) of the F1 score. This treats the performance curve as a mass distribution whose performance value at each point is considered a value of mass. The center of mass is then the value of threshold in which the relative position of the distributed mass sums to zero.

The argmax and center of mass provide bounds (least and most risk averse, respectively) for estimating the optimal similarity threshold. We can adaptively assign an intermediate value between these bounds. When the performance curve is relatively high at t_{opt}^{cog}, then there is incentive for the threshold to be more conservative (closer to t_{opt}^{cog}). When the performance curve is relatively low at t_{opt}^{cog}, then there is more to gain by choosing a threshold value closer to t_{opt}^{argmax}. We can achieve this by first defining a blending weight α as follows:

$$\alpha = F_1\left(t_{opt}^{cog}\right) / F_1\left(t_{opt}^{argmax}\right) \tag{14}$$

Thus, α is the value of the performance curve at the center of mass, normalized by the curve's maximum value. This represents how well the center-of-mass estimate performs (relative to the maximum). We now compute the intermediate value for estimating optimal similarity threshold t_{opt}^{blend} as

$$t_{opt}^{blend} = \alpha^{\beta} \cdot t_{opt}^{cog} + \left(1 - \alpha^{\beta}\right) \cdot t_{opt}^{argmax} \tag{15}$$

Here, the factor β provides a nonlinear bias towards t_{opt}^{argmax} (for $\beta > 1$) or towards t_{opt}^{cog} (for $\beta < 1$). We use $\beta = 3$, which provides good results for all of our experiments.

4 Experimentation

4.1 Evaluating Models

As illustrated in Fig. 1, link prediction in ARES performs the Node Similarity, Node Grouping, and Link Induction functions to induce a set of predicted future links for each of its link prediction models (combination of algorithmic components). In Model Selection, ARES evaluates the quality of the predicted links for each algorithmic combination, to select the combination that has best predictive performance.

We evaluate our method using publicly available network event data (flows) collected from routers within the Los Alamos National Laboratory (LANL) enterprise network [45]. Figure 3 shows the resulting F1 scores for link prediction using the first 50k network events for Day 2 of the LANL dataset (about 27 min of observed traffic) as training data. This plot has an F1 score curve for each combination of link prediction algorithms produced by ARES, as a function of node similarity threshold.

Figure 3 has 12 curves for F1 score. Comparing this to Fig. 1, pairwise grouping (in Node Grouping) is applied for the 5 formulas in Node Similarity (Jaccard, Adamic-Adar, Resource Allocation, Common Neighbor, Preferential Attachment).

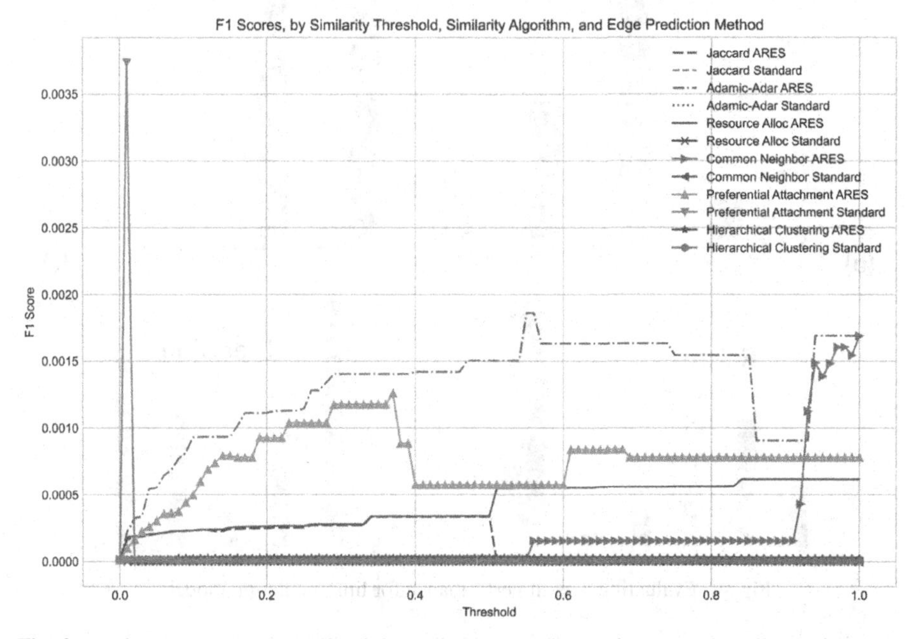

Fig. 3. Performance (F1 scores) for link prediction models as a function of similarity threshold.

In Fig. 3, one of the clustering algorithms in Node Grouping (Hierarchical) also varies as a function of similarity threshold, while the other (HDBSCAN) does not (i.e., it is largely parameter free). For each of those 6 combinations, Fig. 3 has an F1 score curve for ARES (dependency-based) and Standard (affinity-based) semantics for Link Induction. This accounts for the $(5 + 1) \cdot 2 = 12$ performance curves in Fig. 3.

In Fig. 3, a particularly notable F1 score curve is for the Preferential Attachment + Standard model. This model has a high F1 score for a single value of similarity threshold, a score of zero everywhere else. This narrow operationally feasible range suggests that this model would not be robust to changes in the underlying behavior of the network over time, i.e., not suitable for deployment. This highlights the utility performance curve area for selecting models, which requires higher performance over a wider range.

Figure 4(a) shows the areas under each of the F1 score curves in Fig. 3. It also includes the F1 score (for both ARES and Standard link induction) resulting from HDBSCAN clustering (which is independent of similarity threshold and therefore not in Fig. 3), for a total of 14 performance measures.

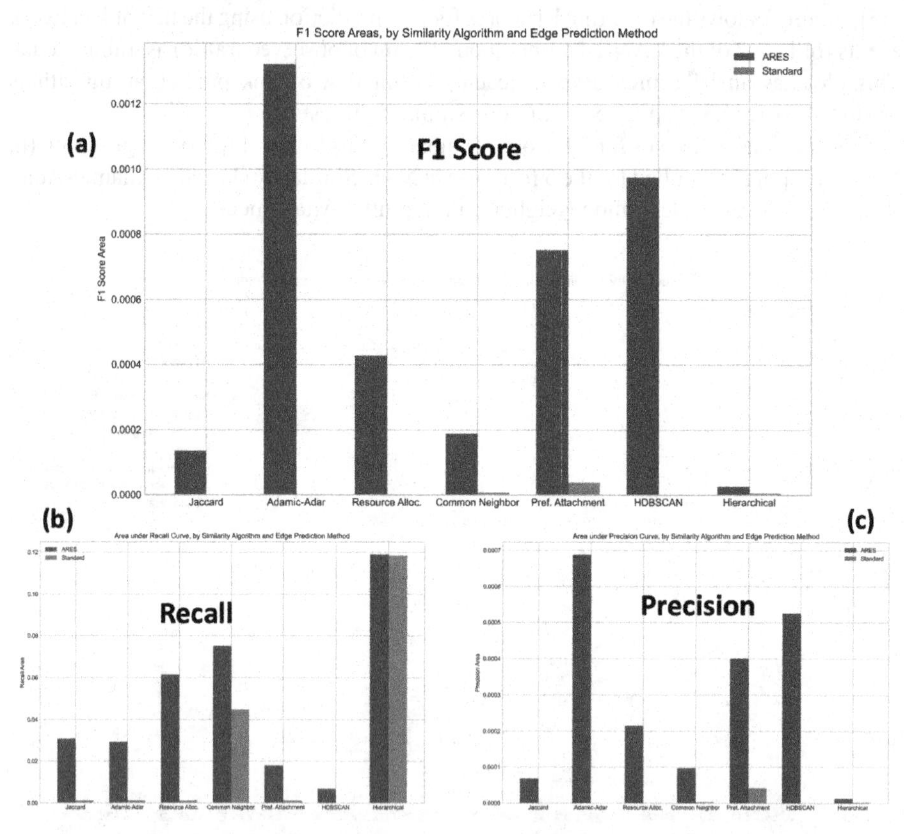

Fig. 4. Evaluating overall performance for link prediction models.

In Fig. 4(a), the application of Adamic-Adar similarity measure with ARES (dependency-based) link induction provides highest area under the F1 curve. As a similarity measure for a node pair, the Adamic-Adar index extends the rewarding of higher numbers of common links (as for Jaccard coefficient) by weighting rarer connections more heavily. The results in Fig. 4(a) demonstrate the value of that approach.

In Fig. 4(a), the second-best model is HDBSCAN (with Jaccard) + ARES. It is interesting that models with both pairwise-based (Adamic-Adar) and partitioning-based

(HDBSCAN with Jaccard) grouping perform well on F1 score, suggesting that HDBSCAN's density-based clustering is a good fit for the semantics of this problem. However, Hierarchical clustering (a greedy algorithm that lacks the adaptive nature of HDBSCAN) has poor F1 score performance.

As shown in Fig. 4(b), performance of the various models is significantly different for recall alone (rather than blended into F1 score). Hierarchical clustering has the highest recall, HDBSCAN has the lowest, and the various models with pairwise grouping lie in between them. This shows that Hierarchical clustering is the most inclusive way of grouping nodes, resulting in a larger portion of true positive link predictions (relative to all the new links that indeed occur). HDBSCAN is the least inclusive in that regard.

As shown in Fig. 4(c), performance of the models for precision (alone) mirrors those for F1 score. Because of the nature of the harmonic mean (having a product rather than sum in its denominator) defining an F1 score, smaller values (for recall versus precision) dominate the score. In this case, since the precision values are significantly lower than the recall values, precision dominates the F1 score.

A particularly interesting result is that across all the link prediction models and performance measures, ARES (dependency-based) link induction significantly outperforms Standard (affinity-based) link induction. The exception is recall for Hierarchical clustering, in which ARES method of link induction only slightly outperforms the Standard one. This validates our intuition that dependency semantics are considerably more applicable to link prediction for cyber networks.

Since the default performance measure for Model Selection in ARES is the area under the F1 curve, for the example in Fig. 4(a), the Adamic-Adar + ARES model is selected for operational deployment. ARES then tunes this selected model, as described in the next section.

4.2 Tuning Selected Model

In most approaches to link prediction, there is a parameter that determines the sensitivity of the predictive algorithm, representing a tradeoff between precision (fraction of predicted links that indeed occurred) and recall (fraction of observed new links that were indeed predicted). Such approaches generally leave that as an adjustable parameter, with little guidance on how to adjust it in an applied setting (in our case, in an operational cyber network). In our experiments, performance curves often exhibit abrupt transitions as the tuning parameter (similarity threshold for node grouping) is adjusted, further complicating the tuning process.

To address this, we apply a low-pass signal filter to the prediction-quality curve for a given link prediction model. This helps handle the abrupt transitions in performance curves, capturing the general trend of the curve as a function of the tuning parameter. We also compute the argmax and center of mass for the filtered curves as bounds for least (argmax) versus most (center of mass) risk averse solutions for optimal threshold, i.e., balancing higher performance against a wider range of feasibility. Overall, these techniques provide robust estimates for optimal tuning values that are expected to perform well given potential variation in future network behavior.

These techniques are illustrated in Fig. 5. Figure 5(a) is the F1 curve for the best performing model (Adamic-Adar + ARES) from the previous section. Figure 5(b) applies a low-pass signal filter to that F1 curve. Figure 5(c) is the unfiltered F1 curve for the Jaccard ARES model, which is particularly challenging for parameter tuning because of its abrupt drop in performance beyond a certain performance threshold value. Figure 5(d) is the filtered version of that more challenging F1 curve.

The curves in Fig. 5 are overlain with vertical lines that show 3 key values of the tuning parameter: (1) argmax, (2) center of mass, and (3) adaptive blend of argmax and center of mass. For the curve in Fig. 5(a), those values are relatively close to the naïve estimate (unfiltered argmax), so that this curve does not present a particularly difficult challenge for estimating the optimal value of the tuning parameter. However, in our experience, that is not the case in general. For example, the curve in Fig. 5(c) is particularly challenging for parameter tuning. Here, the naïve solution (unfiltered argmax) lies close to an abrupt drop in the performance curve. As underlying characteristics change over time for operational networks, shifts in such performance curves could result in poorly tuned predictive models.

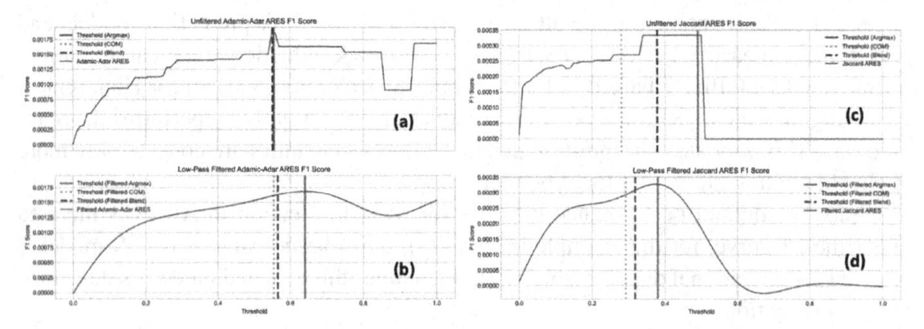

Fig. 5. Optimal tuning of the best performing link prediction model.

On the other hand, our tuning method provides optimal parameter values having high predictive performance over a stable region of the performance curve. In all cases that we have examined, our adaptive blend over the smoothed performance curve avoids regions near abrupt drops in performance. In less challenging cases, our adaptive blend is nearly the same as the naïve (greedy) solution.

Once ARES tunes a selected model, it deploys the model to predict links beyond those that have been observed in a network being protected. The tuned model is then applied for adaptively updating of microsegmentation policy rules, as described in the next section.

4.3 Applying to Cybersecurity Operations

In formulating optimal microsegmentation (network access) policy, the ARES solution first captures a baseline policy to be optimized with respect to cyber resiliency. It then seeks to optimally balance (1) resource availability for organizational mission needs and (2) adversarial threat opportunities. The optimization is carried out by a genetic algorithm, which searches the space of threat mitigation actions (deltas from the policy baseline). The genetic algorithm fitness function includes a variety of factors relevant to cyber resiliency and has a hierarchical structure that allows a desired degree of tradeoff between mission needs and attack risks.

Without link prediction, ARES builds a baseline policy from a historical record of observed traffic for a network. However, that approach is limited since it only considers previous network activity, i.e., the resulting policies overlook potential new mission needs. Link prediction adds an element of adaptivity, by predicting near-term requirements for network access. That is, in formulating the baseline access policy, ARES now includes predicted links along with links derived from observed traffic. Once a link prediction model has been selected (Sect. 4.1) and tuned (Sect. 4.2), it is applied during each instance of ARES policy optimization (e.g., on a recurring schedule).

As denoted in Table 1, any positive prediction of a link (either a true positive or a false positive) represents an allowed access that would not have been allowed without link prediction. Any negative prediction (either a true negative or a false negative) represents no further allowed access beyond policy based on historical records.

Table 1. Link prediction outcomes in the context of cybersecurity operations.

<table>
<tr><td colspan="3" rowspan="3"></td><td colspan="4">Link Predicted</td></tr>
<tr><td colspan="2">Yes</td><td colspan="2">No</td></tr>
<tr><td>Adds access</td><td>Measured via precision</td><td colspan="2">No access added</td></tr>
<tr><td rowspan="4">Link Needed</td><td rowspan="2">Yes</td><td>Potential mission contribution</td><td colspan="2">True Positive</td><td colspan="2">False Negative</td></tr>
<tr><td>Measured via recall</td><td>Supports mission</td><td>Can add attack vector</td><td>Hampers mission</td><td>No attack vector added</td></tr>
<tr><td rowspan="2">No</td><td rowspan="2">Not relevant to mission</td><td colspan="2">False Positive</td><td colspan="2">True Negative</td></tr>
<tr><td>Mission N/A</td><td>Can add attack vector</td><td>Mission N/A</td><td>No attack vector added</td></tr>
</table>

A true positive is a link that is predicted to be needed and is in fact needed. Such a link adds mission value by now being available (i.e., it would not have been available if the baseline policy were based on historical traffic only). On the other hand, if the destination for the allowed access is vulnerable to attack, that provides an additional attack vector (single attack step). A false positive is a link that is predicted to be needed but is not in fact needed. Such a link adds no mission value by being available since it is not used for mission operations. As for a true positive, if the destination is vulnerable, allowing a false-positive link provides an additional attack vector.

A true negative is a link that is not predicted to be needed and is not in fact needed. Thus, no contribution to mission value is possible, since there is inherently no mission need for the link. No additional attack vector is added in this case. A false negative is a link that is not predicted to be needed but is in fact needed. In this case, mission value is reduced (a needed access is blocked), and no additional attack vector is added.

Recall is computed from the elements of the *Link Needed* = *Yes* row of Table 1. It measures the ability to maximize support for the mission, regardless of any associated risks. Precision is computed from the elements of the *Link Predicted* = *Yes* column of Table 1. It measures the ability to minimize unnecessary risks in supporting the mission. F1 score then measures the ability to both maximize mission support and minimize unnecessary risks. True negatives are irrelevant to the organizational mission and do not introduce attack vectors.

ARES includes predicted links along with previously observed links as allowed access rules (deny by default) as a baseline microsegmentation policy. As an illustrative example, reconsider the results described in Sects. 4.1 and 4.2, which are for the first 50k network events for Day 2 of the LANL dataset as training data. There, the model with Adamic-Adar similarity measure and ARES (dependency-based) link induction is the strongest performer in terms of prediction quality alone.

We can assess this model in a security context by computing a baseline resiliency measure resulting from the application of this model's predicted links. This assessment is in terms of (1) threat containment, which is the additive inverse (to be maximized) of an attack score for the predicted links and (2) mission access, which is the relative number of new policy accesses (needed by the mission) introduced by the predicted links. The attack score considers the incremental value (in terms of the number of exploitable paths, up to a maximum path length of 5) that predicted edges give an attacker, from a particular start node to a particular goal node, then finds the mean incremental value over all pairs of nodes.

As shown in Table 2, the Adamic-Adar + ARES model yields a threat containment measure of 0.76 and a mission access measure of 0.006, for a resiliency baseline (mean of threat containment and mission access) measure of 0.39. As a comparison, the Hierarchical Clustering + ARES model yields a threat containment measure of 0.06 and a mission access measure of 0.022, for a resiliency baseline measure of 0.04. This shows that while Hierarchical Clustering + ARES provides significantly more links needed by the mission, it introduces significantly higher risks. A strong bias for preferring mission access over threat containment would need to be applied before the resiliency baseline for Hierarchical Clustering + ARES exceeds that of Adamic-Adar + ARES.

Table 2. Assessing predicted links in a security context.

Link prediction model	Threat containment	Mission access	Resiliency baseline
Adamic-Adar + ARES	0.76	0.006	0.39
Hierarchical clustering + ARES	0.06	0.022	0.04

ARES then considers a variety of actions (alone and in combination) for mitigating risks associated with a baseline policy. Such actions include applying patches to address software weaknesses, clearing password caches, blocking network access, deploying deceptive honeypot computers, or providing redundant resources. Each combination of actions under consideration is scored according to impacts on organizational mission and threats, via simulations under assumed operational scenarios such as likely attack starting points and critical assets to defend. The combination of actions that provides the optimal simulated outcome is then deployed on the defended network.

4.4 Considering Dataset Scale

For the results in Sects. 4.1 and 4.2, with 50k observed network events (~27 min) as input, ARES is configured to apply 75% (37.5k events) for training and 25% (12.5k events) for testing. Further experimentation shows that smaller dataset sizes for testing (or combinations of training and testing) yield better prediction quality, e.g., the results in Fig. 6. This aligns with the assumption that graph link prediction is better suited for near-term predictions, i.e., fewer new links spanning a briefer future time.

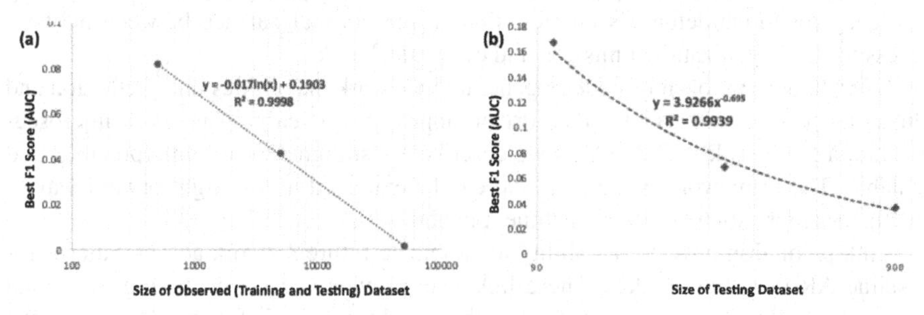

Fig. 6. Link prediction performance as a function of overall dataset size (fixed 75/25% ratio for training versus testing (a) and as a function of the size of dataset for testing (b).

Figure 6(a) is the best F1 curve area for observed datasets of 3 different sizes (0.5k, 5k, and 50k events). Here, the same ratio (75/25%) is used for training versus testing. This shows that smaller datasets (in terms of overall size for both training and testing) yield higher prediction quality. Notably, the performance (area under F1 curve) for link predictions made from the smallest dataset (0.5k events) is over 60 times higher than for the largest dataset (50k events).

The plot in Fig. 6(b) shows the best F1 curve area for varied dataset sizes (100, 300, and 900 events) for testing with a fixed dataset size (300 events) for training. Here, the performance using the smallest dataset (100 events) for testing is over 4 times higher than for using the largest dataset (900 events) for testing.

5 Summary, Conclusions, and Future Work

This paper describes the novel application of link prediction for learning microsegmentation policy in cyber networks. Our approach combines established link prediction formulas for node similarity with dependency-based (rather than traditional affinity-based) link induction, to align with the semantics of cyber networks. Experimental results with real enterprise network data demonstrate that our dependency-based link prediction significantly outperforms traditional link prediction, often by orders of magnitude.

For the practical application of link prediction to operational cyber networks, we address a problem that has been largely ignored in the literature – automatically tuning the sensitivity of link prediction models by finding an optimal value of node similarity threshold. For this, we apply low-pass signal filtering to smooth abrupt changes in the prediction performance curve, extracting the general trend of the performance as a function of tuning parameter. We also adaptively blend the argmax and center of mass of a given model's performance curve to balance risk versus reward in estimating the optimal threshold value. Our tuning method yields parameter values having high predictive performance over a stable region of the performance curve. This provides robust estimates for optimal tuning values that are expected to perform well given potential variation in future network behavior.

Our approach to network microsegmentation enhanced by link prediction is a component of the MITRE ARES capability stack for optimizing cyber network resiliency. ARES applies observed and predicted links as a baseline policy model for optimizing resiliency for threat/defense scenarios, finding the optimal balance between network access for the organizational mission and cyberattack risk.

Overall, our approach provides a general framework that unifies link prediction and clustering as components of a data-driven approach for learning network microsegmentation policy rules. The models we describe are straightforward, interpretable, and scalable. The framework is modular and can be extended in a straightforward way to encompass more sophisticated constituent methods.

Link prediction provides an ability to anticipate future network access patterns for baseline ARES policy models. These links can provide access for the organizational mission as well as introduce new attack paths for adversaries. Future work can examine how ARES defensive responses can best apply these enhanced predictive baselines in balancing attack risks and mission access.

References

1. Rose, S., Borchert, O., Mitchell, S., Connelly, S.: NIST special publication 800-207: zero trust architecture. National Institute of Standards and Technology, Gaithersburg, MD (2020)
2. Executive Order on Improving the Nation's Cybersecurity. https://www.whitehouse.gov/briefing-room/presidential-actions/2021/05/12/executive-order-on-improving-the-nations-cybersecurity/. Accessed 2 Mar 2022
3. Miller, L., Soto, J.: Micro-segmentation for Dummies, VMware special edition. Wiley, Hoboken (2015)
4. Sarker, I.H., Kayes, A.S.M., Badsha, S., Alqahtani, H., Watters, P., Ng, A.: Cybersecurity data science: an overview from machine learning perspective. J. Big Data **7**(1), 1–29 (2020). https://doi.org/10.1186/s40537-020-00318-5
5. Xin, Y., et al.: Machine learning and deep learning methods for cybersecurity. IEEE Access **6**, 35365–35381 (2018)
6. Golnabi, K., Min, R., Khan, L., Al-Shaer, E.: Analysis of firewall policy rules using data mining techniques. In: IEEE/IFIP Network Operations and Management Symposium, Piscataway, NJ, pp. 305–315. Institute of Electrical and Electronics Engineers (2006)
7. Yousefi-Azar, M., Kaafar, M.A., Walker, A.: Unsupervised learning for security of enterprise networks by micro-segmentation. Preprint arXiv:2003.11231v1 (2020)
8. Arifeen, M., Petrovski, A., Petrovski, S.: Automated microsegmentation for lateral movement prevention in industrial internet of things (IIoT). In: International Conference on Security of Information and Networks, Piscataway, NJ, pp. 1–6. Institute of Electrical and Electronics Engineers (2021)

9. Noel, S., Swarup, V., Johnsgard, K.: Optimizing network microsegmentation policy for cyber resilience. J. Defense Model. Simul. Spec. Issue Impact Anal. Cyber Defense Optim. 1–23 (2021)
10. Basta, N., Ikram, M., Kaafar, M.A., Walker, A.: Towards a zero-trust micro-segmentation network security strategy: an evaluation framework. Preprint arXiv:2111.10967v1 (2021)
11. Kumar, A., Singh, S.S., Singh, K., Biswas, B.: Link prediction techniques, applications, and performance: a survey. Phys. A **553**, 1–46 (2020)
12. Wang, P., Xu, B., Wu, Y., Zhou, X.: Link prediction in social networks: the state-of-the-art. Sci. China Inf. Sci. **58**(1), 1–38 (2014). https://doi.org/10.1007/s11432-014-5237-y
13. Mutlu, E.C., Oghaz, T., Rajabi, A., Garibay, I.: Review on learning and extracting graph features for link prediction. Mach. Learn. Knowl. Extr. **2**(4), 672–704 (2020)
14. Marjan, M., Zaki, N., Mohamed, E.A.: Link prediction in dynamic social networks: a literature review. In: IEEE International Congress on Information Science and Technology, Piscataway, NJ, pp. 200–207. Institute of Electrical and Electronics Engineers (2018)
15. Georgousis, S., Kenning, M.P., Xie, X.: Graph deep learning: state of the art and challenges. IEEE Access **9**, 22106–22140 (2021)
16. Katragadda, S., Gottumukkala, R., Pusala, M., Raghavan, V., Wojtkiewicz, J.: Distributed real time link prediction on graph streams. In: IEEE International Conference on Big Data, Piscataway, NJ, pp. 2912–2917. Institute of Electrical and Electronics Engineers (2018)
17. Cai, L., Ji, S.: A multi-scale approach for graph link prediction. In: AAAI Conference on Artificial Intelligence, Palo Alto, CA, pp. 3308–3315. AAAI Press (2020)
18. Aleta, A., Tuninetti, M., Paolotti, D., Moreno, Y., Starnini, M.: Link prediction in multiplex networks via triadic closure. Phys. Rev. Res. **2**, 1–6 (2020)
19. Nassar, H., Benson, A.R., Gleich, D.F.: Pairwise link prediction. In: IEEE/ACM International Conference on Advances in Social Networks Analysis and Mining, New York, pp. 386–393. Association for Computing Machinery (2019)
20. Martínez, V., Berzal, F., Cubero, J.-C.: A survey of link prediction in complex networks. ACM Comput. Surv. **49**(4), 1–33 (2017)
21. Pope, A.S., Tauritz, D.R., Turcotte, M.: Automated design of tailored link prediction heuristics for applications in enterprise network security. In: López-Ibáñez, M. (ed.) Genetic and Evolutionary Computation Conference Companion, pp. 1634–1642. Association for Computing Machinery, New York (2019)
22. Passino, F.S., Bertiger, A.S., Neil, J.C., Heard, N.A.: Link prediction in dynamic networks using random dot product graphs. arXiv:1912.10419 (2021)
23. Eren, M.E., Moore, J.S., Alexandro, B.S.: Multi-dimensional anomalous entity detection via poisson tensor factorization. In: IEEE International Conference on Intelligence and Security Informatics, Piscataway, NJ, pp. 1–6. Institute of Electrical and Electronics Engineers (2020)
24. Bowman, B., Laprade, C., Ji, Y., Huang, H.H.: Detecting lateral movement in enterprise computer networks with unsupervised graph AI. In: International Symposium on Research in Attacks, Intrusions and Defenses, pp. 257–268. USENIX Association, Berkeley (2020)
25. Gupta, M., Fandli, J.G.: Automatically assigning labels to workloads while maintaining security boundaries. United States Patent 11,171,991, 9 November 2021
26. Hamou, C., Brouk, R., McAllister, S.: Micro-segmentation in virtualized computing environments. United States Patent 2017/0374106, 28 December 2017
27. Hui, P., Huang, D., Peylo, C.: Method and system for link prediction in mobile computing. European Patent Office Patent EP 2 911 349, 24 February 2016
28. Choudhury, S., Agarwal, K., Chen, P.-Y., Ray, I.: System and methods for automated detection, reasoning and recommendations for resilient cyber systems. United States Patent 2018/0103052, 1 December 2020
29. Verma, M., et al.: Systems and methods for identifying and mitigating outlier network activity. European Patent Office Patent EP 3 477 906 A1, 31 March 2021

30. Brdiczka, O., Mahadevan, P., Shi, R.: Method and system for thwarting insider attacks through informational network analysis. United States Patent 9,336,388, 10 May 2016
31. Shaashua, T.M., Shaashua, O.: Situation forecast mechanisms for internet of things integration platform. United States Patent 10,990,894, 27 April 2021
32. ATT&CK | cobalt strike. https://attack.mitre.org/software/S0154/. Accessed 3 Mar 2022
33. CALDERA. https://caldera.mitre.org. Accessed 3 Mar 2022
34. Noel, S., Harley, E., Tam, K.H., Limiero, M., Share, M.: CyGraph: graph-based analytics and visualization for cybersecurity. In: Cognitive Computing: Theory and Application, Handbook of Statistics, vol. 35, pp. 117–167. Elsevier, Amsterdam (2016)
35. Noel, S., et al.: Graph analytics and visualization for cyber situational understanding. J. Defense Model. Simul. Impact Anal. Cyber Defense Optim. 1–15 (2021)
36. Noel, S., Harley, E., Tam, K.H., Limiero, M., Share, M.: System and method for visualizing and analyzing cyber-attacks using a graph model. United States Patent 10,313,382, 4 June 2019
37. Liben-Nowell, D., Kleinberg, J.: The link prediction problem for social networks. J. Am. Soc. Inform. Sci. Technol. **58**(7), 1019–1031 (2007)
38. Zhou, T., Lü, L., Zhang, Y.: Predicting missing links via local information. Eur. Phys. J. B **71**, 623–630 (2009)
39. Ahmad, I., Akhtar, M.U., Noor, S., Shahnaz, A.: Missing link prediction using common neighbor and centrality based parameterized algorithm. Sci. Rep. **10**(334), 1–9 (2020)
40. McInnes, L., Healy, J., Astels, S.: HDBSCAN: hierarchical density based clustering. J. Open Source Softw. **2**(11), 205–206 (2017)
41. Murtagh, F., Contreras, P.: Methods of hierarchical clustering. arXiv:1105.0121v1 (2011)
42. Rogers, D.J., Tanimoto, T.T.: A computer program for classifying plants. Science **1115–1118**, 21 (1960)
43. Yang, Y., Lichtenwalter, R.N., Chawla, N.V.: Evaluating link prediction methods. Knowl. Inf. Syst. **45**(3), 751–782 (2014). https://doi.org/10.1007/s10115-014-0789-0
44. Butterworth, S.: On the theory of filter amplifiers. Exper. Wirel. Wirel. Eng. **7**, 536–541 (1930)
45. Turcotte, M.J.M., Kent, A.D., Hash, C.: Unified host and network data set. In: Data Science for Cyber-Security, pp. 1–22. World Scientific, Singapore (2018)

Chuchotage: In-line Software Network Protocol Translation for (D)TLS

Pegah Nikbakht Bideh[1(✉)] and Nicolae Paladi[1,2]

[1] Lund University, Lund, Sweden
{pegah.nikbakht_bideh,nicolae.paladi}@eit.lth.se
[2] CanaryBit, Stockholm, Sweden

Abstract. The growing diversity of connected devices leads to complex network deployments, often made up of endpoints that implement incompatible network application protocols. Communication between heterogeneous network protocols was traditionally enabled by hardware translators or gateways. However, such solutions are increasingly unfit to address the security, scalability, and latency requirements of modern software-driven deployments. To address these shortcomings we propose Chuchotage, a protocol translation architecture for secure and scalable machine-to-machine communication. Chuchotage enables in-line TLS interception and confidential protocol translation for software-defined networks. Translation is done in ephemeral, flow-specific Trusted Execution Environments and scales with the number of network flows. Our evaluation of Chuchotage implementing an HTTP to CoAP translation indicates a minimal transmission and translation overhead, allowing its integration with legacy or outdated deployments.

Keywords: Protocol conversion · IoT · Application layer protocols · Software defined networking · TLS · Cross-layer optimization

1 Introduction

Despite efforts towards standardization and interoperability, many applications use proprietary protocols and incompatible data models for information exchange [25]. This is particularly acute to address in growing density of connected embedded devices or "things". Such devices are increasingly expected to communicate in a machine-to-machine (M2M) pattern. Communication among devices, or between devices and back-end systems that use incompatible protocols can be enabled through *protocol translation*. This is commonly realized either with hardware translators, virtual gateways[1], or distributed software applications [34]. Existing approaches for protocol translation are unfit to address the scalability, latency, and security requirements of current and emerging deployment topologies [7]. Such solutions display at least one of the following limitations.

[1] Communication servers including a virtual gateway to perform protocol translation.

C. Alcaraz et al. (Eds.): ICICS 2022, LNCS 13407, pp. 589–607, 2022.
https://doi.org/10.1007/978-3-031-15777-6_32

1. in-line translation solutions do not support encrypted network traffic;
2. solutions to circumvent limitation (1) rely on deploying trusted certificates to unprotected devices on the network path and increase the attack surface;
3. cloud-based protocol translation solutions support translation over secure network communication by terminating TLS connections in a single centralized component. This increases communication latency between network endpoints and introduces a single point of failure.

Addressing the above challenges is a prerequisite to enable wide-scale device connectivity. This requires support for secure and fast in-line software network protocol translation of encrypted traffic; support for communication over several application layer protocols while maintaining latency requirements; and finally support for distributed protocol translation. Our goal is to enable secure massive M2M communication using protocol translation capable of dynamically adapting to new devices and network topologies. Our **contributions** are as follows:

- we introduce Chuchotage[2], an efficient and secure protocol translator architecture addressing scalability, latency, and security requirements of large-scale networks; Chuchotage builds on earlier work in Software Defined Networking, Trusted Execution Environments, and TLS interception;
- Chuchotage performs in-line protocol translation while supporting secure distributed network communication throughout the network fabric, avoiding translation in a logically or physically centralized back-end;
- we introduce flow-specific, on-demand translator boxes created by software switches on the network path for TLS interception and protocol translation.
- we integrate secure protocol translation in OpenFlow [22] by reusing and extending its signaling. This allows to maintain backward compatibility.

Our solution relies on three principles: (i) secure TLS interception with the use of TEEs; (ii) high-performance confidential protocol translation, and (iii) fault-tolerant distributed architecture with the help of SDN networking. A TEE provides confidentiality and integrity with the use of an isolated execution environment. The loaded code and data to the TEE can be protected from various attacks. In our architecture, we use TEEs to securely decrypt, translate, and re-encrypt data it with a high level of confidentiality and integrity.

In SDN networking, network intelligence is logically centralized, thus abstracting the network infrastructure from network applications [16]. In SDN, the controller has a global view and can decide what suits best for the network. The OpenFlow protocol is usually used in SDN to link the controller and other components, e.g. switches, and routers. OpenFlow is compatible with both hardware and software switches. In Chuchotage, the software switch (Open vSwitch [31] in our implementation) makes informed decisions on application layer protocol translation to provide a high-performance and fault tolerant architecture. To the best of our knowledge, this is the first work that integrates datapath flow matching with secure protocol translation. To improve the performance, we introduce a cross-layer optimization for switch actions described in Sect. 4.

[2] The term *chuchotage* is a form of interpreting where the linguist is near a small target audience and whispers a simultaneous interpretation of what is being said.

The rest of this paper is structured as follows: in Sect. 2 we introduce the relevant background and problem, followed by a review of the related work in Sect. 3. We describe the design of Chuchotage in Sect. 4. We discuss in Sect. 5 the design choices of the Chuchotage implementation, followed by performance and security evaluation in Sect. 6. We conclude in Sect. 7.

2 Background

We define interoperability in IoT networks as the capability of heterogeneous devices and applications to communicate and exchange data or services. Tolk et al. presented interoperability as a layered model with two main layers: *technical* and *semantic* interoperability [30]. *Technical* interoperability enables compatibility of heterogeneous devices through common communication protocols and standards. *Semantic* interoperability enables heterogeneous services and applications to exchange information in a meaningful way [1].

Data or information models used by heterogeneous IoT devices are often incompatible, thus limiting semantic interoperability. Semantic protocol translators are a possible solution; they are able to convert information formats, allowing communication between heterogeneous endpoints. Such translators ingest a standardized way of representing vocabularies of processes or messages. However, despite ongoing efforts for IoT semantic translation, we are yet to see a unified secure platform compatible with most common IoT protocols. We next briefly introduce several interoperability solutions.

Physical Gateways: A traditional way of interoperability is the use of hardware gateways that act as an intermediate component between endpoint devices [25]. Hardware gateways can translate protocols with different standards and specifications, they are commonly one-to-one protocol translators that do not scale (new protocols require adding new hardware); moreover, they require special hardware connectors, thus increasing both the overhead and complexity.

Protocol Translators: Protocol translators replace traditional interoperability solutions, such as gateways; they are intermediate components that perform direct protocol to protocol translation. Depending on where the translation is done, protocol translators are either: a) cloud back-end translators or b) middleboxes. In the first case, the traffic is re-routed to the cloud back-end for translation. In the second case, a middlebox is a hardware component or software network function placed on the communication path between the endpoints.

We review existing protocol translators in Sect. 3.1. These translators either do not consider security or do not scale. Some perform the translation below the application layer, thus adding further network complexity.

For further information about common IoT protocols and different interoperability solutions at different protocol layers refer to Appendix A.1. We propose the Chuchotage architecture to enable protocol interoperability on the application layer. We target the application layer as it has the highest impact on application performance [12].

2.1 Threat Model

Our threat model considers two aspects - security of the network communication, and security of protocol translation. We assume the Dolev-Yao model [9], with an adversary capable of intercepting, and synthesizing any message, being only limited by the constraints of the cryptographic methods used. Considering protocol translation, we assume limited physical access to the platform, akin to the tasks of a legitimate third party user, and excluding physically modifying, probing, or monitoring the system. The adversary is capable of exploiting software vulnerabilities in the host operating system and software network components (network switch and network functions), reloading the switch binary, accessing the host memory, and starting arbitrary processes on the host. The attacker may modify any firmware of software component on the network platforms, including the hypervisor for virtualized set-ups. This threat model is aligned with the threat models of both process-based trusted execution environments (such as Intel SGX [3] or Keystone [27]) as well as virtualization-based trusted execution environments (AMD SEV-SNP [2], Intel TDX [33] and IBM PEF [14]). The Chuchotage architecture may be tuned to use other TEE implementations. Considering the growing diversity of TEE implementations [33] and their various approaches to defending or preventing side-channel attacks, we exclude side-channel attacks. Likewise, we exclude attacks on control-plane components of SDN deployments (such as the SDN controller) or ancillary components (such as the Certificate Authority); these components are trusted and attacks on them can be prevented using best-practice operational security. Translator boxes are not trusted and translation cannot be done securely without a TEE.

3 Related Work

3.1 Protocol Translation

An early work on protocol conversion was presented in 1988 by Lam [17], proposing a formal model to achieve interoperability between processes with different protocols. Its' limitation was that it needs to be implemented as a process or as a low layer protocol in the physical layer, thus adding complexity and overhead to the network.

In [7], the authors proposed a protocol translator for industrial IoT protocols. They proposed the use of an intermediate format in order to translate more than three protocols rather than direct protocol-to-protocol translation. The solution satisfies interoperability features including transparency, scalability, reporting, verifiability, and QoS, however without addressing any security aspects, which Chuchotage explicitly addresses.

Muppet [32] is an edge-based multi-protocol switching architecture that can be used for IoT service automation. Muppet is a P4-based switch which can communicate with IoT devices using different protocols, where switches are managed by an SDN controller. Muppet was designed for translation between Zigbee [26] and Bluetooth low energy (BLE) [11] protocols or translation between

BLE/Zigbee and IP protocols and is therefore complementary to Chuchotage, which works at the application layer. However, similar to [7], Muppet does not support protocol translation over TLS communication.

3.2 TLS Interception

HTTPS interception is implemented for purposes such as content filtering, malware detection, DDoS mitigation, load balancing, etc. [5], and despite the relative maturity of the topic, research on TLS interception proxies gained further attention in recent years. The ME-TLS protocol [20] supports TLS 1.3 and enables endpoints to introduce middleboxes into a session given the consent of both parties. Endpoints can control middlebox access permissions on traffic data, and verify the middlebox service chain. The protocol is based on monitoring handshake messages passively without modifying the handshake of TLS 1.3. An implicit version negotiation mechanism in the ME-TLS handshake protocol enables it to interoperate with TLS endpoints seamlessly. However, ME-TLS requires deploying the Boneh-Franklin identity-based encryption (BF-IBE) [15] instead of the widely adopted Public Key Infrastructure (PKI) approach.

maTLS is an extension to TLS that allows middlebox visibility and auditability by enabling a client to authenticate all middleboxes through a dedicated *middlebox certificate*. The use of middlebox certificates eliminates the insecure practice of installing custom root certificates or servers sharing their private keys with third parties. Furthermore, the middlebox-aware TLS (maTLS) protocol enables auditing the security behaviors of middleboxes [19].

IA2-TLS [4] is an encryption-based approach to enable in-line packet inspection. IA2-TLS is based on binding an *inspection key* to the random nonces that are generated by the endpoints during a TLS handshake. The advantage of this approach is the capacity to introspect traffic both inline and offline, at any location along the network path. This approach requires modifying the client and server TLS implementation. Similar to many other TLS interception approaches, it is not practical considering the lack of backward compatibility.

Considering the properties and backward compatibility of the ME-TLS protocol, we use it for the remainder of this paper as the reference TLS interception protocol. Other approaches to TLS interception are complementary.

4 Chuchotage Protocol Translator

4.1 Architecture

Figure 1 illustrates the Chuchotage architecture, relying on principles introduced in Sect. 1: (i) secure and protocol-compliant TLS interception; (ii) efficient confidential protocol translation; and (iii) fault-tolerant distributed architecture. The proposed architecture assumes that network switches are configured Ⓐ with an action to translate network flows between endpoints that use incompatible application layer network protocols **B** (we use OpenvSwitch for implementation). When invoked, the action triggers the creation of a translator box **C** in a trusted execution environment (Intel SGX in our implementation). The translator box is subsequently attested by a verifier network function and provisioned with credentials for TLS interception **D**. The translator box is network-flow specific, translates subsequent communication between the endpoints **E** and terminates once the network flow is cleared from the switch flow table, as described next.

Dynamic Translator Box Creation. We use TEEs to run *translator boxes* that decrypt the TLS traffic on the respective flow, use application protocol translators to convert it to the target protocol, and re-encrypt it before forwarding. A translator box is instantiated whenever the `translation` action is triggered by a new network flow matching the flow table rule. Depending on the implementation, translator boxes are instantiated either as a child process of the switch daemon (in-switch) or external to the switch. In-switch translator boxes are instantiated by the ovs-vswitch daemon, while external translator boxes are instantiated by the network controller. Translator boxes are deployed in TEEs to ensure execution isolation, confidentiality, and integrity of packet data.

To instantiate a translator box, the parent process first invokes the creation of a TEE and deploys the translation logic configured for the pair of application layer protocols in the respective network flow. Next, a *verifier* network function attests the integrity and authenticity of the translator box [6]. Following a successful attestation, a trusted certificate authority network function provisions the cryptographic artifacts necessary for intercepting the TLS communication between endpoints. The exact artifacts depend on the approach for TLS interception, as described next in Sect. 4.1. The parent process of the translator box terminates it once the respective flow is evicted from the datapath cache.

In our current implementation, we used Intel SGX enclaves to create TEEs. SGX enclaves rely on a trusted computing base of code and data loaded at enclave creation time. Program execution within an enclave is transparent to the underlying operating system and other mutually distrusting enclaves running on the platform. The CPU is an enclave's root of trust; it prevents access to the enclave's memory by the operating system and other enclaves. Library operating systems were used in this context to facilitate both the portability and performance of legacy applications in SGX enclaves [27].

TLS Interception. We focus on the TLS v1.2 [8] and v1.3 [10] for transport security due to their wide adoption. We further use the ME-TLS [20] protocol extension for TLS interception in protocol translator boxes. The use of ME-TLS allows delivering session key materials to translator boxes in-band and does not require additional TLS connections or round-trips. Moreover, this allows retaining backward compatibility with TLS 1.3 [10] through implicit protocol

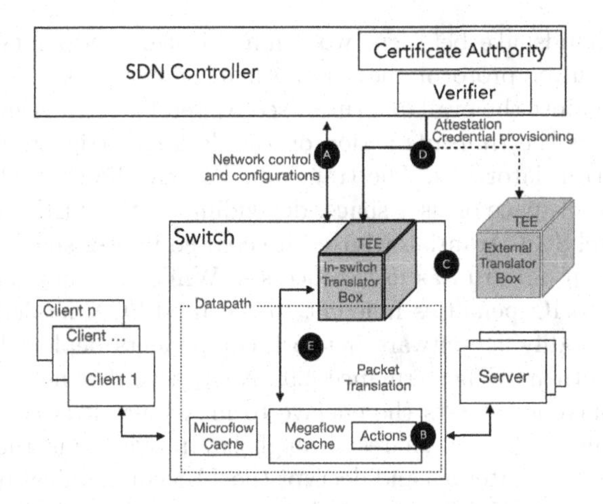

Fig. 1. Conceptual illustration of the Chuchotage architecture

version negotiation. In case one of the endpoints does not support ME-TLS, communication remains encrypted but without protocol translation.

Following the TLS1.3 specification [10], ME-TLS reuses the TLS 1.3 `Finished` message to achieve two additional goals, endpoint authentication and translator box negotiation (agreement between client and server about the translator boxes to be used). For middlebox negotiation, the `ClientFinished` and `ServerFinished` messages each contain two middlebox lists specifying the translator involved in each direction of the network path. Once both client and server endpoints complete the translator box negotiation by including the list of chosen translator boxes to the `ClientFinished` and `ServerFinished` messages, they distribute the necessary session key materials to selected translator boxes. ME-TLS achieves this through an additional `SessionKeyDistribution` message sent by the endpoints to the translator boxes on the communication path. The `SessionKeyDistribution` message is an application data message (not a handshake message); the record field of the message contains a byte sequence, which is an HMAC generated from the shared secret between the client and server (ss_{cs}^{ibe}) and a string constant to differentiate from other application data records, followed by encrypted session key materials for the translator boxes. The ME-TLS protocol uses a property of the BF-IBE scheme [15] that allows endpoints and translator boxes to establish a shared secret between each other through zero-round secret negotiation. In BF-IBE, a trusted authority called a private key generator (PKG) generates private keys for endpoints and translator boxes using their identities and a master key. The endpoints (client and server) can then use the shared secret to encrypt the session key materials communicated to the translator box instances.

Translator Box Integration with OvS. Translator boxes are created following the *translate* action in the flow rules and are instantiated during the transport layer protocol handshake between two communicating endpoints, regardless of the application layer protocol they use. An incoming packet to the switch is first matched against the available rules (see Appendix A.2). A match against a rule that contains the *translate* action on the datapath triggers the creation of a flow-specific translator box. The translator box can be created either on the datapath in kernel space or user space, depending on the TEE implementation. When using Intel SGX, translator boxes are created in user space enclaves, since SGX enclaves can only run as user processes. While this may affect their performance (due to IO penalties inherent to the Intel SGX model), recent work indicates that modifying software network components deployed in TEEs can help to improve their IO performance [29]. Next, a verifier network function of the network controller attests the enclave to make sure it is trustworthy, then the enclave receives the key shares through key provisioning that allows it to compute session key materials and decrypt the TLS communication between the endpoints in the respective flow Fig. 1. Attestation and key provisioning are done *in parallel* with the ongoing transport layer protocol handshake. All subsequent packets in the respective flow will be processed by the translator box.

Protocol to Protocol Translation. Once a translator box inside the enclave receives a packet from the respective flow, it first decrypts the packet using the session key materials computed from the key shared received from the network controller. Next, the translator box parses the decrypted packet, extracts the application data, and formats it into the destination protocol format. Finally, the formatted packet is re-encrypted and returned to the switch data path to be forwarded to its destination.

4.2 Challenges

The design of Chuchotage addresses several important challenges, namely enabling distributed protocol translation and combining TLS interception with attestation primitives of the trusted execution environments. We address distribution and scalability by introducing the concept of ephemeral, flow-specific, on-demand translator boxes created by software switches on the network path. To achieve scalability in high density networks, multiple switches, and SDN controllers can be used in the network depending on the network topology and available resources. Chuchotage combines the ME-TLS protocol for TLS interception [20] with the SGX attestation protocol to provide an uninterrupted chain of trust that includes the communicating endpoints, the translator box, and the certificate authority by the communicating parties.

4.3 Operating Flow

In the following operating flow description, we assume that a network administrator uses a deployment blueprint to define flow rules for the endpoints included in the topology. For the types of devices and communication protocols known beforehand, the network administrator specifies a `translate` action for the flows that require translation. Note that two distinct translation policies will be specified for each source-destination pair in a flow where endpoints implement distinct application layer protocols. In the following operating flow description, we assume the latest version of TLS, version 1.3; while other TLS versions can be made compatible with this operating flow, this requires additional adjustments.

In-line Operating Flow. The sequence diagram in Fig. 2 illustrates how translator boxes instantiated by the switch obtain the session keys negotiated between two endpoints, client and server:

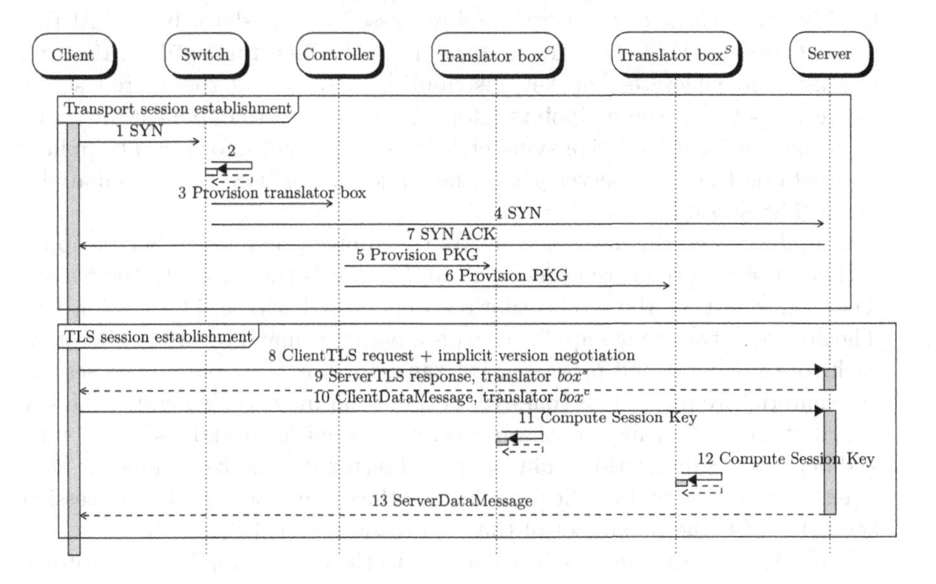

Fig. 2. Chuchotage operating flow

- The Client initiates a communication session by sending a TCP `SYN` packet to Server (step 1).
- A Switch on the network path matches the `SYN` packet against entries in its Microflow cache. Since the Client did not communicate with the Server earlier, the search continues in the Megaflow cache and ultimately in the OpenFlow flow tables, where it matches the translation policy defined by the network administrator (step 2). The results of Megaflow cache lookup will be cached in Microflow cache. The switch triggers the controller to instantiate the translator boxes (step 3).
- The `SYN` packet is immediately forwarded to the destination; this avoids introducing additional latency (step 4).

- The controller instantiates translator boxes for the flows t_c (client-server, step 5) and t_s (server-client, step 6). The controller instantiates the translator box in a TEE, attests it [3,6] and provisions key shares generated by the PKG [15].
- The server returns a `SYN ACK` reply, the transport session is established at this point (step 7).
- The TLS negotiation starts; the negotiation follows the TLS 1.3 with the ME-TLS extensions [20] (step 8). The Client TLS request includes an implicit version negotiation to check that the Server supports the ME-TLS extensions. The Server TLS response follows the TLS 1.3 specification and additionally specifies the identifier of the server translator box (step 9).
- Next the Client starts sending encrypted application data (step 10).
- The `ClientDataMessage` packet containing application data is matched in the Microflow cache of the switch and processed by translator box t_c. At this point, t_c obtains its session key material from the `SessionKeyDistribution` message and generates the key distribution bytes using the shared secret between itself and the endpoints (step 11). It derives the application traffic secrets, allowing it to derive symmetric keys to encrypt and decrypt application data on the client-server path. The session key is used for the remainder of the TLS session.
- Having decrypted the data, t_c converts the application data to Server application protocol format, re-encrypts it, and forwards the packet to the Server;
- The Server returns the application data encrypted with a TLS session key. The `ServerDataMessage` application data packet is matched in the Microflow cache of the Switch and processed by translator box t_s; t_s obtains its session key material from the `SessionKeyDistribution` message, generates the key distribution bytes using the shared secret between itself and the endpoints, and derives the application traffic secrets allowing it to derive symmetric keys to encrypt and decrypt application data on the server-client path. The session key is used for the remainder of the TLS session (step 12);
- t_s converts the decrypted application data to the client's application protocol format, re-encrypts it and forwards it to the client (step 13);
- Translation of application data continues for the remainder of the TLS session; the translator boxes are terminated once the network flow is evicted from the Switch flow cache.

In case of DTLS, the operating flow is modified such that the translator boxes are created after the `ClientHello` message.

5 Implementation

For evaluation purposes, we implemented Chuchotage with two popular IoT protocols, CoAP and HTTP. Our implementation includes the following components. A *client*, an HTTP client representing an IoT device contacts a server with a different protocol, a *Server*, A CoAP server is listening for client connections. *Open vSwitch (OvS)*: endpoints are connected to OvS through the same bridge and OvS is responsible for forwarding incoming client or server packets to the translator box, as well as forwarding outgoing packets from the translator box to their destinations; *SDN controller*: an SDN controller manages the network flows to improve network performance. For that we used Ryu[3], an open source controller. Whenever OvS does not find any matching entry in its flow caches to handle packets in need of translation it contacts the controller, which will trigger a translation. *Translator box*: via the translation process, the controller creates a translator box responsible for translating the traffic between client and server. In the translator box, we used an HTTP to CoAP parser/formatter library[4], capable of parsing and converting HTTP to CoAP messages and vice versa. *TEE*: to ensure execution isolation as well as confidentiality and integrity during packet translation, we ported the protocol translator to an SGX enclave using the Occlum library OS [27]. Occlum[5] is a memory-safe library OS for SGX. Note that for implementing other protocol translation (other than CoAP and HTTP), a new parser/formatter is required but the rest of the components will remain unchanged.

5.1 Implementation Choices

In Chuchotage, the translator box can be instantiated either by the network controller (external) or OvS (in-switch [28]). In our prototype implementation, the SDN controller deploys an SGX enclave with the translator code and attests it, as deploying, managing, and debugging external translators is easier for network administrators. Attestation can be done locally or remotely based on the location of the appraiser and of the target enclave [6]. In our prototype implementation, the SDN controller (appraiser) and translator box (target) both exist on the same platform and hence we used local attestation with a trusted enclave that exists on the SDN controller and keypair provisioning. As mentioned above, the TEE hosting the translator box can be instantiated using several alternative approaches, both virtualization-based [33] or process-based [18,21]. Enterprise deployments should consider remote attestation of translator boxes, or a combination of both as supported by some virtualization-based TEEs [14]. The choice of TEEs depends on constraints on application portability, security, and performance.

For TLS interception, we assume that session key materials are distributed to the involving parties including the client, server, and SDN controller prior to the handshake procedure and ME-TLS overhead is explicitly excluded in our evaluation since it only affects the handshake, not the actual communication.

Our translation policy is defined by using features extracted from the traffic flow, namely a combination of specific source and destination IP addresses and port numbers. When an incoming flow matching these features triggers the translation action and the packets in the matching flow are forwarded to the

[3] https://ryu-sdn.org/.
[4] https://github.com/keith-cullen/FreeCoAP.
[5] https://github.com/occlum/occlum.

translator. After translation, the packets are sent back to the switch to be forwarded to their own destination. While distinct translator boxes can be created for inbound and outbound flows (client to server or server to client, see Sect. 4.3), we use one translator box for both in- and outbound flows.

5.2 Testbed

Our testbed consists of four docker containers representing client, server, a Ryu controller, and a translator box deployed in an SGX enclave (see Fig. 3), as it can be seen in Fig. 3 the testbed is compatible with different pairs of clients and servers. OvS was installed on the host OS and the four docker containers are connected to the OvS via one bridge (`br0` in Fig. 3). Each container is connected to the bridge through its own virtual interface, indicated as `vethp` in Fig. 3. Whenever a flow needs to be translated, Ryu creates and attests an SGX enclave inside container 3. The translation is done inside the enclave and the flow to be translated is afterwards forwarded through container 3.

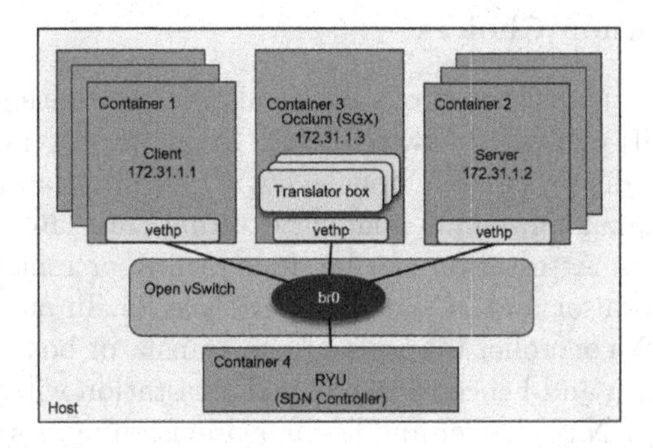

Fig. 3. Testbed overview

6 Evaluation

6.1 Performance Evaluation

We conducted several tests to evaluate the performance of Chuchotage. In the first test, we send packet batches of different sizes (100, 1000, and 10000 packets) from the client to the server and measured the translation time for the entire batch. We also measure the transmission time, i.e. the time between sending the first and last packets excluding the handshake. In this test, the client sends empty `HTTP GET` messages translated to CoAP confirmable `Reset` messages.

We measured translation and transmission time both with and without SGX, to measure the effect of the TEE on the performance (see Fig. 4). Without a TEE, the translator box is created inside container 3 in Fig. 3. As illustrated in Fig. 4, both translation and transmission times slightly increase with the use of a TEE (Intel SGX in this prototype); however, this increase is acceptable in most IoT networks considering the added benefit of protecting network traffic confidentiality. Error bars are based on standard deviation.

We also compared our results with the transmission time of a vanilla CoAP to CoAP communication. Confirmable CoAP Reset messages were sent from a CoAP client to the CoAP server. The transmission time for transferring 100, 1000, and 10000 packets respectively are: 0.00719, 0.07428, and 0.70909 s. We consider these values as a reference point for the added overhead by the translation procedure compared to a vanilla CoAP to CoAP transfer.

In a second test, we send batches of 100 packets of different sizes (128, 256, and 512 Bytes) from client to server and record their translation and transmission time with and without using a TEE. In this test, we send `HTTP POST` requests from the client to the server and they are translated to CoAP confirmable `POST` requests. The results of this test show that using a TEE (Intel SGX in this prototype) results in increasing both the translation and transmission time (see Fig. 5). Packet data length does not affect the translation time.

In a third test, we measured the time to complete a successful handshake. The handshake takes place between the client, server, and translator box; however, the translator box is transparent for the client and server. The overall handshake time (an average of 10 handshakes) including local attestation (0.0164 s), enclave creation (0.80410 s), and additional communication between the Chuchotage components averages 2.83574 s. This is roughly equal to transferring and translating 10000 packets; a vanilla CoAP to CoAP handshake averages to 0.000907 s. However, the handshake is only performed once before translating all subsequent packets in the flow.

The performance of our proposed protocol translator is not comparable to centralized approaches, such as gateway or proxy-based approaches, since they are not suitable for large heterogeneous distribution deployments and often do not consider security of network traffic. Chuchotage is not also comparable to other existing protocol translation solutions, as earlier highlighted in Sect. 3.1.

6.2 Security Evaluation

Reflecting the structure of the threat model (Sect. 2.1) we discuss the security of network communication and of protocol translation.

Network Security. Chuchotage uses TLS 1.3 [10] to implement transport layer security - including key establishment - and inherits its confidentiality and integrity properties. On the other hand, Chuchotage also inherits any potential vulnerabilities yet to be discovered in TLS 1.3; this underscores the importance of following vulnerability management best practices. The security of ME-TLS extensions to TLS 1.3 is reviewed in detail in [20]. There are several types of network based attacks that can target Chuchotage, such as Denial of Service (DoS) or traffic flooding. Similar to other contexts, DoS attacks can be mitigated by DoS prevention techniques including intrusion detection and prevention systems, using load balancers, filtering, etc.

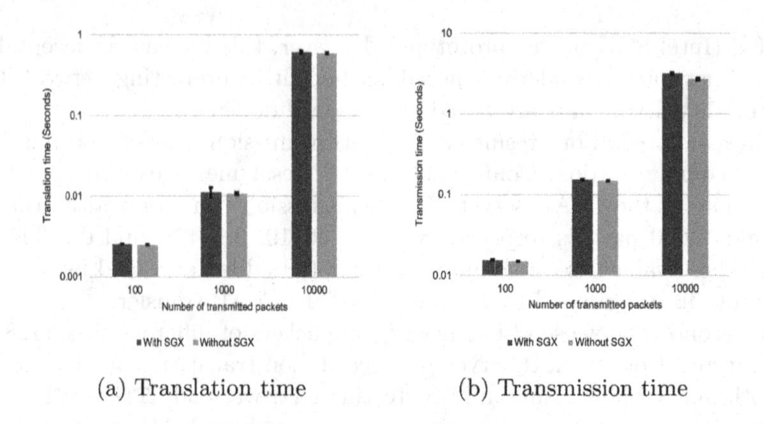

(a) Translation time (b) Transmission time

Fig. 4. Translation and transmission time of translating different number of packets

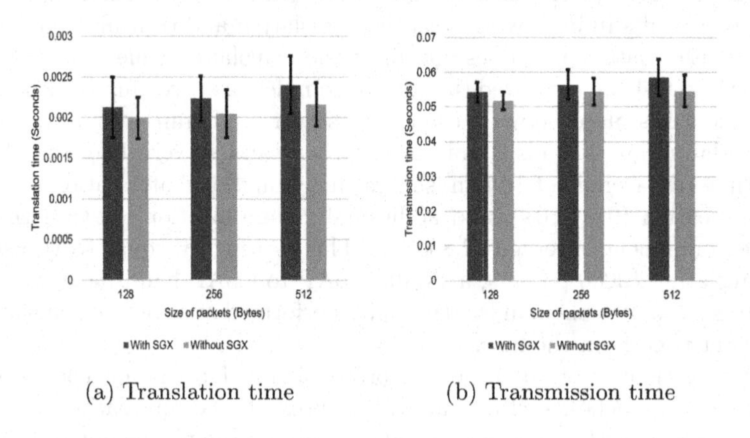

(a) Translation time (b) Transmission time

Fig. 5. Translation and transmission time of translating different packet sizes

Protocol Translation. Availability of a Chuchotage deployment can be ensured through network deployment best practices. High availability is an inherent capability of Chuchotage as translator boxes are instantiated and deployed in TEEs by switches throughout the network topology.

Translator boxes are central to the security of protocol translation and network communication in Chuchotage. Integrity of the protocol translation software deployed in translator boxes is verified through attestation [6]. The chain of trust evaluated through attestation is specific to the platform implementation of the TEE. During protocol translation confidentiality of provisioned cryptographic material and intercepted network traffic is ensured through TEE isolation mechanisms that include memory isolation on hardware or firmware level, run-time memory encryption, and cache flushing upon execution transition [33].

In our current prototype implementation, we use Intel SGX enclaves as a TEE implementation target. SGX is vulnerable to a wide category of attacks reviewed in [24]. Chuchotage can be vulnerable to any attacks applicable to

SGX. However, there are a number of mitigation techniques that can be used to mitigate attacks on realistic applications deployed in SGX enclaves [13].

7 Conclusion

In this paper, we proposed Chuchotage, an in-line application layer protocol translator with transport layer security. Chuchotage relies on secure TLS interception, efficient protocol translation, and fault-tolerant distributed architecture. In Chuchotage we translate, and re-encrypt network flows with minimal latency, on the network path. Scalability is guaranteed by growing the number of translator boxes with the number of flows; translator boxes are instantiated by individual software network switches in the deployment. Depending on the capabilities of the underlying platform and their support for TEEs, Chuchotage allows creating translator boxes either in-switch or external to the switch, in kernel space or user space. We implemented a Chuchotage prototype for HTTP to CoAP translation with Intel SGX enclaves and Open vSwitch. Our evaluation indicates a slight increase in the translation and transmission time. This overhead depends primarily on the choice of TEE in the implementation.

Acknowledgment. This work was financially supported in part by the Swedish Foundation for Strategic Research, with the grant RIT17-0035, and by the Wallenberg AI, Autonomous Systems and Software Program (WASP).

A Appendix

A.1 Common IoT Communication Protocols

In the TCP/IP network model, the physical or data link layer is responsible for physical transmissions; characteristics of applications - such as latency and availability - directly impact traffic characteristics on the link layer. The network layer is responsible for routing and forwarding packets; considering that IoT devices are often resource-constrained, the information necessary for routing should be kept at a minimum. Finally, transport layer protocols (such as TCP and UDP) manage end-to-end communication between network endpoints.

Physical network gateways are commonly used for interoperability in the physical and network layers or transport layer [25]. Gateways have limited scalability [25]: as the number of IoT devices increases, special connectors are required for their interaction, thus adding both cost and complexity to the network.

Application communication between network endpoints is implemented on the application layer. Middleware can perform translation in the application layer; however, connecting middleware components risks further reducing interoperability by locking applications to a specific technology. Interception proxies are an alternative for application layer translation; however, proxies cause delays since all traffic transits through proxies even when translation is unnecessary [7].

Proxies and middleware currently available for application layer protocol translation are increasingly unsuitable for secure, distributed, and transparent application layer protocol translation.

Several application layer protocols - namely HTTP, CoAP, MQTT, and AMQP - have been widely reviewed in academic publications and adopted in large scale deployments. We compare these protocols in Table 1.

Table 1. IoT protocols comparisons

IoT protocols	HTTP	CoAP	MQTT	AMQP
Transport layer	TCP	UDP	TCP	TCP
Security	TLS/SSL	DTLS	TLS/SSL	TLS/SSL
Architecture	Req/Res	Req/Res	Pub/Sub	Pub/Sub
QoS	No	Yes	Yes	Yes
Low power/lossy networks	Fair	Excellent	Fair	Fair
Dynamic discovery	No	Yes	No	No

A.2 Open vSwitch Overview

OpenvSwitch (OvS) is an open source programmable switch [31] that implements packet forwarding on the datapath; it is a flow-based switch, where clients install flows determining forwarding decisions. Flows are installed in a cache level structure that assists the datapath to execute actions on received packets, e.g. allow, drop, etc. For each ingress packet, the datapath consults its cache and forwards the packet to its destination if matching entries exist. For each cache miss, the datapath issues an upcall and forwards the packet to ovs-vswitchd. A datapath can be deployed as a kernel module or in user space with additional firmware support. Packet classification in OvS is computationally expensive, mostly due to the many types of matching fields. Matching is implemented in a hash table of flow rules, with matching fields hashed as keys. OvS uses a modified Tuple Space Search (TSS) algorithm for packet classification. The algorithm searches through the hash map tables based on the maximum entry's priority and terminates after finding the highest priority matching flow rule. Early OvS releases implemented OpenFlow processing exclusively as a kernel module. However, the difficulty of developing and updating kernel modules motivated moving packet classification to user space. A multi-level cache structure kernel implementation compensates the resulting performance impact. The cache structure consists of two levels with increasing lookup costs: a microflow cache (or Exact Match Cache) and a larger megaflow cache. The megaflow cache matches multiple flows with wildcards [23].

Open vSwitch Forwarding. Figure 6 illustrates the OvS internals. An incoming packet reaches the datapath from either a physical or virtual NIC (1). In the datapath, the switch runs a first search based on an exact match (2). If there is a matching entry in the microflow cache, the packet is sent to the specific table in the megaflow cache to retrieve the required actions. Otherwise, the forwarding process performs a second search in the next cache line (3). Failing to find a match, the datapath uses upcalls (4) to inform the ovs-vswitchd that it cannot handle the packet. The ovs-vswitchd uses the classification process (5) to obtain a matching rule via its flow tables. Next, ovs-vswitchd returns to the datapath, inserts the entry in the cache (6), and returns the packet to the kernel (7). Finally, the datapath forwards the packet to the intended destination (8). Failing to find matching information in the flow tables, ovs-vswitchd sends a packet-in request to the network controller to get a matching rule for the unknown packet.

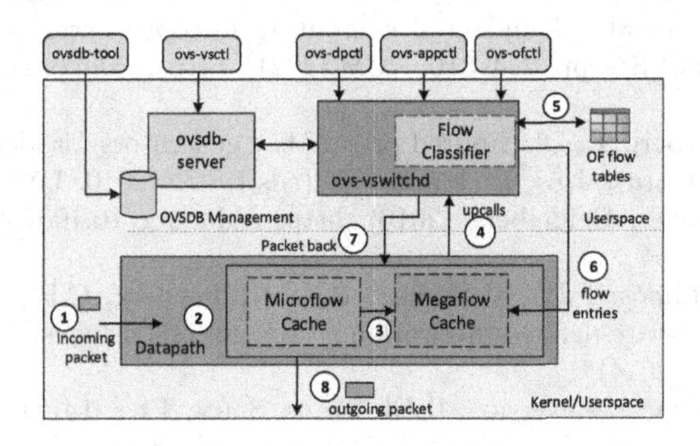

Fig. 6. An overview of Open vSwitch internals

References

1. Semantic Integration & Interoperability Using RDF and OWL (2005). https://www.w3.org/2001/sw/BestPractices/OEP/SemInt/. Accessed 15 Oct 2020
2. AMD SEV-SNP: Strengthening VM isolation with integrity protection and more. White paper, Advanced Micro Devices, January 2020
3. Anati, I., Gueron, S., Johnson, S., Scarlata, V.: Innovative technology for CPU based attestation and sealing. In: Proceedings of the 2nd International Workshop on Hardware and Architectural Support for Security and Privacy, New York, NY, USA, vol. 13, p. 7. ACM (2013)
4. Baek, J., Kim, J., Susilo, W.: Inspecting TLS anytime anywhere: a new approach to TLS interception. In: Proceedings of the 15th ACM Asia Conference on Computer and Communications Security, pp. 116–126 (2020)
5. de Carné de Carnavalet, X., van Oorschot, P.C.: A survey and analysis of TLS interception mechanisms and motivations. arXiv e-prints. arXiv-2010 (2020)
6. Coker, G., et al.: Principles of remote attestation. Int. J. Inf. Secur. **10**(2), 63–81 (2011). https://doi.org/10.1007/s10207-011-0124-7
7. Derhamy, H., Eliasson, J., Delsing, J.: IoT interoperability-on-demand and low latency transparent multiprotocol translator. IEEE Internet Things J. **4**(5), 1754–1763 (2017)
8. Dierks, T., Rescorla, E.: The Transport Layer Security (TLS) Protocol Version 1.2. RFC 5246 (Proposed Standard), August 2008. https://doi.org/10.17487/RFC5246. https://www.rfc-editor.org/rfc/rfc5246.txt. Updated by RFCs 5746, 5878, 6176, 7465, 7507, 7568, 7627, 7685, 7905, 7919

9. Dolev, D., Yao, A.: On the security of public key protocols. IEEE Trans. Inf. Theory **29**(2), 198–208 (1983)
10. Rescorla, E.: The Transport Layer Security (TLS) Protocol Version 1.3. RFC 8446, August 2018. https://doi.org/10.17487/RFC8446
11. Garbelini, M.E., Wang, C., Chattopadhyay, S., Sumei, S., Kurniawan, E.: SweynTooth: unleashing mayhem over bluetooth low energy. In: 2020 USENIX Annual Technical Conference (USENIX ATC 2020), pp. 911–925. USENIX Association, July 2020. https://www.usenix.org/conference/atc20/presentation/garbelini
12. Gregg, B.: Systems Performance, 2nd edn. Pearson, London (2020)
13. Hosseinzadeh, S., Liljestrand, H., Leppänen, V., Paverd, A.: Mitigating branch-shadowing attacks on intel SGX using control flow randomization. In: Proceedings of the 3rd Workshop on System Software for Trusted Execution, pp. 42–47 (2018)
14. Hunt, G.D.H., et al.: Confidential computing for openpower. In: EuroSys 2021, New York, NY, USA, pp. 294–310. ACM (2021). https://doi.org/10.1145/3447786.3456243
15. Kate, A., Goldberg, I.: Distributed private-key generators for identity-based cryptography. In: Garay, J.A., De Prisco, R. (eds.) SCN 2010. LNCS, vol. 6280, pp. 436–453. Springer, Heidelberg (2010). https://doi.org/10.1007/978-3-642-15317-4_27
16. Kreutz, D., Ramos, F.M., Verissimo, P.E., Rothenberg, C.E., Azodolmolky, S., Uhlig, S.: Software-defined networking: a comprehensive survey. Proc. IEEE **103**(1), 14–76 (2014)
17. Lam, S.S.: Protocol conversion. IEEE Trans. Softw. Eng. **14**(3), 353–362 (1988)
18. Lee, D., Kohlbrenner, D., Shinde, S., Asanović, K., Song, D.: Keystone: an open framework for architecting trusted execution environments. In: Proceedings of the Fifteenth European Conference on Computer Systems. EuroSys 2020, New York, NY, USA. ACM (2020). https://doi.org/10.1145/3342195.3387532
19. Lee, H., et al.: maTLS: How to make TLS middlebox-aware? In: NDSS (2019)
20. Li, J., Chen, R., Su, J., Huang, X., Wang, X.: ME-TLS: middlebox-enhanced TLS for internet-of-things devices. IEEE Internet Things J. **7**(2), 1216–1229 (2020). https://doi.org/10.1109/JIOT.2019.2953715
21. McKeen, F., et al.: Innovative instructions and software model for isolated execution. Hasp@ isca **10**(1) (2013)
22. McKeown, N., et al.: Openflow: enabling innovation in campus networks. SIGCOMM Comput. Commun. Rev. **38**(2), 69–74 (2008)
23. Medina, J., Paladi, N., Arlos, P.: Protecting OpenFlow using Intel SGX. In: 2019 IEEE Conference on Network Function Virtualization and Software Defined Networks (NFV-SDN), pp. 1–6. IEEE (2019)
24. Nilsson, A., Bideh, P.N., Brorsson, J.: A survey of published attacks on Intel SGX. arXiv preprint arXiv:2006.13598 (2020)
25. Noura, M., Atiquzzaman, M., Gaedke, M.: Interoperability in internet of things: taxonomies and open challenges. Mob. Netw. Appl. **24**(3), 796–809 (2019)
26. Safaric, S., Malaric, K.: Zigbee wireless standard. In: Proceedings of ELMAR 2006, pp. 259–262 (2006). https://doi.org/10.1109/ELMAR.2006.329562
27. Shen, Y., et al.: Occlum: secure and efficient multitasking inside a single enclave of Intel SGX. In: Proceedings of the Twenty-Fifth International Conference on Architectural Support for Programming Languages and Operating Systems, ASPLOS 2020, New York, NY, USA, pp. 955–970. ACM (2020). https://doi.org/10.1145/3373376.3378469

28. Svenningsson, J., Paladi, N., Vahidi, A.: Faster enclave transitions for IO-intensive network applications. In: Proceedings of the ACM SIGCOMM 2021 Workshop on Secure Programmable Network INfrastructure, SPIN 2021, New York, NY, USA, pp. 1–8. ACM (2021). https://doi.org/10.1145/3472873.3472879
29. Svenningsson, J., Paladi, N., Vahidi, A.: SGX-bundler: speeding up enclave transitions for IO-intensive applications. In: The 22nd IEEE/ACM International Symposium on Cluster, Cloud and Internet Computing. IEEE-Institute of Electrical and Electronics Engineers Inc. (2022)
30. Tolk, A.: Composable mission spaces and M&S repositories-applicability of open standards. In: Spring Simulation Interoperability Workshop, Arlington, VA (2004)
31. Tu, W., Wei, Y.H., Antichi, G., Pfaff, B.: Revisiting the open vswitch dataplane ten years later. In: Proceedings of the 2021 ACM SIGCOMM 2021 Conference, SIGCOMM 2021, New York, NY, USA, pp. 245–257. ACM (2021). https://doi.org/10.1145/3452296.3472914
32. Uddin, M., Mukherjee, S., Chang, H., Lakshman, T.: SDN-based multi-protocol edge switching for IoT service automation. IEEE J. Sel. Areas Commun. **36**(12), 2775–2786 (2018)
33. Yao, J., Zimmer, V.: Virtual Firmware, pp. 459–491. Apress, Berkeley (2020). https://doi.org/10.1007/978-1-4842-6106-4_13
34. Zanella, A., Bui, N., Castellani, A., Vangelista, L., Zorzi, M.: Internet of things for smart cities. IEEE Internet Things J. **1**(1), 22–32 (2014)

Study on the Effect of Face Masks on Forensic Speaker Recognition

Georgiana Bogdanel[1], Nadia Belghazi-Mohamed[1], Hilario Gómez-Moreno[1,2](✉), and Sergio Lafuente-Arroyo[1,2]

[1] Escuela Politécnica Superior, Departamento de Teoría de la Señal y Comunicaciones, Universidad de Alcalá, 28871 Alcalá de Henares, Madrid, Spain
{georgiana.bogdanel,nadia.belghazi}@edu.uah.es,
{hilario.gomez,sergio.lafuente}@uah.es

[2] Instituto Universitario de Investigación en Ciencias Policiales, Facultad de Derecho, Universidad de Alcalá, 28801 Alcalá de Henares, Madrid, Spain

Abstract. The COVID-19 pandemic has led to a dramatic increase in the use of face masks. Face masks can affect both the acoustic properties of the signal and the speech patterns and have undesirable effects on automatic speech recognition systems as well as on forensic speaker recognition and identification systems. This is because the masks introduce both intrinsic and extrinsic variability into the audio signals. Moreover, their filtering effect varies depending on the type of mask used. In this paper we explore the impact of the use of different masks on the performance of an automatic speaker recognition system based on Mel Frequency Cepstral Coefficients to characterise the voices and on Support Vector Machines to perform the classification task. The results show that masks slightly affect the classification results. The effects vary depending on the type of mask used, but not as expected, as the results with FPP2 masks are better than those with surgical masks. An increase in speech intensity has been found with the FPP2 mask, which is related to the increased vocal effort made to counteract the effects of hearing loss.

Keywords: Automatic speaker recognition · Acoustic features · Face mask · Forensic acoustics

1 Introduction

The human voice is the most natural form of communication between people. In addition to words, the speech signal conveys information about the speaker's identity, emotional state, acoustic environment, language and accent. Speaker recognition is the task of identifying the speaker behind an acoustic recording. Forensic speaker recognition involves the identification of a person in any possible speech recording scenario at a crime scene. From this point of view, recognition systems encounter difficulties when dealing with evidence that has some modification or is altered, for example, by the introduction of elements that

C. Alcaraz et al. (Eds.): ICICS 2022, LNCS 13407, pp. 608–621, 2022.
https://doi.org/10.1007/978-3-031-15777-6_33

affect its production as in the case of face masks [22,25,29]. Therefore, speaker recognition techniques cannot produce reliable speaker comparison results under difficult forensic conditions that introduce some alteration in voice production, which limits the admissibility of recorded voice evidence in court.

1.1 Related Work

In recent years, the use of face masks has increased enormously worldwide due to the COVID-19 pandemic. They offer protection against external pathogens, thus preventing human-to-human transmission, but there is no denying the barrier masks pose to the act of communication. For this reason, since the appearance of COVID-19, different studies have appeared in the literature dealing with the possible effects of masks on the acoustic evaluation of the voice.

Several studies point the impact of the use of face masks from a clinical point of view [1,15,19] as well as in the context of the assessment of language ability [5]. It is also worth mentioning the work done by [20] on the comparison of speech intelligibility using different types of masks. This work confirmed the alteration that the mask causes in the intelligibility of speakers by producing variations in relation to the distribution of acoustic power in the frequency bands, especially in the first and second formants[1]. Previous work has also shown that the use of masks produces a significant loss of speech transmission, which varies according to the type of mask used, mainly attenuating sounds above 1 kHz, i.e., there is both attenuation of high frequency sounds and effects on the directivity of the signal [6,18]. However, it is worth noting the scarcity of literature on the effect of masks on speaker recognition. It is worth mentioning some of the works that have served as a basis for the present one [24,25]. It is true that neither of them are contemporary to the period of the COVID-19 pandemic and neither addresses the problems encountered in relation to voice matching. Nevertheless, both present results with English utterances, whereas the present study aims to analyse the effects of masks on Spanish speakers.

1.2 Threat Model

All the mentioned studies in the previous section show that face masks influence voice production and can function as an undeniable acoustic filter for speech. This effect can introduce modifications in the characteristics of the human voice, producing important changes, both consequential and adaptive. For this reason, in the forensic field, there is an undeniable need to know the effect of the use of face masks, due to the possible alteration of the properties of the acoustic signal that they can induce.

In general, we understand forensic field to be the one who has the purpose of collection and analysis of evidence for the clarification of a criminal action.

[1] In Spanish, only the first two formants, F1 and F2, have the characteristics that make the difference between one vowel sound and another. This is due to the relationship between the location of the formants in the spectrogram and the position of the organs involved in articulation [27].

Therefore, in the case of Forensic Speaker Recognition, we will work with spoken broadcasts that are in some way related to the investigation of a criminal act. Furthermore, in most of the cases, a recovered versus a control[2] acoustic emission is confronted, where an expert must carry out an analysis of both voices with the aim of concluding whether the signals correspond to the same person. The possibility of making this comparison lies in the importance of the control sample obtained being generated under the same conditions, or with the highest possible degree of reproducibility, as the recording conditions of the recovered sample. This is the premise under which the different professionals of the State Security Forces and Corps work. However, as a result of the health crisis caused by COVID-19, the conditions for obtaining control samples were slightly altered. This is due to the obligatory use of face masks by those involved, even though the circumstances in which the samples were obtained were not the same.

As stated in several works [8,12], it cannot be denied that voice analysis in forensic conditions involves several drawbacks, since its interpretation and reliability is subject to various factors that generate different types of variability. There are several intrinsic and extrinsic factors that cause variability in acoustic signals. Intrinsic variability refers to the human factors involved in speech production (emotion, speech rate, effort, etc.) causing variation in speech at the time of its generation. On the other hand, extrinsic variability refers to the way in which the acoustic signal reaches the listener, or the analysis system used (background noise, distortions introduced by the transmission or recording channel) producing variations in the signal after it is generated [25]. Covering the face, with face masks in this case, implies both intrinsic variability, i.e., the mask will affect speech production; and extrinsic variability, i.e., it will affect speech production and intelligibility and absorb the signal triggering a loss on speech transmission.

Also, the impact of face masks on speech production does not occur to the same degree with all types of face masks currently on the market but is dependent on the specifications of the individual masks. The higher the level of barrier provided by a face shield, the greater its impact on the acoustic characteristics of speech. During the COVID-19 pandemic, the most marketed face masks in Spain were surgical masks and self-filtering masks. Surgical masks achieve a breathability level of less than $40\,\mathrm{Pa/cm^2}$, which favours the ability to breathe during use; however, in self-filtering masks, the resistance to exhalation is higher. A lower pressure value will indicate that air passes through the mask with less resistance, i.e., it is easier to breathe and talk with the mask on. Therefore, it is presumed that self-filtering masks will have a greater filtering effect than surgical masks, due to the differences in the specifications and materials that make up each mask.

According to the guidelines of the European Network of Forensic Science Institutes (ENFSI), the presence of circumstances that hinder, modify, or alter

[2] According to Delgado-Romero [8], "a control sample is one that belongs to a known subject, while a recovered sample is anonymous, i.e. the identity of the person who carried it out is not known".

evidence during the evaluation stage may limit the admissibility of voice evidence in court and, moreover, state-of-the-art Speech Recognition techniques are not able to produce reliable results under difficult forensic conditions [10]. Therefore, it is undeniable that the use of masks could produce an added challenge for the work of experts by increasing the variability already present in the human voice and by changing the conditions for obtaining control and recovered evidence, making already unsuitable conditions more difficult.

In this paper, we test an Automatic Speaker Recognition (ASR) model based on Support Vector Machine (SVM), with samples recorded under different conditions than those used in the training of the model. Specifically, we tested the strength of the classifier against acoustic signals obtained in the presence of two forensically relevant face masks (surgical IIR and self-filtering FFP2). To support the investigation of the present study, a new corpus of recordings has been generated. Using a recognition system based on acoustic feature extraction, we trained the model using utterances from different speakers under normal conditions, i.e., without the use of a mask, creating a robust identification model. In the test phase, we should find differences in recognition ratios with respect to the use of the mask. All the above has led to the following hypotheses:

1. The use of masks has an impact on the Automatic Speaker Recognition model, decreasing the efficiency of forensic speaker identification. Thus, the voice features of a speaker using a mask are different from those without it.
2. The use of the FFP2 self-filtering respirator will result in a lower hit ratio in speaker identification, due to the higher degree of exhalation resistance in its specifications compared to the IIR surgical mask.

The remainder of this document is organised as follows. Section 2 presents in detail the methodology used for corpus generation, the features extracted and the SVM model used. In Sect. 3, the results generated by the model are presented. Finally, Sect. 4 provides a few concluding remarks and future lines of research.

2 Method

Having set out the objectives, this section will develop the process followed to reach the results and their subsequent conclusions. The methodology followed can be classified as inductive. From the analysis of the collected recordings, a set of interesting vocal characteristics has been extracted from the point of view of forensic speaker identification, in order to feed the voice classifier.

2.1 Voice Recordings

The study involved 30 speakers (15 female and 15 male) with an average age of 26.5 years. All were Spanish speakers and did not report any voice or hearing problems at the time of the study. Informed consent to participate in this study was obtained from all participants.

The recording environment was the same for all participants. The room noise level was measured to take it into account and to avoid possible interferences in the subsequent analysis of the recordings. Voice recordings took place in a laboratory of the Department of Signal Theory and Communications at the University of Alcalá with an average ambient noise of 44.7 dBA.

Participants had to read, in a fluent and normal way, balanced sentences with previously structured linguistic features that facilitate the forensic identification of the speakers. Specifically, three sentences were chosen from the LOCUPOL voice bank, owned by the Spanish National Police Force [27]. This reading task was performed in three different conditions for the speaker: (1) wearing a surgical mask, (2) wearing a FFP2 mask, and (3) not wearing a mask (control samples). During the recording session, participants were required to maintain their habitual voice in terms of pitch, volume and phonation type for each condition, to minimize mask-independent intra-speaker variability in voice production [8].

The entire speech set was captured with an Olympus LS_100 recording device at a constant distance of 30 cm from the speakers, in front of the mouth axis. Recordings were obtained with a sampling frequency of 44.1 kHz and a resolution of 32 bits. The signals were then pre-processed and saved on a laptop in Waveform Audio Format for edition with Audacity [2]. This pre-processing phase consisted of trimming the audios to homogenize their duration, as well as to dispense with those silences before and after the reading of the sentences to avoid possible errors during the extraction of audio features. In this way, a total of 270 voice samples were obtained, with no notable interferences and under controlled conditions. The database generated and the algorithm used is publicly available[3]. Table 1 shows all the speakers involved in the study.

Table 1. Speakers compilation. With each type of mask, 3 sentences were recorded, generating a total of 9 recordings per speaker.

# Speaker	1	2	3	4	5	6	7	8	9	10
Gender	Female	Female	Female	Male	Female	Female	Male	Female	Male	Male
Age	22	22	21	52	18	22	21	18	18	26
# Speaker	11	12	13	14	15	16	17	18	19	20
Gender	Female	Male	Male	Male	Male	Male	Female	Female	Female	Male
Age	22	56	19	18	24	54	40	18	18	59
# Speaker	21	22	23	24	25	26	27	28	29	30
Gender	Male	Female	Male	Female	Female	Male	Female	Male	Male	Female
Age	18	24	42	18	18	19	19	23	25	21

[3] The corpus repository and the ASR system are available at: https://tinyurl.com/8h8dteuu.

2.2 Acoustic Features

As a previous step to any ASR system, the acoustic characteristics of the voice have to be extracted. For this stage, a Python library for audio processing, Librosa [13,14], was used. The features that were extracted from the speech signal with this library correspond to spectral ones. These features are widely used for efficient ASR systems because the spectrum reflects the anatomical structure of the vocal tract. Different speakers will have different spectra. In addition, an advantage of spectral methods is that logarithmic scales, which mimic the functional characteristics of the human ear, could be used to improve the recognition ratio.

There are different features that can be extracted: Mel spectrogram, Mel Frequency Cepstral Coefficients (MFCC), Chroma, Tonnetz and Spectral Contrast [14]. However, many of these features are related and can lead to misclassification errors since the feature vector composition is redundant and confusing for the classification system, as well as inefficient. Therefore, in order to select the most efficient acoustic features for the model, a series of tests were performed to evaluate the performance during training. The best results were obtained using only the MFCCs, specifically 50 per audio. Although in the literature [11,21] it is mentioned that a smaller number of MFCCs is sufficient to represent the characteristics of the voice in a time interval, in our case it has been empirically proven that a larger number of coefficients undoubtedly improves the classification results.

MFCCs are declared by the European Telecommunications Standards Institute (ETSI) as one of the most widely used and well-known features in speech recognition. They are a set of features extracted from the spectral domain representation of the speech signal [9] that concisely describe the general shape of a spectral envelope, represent the vocal timbre and one of their main advantages is the isolation of noise or the removal of irrelevant information from the sound background.

2.3 Automatic Speaker Recognition Model

An ASR system acts as a pattern classifier, each pattern consisting of a set of previously extracted features from the speech signal that allow speaker identification. In this case, the classification is carried out using Support Vector Machines (SVM), with the Radial Basis Function (RBF) kernel which can be described by the following formula:

$$K(x, x') = e^{-\gamma \|x-x'\|^2} \tag{1}$$

This supervised learning modelling technique, consolidated in the field of pattern recognition, allows the resolution of nonlinear classification problems with several classes in an optimal way [7]. It is especially effective when you have a small training set, as in our case. The LibSVM implementation [4] included in the Scikit Learning Python library [17] has been used in all experiments.

In order to build a robust model that correctly assigns speaker identities, it is essential to have a properly trained SVM classifier. For training we have the set of unmasked (NM) audios and a total of 30 classes corresponding to the total number of speakers. The set of audios was divided according to the three sentences uttered by each informant (S1, S2 and S3). Thus, the unmasked audios belonging to two of the sentences were used for training and the audios of the remaining sentence were used to test.

To train the model two hyperparameters must be tuned. C, or regularisation coefficient, is a weighting factor between empirical risk and structural risk, i.e. error tolerance. The adjustment of this parameter may involve a trade-off between margin maximisation and classification violation. A high value of C will imply narrow margins and few observations will be misclassified, which is equivalent to a model well fitted to the data. Conversely, as C decreases the tolerance to errors on the margin will be higher because the margin is wider, this is equivalent to a flexible model [28].

The hyperparameter γ (from the RBF kernel) defines the distance of influence of a single training point. A small value of γ implies a larger distance between the observations separating the classes of the SVM, which makes the estimation more conservative. However, a larger parameter is detrimental because the radius of influence of the support vectors only includes the support vector and the points must be very close to be considered of the same class. Consequently, the model will tend to over-fit despite regularising with C [28]. The estimation of both parameters has been performed on the training data set using a grid search. After evaluating different values for the C and γ parameters, the best results were obtained with $C = 100$ and $\gamma = 0.045$.

The training vectors have been normalised by removing the mean and scaling to a unit variance. The process has been performed for each component of the vector independently. This same scaling has subsequently been applied to the test vectors, but using the mean and variance information obtained by scaling the training vectors. This is done to ensure that the predictors of greater magnitude do not have more influence than the rest.

In the training phase, three different models were used, corresponding to the combinations made with two of the unmasked audio sets for each sentence (S1, S2 or S3). The remaining sentence were used for testing. Therefore, recombining the audios of these three sentences allows training the ASR system with the whole set of unmasked audios, obtaining a robust model for the three sentences in the set. After the training phase, accuracy of 100% have been obtained for all the models tested (see Table 2).

By obtaining the maximum possible accuracy in the three training models, the results can be extrapolated to a single model that uses the whole set of unmasked audios (S1_NM, S2_NM and S3_NM), assuming that its training would also reach an accuracy of 100% when testing. On this model, the set of audios with both surgical masks (MS) and self-filtering masks (MFFP2) will be used to test the classifier. In this way, the predictions obtained for each type of mask can be compared with the values obtained without the mask, in order to appreciate the possible influence of the latter on Forensic Speaker Recognition.

Table 2. Training Results.

Model		Accuracy (%)	#Errors
Train	Test		
S1_NM, S2_NM	S3_NM	100	0
S2_NM, S3_NM	S1_NM	100	0
S1_NM, S3_NM	S2_NM	100	0

3 Results

In this section we present the classification results for the different proposed models. As previously stated, the training was performed with a SVM and adjusting the C and γ parameters, until the highest ratio was achieved. Subsequently, the performance of the models was evaluated considering common evaluation metrics for classification problems.

3.1 ASR Results

The results of the speaker identification on the set of utterances collected using the proposed ASR system are presented in Table 3. In general, a decrease in the accuracy of the automatic classification model employed can be seen when one of the masks is used. When the training and test sentences are pronounced without the use of the mask, the ratio reaches 100% in all test cases. However, when the recorded utterances with the presence of masks are introduced in the testing process, there is a degradation of the performance of the model, introducing errors in the speaker classifications.

Table 3. Results of the performance measures of the no mask models in training.

Model		Accuracy (%)	#Errors
Train	Test		
S1_NM, S2_NM	S3_NM	100	0
S2_NM, S3_NM	S1_NM	100	0
S1_NM, S3_NM	S2_NM	100	0
S1_NM, S2_NM	S1_MS, S2_MS, S3_MS	95.5	4
S2_NM, S3_NM	S1_MS, S2_MS, S3_MS	94.4	5
S1_NM, S3_NM	S1_MS, S2_MS, S3_MS	94.4	4
S1_NM, S2_NM, S3_NM	S1_MS, S2_MS, S3_MS	96.7	3
S1_NM, S2_NM	S1_MFFP2, S2_MFFP2, S3_MFFP2	96.7	3
S2_NM, S3_NM	S1_MFFP2, S2_MFFP2, S3_MFFP2	97.8	2
S1_NM, S3_NM	S1_MFFP2, S2_MFFP2, S3_MFFP2	97.8	1
S1_NM, S2_NM, S3_NM	S1_MFFP2, S2_MFFP2, S3_MFFP2	98.9	1

The results presented in Table 4 are promising. When the training of the models is performed with the speech in the presence of a mask of either of the two types used in this study, the recognition system shows undoubtedly improved identification ratios, reaching 100% accuracy in all cases. As mentioned, the recordings were obtained under the same acoustic conditions, the features extracted in this case were also the 50 MFCCs, and the models have been trained in this case with the same parameters. Therefore, these results seem to confirm the first hypothesis, that the use of masks has some effect on the Automatic Speaker Recognition model, decreasing the efficiency in the identification of speakers.

Table 4. Results of the performance measures of the mask models in training.

Model		Accuracy (%)	#Errors
Train	Test		
S1_MS, S2_MS	S3_MS	100	0
S2_MS, S3_MS	S1_MS	100	0
S1_MS, S3_MS	S2_MS	100	0
S1_MFFP2, S2_MFFP2	S3_MFFP2	100	0
S2_MFFP2, S3_MFFP2	S1_MFFP2	100	0
S1_MFFP2, S3_MFFP2	S2_MFFP2	100	0

Additionally, the confusion matrix was calculated for the model that includes the three sentences without masks. Only one confusion matrix is shown for each of the conditions (with surgical mask and with FFP2 mask). Figure 1 show these graphical representations where it is clear a decrease in the efficiency of the system due to the use of masks. The classification errors identified are consistent with the gender of each speaker, that is, without changing it.

It is worth noting the presence of a repeated error in all the tests carried out. One of the participants (speaker 17), when wearing any face mask, is not correctly identified in many cases. The conditions for obtaining the participant's recordings do not differ from those of the rest of the participants. Therefore, the explanation for this result may lie in a voice characteristic that usually generates problems for speaker recognition systems: the accent. In the speech set used in this study, the participant mentioned is the only one with a highly identifiable accent. Works such as [26] claim that accent is a matter of great interest in the forensic field. In our experiments, there is no problem without a mask but the use of any mask leads to errors in this case.

However, contrary to expectations, the second hypothesis does not hold true. The automatic system when using speech with FFP2 masks gives clearly better results than when using speech with surgical masks. Due to the higher degree of exhalation resistance of FFP2 masks, speech production, intelligibility and signal transmission suffer a considerable loss, which should be translated into lower accuracy in speaker identification. As can be seen in Table 3, the results obtained are not consistent with the initial hypothesis. At this point, further research was carried out in order to find a possible explanation for these data.

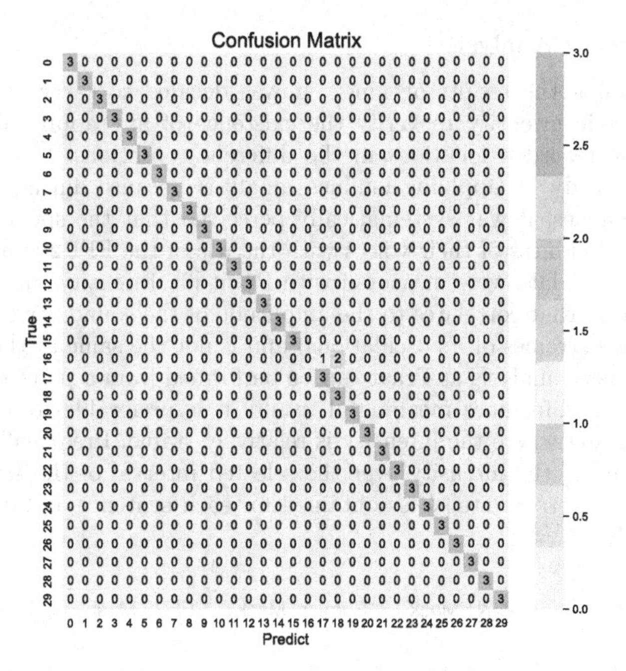

(a) Classifier trained without face mask and tested with surgical mask.

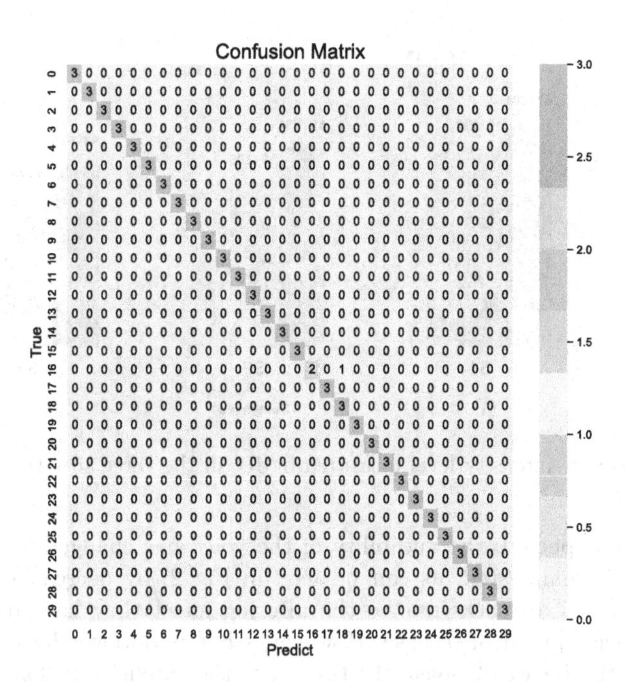

(b) Classifier trained without face mask and tested with FFP2 mask.

Fig. 1. Confusion matrices

3.2 Intensity Analysis

After analysing the results obtained, it was decided to carry out an analysis of the acoustic intensity to verify the existence of any appreciable difference between the recorded locutions in the different conditions. It was decided to extend the study by applying loudness analysis because during the recording of the utterances, it was systematically perceived that the speakers tended to increase the loudness of their voices with the use of the FFP2 mask.

All acoustic data were measured with Praat [3]. Intensity (measured in dB) is a parameter that correlates to the amplitude of the sound wave (the distance between the extremes of its oscillations) and is not the same as the volume of a signal. Intensity analysis at Praat can be performed from a point or over a time interval. To obtain more meaningful results it is preferable to select a region of the recording where the intensity is relatively stable, Praat will generate the average value of the intensity over the selected interval of the signal. Figure 2 compiles the average intensities obtained for each sentence and condition used in this study.

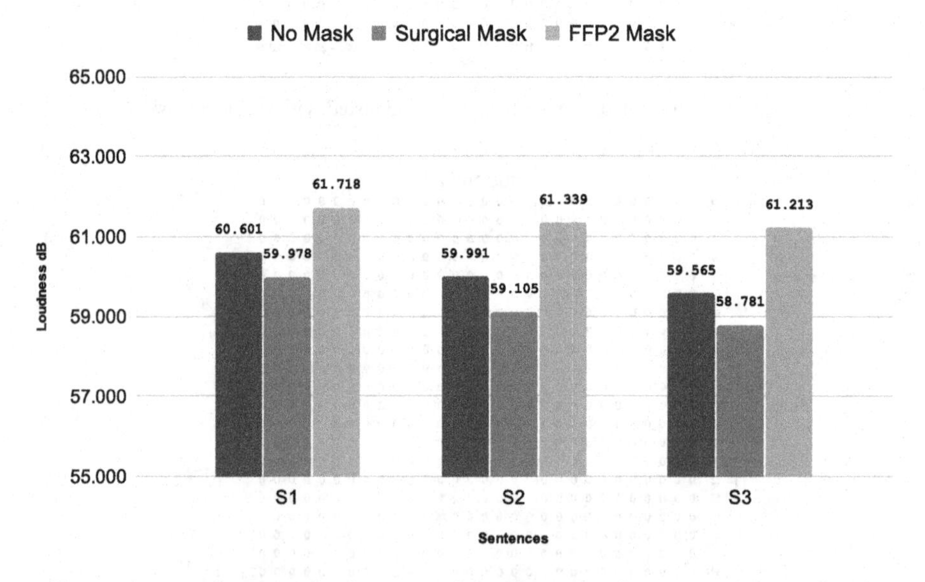

Fig. 2. Average intensity level measured in dB, in the three situations analysed.

As might be expected, the audibility of the voice has changed while wearing a mask. This is manifested, as can be seen in Fig. 2, by decreasing intensity in the case of surgical masks due to the filtering effect they have on exhaled air to prevent the spread of disease. However, what is noteworthy are the results obtained with the FFP2 mask. In this case, the intensity is somewhat higher than in cases where the mouth is less covered or not covered at all.

These results are recurrent for the utterances of all speakers, with an average increase in loudness of approximately 1.6 dB when wearing the FFP2 mask over wearing a surgical mask and an average increase of 1.06 dB over wearing no mask. It is interesting to note the repeated increase in loudness values for all speakers, which leads us to think that this may be the explanation for the non-verification of the second hypothesis.

The masks act as a low-pass acoustic filter, i.e. they reduce the information in high frequencies, which leads to an alteration in the perception of the voice. Therefore, in these unfavourable speech conditions, speakers can adapt their phonation styles to compensate the action of this filter and to improve voice clarity [16]. With the use of the mask we have the sensation that less audibility is produced. With the use of the FFP2 mask, this sensation increases due to its greater resistance to exhalation. Different studies, such as [23], confirm that this may result in an increase in the intensity of the voice, as we believe may have occurred in obtaining the locutions used in this research.

4 Conclusion

The results of this study describe the alterations produced in an ASR system by the use of face masks, due to the changes introduced by them in the acoustic voice characteristics. To support this research, a corpus of recordings of 30 speakers has been collected and a SVM-based classifier has been used after the extraction of MFCCs from the utterances. It has been observed that the system experiences a deterioration of its performance when the set of utterances is used with both types of masks (surgical and FPP2). However, against all odds, a lower percentage of accuracy was obtained for the surgical mask audios than for the FFP2 mask audios. This may be because of the fact that the acoustic filter of these masks produces an alteration of voice perception. Due to the specifications of the FFP2 masks, this alteration is even higher, resulting in an increase in the intensity of the voice to counteract this effect.

The slight degradation of the results indicates the capability of the state-of-art recognition systems to mitigate the face mask mismatch. Within the forensic field, these results underline the importance of obtaining samples under the same conditions to achieve a better identification and that, in some specific cases, wearing a mask could lead to errors. It is necessary to study the acoustic changes in specific parts of speech to obtain more knowledge about the effect of wearing masks in the detected problematic cases. Therefore, although the use of specific software to compare voices can be useful, it should not be the only source of information and forensic expert analysis is desirable.

To conclude, the future work will focus on:

- Increase the number of speakers and the utterances for each one to test the effect of wearing masks.
- Include more linguistically diverse speakers in the training data.
- Add both noise to speech samples and words that hinder intelligibility to test the scalability and robustness of the model.
- Use deep learning algorithms to perform an automatic acoustic features extraction.

Acknowledgement. This research was supported by the Research Grants Program of the Universidad de Alcalá. We acknowledge the valuable counsel and resources provided by G. A. Acha Ruiz, as well as to the Department of Forensic Acoustics of the "Comisaría General de Policía Científica" for the access to the LOCUPOL database sentences.

References

1. Atcherson, S.R., et al.: The effect of conventional and transparent surgical masks on speech understanding in individuals with and without hearing loss. J. Am. Acad. Audiol. **28**, 58–67 (2017)
2. Audacity Team: Audacity (R): Free audio editor and recorder [computer application] (2022). www.audacityteam.org/
3. Boersma, P., Weenink, D.: Praat: doing phonetics by computer [computer program] (version 6.2.10) (2009). www.praat.org. Accessed 17 Mar 2022
4. Chang, C.C., Lin, C.J.: LIBSVM: a library for support vector machines. ACM Trans. Intell. Syst. Technol. **2**, 1–27 (2011). https://doi.org/10.1145/1961189.1961199
5. Coniam, D.: The impact of wearing a face mask in a high-stakes oral examination: an exploratory post-SARS study in Hong Kong. Lang. Assess. Q.: Int. J. **2**, 235–261 (2005)
6. Corey, R.M., Jones, U., Singer, A.C.: Acoustic effects of medical, cloth, and transparent face masks on speech signals. J. Acoust. Soc. Am. **148**, 2371–2375 (2020). https://doi.org/10.1121/10.0002279
7. Cortes, C., Vapnik, V.: Support-vector networks. Mach. Learn. **20**(3), 273–297 (1995). https://doi.org/10.1023/A:1022627411411
8. Delgado-Romero, C.: La Identificación de Locutores en el Ámbito Forense (in Spanish). Ph.D. thesis, Departamento de Comunicación y Publicidad II. Facultad de Ciencias de la Información. Universidad Complutense de Madrid. España (2001)
9. Deller, J.R., Proakis, J.G., Hansen, J.H.L.: Discrete-Time Processing of Speech Signals. Institute of Electrical and Electronics Engineers, New York (2015)
10. ENFSI: Forensic speech and audio analysis working group terms of reference for forensic speaker analysis. European Network of Forensic Science Institutes, pp. 1–4 (2008)
11. Leu, F.Y., Lin, G.L.: An MFCC-based speaker identification system. In: IEEE 31st International Conference on Advanced Information Networking and Applications, AINA, pp. 1055–1062. Institute of Electrical and Electronics Engineers Inc. (2017). https://doi.org/10.1109/AINA.2017.130
12. Maher, R.C.: Principles of Forensic Audio Analysis. Springer, Heidelberg (2018). https://doi.org/10.1007/978-3-319-99453-6
13. McFee, B., et al.: Thassilo: librosa/librosa: 0.9.1 (2022). https://doi.org/10.5281/zenodo.6097378
14. McFee, B., et al.: Librosa: audio and music signal analysis in python. In: Proceedings of the 14th Python in Science Conference, pp. 18–24 (2015). https://doi.org/10.25080/majora-7b98e3ed-003
15. Mendel, L.L., Gardino, J.A., Atcherson, S.R.: Speech understanding using surgical masks: a problem in health care? J. Am. Acad. Audiol. **19**, 686–695 (2008)
16. Nguyen, D.D., et al.: Acoustic voice characteristics with and without wearing a facemask. Sci. Rep. **11**, 1–11 (2021). https://doi.org/10.1038/s41598-021-85130-8

17. Pedregosa, F., et al.: Scikit-learn: machine learning in Python. J. Mach. Learn. Res. **12**, 2825–2830 (2011)
18. Pörschmann, C., Lübeck, T., Arend, J.M.: Impact of face masks on voice radiation. J. Acoust. Soc. Am. **148**, 3663–3670 (2020). https://doi.org/10.1121/10.0002853
19. Radonovich, L.J., Jr., Yanke, R., Cheng, J., Bender, B.: Diminished speech intelligibility associated with certain types of respirators worn by healthcare workers. J. Occup. Environ. Hyg. **7**, 63–70 (2009)
20. Randazzo, M., Koenig, L.L., Priefer, R.: The effect of face masks on the intelligibility of unpredictable sentences. In: Proceedings of Meetings on Acoustics, vol. 42 (2020). https://doi.org/10.1121/2.0001374
21. Rao, K.S., Vuppala, A.K.: Speech Processing in Mobile Environments. SECE, Springer, heidelberg (2014). https://doi.org/10.1007/978-3-319-03116-3
22. Ratha, N.K., Connell, J.H., Bolle, R.M.: Enhancing security and privacy in biometrics-based authentication systems. IBM Syst. J. **40**(3), 614–634 (2001)
23. Ribeiro, V., Dassie-Leite, A.P., Pereira, E.C., Santos, A.D.N., Martins, P., de Irineu, R.: Effect of wearing a face mask on vocal self-perception during a pandemic. J. Voice (2020)
24. Saeidi, R., Huhtakallio, I., Alku, P.: Analysis of face mask effect on speaker recognition. In: Proceedings of the Annual Conference of the International Speech Communication Association, INTERSPEECH, vol. 08, pp. 1800–1804 (2016). https://doi.org/10.21437/Interspeech.2016-518
25. Saeidi, R., Niemi, T., Karppelin, H., Pohjalainen, J., Kinnunen, T., Alku, P.: Speaker recognition for speech under face cover. In: Proceedings of the Annual Conference of the International Speech Communication Association, INTERSPEECH, vol. 2015-January, pp. 1012–1016 (2015). https://doi.org/10.21437/interspeech.2015-275
26. Saleem, S., Subhan, F., Naseer, N., Bais, A., Imtiaz, A.: Forensic speaker recognition: a new method based on extracting accent and language information from short utterances. Forensic Sci. Int.: Digital Invest. **34**, 300982 (2020)
27. Sánchez-López, D.: Análisis acústico y sonográfico de la vocal /a/ para su aplicación en el ámbito de las ciencias forenses (2016). https://tinyurl.com/h5ncwpv. (in Spanish)
28. Wainer, J., Fonseca, P.: How to tune the RBF SVM hyperparameters? An empirical evaluation of 18 search algorithms. Artif. Intell. Rev. **54**, 4771–4797 (2021)
29. Wu, Z., Evans, N., Kinnunen, T., Yamagishi, J., Alegre, F., Li, H.: Spoofing and countermeasures for speaker verification: a survey. Speech Commun. **66**, 130–153 (2015)

Video Forensics for Object Removal Based on Darknet3D

Kejun Zhang[1], Yuhao Wang[1(✉)], and Xinying Yu[2]

[1] Beijing Electronic Science and Technology Institute, Beijing, China
bestiwyh@126.com

[2] School of Cyberspace Security Beijing University of Posts and Telecommunications, Beijing, China

Abstract. To address the problems of insufficient analysis of time-domain information of tampered video by 2D convolutional neural networks and the loss of details in the pooling layer when processing frame images, a 3D video object removal tamper detection and localization model based on Darknet53 optimization network is proposed. While the Darknet53 network can fully retain the detail information in the frame, we try to give the two-dimensional Darknet53 network, which can only process spatial information, the ability to process time-domain information, and extend the two-dimensional convolutional layer into a three-dimensional convolutional layer, and also improve the detection efficiency by adjusting the network structure to reduce feature redundancy, making it more suitable for efficient processing of video tampering detection binary classification tasks. A D3D (Darknet3D) network is constructed to improve feature adequacy representation. Experimental results reveal that the temporal domain classification accuracy of the tamper detection model based on the Darknet3D is 98.9%, and the average Intersection over Union of spatial localization and tamper area labeling is 49.7%, which can effectively detect and locate the object removal tampering.

Keywords: Video object removal tampering · Spatio-temporal localization · Darknet53

1 Introduction

Surveillance video is a crucial basis in legal and criminal investigations. But with the widespread use of cheap and powerful video editing tools, non-professionals can easily tamper with surveillance video without leaving any traces [3]. This has gradually aroused concern about the credibility of digital video content in related fields. Therefore, there is an urgent need for a tampering detection technology that can effectively verify the authenticity and security of video content [1,4,5].

Supported by the National Key Research and Development Program on Cyberspace Security (2018YFB0803601) and the Advanced Discipline Construction Project of Beijing Universities (20210086Z0401).

C. Alcaraz et al. (Eds.): ICICS 2022, LNCS 13407, pp. 622–637, 2022.
https://doi.org/10.1007/978-3-031-15777-6_34

In recent years, video tampering methods based on deep learning technology have developed rapidly. The tamper can use this technology to delete some key content in the original video, making it invisible in subsequent video sequences. This kind of tampered video may cause wrong decisions by law enforcement agencies, such as missed catches. Therefore, the spatio-temporal detection technology for video object removal tamper has significant research value [8]. At present, domestic and foreign scholars have used Convolutional Neural Networks (CNN) with different structures for target detection and positioning, traffic sign recognition, and video action recognition. However, for a target that has a rigid object boundary, the video tamper can remove the boundary of the area where the tampered object is located without showing a trace. Usually, the tamperer initially removes the video object, then fills it with the background scene, and performs transition processing on the adjacent area, so that the tampering is not easy to detect. Therefore, we should not only detect forged videos effectively but also locate the tampered area accurately.

1.1 Related Work

The detection algorithm based on artificially designed features and the detection algorithm based on deep learning are currently the two most widely used object removal detection algorithms. Based on the difference in the correlation between the noise residuals in the tampered area and the original area, Hsu et al. [6] detect and locate the forged area in the video by using Gaussian Mixture Model (GMM) to model the noise residual correlation distribution between the original area and the tampered area. Based on the significant difference between the motion vector distribution of the foreground area in the tampered video and the original video, Li et al. [10] used the correlation of the motion vector before and after the video to detect and locate the tampered area of the video taken by the static camera. In order to solve the problem of low efficiency of tamper detection algorithm, Su et al. [17] propose a method for detecting tampered frames through Energy Factors (EF) and using an Adaptive Parameter-Based Visual Background Extractor (AVIBE) to locate the tamper detection method of tampering area. Saxena et al. [15] proposed a method based on the consistency of optical flow features to detect and locate the tampered area of video inpainting.

Most recently, the rapid development of deep learning technology has injected new vitality into the research of digital video tampering detection [16]. Deep learning technology relies on powerful learning and computing capabilities to extract the features from a large-scale dataset, avoiding the disadvantages of incomplete input information expressed by traditional methods, and greatly improving the accuracy of detection. Many scientific researchers have proposed a variety of tamper detection methods based on deep learning technology, and have made outstanding contributions to the field of video forensics. Yao et al. [23] use the frame difference method to construct the frame difference sequence, extract the frequency signal of the frame difference sequence, and pass the frequency signal through a high-pass filter, and finally use a convolutional neural network to distinguish the tampered frames. Kono et al. [9] consider the temporal and spatial consistency of video by combining CNN and Recurrent Neural Network (RNN). The introduction of the codec structure enables this method to locate the tampered area in the video. Long et al. [11] use propose a method with $3D$ convolutional neural network for frame deletion detection. This method extracts 16 consecutive frames from the video as the input of the $3D$ convolutional neural network. The trained network will be used to detect whether there is frame deletion tampering on the 8th and 9th frames of a 16-frame video clip. Aiming at the high false alarm rate generated when the camera is moving, zooming in or zooming out, Long et al. added confidence processing to the output of the proposed CNN to reduce the impact of the false alarm rate.

1.2 Contributions

Convolutional neural networks have many advantages that traditional methods do not have. However, most of the current tamper detection methods can only process the spatial information in the frame, while ignoring the unique temporal information of the video. In addition, the pooling layer will cause the image to be blurred when pooling the image, resulting in the loss of image details, and it will also ignore the association between the whole and the part. Based on the above problems, this paper proposes a Spatio-temporal localization model for video object removal tamper based on the optimized Darknet53 network. This model can accurately determine whether the input video has been tampered with and locate the tampered area. The main contributions of our work are as follows.

- The Darknet53 feature extraction network with the characteristic of reducing the pooling layer is introduced into video tampering detection to extract the feature of frame images. And the dimension of the data is reduced by continuous convolutional layers and residuals. Thereby reducing the negative effect of gradient caused by pooling, so as to fully retain the detailed information within the frame.
- The temporal domain information is introduced into the Darknet53 network, and construct the three-dimensional Darknet53 feature extraction network using a three-dimensional convolutional layer rather than a two-dimensional one. The features of frame images are jointly extracted from the temporal and spatial perspectives to promote the adequacy of feature expression.
- By compressing the number of residual blocks of the three-dimensional Darknet53 network, a three-dimensional Darknet53 optimized network is designed to reduce network dimensions and feature redundancy.

2 Darknet53

The target detection network YOLOv3 uses Darknet53 fully convolutional network [14] as a feature extraction network to extract image features. The Darknet53 network is integrated with the deep residual network on the basis of YOLOv2's feature extraction network Darknet19 [13], and consists of a series of

1×1 and 3×3 convolutional layers and residual blocks. The batch normalization layer and the LeakyReLu layer together constitute its smallest module. The Darknet53 network structure is shown in Fig. 1.

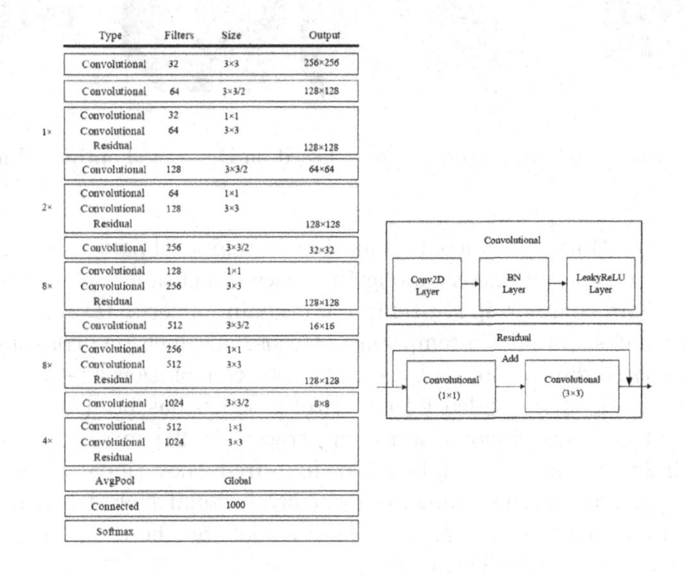

Fig. 1. The network structure of Darknet53.

3 Video Tampering Detection Model Based on Optimized Darknet53

The network model in this paper is shown in Fig. 2, which consists of a time domain classifier and a spatial domain locator. The feature extractor in the model is used to extract the temporal and spatial features of the video; the two-classifier is used to distinguish between the original frame and the tampered frame; the RPN box regressor is used to locate the tampered area in the tampered frame.

3.1 Optimized Darknet53 Feature Extraction Network

In the convolutional neural network, the pooling layer often causes the image to be blurred when pooling the image data, resulting in the loss of detailed information in the image. Therefore, we introduce the full convolutional network Darknet53 as the backbone network, and use its feature of no pooling layer to perform feature extraction on frame images.

The convolutional layer can not only extract the spatial information in the image data, but also extract the temporal information in the data along the

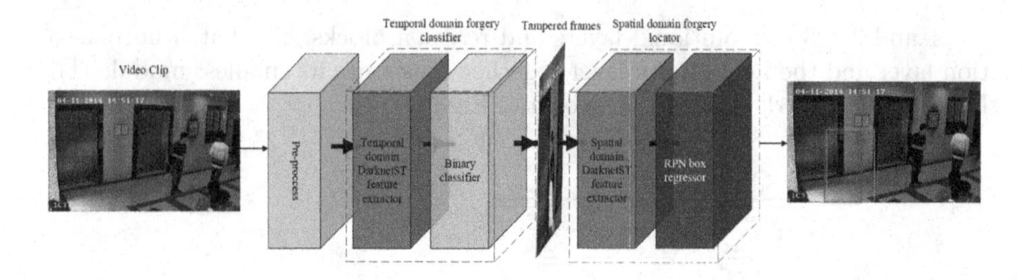

Fig. 2. Video tampering detection model based on Darknet53 optimized network

time dimension. However, video data has both a temporal pattern along the time dimension and a spatial pattern along the space dimension. These two patterns will form a more complex spatio-temporal domain pattern. $3D$ convolution can realize the analysis of spatio-temporal patterns, in which one convolution axis is along the time dimension, and the other two convolution axes are along the spatial dimension of the video frame. After a video has been tampered with, it will definitely leave obvious tampering traces [19,21] in the high-frequency area [20,22,24]. Using the correlation of high-frequency signals in consecutive frames, you can use a three-dimensional convolutional neural network to The tampered frame is processed, and the features left by the tampering operation in the tampered frame are extracted.

Therefore, this article designed the Darknet53 optimized network. On the basis of the original two-dimensional Darknet53 network, the two-dimensional convolutional layer is expanded to the three-dimensional convolutional layer, and the three-dimensional Darknet53 network is constructed, so that the network can process the information of continuous frame images in the temporal domain and the spatial domain at the same time.

In the field of target detection, it is usually necessary to classify and recognize hundreds of objects, and it is necessary to learn the characteristics of each object [12]. However, the detection of tampered video is essentially a two-class problem, so directly using the high-dimensional Darknet53 network for tampering video detection will cause a certain degree of feature redundancy. Therefore, this article adjusts the scale of the three-dimensional Darknet53 network, reduces the number of residual blocks in the network, reduce network dimensions, and designs the Darknet53 optimized network, while maintaining the ability of the three-dimensional Darknet53 network to extract video spatio-temporal features at multiple scales, it also reduces network redundancy and improves training speed. After that, this article applies the Darknet53 optimized network to feature extraction. The five consecutive frames of images first pass through a three-dimensional maximum pooling layer with a sliding window size of $1\times c\times c$ [18] for dimensionality reduction. Subsequently, a high-pass filter SRM layer using three different convolution kernels is used to reduce the influence of the movement of objects in the frame on the frame content, improve the ability of the network model to learn features, and make the traces of tampering more obvious. The three convolution kernels of the high-pass filter are shown in Fig. 3.

$$\frac{1}{4}\begin{pmatrix} 0 & 0 & 0 & 0 & 0 \\ 0 & -1 & 2 & -1 & 0 \\ 0 & 2 & -4 & 2 & 0 \\ 0 & -1 & 2 & -1 & 0 \\ 0 & 0 & 0 & 0 & 0 \end{pmatrix} \quad \frac{1}{12}\begin{pmatrix} -1 & 2 & -2 & 2 & -1 \\ 2 & -6 & 8 & -6 & 2 \\ -2 & 8 & -12 & 8 & -2 \\ 2 & -6 & 8 & -6 & 2 \\ -1 & 2 & -2 & 2 & -1 \end{pmatrix} \quad \frac{1}{4}\begin{pmatrix} 0 & 0 & 0 & 0 & 0 \\ 0 & 0 & -1 & 0 & 0 \\ 0 & -1 & 4 & -1 & 0 \\ 0 & 0 & -1 & 0 & 0 \\ 0 & 0 & 0 & 0 & 0 \end{pmatrix}$$

Fig. 3. Three filter kernels used to extract noise features.

The three convolution kernels of the SRM layer [25] can extract three kinds of high-frequency residual data of the video frame, and then use the extracted high-frequency data as the input of the Darknet53 optimization network to continue to extract the high-frequency features of the frame image in the spatio-temporal domain. The structure of the Darknet53 optimized network is shown in Fig. 4. $R_i (i = 1, 2, 3, ..., 11)$ represents the i-th residual block. In the residual block, each convolutional layer is batch normalized, and the activation function is LeakyReLu.

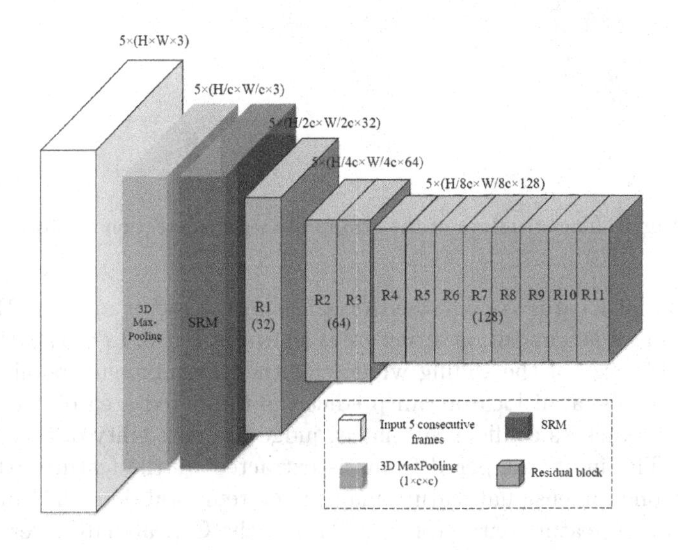

Fig. 4. The structure of the optimized Darknet53 feature extraction network.

3.2 Temporal-Domain Classifier and Spatial-Domain Locator

The temporal-domain classifier is composed of a feature extractor and a two-classifier, which can determine whether the input of five consecutive video frames has been tampered with, and its composition structure is shown in Fig. 5. In the temporal feature extractor, the size of the sliding window of the $3D$ maximum pooling layer is $1 \times 3 \times 3$. The input frame image will be extracted by the feature extractor to generate a feature map. After the feature map passes through the three-dimensional average pooling layer $P3$, it is reduced to 1 dimension in the temporal domain. After passing through two convolutional layers $C4$ and $C5$ with a size of 1×1 convolution kernel, the Global Average Pooling (GAP) layer converts the high-dimensional feature map into a same-dimensional vector. Finally, the high-dimensional features can get the classification result of the input frame through Fully Connected (FC) dimensionality reduction and SoftMax layer regression.

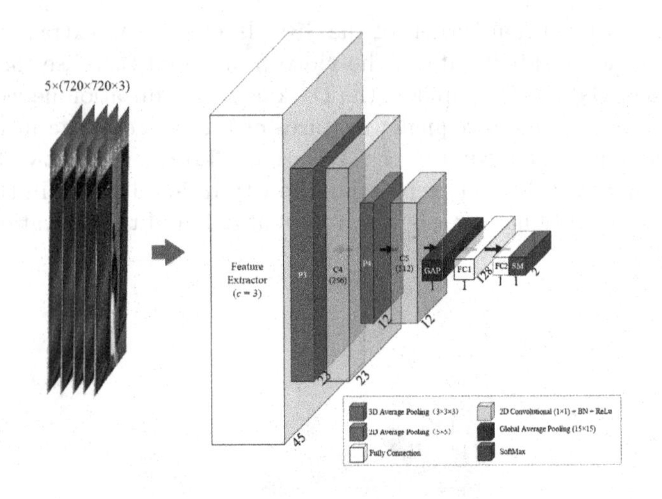

Fig. 5. The structure of the temporal forgery detection network.

The spatial locator is composed of a feature extractor and an RPN frame regressor, and its composition structure is shown in Fig. 6. In the spatial feature extractor, the size of the sliding window of the $3D$ maximum pooling layer is $1 \times 2 \times 2$. The spatial locator can predict the tampered area of the tampered frame and predict its confidence, that is, judge the possibility of the area being tampered. The input tampered frame is extracted by the feature extractor to generate a one-dimensional feature map in the temporal domain. Combine the feature maps of feature extractor and $C4$ into the Concatenate layer [12], then enter the RPN box regressor. $C4$ and $C5$ are convolutional layers with 54 convolution kernels. The upper branch of $C5$ has 36 convolution kernels, and the lower branch has 18 convolution kernels, which correspond to the frame coordinates and their confidence of the 9 size regression boxes at each position of the feature map. While testing the model, Non-Maximum Suppression (NMS) [25] is used to sort the candidate boxes according to the degree of confidence, and the candidate box with the highest degree of confidence is retained as the prediction box.

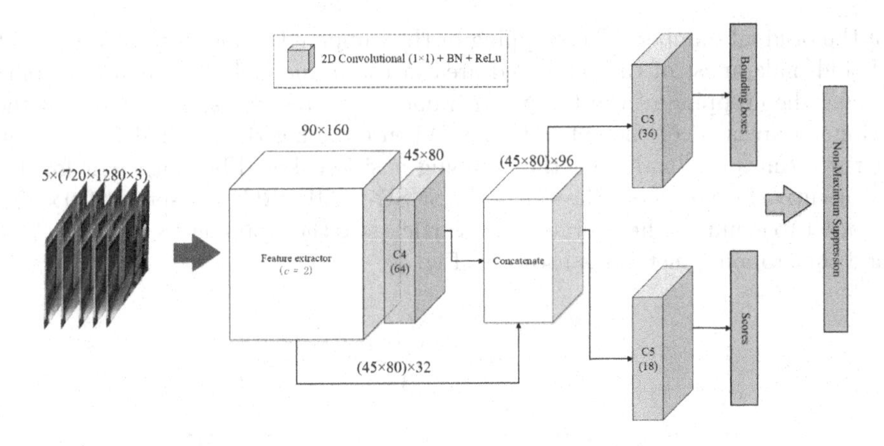

Fig. 6. The structure of the spatial forgery localization network.

4 Experiments

4.1 Dataset

The SYSU-OBJFORG dataset is produced by the video forensics research team of Sun Yat-sen University. It consists of 100 original videos and 100 tampered videos corresponding to the original videos. It is currently the largest object-based forged video dataset. All original videos are directly intercepted from surveillance videos taken by multiple static surveillance cameras. Tampered video is made by the producer after decoding the original video, adding, deleting or changing the original position of the moving objects in the video scene frame by frame, and then encoding and compressing the video. Each video is about 11 s (25 frames/s) long and has a resolution of 1280×720. The SYSU-OBJFORG data set is large in scale and complicated in tampering operations, which can improve the standard of detection and positioning algorithms. Therefore, we choose the SYSU-OBJFORG dataset as the training sample.

In order to make the scale of the selected data set meet the requirements of deep learning training, this article processes the video in the data set into image frames, and then continues to perform cropping and flipping operations on it. Since time-domain classification and spatial-domain positioning are two different network models, this paper adopts two different enhancement operations. Both data enhancement methods operate on five consecutive video frames (the first two frames of the current frame, the current frame, and the last two frames of the current frame total five frames).

The training of the temporal domain classification network requires input of two types of original frames and tampered frames. In the data set, the ratio of the original frame to the tampered frame is 13:3. In order to ensure the training effect, the ratio of the two frames needs to be adjusted to about 1:1. Therefore, this paper adopts the asymmetric data enhancement strategy proposed by Yao et al. [23], that is, the two frames are cropped according to the inverse ratio of the original number. The cropping of the tampered frame in the training set should make most of the tampered area in the frame included in the cropping range; the cropping of the tampered frame in the test set should make all the trimmed areas cover the entire frame. When cropping the original frame, crop it three times randomly, but the cropping position should be consistent for five consecutive frames. Since the size of the SYSU-OBJFORG data set is 1280 × 720, in order to facilitate the operation, this article sets the cropping area to 720 × 720, and the cropping method is shown in Fig. 7.

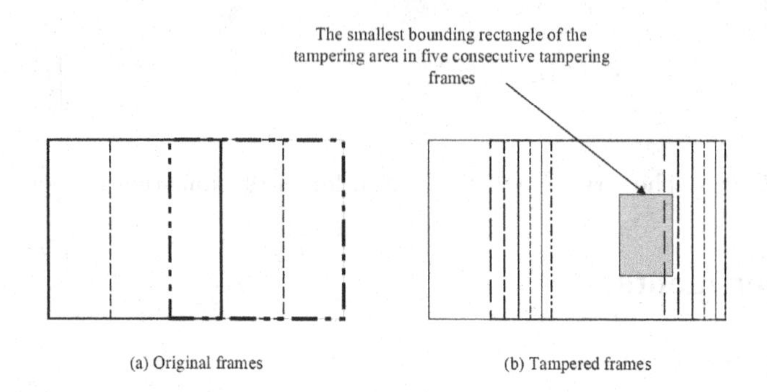

Fig. 7. The cropping method original frame and tampered frame.

The spatial domain localization network only needs to input the tamper frame for training. When performing enhancement operations on the tampered frames in the training set, keep 5 consecutive frames and perform horizontal, vertical, and horizontal and vertical flips at the same time. No action is required on the test set.

4.2 Experimental Settings

This paper verifies the effect of the spatio-temporal detection and localization model based on Darknet3D with the use of the SYSU-OBJFORG dataset. In the training process, 50 pairs, 10 pairs and 40 pairs are randomly selected as the training video set, validation set and test set, using the data enhancement method described in Sect. 4.1. After training the two models, input the test set video into the temporal domain classifier, the classifier will output the detection results frame by frame to obtain preliminary classification data. After recording the frame number detected as a tamper type, take the tamper frame sequence as the input of the spatial locator and make the spatial locator localize the tampered area in the frame. In order to observe whether the model achieves the expected effect, we repeat the above process five times, randomly sort all video sequences each time, and re-divide the training set, validation set, and test set according to the ratio of 5:1:4. And we will calculate the average value of the five test results finally.

The spatio-temporal detection network model in this article is built on the Tensorflow deep learning framework launched by Google, and the training environment is TITAN RTX GPU. The relevant parameter settings in the experiment are as follows: the AdamOptimizer algorithm is used to optimize the loss function, the initial learning rate is set to 0.0001, the $l2$ regularization parameter is 0.0005, and the momentum is 0.9. In the part of training the temporal domain classification network, batchsize is set to 64. In the part of testing, batchsize is set to 3. The input image is the three batches of test data generated by the operation described in Sect. 4.1. The detection and classification rules of the network model in this paper are as follows: if all three images are predicted as original frames, the intermediate frame is detected as the original frame; otherwise, it is detected as a tampered frame. For the evaluation indicators used for time-domain classification and detection, we use the following six metrics defined by Chen et al. [3]:

$$PFACC = \frac{\sum correctly\_classified\_pristine\_frames}{\sum pristine\_frames} \tag{1}$$

$$FFACC = \frac{\sum correctly\_classified\_forged\_frames}{\sum forged\_frames} \tag{2}$$

$$FACC = \frac{\sum correctly\_classified\_frames}{\sum all\_the\_frames} \tag{3}$$

$$Precision = \frac{T_P}{T_P + F_P} \tag{4}$$

$$Recall = \frac{T_P}{T_P + F_N} \tag{5}$$

$$F1Score = \frac{2T_P}{2T_P + F_P + F_N} \tag{6}$$

$PFACC$ is the original frame accuracy rate, $FFACC$ is the tampered frame accuracy rate, and $FACC$ is the accuracy rate of all frames. Precision, recall and $F1$ Score can be calculated by three indicators: TP (number of tampered frames correctly predicted), FP (number of original frames that were incorrectly predicted as tampered frames), and FN (number of tampered frames that were incorrectly predicted as original frames).

The batchsize of the spatial positioning network is set to 1, and the loss function is shown in formula (8).

$$loss = \frac{1}{N_{cls}} \sum_i L_{cls}(i) + \frac{1}{N_{reg}} \sum_j L_{reg}(j) \tag{7}$$

N_{cls} represents the number of tampered frames for all foreground and background frames in a mini-batch. The foreground frame is the frame with the tampered area frame IoU greater than 0.8, the background frame is the frame with the tampered area frame IoU less than 0.2, and i represents the number of boxes, L_{cls} represents the classification (two classification of foreground and background boxes) loss function; N_{reg} is the number of foreground boxes, j is the subscript, and L_{reg} is the box regression loss.

N_{cls} represents the number of tampered frames containing foreground and background frames in each batch, the boxes whose Intersection of Union (IOU) [7] with the border of the tampered area is greater than 0.8 are called the foreground boxes, the boxes whose IOU with the border of the tampered area is less than 0.2 are called the background boxes, and i indicates which box, L_{cls} represents the loss function of classification (two classification of foreground and background boxes); N_{reg} is the number of foreground boxes, j is the subscript, and L_{reg} is the frame regression loss. In this paper, the ratio of the number of foreground boxes to background boxes in each frame is set to 1:5 to balance the number of positive and negative sample frames, strengthen the training of negative samples, and improve the detection rate of the target area.

The constraint formula adopts the definition in literature [2], and the formula is as follows:

$$fg\_num = min(fg\_num, \frac{roi\_num}{\alpha + 1}) \tag{8}$$

$$bg\_num = min(roi\_num - fg\_num, fg\_num \times \alpha) \tag{9}$$

fg_sum is the number of foreground boxes and bg_num is the number of background boxes; the size of roi_num controls the training density of positive and negative samples, which is a constant and is set to 128.

4.3 Experiment Results

Testing of Temporal Forgery Detection. In order to verify the detection effect of the time domain classification network, the algorithm in this paper is compared with the algorithms proposed in [23] and [2], and the measurement standard described in Sect. 4.1 is used. The comparison result is shown in Fig. 8. The solid line in the figure is the test result of the algorithm in this paper, and the dotted line in the figure is the test result of the comparison algorithm. It can be seen from Fig. 8 that compared with the comparison algorithm, the algorithm in this paper has higher accuracy and F1Socre, and can achieve better results in the detection of tampered frames.

Testing of Spatial Forgery Localization. In the field of target detection, the detected object can be directly observed, and it is clearly distinguished from the background or undetected objects. Therefore, when performing spatial localization, the predicted area should be compared with the actual tampered area, rather than simply with the area where the semantic object is located. In the field of video tampering detection, the tampered area generally undergoes careful manipulation by the tamperer. The boundary transition between the tampered area and the original area is natural, and the area of the tampered area is much larger than the area occupied by the semantic object in the frame.

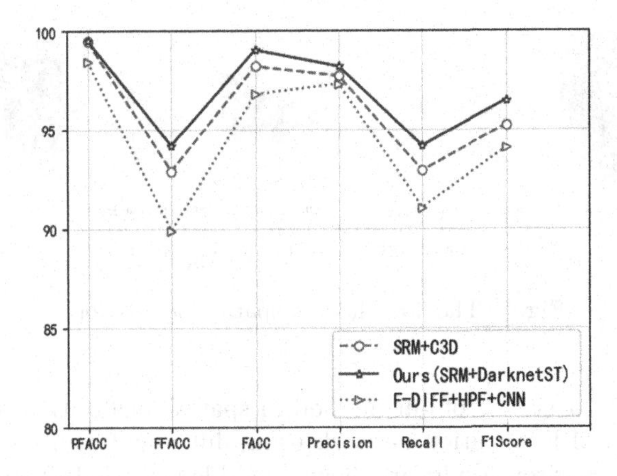

Fig. 8. The results of the temporal detection test.

The index to measure recognition accuracy in target detection is mAP (mean average precision). When calculating mAP, IOU is used as a vital function, and IOU refers to the ratio of intersection and union between the real box and the predicted box, which is used to evaluate the coincidence between the two boxes. The successful detection rate is the percentage of the number of frames where the tampered area is detected to the total number of test frames. Since spatial localization only predicts the tampered area in the tampered frame, so it is more intuitive to use the successful detection rate and the mean Intersection over Union as an evaluation indicator than mAP.

The calculation formula is as follows:

$$Suc\_rate = \frac{\sum F_{suc}}{\sum F_{suc} + \sum F_{mis}} \tag{10}$$

$$IOU\_mean = \frac{1}{N_{suc}} \sum_{i} IOU_i \tag{11}$$

When the IOU of the predicted box and the real tampered box is 0 or the confidence is less than 0.8, it is the missed frame (F_{mis}). Otherwise, it is a successfully detected frame (F_{suc}). $Nsuc$ represents the total number of successfully detected frames, and i is the subscript of the successfully detected frames.

Figure 9 is a flow chart of airspace positioning. After the video clip to be tested passes through the temporal domain classification network, the temporal domain classification network will detect the tampered frame. Then five consecutive frames including the tamper frame (the tamper frame includes two frames before and after it) are input into the spatial localization network, then the prediction boxes on the feature map are sorted according to the size of the corresponding confidence, but only the prediction box with the highest confidence is retained as the prediction area.

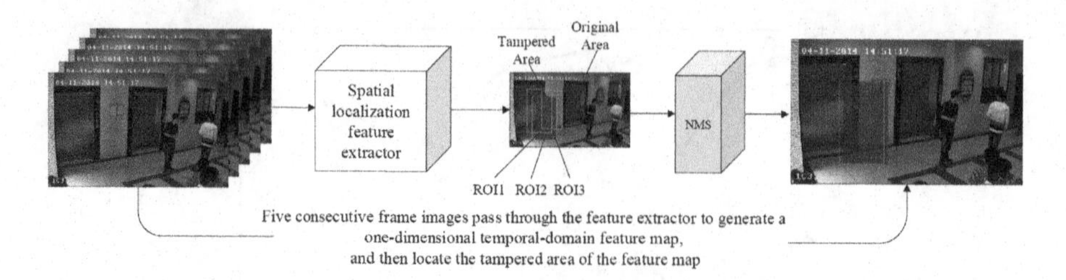

Fig. 9. The flow chart of spatial localization.

To verify the accuracy of our method in spatial localization, we compare it with other spatial localization network using different feature extraction algorithms. The comparison results are shown in Table 1. It can be seen from Table 1, compared with $VGG16$ using $2D$ convolutional layers, both indicators have been greatly improved. It shows that $3D$ convolutional neural network can extract and analyze video features better than $2D$ convolutional neural network. Compared with the feature extraction network using $C3D$, it can be seen that the method proposed in this article which does not use the pooling layer, and the localization effect is also improved.

Table 1. Comparison results of different spatial localization models.

Approaches	Number of test frames	Suc_rate	IOU_mean
SRM + VGG16 + RPN	4557	89.47%	45.65%
C3D + RPN	4594	47.60%	29.80%
SRM + C3D + RPN	4594	95.36%	49.07%
SRM + Darknet3D + RPN	4594	96.87%	49.72%

Complete Test of Spatiotemporal Forgery Localization. When testing the entire spatio-temporal network model, we randomly select 10 tampered videos from the test set for testing. Firstly, the videos will be pre-processed, and the whole video will be divided into a continuous frame image sequence, and then the frame data is input into the temporal domain detection network. Then the frame data passes through the feature extractor and classifier of the temporal network to determine whether each frame has been tampered with, and records the serial number of the tampered frame. Then, the correct tampered frame sequence is turned into the input of the spatial localization network, and the spatial localization is performed after the spatial localization feature extractor and the RPN frame regressor, and finally the overall detection data will be evaluated.

Table 2. The complete test results of our spatiotemporal forgery localization model.

Test Videos	Stage of temporal domain detection				Stage of saptial domain detection		
	Number of video frames	Number of video frames	Number of predicted forged frames	Number of correctly predicted forged frames	FACC	Suc_rate	IOU_mean
Video 1	291	103	81	81	91.75%	97.58%	48.37%
Video 2	284	98	97	97	99.29%	98.69%	72.16%
Video 3	284	71	68	68	98.59%	97.21%	39.64%
Video 4	284	120	118	118	99.29%	95.16%	35.77%
Video 5	285	123	110	107	93.33%	88.26%	49.67%
Video 6	291	180	179	179	99.31%	99.25%	58.35%
Video 7	292	139	136	136	99.31%	99.34%	52.35%
Video 8	292	64	62	62	98.98%	94.36%	38.61%
Video 9	292	130	131	130	99.65%	96.47%	54.63%
Video 10	284	117	116	116	99.65%	100.0%	58.94%

The overall test results are shown in Table 2. $FACC$ represents all frame correct rates and is used to evaluate the results of temporal domain classification; Suc_rate and IOU_mean are used to evaluate the results of spatial localization. It can be seen from Table 2 that whether it is to detect the tampered frame or locate the tampered area, the model proposed in this paper has a good performance when it is used for the complete spatio-temporal test.

5 Conclusion

This paper proposes a removal tampering detection model for video objects based on the optimized Darknet53 network. First, we construct a three-dimensional Darknet53 network. The information of the Spatio-temporal domain is introduced into the two-dimensional Darknet53 network. Then the two-dimensional convolutional layer is replaced by a three-dimensional convolutional layer. The frame image features are jointly extracted from the Spatio-temporal domain to improve the expression of feature sufficiency. Secondly, we design a Darknet53 optimized network by compressing the number of residual blocks in the three-dimensional Darknet53 network to reduce the network dimension and feature redundancy. Finally, the Darknet53 optimized network is applied as a feature extractor to reduce the use of the pooling layer and retain the detailed information within the frame. Experimental results show that the model can effectively extract the Spatio-temporal features in the video data, which is conducive to later detection and localization. Compared with other CNN models, this model can reduce the negative impact of the pooling layer, and has higher detection accuracy and better localization performance. However, the model in this paper still has shortcomings such as high memory consumption and undistinguished transition area of the original frame and the tampered frame. Therefore, we will consider building a tampered video detection network that saves memory and does not depend on the training batch size, further improves the detection effect.

References

1. Battiato, S., Farinella, G.M., Messina, E., Puglisi, G.: Robust image alignment for tampering detection. IEEE Trans. Inf. Forensics Secur. **7**(4), 1105–1117 (2012). https://doi.org/10.1109/TIFS.2012.2194285
2. Chen, L., Yang, Q., Yuan, L.: Passive forensic based on spatio-temporal location of video object removal tampering. J. Commun. (7) (2020)
3. Chen, S., Tan, S., Li, B., Huang, J.: Automatic detection of object-based forgery in advanced video. IEEE Trans. Circ. Syst. Video Technol. **26**(11), 2138–2151 (2016)
4. Chen, W.B., Yang, G.B., Chen, R.C., Zhu, N.B.: Digital video passive forensics for its authenticity and source. J. Commun. **32**(6), 177–183 (2011)
5. Fadl, S.M., Han, Q., Li, Q.: CNN spatiotemporal features and fusion for surveillance video forgery detection. Sign. Process. Image Commun. **90**, 116066 (2020)
6. Hsu, C.C., Hung, T.Y., Lin, C.W., Hsu, C.T.: Video forgery detection using correlation of noise residue. In: 2008 IEEE 10th Workshop on Multimedia Signal Processing, pp. 170–174 (2008). https://doi.org/10.1109/MMSP.2008.4665069
7. Jiang, B., Luo, R., Mao, J., Xiao, T., Jiang, Y.: Acquisition of localization confidence for accurate object detection (2018)
8. Jin, X., He, Z., Xu, J., Wang, Y., Su, Y.: Object-based video forgery detection via dual-stream networks. In: 2021 IEEE International Conference on Multimedia and Expo (ICME), pp. 1–6 (2021). https://doi.org/10.1109/ICME51207.2021.9428319
9. Kono, K., Yoshida, T., Ohshiro, S., Babaguchi, N.: Passive video forgery detection considering spatio-temporal consistency. In: Madureira, A.M., Abraham, A., Gandhi, N., Silva, C., Antunes, M. (eds.) SoCPaR 2018. AISC, vol. 942, pp. 381–391. Springer, Cham (2020). https://doi.org/10.1007/978-3-030-17065-3_38
10. Li, L., Wang, X., Zhang, W., Yang, G., Hu, G.: Detecting removed object from video with stationary background. In: Shi, Y.Q., Kim, H.J., Pérez-González, F. (eds.) The International Workshop on Digital Forensics and Watermarking 2012, pp. 242–252. Springer, Berlin Heidelberg, Berlin, Heidelberg (2013). https://doi.org/10.1007/978-3-642-40099-5_20
11. Long, C., Smith, E., Basharat, A., Hoogs, A.: A C3D-based convolutional neural network for frame dropping detection in a single video shot. In: 2017 IEEE Conference on Computer Vision and Pattern Recognition Workshops (CVPRW) (2017)
12. Mane, S., Mangale, S.: Moving object detection and tracking using convolutional neural networks. In: 2018 Second International Conference on Intelligent Computing and Control Systems (ICICCS), pp. 1809–1813 (2018). https://doi.org/10.1109/ICCONS.2018.8662921
13. Redmon, J., Farhadi, A.: YOLO9000: better, faster, stronger. In: IEEE Conference on Computer Vision and Pattern Recognition, pp. 6517–6525 (2017)
14. Redmon, J., Farhadi, A.: YOLOv3: an incremental improvement. arXiv e-prints (2018)
15. Saxena, S., Subramanyam, A., Ravi, H.: Video inpainting detection and localization using inconsistencies in optical flow. In: 2016 IEEE Region 10 Conference (TENCON), pp. 1361–1365 (2016). https://doi.org/10.1109/TENCON.2016.7848236
16. Su, C., Wei, J.: Hybrid model of vehicle recognition based on convolutional neural network. In: 2020 IEEE 22nd International Conference on High Performance Computing and Communications; IEEE 18th International Conference on Smart City; IEEE 6th International Conference on Data Science and Systems (HPCC/SmartCity/DSS), pp. 1246–1251 (2020). https://doi.org/10.1109/HPCC-SmartCity-DSS50907.2020.00161

17. Su, L., Luo, H., Wang, S.: A novel forgery detection algorithm for video foreground removal. IEEE Access **7**, 109719–109728 (2019). https://doi.org/10.1109/ACCESS.2019.2933871
18. Tran, D., Bourdev, L., Fergus, R., Torresani, L., Paluri, M.: Learning spatiotemporal features with 3D convolutional networks. In: 2015 IEEE International Conference on Computer Vision (ICCV), pp. 4489–4497 (2015). https://doi.org/10.1109/ICCV.2015.510
19. Wang, Q., Zhang, R.: A blind image forensic algorithm based on double quantization mapping relationship of DCT coefficients. J. Electron. Inf. Technol. **36**(009), 2068–2074 (2014)
20. Wang, X., Lu, Z.: Automatic localization of image tampering area based on JPEG block effect difference. Comput. Sci. **37**(002), 269–273 (2010)
21. Wu, W., Zhan, L.: Detection of tampering using color filter array characteristics and fuzzy estimation. Comput. Eng. Des. **28**(21), 5179–5180, 5256 (2007)
22. Yang, H., Zhou, Z., Zhou, C.: Mobile image tampering detection based on pattern noise. J. Comput. Syst. Appl. (2013)
23. Yao, Y., Shi, Y., Weng, S., Guan, B.: Deep learning for detection of object-based forgery in advanced video. Symmetry **10**(1), 3 (2018). https://doi.org/10.3390/sym10010003
24. Zhang, J., Chen, J., Su, Y.: Detection of region-duplication forgery in the video streams. Electron. Meas. Technol. **34**(011), 66–69 (2011)
25. Zhou, P., Han, X., Morariu, V.I., Davis, L.S.: Learning rich features for image manipulation detection. In: 2018 IEEE/CVF Conference on Computer Vision and Pattern Recognition, pp. 1053–1061 (2018). https://doi.org/10.1109/CVPR.2018.00116

Author Index